W. Dyer

SAUNDERS MATHEMATICS BOOKS

Consulting Editor

BERNARD R. GELBAUM, University of California

INTRODUCTION TO THE THEORY OF RANDOM PROCESSES

I. I. GIKHMAN

A. V. SKOROKHOD

Kieve State University

TRANSLATED BY SCRIPTA TECHNICA, INC.

W. B. SAUNDERS COMPANY

Philadelphia • London • Toronto

W. B. Saunders Company: West Washington Square
Philadelphia, Pa. 19105

12 Dyott Street
London, W.C.1

1835 Yonge Street
Toronto 7, Ontario

Originally published in 1965 as Vvedenie v teorige slychainich processov
by the Nauka Press, Moscow

Introduction to the Theory of Random Processes

Preface to the English Translation

The present work of Gikhman and Skorokhod is the first publication since the now classic "Stochastic Processes" by J. Doob to survey in a rigorous way the more modern results in the theory of stochastic processes and to link them to earlier developments in the subject. Not only are the standard classes of stochastic processes discussed, but much of the authors' original work on limit theorems in stochastic processes is also included. A significant feature of the book is a unique and beautiful treatment of processes with independent increments, which assumes no prior knowledge of the theory of infinitely divisible laws.

The book is appropriate for students who have a sound background in probability from a measure-theoretic viewpoint and will, undoubtedly, be welcome as a graduate text. For reference purposes the authors have included a comprehensive discussion of measure theory and of the basic ideas of probability.

The authors take great care to state the topological assumptions underlying each theorem, although occasionally a result may be stated in slightly greater generality than seems warranted by the proof. The book contains a wealth of results, ideas, and techniques, the deepest appreciation of which demands a most careful reading. Certainly, this is not a book for the indolent.

The English translation was reviewed by H. Dym, W. Rosenkrantz, S. Sawyer, D. Stroock, S. R. S. Varadhan, and myself. Although it was our policy not to make major revisions of the manuscript, we corrected many small inadvertent errors.

WARREN M. HIRSCH

Courant Institute of
Mathematical Sciences
New York University

Introduction

The current literature on the theory of random processes is quite extensive. In addition to textbooks and monographs especially devoted to this theory and its various divisions, there are many technical books, for the most part dealing with automation and radio electronics, in which considerable space is given to the theory of random processes.

From the point of view of instruction this literature can be divided into two groups, the first consisting of serious and lengthy monographs whose difficulty hinders their use as beginning textbooks, and the second consisting of books that are either elementary or written for engineering students. There are no books in the [Russian] literature that are designed for rigorous exposition and are at the same time suitable for elementary instruction.

The authors have therefore, decided to write the present book, based on material they have expounded in a number of courses at the University of Kiev. The first five chapters of the book are devoted to general questions in the theory of random processes (including measure theory and axiomatization of probability theory); Chapters 6 through 9 are devoted to more specialized questions (processes with independent increments, Markov processes, and limit theorems for random processes). The book is designed for persons who have had a general course in probability theory and are ready to begin the study of the theory of random processes. The authors hope that it will prove to be useful for students in the universities and also for specialists, other than mathematicians, who wish to familiarize themselves with the fundamental methods and results of the theory in a rigorous though not the most general and exhaustive approach.

The authors have not undertaken to treat all branches of the theory. Certain questions and methods that are well covered in current [Russian] literature are omitted. (These include the semi-group theory of Markov processes, ergodic properties of Markov processes, martingales, and generalized random processes.) On the other hand, questions that, up to the present, have not been included in books on the theory of random processes (such as limit theorems for random processes) but that play an important role in contemporary theory, are considered.

The theory of random processes has recently developed into a separate branch of probability theory. Because the theory of random processes is still so closely related to other divisions of probability theory, the boundaries between this theory and its divisions are often difficult to determine precisely. For example, the theory of random processes is related to the theory of summation of independent random variables by the division of probability theory that studies processes with independent increments, and to mathematical statistics by statistical problems in the theory of random processes.

Let us characterize the problems in the theory of random processes that, from our point of view, may be considered basic.

1. The first problem in the theory of random processes is the construction of a mathematical model that allows a rigorous (formal) definition of a random process, and the investigation of the general properties of that model.

2. A second problem is the classification of random processes. Obviously every classification is arbitrary to some extent. Therefore we need to begin with specific principles that at least indicate the direction the classification will take. The existing classification in the theory of random processes separates from the entire set of random processes certain classes that admit a more or less constructive description. Every class is characterized by the property that only a finite number of functional characteristics need to be specified in order to single out from the entire class an individual random process.

Sometimes, we consider classes of processes that admit a uniform solution to a specified set of problems. In considering such classes, we are usually not interested in the difference between random processes, if the characteristics necessary for the solution of these problems coincide for them.

We might mention the following broad classes of processes: (1) processes with independent increments, (2) Markov processes, (3) Gaussian processes, (4) processes that are stationary in the narrow sense, and (5) processes that are stationary in the broad sense. (We might include in this last group processes with stationary increments.)

3. The third problem, closely associated with the preceding one, consists in finding, for various classes of random processes, an

analytical apparatus that will enable us to calculate the probabilistic characteristics of random processes. Such an apparatus has been constructed for the simplest probabilistic characteristics, and it uses, as a rule, either the theory of differential equations (ordinary and partial) and integrodifferential equations (in the case of Markov processes and processes with independent increments) or the theory of integral equations with symmetric kernel (in the case of Gaussian processes) or Fourier transformations and the theory of functions of a complex variable (for processes with independent increments and for stationary processes).

4. We need to single out a class of problems that has played an important role in the development of certain branches of the theory of random processes and that is of great practical significance. In its general form, the problem consists in the best determination of the value of some functional of a process from the values of other functionals of the same process. An example of such a problem is the problem of prediction: From observation of a process over a certain interval of time, determine the value of the process at some instant of time outside that interval. Under certain restrictions, prediction problems have been solved for processes that are stationary in the broad sense (see Chapter V).

5. An important class of problems in the theory of random processes is the study of various transformations of random processes. These transformations are used to study complicated processes by reducing them to simpler ones.

We might include with the study of transformations of random processes the theory of differential and integral equations involving random processes. This class of problems also includes limit theorems for random processes since the operation of taking the limit is a sort of transformation.

At the present time the principal fields of application of the theory of random processes are electronics (which deals primarily with processes that are stationary in the broad sense and with Gaussian processes) and cybernetics (which deals with processes that are stationary in the narrow sense and with Markov processes).

In mathematical economics and mathematical biology we use Markov processes of a different sort. In the molecular theory of gases we use the process of Brownian motion; in the theory of showers of cosmic particles we apply Markov processes and processes with independent increments.

In general, the methods of the theory of random processes are finding ever new fields of application, and today every one of the natural sciences has felt the influence of this theory, at least to some degree.

Let us characterize briefly the features of the contents of the present book. The first chapter is devoted to random processes in the broad sense. This is the name we have given to the portion of the theory of random processes that deals only with distributions of finite sets of values of a random process. This portion is very close to elementary probability theory, involves no complicated mathematical concepts, and is sufficient for many applications.

For a more profound study of the theory of random processes, a more highly developed theory of measure and integration is necessary. Therefore, following Chapter I we expound all the necessary information in this field (Chapter II), and on the basis of this information we construct an axiomatization of probability theory (Chapter III). We also consider the general questions in the theory of random functions and, after that, specific classes of random processes and special questions in the theory. Among random processes, extensive treatment is given to processes with independent increments (to which one chapter is devoted) and Markov processes (to which two chapters are devoted). Stationary processes are considered to some extent in Chapter I and again in Chapter V, which is devoted to linear transformations of random processes. Chapter V also takes up the problem of linear prediction. An entire chapter is devoted to limit theorems for random processes. In this chapter basic attention is given to processes with independent increments and Markov processes.

Most of the constructions are made for the case in which a random process assumes values belonging to a finite-dimensional Euclidean space. In a few cases we consider complex-valued one-dimensional and multidimensional processes and also processes with values belonging to a complete metric space.

We assume that the reader is familiar with the basic concepts of linear algebra, which is particularly important for the study of Gaussian processes, and the theory of Hilbert spaces, which is used in the study of linear transformations of random processes. The reader should also have some familiarity with functional analysis (complete metric spaces, compact spaces, etc.).

We have not attempted to give a complete bibliography of works on the theory of random processes. In addition to books cited in the text, the bibliography includes only the basic books on the theory of random processes and probability theory that exist in

Russian, as well as articles in which the fundamental results in this field first appeared.

The book is divided into chapters and the chapters into sections. The basic formulas and also the theorems, lemmas, and definitions are numbered afresh in each section. A reference to a theorem or formula in the same section is indicated only by the number of the theorem or formula. If a reference is made to a theorem or formula in another section of the same chapter, the section number is added. If the reference is made to another chapter, the chapter number is added.

The authors express their gratitude to colleagues and students in the Department of Probability Theory and Mathematical Statistics of Kiev State University for the help they have given in the preparation of this book.

Kiev
October 21, 1963 THE AUTHORS.

Contents

4

5

6

7

I

RANDOM PROCESSES IN THE BROAD SENSE

1. DEFINITIONS

The course of a random process, like that of a deterministic process, is described by some function $\xi(\theta)$ (which may assume real, complex, or vector values), where θ assumes values in a reference set Θ. As θ varies, $\xi(\theta)$ describes the evolution of the process. (Of course, the way in which the process evolves is random, and each of the functions $\xi(\cdot)$ describes only one of the possible ways in which the process may develop). These functions $\xi(\cdot)$ are called *sample functions* of the random process.

For each fixed θ, the quantity $\xi(\theta)$ is random. To be able to apply mathematical methods to the questions that we are studying, it is natural to assume that $\xi(\theta)$ is a random variable (possibly vector-valued) in the probabilistic sense. Consequently, by a *random process* we mean a family of random variables $\xi(\theta)$ depending on a parameter θ that assumes values in some set Θ.

If the set Θ is arbitrary, then instead of the term "random process," it is more convenient to use the term "random function" and to reserve the name "random process" for those cases in which the parameter θ is interpreted as time. When the argument of a random function is a spatial variable, this function is also called a *random field*.

This definition of a random process, or a random function as we have just agreed to call it, needs to be made more precise. For the sake of simplicity we shall speak of a random function that assumes real values. First of all we need to make clear just what is meant by "a family of random variables depending on a parameter θ." We recall that in accordance with the principles of probability theory, a finite sequence of random variables

$$\xi_1, \xi_2, \cdots, \xi_n$$

is completely characterized by the *joint distribution function*

$$F(x_1, x_2, \cdots, x_n) = \mathsf{P}\{\xi_1 < x_1, \xi_2 < x_2, \cdots, \xi_n < x_n\} .$$

When we turn to a probabilistic description of a random function, we must consider the question: How should we describe mutual relationships among infinitely many random variables—the values of a random function?

The simplest procedure is to say that the random function $\xi(\theta)$ is given if all possible probabilistic relations among each finite set of values of the random fnnction

$$\xi(\theta_1), \xi(\theta_2), \cdots, \xi(\theta_n), \quad \theta_i \in \Theta, i = 1, 2, \cdots, n; n = 1, 2, \ldots, \tag{1}$$

are defined, that is, if the corresponding joint distribution functions are given. From this point of view the random function $\xi(\theta)$, for $\theta \in \Theta$, is determined by the family of distributions

$$F_{\theta_1,\theta_2,\ldots,\theta_n}(x_1, x_2, \cdots, x_n), \quad \theta_i \in \Theta, i = 1, 2, \cdots, n; n = 1, 2, \cdots, \tag{2}$$

and each distribution function $F_{\theta_1,\theta_2,\ldots,\theta_n}(x_1, x_2, \cdots, x_n)$ is interpreted as the joint distribution function of the sequence of random variables (1).

Of course for such an interpretation to be possible, the family of distributions (2) cannot be completely arbitrary. It must satisfy the following obvious conditions, which we shall call the *compatibility conditions* for the family of distributions (2):

$$\begin{aligned} &F_{\theta_1,\theta_2,\ldots,\theta_n,\theta_{n+1},\ldots,\theta_{n+p}}(x_1, x_2, \cdots, x_n, +\infty, \cdots, +\infty) \\ &\quad = F_{\theta_1,\theta_2,\ldots,\theta_n}(x_1, x_2, \cdots, x_n) , \end{aligned} \tag{3}$$

$$F_{\theta_1,\theta_2,\ldots,\theta_n}(x_1, x_2, \cdots, x_n) = F_{\theta_{i_1},\theta_{i_2},\ldots,\theta_{i_n}}(x_{i_1}, x_{i_2}, \cdots, x_{i_n}) , \tag{4}$$

where $i_1, i_2, \cdots, i_n$ is an arbitrary permutation of the indices $1, 2, \cdots, n$. The necessity of these conditions arises from the following relations:

$$\begin{aligned} &F_{\theta_1,\theta_2,\ldots,\theta_n,\theta_{n+1},\ldots,\theta_{n+p}}(x_1, x_2, \cdots, x_n, +\infty, \cdots, +\infty) \\ &\quad = \mathsf{P}\{\xi(\theta_1) < x_1, \xi(\theta_2) < x_2, \cdots, \xi(\theta_n) < x_n, \\ &\qquad\quad \xi(\theta_{n+1}) < \infty, \cdots, \xi(\theta_{n+p}) < \infty\} \\ &\quad = \mathsf{P}\{\xi(\theta_1) < x_1, \xi(\theta_2) < x_2, \cdots, \xi(\theta_n) < x_n\} \\ &\quad = F_{\theta_1,\theta_2,\ldots,\theta_n}(x_1, x_2, \cdots, x_n) , \\ &F(x_1, x_2, \cdots, x_n) \\ &\quad = \mathsf{P}\{\xi(\theta_1) < x_1, \xi(\theta_2) < x_2, \cdots, \xi(\theta_n) < x_n\} \\ &\quad = \mathsf{P}\{\xi(\theta_{i_1}) < x_{i_1}, \xi(\theta_{i_2}) < x_{i_2}, \cdots, \xi(\theta_{i_n}) < x_{i_n}\} \\ &\quad = F_{\theta_{i_1},\theta_{i_2},\ldots,\theta_{i_n}}(x_{i_1}, x_{i_2}, \cdots, x_{i_n}) . \end{aligned}$$

Just as we can identify a single random variable with its distribution functions, so what has been said above leads us to:

Definition 1. A family of distributions (2) satisfying the compatibility conditions (3) and (4) is said to be a *random function* $\xi(\theta)$ with real values, defined on the set Θ (that is, defined for $\theta \in \Theta$).

The functions belonging to the family $F_{\theta_1,\ldots,\theta_n}(x_1, \cdots, x_n)$ are called *finite-dimensional distributions* of the random function.

This definition of a random function is attractive because of its simplicity, and it is sufficient when we are interested in the values of the random function on a finite set of values of the argument θ. On the other hand, a serious defect in this definition is that it does not enable us to look at the random function in its entirety, that is, to look at all its values simultaneously. In addition, in many experiments the sample function being investigated is defined by means of a graph of some curve. Our definition of a random function not only fails to enable us to construct the graph of this function but also fails to enable us to ask questions regarding such functional properties of the functions $\xi(\theta)$ as their continuity, differentiability, and so forth. Similarly, we cannot immediately ask the question of the probability of the event that the inequality $a < \xi(\theta) < b$, where $a < b$, will be satisfied for all $\theta \in \Theta$.

We can obtain more satisfactory definitions of a random function if we use the axiomatic approach to probability theory. Every probabilistic scheme describes the results of some experiment having random outcomes. If the result of an experiment is described by a number or a finite sequence of numbers, we say that we are observing a random variable or a random vector. On the other hand, if the result of the experiment is described by some function, we say that we are observing a random function. Thus a random function is defined by an arbitrary probabilistic scheme describing experiments whose results are random functions. A more precise analysis of this definition will be given in Chapter IV. Let us agree to call the definition of a random function that we shall use in the present section the definition of a random function in the broad sense.

Up to now, we have been speaking of a single random function. In solving many problems, we need to deal with several distinct random functions. If we are to be able to carry out mathematical operations on them, it is not sufficient that we define each of these functions separately. How can we define a sequence of random functions $\xi_1(\theta), \xi_2(\theta), \cdots, \xi_m(\theta)$, each defined on the single set Θ? Instead of speaking about a sequence of functions, we find that it is simpler to speak of a single vector-valued function $\boldsymbol{\zeta}(\theta)$, whose components are random functions $\xi_1(\theta), \xi_2(\theta), \cdots, \xi_m(\theta)$. Then we can use the preceding definition almost without change. The role of the distribution of the sequence of random variables (1) is played by the joint distribution function of the sequence of vectors $\zeta(\theta_1)$, $\zeta(\theta_2), \cdots, \zeta(\theta_n)$, that is, by the function of nm variables

$$F_{\theta_1, \theta_2, \dots, \theta_n}(x_{11}, x_{12}, \cdots x_{nm})$$
$$= \mathsf{P}\{\xi_1(\theta_1) < x_{11}, \xi_2(\theta_1) < x_{12}, \cdots, \xi_n(\theta_m) < x_{mn}\} .$$

In what follows, the set Θ will usually be a set of real numbers, and the variable θ will be interpreted as the time t. In this case we shall let $\mathfrak{T}$ denote the set Θ, and we shall understand it to mean a finite or infinite interval (closed, open, or semi-open). We shall also consider cases in which $\mathfrak{T}$ is either the set of all nonnegative integers or the set of all real integers. In this case we have a sequence of random variables (vectors) $\{\boldsymbol{\zeta}(k)\}$ (for $k = 0, 1, 2, \cdots$ or $k = 0, \pm 1, \pm 2, \cdots$). We shall call such a process a *random process with discrete time* or a *random sequence*. Processes with discrete time play an important role in the general theory of random processes. There are many probabilistic problems in which time is an essentially discrete quantity. Also, the study of processes with discrete time in most cases makes use of simpler tools. In addition there are a number of cases in which processes with continuous time can be approximated by processes with discrete time.

Another important class of random processes is the class of stationary processes. These are processes whose probabilistic characteristics do not change with displacement of time. More precisely, we have:

Definition 2. A random process $\boldsymbol{\zeta}(t)$ defined on $\mathfrak{T}$ is said to be *stationary* if for arbitrary n, t and $t_1, t_2, \cdots, t_n$, such that $t_i + t \in \mathfrak{T}$ (for $i = 1, 2, \cdots, n$), the joint distribution function of the sequence of random vectors $\boldsymbol{\zeta}(t_1 + t), \boldsymbol{\zeta}(t_2 + t), \cdots, \boldsymbol{\zeta}(t_n + t)$ is independent of t.

As a rule, stationary processes are considered on an infinite (or semi-infinite) interval of the real axis (or the set of real integers). A useful modification in the concept of stationarity is the concept of a process with stationary increments.

Definition 3. A process $\boldsymbol{\zeta}(t)$, for $t \in \mathfrak{T}$, is called a *process with stationary increments* if the joint distribution of the differences,

$$\boldsymbol{\zeta}(t_2 + t) - \boldsymbol{\zeta}(t_1 + t), \boldsymbol{\zeta}(t_3 + t) - \boldsymbol{\zeta}(t_2 + t),$$
$$\cdots, \boldsymbol{\zeta}(t_n + t) - (t_{n-1} + t) ,$$

is independent of t for arbitrary n, t and for $t_1\, t_2, \cdots, t_n$ such that $t_i + t \in \mathfrak{T}$ (for $i = 1, 2, \cdots, n$).

REMARK 1. The definition of a stationary process is equivalent to the following: For an arbitrary bounded continuous function $f(z_1, z_2, \cdots, z_n)$ of n vector-valued variables $z_1, z_2, \cdots, z_n$, the mathematical expectation

$$\mathsf{M}f[\boldsymbol{\zeta}(t_1 + t), \boldsymbol{\zeta}(t_2 + t), \cdots, \boldsymbol{\zeta}(t_n + t)]$$

is independent of t for arbitrary n, t_j, and t such that $t_j + t \in \mathfrak{T}$.

REMARK 2. Stationary processes have the following property: For every continuous function $f(z_1, \cdots, z_n)$, the random process

$$\eta(t) = f[\boldsymbol{\zeta}(t_1 + t), \boldsymbol{\zeta}(t_2 + t), \cdots, \boldsymbol{\zeta}(t_n + t)]$$

is also stationary.

The proof follows from Remark 1.

Definition 4. A sequence of random functions $\xi_n(\theta)$, where $\theta \in \Theta$ and $n = 1, 2, \cdots$, is said to *converge weakly* to $\xi_0(\theta)$, $\theta \in \Theta$, if the finite dimensional distribution functions of the $\xi_n(\theta)$ converge to those of $\xi_0(\theta)$.

2. COVARIANCE FUNCTIONS

The family of joint distributions (2) of Section 1 gives a comprehensive description of a random function in the broad sense. However in many cases we need a more concise description of distributions, one that reflects other important properties of a random function. Furthermore, the solution of many probabilistic problems depends on only a small number of parameters characterizing distributions that come up in the problem. The most important numerical characteristics of distributions are their moments. In the theory of random functions, the role of moments of distributions is played by the moment functions.

Definition 1. The *moment functions*, $m_{j_1, j_2, \cdots, j_s}(\theta_1, \theta_2, \cdots, \theta_s)$, of a random function $\xi(\theta)$, where $\theta \in \Theta$, are defined as the functions

$$m_{j_1, \cdots, j_s}(\theta_1, \theta_2, \cdots, \theta_s) = \mathsf{M}([\xi(\theta_1)]^{j_1}[\xi(\theta_2)]^{j_2} \cdots [\xi(\theta_s)]^{j_s}),$$
$$j_k > 0 \qquad (k = 1, 2, \cdots, s) \tag{1}$$

if the mathematical expectation on the right is meaningful for all $\theta_i \in \Theta$ where $i = 1, \cdots, s$. The quantity $q = j_1 + j_2 + \cdots + j_s$ is called the *order* of the moment function.

Definition 2. A random function $\xi(\theta)$, where $\theta \in \Theta$, is said to belong to the class $L_p(\Theta)$ (and we write $\xi(\theta) \in L_p(\Theta)$) if $\mathsf{M} \mid \xi(\theta) \mid^p < \infty$ for every $\theta \in \Theta$.

Theorem 1. *If* $\xi(\theta) \in L_p(\Theta)$, *the moment functions of order* q *are* *finite for all* $q \leqq p$.

Proof. From the inequality between the geometric and arithmetic means (cf. Hardy, Littlewood, and Polya, Section 2.5)

$$\prod_{k=1}^{s} a_k^{p_k} \leqq \sum_{k=1}^{s} p_k a_k\,, \qquad a_k \geqq 0\,, \qquad p_k \geqq 0\,, \qquad \sum_{k=1}^{s} p_k = 1\,,$$

it follows that

$$\left\{\prod_{k=1}^{s} |\,\xi(\theta_k)\,|^{j_k}\right\}^{1/q} \leqq \left(\sum_{k=1}^{s} \frac{j_k}{q}\,|\,\xi(\theta_k)\,|\right), \qquad q = \sum_{k=1}^{s} j_k\,.$$

Using Hölder's inequality (cf. Hardy, Littlewood, and Polya, Section 2.8), we obtain

$$\sum_{k=1}^{s} \frac{j_k}{q}\,|\,\xi(\theta_k)\,| \leqq \left[\sum_{k=1}^{s}\left(\frac{j_k}{q}\right)^{q'}\right]^{1/q'} \left\{\sum_{k=1}^{s} |\,\xi(\theta_k)\,|^{q}\right\}^{1/q},$$

where $1/q' + 1/q = 1$. Thus

$$\prod_{k=1}^{s} |\,\xi(\theta_k)\,|^{j_k} \leqq s^{q-1}\sum_{k=1}^{s} |\,\xi(\theta_k)\,|^{q}\,,$$

from which the assertion follows.

If we know the characteristic function of the joint distribution of the variables $\xi(\theta_1), \cdots, \xi(\theta_s)$,

$$\psi(\theta_1, \cdots, \theta_s, \lambda_1, \lambda_2, \cdots, \lambda_s) = \mathsf{M}e^{i\sum_{k=1}^{s} \lambda_k \xi(\theta_k)}\,,$$

then the moment functions with integral indices can be found by differentiating:

$$\left.\frac{\partial^q \psi}{\partial \lambda_1^{j_1} \cdots \partial \lambda_s^{j_s}}\right|_{\lambda_1=\lambda_2=\cdots=0} = i^q \mathsf{M}([\xi(\theta_1)]^{j_1} \cdots [\xi(\theta_s)]^{j_s})$$
$$= i^q m_{j_1 \cdots j_s}(\theta_1, \cdots, \theta_s)\,. \qquad (2)$$

Here, differentiation under the mathematical-expectation sign is permissible for all $q = \sum_{k=1}^{s} j_k \leqq p$, if $\xi(\theta) \in L_p(\Theta)$.

A precise converse of this assertion holds for moment functions with even indices. Let Δ_i denote the operation of taking the symmetric finite difference with respect to the variable λ_i and let Δ_i^j denote its jth iterate:

$$\begin{aligned} &\Delta_i \psi(\theta_1, \cdots, \theta_s, \lambda_1, \cdots, \lambda_s) \\ &\quad = \psi(\theta_1, \cdots, \theta_s, \lambda_1, \cdots, \lambda_i + h_i, \cdots, \lambda_s) \\ &\qquad - \psi(\theta_1, \cdots, \theta_s, \lambda_1, \cdots, \lambda_i - h_i, \cdots, \lambda_s)\,, \\ &\Delta_i^j \psi(\theta_1, \cdots, \theta_s, \lambda_1, \cdots, \lambda_s) \\ &\quad = \sum_{r=0}^{j} (-1)^r C_j^r \psi(\theta_1, \cdots, \theta_s, \lambda_1, \cdots, \lambda_i + (j-2r)h_i, \cdots, \lambda_s)\,. \end{aligned}$$

($C_j^r = j!/\{r!(j-r)!\}$ is the binomial coefficient.) Then

$$\begin{aligned}
&\Delta_1^{2j_1}\Delta_2^{2j_2}\cdots\Delta_s^{2j_s}\psi(\theta_1,\cdots,\theta_s,\lambda_1\cdots,\lambda_s)|_{\lambda=0} \\
&= \mathsf{M}\prod_{k=1}^{s}\sum_{r=0}^{2j_k}(-1)^r C_{2j_k}^r e^{i2(j_k-r)h_k\xi(\theta_k)} \\
&= \mathsf{M}\prod_{k=1}^{s}(e^{ih_k\xi(\theta_k)}-e^{-ih_k\xi(\theta_k)})^{2j_k} \\
&= \prod_{k=1}^{s}h_k^{2j_k}(2i)^{2\sum_{k=1}^{s}j_k}\mathsf{M}\prod_{k=1}^{s}\left(\frac{\sin h_k\xi(\theta_k)}{h_k\xi(\theta_k)}\right)^{2j_k}[\xi(\theta_k)]^{2j_k},
\end{aligned}$$

or

$$\prod_{k=1}^{s}\frac{\Delta^{2j_k}}{2h_k^{2j_k}}\psi = (-1)^{\sum_{k=1}^{s}j_k}\mathsf{M}\prod_{k=1}^{s}\left(\frac{\sin h_k\xi(\theta_k)}{h_k\xi(\theta_k)}\right)^{2j_k}[\xi(\theta_k)]^{2j_k}.$$

From this, by using Fatou's lemma (see Chapter II, Section 5), we get

$$\lim_{\substack{h_k\to 0\\ k=1,2,\cdots,s}}\left.\frac{(-1)^{\sum_{k=1}^{s}j_k}\prod_{k=1}^{s}\Delta_k^{2j}\psi}{\prod_{k=1}^{s}(2h_k)^{2j_k}}\right|_{\lambda=0} \geqq \mathsf{M}\prod_{k=1}^{s}[\xi(\theta_k)]^{2j_k}.$$

The expression on the left-hand side of this inequality coincides with the derivative

$$\frac{\partial^{\sum_{k=1}^{s}2j_k}\psi}{\partial\lambda_1^{2j_1},\cdots,\partial\lambda_s^{2j_s}}$$

at the point $\lambda = 0$ if ψ is $2\sum_{k=1}^{s} j_k$ times differentiable.

Thus we have:

Theorem 2. *If the characteristic function $\psi(\theta_1,\cdots,\theta_s\,\lambda,\cdots,\lambda_s)$ is p times differentiable, where p is an even integer, there exist moment functions of order $q \leqq p$ and they can be calculated from formula (2).*

In addition to moment functions, we often consider *central moment functions*

$$\begin{aligned}
&\mu_{j_1,\cdots,j_s}(\theta_1,\cdots,\theta_s) \\
&= \mathsf{M}([\xi(\theta_1)-m_1(\theta_1)]^{j_1}[\xi(\theta_2)-m_1(\theta_2)]^{j_2}\cdots[\xi(\theta_s)-m_1(\theta_s)]^{j_s}), \quad (3)
\end{aligned}$$

which are moment functions of a central random variable $\xi_1(\theta) = \xi(\theta) - m_1(\theta)$ with mathematical expectation equal to 0 for arbitrary $\theta \in \Theta$.

Among the moment functions of special significance are the functions of the first two orders:

$$m(\theta) = m_1(\theta) = \mathsf{M}\xi(\theta), \quad (4)$$

$$R(\theta_1, \theta_2) = \mu_{11}(\theta_1, \theta_2)$$
$$= \mathsf{M}([\xi(\theta_1) - m(\theta_1)][\xi(\theta_2) - m(\theta_2)]) \,. \qquad (5)$$

The function $m(\theta)$ is called the *mean value* and $R(\theta_1, \theta_2)$ is called the *covariance function*. For $\theta_1 = \theta_2 = \theta$, the covariance function gives the variance $\sigma^2(\theta)$ of the random variable $\xi(\theta)$; $R(\theta, \theta) = \sigma^2(\theta)$. For a stationary process ($\Theta = \mathfrak{T}$), it is obvious that

$$m(t) = m = \text{const} \,, \qquad (6)$$

$$R(t_1, t_2) = R(t_1 - t_2, 0) = R(t_1 - t_2) \,, \qquad (7)$$

that is, the covariance function depends only on the difference in the arguments. The function $R(t) = R(t_1 + t, t_1)$ is also called a covariance function of a stationary process. Of course if equations (6) and (7) are satisfied for some process, it does not follow that the process is stationary.

Still, we often encounter problems whose solution depends only on the values of the first two moments of a random function $\xi(t)$. For such problems, the condition that the process be stationary reduces to conditions (6) and (7). Therefore it is natural to consider the following important class of processes (introduced by A. Ya. Khinchin).

Definition 3. A random process $\xi(t)$ is said to be *stationary in the broad sense* if $\mathsf{M}\xi^2(t) < \infty$ and

$$\mathsf{M}\xi(t) = m = \text{const}; \qquad \mathsf{M}([\xi(t_1) - m][\xi(t_2) - m]) = R(t_1 - t_2) \,.$$

We note that for a stationary process in the broad sense the variance σ^2 of a random variable $\xi(t)$ is independent of t:

$$\sigma^2 = R(0) = \mathsf{M}[\xi(t) - m]^2 \,.$$

The quantity

$$r(\theta_1, \theta_2) = \frac{R(\theta_1, \theta_2)}{\sigma(\theta_1)\sigma(\theta_2)} = \frac{R(\theta_1, \theta_2)}{\sqrt{R(\theta_1, \theta_1)R(\theta_2, \theta_2)}}$$

is called the *coefficient of correlation* of the random variables $\xi(\theta_1)$ and $\xi(\theta_2)$.

If $\xi(\theta_1)$ and $\xi(\theta_2)$ are independent, the coefficient of correlation is 0. The converse is not generally true. However in the important particular case in which the random variables $\xi(\theta_1)$ and $\xi(\theta_2)$ have a joint normal distribution, the variables $\xi(\theta_1)$ and $\xi(\theta_2)$ are independent if the coefficient of correlation is 0 or, what amounts to the same thing, if the covariance function $R(\theta_1, \theta_2)$ is identically zero. In the general case, if two random variables ξ and η with finite second-order moments satisfy the condition $R_{\xi,\eta} = \mathsf{M}[(\xi - \mathsf{M}\xi)(\eta - \mathsf{M}\eta)] = 0$, they are said to be *uncorrelated.*

Analogously, we may say that in those branches of the theory that deal only with first- and second-order moments, the concept of uncorrelatedness of random variables replaces the concept of independence of random variables. The coefficient of correlation of a pair of random variables is a measure of linear dependence between them; that is, the coefficient of correlation shows with what accuracy one of the random variables can be linearly expressed in terms of the other. Let us clarify this. For the measure of error δ of the approximate equation $\xi \approx a\eta + b$ where a and b are real numbers, we take the quantity $\delta = \sqrt{\mathsf{M}[\xi - (a\eta + b)]^2}$. Then

$$\delta^2 = \mathsf{M}[(\xi - \mathsf{M}\xi) - a(\eta - \mathsf{M}\eta) + (\mathsf{M}\xi - a\mathsf{M}\eta - b)]^2 ,$$

so that

$$\begin{aligned}\delta^2 &= \mathsf{D}\xi + a^2\mathsf{D}\eta + (\mathsf{M}\xi - a\mathsf{M}\eta - b)^2 - 2aR_{\xi\eta} \\ &= \left(a\sigma_\eta - \frac{R_{\xi\eta}}{\sigma_\eta}\right)^2 + \sigma_\xi^2(1 - r_{\xi\eta}^2) + (\mathsf{M}\xi - a\mathsf{M}\eta - b)^2 ,\end{aligned}$$

where $\sigma_\xi^2 = \mathsf{D}\xi$ and $\sigma_\eta^2 = \mathsf{D}\eta$ are the variances of the variables ξ and η respectively. This expression attains its minimum when

$$a = \frac{R_{\xi\eta}}{\sigma_\eta^2} = \frac{\sigma_\xi}{\sigma_\eta} r_{\xi\eta} , \qquad b = m_\xi - am_\eta = m_\xi - \frac{\sigma_\xi}{\sigma_\eta} r_{\xi\eta} m_\eta ,$$

and this minimum is equal to $\min \delta^2 = \sigma_\xi^2(1 - r_{\xi\eta}^2)$. Thus the greater in absolute value the coefficient of correlation between two random variables, the greater the accuracy with which one of them can be represented as a linear function of the other.

We often consider complex-valued random functions $\zeta(\theta)$, which can be represented in the form $\zeta(\theta) = \xi(\theta) + i\eta(\theta)$; we can also regard them as two-dimensional vector-valued random functions.

For a complex-valued function, the relation $\zeta(\theta) \in L_2(\theta)$ means that $\mathsf{M} \mid \zeta(\theta) \mid^2 < \infty$, where $\theta \in \Theta$, that is, $\xi(\theta) \in L_2(\Theta)$ and $\eta(\theta) \in L_2(\Theta)$.

The covariance function of a complex random function is defined by the equation $R(\theta_1, \theta_2) = \mathsf{M}([\zeta(\theta_1) - \mathsf{M}\zeta(\theta_1)]\overline{[\zeta(\theta_2) - \mathsf{M}\zeta(\theta_2)]}$, where the vinculum denotes the complex conjugate.

Let us note certain properties of covariance functions:

1. $R(\theta, \theta) \geqq 0$ with equality holding if and only if $\zeta(\theta)$ is constant with probability 1;

2. $$R(\theta_1, \theta_2) = \overline{R(\theta_2, \theta_1)} ; \tag{8}$$

3. $$\mid R(\theta_1, \theta_2) \mid^2 \leqq R(\theta_1, \theta_1)R(\theta_2, \theta_2) ; \tag{9}$$

4. For every n, $\theta_1, \theta_2, \cdots, \theta_n$ and complex numbers $\lambda_1, \lambda_2, \cdots, \lambda_n$,

$$\sum_{j,k=1}^{n} R(\theta_j, \theta_k)\lambda_j\bar{\lambda}_k \geqq 0 . \tag{10}$$

The first two assertions are obvious. The third is obtained as a consequence of the Cauchy-Schwarz inequality

$$(\mathsf{M}\,|\,\xi\eta\,|)^2 \leqq \mathsf{M}\,|\,\xi\,|^2\mathsf{M}\,|\,\eta\,|^2 .$$

To prove 4, it suffices to note that

$$\sum_{j,k=1}^{n} R(\theta_j, \theta_k)\lambda_j\bar{\lambda}_k = \mathsf{M}\sum_{j,k=1}^{n} \zeta(\theta_j)\overline{\zeta(\theta_k)}\lambda_j\bar{\lambda}_k = \mathsf{M}\left|\sum_{j=1}^{n} \zeta(\theta_j)\lambda_j\right|^2 \geqq 0 .$$

We note that properties 1, 2, and 3 follow from property 4.

A function $R(\theta_1, \theta_2)$ that satisfies property 4 is called the *non-negative-definite kernel* on Θ. A complex-valued process $\zeta(t)$ is said to be *stationary in the broad sense* if $m(t) = \mathsf{M}\zeta(t) = \text{const}$, $R(t_1, t_2) = R(t_1 - t_2)$. For processes that are stationary in the broad sense, properties 1 to 4 of a correlation function take the forms

1′. $$R(0) \geqq 0 ,$$

2′. $$\overline{R(t)} = R(-t) , \tag{11}$$

3′. $$|\,R(t)\,| \leqq R(0) , \tag{12}$$

4′. $$\sum_{j,k=1}^{n} R(t_j - t_k)\lambda_j\bar{\lambda}_k \geqq 0 . \tag{13}$$

Let $\zeta_1(\theta)$ and $\zeta_2(\theta)$ denote two random functions belonging to $L_2(\Theta)$. To characterize the degree of linear dependence between two such functions, we introduce the joint covariance function.

Definition 4. The *joint covariance function* of two random functions $\zeta_1(\theta)$ and $\zeta_2(\theta)$ in $L_2(\Theta)$ is defined as the quantity

$$R_{\zeta_1\zeta_2}(\theta_1, \theta_2) = \mathsf{M}([\zeta_1(\theta_1) - \mathsf{M}\zeta_1(\theta_1)]\overline{[\zeta_2(\theta_2) - \mathsf{M}\zeta_2(\theta_2)]}) .$$

Suppose that Θ is an interval of the real axis $\mathfrak{T}$. Then two processes $\zeta_1(t)$ and $\zeta_2(t)$ are said to be *stationarily connected* if each is stationary in the broad sense and $R_{\zeta_1\zeta_2}(\theta_1, \theta_2) = R(\theta_1 - \theta_2)$. Suppose that we are given a sequence of complex-valued random functions $\zeta_1(\theta), \zeta_2(\theta), \cdots, \zeta_r(\theta)$, $\zeta_i(\theta) \in L_2(\Theta)$, $i = 1, 2, \cdots, r$. Let us agree to treat this sequence as a single r-dimensional complex-valued random function $\boldsymbol{\zeta}(\theta) = \{\zeta_1(\theta), \zeta_2(\theta), \cdots, \zeta_r(\theta)\}$, $\theta \in \Theta$. If $\boldsymbol{\xi}$ and $\boldsymbol{\eta}$ are two vectors $\boldsymbol{\xi} = (\xi_1, \xi_2, \cdots, \xi_r)$, $\boldsymbol{\eta} = (\eta_1, \eta_2 \cdots, \eta_r)$, we shall let $\boldsymbol{\xi\eta}^*$ denote the matrix

$$\boldsymbol{\zeta\eta}^* = \begin{pmatrix} \xi_1\bar{\eta}_1 & \xi_1\bar{\eta}_2 & \cdots & \xi_1\bar{\eta}_r \\ \xi_2\bar{\eta}_1 & \xi_2\bar{\eta}_2 & \cdots & \xi_2\bar{\eta}_r \\ \cdots & \cdots & \cdots & \cdots \\ \xi_r\bar{\eta}_1 & \xi_r\bar{\eta}_2 & \cdots & \xi_r\bar{\eta}_r \end{pmatrix} = (\xi_i\bar{\eta}_i) .$$

We set

$$m(\theta) = \mathsf{M}\boldsymbol{\zeta}(\theta) = \{\mathsf{M}\zeta_1(\theta), \mathsf{M}\zeta_2(\theta), \cdots, \mathsf{M}\zeta_r(\theta)\} ,$$

$$\begin{aligned} R(\theta_1, \theta_2) &= (R_{ij}(\theta_1, \theta_2)) = \mathsf{M}([\boldsymbol{\zeta}(\theta_1) - m(\theta_1)][\boldsymbol{\zeta}(\theta_2) - m(\theta_2)]^*) \\ &= (\mathsf{M}\{[\zeta_i(\theta_1) - m_i(\theta_1)]\overline{[\zeta_j(\theta_2) - m_j(\theta_2)]}\}) , \qquad i, j = 1, \cdots, r . \end{aligned}$$

The function $m(\theta)$ is an r-dimensional complex vector-valued function. It is called the *mean value* of the vector-valued random function $\boldsymbol{\zeta}(\theta)$. The matrix $R(\theta_1, \theta_2)$ is called the *covariance matrix* of $\boldsymbol{\zeta}(\theta)$. If $\Theta = \mathfrak{T}$ and $m(t) = m = \text{const}$, $R(t_1, t_2) = R(t_1 - t_2)$, the process $\boldsymbol{\zeta}(t)$ is said to be *stationary in the broad sense.*

Corresponding to properties 1 to 4 of covariance functions are the following properties of the covariance matrix of a random function:

1. $R(\theta, \theta)$ is a nonnegative-definite matrix

$$\sum_{j,k=1}^{n} R_{jk}(\theta, \theta)\lambda_j\bar{\lambda}_k = \mathsf{M}\left|\sum_{j=1}^{r} \lambda_j\zeta_j(\theta)\right|^2 \geqq 0 ; \tag{14}$$

2. $$R(\theta_1, \theta_2)^* = R(\theta_2, \theta_1) , \tag{15}$$

where the asterisk denotes the complex conjugate matrix;

3. $$| R_{jk}(\theta_1, \theta_2) |^2 \leqq R_{jj}(\theta_1, \theta_1)R_{kk}(\theta_2, \theta_2) , \quad j, k = 1, \cdots, r ; \tag{16}$$

4. For arbitrary $n, \theta_1, \cdots, \theta_n$ and a sequence of complex vectors $A_1, A_2, \cdots, A_n$,

$$\sum_{j,k=1}^{n} (R(\theta_j, \theta_k)A_k, A_j) \geqq 0 . \tag{17}$$

This last condition is equivalent to:

4′. For an arbitrary sequence of matrices $\Lambda_1, \cdots, \Lambda_n$, the matrix $\sum_{j,k=1}^{n} \Lambda_j R(\theta_j, \theta_k)\Lambda_k^*$ is nonnegative-definite.

Properties 1 and 2 are obvious. To prove property 3, let us use the Cauchy-Schwarz inequality for the mathematical expectation:

$$\begin{aligned} &| R_{jk}(\theta_1, \theta_2) |^2 \\ &= | \mathsf{M}[(\zeta_j(\theta_1) - m_j(\theta_1))\overline{(\zeta_k(\theta_2) - m_k(\theta_2))}] |^2 \leqq R_{jj}(\theta_1, \theta_1)R_{kk}(\theta_2, \theta_2) . \end{aligned}$$

To prove property 4, let us set $A_k = (a_{k1}, \cdots, a_{kr})$. Then

$$\begin{aligned} \sum_{j,k=1}^{n} (R(\theta_j, \theta_k)A_k, A_j) &= \sum_{j,k=1}^{n} \sum_{p,q=1}^{r} R_{pq}(\theta_j, \theta_k)a_{kq}\bar{a}_{jp} \\ &= \mathsf{M}\left|\sum_{j=1}^{n} \sum_{p=1}^{r} (\zeta_p(\theta_j) - m_p(\theta_j))\bar{a}_{jp}\right|^2 \geqq 0 . \end{aligned}$$

In conclusion let us look at some modifications in the preceding definitions.

Let us agree to call a process $\zeta(t)$, for $t \in \mathfrak{T}$ *a process belonging to the class* L_2^Δ or $L_2^\Delta(\mathfrak{T})$ if for arbitrary $t_1, t_2 \in \mathfrak{T}$, $\mathsf{M}\,|\boldsymbol{\zeta}(t_2) - \boldsymbol{\zeta}(t_1)|^2 < \infty$, where $|\boldsymbol{\zeta}(t)|$ denotes the norm of the vector. For processes belonging to the class L_2^Δ, we introduce the vector-valued function

$$\mathsf{M}\{\boldsymbol{\zeta}(t_2) - \boldsymbol{\zeta}(t_1)\} = \boldsymbol{m}(t_1, t_2) = \{m_1(t_1, t_2), m_2(t_1, t_2), \cdots, m_r(t_1, t_2)\} ,$$

which we call the *mean value* of the increment of the process, and we introduce the matrix

$$\begin{aligned} \boldsymbol{D}&(t_1, t_2, t_3, t_4) \\ &= (D_{jk}(t_1, t_2, t_3, t_4)) \\ &= (\mathsf{M}\{[\zeta_j(t_2) - \zeta_j(t_1) - m_j(t_1, t_2)]\overline{[\zeta_k(t_4) - \zeta_k(t_3) - m_k(t_3, t_4)]}\}) \\ &\quad (j, k = 1, \cdots, r) \end{aligned}$$

which we call the *structural matrix* of the process $\boldsymbol{\zeta}(t)$. If the functions $\boldsymbol{m}(t_1, t_2)$ and $\boldsymbol{D}(t_1, t_2, t_3, t_4)$ are independent of the displacement of the arguments, that is, if

$$\boldsymbol{m}(t_1 + h, t_2 + h) = \boldsymbol{m}(t_1, t_2) ,$$

$$\boldsymbol{D}(t_1 + h, t_2 + h, t_3 + h, t_4 + h) = \boldsymbol{D}(t_1, t_2, t_3, t_4) ,$$

where $t_i \in \mathfrak{T}$, $t_i + h \in \mathfrak{T}$, $i = 1, \cdots, 4$, then the $\boldsymbol{\zeta}(t)$ is called a *process with stationary increments in the broad sense*.

For a process with stationary increments, $\boldsymbol{m}(h) = \boldsymbol{m}(t, t + h)$ is an additive function of h. If we make the additional requirement that the function $\boldsymbol{m}(h)$ be continuous or bounded on some interval, it then follows that $\boldsymbol{m}(h)$ is linear, that is, $\boldsymbol{m}(h) = (m_1 h, m_2 h, \cdots, m_r h)$.

For real processes the structural function $\boldsymbol{D}(t_1, t_2, t_3, t_4)$ can be expressed in terms of the simpler functions $\boldsymbol{D}(t_1, t_2) = \boldsymbol{D}(t_1, t_2, t_1, t_2)$, which is also called the structural function of the process. Indeed,

$$\begin{aligned} \boldsymbol{D}(t_1, t_4) &= \mathsf{M}([\boldsymbol{\zeta}(t_4) - \boldsymbol{\zeta}(t_3) - \boldsymbol{m}(t_3, t_4) + \boldsymbol{\zeta}(t_3) \\ &\quad - \boldsymbol{\zeta}(t_1) - \boldsymbol{m}(t_1, t_3)][\boldsymbol{\zeta}(t_4) - \boldsymbol{\zeta}(t_3) - \boldsymbol{m}(t_3, t_4) + \boldsymbol{\zeta}(t_3) \\ &\quad - \boldsymbol{\zeta}(t_1) - \boldsymbol{m}(t_1, t_3)]^*) \\ &= \boldsymbol{D}(t_3, t_4) + 2\boldsymbol{D}(t_1, t_3, t_3, t_4) + \boldsymbol{D}(t_1, t_3) , \end{aligned}$$

so that

$$\boldsymbol{D}(t_1, t_3, t_3, t_4) = \frac{1}{2}[\boldsymbol{D}(t_1, t_4) - \boldsymbol{D}(t_1, t_3) - \boldsymbol{D}(t_3, t_4)] .$$

Furthermore,

$$\boldsymbol{D}(t_1, t_2, t_3, t_4) = \boldsymbol{D}(t_1, t_2, t_2, t_4) - \boldsymbol{D}(t_1, t_2, t_2, t_3) .$$

The last two formulas together express the function $\boldsymbol{D}(t_1, t_2, t_3, t_4)$ in terms of the $\boldsymbol{D}(t_1, t_2)$.

3. GAUSSIAN RANDOM FUNCTIONS

In many practical problems an important role is played by random functions for which the family of joint distributions defining a random function consists of Gaussian (normal) distributions. First we shall give the definition and basic properties of a multi-dimensional Gaussian distribution.

Definition 1. A random vector $\boldsymbol{\xi} = (\xi_1, \xi_2, \cdots, \xi_n)$ is said to have a *Gaussian* (*normal*) *distribution* if the characteristic function can be written in the form

$$\psi(t_1, t_2, \cdots, t_n) = \mathsf{M} \exp\{i(\boldsymbol{t}, \boldsymbol{\xi})\} = \exp\left(i(\boldsymbol{m}, \boldsymbol{t}) - \frac{1}{2}(\Lambda \boldsymbol{t}, \boldsymbol{t})\right), \quad (1)$$

where $\boldsymbol{m} = (m_1, m_2, \cdots, m_n)$ and $\boldsymbol{t} = (t_1, t_2, \cdots, t_n)$ are vectors and $\boldsymbol{\Lambda} = (\lambda_{ik})$ for $i, k = 1, \cdots, n$, is a nonnegative-definite real symmetric matrix. Here $(\boldsymbol{\alpha}, \boldsymbol{\beta})$ denotes the scalar product of the vectors $\boldsymbol{\alpha}$ and $\boldsymbol{\beta}$, so that

$$(\boldsymbol{m}, \boldsymbol{t}) = \sum_{k=1}^{n} m_k t_k, \qquad (\boldsymbol{\Lambda t}, \boldsymbol{t}) = \sum_{j,k=1}^{n} \lambda_{jk} t_j t_k .$$

The following theorem serves as a formal justification of the definition that we have just given.

Theorem 1. *For a function* $\psi(\boldsymbol{t}) = \exp[i(\boldsymbol{m}, \boldsymbol{t}) - 1/2(\boldsymbol{\Lambda t}, \boldsymbol{t})]$ *to be the characteristic function of an n-dimensional random vector* $\boldsymbol{\xi}$, *it is necessary and sufficient that the real matrix* $\boldsymbol{\Lambda}$ *be nonnegative-definite and symmetric. The rank of the matrix* $\boldsymbol{\Lambda}$ *is equal to the dimension of the subspace in which the distribution of the vector* $\boldsymbol{\xi}$ *can be concentrated.*

Proof of the Necessity. Suppose that the characteristic function $\psi(\boldsymbol{t})$ of a random vector is given by formula (1). If we differentiate it first with respect to t_j and then with respect to t_k and then set $\boldsymbol{t} = 0$, we see that the distribution has finite moments (Theorem 2, Section 2) and that

$$\left.\frac{\partial \psi}{\partial t_j}\right|_{\boldsymbol{t}=0} = i\mathsf{M}\xi_j = im_j , \quad (2)$$

$$\left.\frac{\partial^2 \psi}{\partial t_j \partial t_k}\right|_{\boldsymbol{t}=0} = -\mathsf{M}\xi_j\xi_k = -m_j m_k - \lambda_{ij} . \quad (3)$$

It follows from these formulas that the matrix $\boldsymbol{\Lambda}$ is real, symmetric, and nonnegative-definite:

$$(\boldsymbol{\Lambda t}, \boldsymbol{t}) = \mathsf{M}\left(\sum_{j=1}^{n} (\xi_j - m_j) t_j\right)^2 = D(\boldsymbol{\xi}, \boldsymbol{t}) \geqq 0 . \quad (4)$$

If the rank of the matrix Λ is equal to $r(\leq n)$, then by making a suitable change of variables $t_j = \sum_{k=1}^{n} \alpha_{jk}\tau_k$ we can reduce the quadratic form to principal axes:

$$(\Lambda t, t) = \sum_{k=1}^{r} \lambda_k \tau_k^2 = \mathsf{M}\left[\sum_{k=1}^{u} \sum_{j=1}^{n} (\xi_j - m_j)\alpha_{jk}\tau_k\right]^2 .$$

Thus $\sum_{j=1}^{n} (\xi_j - m_j)\alpha_{jk} = 0$ for $k = r + 1, \cdots, n$ with probability 1. These relations show that there exist with probability 1, $n - r$ linearly independent relationships among the components of the vector $\boldsymbol{\xi}$ and hence that its distribution is concentrated in the r-dimensional hyperplane defined by $\sum_{j=1}^{n} (x_j - m_j)\alpha_{jk} = 0, k = r + 1, \cdots, n$.

Proof of the Sufficiency. Let us suppose that Λ is a positive-definite symmetric matrix. The function

$$\psi(t) = \exp\left\{i(m, t) - \frac{1}{2}(\Lambda t, t)\right\}$$

is absolutely integrable and differentiable. Consequently we can apply Fourier's integral formula to it:

$$\psi(t) = \int_{-\infty}^{\infty} f(x)\, e^{i(x, t)} dx\,, \qquad f(x) = \frac{1}{(2\pi)^n}\int_{-\infty}^{\infty} \psi(t) e^{-i(t, x)} dt\,. \quad (5)$$

These integrals are n-dimensional, and dx and dt denote n-dimensional elements of volume. Let C denote an orthogonal matrix that reduces Λ to diagonal form, so that $C^*\Lambda C = D$, where $D = (\lambda_i\delta_{ik})$ for $i, k = 1, \cdots, n$ and $\lambda_i > 0$, and where C^* is the adjoint of C (Here, we note that C is real and orthogonal and thus C^* coincides both with the transpose and with the inverse of C (that is, $C^* = C' = C^{-1}$).) Let us make a change of variables of integration by setting

$$t = Cu \text{ or } u = C^*t, \text{ where } u = (u_1, u_2, \cdots, u_n)\,.$$

Since the element of volume is not changed under an orthogonal transformation, it follows that

$$f(x) = \frac{1}{(2\pi)^n}\int_{-\infty}^{\infty} \exp\left\{-i(x - m, Cu) - \frac{1}{2}(\Lambda Cu, Cu)\right\} du\,.$$

We have

$$(\Lambda Cu, Cu) = (C^*\Lambda Cu, u) = \sum_{k=1}^{n} \lambda_k u_k^2,$$

$$(x - m, Cu) = (C^*(x - m), u) = \sum_{k=1}^{n} x_k^* u_k\,,$$

where x_k^* is the kth component of the vector $x^* = C^*(x - m)$. Therefore,

$$f(x) = \frac{1}{(2\pi)^n}\int_{-\infty}^{\infty} \exp\left\{-i\sum_{k=2}^{n} x_k^* u_k - \frac{1}{2}\sum_{n=1}^{n} \lambda_k u_k^2\right\} du$$

$$= \prod_{k=1}^{n} \frac{1}{2\pi}\int_{-\infty}^{\infty} \exp\left\{-ix_k^* u_k - \frac{1}{2}\lambda_k u_k^2\right\} du_k$$

$$\times \prod_{k=1}^{n} \frac{1}{\sqrt{2\pi\lambda_k}} e^{-x_k^{*2}/2\lambda_k} = (2\pi)^{-\pi/2}\left(\prod_{k=1}^{n} \lambda_k\right)^{-1/2} e^{-1/2(D^{-1}x^*, x^*)} .$$

Furthermore, $\prod_{k=1}^{n} \lambda_k = \Delta$, where Δ is the determinant of the matrix $\mathbf{\Lambda}$, and

$$\begin{aligned}(\mathbf{D}^{-1}\mathbf{x}^*, \mathbf{x}^*) &= (\mathbf{D}^{-1}\mathbf{C}^*(\mathbf{x} - \mathbf{m}), \mathbf{C}^*(\mathbf{x} - \mathbf{m}))\\ &= (\mathbf{C}\mathbf{D}^{-1}\mathbf{C}^*(\mathbf{x} - \mathbf{m}), (\mathbf{x} - \mathbf{m}))\\ &= ((\mathbf{C}\mathbf{D}\mathbf{C}^*)^{-1}(\mathbf{x} - \mathbf{m}), (\mathbf{x} - \mathbf{m}))\\ &= \mathbf{\Lambda}^{-1}(\mathbf{x} - \mathbf{m}), (\mathbf{x} - \mathbf{m})) ,\end{aligned}$$

where $\mathbf{\Lambda}^{-1}$ is the inverse of the matrix $\mathbf{\Lambda}$. Finally we obtain

$$(\mathbf{x})f = \frac{1}{\sqrt{(2\pi)^n\Delta}} \exp\left\{-\frac{1}{2}(\mathbf{\Lambda}^{-1}(\mathbf{x} - \mathbf{m}), (\mathbf{x} - \mathbf{m})\right\}$$

$$= \frac{1}{\sqrt{(2\pi)^n\Delta}} \exp\left\{-\frac{1}{2}\sum_{j,k=1}^{n} \frac{\Delta_{kj}(x_j - m_j)(x_k - m_k)}{\Delta}\right\}, \quad (6)$$

where the Δ_{kj} are the cofactors of the elements of the matrix $\mathbf{\Lambda}$. It follows from (6) that $f(\mathbf{x}) \geqq 0$ and it follows from (5) that

$$\int_{-\infty}^{\infty} f(\mathbf{x})d\mathbf{x} = \psi(0) = 1 .$$

Thus the function $f(\mathbf{x})$ can be regarded as an n-dimensional distribution density, and $\psi(\mathbf{t})$ is its characteristic function.

Turning to the general case, let us suppose that the matrix $\mathbf{\Lambda}$ is a matrix of rank r (where $r < n$) and that $\mathbf{C}$ is an orthogonal transformation that reduces $\mathbf{\Lambda}$ to diagonal form: $\mathbf{C}^*\mathbf{\Lambda}\mathbf{C} = \mathbf{D}_r$, where $\mathbf{D}_r$ is a diagonal matrix whose diagonal elements λ_k are zero for $k = r + 1, \cdots, n$ and positive for $k = 1, 2, \cdots, r$. Suppose that $\lambda_j^\varepsilon = \lambda_j$ for $j = 1, \cdots, r$ but that $\lambda_j^\varepsilon = \varepsilon$ for $j = r + 1, \cdots, n$. Then, $\mathbf{\Lambda}_\varepsilon = \mathbf{C}\mathbf{D}_\varepsilon\mathbf{C}^*$ is a positive-definite matrix and

$$\psi_\varepsilon(\mathbf{t}) = \exp\left\{i(\mathbf{m}, \mathbf{t}) - \frac{1}{2}(\mathbf{\Lambda}_\varepsilon\mathbf{t}, \mathbf{t})\right\}$$

is the characteristic function of some distribution. As $\varepsilon \longrightarrow 0$, the function $\psi_\varepsilon(\mathbf{t})$ converges uniformly to $\psi(\mathbf{t})$. Hence $\psi(\mathbf{t})$ is the characteristic function of some distribution. As shown above, this distribution is concentrated in an r-dimensional hyperplane, so that it has no density. Such a distribution is called an *improper Gaussian distribution.*

Corollary 1. *In the expression* (1) *for the characteristic function of a Gaussian distribution,* $\boldsymbol{m} = (m_1, m_2, \cdots, m_n)$ *is the vector of the mathematical expectation and* $\boldsymbol{\Lambda}$ *is the covariance matrix*:

$$\boldsymbol{m} = \mathsf{M}\boldsymbol{\xi}\,, \qquad \lambda_{jk} = \mathsf{M}[(\xi_j - m_j)(\xi_k - m_k)]\,.$$

This corollary follows immediately from formulas (2) and (3).

Corollary 2. *If the covariance matrix* $\boldsymbol{\Lambda}$ *of a Gaussian random vector* $\boldsymbol{x}$ *is nondegenerate, there exists an n-dimensional distribution density* $f(\boldsymbol{x})$ *defined by formula* (6).

Corollary 3. *The joint distribution of an arbitrary group of components of a Gaussian random vector is Gaussian.*

Theorem 2. *If a random vector* $\boldsymbol{\xi} = (\xi_1, \xi_2, \cdots, \xi_n)$ *has a Gaussian distribution and if the random vectors* $\boldsymbol{\xi}' = (\xi_1, \cdots, \xi_r)$ *and*

$$\boldsymbol{\xi}'' = (\xi_{r+1}, \cdots, \xi_n)$$

(*for* $r < n$) *are uncorrelated, then the vectors* $\boldsymbol{\xi}'$ *and* $\boldsymbol{\xi}''$ *are independent.*

Proof. The fact that $\boldsymbol{\xi}'$ and $\boldsymbol{\xi}''$ are uncorrelated implies that $\mathsf{M}\xi_i\xi_j - \mathsf{M}\xi_i\mathsf{M}\xi_j = 0$, $i = 1, \cdots, r$, $j = r + 1, \cdots, n$. Therefore

$$\psi(\boldsymbol{t}) = \exp\left\{i(\boldsymbol{m}', \boldsymbol{t}') + i(\boldsymbol{m}'', \boldsymbol{t}'') - \frac{1}{2}\sum_{j,k=1}^{r}\lambda_{jk}t_jt_k - \frac{1}{2}\sum_{j,k=r+1}^{n}\lambda_{jk}t_jt_k\right\},$$

where

$$\boldsymbol{m}' = (m_1, m_2, \cdots, m_r),\ \boldsymbol{m}'' = (m_{r+1}, \cdots, m_n)\,,$$
$$\boldsymbol{t}' = (t_1, \cdots, t_r),\ \boldsymbol{t}'' = (t_{r+1}, \cdots, t_n)\,.$$

The preceding formula can be rewritten in the form

$$\begin{aligned}\psi(\boldsymbol{t}) &= \mathsf{M}\exp\{i(\boldsymbol{t}', \boldsymbol{\xi}') + i(\boldsymbol{t}'', \boldsymbol{\xi}'')\}\\ &= \mathsf{M}\exp i(\boldsymbol{t}', \boldsymbol{\xi}')\mathsf{M}\exp i(\boldsymbol{t}'', \boldsymbol{\xi}'') = \psi'(\boldsymbol{t}')\psi''(\boldsymbol{t}'')\,,\end{aligned}$$

where $\psi'(\boldsymbol{t}')$ and $\psi''(\boldsymbol{t}'')$ are the characteristic functions of the vectors $\boldsymbol{\xi}'$ and $\boldsymbol{\xi}''$. This relation proves the independence of $\boldsymbol{\xi}'$ and $\boldsymbol{\xi}''$.

Let $\mathfrak{A} = \|a_{jk}\|$ (for $j = 1, \cdots, h$ and $k = 1, \cdots, n$) denote an arbitrary rectangular matrix, and set $\boldsymbol{\eta} = \mathfrak{A}\boldsymbol{\xi}$; that is

$$\boldsymbol{\eta} = \{\eta_1, \cdots, \eta_h\},\ \eta_j = \sum_{k=1}^{n} a_{jk}\xi_k\,, \qquad j = 1, \cdots, h\,.$$

The vector $\boldsymbol{\eta}$ is a *linear transform* of the vector $\boldsymbol{\xi}$.

Theorem 3. *Linear transformations of random vectors map Gaussian distributions into Gaussian distributions.*

Proof. Let $\psi_{\boldsymbol{\eta}}(t_1, \cdots, t_h)$ denote the characteristic function of the vector $\boldsymbol{\eta}$. Then

$$\begin{aligned}\psi_\eta(t_1, \cdots, t_h) &= \mathsf{M} \exp\left\{i \sum_{j=1}^{h} t_j \eta_j\right\} \\ &= \mathsf{M} \exp\left\{i \sum_{k=1}^{n} \left(\sum_{j=1}^{h} t_j a_{jk}\right) \xi_k\right\} \\ &= \exp\left\{i(\boldsymbol{t}, \mathfrak{A}\boldsymbol{m}) - \frac{1}{2}(\mathfrak{A}\boldsymbol{\Lambda}\mathfrak{A}'\boldsymbol{t}, \boldsymbol{t})\right\}, \qquad (7)\end{aligned}$$

that is, $\boldsymbol{\eta}$ has a Gaussian distribution with mathematical expectation $\mathfrak{A}\boldsymbol{m}$ and with variance-covariance matrix $\boldsymbol{\Lambda}_\eta = \mathfrak{A}\boldsymbol{\Lambda}\mathfrak{A}'$.

Theorem 4. *Let $\boldsymbol{\xi}^{(\alpha)}$ (where $\alpha = 1, 2, \cdots, r, \cdots$) denote a sequence of n-dimensional vectors having Gaussian distributions with parameters $(\boldsymbol{m}^{(\alpha)}, \boldsymbol{\Lambda}^{(\alpha)})$. The sequence of distributions of the vectors $\boldsymbol{\xi}^{(\alpha)}$ converges weakly (converges in distribution) to some limiting distribution if and only if*

$$\boldsymbol{m}^{(\alpha)} \longrightarrow \boldsymbol{m}, \qquad \boldsymbol{\Lambda}^{(\alpha)} \longrightarrow \boldsymbol{\Lambda}. \qquad (8)$$

Then the limiting distribution is also a Gaussian distribution with parameters $\boldsymbol{m}$ and $\boldsymbol{\Lambda}$.

For a sequence of distributions of random vectors $\boldsymbol{\xi}^{(\alpha)}$ to converge weakly to a limit, it is necessary and sufficient that the sequence of their characteristic functions $\psi^{(\alpha)}(\boldsymbol{t})$ converge to a continuous function. Let us consider the sequence $\{\ln \psi^{(\alpha)}(\boldsymbol{t})\}$, where $\ln \psi^{(\alpha)}(\boldsymbol{t}) = i(\boldsymbol{m}^{(\alpha)}, \boldsymbol{t}) - 1/2(\boldsymbol{\Lambda}^{(\alpha)}\boldsymbol{t}, \boldsymbol{t})$ in some neighborhood of the point $\boldsymbol{t} = 0$. For this sequence to converge it is necessary and sufficient that conditions (8) be satisfied. If conditions (8) are satisfied, then $\psi^{(\alpha)}(\boldsymbol{t}) \longrightarrow \psi(\boldsymbol{t}) = \exp\{i(\boldsymbol{m}, \boldsymbol{t}) - 1/2(\boldsymbol{\Lambda t}, \boldsymbol{t})\}$ for all $\boldsymbol{t}$; that is, a limiting distribution exists and is Gaussian.

Let us turn now to random functions. A real r-dimensional random function $\boldsymbol{\xi}(\theta) = \{\xi_1(\theta), \cdots, \xi_r(\theta)\}$ is said to be *Gaussian* if for every n the joint distribution of all components of the random vectors

$$\boldsymbol{\xi}(\theta_1), \boldsymbol{\xi}(\theta_2), \cdots, \boldsymbol{\xi}(\theta_n) \qquad (9)$$

are Gaussian.

The covariance matrix $\boldsymbol{R}$ of the joint distribution of a sequence of random vectors (9) is $rn \times rn$ and it can be partitioned into square $r \times r$ cells as follows:

$$\boldsymbol{R} = \begin{pmatrix} \boldsymbol{R}(\theta_1, \theta_1) & \boldsymbol{R}(\theta_1, \theta_2) & \cdots & \boldsymbol{R}(\theta_1, \theta_n) \\ \boldsymbol{R}(\theta_2, \theta_1) & \boldsymbol{R}(\theta_2, \theta_2) & \cdots & \boldsymbol{R}(\theta_2, \theta_n) \\ \cdots & \cdots & \cdots & \cdots \\ \boldsymbol{R}(\theta_n, \theta_1) & \boldsymbol{R}(\theta_n, \theta_2) & \cdots & \boldsymbol{R}(\theta_n, \theta_n) \end{pmatrix},$$

where $\boldsymbol{R}(\theta_1, \theta_2)$ is the covariance matrix of the function $\boldsymbol{\xi}(\theta)$. The matrix $\boldsymbol{R}$ is real and nonnegative-definite.

The converse is obvious. Specifically, for any real-vector-valued function $\boldsymbol{m}(\theta)$ and any real-vector-valued nonnegative-definite matrix function $\boldsymbol{R}(\theta_1, \theta_2)$, where $\theta_i \in \Theta$ (for $i = 1, 2$), there exists an r-dimensional Gaussian random function (in the broad sense) for which $\boldsymbol{m}(\theta)$ is the mathematical-expectation vector and $\boldsymbol{R}(\theta_1, \theta_2)$ is the covariance matrix.

The moments of a Gaussian real valued random function can be obtained from the decomposition of the characteristic function. Confining ourselves to the case of central moments, we set $\boldsymbol{m}(\theta) = 0$. Then

$$\psi(\theta_1, \cdots, \theta_s, t_1, \cdots, t_s)$$
$$= e^{(-1/2)(\boldsymbol{\Lambda t}, \boldsymbol{t})} = 1 - \frac{1}{2}(\boldsymbol{\Lambda t}, \boldsymbol{t})$$
$$+ \frac{1}{2!\,2^2}(\boldsymbol{\Lambda t}, \boldsymbol{t})^2 - \cdots + (-1)^n \frac{1}{2^n n!}(\boldsymbol{\Lambda t}, \boldsymbol{t})^n + \cdots ,$$

where $\boldsymbol{\Lambda} = (R(\theta_j, \theta_k))$ for $j, k = 1, \cdots, s$. From this we obtain for an arbitrary moment function of odd order, $\mu_{j_1 \cdots j_s}(\theta_1, \cdots, \theta_s) = 0$ if $\sum_{k=1}^{s} j_k = 2n + 1$. For central moments of functions of even order,

$$\mu_{j_1 \cdots j_s}(\theta_1, \cdots, \theta_s) = \frac{\partial^{2n}}{\partial t_1^{j_1} \cdots \partial t_s^{j_s}} \frac{1}{2^n n!}(\boldsymbol{\Lambda t}, \boldsymbol{t})^n \bigg|_{t=0} , \qquad \sum_{k=1}^{s} j_k = 2n . \quad (10)$$

For example, for fourth-order moment functions we have the following formulas:

$$\mu_4(\theta) = 3R^2(\theta, \theta),\ \mu_{31}(\theta_1, \theta_2) = 3R(\theta_1, \theta_1)R(\theta_1, \theta_2) ,$$
$$\mu_{21}(\theta_1, \theta_2, \theta_3) = R(\theta_1, \theta_1)R(\theta_2, \theta_3) + 2R(\theta_1, \theta_2)R(\theta_1, \theta_3) ,$$
$$\mu_{1111}(\theta_1, \theta_2, \theta_3, \theta_4) = R(\theta_1, \theta_2)R(\theta_3, \theta_4)$$
$$+ R(\theta_1, \theta_3)R(\theta_2, \theta_4) + R(\theta_1, \theta_4)R(\theta_2, \theta_3) .$$

In the general case,

$$\mu_{j_1 \cdots j_s}(\theta_1, \cdots, \theta_s) = \Sigma\Pi R(\theta_p, \theta_q) , \quad (11)$$

the structure of which can be described as follows: We write the points $\theta_1, \cdots, \theta_s$ in order, where θ_k is repeated until it appears j_k times. We partition this sequence into arbitrary pairs. Then we take the product on the right side of formula (11) over all pairs of this partition, and we take the sum over all partitions (pairs that differ only by a permutation of the elements are considered as a single pair). The assertion follows immediately from formula (10).

The fact that Gaussian random functions play an important role in practical problems can be explained in part as follows: Under quite broad conditions, the sum of a large number of independent small (in absolute value) random functions is apporoximately a Gaussian random function, regardless of the probabilistic nature of the individual terms. This so-called theorem on the normal correlation is a multi-dimensional generalization of the central limit theorem. Here is one of its simpler formulations:

Theorem 5. *Let* $\{\eta_n\}$ *denote a sequence of sums of random functions* $\eta_n(\theta) = \sum_{k=1}^{m_n} \alpha_{nk}(\theta)$, $\theta \in \Theta$, $n = 1, 2, \cdots$. *Suppose that the following three conditions are satisfied:*

a. *For fixed n, the random variables* $\alpha_{n1}(\theta_1), \alpha_{n2}(\theta_2), \cdots, \alpha_{nm_n}(\theta_{m_n})$ *are mutually independent for arbitrary* $\theta_1, \theta_2, \cdots, \theta_{m_n}$, *possess second order moments, and* $\mathsf{M}\alpha_{nk}(\theta) = 0$, $\mathsf{M}\alpha_{nk}^2(\theta) = b_{nk}^2(\theta)$, $\max_k b_{nk}^2(\theta) \longrightarrow 0$ *as* $n \longrightarrow \infty$;

b. *The sequence of covariance functions* $R_n(\theta_1, \theta_2) = \mathsf{M}[\eta_n(\theta_1)\eta_n(\theta_2)]$ *converges as* $n \longrightarrow \infty$ *to some limit* $\lim_{n\to\infty} R_n(\theta_1, \theta_2) = R(\theta_1, \theta_2)$;

c. *For every* θ, *the sums* $\eta_n(\theta) = \sum_{k=1}^{m_n} \alpha_{nk}(\theta)$ *satisfy Lindeberg's condition: For arbitrary positive* τ,

$$\frac{1}{B_n^2} \sum_{k=1}^{m_n} \int_{|x|>\tau B_n} x^2 d\Pi_{nk}(\theta, x) \longrightarrow 0 ,$$

where $\Pi_{nk}(\theta, x)$ *is the distribution function of the random variable* $\alpha_{nk}(\theta)$, *and* $B_n^2 = \sum_{k=1}^{m_n} b_{nk}^2(\theta) = R_n(\theta, \theta)$. *Then the sequence* $\{\eta_n(\theta)\}$ *converges weakly as* $n \longrightarrow \infty$ *to a Gaussian random function with mathematical expectation zero and covariance function* $R(\theta_1, \theta_2)$.

Proof of this theorem reduces to one of the variants of the one-dimensional central limit theorem. We recall the necessary formulation:

Theorem 6. *Suppose that* $\zeta_n = \sum_{k=1}^{m_n} \xi_{nk}$, *for* $n = 1, 2, \cdots$, *where the random variables* ξ_{nk} *satisfy the following conditions:*

a. *For fixed n, the random variables* $\xi_{n1}, \xi_{n2}, \cdots, \xi_{nm_n}$ *are nondegenerate and mutually independent,* $\mathsf{M}\xi_{nk} = 0$ *and* $\mathsf{M}\xi_{nk}^2 = b_{nk}^2$;

b. *For arbitrary* $\tau > 0$, *Lindeberg's condition is satisfied:*

$$B_n^{-2} \sum_{k=1}^{m_n} \int_{|x|>\tau B_n} x^2 d\Pi_{nk}(x) \longrightarrow 0 \quad as \quad n \longrightarrow \infty ,$$

where $B_n^2 = \sum_{k=1}^{m_n} b_{nk}^2$, *and the* $\Pi_{nk}(x)$ *are distribution functions of the variables* ξ_{nk}.

Then the sequence $\{\zeta_n/B_n\}$ *of the distributions of the variable* ζ_n/B_n *converges as* $n \longrightarrow \infty$ *to a Gaussian distribution with parameters* 0 *and* 1 (cf. Gnedenko [1963], p. 306).

Consider the characteristic function

$$\psi_n(\theta_1, \cdots, \theta_s, tt_1, \cdots, tt_s) = \mathsf{M}e^{it\sum_{j=1}^{s} t_j\eta_n(\theta_j)} ,$$

which for $t = 1$ is the characteristic function of the joint distribution of the variables $\eta_n(\theta_1), \cdots, \eta_n(\theta_s)$, and for fixed $t_1, \cdots, t_s$ is the characteristic function of the random variable $\sum_{j=1}^{s} t_j\eta_n(\theta_j) = \zeta_n$. The quantity ζ_n can be written in the form

$$\zeta_n = \sum_{k=1}^{m_n} \beta_{nk}, \qquad \beta_{nk} = \sum_{j=1}^{s} t_j\alpha_{nk}(\theta_j), \qquad k = 1, \cdots, m_n .$$

Here,

$$\mathsf{M}\beta_{nk} = 0 , \ \tilde{b}_{nk}^2 = \mathsf{M}\beta_{nk}^2 = \sum_{j,p=1}^{s} t_jt_p\mathsf{M}[\alpha_{nk}(\theta_j)\alpha_{nk}(\theta_p)] ,$$

$$\tilde{B}_n^2 = \sum_{k=1}^{m_n} \tilde{b}_{nk}^2 = \sum_{j,p=1}^{s} t_jt_pR_n(\theta_j, \theta_p) .$$

From this we see that

$$\max_k \tilde{b}_{nk}^2 \leqq \sum_{j,p=1}^{s} t_jt_p \max_k b_{nk}(\theta_j) \max_k b_{nk}(\theta_p) \longrightarrow 0 \quad \text{as} \quad n \longrightarrow \infty ,$$

and $\tilde{B}_n^2 \longrightarrow \tilde{B}^2 = \sum_{j,p=1}^{s} t_jt_pR(\theta_j, \theta_p)$. If $\tilde{B}^2 = 0$, then $\{\zeta_n\}$ converges weakly to zero and $\psi_n(\theta_1, \cdots, \theta_s, t_1, \cdots, t_s) \longrightarrow 1$, which is a special case of the assertion of the theorem. For $\tilde{B}^2 > 0$ we verify the satisfaction of Lindeberg's conditions for the variables β_{nk}. In this case it is sufficient to show that

$$\sum_{k=1}^{m_n} \mathsf{M}\Big[g_\tau\Big(\sum_{j=1}^{s} t_j\alpha_{nk}(\theta_j)\Big)\Big(\sum_{j=1}^{s} t_j\alpha_{nk}(\theta_j)\Big)^2\Big] \longrightarrow 0 \quad \text{as} \quad n \longrightarrow \infty$$

for arbitrary $\tau > 0$, where $g_\tau(x) = 0$ for $|x| \leqq \tau$ and $g_\tau(x) = 1$ for $|x| > \tau$. If $t_p\alpha_{nk}(\theta_p)$ is the greatest in absolute value of the $t_j\alpha_{nk}(\theta_j)$ (for $j = 1, 2, \cdots, s$), then

$$g_\tau\Big(\sum_{j=1}^{s} t_j\alpha_{nk}(\theta_j)\Big)\Big(\sum_{j=1}^{s} t_j\alpha_{nk}(\theta_j)\Big)^2 \leqq g_{\tau/s}(t_p\alpha_{nk}(\theta_p))s^2(t_p\alpha_{nk}(\theta_p))^2 .$$

Therefore we always have

$$g_\tau\Big(\sum_{j=1}^{s} t_j\alpha_{nk}(\theta_j)\Big)\Big(\sum_{j=1}^{s} t_j\alpha_{nk}(\theta_j)\Big)^2 \leqq s^2 \sum_{j=1}^{s} g_{\tau/s}(t_j\alpha_{nk}(\theta_j))(t_j\alpha_{nk}(\theta_j))^2$$

and

$$\sum_{k=1}^{m_n} \mathsf{M}\Big[g_\tau\Big(\sum_{j=1}^{s} t_j\alpha_{nk}(\theta_j)\Big)\Big(\sum_{j=1}^{s} t_j\alpha_{nk}(\theta_j)\Big)^2\Big]$$
$$\leqq s^2 \sum_{j=1}^{s} t_j^2 \sum_{k=1}^{m_n} \int_{|x| \geqq \tau/s|t_j|} x^2 d\Pi_{nk}(\theta_j, x) \longrightarrow 0 \quad \text{as} \quad n \longrightarrow \infty .$$

Thus the central limit theorem is applicable to the quantities β_{nk}. By virtue of this theorem,

$$\psi_n(\theta_1, \cdots, \theta_s, tt_1, \cdots, tt_s) \longrightarrow e^{-(\tilde{B}^2t^2)/2} = e^{-t^2/2 \sum_{j,p=1}^{s} t_jt_pR(\theta_j,\theta_p)} .$$

Here, if we set $t = 1$, we see that the sequence of characteristic functions of the joint distributions of the quantities $\eta_n(\theta_j)$, for $j = 1, \cdots, s$, converges as $n \to \infty$ to the characteristic function of a Gaussian distribution. The continuity of the correspondence between the distributions and the characteristic functions implies the conclusion of the theorem.

A vector process $\boldsymbol{\xi}(t)$, for $t \in [0, T]$, is called a *process with independent increments* if, for arbitrary $t_1, t_2, \cdots, t_n$, where

$$0 < t_1 < t_2 < \cdots < t_n \leqq T ,$$

the random variables $\boldsymbol{\xi}(0), \boldsymbol{\xi}(t_1) - \boldsymbol{\xi}(0), \boldsymbol{\xi}(t_2) - \boldsymbol{\xi}(t_1), \cdots, \boldsymbol{\xi}(t_n) - \boldsymbol{\xi}(t_{n-1})$ are mutually independent. Chapter VI is devoted to the theory of these processes.

A Gaussian process with independent increments is called a *Wiener process or a process of Brownian motion.* If we observe through a high-powered microscope a small particle suspended in a liquid, we see that the particle is in constant motion and that its path is an extremely complicated broken line with chaotically directed links. This phenomenon is explained by the collisions of molecules of the liquid with the particle. The particle is considerably larger than the molecules of the liquid, and in a single second it experiences an enormous number of such collisions. It is impossible to follow the result of each collision of the particle with the molecules. The visible motion of the particle is called *Brownian motion.* As a first approximation we may assume that the displacements of the particle under the influence of collisions with the molecules of the medium are independent of each other and thus we may consider Brownian motion as a process with independent increments. Since each individual displacement is small, we may assume that the central limit theorem in probability theory is applicable to their sum, and we may treat Brownian motion as a Gaussian process.

Let us make some remarks regarding the structural function of a process of Brownian motion. If $t_1 < t_2 < t_3 < t_4$, we have, from the independence of the increments $\boldsymbol{\xi}(t)$, the obvious equation for the structural function $\boldsymbol{D}(t_1, t_2, t_3, t_4)$:

$$\boldsymbol{D}(t_1, t_2, t_3, t_4) = 0 .$$

Let us define $\boldsymbol{D}(s, t) = \boldsymbol{D}(s, t, s, t)$ and let us suppose that $s < u < t$. Then

$$\boldsymbol{D}(s, t) = \boldsymbol{D}(s, u) + \boldsymbol{D}(u, t) , \tag{12}$$

that is, the matrix $\boldsymbol{D}(s, t)$ is an additive function of the interval. If $\boldsymbol{\xi}(t)$ is a process with stationary increments, then $\boldsymbol{D}(s, t) = \boldsymbol{D}(t - s)$ and the preceding formula can be written

$$\boldsymbol{D}(s_1) + \boldsymbol{D}(s_2) = \boldsymbol{D}(s_1 + s_2) , \qquad s_i > 0 , \qquad i = 1, 2 . \tag{13}$$

If $\boldsymbol{D}(s)$ is continuous, it is easy to see that solutions of equation (13) are of the form

$$\boldsymbol{D}(s) = \boldsymbol{D}s , \tag{14}$$

where $\boldsymbol{D}$ is a nonnegative-definite matrix. Formula (14) completely describes the structural function of a process of Brownian motion with stationary increments. If the mathematical-expectation vector $\boldsymbol{m}(t) = \mathsf{M}(\boldsymbol{\xi}(t) - \boldsymbol{\xi}(0))$ is continuous with respect to t, then $\boldsymbol{m}(t) = \boldsymbol{m}t$. In the one-dimensional case, Brownian motion with stationary increments is completely determined by the distribution of the quantity $\xi(0)$ and the parameters m and σ, where $\sigma > 0$. Here,

$$\mathsf{M}[\xi(t) - \xi(0)] = mt$$

and $D[\xi(t) - \xi(0)] = \sigma^2 t$. It was just this process that N. Wiener first studied in detail.

4. OSCILLATIONS WITH RANDOM PARAMETERS

We now introduce some terminology that will be useful for a physical interpretation of random processes. If $\xi(t)$ denotes an electric current at an instant t and if we are interested in the energy dissipated by this current per unit of resistance, we have the following natural definitions. The quantity $\int_{t_1}^{t_2} \xi^2(t)dt$ is called the energy of the random process $\xi(t)$ in the course of an interval $[t_1, t_2]$. The intergral $\int_{-\infty}^{\infty} \xi^2(t)dt$, if it converges, is called the total energy of the process $\xi(t)$ (for $-\infty < t < \infty$). The limit

$$\lim_{T\to\infty} \frac{1}{2T} \int_{-T}^{T} \xi^2(t)dt$$

is called the mean power of the random process. If the process $\xi(t)$ is complex we should write $|\xi(t)|^2$ in place of $\xi^2(t)$ in the two preceding expressions.

In the case in which the process has a different physical interpretation, this terminology may prove contrary to the usual representations. Nonetheless, we shall use it in what follows.

In many questions random processes are simulated by the sum of harmonics with given frequencies and random amplitudes and phases. In other words, one considers processes of the form

$$\xi(t) = \sum_{k=1}^{n} \alpha_k \cos(u_k t + \varphi_k) \tag{1}$$

where the u_k are given numbers and the quantities α_k and φ_k are random. The probabilistic structure of this process is completely determined by the joint distribution of the random variables α_k and φ_k (for $k = 1, 2, \cdots, n$).*

If a random process $\xi(t)$ is the sum of n harmonic oscillations with amplitudes $|\alpha_k|$ and frequencies u_k the set of all frequencies $\{u_k\}$, for $k = 1, 2, \cdots, n$, considered as a set of points on the line $(-\infty < u < \infty)$, is called the *spectrum* of the random function $\xi(t)$.

In many cases it is convenient to consider complex-valued random processes of oscillatory nature,

$$\zeta(t) = \sum_{k=1}^{n} \gamma_k e^{iu_k t} , \tag{2}$$

where the complex amplitudes γ_k are random variables:

$$\gamma_k = \alpha_k + i\beta_k .$$

Here the α_k and the β_k, for $k = 1, 2, \cdots, n$, are real.

The process $\zeta(t)$ can be split into real and imaginary parts:

$$\zeta(t) = \xi(t) + i\eta(t) ,$$

where

$$\left.\begin{aligned} \xi(t) &= \sum_{k=1}^{n} \alpha_k \cos u_k t - \beta_k \sin u_k t , \\ \eta(t) &= \sum_{k=1}^{n} \alpha_k \sin u_k t + \beta_k \cos u_k t . \end{aligned}\right\} \tag{3}$$

If we set $\alpha_k = |\gamma_k| \cos \varphi_k$ and $\beta_k = |\gamma_k| \sin \varphi_k$, we obtain an expression for the functions $\xi(t)$ and $\eta(t)$ in the form

$$\xi(t) = \sum_{k=1}^{n} |\gamma_k| \cos (u_k t + \varphi_k) ,$$

$$\eta(t) = \sum_{k=1}^{n} |\gamma_k| \sin (u_k t + \varphi_k) . \tag{4}$$

It is easy to compute the mean power of the process $\zeta(t)$. We have

$$\frac{1}{2T}\int_{-T}^{T} |\zeta(t)|^2 dt = \frac{1}{2T}\int_{-T}^{T} \sum_{k,r=1}^{n} \gamma_k \gamma_r e^{it(u_k - u_r)} dt$$

$$= \sum_{k=1}^{n} |\gamma_k|^2 + \sum_{\substack{k,r=1 \\ k \neq r}}^{n} \gamma_k \gamma_r \frac{\sin T(u_k - u_r)}{T(u_k - u_r)} .$$

We obtain, with probability 1,

* In defining a random process by means of formula (1), we have in mind a random process whose finite-dimensional distributions can be calculated by beginning with formula (1) and the joint distribution of the quantities $\alpha_1, \cdots, \alpha_n, \varphi_1, \cdots, \varphi_n$.

$$\lim_{T\to\infty} \frac{1}{2T} \int_{-T}^{T} |\zeta(t)|^2 dt = \sum_{k=1}^{n} |\gamma_k|^2 .$$

Thus the mean power of the oscillatory random process $\zeta(t)$ is equal to the sum of the mean powers of the harmonic components of the process.

We compute in analogous fashion the mean value of the random function $\zeta(t)$ in an infinite interval of time. We have

$$\frac{1}{2T} \int_{-T}^{T} \zeta(t) dt = \sum_{k=1}^{n} \gamma_k \frac{\sin Tu_k}{Tu_k} .$$

The value of $(\sin u)/u$ is taken to be 1 at $u = 0$. Thus if 0 is a point of the spectrum, then

$$\lim_{T\to\infty} \frac{1}{2T} \int_{-T}^{T} \zeta(t) dt = \gamma_0 ,$$

where γ_0 is the amplitude corresponding to the frequency $u = 0$.

Even under special assumptions regarding the distribution of the variables γ_k, the distribution of the variable $\zeta(t)$ is extremely complicated. However, the simplest characteristics of the distribution of the random variable $\zeta(t)$ can be obtained without difficulty. Let us suppose that the complex amplitudes γ_k have mathematical expectations equal to 0 and that they are uncorrelated, that is, that

$$\mathsf{M}\gamma_k = 0, \ \mathsf{M}\gamma_k\gamma_r = 0 , \qquad k \neq r . \tag{5}$$

Then $\mathsf{M}\zeta(t) = 0$. We note that if 0 is not a point of the spectrum of a random process the mathematical expectation of the function $\zeta(t)$, that is, its mean value in the probabilistic sense, coincides with the mean value over the infinite time interval $(-\infty, \infty)$. On the other hand, if 0 is a point of the spectrum of the process, then the mean value of the sample function over the time is a random variable. We have the following expression for the covariance function of the process $\zeta(t)$:

$$R(t_1, t_2) = \mathsf{M}[\zeta(t_1)\overline{\zeta(t_2)}] = \mathsf{M}\Big[\sum_{k,r=1}^{n} \gamma_k \overline{\gamma}_r e^{i(u_k t_1 - u_r t_2)}\Big] = \sum_{k=1}^{n} c_k^2 e^{iu_k(t_1-t_2)} ,$$

where $c_k^2 = \mathsf{M}|\gamma_k|^2$. Thus the covariance function of the process $\zeta(t)$ depends only on the difference $t_1 - t_2$ of the arguments:

$$R(t_1, t_2) = R(t_1 - t_2) , \tag{6}$$

$$R(t) = \sum_{k=1}^{n} c_k^2 e^{iu_k t} . \tag{7}$$

Consequently, if the quantities γ_k in the process (2) are uncorrelated and if they have mean values equal to 0, then the process $\zeta(t)$ is a stationary process in the broad sense and its covariance function is given by formula (7). This formula is called the *spectral distribution* of the covariance function. It determines the spectrum of the random process (that is, the set of frequencies $\{u_k\}$, for $k = 1, 2, \cdots, n$, of the harmonic oscillations constituting the process $\zeta(t)$) and the mathematical expectations c_k^2 of the mean powers of the corresponding harmonic components of the process in a unit of time. The quantity c_k^2 can be called simply the mean value of the power of the harmonic component of the process with frequency u_k. It is obtained by first averaging the power over time and then averaging it in the probabilistic sense, that is, by taking the mathematical expectation. In connection with these energy representations we introduce the following important characteristic of a random process, the spectral function of the process.

The *spectral function* $F(u)$ of the process (2) is defined by

$$F(u) = \sum_{\substack{k \\ u_k < u}} c_k^2 \ .$$

This means that $F(u)$ is equal to the mean power of the harmonic components of the process $\zeta(t)$ (whose frequencies are less than a given quantity u) in a unit of time. It completely characterizes both the mean power of each harmonic component of the process $\zeta(t)$ and the overall mean power of the harmonic components of the process, whose frequencies lie in a given interval. Specifically,

$$c_k^2 = F(u_k + 0) - F(u_k) \ , \qquad \sum_{u_1 \leq u_k < u_2} c_k^2 = F(u_2) - F(u_1) \ .$$

By means of the spectral function, the covariance function of the process $\zeta(t)$ can be written in the form

$$R(t) = \int_{-\infty}^{\infty} e^{iut} dF(u) \ . \tag{8}$$

From a mathematical point of view, the spectral function is a nonnegative nondecreasing left-continuous function that is piecewise-constant with finitely many jumps of magnitudes c_k^2. We shall find that the concept of a spectral function can be introduced for arbitrary processes that are stationary in the broad sense. This question, like the question of generalizing the representation (8) to arbitrary stationary processes, is considered in Section 5.

Of particular interest are random processes that can be obtained from processes of the form

$$\zeta(t) = \sum_{k=1}^{n} \gamma_k e^{iu_k t} \tag{9}$$

by passage to the limit. The basis of this procedure is the fact that the number of terms in the sum (9) increases without bound as we decrease the complex amplitudes γ_k whereas the spectrum of the process (that is, the set of all frequencies u_k) fills the line $(-\infty, \infty)$ ever more densely. In passing to the limit, we can obtain a random process that has a continuous spectrum. The more precise meaning of this term and the question of the analogue of the representation (9) for the random processes obtained in this manner will be considered in Section 5 and later in Chapter V. Here we confine ourselves to a consideration of the limiting distributions.

Theorem 1. *Let* $\{\alpha_{nk}\}$ *and* $\{\beta_{nk}\}$ *(for* $k = 1, \cdots, n$ *and* $n = 1, 2, \cdots$*) denote two sequences of sequences of mutually independent (in each sequence) random variables*

$$\mathsf{M}\alpha_{nk} = \mathsf{M}\beta_{nk} = 0 , \qquad \mathsf{D}\alpha_{nk} = \mathsf{D}\beta_{nk} = b_{nk}^2 < \infty$$

and $\zeta_n(t) = \sum_{k=1}^n \gamma_{nk} e^{iu_{nk}t} = \xi_n(t) + i\eta_n(t)$, $\gamma_{nk} = \alpha_{nk} + i\beta_{nk}$. *Suppose that the following conditions are satisfied as* $n \longrightarrow \infty$:

a. *The sequence of the spectral functions* $F_n(u)$ *of the processes* $\xi_n(t)$ *converges on some everywhere-dense set of values on the line* $(-\infty, \infty)$ *to a bounded function* $F(u)$, *such that*

$$F(-\infty) = 0, \; F(+\infty) = \sigma^2 = \lim_{n\to\infty} 2 \sum_{k=1}^{n} b_{nk}^2 ,$$

b. *The random variables* α_{nk} *and* β_{nk} *satisfy Lindeberg's condition* (cf. Theorem 5 of Section 3).

Then the sequence of random processes $\zeta(t)$ *converges weakly to the stationary process* $\zeta(t)$, *where* $\zeta(t) = \xi(t) + i\eta(t)$. *The characteristic function of the joint distribution of the variables*

$$\xi(t_1), \cdots, \xi(t_s), \; \eta(t_1), \cdots, \eta(t_s) \tag{10}$$

is given by the expression

$$\psi(\lambda_1, \cdots, \lambda_s, \; \mu_1, \cdots, \mu_s) = \exp\left\{-\frac{1}{2}B^2\right\} . \tag{11}$$

where

$$B^2 = \sum_{j,k=1}^{s} R(t_j - t_k) z_j z_k, \; z_j = \lambda_j - i\mu_j , \qquad j = 1, \cdots, s , \tag{12}$$

$$R(t) = \int_{-\infty}^{\infty} e^{iut} dF(u) . \tag{13}$$

We note that under the conditions of this theorem we can use an arbitrary bounded nondecreasing function for the function $F(u)$. This theorem is a special case of Theorem 5 of Section 3. In this application we take Θ to be the set of points $\theta = (t, q)$, $-\infty < t < \infty$ and $q = 1, 2$, and we define $\alpha_{n_k}(\theta)$ by

$$\alpha_{nk}(t, 1) = \operatorname{Re} \gamma_{nk} e^{iu_{nk}t}, \quad \alpha_{nk}(t, 2) = \operatorname{Im} \gamma_{nk} e^{iu_{nk}t} .$$

It is easy to verify that if the α_{nk} and β_{nk} satisfy Lindeberg's condition, the variables $\alpha_{nk}(\theta)$ also satisfy it for arbitrary θ. Using the notation of Theorem 5 of Section 3, we have $R^{(n)}(\theta_1, \theta_2) = R^{(n)}_{q_1 q_2}(t_1, t_2)$ for $\theta_i = (t_i, q_i)$, where

$$R^{(n)}_{11}(t_1, t_2) = \mathsf{M}\xi_n(t_1)\xi_n(t_2) ,$$

$$R^{(n)}_{22}(t_1, t_2) = \mathsf{M}\eta_n(t_1)\eta_n(t_2), \quad R^{(n)}_{12}(t_1, t_2) = \mathsf{M}\xi_n(t_1)\eta_n(t_2) .$$

On the basis of formulas (3),

$$R^{(n)}_{11}(t_1, t_2) = R^{(n)}_{22}(t_1, t_2) = \frac{1}{2}\int_{-\infty}^{\infty} \cos (t_2 - t_1)u \, dF_n(u) ,$$

$$R^{(n)}_{12}(t_1, t_2) = \frac{1}{2}\int_{-\infty}^{\infty} \sin (t_2 - t_1)u \, dF_n(u) .$$

Keeping Helly's theorem in mind, by letting n approach ∞ we obtain from condition *a.*, the existence of the limits

$$R_{ij}(t) = \lim_{n\to\infty} R^{(n)}_{ij}(\tau, \tau + t), \qquad i, j = 1, 2 ,$$

$$R_{11}(t) = R_{22}(t) = \frac{1}{2}\int_{-\infty}^{\infty} \cos tu \, dF(u) ,$$

$$R_{12}(t) = \frac{1}{2}\int_{-\infty}^{\infty} \sin tu \, dF(u) .$$

Thus the conditions of Theorem 5 of Section 3 are satisfied. It follows from that theorem that the characteristic function of the joint distribution of the variables (10) is $\psi(\lambda_1, \cdots, \lambda_s, \mu_1, \cdots, \mu_s) = e^{-B^2/2}$, where

$$B^2 = \sum_{r,k=1}^{s} R_{11}(t_r - t_k)\lambda_k\lambda_r + 2R_{12}(t_r - t_k)\lambda_k\mu_r + R_{22}(t_r - t_k)\mu_k\mu_r$$

$$= \frac{1}{2}\sum_{k,r=1}^{s} R(t_r - t_k)z_r z_k, \quad z_r = \lambda_r - i\mu_r, \qquad r = 1, \cdots, s .$$

This completes the proof of the theorem.

5. THE SPECTRAL REPRESENTATIONS OF THE COVARIANCE FUNCTION OF A STATIONARY PROCESS AND OF THE STRUCTURAL FUNCTION OF A PROCESS WITH STATIONARY INCREMENTS

In the preceding section we constructed an example of a stationary process whose covariance function was given by the formula

$$R(t) = \int_{-\infty}^{\infty} e^{itu} dF(u) \tag{1}$$

where $F(u)$ was an arbitrary nonnegative bounded nondecreasing function that was everywhere continuous on the left.

We find that equation (1) applies for all processes that are stationary in the broad sense. Specifically, we have:

Theorem 1 (Khinchin). *For a function $R(t)$ (defined for $-\infty < t < \infty$) to be the covariance function of a process that is stationary in the broad sense and that satisfies the condition*

$$\mathsf{M}|\zeta(t+\tau)-\zeta(t)|^2 \to 0 \quad \text{as} \quad \tau \to 0\,, \tag{2}$$

it is necessary and sufficient that it have a representation of the form (1).

Proof of the Necessity. Let us set $\alpha^2 = \mathsf{M}|\zeta(0)|^2$. Then

$$\begin{aligned}|R(t+\tau)-R(t)| &= \mathsf{M}([\zeta(t+\tau)-\zeta(t)]\overline{\zeta(0)})|\\ &\leqq \alpha\{\mathsf{M}|\zeta(t+\tau)-\zeta(t)|^2\}^{1/2}\,.\end{aligned}$$

It follows from condition (2) that the function $R(t)$ is continuous. Furthermore, it is a positive-definite function: that is,

$$\sum_{k,j=1}^{n} R(t_k - t_j)\lambda_k\overline{\lambda}_j \geqq 0$$

for an arbitrary choice of numbers $n, t_1, t_2, \cdots, t_n$. It follows from the Bochner-Khinchin theorem (see, for example, B. V. Gnedenko, 1967, §38,) that the function has a representation of the form (1).

The sufficiency of the condition of the theorem follows from the preceding section in which it was shown that for an arbitrary nonnegative nondecreasing function $F(u)$ one can construct a normal complex-valued stationary (even in the narrow sense) process whose correlation function is given by formula (1).

Actually, one can construct in a much simpler manner a process that is stationary in the broad sense with covariance function given by formula (1) for a given spectral function $F(u)$. Suppose that $F(+\infty) = \sigma^2$. Let ξ denote a random variable with distribution function $(1/\sigma^2)F(x)$. Let $\zeta(t) = \sigma e^{i(t\xi+\varphi)}$, where φ and ξ are independent and φ is uniformly distributed on the interval $(0, 2\pi)$. Then

$$\mathsf{M}\zeta(t) = \sigma\int_{-\infty}^{\infty}\int_0^{2\pi} e^{i(tu+\varphi)}\frac{1}{\sigma^2}dF(u)\frac{d\varphi}{2\pi} = 0$$

and

$$\begin{aligned}R(t+h, h) &= \mathsf{M}[\zeta(t+h)\overline{\zeta(h)}]\\ &= \sigma^2\int_{-\infty}^{\infty}\int_0^{2\pi} e^{itu}\frac{1}{\sigma^2}dF(u)\frac{d\varphi}{2\pi} = \int_{-\infty}^{\infty} e^{itu}dF(u)\,.\end{aligned}$$

Definition 1. The function $F(u)$ in the representation (1) of the covariance function of a process that is stationary in the broad sense is called a *spectral function*. If $F(u)$ is absolutely continuous, the function f defined by

$$F(u) = \int_{-\infty}^{u} f(v)dv$$

is called the *spectral density* of the process.

Let us recall the physical meaning of the spectral function. If the random function $\xi(t)$ represents an electric current, then the function $F(u)$ can be interpreted as follows: The process $\xi(t)$ can be represented as a "continuous sum" of simple harmonic oscillations with random amplitudes, and the increment $F(u_2) - F(u_1)$, where $u_1 < u_2$, is equal to the mean power of the harmonic components of the process whose frequencies lie in the half-closed half-open $[u_1, u_2)$.

The spectral function can be obtained from the covariance function. If a and b are points of continuity of the distribution $F(u)$, then as one can see from the theory of characteristic functions (cf. Gnedenko, 1967, §35),

$$F(b) - F(a) = \frac{1}{2\pi}\int_{-\infty}^{\infty} \frac{e^{-ita} - e^{itb}}{it} R(t)dt \; , \tag{3}$$

where the integral on the right should be understood in the sense of the principal value. At points of discontinuity of the function $F(u)$, formula (3) remains valid if we replace F on the left with the function G defined by $G(u) = [F(u + 0) - F(u)]/2$.

If the covariance function $R(t)$ is absolutely integrable on the interval $(-\infty, \infty)$, that is, if $\int_{-\infty}^{\infty} |R(t)| dt < \infty$, then the right-hand member of equation (3) is differentiable with respect to the parameter b. Thus the absolute integrability of the function $R(t)$ implies the existence of a spectral density equal to

$$f(u) = \frac{1}{2\pi}\int_{-\infty}^{\infty} e^{-iut} R(t)dt \; ,$$

that is, $f(u)$ is the inverse Fourier transform of the correlation function $R(t)$.

This situation is analogous with multidimensional processes, in which case the role of the covariance function is played by a covariance matrix function.

We introduce the following notation: The *trace* of a square matrix $\mathbf{\Phi} = (\Phi_{jk})$, for $j, k = 1, \cdots, n$, is defined as the sum of the diagonal elements of that matrix and is denoted by tr $\mathbf{\Phi}$: tr $\mathbf{\Phi} = \sum_{j=1}^{n} \Phi_{jj}$.

Let $\mathbf{\Phi}$ and $\mathbf{\Psi}$ denote two $m \times n$ matrices

$$\mathbf{\Phi} = (\Phi_{jk}),\ j = 1, \cdots, m\,; \qquad k = 1, \cdots, n\,,$$
$$\mathbf{\Psi} = (\Psi_{jk}),\ j = 1, \cdots, m\,; \qquad k = 1, \cdots, n\,.$$

The scalar product of the matrices $\mathbf{\Phi}$ and $\mathbf{\Psi}$ is defined by

$$(\mathbf{\Phi}, \mathbf{\Psi}) = \sum_{\substack{j=1,\cdots,m\\k=1,\cdots,n}} \Phi_{jk}\overline{\Psi}_{jk} = \operatorname{tr} \mathbf{\Phi}\mathbf{\Psi}^*\,,$$

where $\mathbf{\Psi}^*$ is the adjoint of the matrix $\mathbf{\Psi}$; that is $\mathbf{\Psi}^* = (\Psi^*_{jk})$ for $j = 1, \cdots, m$ and $k = 1, \cdots, n$; $\Psi^*_{jk} = \overline{\Psi}_{kj}$. The *norm* $\| \mathbf{\Phi} \|$ of the matrix $\mathbf{\Phi}$ is defined as

$$\| \mathbf{\Phi} \| = \sqrt{(\mathbf{\Phi}, \mathbf{\Phi})} = \sqrt{\sum_{\substack{j=1,\cdots,m\\k=1,\cdots,n}} | \Phi_{jk} |^2}\,.$$

Theorem 2. *For a matrix function $\boldsymbol{R}(t) = (R_{jk}(t))$, where $j, k = 1, \cdots, n$, to be the covariance matrix of some vector-valued process $\boldsymbol{\zeta}(t)$ that is stationary in the broad sense and satisfies the condition*

$$\lim_{\tau\to 0} \mathsf{M} \| \boldsymbol{\zeta}(t + \tau) - (t) \|^2 = 0\,, \tag{4}$$

it is necessary and sufficient that it have a representation of the form

$$\boldsymbol{R}(t) = \int_{-\infty}^{\infty} e^{iut} d\boldsymbol{F}(u)\,, \tag{5}$$

where $\boldsymbol{F}(u) = (F_{jk}(u))$, for $j, k = 1, \cdots, n$, is a matrix function satisfying the following two conditions: (a) For arbitrary $u_1 < u_2$, the matrix $\Delta\boldsymbol{F}(u) = \boldsymbol{F}(u_2) - \boldsymbol{F}(u_1)$ is nonnegative-definite, and (b)

$$\operatorname{tr} \{\boldsymbol{F}(+\infty) - \boldsymbol{F}(-\infty)\} < \infty\,.$$

With regard to condition (b) we note that by virtue of condition (a) the diagonal elements of the matrix $\boldsymbol{F}(u)$ are nondecreasing functions of u. Thus condition (b) is equivalent to the requirement that each diagonal element $F_{ii}(u)$ of the matrix $\boldsymbol{F}(u)$ be of bounded variations on the line $(-\infty, \infty)$.

Furthermore, the positive-definiteness of the matrix $\Delta\boldsymbol{F}(u)$ implies that $| \Delta F_{jk}(u) |^2 \leqq \Delta F_{jj}(u)\Delta F_{kk}(u)$, so that

$$\sum_{p=1}^{s} | \Delta F_{jk}(u_p) | \leqq \left[\sum_{p=1}^{s} \Delta F_{jj}(u_p)\right]^{1/2} \left[\sum_{p=1}^{s} \Delta F_{kk}(u_p)\right]^{1/2}\,,$$

where $\Delta\boldsymbol{F}(u_p) = \boldsymbol{F}(u_p) - \boldsymbol{F}(u_{p-1})$, for $p = 1, 2, \cdots, s$, and where $u_0 < u_1 < \cdots u_s$.

From this it follows that the offdiagonal elements $F_{jk}(u)$ of the matrix $\boldsymbol{F}(u)$ are also functions of bounded variation. Without loss of generality we can assume that $\boldsymbol{F}(-\infty) = 0$.

Let us now prove the theorem.

Proof of the Necessity. Let $\boldsymbol{R}(t)$ denote the covariance matrix function of an r-dimensional vector-valued process $\boldsymbol{\zeta}(t)$ that is stationary in the broad sense and that satisfies condition (4). For an arbitrary r-dimensional vector $\boldsymbol{c}$, we introduce the random variable $\zeta_c(t) = (\boldsymbol{\zeta}(t), \boldsymbol{c})$. Then

$$\mathsf{M}\,|\,\zeta_c(t+\tau) - \zeta_c(t)\,|^2 \leqq \|\,\boldsymbol{c}\,\|\,\{\mathsf{M}\,\|\,\boldsymbol{\zeta}(t+\tau) - \boldsymbol{\zeta}(t)\,\|^2\}^{1/2} ,$$

so that $\lim_{\tau\to 0} \mathsf{M}\,|\,\zeta_c(t+\tau) - \zeta_c(t)\,|^2 = 0$. In addition, $\zeta_c(t)$ is a random process that is stationary in the broad sense:

$$m_c = \mathsf{M}\zeta_c(t) = (\mathsf{M}\boldsymbol{\zeta}(t), \boldsymbol{c}) = \text{const} ,$$

$$\mathsf{M}([\zeta_c(t+\tau) - m_c]\overline{[\zeta_c(\tau) - m_c]}) = (\boldsymbol{c}, \boldsymbol{R}(t)\boldsymbol{c}) .$$

Let $R_c(t)$ denote the covariance function of the process $\zeta_c(t)$. We have

$$R_c(t) = (\boldsymbol{c}, \boldsymbol{R}(t)\boldsymbol{c}) . \tag{6}$$

On the basis of the preceding theorem, the covariance function $R_c(t)$ can be represented in the form

$$R_c(t) = \int_{-\infty}^{\infty} e^{itu} dF_c(u) , \tag{7}$$

where $F_c(u)$ is a nondecreasing function of u and $F_c(u) < \infty$.

Let $\boldsymbol{e}^{(k)}$ denote the r-dimensional vector whose kth component is equal to one and whose other components are equal to zero,

$$e_p^{(k)} = \begin{cases} 0 , & p \neq k ; \\ 1 , & p = k . \end{cases}$$

It follows from (6) that $R_{e^{(k)}}(t) = R_{kk}(t)$ is the covariance function of the kth component of the vector-valued process $\boldsymbol{\zeta}(t)$. We set $\boldsymbol{e}^{(k,j)} = \boldsymbol{e}^{(k)} + \boldsymbol{e}^{(j)}$, $\tilde{\boldsymbol{e}}^{(k,j)} = i\boldsymbol{e}^{(k)} + \boldsymbol{e}^{(j)}$. From (6), we obtain

$$R_{e^{(k,j)}}(t) = R_{kk}(t) + R_{kj}(t) + R_{jk}(t) + R_{jj}(t) , \qquad k \neq j ,$$

$$R_{\tilde{e}^{(k,j)}}(t) = R_{kk}(t) + iR_{kj}(t) - iR_{jk}(t) + R_{jj}(t) ,$$

from which we get

$$R_{kj}(t) = \frac{R_{e^{(k,j)}}(t) - iR_{\tilde{e}^{(k,j)}}(t)}{2} - \frac{1-i}{2}(R_{e^{(k)}} - R_{e^{(j)}}) .$$

If we set

$$F_{kj}(u) = \frac{F_{e^{(k,j)}}(u) - iF_{\tilde{e}^{(k,j)}}(u)}{2} - \frac{1-i}{2}(F_{e^{(k)}}(u) - F_{e^{(j)}}(u)) , \qquad k \neq j ,$$

$$F_{kk}(u) = F_{e^{(k)}}(u) ,$$

we obtain

$$R_{kj}(t) = \int_{-\infty}^{\infty} e^{itu} dF_{kj}(u) , \qquad k, j = 1, 2, \cdots, r .$$

Furthermore,

$$R_c(t) = \sum_{k,j=1}^{r} \overline{c_k} R_{kj}(t) c_j = \int_{-\infty}^{\infty} e^{itu} d\Big(\sum_{k,j=1}^{r} \overline{c_k} F_{kj}(u) c_j \Big) .$$

By virtue of the uniqueness of the function $F_c(u)$, we obtain

$$F_c(u) = \sum_{k,j=1}^{r} \overline{c_k} F_{kj}(u) c_j .$$

From this it follows that the matrix $\Delta \boldsymbol{F}(u) = (\Delta F_{kj}(u))$, for $k, j = 1, \cdots, r$, is nonnegative-definite and $F_{kk}(+\infty) < \infty$.

Proof of the Sufficiency. Consider the matrix function of two variables $\boldsymbol{R}(t_1, t_2) = \boldsymbol{R}(t_1 - t_2)$, where $\boldsymbol{R}(t)$ is defined by formula (5). Let us show that this function is nonnegative-definite. We have

$$\sum_{\substack{k,j=1,\cdots,r \\ p,q=1,\cdots,n}} R_{kj}(t_p, t_q) z_k^{(p)} \overline{z_j^{(q)}} = \int_{-\infty}^{\infty} d(\boldsymbol{W}, \boldsymbol{F}(u) \boldsymbol{W}) ,$$

where

$$\boldsymbol{W} = \sum_{p=1}^{n} e^{iut_p} \boldsymbol{z}^{(p)}, \quad \boldsymbol{z}^{(p)} = (z_1^{(p)}, \cdots, z_r^{(p)}) .$$

It follows from condition (a) of the Theorem 2 that $\boldsymbol{W}, \boldsymbol{F}(u)\boldsymbol{W}$ is a nondecreasing function of u. This means that $\boldsymbol{R}(t_1 - t_2)$ is a nonnegative-definite matrix. Let us construct an r-dimensional complex Gaussian process with vector-valued mathematical expectation equal to 0 and with covariance matrix function $\boldsymbol{R}(t_1 - t_2)$ (cf. Section 3). We obtain a process $\boldsymbol{\zeta}(t)$ that is stationary in the strict sense. Here,

$$\mathsf{M} \| \boldsymbol{\zeta}(t + \tau) - \boldsymbol{\zeta}(t) \|^2 = 2 \mathrm{tr} \, [\boldsymbol{R}(0) - \mathrm{Re} \, \boldsymbol{R}(\tau)] .$$

Since the function $\boldsymbol{R}(t)$ defined by equation 5 is continuous, this last formula shows that the process $\boldsymbol{\zeta}(t)$ satisfies condition 4 of Theorem 2. This completes the proof of the theorem.

Let us look at the question of the spectral distribution of the covariance function of a stationary process with discrete time.

Theorem 3. *The matrix function* $\boldsymbol{R}(t) = (R_{kj}(t))$, *for*

$$k, j = 1, 2, \cdots, r \quad \textit{and} \quad t = 0, \pm 1, \pm 2, \cdots ,$$

is the covariance function of some stationary (in the broad sense) vector-valued process $\boldsymbol{\zeta}(t)$ *with discrete time if and only if it can be represented in the form*

$$\boldsymbol{R}(t) = \int_0^{2\pi} e^{itu} d\boldsymbol{F}(u) ,$$

where $\boldsymbol{F}(u) = (F_{kj}(u))$, *for* $k, j = 1, \cdots, r$, *is a matrix function defined on* $[0, 2\pi)$ *and enjoying the following properties:* (*a*) *For arbitrary* u_1 *and* u_2 *such that* $0 \leqq u_1 < u_2 < 2\pi$, *the matrix* $\Delta \boldsymbol{F}(u) = \boldsymbol{F}(u_2) - \boldsymbol{F}(u_1)$ *is nonnegative-definite*; (*b*) $\mathrm{tr}\,[\boldsymbol{F}(2\pi - 0] - \boldsymbol{F}(0)] < \infty$.

Proof of this theorem is analogous to that of Theorem 2. One first considers the one-dimensional case, just as in Theorem 1, except that instead of referring to the Bochner-Khinchin theorem, one needs to refer here to a theorem of Herglotz (see Gnedenko, 1967, p. 259) on the representation of a nonnegative-definite sequence of numbers. Then as in the proof of Theorem 2, one shifts from the one-dimensional to the multidimensional case.

Let us now look at processes with stationary increments (in the broad sense). For structural functions of such processes, we obtain a formula analogous to the spectral representation of the covariance function of a stationary process.

Theorem 4. *Let* $\boldsymbol{\zeta}(t)$ *denote an r-dimensional process with stationary increments (in the broad sense). The function* $\boldsymbol{D}(t_1, t_2, t_3, t_4)$ *is the structural function of the process* $\boldsymbol{\zeta}(t)$ *satisfying the relation*

$$\mathsf{M}\,\|\boldsymbol{\zeta}(t+\tau) - \boldsymbol{\zeta}(t)\|^2 \to 0 \quad as \quad \tau \to 0\,, \tag{8}$$

if and only if this function possesses a representation of the form

$$\boldsymbol{D}(t_1, t_2, t_3, t_4) = -\int_{-\infty}^{\infty} \frac{e^{it_2u} - e^{it_1u}}{iu} \frac{e^{-it_4u} - e^{-it_3u}}{iu} d\boldsymbol{F}(u)\,, \tag{9}$$

where $\boldsymbol{F}(u)$ *is a matrix function* $(-\infty < u < \infty)$ *with the following properties:* (*a*) *For every* u_1 *and* u_2 *such that* $u_2 > u_1$, *the matrix* $\boldsymbol{F}(u_2) - \boldsymbol{F}(u_1)$ *is nonnegative-definite*; (*b*)

$$\int_{-\infty}^{\infty} k(u)^{d[\mathrm{tr}\,\boldsymbol{F}(u)]} < \infty\,, \tag{10}$$

where

$$k(u) = \begin{cases} 1, & |u| \leqq 1\,, \\ \dfrac{1}{u^2}, & |u| > 1\,. \end{cases}$$

Proof. Let us set

$$\bar{\boldsymbol{\zeta}}_m(k) = \boldsymbol{\zeta}\Big(\frac{k+1}{2^m}\Big) - \boldsymbol{\zeta}\Big(\frac{k}{2^m}\Big)\Big\}\,,$$

$$m = 1, 2, \cdots, \qquad k = 0, \pm 1, \pm 2, \cdots \tag{11}$$

For arbitrary m, the sequence $\bar{\boldsymbol{\zeta}}_m(k)$, for $k = 0, \pm 1, \pm 2, \cdots$, is a process with discrete time that is stationary in the broad sense. On the basis of Theorem 3, the covariance matrix $\tilde{\boldsymbol{R}}_m(k)$ of the sequence (11) has the following representation

$$\tilde{R}_m(k) = \int_{-\pi}^{\pi} e^{iku} dF_m(u) , \tag{12}$$

where the $F_m(u)$ satisfy condition (a) and $\operatorname{tr}\{F_m(\pi) - F_m(-\pi - 0)\} < \infty$. It is more convenient to write formula (12) in the form

$$\tilde{R}_m(k) = \int_{-\infty}^{\infty} e^{i(k/2^m)u}\, d\tilde{F}_m(u) , \tag{13}$$

where

$$\tilde{F}_m(u) = 0 \quad \text{for} \quad u < -2^m\pi ,$$

$$\tilde{F}_m(u) = F_m\left(\frac{u}{2^m}\right) - F_m\left(\frac{-\pi}{2^m}\right) \quad \text{for} \quad -2^m\pi \leqq u \leqq 2^m\pi$$

$$\tilde{F}_m(u) = F_m(\pi) \quad \text{for} \quad u > 2^m\pi .$$

Since

$$D\left(t + \frac{k}{2^m},\ t + \frac{s+k+1}{2^m},\ t,\ t + \frac{r+1}{2^m}\right)$$

$$= \sum_{j=1}^{r+1} \sum_{l=1}^{s+1} D\left(t + \frac{k+l-1}{2^m},\ t + \frac{k+l}{2^m},\ t + \frac{j-1}{2^m},\ t + \frac{j}{2^m}\right)$$

$$= \sum_{j=1}^{r} \sum_{l=1}^{s} \tilde{R}_m(k + l - j) ,$$

it follows that for numbers t_1, t_2, t_3, and t_4 of the form $k/2^m$ (where $k = 0, \pm 1, \pm 2, \cdots$), we obtain

$$D(t_1, t_2, t_3, t_4) = \int_{-\infty}^{\infty} \sum_{j=1}^{(t_4-t_3)2^m-1} \sum_{l=1}^{(t_2-t_1)2^m-1} \exp\left\{\frac{iu(k+l-j)}{2^m}\right\} d\tilde{F}_m(u)$$

$$= \int_{-\infty}^{\infty} \frac{e^{iut_2} - e^{iut_1}}{e^{i(u/2^m)} - 1} \frac{e^{-iut_4} - e^{-iut_3}}{e^{-i(u/2^m)} - 1} d\tilde{F}_m(u)$$

which can also be written in the form

$$D(t_1, t_2, t_3, t_4) = -\int_{-\infty}^{\infty} \frac{e^{iut_2} - e^{iut_1}}{iu} \frac{e^{-iut_4} - e^{-iut_3}}{iu} \frac{1}{k(u)} dH_m(u) , \tag{14}$$

where

$$H_m(u) = \int_{-\infty}^{u} \frac{\alpha^2 k(\alpha)}{4 \sin^2 \dfrac{\alpha}{2^{m+1}}} d\tilde{F}_m(u) . \tag{15}$$

Let us now show that the sequence of matrix functions $H_m(u)$ is weakly compact. This means that the sequence $\{H_m(u)\}$ contains a subsequence $\{H_{m_k}(u)\}$ such that for an arbitrary $f(u)$ that is bounded and continuous on $(-\infty, \infty)$,

$$\lim_{k\to\infty} \int_{-\infty}^{\infty} f(u) dH_{m_k}(u) = \int_{-\infty}^{\infty} f(u) dH(u) ,$$

where $\boldsymbol{H}(u)$ is some matrix whose elements are functions of bounded variation. The matrix $\boldsymbol{H}(u)$ is then the weak limit of the sequence of the matrices $\boldsymbol{H}_{m_k}(u)$. On the basis of Helly's theorem the sequence $\{\boldsymbol{H}_m(u)\}$ is weakly compact if the norms of the matrices $\boldsymbol{H}_m(u)$ are uniformly bounded (with respect to m) and $\left\|\int_{|u|>A} d\boldsymbol{H}_m(u)\right\| \to 0$ as $A \to \infty$ uniformly with respect to m. Let us show that these conditions can be satisfied. We have

$$\boldsymbol{D}(t, t+h, t, t+h) = \mathsf{M}([\boldsymbol{\zeta}(t+h) - \boldsymbol{\zeta}(t)][\boldsymbol{\zeta}(t+h) - \boldsymbol{\zeta}(t)]^*)$$

$$= \int_{-\infty}^{\infty} \frac{4\sin^2 \frac{uh}{2}}{u^2 k(u)} d\boldsymbol{H}_m(u), \quad t = \frac{k}{2^m}, \quad h = \frac{j}{2_m}. \tag{16}$$

Let us set $\psi(h) = \mathsf{M}\|\boldsymbol{\zeta}(t+h) - \boldsymbol{\zeta}(t)\|^2$. By the hypothesis of the theorem, $\psi(h) \to 0$ as $h \to 0$. From this it follows that the function $\psi(h)$ is continuous. This is true because

$$|\psi(h'') - \psi(h')| \leqq \mathsf{M}(\|\boldsymbol{\zeta}(t+h') - \boldsymbol{\zeta}(t+h'')\|$$
$$\times [\|\boldsymbol{\zeta}(t+h'') - \boldsymbol{\zeta}(t)\| + \|\boldsymbol{\zeta}(t+h') - \boldsymbol{\zeta}(t)\|])$$
$$\leqq \sqrt{\psi(|h'-h''|)\psi(h'')} + \sqrt{\psi(|h'-h''|)\psi(h')}.$$

It follows from (16) that for $A > 0$,

$$\psi(h) \geqq \|\boldsymbol{D}(t, t, +h, t, t, +h)\| \geqq \int_{-A}^{A} \frac{4\sin^2 \frac{uh}{2}}{u^2} d\boldsymbol{H}_m^{(k,k)}(u),$$

where $\boldsymbol{H}_m(u) = (H_m^{(r,s)}(u))$ for $r, s = 1, \cdots, h$. If $A|h| < \pi$ we have

$$\frac{4\sin^2 \frac{uh}{2}}{u^2} \geqq \frac{4}{\pi}h^2$$

for $|u| \leqq 1$. Consequently

$$\psi(h) \geqq \frac{4h^2}{\pi^2}\int_{-A}^{A} dH_m^{(k,k)}(u). \tag{17}$$

Furthermore, for $A \geqq 1$ we have

$$\psi(h) \geqq 4\int_{|u|>A} \sin^2 \frac{uh}{2} dH_m^{(k,k)}(u)$$

$$= 2\int_{|u|>A} (4 - \cos uh) dH_m^{(k,k)}(u). \tag{18}$$

Integrating this inequality with respect to h, we obtain

$$\frac{1}{h}\int_0^h \psi(\alpha)d\alpha \geqq 2\int_{|u|>A}\left(1 - \frac{\sin uh}{uh}\right) dH_m^{(k,k)}(u)$$

$$\geqq 2\left(1 - \frac{1}{Ah}\right)\int_{|u|>A} dH_m^{(k,k)}(u).$$

Since the left-hand member of this inequality approaches 0 independently of m as $h \to 0$, it follows that

$$\int_{|u|>A} dH_m^{(k,k)}(u) \to 0$$

uniformly with respect to m as $A \to \infty$. From the positive-definiteness of the matrix $\Delta \boldsymbol{H}_m$, we have

$$\| \Delta \boldsymbol{H}_m(u) \| \leqq \operatorname{tr} \Delta \boldsymbol{H}_m ,$$

so that

$$\left\| \int_{|u|>A} d\boldsymbol{H}_m(u) \right\| \leqq \operatorname{tr} \int_{|u|>A} d\boldsymbol{H}_m(u) \to 0 \quad \text{as} \quad A \to \infty$$

uniformly with respect to m, and by virtue of (17),

$$\left\| \int_{|u| \leqq A} d\boldsymbol{H}_m(u) \right\| \leqq \frac{\pi^2 r}{4h^2} \psi(h) , \qquad |h| \leqq \frac{\pi}{A} .$$

These inequalities prove the weak compactness of the sequence of matrices $\boldsymbol{H}_m(u)$.

If we now take the limit in (14) with respect to the subsequence of indices m_k such that $\{\boldsymbol{H}_{m_k}(u)\}$ converges weakly to $\boldsymbol{H}(u)$, we obtain

$$\boldsymbol{D}(t_1, t_2, t_3, t_4) = \int_{-\infty}^{\infty} \frac{e^{iut_2} - e^{iut_1}}{iu} \frac{e^{-iut_4} - e^{-iut_3}}{-iu} \frac{d\boldsymbol{H}(u)}{k(u)} .$$

This equation is valid for all dyadic t_1, t_2, t_3, t_4. Obviously

$$\boldsymbol{H}(u_2) - \boldsymbol{H}(u_1)$$

is a nonnegative-definite matrix. Since the left and right members of the formula are continuous functions and since these two members coincide on an everywhere-dense subset of the values of t_1, t_2, t_3, t_4, they are equal for *all* values of these variables. It only remains for us to set $\boldsymbol{F}(u) = \int_{-\infty}^{u} \{d\boldsymbol{H}(u)/k(u)\}$ to obtain the desired result.

Let us now look at a scalar random field $\xi(\boldsymbol{x})$ in n-dimensional space E_n: $\boldsymbol{x} = (x^1, x^2, \cdots, x^n)$ where $-\infty < x^i < \infty$. This field is said to be *homogeneous* if $\mathsf{M}\xi(\boldsymbol{x}) = a = \text{const}$,

$$R[\boldsymbol{x}_1 + \boldsymbol{z}, \boldsymbol{x}_2 + \boldsymbol{z}] = \mathsf{M}([\boldsymbol{\xi}(\boldsymbol{x}_2 + \boldsymbol{z}) - a][\xi(\boldsymbol{x}_1 + \boldsymbol{z}) - a]) = R(\boldsymbol{x}_1, \boldsymbol{x}_2) .$$

If we set $\boldsymbol{z} = -\boldsymbol{x}_1$ in this last condition, we obtain

$$R(\boldsymbol{x}_1, \boldsymbol{x}_2) = R(0, \boldsymbol{x}_2 - \boldsymbol{x}_1) .$$

The last equation means that the covariance between the random variables $\xi(\boldsymbol{x}_1)$ and $\xi(\boldsymbol{x}_2)$ depends only on the vector $\boldsymbol{x}_2 - \boldsymbol{x}_1$ connecting the points $\boldsymbol{x}_1$ and $\boldsymbol{x}_2$. The function $R(\boldsymbol{x}) = R(\boldsymbol{z}, \boldsymbol{z} + \boldsymbol{x})$ is also called the covariance function of the homogeneous random field.

It is nonnegative-definite function of n variables; that is, the quadratic form

$$\sum_{i,k=1}^{N} R(x_i - x_k)\lambda_i\lambda_k$$

is nonnegative-definite for arbitrary choice of N and points

$$x_1, x_2, \cdots, x_N .$$

The function $R(x)$ is continuous if and only if the random field $\xi(x)$ satisfies the condition

$$\mathsf{M}\,|\,\xi(x + z) - \xi(x)\,|^2 \longrightarrow 0 \quad \text{as} \quad z \longrightarrow 0 . \tag{19}$$

The Bochner-Khinchin theorem for nonnegative-definite functions of a single variable can be carried over almost without change in the course of the proof to functions of several variables. Thus the covariance function of a homogeneous field satisfying condition (9) has a representation of the form

$$R(x) = \int_{E_n} e^{i(x,\nu)} d\sigma(\nu) ,$$

where (x, ν) denotes the scalar product of the n-dimensional vectors x and ν and $\sigma(A)$ is a finite measure in E_n.

A random field is said to be *isotropic* if the covariance function $R(x_1, x_2)$ depends only on x_1 and the distance between the points x_1 and x_2. If in addition it is homogeneous, then $R(x_1, x_2) = R(\rho)$, where ρ is the distance between the points x_1 and x_2:

$$\rho = \sqrt{\sum_{j=1}^{n} (x_1^j - x_2^j)^2} .$$

We will find a representation of the covariance function of a homogeneous isotropic field. Let us look at the expression for the covariance function of a homogeneous field

$$R(\rho) = \int_{E_n} e^{i(x,\nu)} d\sigma(\nu)$$

and let us integrate this expression over the surface of a sphere S_ρ of radius ρ. Reversing the order of integration, we obtain

$$R(\rho) = \frac{\Gamma\left(\frac{n}{2}\right)}{2\pi^{n/2}\rho^{n-1}} \int_{E_n} \left\{ \int_{S_\rho} e^{i(x,\nu)} ds \right\} d\sigma(\nu) . \tag{20}$$

Let $f(x)$ denote an arbitrary integrable function in E_n and let V_ρ denote the ball of radius ρ with center at a fixed point. Then

$$\frac{d}{d\rho} \int_{V_\rho} f(x) dx^1 \cdots dx^n = \int_{S_\rho} f(x) ds ,$$

where the integral on the right is over the surface S_ρ of the ball V_ρ. Let us use this formula to evaluate the inner integral on the right side of formula (20). Shifting to spherical coordinates in n-dimensional space (cf. G. M. Fikhtengol'ts, vol. III, p. 401) and taking for φ_1 the angle between the vector x and ν, we obtain

$$\begin{aligned}\int_{V_\rho} e^{i(x,\nu)}dx^1 \cdots dx &= \int_0^\rho\int_0^\pi \cdots \int_0^\pi\int_0^{2\pi} e^{ir|\nu|\cos\varphi_1} r^{n-1}\sin^{n-2}\varphi_1 \\ &\quad \times \sin^{n-3}\varphi_2 \cdots \sin\varphi_{n-2}\, dr\, d\varphi_1 \cdots d\varphi_{n-2} \\ &= \frac{2\pi^{(n-1)/2}}{\Gamma\left(\frac{n-1}{2}\right)}\int_0^\rho\int_0^\pi e^{ir|\nu|\cos\varphi_1} r^{n-1}\sin^{n-2}\varphi_1\, dr\, d\varphi_1 .\end{aligned}$$

Furthermore,

$$\begin{aligned}\int e^{ir|\nu|\cos\varphi_1}\sin^{n-2}\varphi_1 d\varphi_1 &= \sum_{k=0}^{\infty}\int_0^\pi \frac{(ir|\nu|\cos\varphi_1)^k}{k!}\sin^{n-2}\varphi_1 d\varphi_1 \\ &= \sum_{k=0}^{\infty}(-1)^k \frac{r^{2k}|\nu|^{2k}}{(2k)!}\frac{\Gamma\left(\frac{2k+1}{2}\right)\Gamma\left(\frac{n-1}{2}\right)}{\Gamma\left(\frac{2k+n}{2}\right)},\end{aligned}$$

and

$$\begin{aligned}I &= \int_0^\rho\int_0^\pi e^{ir|\nu|\cos\varphi_1} r^{n-1}\sin^{n-2}\varphi_1 d\varphi_1 \\ &= \sum_{k=0}^{\infty}(-1)^k \frac{\rho^{2k+n}|\nu|^{2k}}{(2k)!(2k+n)}\frac{\Gamma\left(\frac{2k+1}{2}\right)\Gamma\left(\frac{n-1}{2}\right)}{\Gamma\left(\frac{2k+n}{2}\right)}.\end{aligned}$$

Using the formula for the gamma function for half-integral values

$$\Gamma\left(k+\frac{1}{2}\right) = \frac{\sqrt{\pi}}{2^{2k-1}}\frac{\Gamma(2k)}{\Gamma(k)},$$

we obtain

$$\begin{aligned}I &= \frac{\sqrt{\pi}\,\Gamma\left(\frac{n-1}{2}\right)2^{(n/2)-1}\rho^{n/2}}{|\nu|^{n/2}}\sum_{k=0}^{\infty}(-1)^k\frac{\left(\frac{\rho|\nu|}{2}\right)^{2k+(n/2)}}{k!\,\Gamma\left(k+\frac{n}{2}+1\right)} \\ &= \frac{\sqrt{\pi}}{2}\Gamma\left(\frac{n-1}{2}\right)\left(\frac{2\rho}{|\nu|}\right)^{n/2} J_{n/2}(\rho|\nu|)] ,\end{aligned}$$

where $J_m(x)$ is the Bessel function of the first kind of order m. Consequently,

$$\int_{V_\rho} e^{i(x,\nu)}dx^1 \cdots dx^n = \left(\frac{2\pi\rho}{|\nu|}\right)^{n/2} J_{n/2}(\rho|\nu|) .$$

From this it follows that

$$\int_{S_\rho} e^{i(x,\nu)}ds = \left(\frac{2\pi\rho}{|\nu|}\right)^{n/2} |\nu| J_{(n-2)/2}(\rho|\nu|) .$$

In particular, the integral depends on $|\nu|$. We introduce the positive parameter λ and we set $g(\lambda) = \sigma(V_\lambda)$, $\lambda > 0$. This last formula and formula (20) yield

$$R(\rho) = 2^{(n-2)/2}\,\Gamma\left(\frac{n}{2}\right)\int_0^\infty \frac{J_{(n-2)/2}(\lambda\rho)}{(\lambda\rho)^{(n-2)/2}}\,dg(\lambda) , \tag{21}$$

where $g(\lambda)$ is an increasing function, $g(0) = 0$, and

$$g(+\infty) = \sigma(E_n) = R(0) < \infty .$$

Thus we have obtained

Theorem 5. *For $R(\rho)(0 \leqq \rho < \infty)$ to be the covariance function of a homogeneous isotropic n-dimensional random field satisfying condition* (19), *it is necessary and sufficient that it have a representation of the form* (21), *where $g(\lambda)$ is a bounded nondecreasing function.*

For $n = 2$ the formula takes the following simple form

$$R(\rho) = \int_0^\infty J_0(\lambda\rho)dg(\lambda) , \tag{22}$$

and for $n = 3$,

$$R(\rho) = 2\int_0^\infty \frac{\sin \lambda\rho}{\lambda\sigma}\,dg(\lambda) . \tag{23}$$

II

MEASURE THEORY

We assume that the reader is familiar with the elements of the set-theoretic construction of probability theory. Therefore, in the present chapter we have omitted elementary examples and details that give the intuitive basis underlying formal definitions (see, for example, Gnedenko, 1967).

The starting point in probability theory is the assumption that one can define a set U and a class $\mathfrak{S}$ of subsets of U so that every event A, for which it is meaningful to speak of its probability within the framework of a particular problem, can be interpreted as some subset of the set U belonging to $\mathfrak{S}$. Since an arbitrary event A interpreted in this manner can be regarded as the union of elements of U that belong to A, we call the points in the set U *elementary events* and the set U itself the *space of elementary events.* For example, if an experiment consists of drawing the graph of a continuous random function in the course of a fixed interval of time $[a, b]$, then U can be understood as the space of continuous functions on the interval $[a, b]$.

In what follows the events are identified with the sets corresponding to them. Obviously a *certain* event then coincides with the set U and an *impossible* event coincides with the empty set; the union, coincidence, and difference of two or more events then coincide with the set-theoretic union, intersection, and difference of sets. The incompatibility of a class of events means that the intersection of the corresponding sets is empty. If A is an event, the complementary event A is the set-theoretic complement of A in U. Furthermore, to every event $A \in \mathfrak{S}$ is assigned a nonnegative number $\mathsf{P}(A)$ called the *probability* of the event A. It is natural to require that the class $\mathfrak{S}$ of events and their probabilities (which are defined) enjoy the following properties (with which we are familiar from elementary probability theory): (a) The difference of two events or the union of an arbitrary sequence of events in the class $\mathfrak{S}$ are events (that is, they belong to $\mathfrak{S}$); (b) the probability of the union of an arbitrary sequence of incompatible events is equal to the sum of the probabilities of the events of the given sequence; and (c) the probability of a certain event is equal to 1.

The mathematical apparatus with which we formulate the basic assumptions and concepts of probability theory and derive the general theorems is the abstract theory of measure and integration. The material from this theory that we shall need in this book is expounded in the present chapter.

1. MEASURE

Let U denote an abstract set, which we shall call a *space*. We shall indicate subsets of U by italic letters and classes of subsets of U by German letters (capitals in both cases).

We assume that the definitions and simplest properties of the algebraic operations on sets are known and we mention only the frequently used duality relationships

$$\overline{\bigcap_k A_k} = \bigcup_k \bar{A}_k \tag{1}$$

and

$$\overline{\bigcup_k A_k} = \bigcap_k \bar{A}_k \tag{2}$$

where the index k ranges over an arbitrary (finite or infinite) set of values (cf. for example, Kolmogorov and Fomin).

Definition 1. A nonempty class $\mathfrak{R}$ of subsets of U is called an *algebra of sets* of U if it enjoys the following properties:

a. $A \in \mathfrak{R}$ and $B \in \mathfrak{R}$ imply $A \cup B \in \mathfrak{R}$,

b. $A \in \mathfrak{R}$ implies $\bar{A} \in \mathfrak{R}$.

Let us give some of the simpler consequences of this definition. Since $A \cup \bar{A} = U$, the relation $A \in \mathfrak{R}$ implies $U \in \mathfrak{R}$. This in turn implies that the empty set belongs to the algebra of sets. Furthermore, if $A \in \mathfrak{R}$ and $B \in \mathfrak{R}$, then on the basis of relationships (1) and (2), $A \cap B = \overline{(\bar{A} \cup \bar{B})} \in \mathfrak{R}$, and $A \backslash B = A \cap \bar{B} \in \mathfrak{R}$, that is, the intersection and difference of two sets belonging to the algebra $\mathfrak{R}$ also belong to $\mathfrak{R}$. From this it follows by induction that the union and intersection of an arbitrary finite number of sets belonging to the algebra $\mathfrak{R}$ also belong to $\mathfrak{R}$. With respect to the unions and intersections of a countably infinite collection of sets in $\mathfrak{R}$, the latter assumption generally ceases to be valid. Therefore we introduce the following definition, which plays a fundamental role.

Definition 2. An algebra of sets $\mathfrak{S}$ is called a *σ-algebra* if for an arbitrary sequence of sets $A_k \in \mathfrak{S}$, where $k = 1, 2, \cdots, \bigcup_{k=1}^{\infty} A_k \in \mathfrak{S}$. The sets $A \in \mathfrak{S}$ are said to be *$\mathfrak{S}$-measurable*. (Since $\bigcap_{k=1}^{\infty} A_k = \overline{\bigcup_{k=1}^{\infty} \bar{A}_k}$, the intersection of an arbitrary countable collection of sets belonging to $\mathfrak{S}$ also belongs to $\mathfrak{S}$.)

Theorem 1. *For every class of sets $\mathfrak{A}$ there exists a smallest σ-algebra $\mathfrak{S}$ containing $\mathfrak{A}$.*

This σ-algebra is called the *σ-algebra generated by the class* $\mathfrak{A}$, and is denoted by $\sigma\{\mathfrak{A}\}$. It is easy to prove the existence of such a σ-algebra. There exist σ-algebras containing $\mathfrak{A}$. To exhibit one, it suffices to take the class of all subsets of the set U. Noting that the intersection of an arbitrary set of σ-algebra is again a σ-algebra, we see that the intersection of all σ-algebras containing $\mathfrak{A}$ is the minimal σ-algebra containing $\mathfrak{A}$.

Definition 3. In a metric space, the σ-algebra of sets generated by the class $\mathfrak{G}$ of open sets is called the *σ-algebra of Borel sets* and its elements are called *Borel sets*.

Obviously the σ-algebra generated by the closed sets of a metric space coincides with the σ-algebra of Borel sets.

We can easily see that in a separable metric space the σ-algebra of Borel sets is the σ-algebra generated by the set of open (or closed) spheres. On the real line the σ-algebras generated by open or closed (or even half-open half-closed) intervals coincide with the σ-algebra of Borel sets.

In n-dimensional Euclidean space E_n we choose for a system of sets generating the σ-algebra of Borel sets the systems of closed, open, or half-open half-closed parallelepipeds (or intervals) $J[\boldsymbol{a}, \boldsymbol{b}]$, $J(\boldsymbol{a}, \boldsymbol{b})$, $J[\boldsymbol{a}, \boldsymbol{b})$, $J(\boldsymbol{a}, \boldsymbol{b}]$. (If $\boldsymbol{a} = (a_1, a_2, \cdots, a_n)$ and $\boldsymbol{b} = (b_1, b_2, \cdots, b_n)$, then $J[\boldsymbol{a}, \boldsymbol{b}) = \{(x_1, x_2, \cdots, x_n); a_i \leqq x_i < b_i, i = 1, 2, \cdots, n\}$. The other intervals are defined in analogous fashion.)

Suppose that to every set A in a certain class of sets $\mathfrak{A}$ we assign a definite number $W = W(A)$, which may be $+\infty$ or $-\infty$. This defines a set function W on $\mathfrak{A}$ into the set of real numbers, $A \rightarrow W = W(A)$.

Definition 4. *A* set function W is said to be *additive* (*or finitely additive*) if it assumes infinite values of only one sign and if, for an arbitrary finite sequence of sets $A_k \in \mathfrak{A}$ (for $k = 1, 2, \cdots, n$) that are pairwise disjoint (that is, $A_k \cap A_r = \emptyset$, for $k \neq r$ where $k, r = 1, 2, \cdots, n$ and $\emptyset$ denotes the empty set) such that

$$\bigcup_{k=1}^{n} A_k \in \mathfrak{A} ,$$

we find that

$$W\left(\bigcup_{k=1}^{n} A_k\right) = \sum_{k=1}^{n} W(A_k) .$$

If this equation holds for an arbitrary countable collection of sets, that is, if

$$W\left(\bigcup_{k=1}^{\infty} A_k\right) = \sum_{k=1}^{\infty} W(A_k)$$

for an arbitrary sequence of sets $A_k \in \mathfrak{A}$, where $A_k \cap A_r = \varnothing$ whenever $k \neq r$, for $k, r = 1, 2, \cdots, n, \cdots$, such that

$$\bigcup_{k=1}^{\infty} A_k \in \mathfrak{A} ,$$

then the set function $W = W(A)$ is said to be *countably additive* (or *completely additive*).

Definition 5. A countably additive nonnegative set function $\mu = \mu(A)$ defined on a σ-algebra of sets $\mathfrak{S}$ and satisfying the equation $\mu(\varnothing) = 0$ is called a *measure*.

If a σ-algebra of sets $\mathfrak{S}$ is defined on a set U and a measure μ is defined on $\mathfrak{S}$, then the set U is called a *space with measure* $\{U, \mathfrak{S}, \mu\}$ or a *measurable space*. The latter term will be applied for a set U with a fixed σ-algebra of sets $\mathfrak{S}$ even when the measure μ is not given. We can easily see that the condition $\mu(\varnothing) = 0$ is equivalent to the condition that $\mu(A)$ is not identically equal to $+\infty$ for all $A \in \mathfrak{S}$.

An arbitrary set $A \in \mathfrak{S}$ of a space with measure $\{U, \mathfrak{S}, \mu\}$ can itself be regarded as a space with measure $\{A, \mathfrak{S}_A, \mu_A\}$, where $\mathfrak{S}_A$ is the σ-algebra of subsets A of the form $A \cap B$ for an arbitrary subset B of $\mathfrak{S}$ and $\mu_A(C) = \mu(C)$ for every $C \in \mathfrak{S}_A$.

We now present a few properties of measures.

Theorem 2. a. *If A and $B \supset A$ belong to $\mathfrak{S}$, then $\mu(A) \leqq \mu(B)$, and if $\mu(A) \neq \infty$, then $\mu(B \backslash A) = \mu(B) - \mu(A)$.*

b. *If $\{A_n\}$ is a countable or infinite sequences of sets belonging to $\mathfrak{S}$, then $\mu(\bigcup_n A_n) \leqq \sum_n \mu(A_n)$.*

c. *If $\{A_n\}$ is an increasing sequence of sets in $\mathfrak{S}$, that is, if $A_{n+1} \supset A_n$ for $n = 1, 2, \cdots$, then*

$$\lim_{n \to \infty} \mu(A_n) = \mu\left(\bigcup_{n=1}^{\infty} A_n\right) . \tag{3}$$

d. *If $\{A_n\}$, for $n = 1, 2, \cdots$, is a decreasing sequence of sets* *in* $\mathfrak{S}$ *and if* $\mu(A_1) < \infty$, *then*

$$\lim_{n\to\infty} \mu(A_n) = \mu\left(\bigcap_{n=1}^{\infty} A_n\right) . \tag{4}$$

Proof. a. Since $B \backslash A \in \mathfrak{S}$ and $B = A \cup (B \backslash A)$ (for $A \subset B$), we have $\mu(B) = \mu(A) + \mu(B \backslash A)$.

b. Let us set $C_1 = A_1$ and $C_n = A_n \backslash (\bigcup_{k=1}^{n-1} A_k)$ for $n = 2, 3, \cdots$. Then the sets C_n belong to $\mathfrak{S}$ and they are pairwise disjoint (that is, $C_n \cap C_r = \varnothing$ for $n \neq r$). Furthermore, $\bigcup_{n=1}^{\infty} C_n = \bigcup_{n=1}^{\infty} A_n$ and $\mu(C_n) \leqq \mu(A_n)$. Therefore,

$$\mu\left(\bigcup_{n=1}^{\infty} A_n\right) = \mu\left(\bigcup_{n=1}^{\infty} C_n\right) = \sum_{n=1}^{\infty} \mu(C_n) \leqq \sum_{n=1}^{\infty} \mu(A_n) .$$

c. If $A_n \subset A_{n+1}$ for $n = 1, 2, \cdots$, we obtain, in the notation used above, $C_n = A_n \backslash A_{n-1}$ and $\mu(C_n) = \mu(A_n) - \mu(A_{n-1})$ if $\mu(A_{n-1}) \neq \infty$. Let us suppose that $\mu(A_n) \neq \infty$ for every n and $A_0 = \varnothing$. Then

$$\mu\left(\sum_{n=1}^{\infty} A_n\right) = \sum_{n=1}^{\infty} \mu(C_n) = \sum_{n=1}^{\infty} [\mu(A_n) - \mu(A_{n-1})] = \lim_{n\to\infty} \mu(A_n) .$$

On the other hand, if $\mu(A_{n_0}) = \infty$ for some $n = n_0$, then for $n > n_0$ we have *a fortiori* $\mu(A_n) = \infty$ and $\mu(\bigcup_{n=1}^{\infty} A_n) = \infty$.

d. Let us set $B_n = A_1 \backslash A_n$ for $n = 1, 2, \cdots$. The sets B_n belong to the σ-algebra $\mathfrak{S}$, they increase monotonically (that is, $B_n \subset B_{n+1}$), and from (c), $\mu(\bigcup_{n=1}^{\infty} B_n) = \lim_{n\to\infty} \mu(B_n)$. On the other hand, $\bigcap_{n=1}^{\infty} A_n = A_1 \backslash \bigcup_{n=1}^{\infty} B_n$. Therefore

$$\begin{aligned} \mu\left(\bigcap_{n=1}^{\infty} A_n\right) &= \mu(A_1) - \mu\left(\bigcup_{n=1}^{\infty} B_n\right) = \mu(A_1) - \lim_{n\to\infty} \mu(B_n) \\ &= \mu(A_1) - \lim_{n\to\infty}[\mu(A_1) - \mu(A_n)] = \lim_{n\to\infty} \mu(A_n) . \end{aligned}$$

Definition 6. Let $\{A_n\}$, for $n = 1, 2, \cdots$, denote an infinite sequence of sets. The *limit superior* $\overline{\lim}\, A_n$ of the sequence $\{A_n\}$ is defined as the set consisting of those points of U that belong to infinitely many of the sets A_n. The *limit inferior* $\underline{\lim}\, A_n$ of the sequence $\{A_n\}$ is defined as the set of those points of the space U that belong to all except possibly finitely many of the sets A_n for $n = 1, 2, \cdots$.

Thus

$$\overline{\lim}\, A_n = \bigcap_{n=1}^{\infty} \bigcup_{k=n}^{\infty} A_k , \tag{5}$$

$$\underline{\lim}\, A_n = \bigcup_{n=1}^{\infty} \bigcap_{k=n}^{\infty} A_k . \tag{6}$$

If $\{A_n\}$, for $n = 1, 2, \cdots$, is an increasing sequence, then $\overline{\lim}\, A_n =$

$\underline{\lim}\, A_n = \bigcup_{n=1}^{\infty} A_n$. On the other hand, if $\{A_n\}$ is a decreasing sequence, then $\overline{\lim}\, A_n = \underline{\lim}\, A_n = \bigcap_{n=1}^{\infty} A_n$. It follows from (5) and (6) that the limits superior and inferior of a sequence of sets belonging to a σ-algebra $\mathfrak{S}$ also belongs to $\mathfrak{S}$. If μ denotes a measure on $\mathfrak{S}$, then it follows from assertions c and d of Theorem 2 that

$$\mu(\overline{\lim}\, A_n) = \lim_{n\to\infty} \mu\Big(\bigcup_{k=n}^{\infty} A_k\Big), \tag{7}$$

$$\mu(\underline{\lim}\, A_n) = \lim_{n\to\infty} \mu\Big(\bigcap_{k=n}^{\infty} A_k\Big), \tag{8}$$

with equation 7 holding if the measure μ is finite.

Definition 7. A sequence of sets $\{A_n\}$, for $n = 1, 2, \cdots$, is said to be *convergent* if $\overline{\lim}\, A_n = \underline{\lim}\, A_n$. In this case the common value of the limits superior and inferior of the sequence $\{A_n\}$ is called the *limit* of the sequence $\{A_n\}$: $\lim A_n = \overline{\lim}\, A_n = \underline{\lim}\, A_n$.

It follows from our definition of convergence of a sequence of sets that every point $u \in U$ either belongs to only a finite number of the sets A_n or belongs to all the A_n from some n on. It follows from what was said above that every monotonic sequence is convergent. Since

$$\mu\Big(\bigcap_{k=n}^{\infty} A_k\Big) \leqq \mu(A_n) \leqq \mu\Big(\bigcup_{k=n}^{\infty} A_k\Big),$$

it follows on the basis of formulas (3) and (4) that for every convergent sequence $\{A_n\}$ of sets A_n and every finite measure μ,

$$\mu(\lim A_n) = \lim \mu(A_n). \tag{9}$$

We now introduce the following useful concept:

Definition 8. A class $\mathfrak{M}$ of sets is said to be *monotonic* if the convergence of an arbitrary monotonic sequence of sets $A_n \in \mathfrak{M}$, for $n = 1, 2, \cdots$, implies tnat the limit of such a sequence belongs to $\mathfrak{M}$.

Since the intersection of monotonic classes is a monotonic class, it follows that corresponding to an arbitrary class $\mathfrak{A}$ of sets there is a minimal monotonic class $m\{\mathfrak{A}\}$ containing $\mathfrak{A}$.

Obviously, every σ-algebra is a monotonic class and every algebra that is a monotonic class is a σ-algebra: $\bigcup_{n=1}^{\infty} A_n = \lim \bigcup_{k=1}^{n} A_k$. In many cases we need to show that a particular class of sets contains a minimal σ-algebra generated by a given algebra. For which the following theorem is useful.

Theorem 3. *The minimal monotonic class $m\{\mathfrak{A}\}$ containing the algebra $\mathfrak{A}$ coincides with the minimal σ-algebra $\sigma\{\mathfrak{A}\}$.*

On the basis of the remark made above it is sufficient to show that $m\{\mathfrak{A}\} \supset \sigma\{\mathfrak{A}\}$, and to do this it is sufficient to show that $m\{\mathfrak{A}\}$ is an algebra. Let $\mathfrak{K}\{A\}$ denote the class of all sets B such that

$$A \cup B \in m\{\mathfrak{A}\},\ A \backslash B \in m\{\mathfrak{A}\};\ B \backslash A \in m\{\mathfrak{A}\}\ . \tag{10}$$

The class $\mathfrak{K}\{A\}$ is monotonic: if $\{B_n\}$ is a monotonic sequence of sets and each B_n satisfies conditions (10), then $A \cup B_n$, $A \backslash B_n$, $B_n \backslash A$, are also monotonic sequences of sets and

$$\lim (A \cup B_n) = A \cup \lim B_n \in m\{\mathfrak{A}\}\ ,$$
$$\lim (A \backslash B_n) = A \backslash \lim B_n \in m\{\mathfrak{A}\}\ ,$$
$$\lim (B_n \backslash A) = \lim B_n \backslash A \in m\{\mathfrak{A}\}\ ;$$

that is, $\lim B_n \in m\{\mathfrak{A}\}$. If $A \in \mathfrak{A}$, then $\mathfrak{K}\{A\} \supset \mathfrak{A}$. Consequently, $m\{\mathfrak{A}\} \subset \mathfrak{K}\{A\}$; that is, for every $F \in m\{\mathfrak{A}\}$ we have $F \in \mathfrak{K}\{A\}$. It then follows from the definition of $\mathfrak{K}\{F\}$ that $A \in \mathfrak{K}\{F\}$. Then just as above, it follows from the monotonicity of $\mathfrak{K}\{F\}$ that $m\{\mathfrak{A}\} \subset \mathfrak{K}\{F\}$. This means that relations (10) hold for arbitrary A and B in $m\{\mathfrak{A}\}$; that is, $m\{\mathfrak{A}\}$ is an algebra of sets.

Let us pause to look at arbitrary countably additive functions defined on a σ-algebra of $\mathfrak{S}$. We shall call them *charges*. Since every charge is the difference between two measures the study of charges reduces to the study of measures. This follows immediately from Theorem 4 below.

Definition 9. For arbitrary $A \in \mathfrak{S}$, the quantities

$$W^+(A) = \sup_{A' \subset A,\, A' \in \mathfrak{S}} W(A');\quad W^-(A) = -\inf_{A' \subset A,\, A' \in \mathfrak{S}} W(A') \tag{11}$$

are called respectively the *positive* and *negative variations* of the charge W on the set A, and the quantity

$$|\,W\,|\,(A) = W^+(A) + W^-(A) \tag{12}$$

is called the *absolute variation.*

We note that for arbitrary $A \in \mathfrak{S}$,

$$|\,W\,|\,(A) \geqq |\,W(A)\,|\ . \tag{13}$$

It follows immediately from the definition that W^+ and W^- are nonnegative and nondecreasing set functions: If $A \subset B$, then

$$0 \leqq W^{\pm}(A) \leqq W^{\pm}(B)\ . \tag{14}$$

Furthermore,

$$W^{\pm}(A_1 \cup A_2) \leqq W^{\pm}(A_1) + W^{\pm}(A_2)\ . \tag{15}$$

Throughout the remaining portion of this section we shall assume that the space U and a σ-algebra $\mathfrak{S}$ on it are fixed. All the sets that we shall consider are assumed to be $\mathfrak{S}$-measurable.

Lemma 1. *If $W(A) < \infty$ for every A, then $W^+(U) < \infty$.*

To prove this, let us assume the opposite, namely that $W^+(U) = \infty$. Let us show that in this case there exists for arbitrary $c > 0$, a set A such that $W(A) > c$ and $W^+(A) = \infty$.

We also prove this assertion by contradiction. If it is not valid, there exists an A_1 such that $W(A_1) > c$ and $W^+(A_1) < \infty$. If we set $A_2 = U \backslash A_1$ in inequality (15) we obtain the result that $W^+(U \backslash A_1) = \infty$. Repeating the above reasoning with U replaced by $U \backslash A_1$, we see that there exists an $A_2 \subset U \backslash A_1$ such that $W(A_2) > c$ and $W^+(A_2) < \infty$. We obtain by induction an infinite sequence of sets $A_1, A_2, \cdots$, belonging to $\mathfrak{S}$, pairwise disjoint, and such that $W(A_n) > c$, so that

$$W\left(\bigcup_{n=1}^{\infty} A_n\right) = \sum_{n=1}^{\infty} W(A_n) = \infty ,$$

which contradicts the hypothesis of the theorem. Thus for every c there exists an A such that $W(A) > c$ and $W^+(A) = \infty$. Let us take successively $c = 1, 2, \cdots, n \cdots$ and then apply what we have just proven to construct a sequence of sets B_n, such that for each n, $W(B_n) > n$, $W^+(B_n) = \infty$ and $B_{n+1} \subset B_n$. For this we first use U to find B_1, then B_1 to find $B_2 \subset B_1$, and so forth. Let us set $D = \bigcap_{n=1}^{\infty} B_n$. Then $W(B_1 \backslash D) = \lim[W(B_1) - W(B_n)] = -\infty$ so that $W(D) = +\infty$, which is impossible. This contradiction completes the proof of the lemma.

Theorem 4 (Hahn). *Let W denote an arbitrary charge on a σ-algebra $\mathfrak{S}$. Then U can be partitioned into two sets P and N such that $U = P \cup N$, $P \cap N = \varnothing$, and for every $A \in \mathfrak{S}$, $W(A \cap P) \geqq 0$, $W(A \cap N) \leqq 0$.*

Proof. Since the charge can assume infinite values of only one sign, we may suppose that W does not assume the value $+\infty$. On the basis of Lemma 1,

$$\beta = W^+(U) = \sup_{A \in \mathfrak{S}} W(A) < +\infty$$

Let $\{C_n\}$ denote a sequence of sets such that $W(C_n) > \beta - 2^{-n}$ and

$$P = \overline{\lim}\, C_n = \bigcap_{n=1}^{\infty} \bigcup_{k=n}^{\infty} C_k .$$

If $A \subset C_n$, then

$$W(A) = W(C_n) - W(C_n \backslash A) > -\frac{1}{2^n} .$$

Therefore, for arbitrary $A \subset P$ we obtain from the relation

$A \subset \bigcup_{k=n}^{\infty} A \cap C_k$ the result

$$W(A) = \sum_{k=n}^{\infty} W\Big(A \cap \Big(C_k \setminus \bigcup_{j=1}^{k-1} C_j\Big)\Big) \geqq - \sum_{k=n}^{\infty} \frac{1}{2^k} = - \frac{1}{2^{n-1}},$$

or taking the limit as $n \longrightarrow \infty$, we find

$$W(A) \geqq 0 . \tag{16}$$

Let us now set

$$N = U \setminus P = \bigcup_{n=1}^{\infty} D_n ,$$

where

$$D_n = \bigcap_{k=n}^{\infty} (U \setminus C_k) .$$

If $A \cap C_n = \varnothing$, then $W(A) \leqq 2^{-n}$. (This is true because the inequality $W(A) > 2^{-n}$ implies that $W(A \cup C_n) = W(A) + W(C_n) > \beta$, which is impossible.) From this it follows that the relation

$$A \cap \Big(\bigcup_{k=n}^{\infty} C_k\Big) = \varnothing$$

implies that $W(A) \leqq 0$. Therefore if $A \subset N$,

$$A = \bigcup_{n=1}^{\infty} (A \cap D_n), \quad W(A) = \sum_{n=1}^{\infty} W(A \cap (D_n \setminus D_{n-1})) \leqq 0 .$$

Thus

$$W(A) \leqq 0 \text{ for every } A \subset N . \tag{17}$$

This completes the proof of the theorem.

Corollary 1. *The positive, negative, and absolute variations of a charge are measures and*

$$W^+(A) = W(A \cap P), \quad W^-(A) = - W(A \cap N) , \tag{18}$$

$$|W|(A) = W(A \cap P) - W(A \cap N) . \tag{19}$$

Corollary 2. *An arbitrary charge can be represented as the difference of two measures:*

$$W(A) = W^+(A) - W^-(A) . \tag{20}$$

Corollary 3.

$$\sup_{\substack{A' \subset A \\ A' \in \mathfrak{S}}} |W(A')| \leqq |W|(A) \leqq 2 \sup_{\substack{A' \subset A \\ A' \in \mathfrak{S}}} |W(A')| . \tag{21}$$

Proof. Formula (20) follows from (18) since

$$W(A) = W(A \cap P) + W(A \cap N) = W^+(A) - W^-(A) ,$$

and inequalities (21) follow from (20) and (19).

Let W denote the set of all finite charges on a σ-algebra $\mathfrak{S}$. This set is a linear space if the sum of two charges and the product of a charge and a number are defined in the natural manner

$$(W_1 + W_2)(A) = W_1(A) + W_2(A)\,, \qquad (tW)(A) = tW(A)\,.$$

We now define a norm on W:

$$\| W \| = | W |(U)\,.$$

It follows from formulas (18) that

$$\| tW \| = | t | \, \| W \|$$

and from (15) that

$$\| W_1 + W_2 \| \leqq \| W_1 \| + \| W_2 \|\,.$$

Thus W becomes a normed space. Convergence in W is called *convergence in variation*. If $\{W_n\}$, for $n = 1, 2, \cdots$, converges to W in variation, that is, if $\| W - W_n \| \to 0$ as $n \to \infty$, then $W_n(A) \to W(A)$ uniformly over all sets $A \in \mathfrak{S}$:

$$\sup_{A \in \mathfrak{S}} | W(A) - W_n(A) | \leqq \| W - W_n \|$$

(cf. inequality (21)).

Theorem 5. *The space W with norm $\| W \| = | W |(U)$ is a Banach space* (that is, a complete normed linear space).

We need prove only the completeness of the space. Suppose that $\| W_{n'} - W_n \| \to 0$ as $n, n' \to \infty$. For every $A \in \mathfrak{S}$, the sequence of the numbers $W_n(A)$ is a fundamental sequence and it converges to a finite limit. Let us set $W_0(A) = \lim_{n\to\infty} W_n(A)$. The set function $W_0(A)$ is defined on $\mathfrak{S}$ and is finite and additive. Let us show that it is countably additive. Let $\{A_k\}$, for $k = 1, 2, \cdots$, denote a sequence of disjoint sets in $\mathfrak{S}$. Then

$$\begin{aligned}
&\left| W_0\Big(\bigcup_{k=1}^{\infty} A_k\Big) - W_0\Big(\bigcup_{k=1}^{n} A_k\Big) \right| \\
&\qquad = \left| W_0\Big(\bigcup_{k=n+1}^{\infty} A_k\Big) \right| = \lim_{p\to\infty} \left| W_{m+p}\Big(\bigcup_{k=n+1}^{\infty} A_k\Big) \right| \\
&\qquad \leqq \left| W_m\Big(\bigcup_{k=n+1}^{\infty} A_k\Big) \right| + \overline{\lim_{p\to\infty}} \| W_{m+p} - W_m \|\,. \qquad (22)
\end{aligned}$$

The right-hand member of this inequality can be made arbitrarily small by the suitable choice of m and n. It follows from (21) that $\{W_n\}$ converges in variation to W_0. This completes the proof of the theorem.

2. MEASURABLE FUNCTIONS

From an intuitive point of view, a random variable ξ is a (variable) number that corresponds to each possible outcome of an experiment. Since the outcomes of an experiment are described by elementary events, a random variable can be regarded as a function of an elementary event, $\xi = f(u)$ for $u \in U$. On the other hand, in elementary probability theory a random variable ξ is completely characterized by its distribution function $F(x) = \mathsf{P}\{\xi < x\}$. Corresponding to the event $\{\xi < x\}$ is the set of elementary events $\{u, f(u) < x\}$. Therefore, for it to be meaningful to speak of a distribution function of a random variable, the set $\{u, f(u) < x\}$ must for arbitrary real x belong to $\mathfrak{S}$. In this section we shall study the class of functions defined on a measurable space $\{U, \mathfrak{S}, \mu\}$ which enjoy this property.

Definition 1. Let $\mathfrak{S}$ denote a σ-algebra of sets of the space U. Let $f(u)$ denote a function defined on an $\mathfrak{S}$-measurable set M and assuming real values (and possibly the values $\pm\infty$). Such a function $f(u)$ is said to be *$\mathfrak{S}$-measurable* if for every real x the set $\{u; f(u) < x\}$ is $\mathfrak{S}$-measurable.

We note a few properties of measurable functions.

Theorem 1. *Let A denote an arbitrary Borel set in the n-dimensional space E_n and let $f_1(u), \cdots, f_n(u)$ denote $\mathfrak{S}$-measurable functions all defined on the same set $M \in \mathfrak{S}$. Then the set*

$$\{u; u \in M, (f_1(u), f_2(u), \cdots, f_n(u)) \in A\}$$

is $\mathfrak{S}$-measurable.

Proof. Since

$$\begin{aligned}\{u; u \in M, (f_1(u), \cdots, f_n(u)) \in A'\backslash A''\} \\ = \{u; u \in M, (f_1(u), \cdots, f_n(u)) \in A']\backslash\{u; u \in M, \\ (f_1(u), \cdots, f_n(u)) \in A''\},\end{aligned}$$

$$\begin{aligned}\left\{u; u \in M, (f_1(u), \cdots, f_n(u)) \in \bigcup_{k=1}^{\infty} A^{(k)}\right\} \\ = \bigcup_{k=1}^{\infty} \{u; u \in M, (f_1(u), \cdots, f_n(u)) \in A^{(k)}\},\end{aligned}$$

the class $\mathfrak{A}$ of sets A contained in E_n such that the set

$$\{u; u \in M, (f_1(u), \cdots, f_n(u)) \in A\}$$

is $\mathfrak{S}$-measurable constitutes a σ-algebra. Furthermore, $\mathfrak{A}$ contains the n-dimensional infinite intervals

$$I_{a_1 \cdots a_n} = \{(x_1, \cdots, x_n); x_1 < a_1, \cdots, x_n < a_n\}$$

since

$$\{u;\ u \in M,\ (f_1(u), \cdots, f_n(u)) \in I_{a_1\cdots a_n}\} = \bigcap_{k=1}^{n} \{u;\ u \in M,\ f_k(u) < a_k\} .$$

Consequently $\mathfrak{A}$ contains all Borel sets in E_n.

Corollary 1. *If $f(u)$ is an $\mathfrak{S}$-measurable function, then for every x the sets*

$$\{u;\ u \in M, f(u) \leqq x\};\ \{u;\ u \in M,\ f(u) > x\} ,$$
$$\{u;\ u \in M, f(u) \geqq x\};\ \{u;\ u \in M,\ f(u) = x\} ,$$
$$\{u;\ u \in M,\ a \leqq f(x) < b\} ,\quad \text{etc.},$$

are $\mathfrak{S}$-measurable.

REMARK 1. As one can see from the proof of Theorem 1, the assertion in that theorem holds for an arbitrary function $f(u)$ defined on an $\mathfrak{S}$-measurable set M and satisfying the condition $\{u,\ u \in M,\ f(u) \in K\} \in \mathfrak{S}$ for a certain class of sets K that generates a σ-algebra containing $\mathfrak{B}_1$ (the σ-algebra of Borel sets in E_1). In particular the function $f(u)$ defined on $M \in \mathfrak{S}$ is $\mathfrak{S}$-measurable if for arbitrary real x one of the following systems of sets

$$\{u;\ u \in M,\ f(u) \leqq x\}$$
$$\{u;\ u \in M,\ x \leqq f(u)\}$$
$$\{u;\ u \in M,\ x < f(u)\}$$

is $\mathfrak{S}$-measurable (x may range only over an arbitrary everywhere-dense set).

Theorem 2. *Let $\{f_n(u), n = 1, 2, \cdots, u \in M\}$ denote a sequence of $\mathfrak{S}$-measurable functions. Then the functions*

$$\sup_n f_n(u),\ \inf_n f_n(u),\ \overline{\lim_n} f_n(u),\ \underline{\lim_n} f_n(u)$$

are $\mathfrak{S}$-measurable.

The proof follows from the relations:

$$\{u;\ u \in M,\ \sup_n f_n(u) > x\} = \bigcup_{n=1}^{\infty} \{u;\ u \in M,\ f_n(u) > x\} ,$$

$$\{u;\ u \in M,\ \inf_n f_n(u) < x\} = \bigcup_{n=1}^{\infty} \{u;\ u \in M,\ f_n(u) < x\} ,$$

$$\{u;\ u \in M, \overline{\lim} f_n(u) < x\} = \bigcup_{k=1}^{\infty} \bigcup_{n=1}^{\infty} \bigcap_{j=n}^{\infty} \left\{u;\ u \in M,\ f_j(u) < x - \frac{1}{k}\right\} ,$$

$$\{u;\ u \in M,\ \underline{\lim_n}\ f_n(u) > x\} = \bigcup_{k=1}^{\infty} \bigcup_{n=1}^{\infty} \bigcap_{j=n}^{\infty} \left\{u;\ u \in M,\ f_j(u) > x + \frac{1}{k}\right\} .$$

Definition 2. The *characteristic function* $\chi_A(u)$ of a set A is defined as the function that is equal to 1 for $u \in A$ and equal to 0 for $u \notin A$.

Note the following obvious relations:

$$\chi_{A \cap B}(u) = \chi_A(u)\chi_B(u); \tag{1}$$

$$\chi_{A \cup B}(u) = \chi_A(u) + \chi_B(u) \qquad (A \cap B = \varnothing); \tag{2}$$

$$\chi_{\bar{A}}(u) = 1 - \chi_A(u); \tag{3}$$

$$\chi_{\overline{\lim} A_n}(u) = \overline{\lim}\, \chi_{A_n}(u); \tag{4}$$

$$\chi_{\underline{\lim} A_n}(u) = \underline{\lim}\, \chi_{A_n}(u)\ . \tag{5}$$

Definition 3. An $\mathfrak{S}$-measurable function $f(u)$ is called a *simple function* if there exists a finite collection of sets, each contained in the domain of definition of f and together covering this domain of definition, such that f assumes a constant finite value on each member of the collection (though possibly differing from member to member).

Suppose that a simple function $f(u)$ is defined on a set $M \in \mathfrak{S}$ and assumes the values $a_1, a_2, \cdots, a_n$ (where $a_i \neq a_j$ if $i \neq j$, for $i, j = 1, \cdots, n$).

Let us set $A_j = \{u; u \in M, f(u) = a_j\}$ for $j = 1, \cdots, n$. Then the A_j are $\mathfrak{S}$-measurable and

$$f(u) = \sum_{j=1}^{n} a_j \chi_{A_j}(u), \quad u \in M \tag{6}$$

where $\chi_{A_j}(u)$ is the characteristic function of the set A_j. On the other hand, every function that can be represented in the form (6) is a simple function defined on M.

Theorem 3. *For a function $f(u)$ (where $u \in M \in \mathfrak{S}$) to be $\mathfrak{S}$-measurable, it is necessary and sufficient that it be the limit of a sequence of simple functions that converges everywhere on M.*

Proof. The sufficiency follows from Theorem 2. To prove the necessity we set

$$A_{N,-2^N N} = \{u; u \in M, f(u) < -N\}\ ;$$

$$A_{N,k} = \left\{u; n \in M, \frac{k-1}{2^N} \leqq f(u) < \frac{k}{2^N}\right\},$$

$$k = -2^N N + 1,\ -2^N N + 2, \cdots, 2^N N\ ,$$

$$A_{N,2^N N+1} = \{u; u \in M, f(u) \geqq N\}\ ,$$

$$f_N(u) = \sum_{k=-2^N N}^{2^N N+1} \frac{k-1}{2^N} \chi_{A_{N,k}}(u), \quad u \in M\ .$$

Then $|f_N(u) - f(u)| < 2^{-N}$ if $|f(u)| \leqq N$, $f_N(u) = N$ if $f(u) \geqq N$, and $f_N(u) < -N$ if $f(u) < -N$. Consequently, $\lim f_N(u) = f(u)$, $u \in M$.

This completes the proof of the theorem.

REMARK 2. If $f(u)$ is nonnegative (or at least bounded below) and $\mathfrak{S}$-measurable, it is the limit of an everywhere-convergent nondecreasing sequence of simple functions. To see this, we note that the functions $f_N(u)$ constitute in this case a nondecreasing sequence beginning with some number N.

Let us now look at functions $g(x)$ defined on some metric space R into the extended real line (that is, the set of real numbers with the values $\pm\infty$ included). Let $\mathfrak{B}$ denote the σ-algebra of Borel sets contained in R.

Definition 4. A function $f(x)$ for $x \in R$ is called a *Borel function* if for arbitrary real a the set $\{x; f(x) < a\}$ is a Borel set.

Definition 5. A *Baire function* is defined as a function belonging to the smallest class B of functions defined on R that satisfies the following two conditions: (a) B contains all continuous functions; (b) B is closed under passage to the limit; that is, if $\{f_n(x)\}$, for $n = 1, 2, \cdots$, is an arbitrary sequence of functions $f_n(x) \in B$ that converges in R, then $\lim_{n\to\infty} f_n(x) \in B$.

Theorem 4. *The classes of Borel and Baire functions coincide.*

Proof. Let Q denote the class of Borel functions. Q contains all continuous functions and is closed under passage to the limit:

$$\{x; \lim f_n(x) < a\} = \{x; \overline{\lim}\, f_n(x) < a\}$$
$$= \bigcup_{k=1}^{\infty} \bigcup_{n=1}^{\infty} \bigcap_{j=n}^{\infty} \left\{x; f_j(x) < a - \frac{1}{k}\right\} \in \mathfrak{B}\,,$$
$$\text{if } f_n(x) \in Q \,(\text{for } n = 1, 2, \cdots)\,.$$

Thus Q contains all Baire functions. Let us now show that every Borel function is also a Baire function. Let us show first that the class of Baire functions is linear. Let $K(g)$ denote the class of all functions $f(x)$ defined on R such that for arbitrary real α and β the function $\alpha f + \beta g$ is a Baire function. $K(g) = \{f(x); \alpha f(x) + \beta g(x) \in B$ for arbitrary α and $\beta\}$. If g and f are continuous, then $f(x) \in K(g)$. Furthermore, the class $K(g)$ is closed under passage to the limit: If $f_n(x) \in K(g)$, for $n = 1, 2, \cdots$, and if $\lim f_n(x) = f_0(x)$, then $f_0(x) \in K(g)$. Consequently $B \subset K(g)$ for an arbitrary continuous function $g(x)$. From the definition of the class $K(g)$ it follows that the relation $f \in K(g)$ implies $g \in K(f)$. Thus, the relation $B \subset K(g)$ for $g \in C$ implies that $C \subset K(f)$ for arbitrary $f \in B$, where C is the class of continuous functions. Since $K(f)$ is closed under passage to the limit and contains C it follows that $B \subset K(f)$ for arbitrary $f \in B$. This means that for arbitrary Baire functions f and g and arbitrary real numbers α and β, $\alpha f(x) + \beta g(x) \in B$; that is, the class B constitutes a linear space.

Let us now look at the class of sets $\mathfrak{A}$ whose characteristic functions are Baire functions. We have the following:

a. $\mathfrak{A}$ contains all closed sets. Let F denote a closed set, let $\chi_F(x)$ denote the characteristic function of F, and let $\rho(x, F)$ denote the distance from the point x to the set F. Since $|\rho(x, F) - \rho(y, F)| \leqq \rho(x, y)$, the function $\rho(x, F)$ is continuous. Let us set

$$f_\varepsilon(x) = \begin{cases} 1 - \dfrac{\rho(x, F)}{\varepsilon} & \text{if } \rho(x, F) < \varepsilon \,, \\ 0 & \text{if } \rho(x, F) \geqq \varepsilon \,. \end{cases}$$

The function $f_\varepsilon(x)$ is continuous and $\lim_{\varepsilon \to 0} f_\varepsilon(x) = \chi_F(x)$. Consequently $F \in \mathfrak{A}$.

b. The class $\mathfrak{A}$ is monotonic. Let A_n denote a monotonic sequence of sets $A_n \in \mathfrak{A}$ and define $A_0 = \lim A_n$. Then $x_{A_0}(x) = \lim \chi_{A_n}(x) \in B$. It follows from a and b that $\mathfrak{A}$ contains all Borel sets in the space (Theorem 3, Section 1).

c. Let $f(x)$ denote an arbitrary Borel function. On the basis of Theorem 3, there exists a sequence of simple Borel functions $f_N(x)$ such that

$$f(x) = \lim f_N(x) \,. \tag{7}$$

The simple functions $f_N(x)$ admit a representation of the form (6) in which the A_j are Borel sets. Since the class B constitutes a linear space, simple Borel functions belong to B. Since B is closed under passage to the limit, on the basis of equation (7) an arbitrary Borel function is a Baire function. This completes the proof of the theorem.

Let us now look at the properties of measurable functions.

Theorem 5. *Let $f_1(u), \cdots, f_n(u)$ denote a sequence of finite $\mathfrak{S}$-measurable functions defined on an $\mathfrak{S}$-measurable set M and let $\varphi(t_1, \cdots, t_n)$ denote a Borel function in n-dimensional space E_n. Then the function $\varphi(f_1(u), \cdots, f_n(u))$ for $u \in M$ is $\mathfrak{S}$-measurable.*

Proof. For arbitrary real a, the set

$$B_a = \{(t_1, \cdots, t_n);\ \varphi(t_1, \cdots, t_n) < a\}$$

is a Borel subset of E_n. The set

$$\{u;\ u \in M,\ \varphi(f_1(u),\ f_2(u), \cdots, f_n(u)) < a\} = \{u;\ u \in M,\ (f_1(u), \cdots, f_n(u)) \in B_a\}$$

is $\mathfrak{S}$-measurable on the basis of Theorem 1. This completes the proof of the theorem.

Corollary 1. *If f and g are $\mathfrak{S}$-measurable finite functions, then the functions $f \pm g$, fg, and $1/g$ are also $\mathfrak{S}$-measurable. Here $1/x$ must be assigned some value unique for $x = 0$.*

This follows from the fact that the functions $x \pm y$, xy, and $1/x$ are Borel functions.

Corollary 2. *For any two $\mathfrak{S}$-measurable functions $f(u)$ and $g(u)$, where $u \in M$, the sets $\{u; u \in M, f(u) < g(u)\}$; $\{u; u \in M, f(u) = g(u)\}$ are measurable.*

The proof follows from the measurability of the function $f(u) - g(u)$.

Definition 6. Let μ denote a measure with domain of definition $\mathfrak{S}$. Two functions f and g are said to be *equivalent* (more precisely, *μ-equivalent*) on a set $M \in \mathfrak{S}$ if the set $A = \{u; u \in M, f(u) \neq g(u)\}$ is $\mathfrak{S}$-measurable and $\mu(A) = 0$.

Definition 7. A σ-algebra of sets $\mathfrak{S}$ is said to be *complete* (or *μ-complete* or *complete with respect to the measure μ*) if an arbitrary subset N' of a set N of μ-measure 0 is $\mathfrak{S}$-measurable; that is, if the relations $N' \subset N$, $N \in \mathfrak{S}$, and $\mu(N) = 0$ imply

$$N' \in \mathfrak{S} . \tag{8}$$

The measure μ defined on a μ-complete σ-algebra of sets is also said to be complete.

Of course relation (8) implies that $\mu(N') = 0$.

Theorem 6. *If $\mathfrak{S}$ is a μ-complete σ-algebra, if $f(u)$, for $u \in M$, is a $\mathfrak{S}$-measurable function, and if the functions $f(u)$ and $g(u)$ are equivalent on M, then $g(u)$ for $u \in M$ is also $\mathfrak{S}$-measurable.*

It follows from the hypothesis of the theorem that for arbitrary real a, $\{u; u \in M, g(u) < a\} = \{u; u \in M, f(u) < a\} \backslash N' \cup N''$, where N' and N'' are subsets of the set $N = \{u; u \in M, f(u) \neq g(u)\}$ of μ-measure 0. By virtue of the completeness of the measure, the set $\{u; u \in M, g(u) < a\}$ is $\mathfrak{S}$-measurable.

The set of all $\mathfrak{S}$-measurable functions defined on M and equivalent to a given function $f(u)$ is obviously some complete equivalence class of functions. In many cases there is no point in distinguishing among equivalent functions. Then the word "function" actually refers to an entire class of $\mathfrak{S}$-measurable functions that are equivalent to each other. In what follows we shall often proceed from this point of view.

Let us make some remarks concerning terminology. A certain property is said to *hold μ-almost-everywhere* on M if the μ-measure of the set of points at which this property does not hold is equal to 0. For example, if two functions f and g are equivalent on M, we may say that f and g coincide μ-almost-everywhere on M. A sequence of functions $f_n(u)$, for $n = 1, 2, \cdots$, and where $u \in M$, is said to *converge μ-almost-everywhere* to the function $f(u)$ on M if the μ-measure of the set of those points $u \in M$ at which $\lim_{n\to\infty} f_n(u)$ does not exist or does not coincide with $f(u)$ is equal to 0.

If a property holds μ-almost-everywhere on a set M, let us indicate this property with the expression (mod μ) instead of the more cumbrous "μ-almost-everywhere." Thus if f and g are equivalent on a set M, we can write $f(u) = g(u)$, $u \in M(\text{mod } \mu)$. Similarly, if $\{f_n(u)\}$ converges to $f(u)$ μ-almost-everywhere on M, we can write simply $\lim_{n\to\infty} f_n(u) = f(u)$, $u \in M(\text{mod } \mu)$.

3. CONVERGENCE IN MEASURE

A sequence of random variables ξ_n is said to *converge in probability* to a random variable ξ if for arbitrary $\varepsilon > 0$ $P\{|\xi_n - \xi| > \varepsilon\} \to 0$ as $n \to \infty$, and we indicate this fact by writing $\xi = P - \lim \xi_n$.

Corresponding to this definition in the general theory of functions is the following: Let $\{U, \mathfrak{S}, \mu\}$ denote a space with a measure and let $\{f_n(u)\}$ for $n = 1, 2, \cdots$ denote a sequence of μ-almost-everywhere finite $\mathfrak{S}$-measurable functions on U.

Definition 1. A sequence $\{f_n(u)\}$ is said to *converge in μ-measure* to a $\mathfrak{S}$-measurable function $f(u)$ if for arbitrary $\varepsilon > 0$,

$$\mu\{u; |f_n(u) - f(u)| > \varepsilon\} \to 0 \text{ as } n \to \infty .$$

We indicate this by writing

$$f(u) = \mu\text{-lim } f_n(u) .$$

In this section we shall consider those properties of sequences of functions that are related to convergence in measure and we shall look at the relation between convergence in measure and ordinary convergence (convergence at each point of some set). Some measurable space $\{U, \mathfrak{S}, \mu\}$ is considered fixed. All the functions in question will be assumed $\mathfrak{S}$-measurable and (mod μ) finite even though this may not be explicitly stated.

Let $\{f_n(u)\}$, for $n = 1, 2, \cdots$, denote a given sequence of $\mathfrak{S}$-measurable finite (mod μ) functions on U. Let S denote the set of points of U at which the sequence $\{f_n(u)\}$ converges to a finite limit and let D denote the set of points at which this sequence diverges. Then

$$S = \bigcap_{k=1}^{\infty} \bigcup_{n=1}^{\infty} \bigcap_{m=1}^{\infty} \left\{u; \mid f_n(u) - f_{n+m}(u) \mid < \frac{1}{k}\right\}, \tag{1}$$

$$D = \bigcup_{k=1}^{\infty} \bigcap_{n=1}^{\infty} \bigcup_{m=1}^{\infty} \left\{u; \mid f_n(u) - f_{n+m}(u) \mid \geqq \frac{1}{k}\right\}, \tag{2}$$

that is, the sets S and D are $\mathfrak{S}$-measurable, so that it is always meaningful to speak of the measure of the set on which a sequence $\{f_n(u)\}$ converges or diverges. If $\mu(D) = 0$, the sequence of functions converges μ-almost-everywhere. We set $f(u) = \lim f_n(u)$, $u \in S$, and we extend the definition of f_n to the set $U \backslash S = D$ by setting $f(u) = 0$ for $u \in D$. Then $\lim f_n(u) = f(u) \pmod{\mu}$, and the function $f(u)$ is the finite limit (mod μ) of the sequence $\{f_n(u)\}$ on U.

Theorem 1. *Let μ denote a finite measure. If a sequence $\{f_n(u)\}$ of functions $f_n(u)$ converges* (mod μ) *to a finite* (mod μ) *function $f(u)$ on U, then $\{f_n(u)\}$ converges to $f(u)$ in μ-measure.*

Proof. Let D denote the subset of U on which the sequence $\{f_n(u)\}$ does not converge to $f(u)$. Define

$$D_{km} = \bigcup_{n=m}^{\infty} \left\{u; \mid f_n(u) - f(u) \mid \geqq \frac{1}{k}\right\},$$

$$D_k = \bigcap_{m=1}^{\infty} D_{km}, \quad D = \bigcup_{k=1}^{\infty} D_k .$$

The sets D_k constitute an increasing sequence and the sets D_{km} constitute for fixed k a decreasing sequence. Consequently, $\mu(D) = \lim_{k\to\infty} \mu(D_k) = 0$, so that $\mu(D_k) = 0$ and $\mu(D_k) = \lim \mu(D_{km}) = 0$. Thus for every k and $\varepsilon > 0$ there exists an m such that

$$\mu\left\{u; \mid f_n(u) - f(u) \mid \geqq \frac{1}{k}\right\} < \varepsilon$$

for all $n \geqq m$, which means that the sequence $\{f_n(u)\}$ converges to $f(u)$ in μ-measure.

The following theorem asserts that a sequence of functions that converges in μ-measure has no more than one limit (mod μ).

Theorem 2. *If $f(u) = \mu\text{-}\lim f_n(u)$ and $g(u) = \mu\text{-}\lim f_n(u)$ (for $u \in M$), then the functions f and g are equivalent on M.*

This is true because

$$\mu\{u; u \in M, f \neq g\}$$
$$\leqq \lim_{k\to\infty} \mu\left\{u; u \in M, \mid f(u) - g(u) \mid > \frac{2}{k}\right\},$$

$$\mu\left\{u;\, u \in M,\, |f(u) - g(u)| > \frac{2}{k}\right\}$$
$$\leqq \mu\left\{u;\, u \in M,\, |f(u) - f_n(u)| + |f_n(u) - g(u)| > \frac{2}{k}\right\}$$
$$\leqq \mu\left\{u;\, u \in M,\, |f(u) - f_n(u)| > \frac{1}{k}\right\}$$
$$+ \mu\left\{u;\, u \in M,\, |f_n(u) - g(u)| > \frac{1}{k}\right\} > 0 \text{ as } k \longrightarrow \infty\ .$$

Definition 2. A sequence $\{f_n(u)\}$, for $n = 1, 2, \cdots$, is said to be *bounded with respect to a measure* μ on a set M if

$$\sup_n \mu\{u;\, u \in M,\, |f_n(u)| > L\} \longrightarrow 0 \text{ as } L \longrightarrow \infty\ .$$

Theorem 3. *If a sequence* $\{f_n(u)\}$, *for* $n = 1, 2, \cdots$, *converges in* μ*-measure to a finite* (mod μ) *function* $f(u)$ *on* U, *it is bounded on* U *with respect to the measure* μ.

Proof. For arbitrary L,

$$\mu\{u;\, u \in M,\, |f_n(u)| > L\} \leqq \mu\left\{u;\, u \in M,\, |f_n(u) - f(u)| > \frac{L}{2}\right\}$$
$$+ \mu\left\{u;\, u \in M,\, |f(u)| > \frac{L}{2}\right\} \tag{3}$$

The sets $\{u;\, u \in M,\, |f(u)| > L/2\}$ constitute a decreasing sequence as $L \longrightarrow \infty$ and

$$\bigcap_{L=1}^{\infty} \left\{u;\, |f(u)| > \frac{L}{2}\right\} = \{u;\, |f(u)| = \infty\}\ .$$

Therefore the second term on the right side of formula (3) approaches 0. That the same is true of the first term follows from the convergence in measure of $\{f_n\}$ to f.

Let us find a necessary and sufficient condition for convergence in measure of a sequence of functions. We introduce a useful auxiliary concept:

Definition 3. A sequence $\{f_n(u)\}$, for $n = 1, 2, \cdots$, is said to be *almost uniformly convergent* on U if for arbitrary $\varepsilon > 0$ there exists a set H such that $\mu(H) < \varepsilon$ and the sequence $\{f_n(u)\}$ converges uniformly on $U \backslash H$.

The concept of an almost uniformly convergent sequence should not be confused with the concept of a sequence that converges uniformly almost everywhere on U.

Lemma 1. *If a sequence* $\{f_n(u)\}$, *for* $n = 1, 2, \cdots$, *of* $\mathfrak{S}$*-measurable finite* (mod μ) *functions converges almost uniformly on* U, *it converges* *almost everywhere on* U.

Proof. For every integer k, there exists a set H_k such that $\mu(H_k) < 2^{-k}$, and the sequence $\{f_n(u)\}$ converges uniformly on $U \backslash H_k$. Then the sequence $\{f_n(u)\}$ converges on an arbitrary set $M_p = U \backslash \bigcup_{k=p}^{\infty} H_k$ and consequently converges on the set $\bigcup_{p=1}^{\infty} M_p$. The set

$$H = U \backslash \bigcup_{p=1}^{\infty} M_p = \bigcap_{p=1}^{\infty} \bigcup_{k=p}^{\infty} H_k ,$$

where the sequence $\{f_n(u)\}$ (for $n = 1, 2, \cdots$) may diverge, has measure

$$\mu(H) = \lim_{p \to \infty} \mu\left(\bigcup_{k=p}^{\infty} H_k\right) = 0 .$$

Definition 4. A sequence $\{f_n(u)\}$ of $\mathfrak{S}$-measurable finite (mod μ) functions is said to be *fundamental* in μ-measure if for arbitrary $\varepsilon > 0$, $\mu\{u; | f_n(u) - f_m(u) | > \varepsilon\} \to 0$ as $n, m \to \infty$.

Theorem 4. *If a sequence* $\{f_n(u)\}$ *of* $\mathfrak{S}$*-measurable finite* (mod μ) *functions is fundamental in* μ*-measure it contains an almost uniformly convergent subsequence* $\{f_{n_k}(u)\}$, *for* $k = 1, 2$.

Proof. Let us find an n_k such that

$$\mu\left\{u; | f_n(u) - f_m(u) | > \frac{1}{2^k}\right\} < \frac{1}{2^k}$$

for $n, m \geqq n_k$. Without loss of generality we may assume that the sequence $\{n_k\}$ is an increasing sequence. Suppose that

$$E_k = \left\{u; | f_{n_k}(u) - f_{n_{k+1}}(u) | > \frac{1}{2^k}\right\} .$$

Then if $u \notin \bigcup_{j=k}^{\infty} E_j$ and $i, j \geqq k$ (for $i < j$), it follows that

$$\begin{aligned} &| f_{n_i}(u) - f_{n_j}(u) | \\ &\leqq | f_{n_i}(u) - f_{n_{i+1}}(u) | + | f_{n_{i+1}}(u) - f_{n_{i+2}}(u) | + \cdots + | f_{n_{j-1}}(u) - f_{n_j}(u) | \\ &\leqq \frac{1}{2^i} + \frac{1}{2^{i+1}} + \cdots + \frac{1}{2^{j-1}} \leqq \frac{1}{2^{i-1}} ; \end{aligned}$$

that is, the sequence $\{f_{n_k}(u)\}$ converges uniformly on the set $U \backslash H_k$, where

$$H_k = \bigcup_{j=k}^{\infty} E_k \quad \text{and} \quad \mu(H_k) \leqq \sum_{j=k}^{\infty} \mu(E_j) \leqq 2^{1-k} ;$$

in other words, the sequence $\{f_{n_k}(u)\}$ converges almost uniformly on U.

Theorem 5. *For a sequence* $\{f_n(u)\}$ *of functions to converge in*

μ-measure, it is necessary and sufficient that it be a fundamental sequence in μ-measure.

Proof of the Necessity. If $\{f_n(u)\}$ converges in measure to the function $f(u)$, then

$$\mu\{u; | f_n(u) - f_m(u) | > \varepsilon\}$$
$$\leqq \mu\left\{u; | f_n(u) - f(u) | > \frac{\varepsilon}{2}\right\} + \mu\left\{u; | f(u) - f_m(u) | > \frac{\varepsilon}{2}\right\} \longrightarrow 0$$

as $n, m \longrightarrow \infty$.

Proof of the Sufficiency. If a sequence $\{f_n(u)\}$ is fundamental in measure, it contains by virtue of Theorem 4 a subsequence $\{f_{n_k}(u)\}$ that converges almost uniformly and hence in measure to some finite (mod μ) $\mathfrak{S}$-measurable function $f(u)$. Then

$$\mu\{u; | f(u) - f_n(u) | > \varepsilon\}$$
$$\leqq \mu\left\{u; | f(u) - f_{n_k}(u) | > \frac{\varepsilon}{2}\right\} + \mu\left\{u; | f_{n_k}(u) - f_n(u) | > \frac{\varepsilon}{2}\right\}.$$

On the basis of the choice of sequence $\{f_{n_k}(u)\}$, the first term on the right side of the inequality approaches zero as $k \longrightarrow \infty$ and the second term approaches zero as $k, n \longrightarrow \infty$ by virtue of the fact that the sequence $\{f_n(u)\}$ is fundamental in measure. This completes the proof of the theorem.

Corollary. *For a sequence $\{f_n(u)\}$ to converge in μ-measure to $f(u)$, it is necessary and sufficient that every subsequence of the functions $f_n(u)$ contain another subsequence that converges to $f(u)$ almost uniformly.*

The necessity follows from Theorems 4 and 5. To prove the sufficiency, we note that if the sequence $\{f_n(u)\}$ does not converge to $f(u)$ in measure there exists a sequence of indices n_k such that $\mu\{u; | f_{n_k}(u) - f(u) | > \varepsilon\} > \delta$ for some $\varepsilon > 0$ and $\delta > 0$. But this contradicts the assumption that the sequence $\{f_{n_k}(u)\}$ contains a subsequence that converges almost uniformly to $f(u)$.

If $\mu(U) < \infty$, we can replace almost uniform convergence in the statement of the corollary with convergence (mod μ).

Theorem 6. *If $\mu(U) < \infty$ and if the sequences $\{f_n^{(k)}(u)\}$, for $k = 1, 2, \cdots, s$, of finite* (mod μ) *functions converge (in μ-measure) as $n \longrightarrow \infty$ to functions $g_k(u) = \mu\text{-lim} f_n^{(k)}(u)$, and if $\varphi(t_1, \cdots, t_s)$ is an arbitrary continuous function of s variables t_j, where $-\infty < t_j < \infty$, for $j = 1, \cdots, s$, then $\varphi(g_1(u), g_2(u), \cdots, g_s(u)) = \mu\text{-lim}\, \varphi(f_n^{(1)}(u), f_n^{(2)}(u), \cdots, f_n^{(s)}(u))$.*

Proof. Let us set $F_n(u) = \varphi(f_n^{(1)}(u), \cdots, f_n^{(s)}(u))$. An arbitrary sequence of indices n_k contains a subsequence $\{n_{k_r}\}$ such that the sequence $\{f_{n_{k_r}}^{(j)}(u)\}$ converges μ-almost-everywhere to $g_j(u)$ (for $j = 1, \cdots, s$) as $n \longrightarrow \infty$. It follows from the continuity of the function $\varphi(t_1, \cdots, t_s)$ that an arbitrary subsequence $\{F_{n_k}(u)\}$ contains in turn a subsequence that converges to $F(u) = \varphi(g_1(u), \cdots, g_s(u))$ μ-almost-everywhere as $n \longrightarrow \infty$. This, together with the above corollary, yields the desired result.

Corollary. *If* $\mu\text{-lim } f_n(u) = f(u)$ *and* $\mu\text{-lim } g_n(u) = g(u)$, *then*

$$\mu\text{-lim } (\alpha f_n(u) + \beta g_n(u)) = \alpha f(u) + \beta g(u) ,$$

$$\mu\text{-lim } f_n(u) g_n(u) = f(u)\, g(u) .$$

REMARK. It follows from the proof of Theorem 6 that the conclusion of that theorem remains in force even when the function $\varphi(t_1, \cdots, t_s)$ is not necessarily continuous but has the property $\mu\{u; (g_1(u), \cdots, g_s(u)) \in A\} = 0$ where A is the set of points of discontinuity of φ. In particular, if $\mu\text{-lim } g_n(u) = g(u)$ and $g(u) \neq 0$ (mod μ), then the sequence $\{1/g_n(u)\}$ converges in measure to $1/g(u)$.

4. INTEGRALS

In probability theory, we assign to a random variable ξ a specific number $\mathsf{M}\xi$ known as the *mathematical expectation* of ξ. If the random variable ξ assumes finitely many values $x_1, x_2, \cdots, x_s$, the mathematical expectation is given by the formula

$$\mathsf{M}\xi = \sum_{i=1}^{s} x_i \mathsf{P}\{\zeta = x_i\} \tag{1}$$

and it enjoys the following properties:

$$\mathsf{M}(a\xi + b\eta) = a\mathsf{M}\zeta + b\mathsf{M}\eta ,$$

and the inequality $\xi \leqq \eta$ implies $\mathsf{M}\xi \leqq \mathsf{M}\eta$.

We now define the mathematical exception for more general cases. For an arbitrary random variable ξ, we construct a sequence $\{\xi_n\}$ of random variables, each assuming finitely many values, that converge to ξ. We set $\mathsf{M}\xi = \lim \mathsf{M}\xi_n$. This definition will be meaningful if (a) $\lim \mathsf{M}\xi_n$ exists, (b) $\lim \mathsf{M}\xi_n$ depends only on ξ and not on the particular choice of sequence $\{\xi_n\}$ of random variables ξ_n approximating ξ, and (c) the extended definition of the mathematical expectation has "good" analytical properties. It has proven impossible to extend the concept of mathematical expectation to all random variables and at the same time to satisfy these three conditions. However we can do this for a rather broad class of random variables (for example, for all random variables that are bounded either above or below). Difficulties arise here because in taking the limit in a sequence of sums of the form (1), the parts corresponding to both positive and negative terms can approach $\pm\infty$. Therefore it is expedient to consider first random variables that assume values of a single sign.

When we shift from random variables to arbitrary functions defined on spaces with measure, the concept of mathematical expectation becomes the concept of an integral. The present section is devoted to carrying this out in the general case. We assume that some measurable space $\{U, \mathfrak{S}, \mu\}$ is fixed and that all the functions in question are $\mathfrak{S}$-measurable.

Let us first look at simple functions. Let $f(u)$ denote a simple function defined on U and assuming the values $c_1, c_2, \cdots, c_n$. The quantity

$$\int_U f(u)\mu(du) = \sum_{k=1}^{\infty} c_k \mu\{u; f(u) = c_k\} \tag{1'}$$

is called the *integral* of the function $f(u)$ over U.

Let us note the simpler properties of integrals of simple functions:

$$f(u) \geqq 0 \text{ implies } \int_U f(u)\mu(du) \geqq 0 \,, \tag{2}$$

$$\int_U kf(u)\mu(du) = k\int_U f(u)\mu(du) \,, \tag{3}$$

where k is an arbitrary constant;

$$\int_U [f_1(u) + f_2(u)]\mu(du) = \int_U f_1(u)\mu(du) + \int_U f_2(u)\mu(du) \,. \tag{4}$$

Only (4) needs to be proved. Suppose that $f_i(u)$, for $i = 0, 1, 2$, assumes the values $c_k^{(i)}$ (for $k = 1, 2, \cdots, m_i$), that $f_0(u) = f_1(u) + f_2(u)$ and that $A_k^{(i)} = \{u; f^{(i)}(u) = c_k^{(i)}\}$. Then

$$\bigcup_{k=1}^{m_i} A_k^{(i)} = U$$

and either $A_k^{(0)} \cap A_r^{(1)} \cap A_s^{(2)} = \varnothing$ or $c_k^{(0)} = c_r^{(1)} + c_s^{(2)}$. Consequently,

$$\begin{aligned}\int_U f_0(u)\mu(du) &= \sum_{k=1}^{m_0} c_k^{(0)}\mu(A_k^{(0)}) = \sum_{k=1}^{m_0} c_k^{(0)} \sum_{r,s} \mu(A_k^{(0)} \cap A_r^{(1)} \cap A_s^{(2)}) \\ &= \sum_{k=1}^{m_0} \sum_{r,s} c_r^{(1)}\mu(A_k^{(0)} \cap A_r^{(1)} \cap A_s^{(2)}) \\ &\quad + \sum_{k=1}^{m_0} \sum_{r,s} c_s^{(2)}\mu(A_k^{(0)} \cap A_r^{(1)} \cap A_s^{(2)}) \\ &= \sum_r c_r^{(1)}\mu(A_r^{(1)}) + \sum_s c_s^{(2)}\mu(A_s^{(2)}) \\ &= \int_U f^{(1)}(u)\mu(du) + \int_U f^{(2)}(u)\mu(du) ,\end{aligned}$$

which proves (4).

Let us now define the integral of a simple function $f(u)$ over an arbitrary set $M \in \mathfrak{S}$ by means of the formula

$$\int_M f(u)\mu(du) = \int_U f(u)\chi_M(u)\mu(du) , \tag{5}$$

where $\chi_M(u)$ is the characteristic function of the set M.

If M_1 and M_2 are $\mathfrak{S}$-measurable and have no points in common, then

$$\chi_{M_1 \cup M_2}(u) = \chi_{M_1}(u) + \chi_{M_2}(u) .$$

Consequently,

$$\int_{M_1 \cup M_2} f(u)\mu(du) = \int_{M_1} f(u)\mu(du) + \int_{M_2} f(u)\mu(du) . \tag{6}$$

It follows from (2) and (4) that the inequality $f(u) \geqq g(u)$ for $u \in M$ implies

$$\int_M f(u)\mu(du) \geqq \int_M g(u)\mu(du) . \tag{7}$$

Before generalizing the definition of an integral, let us prove:

Lemma 1. *Let $\{g_n(u)\}$, for $n = 1, 2, \cdots$, denote a nondecreasing sequence of nonnegative simple functions and suppose that* $\lim g_n(u) \geqq h(u)$ *for $u \in M$, where $h(u)$ is a nonnegative simple function. Then,*

$$\lim \int_M g_n(u)\mu(du) \geqq \int_M h(u)\mu(du) . \tag{8}$$

Proof. Let $0 \leqq h_1 < h_2 < \cdots < h_s$ denote values assumed by the function $h(u)$. To prove inequality (8), let us confine ourselves to the case $h_1 > 0$.

In the opposite case, it would suffice to prove inequality (8) for the set $M \backslash \{u; h(u) = 0\}$.

Let ε denote an arbitrary positive number and let Q_m, for each $m = 1, 2, \cdots$, denote the set $\{u; u \in M, g_n(u) \geqq h(u) - \varepsilon, n \geqq m\}$.

These Q_m constitute a nondecreasing sequence of $\mathfrak{S}$-measurable sets, and $\bigcup_{m=1}^{\infty} Q_m = M$. Suppose that $\mu(M) < \infty$. Then $\lim \mu(Q_m) = \mu(M)$ and $\lim \mu(M \backslash Q_m) = 0$. Furthermore, for $n \geqq m$,

$$\begin{aligned}\int_M g_n(u)\,\mu(du) &\geqq \int_{Q_m} g_n(u)\,\mu(du) \geqq \int_{Q_m} (h(u) - \varepsilon)\mu(du) \\ &= \int_{Q_m} h(u)\mu(du) - \varepsilon\mu(Q_m) \\ &\geqq \int_M h(u)\mu(du) - \varepsilon\mu(M) - h_s\mu(M\backslash Q_m)\ .\end{aligned}$$

If we let ε approach 0 and m approach ∞, this inequality becomes inequality (8). If $\mu(M) = \infty$, let us take $\varepsilon = h_1/2$. Then for $n \geqq m$,

$$\int_M g_n(u)\mu(du) \geqq \frac{h_1}{2}\,\mu(Q_m) \longrightarrow \infty\ ,$$

as $m \longrightarrow \infty$, which again yields (8).

Lemma 2. *Let $\{f_n(u)\}$ and $\{g_n(u)\}$ denote two nondecreasing sequences of nonnegative simple functions and suppose that* $\lim f_n(u) = \lim g_n(u)$ *for $u \in M \in \mathfrak{S}$. Then*

$$\lim \int_M f_n(u)\mu(du) = \lim \int_M g_n(u)\mu(du)\ .$$

Proof. It follows from the hypothesis of the theorem that

$$\lim f_n(u) \geqq g_m(u)\ .$$

On the basis of Lemma 1,

$$\lim \int_M f_n(u)\mu(du) \geqq \int_M g_m(u)\mu(du)\ .$$

If we now let m approach ∞ we obtain

$$\lim \int_M f_n(u)\mu(du) \geqq \lim \int_M g_n(u)\mu(du)\ .$$

By reversing the roles of the sequences $\{f_n(u)\}$ and $\{g_n(u)\}$ in this last inequality, we obtain the desired result.

Let us now define an integral in the general case. Let $f(u)$ denote an arbitrary $\mathfrak{S}$-measurable function on M. We define

$$\begin{aligned}f^+(u) &= \max\{f(u), 0\}\ ,\\ f^-(u) &= -\min\{f(u), 0\},\quad u \in M\ .\end{aligned}$$

Then

$$f(u) = f^+(u) - f^-(u)\ , \tag{9}$$

and the functions f^+ and f^- are nonnegative and $\mathfrak{S}$-measurable.

Definition 1. The quantity

$$\int_M f(u)\mu(du) = \lim \int_M f_n(u)\mu(du) , \tag{10}$$

where $\{f_n(u)\}$ is an arbitrary nondecreasing sequence of nonnegative simple functions that converges to $f(u)$, is called the *integral* of the $\mathfrak{S}$-measurable nonnegative function $f(u)$ over the set M. If $f(u)$ is an arbitrary $\mathfrak{S}$-measurable function and if one of the integrals

$$\int_M f^+(u)\mu(du), \int_M f^-(u)\mu(du) \tag{11}$$

is finite the integral of the function $f(u)$ is defined by

$$\int_M f(u)\mu(du) = \int_M f^+(u)\mu(du) - \int_M f^-(u)\mu(du) . \tag{12}$$

If both integrals (11) are finite, the integral of the function f is also finite and the function f is said to be *integrable* over the set M.

This definition is meaningful: On the basis of Lemma 2, the definition of an integral of a nonnegative function is independent of the particular choice of approximating sequence in formula (10). In particular, for a nonnegative simple function the definition of the integral (10) coincides with the original definition. Furthermore, for an arbitrary simple function, the functions f^+ and f^- corresponding to it are simple, and on the basis of property (4), which asserts the linearity of the integral of a simple function, formula (12) for simple functions gives the originally determined value of the integral.

Let us now look at the basic properties of an integral.

Theorem 1. *If $\mu(M) = 0$ and f is an arbitrary function defined on M, then*

$$\int_M f(u)\mu(du) = 0 . \tag{13}$$

This follows immediately from formulas (12), (10), and (1′),

Theorem 2.

$$\int_{M_1 \cup M_2} f(u)\mu(du) = \int_{M_1} f(u)\mu(du) + \int_{M_2} f(u)\mu(du) , \tag{14}$$

where $M_1 \cap M_2 = \varnothing$, where M_1 and M_2 belong to $\mathfrak{S}$, and where one side or the other of the equation is meaningful.

It suffices to consider (14) for nonnegative functions. But for these functions, equation (14) follows from the definition of an integral and the fact that this equation holds for simple functions.

Theorem 3.

$$\int_M \alpha f(u)\,\mu(du) = \alpha \int_M f(u)\mu(du)\ . \tag{15}$$

Equation (15) is valid for simple functions. By taking limits, we extend this equation to arbitrary nonnegative functions, and by virtue of

$$(\alpha f)^+ = \alpha f^+, \quad (\alpha f)^- = \alpha f^- \quad \text{for } \alpha \geqq 0\ ,$$
$$(\alpha f)^+ = -\alpha f^-, \ (\alpha f)^- = -\alpha f^+ \ \text{for } \alpha < 0$$

and formula (12), the result is carried over for arbitrary $\mathfrak{S}$-measurable functions whose integrals over M are meaningful.

Theorem 4.

$$\int_M [f(u) + g(u)]\mu(du) = \int_M f(u)\mu(du) + \int_M g(u)\mu(du)\ , \tag{16}$$

if the the integrals on the right are not infinite of different signs.

If the functions $f(u)$ and $g(u)$ are nonnegative on M, then (16) is valid because this equation holds for simple functions. The case in which $f(u)$ and $g(u)$ are both nonpositive reduces to this case by means of (15) with $\alpha = -1$. Suppose now that $f(u) \geqq 0$ and $g(u) \leqq 0$ everywhere on M, that $f(u) + g(u) \geqq 0$ on $M_1 \subset M$, and that $f(u) + g(u) < 0$ on $M_2 \subset M$, where $M_1 \cup M_2 = M$. Then by setting $f(u) + g(u) = h(u)$, we obtain $f(u) = h(u) + (-g(u))$, so that

$$\int_{M_1} f(u)\,\mu(du) = \int_{M_1} h(u)\,\mu(du) + \int_{M_1} -g(u)\,\mu(du)\ ,$$

that is,

$$\int_{M_1} [f(u) + g(u)]\mu(du) = \int_{M_1} f(u)\,\mu(du) - \int_{M_1} g(u)\mu(du)$$

if the right-hand member of this equation is not an indeterminate form of the form $\infty - \infty$. Analogously, $-g(u) = f(u) + (-h(u))$. Therefore,

$$\begin{aligned}\int_{M2} -g(u)\,\mu(du) &= -\int_{M2} g(u)\mu(du)\\ &= \int_{M_2} f(u)\,\mu(du) + \int_{M_2} -h(u)\mu(du)\\ &= \int_{M_2} f(u)\mu(du) - \int_{M_2} h(u)\mu(du)\ ,\end{aligned}$$

so that, again,

$$\int_{M_2} (f(u) + g(u))\mu(du) = \int_{M_2} f(u)\,\mu(du) + \int_{M_2} g(u)\mu(du)$$

if the right-hand member of this equation is not an indeterminate form of the form $\infty - \infty$. By combining the integrals over the sets M_1 and M_2, we obtain formula (16), by virtue of (2).

Let $f(u)$ and $g(u)$ denote any two functions defined on M. Let us set $M_1 = \{u; f(u) \geqq 0, g(u) \geqq 0\}$, $M_2 = \{u; f(u) < 0, g(u) < 0\}$, $M_3 = \{u; f(u) \geqq 0, g(u) < 0\}$, $M_4 = \{u; f(u) < 0, g(u) \geqq 0\}$. Then

$$M = \bigcup_{i=1}^{4} M_i$$

where $M_i \cap M_j = \varnothing$ for $i \neq j$, and formula (16) is proved for each of the sets M_i for $i = 1, 2, 3, 4$. If we now replace M in formula (16) with the set M_i for $i = 1, 2, 3, 4$ and add the equations obtained, we get, in view of (14), equation (16) in the general case.

Theorem 5. *If $f(u)$ and $g(u)$ are equivalent on M, then*

$$\int_M f(u)\mu(du) = \int_M g(u)\mu(du) .$$

Indeed,

$$\int_M f(u)\,\mu(du) = \int_M g(u)\,\mu(du) + \int_M [f(u) - g(u)]\mu(du) ,$$

and

$$\int_M [f(u) - g(u)]\mu(du) = \int_{\{u; f(u) \neq g(u)\}} [f(u) - g(u)]\mu(du) = 0$$

(by virtue of Theorem 1).

Theorem 6. *If $f(u) \geqq g(u)$ for $u \in M$, we have*

$$\int_M f(u)\mu(du) \geqq \int_M g(u)\mu(du) \tag{17}$$

if both integrals exist.

Proof. The function $\varphi(u) = f(u) - g(u)$ is nonnegative for $u \in M$. It follows from the definition of an integral that $\int \varphi(u)\mu(du) \geqq 0$ and hence

$$\int_M f(u)\mu(du) - \int_M g(u)\mu(du) \geqq 0 ,$$

if the difference on the left is defined. On the other hand, if the integrals are infinite but of the same sign, (17) is obvious.

Theorem 7. *If*

$$\int_M f(u)\mu(du) \neq +\infty ,$$

then $\mu\{u; u \in M, f(u) = +\infty\} = 0$. In particular, if the function $f(u)$ is integrable, it is finite almost everywhere.

Proof.

$$\int_M f^+(u)\mu(du) \geqq \int_M h(u)\mu(du) ,$$

where $h(u) = 0$ if $f^+(u) < +\infty$ and $h(u) = h$ if $f^+(u) = +\infty$, where h is an arbitrary positive constant. Consequently,

$$\int_M f^+(u)\mu(du) \geqq h\mu\{u;\, u \in M,\ f(u) = +\infty\} ,$$

which can be bounded as $h \longrightarrow \infty$ only when $f(u) < +\infty$ almost everywhere.

Theorem 8. *If $f(u)$ is integrable on M, then $|f(u)|$ is also integrable on M.*

Proof. If $f(u)$ is integrable on M, then

$$\int_M f^+(u)\mu(du) < \infty, \quad \int_M f^-(u)\mu(du) < \infty .$$

Since

$$|f(u)| = f^+(u) + f^-(u) ,$$

it follows that

$$\int_M |f(u)|\, \mu(du) = \int_M f^+(u)\mu(du) + \int_M f^-(u)\mu(du) < \infty .$$

Theorem 9. *If $|f(u)| \leqq h(u)$ for $u \in M$ and if $h(u)$ is integrable on M, then $f(u)$ is also integrable on M.*

This follows from the fact that $f^+(u) \leqq h(u)$ and $f^-(u) \leqq h(u)$ and from equation (16).

5. INTERCHANGING LIMITS AND INTEGRATIONS. L_p SPACES

Let us now look at the possibility of taking the limit under the integral sign. What conditions do we need to impose on a sequence of functions $f_n(u)$ to ensure that

$$\lim_{n\to\infty} \int_U f_n(u)\mu(du) = \int_U f(u)\mu(du)$$

where $f(u)$ is the limit (in a specified sense) of the sequence $\{f_n(u)\}$?

Theorem 1 (Lebesgue). *Suppose that $\{f_n(u)\}$ is a nondecreasing sequence of nonnegative $\mathfrak{S}$-measurable functions. Define* $f(u) = \lim f_n(u) \pmod{\mu}$. *Then*

$$\lim \int_U f_n(u)\mu(du) = \int_U f(u)\mu(du) .$$

Proof. For each n, let $\{g_{nk}(u)\}$ denote a nondecreasing sequence of nonnegative simple functions that converges to $f_n(u)$,

$$\lim_{k\to\infty} g_{nk}(u) = f_n(u) ,$$

and define $h_n(u) = \max_{i\leqq n} g_{in}(u)$. Since $g_{in}(u) \leqq f_i(u)$, we have

$$h_n(u) \leqq \max_{i\leqq n} f_i(u) = f_n(u)$$

and

$$\lim h_n(u) \leqq \lim f_n(u) = f(u) . \tag{1}$$

The sequence $\{h_n(u)\}$ is nondecreasing, it consists of simple functions, and for all k,

$$\lim_{n\to\infty} h_n(u) \geqq \lim_{n\to\infty} g_{kn}(u) = f_k(u) .$$

Consequently,

$$\lim_{n\to\infty} h_n(u) \geqq \lim_{k\to\infty} f_k(u) = f(u) .$$

Comparing this with (1), we obtain

$$f(u) = \lim_{n\to\infty} h_n(u) .$$

It follows from the definition of an integral and formula (1) that

$$\int_U f(u)\mu(du) = \lim \int_U h_n(u)\mu(du) \leqq \lim \int_U f_n(u)\mu(du) .$$

Inequality in the opposite direction follows from the fact that $f_n(u) \leqq f(u)$.

Theorem 2 (Fatou's lemma). *Let $\{f_n(u)\}$ denote a sequence of nonnegative $\mathfrak{S}$-measurable functions. Then*

$$\int_U \underline{\lim}\, f_n(u)\mu(du) \leqq \underline{\lim} \int_U f_n(u)\mu(du) . \tag{2}$$

Proof. We have

$$\underline{\lim}\, f_n(u) = \lim g_n(u) ,$$

where

$$g_n(u) = \inf \{f_n(u), f_{n+1}(u), \cdots\}$$

is a nondecreasing sequence of nonnegative functions. It follows from Lebesgue's theorem (1) that

$$\int_U \underline{\lim}\, f_n(u)\,\mu(du) = \lim \int_U g_n(u)\,\mu(du) \leqq \underline{\lim} \int_U f_n(u)\mu(du) ,$$

which completes the proof of inequality (2).

Theorem 3 (Lebesgue). *Let $\{f_n(u)\}$ denote a sequence of meas-*

urable functions. Suppose that there exists a nonnegative integrable function $s(u)$ *such that* $|f_n(u)| \leqq s(u)$ *for* $n = 1, 2, \cdots$. *Then*

$$\int_U \underline{\lim}\, f_n(u)\mu(du) \leqq \underline{\lim} \int_U f_n(u)\mu(du)$$

$$\leqq \overline{\lim} \int_U f_n(u)\mu(du) \leqq \int_U \overline{\lim}\, f_n(u)\mu(du) . \tag{3}$$

In particular, if $\{f_n(u)\}$ *converges almost everywhere in* U *to* $f(u)$, *then*

$$\lim \int_U f_n(u)\mu(du) = \int_U f(u)\mu(du) . \tag{4}$$

Proof. Applying Fatou's lemma to the sequence $\{s(u) + f_n(u)\}$, we obtain

$$\underline{\lim} \int_U [s(u) + f_n(u)]\mu(du) \geqq \int_U \underline{\lim}\, [s(u) + f_n(u)]\mu(du)$$

$$= \int_U [s(u) + \underline{\lim}\, f_n(u)]\mu(du)$$

$$= \int_U s(u)\mu(du) + \int_U \underline{\lim}\, f_n(u)\, \mu(du) . \tag{5}$$

Furthermore,

$$\underline{\lim} \int_U [s(u) + f_n(u)]\mu(du) = \underline{\lim} \left\{ \int_U s(u)\mu(du) + \int_U f_n(u)\, \mu(du) \right\}$$

$$= \int_U s(u)\mu(du) + \underline{\lim} \int_U f_n(u)\mu(du) . \tag{6}$$

It follows from (5) and (6), by virtue of the integrability of the function $s(u)$, that

$$\underline{\lim} \int_U f_n(u)\mu(du) \geqq \int_U \underline{\lim}\, f_n(u)\mu(du) . \tag{7}$$

Applying this last inequality to the sequence $\{- f_n(u)\}$, we obtain

$$\underline{\lim} \int_U - f_n(u)\mu(du) = - \overline{\lim} \int_U f_n(u)\mu(du)$$

$$\geqq \int_U \underline{\lim}(- f_n(u))\mu(du) = - \int_U \overline{\lim}\, f_n(u)\mu(du) ,$$

so that

$$\int_U \overline{\lim}\, f_n(u)\mu(du) \geqq \overline{\lim} \int_U f_n(u)\mu(du) . \tag{8}$$

Inequalities (7) and (8) prove the theorem.

Corollary. *Under the conditions of the preceding theorem, if* $\lim f_n(u) = f(u)$ (mod μ), *where* $u \in U$, *then*

$$\int_U | f(u) - f_n(u) | \, \mu(du) \longrightarrow 0 \quad as \quad n \longrightarrow \infty . \tag{9}$$

Proof.

$$\lim_{n\to\infty} | f(u) - f_n(u) | = 0$$

and

$$| f(u) - f_n(u) | \leqq 2s(u) .$$

Applying equation (4) to the sequence $\{| f(u) - f_n(u) |\}$, we obtain assertion (9),

Let $f(u)$ denote an arbitrary function for which the integral

$$\int_U f(u)\mu(du) \tag{10}$$

is meaningful. Then the formula

$$\varphi(A) = \int_A f(u)\mu(du), \quad A \in \mathfrak{S}$$

defines a set function on the σ-algebra $\mathfrak{S}$.

Theorem 4. *If the integral* (10) *is meaningful, then the set function $\varphi(A)$ is countably additive on $\mathfrak{S}$.*

Proof. Since $f(u) = f^+(u) - f^-(u)$ and one of the functions $f^+(u)$ or $f^-(u)$ is integrable, it will be sufficient for us to prove the theorem for a nonnegative function. Define $A = \bigcup_{n=1}^{\infty} A_n$, where and $A_n \in \mathfrak{S}$, $A_n \cap A_m = \varnothing$ for $n \neq m$. Let $\chi_0(u)$ and $\chi_n(u)$, denote the characteristic functions of the sets A and A_n for $n = 1, 2, \cdots$. Then

$$\chi_0(u) = \sum_{n=1}^{\infty} \chi_n(u)$$

and by virtue of Theorem 1,

$$\varphi(A) = \int_A f(u)\mu(du) = \int_U \chi_0(u)f(u)\mu(du) = \lim_{n\to\infty} \int_U f(u) \times \sum_{k=1}^{n} \chi_k(u)\mu(du)$$
$$= \lim_{n\to\infty} \sum_{k=1}^{n} \int_{A_k} f(u)\mu(du) = \sum_{k=1}^{\infty} \varphi(A_k) .$$

Corollary. *If a function $f(u)$ is integrable on U, then for arbitrary $\varepsilon > 0$ there exists a $\delta > 0$ such that the inequality $\mu(A) < \delta$ implies*

$$\int_A | f(u) | \mu(du) < \varepsilon . \tag{11}$$

Proof. Since integrability of the function $f(u)$ implies integrability of $| f(u) |$, it will be sufficient to prove the corollary for nonnegative integrable functions. We note first of all that $\mu\{u; f(u) = \infty\} = 0$. Therefore,

$$\lim_{N\to\infty} \varphi\{u; f(u) > N\} = \varphi\{u, f(u) = \infty\} = 0$$

(cf. Theorem 1, section 4). Consequently, for some N_0, $\varphi\{u; f(u) > N_0\} < \varepsilon/2$. Furthermore, for arbitrary N,

$$\varphi(A) = \int_A f(u)\mu(du) \leqq \int_{A\cap\{u; f\leqq N\}} f(u)\mu(du) + \int_{A\cap\{u; f>N\}} f(u)\mu(du)$$
$$\leqq N\mu(A) + \varphi\{u; f > N\} .$$

Taking $N = N_0$ and $\mu(A) \leqq \delta = \varepsilon/2N_0$, we obtain the desired result.

Theorem 5. *Let $\{f_n(u)\}$ denote a sequence of measurable functions that converges in measure to a function $f(u)$ on U. Suppose that $|f_n(u)| \leqq s(u)$ (mod μ), $n = 1, 2, \cdots$, where $s(u)$ is an integrable function. Then*

$$\lim \int_U f_n(u)\mu(du) = \int_U f(u)\mu(du) .$$

Proof. In accordance with Theorem 4 of Section 3, an arbitrary sequence $\{f_{n_k}(u)\}$ contains a subsequence $\{f_{n_{k_j}}(u)\}$ that converges almost uniformly to $f(u)$. It follows from Theorem 3 that

$$\lim_{j\to\infty} \int_U f_{n_{k_j}}(u)\mu(du) = \int_U f(u)\mu(du) .$$

Thus the bounded sequence of the integrals $\int_U f_n(u)\mu(du)$ has a unique point of accumulation $\int_U f(u)\mu(du)$. This completes the proof of the theorem.

We present one more theorem on limits. Here the requirement that an integrable majorant exist is replaced with another requirement whose satisfaction is in many problems easier to verify.

Theorem 6. *If a sequence $\{f_n(u)\}$ converges in measure to a function $f(u)$ for $u \in U$ where $\mu U < \infty$, and if for some $p > 1$,*

$$\int_U | f_n(u) |^p \mu(du) \leqq c, \quad n = 1, 2 \cdots,$$

then

$$\lim_{n\to\infty} \int_U f_n(u)\mu(du) = \int_U f(u)\mu(du) .$$

Proof. Let ε denote an arbitrary positive number. Define

$$A_n(\varepsilon) = \{u; | f(u) - f_n(u) | > \varepsilon\} .$$

Then for $N > 1/\varepsilon$,

$$\left|\int_U f(u)\mu(du) - \int_U f_n(u)\,\mu(du)\right| \leqq \int_{U\setminus A_n(\varepsilon)} |\,f(u) - f_n(u)\,|\,\mu(du)$$
$$+ \int_{A_n(\varepsilon)\setminus A_n(N)} |\,f(u) - f_n(u)\,|\,\mu(du)$$
$$+ \int_{A_n(N)} |\,f(u) - f_n(u)\,|\,\mu(du) \leqq \varepsilon\mu(U)$$
$$+ N\mu(A_n(\varepsilon)) + \frac{1}{N^{p-1}}\int_{A_n(N)} |\,f(u) - f_n(u)\,|^p\,\mu(du)\,.$$

It follows from the inequality $\left(\frac{a+b}{2}\right)^p \leqq \frac{|\,a\,|^p + |\,b\,|^p}{2}$ for $p > 1$ that

$$\int_{A_n(N)} |\,f(u) - f_n(u)\,|^p\,\mu(du)$$
$$\leqq 2^{p-1}\left\{\int_{A_n(N)} [|\,f(u)\,|^p + |\,f_n(u)\,|^p]\,\mu(du)\right.$$
$$\leqq 2^{p-1}\left\{\int_U [|\,f(u)\,|^p + |\,f_n(u)\,|^p]\,\mu(du)\right\}.$$

On the other hand, since $f(u) = \mu\text{-}\lim f_n(u)$, there exists a sequence $\{f_{n_k}(u)\}$ that converges to $f(u)$ (mod μ). By virtue of Fatou's lemma and the hypothesis of the theorem,

$$\int_U |\,f(u)\,|^p\,\mu(du) = \int_U \underline{\lim}\,|\,f_{n_k}(u)\,|^p\,\mu(du)$$
$$\leqq \underline{\lim}\int_U |\,f_{n_k}(u)\,|^p\,\mu(du) \leqq c\,.$$

Thus

$$\overline{\lim_{n\to\infty}}\left|\int_U f(u)\mu(du) - \int_U f_n(u)\mu(du)\right|$$
$$\leqq \varepsilon\mu(U) + N\,\overline{\lim_{n\to\infty}}\,\mu(A_n(\varepsilon)) + \frac{2^p c}{N^{p-1}} = \varepsilon\mu(U) + \frac{2^p c}{N^{p-1}}\,.$$

Since ε is an arbitrary positive number and since $N > 1/\varepsilon$, the proof is completed.

Let us now look at complex-valued functions $f(u)$ defined on U. The function $f(u) = f_1(u) + if_2(u)$ where $f_1(u)$ and $f_2(u)$ are real, is said to be $\mathfrak{S}$-measurable if $f_1(u)$ and $f_2(u)$ are $\mathfrak{S}$-measurable. The function $f(u)$ is said to be integrable if $f_1(u)$ and $f_2(u)$ are integrable. The integral of the function $f(u)$ over U is defined by the equation

$$\int_U f(u)\mu(du) = \int_U f_1(u)\mu(du) + i\int_U f_2(u)\mu(du)\,.$$

The properties of an integral that were established for real functions are easily carried over to complex-valued functions. To the end of the present section, equivalent functions are assumed equal, so that the word "function" actually means an entire class of functions that are equivalent to each other.

Let $L_p = L_p\{U, \mathfrak{S}, \mu\}$ (for $p \geqq 1$) denote the class of all $\mathfrak{S}$-measurable functions defined on U into the set of complex numbers and satisfying the inequality $\int_U | f(u) |^p \mu(du) < \infty$.

Minkowski's inequality, which is proven for integrals in abstract spaces in the same way as for ordinary spaces (see Hardy, Littlewood, and Polya; Heider & Simpson):

$$\left\{\int | f(u) + g(u) |^p \mu(du)\right\}^{1/p}$$
$$\leqq \left\{\int_U | f(u) |^p \mu(du)\right\}^{1/p} + \left\{\int_U | g(u) |^p \mu(du)\right\}^{1/p},$$

implies that the sum of two functions belonging to L_p also belongs to L_p. Obviously if f belongs to L_p, so does αf, where α is a complex number. Thus L_p is a linear space. Defining the norm of an element f of L_p by

$$\| f \| = \left\{\int_U | f(u) |^p \mu(du)\right\}^{1/p},$$

we make L_p a normed space. Obviously

$$\| \alpha f \| = | \alpha | \, \| f \| , \qquad \| f + g \| \leqq \| f \| + \| g \| ,$$

so that the axioms of a normed space are satisfied. The distance between two functions f, $g \in L_p$ is then defined by

$$\rho(f, g) = \| f - g \| = \left\{\int_U | f(u) - g(u) |^p d\mu\right\}^{1/p}.$$

If a sequence $\{f_n(u)\}$ of functions in L_p converges to $f(u)$ in the sense of convergence in L_p, that is, if

$$\| f(u) - f_n(u) \| \longrightarrow 0 \quad \text{as} \quad n \longrightarrow \infty ,$$

we say that $\{f_n(u)\}$ converges in mean of order p to $f(u)$. A sequence $\{f_n(u)\}$ is said to be *fundamental* in L_p if $\| f_n(u) - f_{n'}(u) \| \to 0$ as $n, n' \to \infty$. It follows from the inequality

$$\varepsilon^p \mu\{u; | f(u) - g(u) | > \varepsilon\} \leqq \int_U | f(u) - g(u) |^p \mu(du)$$

that if a sequence is convergent (resp. fundamental) in L_p, it is convergent (resp. fundamental) in measure.

Theorem 7. *For a sequence $\{f_n\}$ of functions in L_p to converge* *in L_p to some limit, it is necessary and sufficient that the sequence $\{f_n\}$ be fundamental. In other words, the space L_p is complete.*

Proof. The necessity of the hypothesis of the theorem is general for metric spaces as a consequence of the triangle inequality:

$$\| f_n(u) - f_{n'}(u) \| \leqq \| f - f_{n'} \| + \| f_n - f \| .$$

Let us prove the sufficiency. Since $| \| f_n \| - \| f_{n'} \| | \leqq \| f_n - f_{n'} \|$, if a sequence $\{f_n\}$ is a fundamental sequence, the sequence $\{\| f_n \|\}$ is bounded; thus $\| f_n \| \leqq c$. Furthermore, the fact that the sequence is fundamental in L_p implies that it is fundamental in measure. Consequently (see Theorem 4, Section 3) the sequence $\{f_n\}$ contains a subsequence $\{f_{n_k}\}$ that converges almost uniformly in U. Using Fatou's lemma, we obtain

$$c^p \geqq \underline{\lim} \int_U | f_{n_k}(u) |^p \mu(du) \geqq \int_U \underline{\lim} \, | f_{n_k}(u) |^p \mu(du) = \int_U | f(u) |^p \mu(du) ,$$

from which it follows that $f \in L_p$. Let ε denote an arbitrary positive number. Since the sequence $\{f_n\}$ is fundamental, it follows that for all $n \geqq n_0$,

$$\varepsilon \geqq \underline{\lim}_{k\to\infty} \int_U | f_{n_k}(u) - f_n(u) |^p \mu(du)$$

$$\geqq \int_U \underline{\lim}_{k\to\infty} | f_{n_k}(u) - f_n(u) |^p \mu(du) = \int_U | f(u) - f_n(u) |^p \mu(du) ,$$

which proves the convergence of the sequence $\{f_n\}$ to f in L_p.

In what follows, the L_p space that we shall be primarily interested in is the space L_2, that is, the space of functions $f(u)$ such that

$$\int_U | f(u) |^2 \mu(du) < \infty .$$

Since $| f\bar{g} | \leqq (| f |^2 + | g |^2)/2$, it follows that the function fg is integrable for all $f, g \in L_2$. Consequently we can define a functional on the set of pairs of functions f and g in L_2:

$$(f, g) = \int_U f(u) \overline{g(u)} \mu(du) .$$

This functional is known as the *scalar product* of the functions f and g. The following properties of the scalar product follow directly from the definition:

1. $(f, f) \geqq 0$ with equality holding if and only if f is equivalent to 0;

2. $(f, g) = \overline{(g, f)}$;

3. $(\alpha f_1 + \beta f_2, g) = \alpha(f_1, g) + \beta(f_2, g)$;

that is, the scalar product is a positive-definite bilinear form in the two variables f and g. A metric in the space L_2 is connected with the scalar product by the formula

$$\rho(f, g) = \| f - g \| = \sqrt{(f - g, f - g)} .$$

Since the space L_2 is complete, it is a particular case of an abstract Hilbert space. We assume that the reader is familiar with the basic properties and the simpler theorems having to do with Hilbert-space theory (see for example Kantorovich and Akilov, Heider and Simpson, and also Section 1 of Chapter V of this book). We mention here only that the scalar product satisfies the Cauchy-Bunyakovskiy-Schwarz inequality:

$$| (f, g) |^2 \leqq (f, f)\,(g, g) \tag{12}$$

or

$$\left| \int_U f(u)\overline{g(u)}\,\mu(du) \right|^2 \leqq \int_U | f(u) |^2 \mu(du) \int_U | g(u) |^2 \mu(du) . \tag{13}$$

6. ABSOLUTE CONTINUITY OF MEASURES. MAPPINGS

In the preceding section, for an arbitrary $\mathfrak{S}$-measurable function $f(x)$ such that the integral $\int_U f(u)\mu(du)$ is meaningful, we defined a countably additive set function

$$\varphi(A) = \int_A f(u)\mu(du) , \tag{1}$$

on $\mathfrak{S}$. The function $\varphi(A)$ is called the *indefinite integral* of the function $f(u)$ (with respect to the integrating measure μ). Corresponding to this, the function $f(u)$ in formula (1) can be regarded as the derivative of the countably additive function φ with respect to the measure μ. By definition we set $f(u) = \dfrac{d\varphi}{d\mu}$.

How can we characterize the class of countably additive functions φ having a representation of the form (1)? Is the function $f(u)$ defined naturally by formula (1)? If $\varphi(A)$ is the indefinite integral of the function $f(u)$, then by virtue of one of the simplest properties of an integral (see Section 4, Theorem 1), $\varphi(A) = 0$ whenever $\mu(A) = 0$. We find that this condition is not only necessary but in the most important cases is also sufficient if the function $\varphi(A)$ is to have a representation of the form (1). In what follows we shall consistently assume that μ and φ are finite measures defined on some fixed σ-algebra $\mathfrak{S}$.

Definition 1. A measure φ is said to be *absolutely continuous* with respect to a measure μ (we write $\varphi \ll \mu$) if $\varphi(A) = 0$ whenever $A \in \mathfrak{S}$ and $\mu(A) = 0$. The measure φ and μ are said to be *singular* with respect to each other (and we write $\varphi \perp \mu$) if U can be partitioned into two disjoint subsets (that is, $U = C_\mu \cup C_\varphi$, where C_μ, $C_\varphi \in \mathfrak{S}$ and $C_\mu \cap C_\varphi = \emptyset$) such that $\varphi(C_\mu) = 0$ and $\mu(C_\varphi) = 0$. The singularity of the measures means that the measure φ is nonzero only on subsets of some set of μ-measure 0 and vice versa.

Theorem 1. *Suppose that φ and μ are two finite measures. The measure φ can be represented in the form*

$$\varphi(A) = \int_A f(u)\mu(du) + \psi(A) , \tag{2}$$

where $f(u)$ is a nonnegative μ-integrable function on U and ψ and μ are singular.

Proof. Define $\lambda = \varphi + \mu$. We recall that $L_2(\lambda)$ is the space of $\mathfrak{S}$-measurable functions such that $\int_U | f(u) |^2 \lambda(du) < \infty$. In this space, let us look at the functional $\alpha(f) = \int_U f(u)\varphi(du)$. It is defined for every function $f \in L_2(\lambda)$ since $| f | \leqq 1 + | f |^2$ and $\int_U | f |^2 (u)\varphi(du) \leqq \int_U | f |^2(u)\lambda(du)$, and it is linear and bounded:

$$\sup_{||f|| \leqq 1} | \alpha(f) | \leqq \sup_{||f|| \leqq 1} \left\{\varphi(U)\int_U | f(u) |^2 \varphi(du)\right\}^{1/2} \leqq \sqrt{\varphi(U)}|| f || .$$

Thus $|| \alpha || \leqq \sqrt{\varphi(U)}$. Every bounded linear functional $\alpha(f)$ in a Hilbert space H can be represented as a scalar product $\alpha(f) = (f, g)$, where g is some element of H (see Kantorovich and Akilov; Heider and Simpson). Consequently,

$$\int_U f(u)\varphi(du) = \int f(u)g(u)\lambda(du), \; g \in L_2(\lambda) . \tag{3}$$

Here, if we set $f(u) = \chi_A(u)$ where χ_A is the characteristic function of the set $A \in \mathfrak{S}$, we obtain

$$\varphi(A) = \int_A g(u)\lambda(du) \leqq \lambda(A) .$$

From this it follows that $0 \leqq g(u) \leqq 1 \pmod{\lambda}$. Define $N = \{u; g(u) = 1\}$. Then $\varphi(N) = \lambda(N) = \mu(N) + \varphi(N)$. Consequently $\mu(N) = 0$. We set $\varphi(A) = \varphi_1(A) + \psi(A)$ where $\varphi_1(A) = \varphi(A \backslash N)$ and $\psi(A) = \varphi(A \cap N)$. It follows from the definition of $\psi(A)$ that ψ and μ are singular. Remembering that $\lambda = \varphi + \mu$, we obtain from (3),

$$\int_U f(u) [1 - g(u)] \varphi(du) = \int_U f(u)g(u)\mu(du)$$

for an arbitrary function f that is square-integrable with respect to the measure λ. But this equation then holds for an arbitrary nonnegative $\mathfrak{S}$-measurable function. If we set $f = \chi_A/(1 - g)$ for $u \notin N$ and $f = 0$ for $u \in N$, we obtain

$$\varphi(A \backslash N) = \varphi_1(A) = \int_{A \backslash N} \bar{g}(u)\mu(du) = \int_A \bar{g}(u)\mu(du) ,$$

where $\bar{g}(u) = g(u)/[1 - g(u)]$ for $u \notin N$ and $\bar{g}(u) = 0$ for $u \in N$. This completes the proof of the theorem.

Corollary 1 (The Radon-Nikodym theorem). *If φ and μ are finite measures and φ is absolutely continuous with respect to μ, then φ is the indefinite integral of some μ-integrable function*

$$\varphi(A) = \int_A f(u)\mu(du) . \qquad (4)$$

The function $f(u)$ is uniquely defined (mod μ) *by formula* (4).

Formula (4) follows from (2) since $\psi(A) = \varphi(A \cap N) = 0$ for $\varphi \ll \mu$. If

$$\int_A f_1(u)\mu(du) = \int_A f_2(u)\mu(du)$$

for arbitrary $A \in \mathfrak{S}$, then

$$\int_A g(u)\mu(du) = 0$$

for arbitrary A where $g(u) = f_1(u) - f_2(u)$. In particular, for arbitrary positive ε,

$$0 = \int_{\{u;g(u)>\varepsilon\}} g(u)\mu(du) > \varepsilon\mu\{u; g(u) > \varepsilon\} ,$$

that is, $\mu\{(u; g(u) > \varepsilon\} = 0$. It follows from this that $\mu\{u; g(u) > 0\} = 0$. Analogously, $\mu\{u; g(u) < 0\} = 0$ and $g = 0$ (mod μ). This proves the uniqueness (mod μ) of the function $f(u)$ in formula (4).

Corollary 2. *If φ and μ are finite measures on $\mathfrak{S}$, then φ can be represented as a sum of the form $\varphi = \varphi_1 + \psi$, where $\varphi_1 \ll \mu$ and $\psi \perp \mu$. This representation is unique.*

Corollary 3. *For the measure μ to be absolutely continuous with respect to μ (where $\varphi < \infty$ and $\mu < \infty$), it is necessary and sufficient that for arbitrary $\varepsilon > 0$, there exists a $\delta > 0$ such that the inequality $\mu(A) < \delta$ implies the inequality $\varphi(A) < \varepsilon$.*

The proof of this condition is obvious. The necessity follows from a property of an integral (cf. corollary to theorem 4 of Section 5).

Theorem 2. *If μ and φ are finite measures and $\varphi \ll \mu$, then for every $\mathfrak{S}$-measurable function $f(u)$,*

$$\int_U f(u)\varphi(du) = \int_U f(u)\frac{d\varphi}{d\mu}\,\mu(du)\,, \tag{5}$$

if one side of this equation is meaningful.

Proof. The class K of functions for which equation (5) holds enjoys the following properties: (a) It contains the characteristic functions of arbitrary $\mathfrak{S}$-measurable sets by virtue of the definition of the derivative $d\varphi/d\mu$; (b) it is linear and hence contains all simple functions; (c) it is closed under passage to the limit for an arbitrary nondecreasing sequence of nonnegative (or nonpositive) functions (see Section 5, Theorem 1). It follows from these three properties that K contains all $\mathfrak{S}$-measurable functions for which one side of equation (5) is meaningful.

Let X and Y denote two sets and let $y = f(x)$ denote a mapping of X onto Y (which we denote by $X \to Y$). This means that to every $x \in X$ is assigned an element $y = f(x)$ of the set Y and that for every element $y \in Y$ there exists an $x \in X$ such that $y = f(x)$. In the relation $y = f(x)$, the element y (of the set Y) is called the *image* of the element x (of the set X) and the element x is called the *preimage* of the element y. If A is a subset of X, the image $f(A)$ of the set A is the subset of Y consisting of the images for all elements of the set A:

$$f(A) = \{y; y = f(x), x \in A\}\,.$$

The set of all x whose images belong to B is called the *preimage* of the set $B \subset Y$ and is denoted by $f^{-1}(B)$:

$$f^{-1}(B) = \{x; f(x) \in B\}\,.$$

It is easy to see that for an arbitrary class of sets $\{B_\alpha\}$,

$$f^{-1}(\bigcup_\alpha B_\alpha) = \bigcup_\alpha f^{-1}(B_\alpha)\,, \qquad f^{-1}(\bigcap_\alpha B_\alpha) = \bigcap_\alpha f^{-1}(B_\alpha)\,.$$

In addition,

$$f^{-1}(\bar{B}) = \overline{f^{-1}(B)}\,.$$

Suppose that a σ-algebra of sets $\mathfrak{S}$ is defined on X. The mapping $y = f(x)$ of X onto Y induces in Y a unique new σ-algebra of sets $\mathfrak{S}_f$ consisting of those sets $B \subset Y$ whose preimages are sets of the σ-algebra $\mathfrak{S}$: $\mathfrak{S}_f = \{B, f^{-1}(B) \in \mathfrak{S}\}$. We shall say that the σ-algebra $\mathfrak{S}_f$ is the *image* of $\mathfrak{S}$ under the mapping $f(x)$ and we shall let it be denoted by $f(\mathfrak{S})$.

Suppose that a measure μ is defined on $\mathfrak{S}$. By means of the mapping $y = f(x)$, we can transfer the measure from $\mathfrak{S}$ to $f(\mathfrak{S})$. To do this, we set $\mu'(B) = \mu[f^{-1}(B)]$ for every $B \in f(\mathfrak{S})$. Obviously μ' is a measure on $f(\mathfrak{S})$. Let us agree in this case to write $\mu' = \mu f^{-1}$ and to call the measure μ' the image of the measure μ.

Let Y denote a metric space. The mapping $y = f(x)$ is said to be *$\mathfrak{S}$-measurable* if $f(\mathfrak{S})$ contains all Borel sets in Y. In this definition, it is sufficient to require that $f(\mathfrak{S})$ contain some class of sets $\mathfrak{M}$ that generates the σ-algebra of Borel sets $\mathfrak{B}$ since it will then follow that $f(\mathfrak{S}) \supset \mathfrak{B}$.

The definition of a measurable mapping can be generalized as follows: Let X and Y denote two spaces with σ-algebras $\mathfrak{S}_x$ and $\mathfrak{S}_y$ defined for X and Y respectively. A mapping $y = f(x)$ defined on X onto Y is said to be measurable (for given $\mathfrak{S}_x$ and $\mathfrak{S}_y$) if $f(\mathfrak{S}_x) \supset \mathfrak{S}_y$. The usefulness of this definition is related to the fact that a measurable mapping $y = f(x)$ enables us to transfer the measure μ from the σ-algebra $\mathfrak{S}_x$ on which it is defined to a given σ-algebra $\mathfrak{S}_y$ by using the device just described. We set $\mu'(B) = \mu f^{-1}(B) = \mu[f^{-1}(B)]$ for $B \in \mathfrak{S}_y$. The following theorem corresponds to the rule for change of variable in evaluating an integral.

Theorem 3. *Let $y = f(x)$ denote a mapping of a space with measure $\{X, \mathfrak{S}, \mu\}$ onto Y and let $g(y)$ denote an $f(\mathfrak{S})$-measurable real function defined on Y. Then*

$$\int_Y g(y)\mu f^{-1}(dy) = \int_X g(f(x))\mu(dx) , \qquad (6)$$

if at least one of these integrals is meaningful.

Since the class of functions $g(y)$ for which equation 6 holds is linear and closed under passage to the limit for nonnegative nondecreasing sequences, it will be sufficient to verify this equation for the characteristic functions $\chi_B(y)$ of $f(\mathfrak{S})$-measurable sets. Let B denote a member of $f(\mathfrak{S})$. Then

$$\begin{aligned}\int_Y \chi_B(y)\mu f^{-1}(dy) &= \int_B \mu f^{-1}(dy) = \mu(f^{-1}(B)) \\ &= \int_{f^{-1}(B)} \mu(dx) = \int_X \chi_B[f(x)]\mu(dx) ,\end{aligned}$$

which yields the desired result.

7. EXTENSION OF MEASURES

At the heart of the axiomatization of probability theory is the concept of a probability space, that is, the space of elements and a probability defined on it. A probability space is not a simple mathematical object, however, and it is by no means always defined in elementary probability-theoretic problems and constructions. For example, in an elementary exposition of probability theory a finite collection of random variables is characterized by a joint distribution function

$$F(x_1, x_2, \cdots, x_n) = \mathsf{P}\{\xi_1 < x_1, \xi_2 < x_2, \cdots, \xi_n < x_n\} .$$

There is no mention here of any σ-algebra of sets or of countably additive probability. This section is devoted to the question: How can we construct probability spaces from more elementary objects?

Definition 1. A nonnegative set function λ defined on all subsets of a set U is called an *outer measure* on U if

$$\left.\begin{array}{ll} \text{a.} & \lambda(\phi) = 0 , \\ \text{b.} & \lambda(A) \leqq \sum_k \lambda(A_k) \end{array}\right\} \tag{1}$$

for an arbitrary finite or countably infinite collection of sets A_k covering A (that is, $A \subset \bigcup_k A_k$).

The property expressed by inequality b is called the *subadditivity* of the set function.

We find that we can use the outer measure to define a nonempty σ-algebra $\mathfrak{S}$ of the sets such that the outer measure λ will be a countably additive set function on $\mathfrak{S}$.

Definition 2. A set E is said to be *Caratheodory-measurable* if for arbitrary $A \subset U$,

$$\lambda(A) = \lambda(A \cap E) + \lambda(A \backslash E) . \tag{2}$$

Theorem 1 (Caratheodory). *The nonempty class $\mathfrak{S}$ of all Caratheodory-measurable sets is a complete (with respect to λ) σ-algebra. The outer measure λ on $\mathfrak{S}$ is a measure.*

The following propositions are steps in the proof of this theorem. Let us agree throughout this section to refer to Caratheodory-measurable sets simply as "measurable sets."

Lemma 1. *The sets $\varnothing$ and U are measurable.*

Proof. By virtue of property a of an outer measure, the measurability condition (2) is trivially satisfied for the empty set and the entire space U since

$$A \cap \varnothing = \varnothing , \qquad A \backslash \varnothing = A ,$$
$$A \cap U = A , \qquad A \backslash U = \varnothing .$$

Lemma 2. *The complement of a measurable set is measurable.*

Proof. If E is measurable, it follows from the equations $A \cap \bar{E} = A \backslash E$ and $A \backslash \bar{E} = A \cap E$, where A is arbitrary, that

$$\lambda(A \cap \bar{E}) + \lambda(A \backslash \bar{E}) = \lambda(A \cap E) + \lambda(A \backslash E) = \lambda(A) ;$$

that is, $\bar{E}$ is measurable.

Lemma 3. *If E is measurable, if A and B are subsets of U, and if B and E are disjoint, then*

$$\lambda[A \cap (E \cup B)] = \lambda(A \cap E) + \lambda(A \cap B) . \tag{3}$$

Proof. Let us apply condition (2) with A represented here by the set $A \cap (E \cup B)$. Since

$$[A \cap (E \cap B)] \cap E = A \cap E \quad \text{and} \quad [A \cap (E \cup B)] \backslash E = A \cap B ,$$

equation (2) reduces to equation (3).

Lemma 4. *If E_k, for $k = 1, 2, \cdots, n$, are measurable, if A and B are arbitrary subsets of U, if $B \cap E_k = \varnothing$, and if $E_k \cap E_r = \varnothing$ (for $r, k = 1, 2, \cdots, n$) whenever $r \neq k$, then*

$$\begin{aligned} &\lambda[A \cap (E_1 \cup \cdots \cup E_n \cup B)] \\ &\qquad = \lambda(A \cap E_1) + \cdots + \lambda(A \cap E_n) + \lambda(E \cap B) . \end{aligned} \tag{4}$$

The proof follows from (3) by induction.

Lemma 5. *The collection of all measurable sets constitutes an algebra of sets.*

Proof. Since $E_1 \backslash E_2 = \overline{(\bar{E}_1 \cup E_2)}$ by virtue of Lemma 2 it will be sufficient to show that measurability of E_1 and E_2 implies measurability of $E_1 \cup E_2$. Let us apply condition (2), twice, once with E replaced by E_1:

$$\lambda(A) = \lambda(A \cap E_1) + \lambda(A \backslash E_1) ,$$

and once with E replaced by E_2 and A replaced by $A \backslash E_1$:

$$\lambda(A \backslash E_1) = \lambda[(A \backslash E_1) \cap E_2] + \lambda[(A \backslash E_1) \backslash E_2] .$$

Since $(A \backslash E_1) \cap E_2$ and $A \cap E_1$ are disjoint and their sum is $A \cap (E_1 \cup E_2)$, by using Lemma 3 we obtain

$$\begin{aligned} \lambda(A) &= \lambda(A \cap E_1) + \lambda[(A \backslash E_1) \cap E_2] + \lambda[(A \backslash (E_1 \cup E_2)] \\ &= \lambda[A \cap (E_1 \cup E_2)] + \lambda[A \backslash (E_1 \cup E_2)] , \end{aligned}$$

which proves the measurability of the set $E_1 \cup E_2$.

It follows from Lemma 4 and 5 that the outer measure λ is an additive function defined on the algebra of measurable sets $\mathfrak{S}$.

Lemma 6. *If $\{E_k\}$, for $k = 1, 2, \cdots$, is a sequence of measurable pairwise disjoint sets (that is, $E_k \cap E_s = \varnothing$ for $k \neq s$), then*

$$\lambda\left(A \cap \bigcup_{k=1}^{\infty} E_k\right) = \sum_{k=1}^{\infty} \lambda(A \cap E_k) \tag{5}$$

Proof. By virtue of the subadditivity of the outer measure,

$$\lambda\left[A \cap \left(\bigcup_{k=1}^{\infty} E_k\right)\right] \leqq \sum_{k=1}^{\infty} \lambda(A \cap E_k) \,. \tag{6}$$

On the other hand,

$$\lambda\left[A \cap \left(\bigcup_{k=1}^{\infty} E_k\right)\right] \geqq \lambda\left[A \cap \left(\bigcup_{k=1}^{n} E_k\right)\right] = \sum_{k=1}^{n} \lambda(A \cap E_k) \,.$$

Taking the limit as $n \longrightarrow \infty$ in the last inequality, we obtain

$$\lambda\left[A \cap \left(\bigcup_{k=1}^{\infty} E_k\right)\right] \geqq \sum_{k=1}^{\infty} \lambda(A \cap E_k) \,. \tag{7}$$

Inequalities (6) and (7) prove the lemma.

Lemma 7. *The class of all measurable sets constitutes a σ-algebra.*

Proof. Let $\{E_k\}$ denote a sequence of measurable sets. We need to show that $\bigcup_{k=1}^{\infty} E_k \in \mathfrak{S}$. Since

$$\bigcup_{k=1}^{\infty} E_k = E_1 \cup (E_2 \backslash E_1) \cup \cdots \cup \left(E_n \Big\backslash \bigcup_{k=1}^{n-1} E_k\right) \cup \cdots,$$

we may confine ourselves to the case in which the sets E_k are pairwise disjoint. Suppose that $\bigcup_{k=1}^{\infty} E_k = (\bigcup_{k=1}^{n} E_k) \cup B_n$. The set $\bigcup_{k=1}^{n} E_k$ is measurable and therefore

$$\begin{aligned} \lambda(A) &= \lambda\left[A \cap \left(\bigcup_{k=1}^{n} E_k\right)\right] + \lambda\left[A \Big\backslash \bigcup_{k=1}^{n} E_k\right] \\ &\geqq \sum_{k=1}^{n} \lambda(A \cap E_k) + \lambda\left[A \Big\backslash \bigcup_{k=1}^{\infty} E_k\right] . \end{aligned}$$

Taking the limit as $n \longrightarrow \infty$, we obtain (using Lemma 6)

$$\lambda(A) \geqq \sum_{k=1}^{\infty} \lambda(A \cap E_k) + \lambda\left(A \Big\backslash \bigcup_{k=1}^{\infty} E_k\right) .$$

Since inequality in the opposite direction always holds because of the subadditivity of the outer measure, the assertion is proved.

Lemma 8. *Every subset N' of a set N of outer measure zero ($\lambda(N) = 0$) has outer measure zero and is measurable.*

Proof. If $N' \subset N$ and $\lambda(N) = 0$, then $\lambda(N') = 0$ because of the subadditivity. Furthermore,

$$\lambda(A \backslash N') \leqq \lambda(A) \leqq \lambda(A \cap N') + \lambda(A \backslash N') \,, \qquad \lambda(A \cap N') = 0 \,.$$

and consequently,

$$\lambda(A) = \lambda(A \cap N') + \lambda(A \backslash N') = \lambda(A \backslash N') \,,$$

which proves the measurability of the set N'.

Theorem 1 is an immediate consequence of Lemmas 6–8.

Theorem 1 shows how we can use the outer measure of sets to construct a measure. However, the possibility is not ruled out that the σ-algebra of the sets that are measurable with respect to this measure is trivial, that is, that it consists only of the empty set and the entire space U. On the other hand, in most cases we are interested not in arbitrary measures but in measures that coincide with a given set function on some class of sets $\mathfrak{A}$ that does not constitute a σ-algebra. For example, let us look at the particular problem characterizing the statement of the question in the general case. Let $F(x)$ denote a given distribution function of some random variable ξ. We wish to construct a measurable space $\{U, \mathfrak{S}, \mathsf{P}\}$ in which U is the real line $(-\infty, \infty)$, $\mathfrak{S}$ is the complete σ-algebra of sets containing all Borel subsets of the real line, and P is a probability measure on $\mathfrak{S}$ that is consistent with the given distribution function $F(x)$, that is, such that

$$\mathsf{P}\{\xi < x\} = \mathsf{P}\{(-\infty, x)\} = F(x) .$$

The distribution function $F(x)$ determines the probability that the value of the random variable ξ will fall in the left-closed right-open interval $[a, b)$:

$$\mathsf{P}\{a \leqq \xi < b\} = F(b) - F(a) = F[a, b) .$$

The class $\mathfrak{A}$ of all finite left-closed right-open intervals satisfies the following two conditions (we treat the empty set as belonging to $\mathfrak{A}$): (a) The intersection of two left-closed right-open intervals is such an interval. (b) The difference of two such intervals is either such an interval or the union of two such intervals.

Note that the interval function $F[a, b)$, is additive; that is, if $a_0 < a_1 < \cdots < a_n$, then

$$F[a_0, a_n) = F[a_0, a_1) + F(a_1, a_2) + \cdots + F[a_{n-1}, a_n) .$$

The problem now is to find a measurable space $\{U, \mathfrak{S}, \mathsf{P}\}$ such that $\mathfrak{S}$ contains all left-closed right-open intervals $[a, b)$ and $\mathsf{P}[a, b) = F(b) - F(a)$. Let us consider this problem in the general case.

Definition 3. Let $\{m, \mathfrak{M}\}$ and $\{m', \mathfrak{M}'\}$ denote two set functions such that $\mathfrak{M} \subset \mathfrak{M}'$ and $m(\Delta) = m'(\Delta)$ for $\Delta \in \mathfrak{M}$. The set function $\{m', \mathfrak{M}'\}$ is then said to be an *extension* of the set function $\{m, \mathfrak{M}\}$.

An extension of a given nonnegative set function as a measure can often be obtained by means of a device proposed by Lebesgue: First we construct from m an outer measure defined on all subsets of the space; then we use Theorem 1.

Let $\{m, \mathfrak{M}\}$ denote an arbitrary nonnegative set function satisfying the conditions $\varnothing \in \mathfrak{M}$ and $m(\varnothing) = 0$. For arbitrary A we define

$$\lambda(A) = \inf\left\{\sum_{k=1}^{\infty} m(\Delta_k);\ \Delta_k \in \mathfrak{M},\ \bigcup_{k=1}^{\infty} \Delta_k \supset A\right\}, \tag{8}$$

where the infimum is taken over all finite or countably infinite coverings of the set A by the sets $\Delta_k \in \mathfrak{M}$, for $k = 1, 2, \cdots$. If A is not covered by a countable union of sets belonging to $\mathfrak{M}$, we define

$$\lambda(A) = \infty . \tag{9}$$

The function λ is defined on the class $\mathfrak{U}$ of all subsets of the set U.

Lemma 9. *The set function* $\{\lambda, \mathfrak{U}\}$ *is an outer measure.*

Proof. Since $m(\varnothing) = 0$, it follows that $\lambda(\varnothing) = 0$. We must now show that the function λ is subadditive. Suppose that

$$A \subset \bigcup_{k=1}^{\infty} A_k .$$

If $\lambda(A_k) = \infty$ for at least one k, the subadditivity is obvious. Suppose that $\lambda(A_k) < \infty$ for every k. Then for arbitrary $\varepsilon > 0$ we construct for every k a sequence $\{\Delta_{kr};\ r = 1, 2, \cdots\}$ of sets belonging to $\mathfrak{M}$ such that

$$\sum_r m(\Delta_{kr}) \leqq \lambda(A_k) + \frac{\varepsilon}{2^k},\quad \bigcup_r \Delta_{kr} \supset A_k .$$

Then the set $\{\Delta_{kr};\ k = 1, 2, \cdots,\ r = 1, 2, \cdots\}$ is a finite or countably infinite collection of sets that covers A, and we have

$$\lambda(A) \leqq \sum_k \sum_r m(\Delta_{kr}) \leqq \sum_k \lambda(A_k) + \varepsilon .$$

Taking the limit as $\varepsilon \to 0$, we obtain the desired result.

Let us call the set function $\{\lambda, \mathfrak{U}\}$ defined by equations (8) and (9) the *Lebesgue outer measure* corresponding to $\{m, \mathfrak{M}\}$. On the σ-algebra $\mathfrak{L}$ of all λ-measurable (in the sense of Caratheodory) sets, λ is a measure. Let us agree to call it the *Lebesgue measure* in the present section. We now consider the question: When is $\{\lambda, \mathfrak{L}\}$ the extension of a set function $\{m, \mathfrak{M}\}$? We note that if $\{m, \mathfrak{M}\}$ has an extension as a measure, m must be an additive function on $\mathfrak{M}$. In many cases $\mathfrak{M}$ has the following structure:

Definition 4. A class of sets $\mathfrak{M}$ is said to be *decomposable* if for arbitrary Δ_1 and Δ_2 in $\mathfrak{M}$,

$$\Delta_1 \cap \Delta_2 \in \mathfrak{M} ,$$

$$\Delta_1 \backslash \Delta_2 = \bigcup_{k=1}^{s} \Delta_k^*, \text{ where } \Delta_k^* \in \mathfrak{M},\ \Delta_k^* \cap \Delta_r^* = \varnothing \text{ for } k \neq r .$$

REMARK 1. The concept of a decomposable class of sets is close to the concept of an algebra of sets. Specifically, a class of sets of the form

$$B = \left(\bigcup_{k=1}^{r} \Delta_k\right) \cap \Delta ,$$

where r is an arbitrary integer and the Δ_k are arbitrary sets in $\mathfrak{M}$, is an algebra of the subsets Δ.

Definition 5. A nonnegative additive set function not identically $+\infty$ that is defined on a decomposable class of sets $\mathfrak{M}$ is called an *elementary measure*.

Theorem 2. *For a Lebesgue measure* $\{\lambda, \mathfrak{L}\}$ *constructed from an elementary measure* $\{m, \mathfrak{M}\}$ *to be its extension, it is necessary and sufficient that for arbitrary* $\Delta \in \mathfrak{M}$ *the relations* $\bigcup_{k=1}^{\infty} \Delta_k \supset \Delta$, *where each* $\Delta_k \in \mathfrak{M}$, *imply*

$$m(\Delta) \leqq \sum_{k=1}^{\infty} m(\Delta_k) . \tag{10}$$

Proof of the Necessity. If $\{\lambda, \mathfrak{L}\}$ is the extension of $\{m, \mathfrak{M}\}$, then $\lambda(\Delta) = m(\Delta)$ and inequality (10) follows from Lemma 9.

Proof of the Sufficiency. It follows from the definition of outer measure that $\lambda(\Delta) \leqq m(\Delta)$, if $\Delta \in \mathfrak{M}$. On the other hand if condition (10) is satisfied, then

$$\lambda(\Delta) = \inf \sum_{k=1}^{\infty} m(\Delta_k) \geqq m(\Delta) ,$$

from which we get

$$\lambda(\Delta) = m(\Delta) \quad \text{for every} \quad \Delta \in \mathfrak{M} . \tag{11}$$

We must now show that all the $\Delta \in \mathfrak{M}$ are measurable. From the subadditivity of the outer measure, we have for arbitrary $\Delta^* \in \mathfrak{M}$

$$\lambda(\Delta^*) \leqq \lambda(\Delta^* \cap \Delta) + \lambda(\Delta^* \backslash \Delta) . \tag{12}$$

Since $\Delta^* \cap \Delta = \Delta_0$ and $\Delta^* \backslash \Delta = \bigcup_{k=1}^{s} \Delta_k$, where $\Delta_j \in \mathfrak{M}$ (for $j = 0, 1, \cdots, s$), and since we may assume that the Δ_j are disjoint, it follows that

$$\begin{aligned} \lambda(\Delta^*) &= m(\Delta^*) = \sum_{k=0}^{s} m(\Delta_k) \\ &= \lambda(\Delta^* \cap \Delta) + \sum_{k=1}^{s} m(\Delta_k) \geqq \lambda(\Delta^* \cap \Delta) + \lambda(\Delta^* \backslash \Delta) . \end{aligned} \tag{13}$$

Comparing (12) and (13) we obtain

$$\lambda(\Delta^*) = \lambda(\Delta^* \cap \Delta) + \lambda(\Delta^* \backslash \Delta) . \tag{14}$$

Let A denote an arbitrary subset of U. If $\lambda(A) = \infty$ we have

$$\lambda(A) \geqq \lambda(A \cap \Delta) + \lambda(A \backslash \Delta) .$$

On the other hand, if $\lambda(A) < \infty$ there exists a sequence of sets $\Delta_k \in \mathfrak{M}$ such that $A \subset \bigcup_{k=1}^{\infty} \Delta_k$ and $\lambda(A) + \varepsilon > \sum_k \lambda(\Delta_k)$, for $\varepsilon > 0$. On the basis of (14) we have

$$\lambda(\Delta_k) = \lambda(\Delta_k \cap \Delta) + \lambda(\Delta_k \backslash \Delta) ,$$

from which we get

$$\lambda(A) + \varepsilon > \sum_{k=1}^{\infty} \lambda(\Delta_k \cap \Delta) + \sum_{k=1}^{\infty} \lambda(\Delta_k \backslash \Delta) \geqq \lambda(A \cap \Delta) + \lambda(A \backslash \Delta) ,$$

since

$$A \cap \Delta \subset \bigcup_{k=1}^{\infty} (\Delta_k \cap \Delta) \quad \text{and} \quad A \backslash \Delta \subset \bigcup_{k=1}^{\infty} (\Delta_k \backslash \Delta) .$$

Thus since ε is arbitrary, we have in all cases:

$$\lambda(A) \geqq \lambda(A \cap \Delta) + \lambda(A \backslash \Delta) .$$

Inequality in the opposite direction follows from the subadditivity of the outer measure. Consequently for arbitrary $A \subset U$ and arbitrary $\Delta \in \mathfrak{M}$,

$$\lambda(A) = \lambda(A \cap \Delta) + \lambda(A \backslash \Delta) ,$$

that is, Δ is measurable. Thus $\mathfrak{M} \subset \mathfrak{L}$ and $m(\Delta) = \lambda(\Delta)$ on $\mathfrak{M}$, that is, the Lebesgue measure $\{\lambda, \mathfrak{L}\}$ is the extension of the elementary measure $\{m, \mathfrak{M}\}$.

In connection with the solution obtained in Theorem 2 to the problem of the extension of an additive set function $\{m, \mathfrak{M}\}$, the question naturally arises: Is it possible to extend an elementary measure $\{m, \mathfrak{M}\}$ in another manner and obtain a measure distinct from the Lebesgue measure $\{\lambda, \mathfrak{L}\}$? In a certain sense the answer to this question is negative. We find that extension of an elementary measure to the smallest σ-algebra $\sigma\{\mathfrak{M}\}$ containing $\mathfrak{M}$ is unique if it exists. We shall say that an elementary measure $\{m, \mathfrak{M}\}$ is σ-finite if U can be represented as the sum of countably many elementary sets each of finite measure.

Theorem 3. *The extension of an elementary σ-finite measure $\{m, \mathfrak{M}\}$ as a measure $\{\mu, \sigma\{\mathfrak{M}\}\}$ is unique.*

Proof. Let Δ denote an arbitrary fixed set in $\mathfrak{M}$ such that $m(\Delta) < \infty$. Let $\mathfrak{A}_\Delta$ denote the class of sets of the form

$$B = \bigcup_{k=1}^{s} A_k \cap \Delta, \; A_k \in \mathfrak{M}, \; A_k \cap A_j = \varnothing \quad \text{for} \quad k \neq j \,.$$

Then $\mathfrak{A}_\Delta$ is the algebra of subsets of Δ. If $\{\mu_1, \sigma\{\mathfrak{M}\}\}$ and $\{\mu_2, \sigma\{\mathfrak{M}\}\}$ are two extensions of an elementary measure $\{m, \mathfrak{M}\}$, then by virtue of the additivity of a measure,

$$\mu_1(B) = \sum_{k=1}^{s} m(A_k \cap \Delta) = \mu_2(B) \,.$$

Let $\mathfrak{K}$ denote the class of sets C such that $\mu_1(C) = \mu_2(C)$. Obviously, $\mathfrak{K}$ is a monotonic class and $\mathfrak{K} \supset \mathfrak{A}_\Delta$. Consequently, $\mathfrak{K} \supset \sigma\{\mathfrak{A}_\Delta\}$. Now let E denote an arbitrary set belonging to $\sigma\{\mathfrak{M}\}$. The set E is covered by a countable disjoint collection of sets $\Delta_j \in \mathfrak{M}$ each of finite m-measure. Consequently, for arbitrary i we have $\mu_1(E \cap \Delta_i) = \mu_2(E \cap \Delta_i)$, so that $\mu_1(E) = \mu_2(E)$.

We shall now show the relation between the extension of an elementary measure on $\sigma\{\mathfrak{M}\}$ and the σ-algebra $\mathfrak{L}$ of all measurable sets.

Let $\mathscr{E}_\sigma$ (resp. $\mathscr{E}_\delta$) denote the class of sets that can be represented as the union (resp. intersection) of a countable collection of sets belonging to $\mathfrak{M}$ and let $\mathscr{E}_{\sigma\delta}$ (resp. $\mathscr{E}_{\delta\sigma}$) denote the class of sets that can be represented as the intersection (resp. union) of a countable collection of sets belonging to $\mathscr{E}_\sigma$ (resp. $\mathscr{E}_\delta$). The classes $\mathscr{E}_\sigma$, $\mathscr{E}_\delta$, $\mathscr{E}_{\sigma\delta}$, $\mathscr{E}_{\delta\sigma}$ all belong to $\sigma\{\mathfrak{M}\}$.

Theorem 4. *If* $(\lambda, \mathfrak{L})$ *is the Lebesgue measure that is the extension of a σ-finite elementary measure* $\{m, \mathfrak{M}\}$ *satisfying condition* (10), *then for every set* $A \in \mathfrak{L}$, *there exists a measurable set* E *in* $\mathscr{E}_{\sigma\delta}$ *such that* $A \subset E$ *and* $\lambda(A) = \lambda(E)$.

Proof. Since

$$\lambda(A) = \inf_{\cup\Delta_k \supset A} \sum m(\Delta_k) = \inf_{\cup\Delta_k \supset A} \sum \lambda(\Delta_k) \,,$$

there exists a sequence $\{\Delta_k^{(n)}\}$ of coverings of the set A (that is, $\bigcup_{k=1}^\infty \Delta_k^{(n)} \supset A$, for each n) such that

$$\lambda(A) = \lim_{n\to\infty} \sum_k \lambda(\Delta_k^{(n)}) \,.$$

Suppose that $E^{(n)} = \bigcup_k \Delta_k^{(n)}$ and $\widetilde{E}^{(n)} = \bigcap_{k=1}^n E^{(k)}$. Then

$$\lambda(A) \leqq \lambda(\widetilde{E}^{(n)}) \leqq \lambda(E^{(n)}) \leqq \sum_{k=1}^\infty \lambda(\Delta_k^{(n)})$$

and

$$\lambda(A) = \lim \lambda(\widetilde{E}^{(n)}) = \lambda(E) \,,$$

where $E = \bigcap_{k=1}^\infty \widetilde{E}^{(k)}$ is the limit of a decreasing sequence of sets and $E \in \mathscr{E}_{\sigma\delta}$. This completes the proof of the theorem.

Let $\mathfrak{N}$ denote the class of all subsets of sets in $\sigma\{\mathfrak{M}\}$ that are of μ-measure 0, where μ is some measure defined on $\sigma\{\mathfrak{M}\}$:

$$\mathfrak{N} = \{N; N \subset E, \ E \in \sigma\{\mathfrak{M}\}, \ \mu(E) = 0\} ,$$

Let $\mathfrak{S}$ denote the class of all sets of the form $A = (E \cup N')\backslash N''$, where $E \in \sigma\{\mathfrak{M}\}$, $N' \in \mathfrak{N}$, and $N'' \in \mathfrak{N}$. In short, $\mathfrak{S}$ is the class of sets that differ from the sets in $\sigma\{\mathfrak{M}\}$ by a subset of μ-measure 0. Let us set $\tilde{\mu}(A) = \mu(E)$. This definition is unambiguous: If $A = (E_1 \cup N_1')\backslash N_1'' = (E_2 \cup N_2')\backslash N_2''$, then $E_1\backslash E_2 \subset N_1'' \cup N_2'$ and $E_2\backslash E_1 \subset N_1' \cup N_2''$. Since $E_1\backslash E_2$ and $E_2\backslash E_1$ belong to $\sigma\{\mathfrak{M}\}$ we have $\mu(E_2\backslash E_1) = \mu(E_1\backslash E_2) = 0$ so that $\tilde{\mu}(E_2) = \tilde{\mu}(E_1)$. One can easily show that $\mathfrak{S}$ is a σ-algebra and that $\tilde{\mu}$ is a measure on it.

Definition 6. The measure $\{\tilde{\mu}, \mathfrak{S}\}$ is called the *completion* of the measure $\{\mu, \sigma\{\mathfrak{M}\}\}$.

Theorem 5. *If a σ-finite elementary measure $\{m, \mathfrak{M}\}$ satisfies condition* (10), *the completion $\{\lambda, \mathfrak{S}\}$ of its extension $\{\lambda, \sigma\{\mathfrak{M}\}\}$ onto the minimal σ-algebra coincides with the Lebesgue measure $\{\lambda, \mathfrak{L}\}$.*

Proof. Since the σ-algebra $\mathfrak{L}$ is λ-complete it follows that for $E \in \sigma(\mathfrak{M})$, $N', N'' \in \mathfrak{N}$, and $A = E \cup N'\backslash N''$, we have $N', N'' \in \mathfrak{L}$ and $\lambda(A) = \lambda(E)$. In other words $\mathfrak{S} \subset \mathfrak{L}$. Suppose now that A is an arbitrary set of finite measure in $\mathfrak{L}$. Then there exists an $E \supset A$ belonging to $\sigma\{\mathfrak{M}\}$ (in accordance with Theorem 4) and $\lambda(E\backslash A) = 0$. Let us set $E\backslash A = F$. Then there exists an $E' \supset F$ belonging to $\sigma\{\mathfrak{M}\}$ such that $\lambda(E') = \lambda(F) = 0$. Thus

$$A = E\backslash F, \ F \subset E', \ E' \in \sigma\{\mathfrak{M}\}, \ \lambda(E') = 0 ,$$

that is, $F \in \mathfrak{N}$. In the general case in which A is an arbitrary set belonging to $\mathfrak{L}$ we apply the last relation to the set $A \cap \Delta_k$. This completes the proof of the theorem.

These last theorems show that in a certain sense, measurable sets do not differ greatly from sets belonging to the minimal σ-algebra $\sigma\{\mathfrak{M}\}$. Specifically, a measurable set A differs from some $E \in \sigma\{\mathfrak{M}\}$ by a subset of a set in $\sigma\{\mathfrak{M}\}$ that is of measure 0.

The following result is a particular case, but it will be used in what follows.

Theorem 6. *Let U denote a complete separable metric space, let $\mathfrak{B}$ denote the σ-algebra of Borel subsets of U, and let μ denote a finite measure on $\mathfrak{B}$. Then for arbitrary $A \in \mathfrak{B}$ and arbitrary $\varepsilon > 0$ there exists a compact set K_ε contained in A such that $\mu(A\backslash K_\varepsilon) < \varepsilon$.*

Proof. Let us first prove the theorem for the particular case of $A = U$. Let $\{x_n\}$ denote a countable everywhere-dense set in

U. For each $n = 1, 2, \cdots$, let $\{\bar{S}_k(x_n)\}$ denote the sequence of closed spheres of radius $1/k$ (for $k = 1, 2, \cdots$) with center at the point x_n. For each k let us choose n_k so that

$$\mu\Big(\bigcup_{n=1}^{n_k} \bar{S}_k(x_n)\Big) \geqq \mu(U) - \frac{\varepsilon}{2^k}.$$

Define

$$\bar{S}_k = \bigcup_{n=1}^{n_k} \bar{S}_k(x_n).$$

The set $\bar{S}_k$ is closed and admits a finite $(1/k)$-net. Let us set

$$K_\varepsilon = \bigcap_{k=1}^{\infty} \bar{S}_k.$$

Then K_ε is closed and it admits a finite $(1/k)$-net for arbitrary integral k; that is, K_ε is compact. On the other hand,

$$\mu(U \backslash K_\varepsilon) \leqq \sum_{k=1}^{\infty} \mu(U \backslash \bar{S}_k) \leqq \varepsilon.$$

Thus the assertion of the theorem holds when A coincides with the entire space U. Furthermore, an arbitrary closed set F contained in a complete separable metric space is itself complete and separable. Consequently the theorem holds for an arbitrary closed set $F = A$. Let $\mathfrak{K}$ denote the class of sets B that can be represented simultaneously in the form

$$B = \bigcup_{n=1}^{\infty} F_n = \bigcap_{n=1}^{\infty} G_n$$

where the F_n are closed and the G_n are open. The class $\mathfrak{K}$ contains all closed sets and is an algebra of sets. Therefore $\sigma\{\mathfrak{K}\} = \mathfrak{B}$.

Let $\mathfrak{K}^*$ denote the class of sets for which the assertion of the theorem is valid. Let $\{B_n\}$, for $n = 1, 2, \cdots$, denote a nondecreasing sequence of sets belonging to $\mathfrak{R}^*$. Define

$$B_0 = \bigcup_{n=1}^{\infty} B_n,$$

Let $\{K_n\}$ denote a sequence of compact sets such that $K_n \subset B_n$, $\mu(B_n \backslash K_n) < \varepsilon/2^{n+1}$ and $\mu(B_0 \backslash B_{n_0}) < \varepsilon/2$. Then

$$\mu\Big(B_0 \Big\backslash \bigcup_{n=1}^{n_0} K_n\Big) \leqq \mu(B_0 \backslash B_{n_0}) + \sum_{n=1}^{n_0} \mu(B_n \backslash K_n) < \varepsilon,$$

so that $B_0 \in \mathfrak{K}^*$. In analogous fashion one can show that

$$B_0 = \bigcap_{n=1}^{\infty} B_n \in \mathfrak{K}^*$$

if $B_n \in \mathfrak{K}^*$ and if $\{B_n\}$ is a decreasing sequence of sets. Thus $\mathfrak{K}^*$

is a monotonic class. Since it contains closed sets, $\mathfrak{K} \subset \mathfrak{K}^*$. Therefore $\mathfrak{B} = \sigma\{\mathfrak{K}\} \subset \mathfrak{K}^*$. This completes the proof of the theorem.

Let us prove a frequently used result dealing with the possibility of approximating an arbitrary measurable function by simple ones.

Theorem 7. *Let $\mathfrak{M}$ denote a decomposable class of sets, let μ denote a finite elementary measure admitting an extension $\tilde{\mu}$ to the complete σ-algebra $\mathfrak{S}$, let $f(u)$ denote a $\mathfrak{S}$-measurable function, and let $g_\varepsilon(u)$ denote a function that assumes only finitely many distinct values and that is measurable with respect to $\mathfrak{M}$. Then for arbitrary $\varepsilon > 0$,* (a) *if $f(u)$ is finite* (mod $\tilde{\mu}$), *there exists a $g_\varepsilon(u)$ such that* $\tilde{\mu}\{x; |f(u) - g_\varepsilon(u)| > \varepsilon\} < \varepsilon$, and (b) *if $f(x) \in L_r\{U, \mathfrak{S}, \tilde{\mu}\}$, where $r \geqq 1$, there exists a $g_\varepsilon(x)$ such that* $\int_U |f(u) - g_\varepsilon(u)|^r \tilde{\mu}(du) < \varepsilon$.

Proof. We note that the class of functions for which the theorem is valid is linear. Since every measurable function $f(x)$ can be approximated in the sense of convergence in measure and since every $f(u) \in L_r\{U, \mathfrak{S}, \tilde{\mu}\}$ can be approximated in the sense of the metric of the space L_r by a simple $\mathfrak{S}$-measurable function, it will be sufficient for us to prove the theorem for the characteristic functions of $\mathfrak{S}$-measurable sets. Let A denote a member of $\mathfrak{S}$ and let $\chi_A(u)$ denote its characteristic function. Since

$$\tilde{\mu}(A) = \inf_{\cup \Delta_k \supset A} \sum_k \mu(\Delta_k) \quad \text{for} \quad \Delta_k \in \mathfrak{M} ,$$

there exists for given positive ε a sequence $\{\Delta_k\}$ such that

$$\sum_{k=1}^{\infty} \mu(\Delta)_k - \tilde{\mu}(A) < \frac{\varepsilon}{2} .$$

Let us now find an n such that

$$\sum_{k=n+1}^{\infty} \mu(\Delta_k) < \frac{\varepsilon}{2} .$$

Let $g_\varepsilon(u)$ denote the characteristic function of the set $\bigcup_{k=1}^n \Delta_k$. Then

$$\{u; |\chi_A(u) - g_\varepsilon(u)| > 0\} \subset \left\{\bigcup_{k=1}^{\infty} \Delta_k \setminus A\right\} \cup \left\{\bigcup_{k=n+1}^{\infty} \Delta_k\right\} ,$$

$$\tilde{\mu}\{x; |\chi_A(u) - g_\varepsilon(u)| > 0\} < \varepsilon$$

and

$$\int_U |\chi_A(u) - g_\varepsilon(u)|^r \tilde{\mu}(du) < \varepsilon ,$$

which completes the proof of the theorem.

REMARK 2. Assertion (*b*) of Theorem 7 remains valid even without the assumption that the measure $\tilde{\mu}$ is finite. To see this, we note that if $N_\varepsilon = \{u; \mid f(u) \mid > \varepsilon\}$, $\tilde{\mu}(N_\varepsilon) < \infty$ and

$$\int_U \mid f(u) \mid^r \tilde{\mu}(du) = \lim_{\varepsilon \to 0} \int_{N_\varepsilon} \mid f(u) \mid^r \tilde{\mu}(du) .$$

Therefore it suffices to approximate the function $f(u)$ (in the sense of the metric of L_r) on the set N_ε.

Let us now look at the extension of an arbitrary additive set function.

On a decomposable class $\mathfrak{M}$ of sets consider the real set function $\varphi(\Delta)$ satisfying the conditions

1. $\varphi(\varnothing) = 0$,
2. $\varphi\left(\bigcup_{k=1}^{n} \Delta_k\right) = \sum_{k=1}^{n} \varphi(\Delta_k)$, if $\Delta_i \cap \Delta_j = \varnothing (i \neq j)$, $\Delta_i \in \mathfrak{M}$.

Let $R(\Delta)$ denote the class of all decompositions of the set Δ as the union of finitely many disjoint sets belonging to $\mathfrak{M}$ and set $\mu(\Delta) = \sup_{R(\Delta)} \sum \mid \varphi(\Delta_k) \mid$. The function $\mu(\Delta)$ is called the *absolute variation* of the set function. The absolute variation enjoys the following properties:

1′. $\mu(\Delta) \geqq 0$, $\mu(\varnothing) = 0$,

2′. $\mu\left(\bigcup_{k=1}^{n} \Delta_k\right) = \sum_{k=1}^{n} \mu(\Delta_k)$, $\Delta_i \cap \Delta_j = \varnothing (i \neq j)$,

3′. $\mu(\Delta) \geqq \mid \varphi(\Delta) \mid$;

that is, it is an elementary measure on $\mathfrak{M}$.

A function $\varphi(\Delta)$, where $\Delta \in \mathfrak{M}$, is called a *function of bounded variation* on $\mathfrak{M}$ if $\mu(\Delta) < \infty$ for arbitrary $\Delta \in \mathfrak{M}$.

Theorem 8. *Let* $\{\varphi, \mathfrak{M}\}$ *denote an additive real set function satisfying the following conditions*: (a) *it is of bounded variation on* $\mathfrak{M}$, (b) *the absolute variation* $\mu(\Delta)$ *of the function* $\varphi(\Delta)$ *is subadditive*:

$$\mu(\Delta) \leqq \sum_k \mu(\Delta_k) , \quad \text{if} \quad \bigcup_k \Delta_k \supset \Delta .$$

Then there exist two measures $\{\tilde{\nu}_1, \mathfrak{S}\}$ *and* $\{\tilde{\nu}_2, \mathfrak{S}\}$ *such that* $\mathfrak{M} \subset \mathfrak{S}$ *and*

$$\varphi(\Delta) = \tilde{\nu}_1(\Delta) - \tilde{\nu}_2(\Delta) , \qquad \mu(\Delta) = \tilde{\nu}_1(\Delta) + \tilde{\nu}_2(\Delta) .$$

Proof. On the basis of the hypotheses of the theorem, the function $\mu(\Delta)$ can be extended as a complete measure $\{\tilde{\mu}, \mathfrak{S}\}$. We note that condition (b) stipulating the subadditivity of μ is equivalent to the condition that μ is countably additive on $\mathfrak{M}$. From this it follows that the function $\varphi(\Delta)$ is also countably additive on $\mathfrak{M}$. This is true because $\Delta = \bigcup_{k=1}^{\infty} \Delta_k$ and $\Delta_n \cap \Delta_r = \varnothing$ (for $n \neq r$). Let us set

$$\Delta \setminus \bigcup_{k=1}^{N} \Delta_k = \bigcup_{k=1}^{m_N} \Delta_k^N$$

where the Δ_k^N are pairwise disjoint members of $\mathfrak{M}$. Then

$$\left| \varphi(\Delta) - \sum_{k=1}^{N} \varphi(\Delta_k) \right| \leqq \sum_{k=1}^{m_N} \mu(\Delta_k^N) = \mu\left(\Delta \setminus \bigcup_{n=1}^{N} \Delta_n\right) \longrightarrow 0$$

as $N \longrightarrow \infty$. We define on $\mathfrak{M}$ the two nonnegative set functions

$$\nu_1(\Delta) = \frac{\mu(\Delta) + \varphi(\Delta)}{2}, \ \nu_2(\Delta) = \frac{\mu(\Delta) - \varphi(\Delta)}{2}.$$

It follows from the definition that

$$\mu(\Delta) = \nu_1(\Delta) + \nu_2(\Delta), \ \varphi(\Delta) = \nu_1(\Delta) - \nu_2(\Delta).$$

Since μ and φ are countably additive on $\mathfrak{M}$, so are ν_1 and ν_2. Therefore ν_1 and ν_2 have Lebesgue extensions as measures $\{\tilde{\nu}_1, \mathfrak{S}_1\}$ and $\{\tilde{\nu}_2, \mathfrak{S}_2\}$ respectively. Since $\nu_i(\Delta) \leqq \mu(\Delta)$ we have $\mathfrak{S} \subset \mathfrak{S}_i$ for $i = 1, 2$. This completes the proof of the theorem.

REMARK 3. We cannot assert that satisfaction of the conditions of Theorem 8 implies that a set function φ has an extension $\tilde{\varphi}$ onto the σ-algebra $\mathfrak{S}$ of the sets. The function $\tilde{\varphi} = \tilde{\nu}_1 - \tilde{\nu}_2$ is defined only on some class of sets $\mathfrak{S}_0 \subset \mathfrak{S}$ consisting of those $A \in \mathfrak{S}$ such that $\tilde{\nu}_i(A) < \infty$ for $i = 1, 2$. However we can define the integral with respect to the set function $\tilde{\varphi}$ for every $\mathfrak{S}$-measurable and $\tilde{\mu}$-integrable function $f(u)$. We make this definition by means of the relation

$$\int_U f(u)\tilde{\varphi}(du) = \int_U f(u)\tilde{\nu}_1(du) - \int_U f(u)\tilde{\nu}_2(du). \tag{15}$$

Here we should remember that if $f(u)$ is a $\tilde{\mu}$-integrable function, it is also $\tilde{\nu}_i$-integrable (for $i = 1, 2$). The integral with respect to the set function $\tilde{\varphi}$ defined by equation (15) is called the *Lebesgue-Stieltjes integral*.

8. THE PRODUCT OF TWO MEASURES

Let U and V denote two arbitrary sets. The set of ordered pairs (u, v), where $u \in U$ and $v \in V$, is called the *Cartesian product* of the sets U and V and is denoted by $U \times V$. Here $(u, v) = (u_1, v_1)$ if and only if $u = u_1$ and $v = v_1$. If $V = U$, instead of writing $U \times U$ we may write U^2. The simplest example of the Cartesian product of two sets is the coordinate plane, which can be regarded as the Cartesian product of the real axis with itself. If $A \subset U$ and $B \subset V$ then the set $A \times B$ is a subset of $U \times V$. It is called a *rectangle* with sides A and B.

If $\mathfrak{M}$ denotes a class of subsets of U and $\mathfrak{N}$ denotes a class of subsets of V, the product $\mathfrak{M} \times \mathfrak{N}$ of these classes is defined as the family of all sets of the form $M \times N$ for $M \in \mathfrak{M}$ and $N \in \mathfrak{N}$. If $\mathfrak{M} = \mathfrak{S}_1$ and $\mathfrak{N} = \mathfrak{S}_2$ are two σ-algebras of sets, their product $\mathfrak{S}_1 \times \mathfrak{S}_2$ is not an algebra. However, it is a decomposable class of sets (cf. Definition 4 of Section 7). To see this, note that

$$(M_1 \times N_1) \cap (M_2 \times N_2) = (M_1 \cap M_2) \times (N_1 \cap N_2) ;$$

$$(M_1 \times N_1)\backslash(M_2 \times N_2) = [(M_1\backslash M_2) \times N_1] \cup [(M_1 \cap M_2) \times (N_1\backslash N_2)] . \tag{1}$$

Let μ_1 denote a measure defined on $\mathfrak{S}_1$ and let μ_2 denote a measure defined on $\mathfrak{S}_2$. Let us define on $\mathfrak{S}_1 \times \mathfrak{S}_2$ a set function $m(\Delta)$,

$$m(\Delta) = \mu_1(A)\mu_2(B) \quad \text{if} \quad \Delta = A \times B,\ A \in \mathfrak{S}_1,\ B \in \mathfrak{S}_2 . \tag{2}$$

In the case of an indeterminate form of the type $0 \cdot \infty$, we define the value of m to be 0.

Lemma 1. *The function $m(\Delta)$ defined by formula (2) is an elementary measure on $\mathfrak{S}_1 \times \mathfrak{S}_2$ satisfying the subadditivity condition*

$$m(\Delta_0) \leqq \sum_k m(\Delta_k) , \quad \textit{if} \quad \Delta_0 \subset \bigcup_k \Delta_k , \tag{3}$$

where

$$\Delta_k \in \mathfrak{S}_1 \times \mathfrak{S}_2 \quad \textit{for} \quad k = 0, 1, \cdots .$$

Proof. We note that if $\Delta_2 \subset \Delta_1$ where $\Delta_i = M_i \times N_i$ for $i = 1, 2$, by virtue of formula (1) we have

$$\Delta_1 = \Delta_2 \cup \Delta' \cup \Delta'', \ \Delta' = (M_1\backslash M_2) \times N_1 ,$$
$$\Delta'' = M_2 \times (N_1\backslash N_2) \tag{4}$$

where Δ_2, Δ', and Δ'' are pairwise disjoint. Furthermore,

$$\begin{aligned} m(\Delta_1) &= \mu_1(M_1)\mu_2(N_1) \\ &= \mu_1(M_2)\mu_2(N_2) + \mu_1(M_1\backslash M_2)\mu_2(N_1) + \mu_1(M_2)\mu_2(N_1\backslash N_2) \\ &= m(\Delta_2) + m(\Delta') + m(\Delta'') . \end{aligned}$$

Thus the function $m(\Delta)$ is additive with respect to these particular decompositions of the set Δ_1. In particular, if

$$\Delta_1 = \Delta_2 \cup \Delta_3 (\Delta_2 \cap \Delta_3 = \varnothing) ,$$

then

$$m(\Delta_1) = m(\Delta_2) + m(\Delta_3) .$$

Additivity in the general case is obtained by induction. If

$$\Delta = \bigcup_{k=1}^{n} \Delta_k ,$$

where the rectangles Δ_k are pairwise disjoint, then $\Delta \backslash \Delta_n = \Delta' \cup \Delta''$, where Δ' and Δ'' are determined from formula (1). From what we have shown,

$$m(\Delta) = m(\Delta_n) + m(\Delta') + m(\Delta'') .$$

From the induction hypothesis,

$$m(\Delta') = m\left(\bigcup_{k=1}^{n-1} (\Delta' \cap \Delta_k)\right) = \sum_{k=1}^{n-1} m(\Delta' \cap \Delta_k) .$$

It then follows that

$$\begin{aligned} m(\Delta) &= m(\Delta_n) + \sum_{k=1}^{n-1} m(\Delta' \cap \Delta_k) + m(\Delta'' \cap \Delta_k) \\ &= m(\Delta_n) + \sum_{k=1}^{n-1} m([\Delta' \cup \Delta''] \cap \Delta_k) \\ &= m(\Delta_n) + \sum_{k=1}^{n-1} m(\Delta_k) = \sum_{k=1}^{n} m(\Delta_k) . \end{aligned}$$

This proves the additivity of the function $m(\Delta)$.

Let us now prove the subadditivity property (3). Suppose that $\Delta_0 \subset \bigcup_{k=1}^{\infty} \Delta_k$ where $\Delta_k = M_k \times N_k$ for $k = 0, 1, \cdots$, and suppose that $\chi_k(u, v) = \chi_k'(u)\chi_k''(v)$ where $\chi_k(u, v)$, $\chi_k'(u)$, and $\chi_k''(v)$ are the characteristic functions of the sets Δ_k, M_k, and N_k respectively. Then

$$\chi_0(u, v) \leqq \sum_{k=1}^{\infty} \chi_k(u, v) = \sum_{k=1}^{\infty} \chi_k'(u)\chi_k''(v) .$$

If we integrate this inequality with respect to μ_1 we obtain

$$\chi_0''(v)\mu_1(M_0) \leqq \sum_{k=1}^{\infty} \chi_k''(v)\mu_1(M_k) .$$

Then integrating with respect to μ_2, we obtain

$$m(\Delta_0) = \mu_2(N_0)\mu_1(M_0) \leqq \sum_{k=1}^{\infty} \mu_2(N_k)\mu_1(M_k) = \sum_{k=1}^{\infty} m(\Delta_k)$$

which completes the proof of the lemma.

Theorem 1. *Let $\{U, \mathfrak{S}_1, \mu_1\}$ and $\{V, \mathfrak{S}_2, \mu_2\}$ denote two measurable spaces and let $\mathfrak{S}_3$ denote the smallest σ-algebra containing $\mathfrak{S}_1 \times \mathfrak{S}_2$. There exists a unique measure ν defined on $\mathfrak{S}_3$ that satisfies the condition*

$$\nu(M \times N) = \mu_1(M)\mu_2(N), \quad M \in \mathfrak{S}_1, \quad N \in \mathfrak{S}_2 . \tag{5}$$

Proof. The proof follows from Theorems 2 and 3 of Section 7 and the preceding lemma.

Definition 1. The measurable space $\{U \times V, \mathfrak{S}_3, \nu\}$ is called the *product* of the measurable spaces $\{U, \mathfrak{S}_1, \mu_1\}$ and $\{V, \mathfrak{S}_2, \mu_2\}$, the σ-algebra $\mathfrak{S}_3 = \sigma\{\mathfrak{S}_1 \times \mathfrak{S}_2\}$ is called the product of the σ-algebras $\mathfrak{S}_1$ and $\mathfrak{S}_2$; and the measure ν is called the product of the measures μ_1 and μ_2 (and we write $\nu = \mu_1 \times \mu_2$).

Let the completion of the σ-algebra $\mathfrak{S}_3$ (with respect to the measure ν) and the completed measure be denoted by $\tilde{\mathfrak{S}}_3$ and $\tilde{\nu}$ respectively.

Let E denote an arbitrary set contained in $U \times V$.

Definition 2. The set of all $v \in V$ for which the point (u, v) belongs to E for fixed $u \in U$ is called the *u-section* of the set E and is denoted by $E_{u\cdot}$. We write $E_{u\cdot} = \{v; u, v) \in E, \ u \text{ fixed}\}$.

The v-section of the set E is defined analogously. We indicate it by $E_{\cdot v}$ and we write $E_{\cdot v} = \{u; (u, v) \in E, \ v \text{ fixed}\}$.

Now consider a function $f(u, v)$ defined on $U \times V$.

Definition 3. A function of a single variable ($v \in V$) defined by the equation $f_{u\cdot}(v) = f(u, v)$ where u is fixed is called the *u-section* of the function $f(u, v)$. We write $f_{u\cdot}(u, v) = f_{u\cdot}(v)$.

The v-section $f_{\cdot v}(u, v) = f_{\cdot v}(u)$ of a function $f(u, v)$ is defined analogously.

Lemma 2. *Let $\{U, \mathfrak{S}_1, \mu_1\}$ and $\{V, \mathfrak{S}_2, \mu_2\}$ denote two measurable spaces and let $\{U \times V, \mathfrak{S}_3, \nu\}$ denote their product. Suppose that the measures μ_1 and μ_2 are finite. Then for any $\mathfrak{S}_3$-measurable set E*

a. *for arbitrary u the set $E_{u\cdot}$ is $\mathfrak{S}_2$-measurable and for arbitrary v the set $E_{\cdot v}$ is $\mathfrak{S}_1$-measurable,*

b. *$\mu_2(E_{u\cdot})$ is an $\mathfrak{S}_1$-measurable function of u and $\mu_1(E_{\cdot v})$ is an $\mathfrak{S}$-measurable function of v,*

c. $$\nu(E) = \int_U \mu_2(E_{u\cdot})\mu_1(du) = \int_U \mu_1(E_{\cdot v})\mu_2(dv) \,. \qquad (6)$$

Proof. Let $\mathfrak{R}$ denote the class of sets for which the theorem is valid. Obviously, $\mathfrak{R}$ contains rectangles and finite unions of them and $\mathfrak{R}$ is a monotonic class of sets (on the basis of Lebesgue's theorem on the integration of monotonic sequences of functions, Theorem 2 of Section 5). Since finite sums of rectangles constitute an algebra of sets (cf. Remark 1, Definition 4, Section 7), it follows on the basis of Theorem 3 of Section 1 that $\mathfrak{R}$ contains $\mathfrak{S}_3$. This completes the proof of the lemma.

Lemma 3. *Suppose that $\{U, \mathfrak{S}_1, \mu_1\}$ and $\{V, \mathfrak{S}_2, \mu_2\}$ are measurable sets with complete finite measures. Let $\{\tilde{\mathfrak{S}}_3, \tilde{\nu}\}$ denote the completion of the measure $\{\mathfrak{S}_3, \nu\}$. The assertion of Lemma 2 remains in force for μ_1-almost-all u and μ_2-almost-all v and for an arbitrary $\tilde{\mathfrak{S}}$-measurable set E.*

Proof. By virtue of Theorems 4 and 5 of Section 7, for a set $E \in \tilde{\mathfrak{S}}_3$ there exists a set $E' \in \mathfrak{S}_3$ such that $E \subset E'$ and $\nu(E) = \nu(E')$. Applying this conclusion to the set $\bar{E} = (U \times V)\backslash E$, we obtain a set $E_1 \in \mathfrak{S}_3$ such that $E_1 \supset \bar{E}$ and $\nu(E_1) = \nu(\bar{E})$. Setting $E'' = (U \times V)\backslash E_1$, we see that $E'' \subset E$, $E'' \in \mathfrak{S}_3$, and $\nu(E'') = \nu(E)$. Thus

$$E'' \subset E \subset E', \; E', \; E'' \in \mathfrak{S}_3, \; \nu(E'') = \nu(E) = \nu(E') .$$

Let us apply equation (6) to the sets E' and E''. We obtain

$$\int_U [\mu_2(E'_{u\cdot}) - \mu_2(E''_{u\cdot})]\mu_1(du) = 0, \; \mu_2(E'_{u\cdot}) \geqq \mu_2(E''_{u\cdot}) ,$$

from which we see that $\mu_2(E''_{u\cdot}) = \mu_2(E'_{u\cdot})$ for μ_1-almost-all u. Since

$$\mu_2(E''_{u\cdot}) \leqq \mu_2(E_{u\cdot}) \leqq \mu_2(E'_{u\cdot}) ,$$

this triple of functions is μ_1-equivalent. Thus

$$\nu(E) = \nu(E'') = \int_U \mu_2(E''_{u\cdot})\mu_1(du) = \int_U \mu_2(E_{u\cdot})\mu_1(du) .$$

Reversing the roles of the measures μ_1 and μ_2, we obtain proof of the lemma.

Theorem 2 (Fubini). *Suppose that $\{U, \mathfrak{S}_1, \mu_1\}$ and $\{V, \mathfrak{S}_2, \mu_2\}$ are measurable spaces. Let $\{U \times V, \mathfrak{S}_3, \nu\}$ denote their product. Suppose that the measures μ_1 and μ_2 are finite. Let $f(u, v)$ denote an arbitrary $\mathfrak{S}_3$-measurable ν-integrable function. Then for all $u(v)$, the section $f_{u\cdot}(u, v)$ (resp. $f_{\cdot v}(u, v)$) of the function $f(u, v)$ is $\mathfrak{S}_2$-measurable (resp. $\mathfrak{S}_1$-measurable) and*

$$\int_{U\times V} f(u, v)\,\nu(du \times dv) = \int_U \mu_1(du) \int_V f_{u\cdot}(u,v)\mu_2(dv)$$
$$= \int_V \mu_2(dv) \int_U f_{\cdot v}(u, v)\mu_1(du) . \tag{7}$$

If the measures $\{\mathfrak{S}_1, \mu_1\}$ and $\{\mathfrak{S}_2, \mu_2\}$ are complete, the assertion of the theorem remains valid for an arbitrary $\tilde{\mathfrak{S}}_3$-measurable function for μ_1-almost-all u and for μ_2-almost-all v.

Proof. It is sufficient to consider the case in which the function $f(u, v)$ is nonnegative. Proofs of the general assertions of the theorem are carried out in the same way except that in proving the first assertion we use Lemma 2 and in proving the second we use Lemma 3. Let us confine ourselves to the first assertion. Formula (7) is valid (a) for the characteristic functions of $\mathfrak{S}_3$-measurable sets (according to Lemma 2), (b) for linear combinations of them, and (c) for the limits of nondecreasing sequences of nonnegative simple $\mathfrak{S}_3$-measurable functions. On the basis of Remark 2 following Theorem 3 of Section 2, formula (7) is valid for an arbitrary $\mathfrak{S}_3$-measurable function. This completes the proof of the theorem.

REMARK 1. Formula (7) holds for nonnegative functions $f(u, v)$ without the assumption of integrability when the other conditions of Theorem 2 are satisfied.

REMARK 2. Theorem 2 remains valid for σ-finite measures μ_1 and μ_2, that is, for all cases in which the spaces U and V are unions of a countable collection of measurable sets of finite measure.

The concepts of the product of σ-algebras and measures can be generalized to finite or even countable products. Let $U_1, U_2, \cdots, U_n$ denote n sets. Their Cartesian product, denoted by

$$\prod_{k=1}^{n} U_k = U_1 \times U_2 \times \cdots \times U_n ,$$

is defined as the set of all ordered n-tuples $(u_1, u_2, \cdots, u_n)$ where u_k is an arbitrary element of U_k for $k = 1, 2, \cdots, n$. Strictly speaking, the formation of Cartesian products is not an associative operation. For example, $(U_1 \times U_2) \times U_3$ is the set of all possible pairs (z, y_3) where z is a pair of the form $z = (y_1, y_2)$. This set is different from the product $U_1 \times U_2 \times U_3$ and from the product $U_1 \times (U_2 \times U_3)$. However between these three spaces there exists a natural isomorphism

$$((y_1, y_2), y_3) \leftrightarrow (y_1, y_2, y_3) \leftrightarrow (y_1, (y_2, y_3)) ,$$

which enables us to identify them. Keeping this in mind, let us agree always to consider products of sets as defined up to this isomorphism and to treat the Cartesian product of sets as an associative operation.

If n sets $U_k = U$ for $k = 1, \cdots, n$ coincide, the Cartesian product

$$\prod_{k=1}^{n} U_k = U^n$$

is called the nth *power* of the set U.

If measures $\{\mu_k, \mathfrak{S}_k\}$ are defined on U_k, measures $\{\mu^{(k,\cdots,n)}, \mathfrak{S}^{(k,\cdots,n)}\}$ and their completions $\{\tilde{\mu}^{(k,\cdots,n)}, \tilde{\mathfrak{S}}^{(k,\cdots,n)}\}$ can be defined inductively on the space

$$\prod_{r=k}^{n} U_r .$$

Here $\mathfrak{S}^{(k,\cdots,n)}$ denotes the minimal σ-algebra containing sets of the form $A_k \times A_{k+1} \times \cdots \times A_n$ where $A_j \in \mathfrak{S}_j$. This construction is associative, and

$$\mu^{(1,2,\cdots,n)} = \mu^{(1,2,\cdots,k)} \times \mu^{(k+1,k+2,\cdots,n)} . \tag{8}$$

Formula (8) follows from the uniqueness of the extension of an elementary measure as a minimal σ-algebra.

Let us suppose that the function $f(u_1, u_2, \cdots, u_n)$ defined on $U_1 \times U_2 \times \cdots \times U_n$ is $\tilde{\mathfrak{S}}^{(1,\cdots,n)}$-measurable and nonnegative. On the basis of Fubini's theorem it follows that with fixed $(u_1, \cdots, u_k)$, for $\tilde{\mu}^{(1,\cdots,k)}$-almost-all $(u_1, \cdots, u_k)$, the function

$$f(u_1, \cdots, u_k, u_{k+1}, \cdots, u_n) = f_{u_1,u_2,\cdots,u_k}(u_{k+1}, u_{k+2}, \cdots, u_n)$$

is $\tilde{\mathfrak{S}}^{(k+1,\cdots,n)}$-measurable. This follows from the fact that

$$\tilde{\mathfrak{S}}^{(1,\cdots,n)} \supset (\tilde{\mathfrak{S}}^{(1,\cdots,k)} \times \tilde{\mathfrak{S}}^{(k+1,\cdots,n)})$$

and

$$\int_{U_1 \times U_2 \times \cdots \times U_n} f(u_1, \cdots, u_n)\, \mu^{(1,\cdots,n)}(d(u_1, u_2, \cdots, u_n))$$
$$= \int_{U_1 \times \cdots \times U_k} \Big[\int_{U_{k+1} \times \cdots \times U_n} f_{u_1 \cdots u_k}(u_{k+1}, \cdots, u_n) \mu^{(k+1,\cdots,n)}$$
$$\times\, (d(u_{k+1}, \cdots, u_n)) \Big] \mu^{(1,\cdots,k)}(d(u_1, \cdots, u_k)) .$$

III

AXIOMATIZATION OF PROBABILITY THEORY

1. PROBABILITY SPACES

Definition 1. Let U denote a set and let $\mathfrak{S}$ denote a σ-algebra defined on U. Let P denote a complete measure defined on $\mathfrak{S}$ such that $\mathsf{P}(U) = 1$. Then the triple $\{U, \mathfrak{S}, \mathsf{P}\}$ is called a *probability space*. Here the elements of the set U are called *elementary events*, the set U itself is called the *space of the elementary events*, and the subsets of U that belong to $\mathfrak{S}$ are called *events*.

The space U, treated as an event, is called the *certain event*; the empty set $\varnothing$ is called the *impossible event*. If A and B are two events such that $A \subset B$, we say that event A implies event B. If $A \cap B = \varnothing$, the events A and B are said to be *incompatible*. The event $\bar{A} = U \backslash A$ is called the *opposite* of the event A. For every $A \in \mathfrak{S}$, the quantity $\mathsf{P}(A)$ is called the *probability* of the event A.

Let us enumerate a few of the simpler properties of events and their probabilities that follow immediately from the definitions and the properties of measures that we have already proven.

1. The sum and intersection of finitely or countably many events are events.

2. The probability of the certain event is 1; the probability of the impossible event is 0; the relation $A \subset B$ implies $\mathsf{P}(A) \leqq \mathsf{P}(B)$.

3. If A is an event and $\bar{A}$ is the opposite event, then $\mathsf{P}(\bar{A}) = 1 - \mathsf{P}(A)$.

4. If $\{A_n\}$ is a finite or countable set of pairwise incompatible events, then $\mathsf{P}(\bigcup_n A_n) = \sum_n \mathsf{P}(A_n)$.

5. If $\{A_n\}$ is an increasing (resp. decreasing) sequence of events, that is, if A_n implies A_{n+1} (resp. of A_{n+1} implies A_n), then

$$\lim_{n\to\infty} \mathsf{P}(A_n) = \mathsf{P}\Big(\bigcup_{n=1}^{\infty} A_n\Big) \quad \Big(\text{resp. } \lim_{n\to\infty} \mathsf{P}(A_n) = \mathsf{P}\Big(\bigcap_{n=1}^{\infty} A_n\Big)\Big).$$

Definition 2. An arbitrary finite (mod P) $\mathfrak{S}$-measurable real function is called a *random variable*. Two random variables $\xi_1 = f_1(u)$ and $\xi_2 = f_2(u)$ are considered equal if they are equivalent: $\xi_1 = \xi_2$ if $f_1(u) = f_2(u)$ (mod P).

For every random variable ξ, a function $F(x)$ of a real argument x, known as the *distribution function* of the random variable ξ, is unambiguously defined by $F(x) = \mathsf{P}\{\xi < x\} = \mathsf{P}\{u; f(u) < x\}$. The distribution function has the following properties:

6. The distribution function $F(x)$ of a random variable ξ is nonnegative, nondecreasing, and continuous from the left; also $F(-\infty) = 0$ and $F(+\infty) = 1$.

Analogously, if $\{\xi_k = f_k(u), k = 1, 2, \cdots, n\}$ is an n-tuple of random variables, the function (of n real variables)

$$F(x_1, x_2, \cdots, x_n) = \mathsf{P}\{\xi_1 < x_1, \cdots, \xi_n < x_n\} = \mathsf{P}\left[\bigcap_{k=1}^{n} \{u, f_k(u) < x_{kk}\}\right]$$

is called the *joint distribution function* of the variables ξ_k for $k = 1\ 2, \cdots, n$. Obviously the function $F(x_1, x_2, \cdots, x_n)$ is uniquely defined in n-dimensional Euclidean space E_n, is nondecreasing, and is continuous from the left with respect to each variable. Furthermore,

$$F(x_1, x_2, \cdots, x_k, -\infty, x_{k+2}, \cdots, x_n) = 0$$

and

$$F(x_1, \cdots, x_k, +\infty, \cdots, +\infty) = F^{(k)}(x_1, \cdots, x_k)\ ,$$

where $F^{(k)}(x_1, \cdots, x_k)$ denotes the distribution function of the k-tuple of random variables $\xi_1, \xi_2, \cdots, \xi_k$.

All of the results obtained in Chapter II can now be repeated in probability-theoretic terms. We confine ourselves to a few examples.

Let $\varphi(t_1, t_2, \cdots, t_n)$ denote a Borel function in n-dimensional Euclidean space and let $\xi_1, \xi_2, \cdots, \xi_n$ denote n random variables. Let us define

$$\varphi(\xi_1, \xi_2, \cdots, \xi_n) = \varphi(f_1(u), \cdots, f_n(u)) = F(u)$$

where $\xi_k = f_k(u)$ for $k = 1, 2, \cdots, n$. Note that if we replace the function $f_k(u)$ by an equivalent (mod P) function $f_k(u)$, the function $F(u)$ is replaced by an equivalent function (mod P). On the basis of Theorem 5, Section 2, Chapter II, we have

7. If a Borel function $\varphi(\xi_1, \cdots, \xi_n)$ of random variables $\xi_1, \cdots, \xi_n$ is finite (mod P), then it too is a random variable.

In particular, the sum and product of finitely many random variables is a random variable. If a random variable $\xi \neq 0$

(mod P), then $1/\xi$ is also a random variable.

Let $\{\xi_n\}$ denote a sequence of random variables. It is always meaningful to speak of the event S that the sequence $\{\xi_n\}$ will converge to a finite limit, and to speak of the probability $\mathsf{P}(S)$ of that event (see Section 3, Chapter II).

Definition 3. If $\mathsf{P}(S) = 1$, that is, if the sequence of functions $\xi_n = f_n(u)$ converges P-almost-everywhere, we say that the sequence $\{\xi_n\}$ converges *almost certainly* (or that it *converges with probability* 1).

Definition 4. If there exists a random variable ξ such that $\mathsf{P}\{|\xi_n - \xi| > \varepsilon\} \longrightarrow 0$ as $n \longrightarrow \infty$ for arbitrary $\varepsilon > 0$, the sequence $\{\xi_n\}$ is said to *converge in probability* to the random variable ξ, that is, $\xi = \mathsf{P}\text{-lim}\, \xi_n$.

If a sequence $\{\xi_n\}$ converges almost certainly, it converges in probability (cf. Theorem 1, Section 3, Chapter II). The converse is not generally true. However a sequence $\{\xi_n\}$ of random variables that converges in probability contains a subsequence that converges almost certainly (cf. Theorem 4, Section 3, Chapter II).

8. A necessary and sufficient condition for a sequence of random variables to converge in probability is that for arbitrary $\varepsilon > 0$ and $\delta > 0$ there exists an $n_0 = n_0(\varepsilon, \delta)$ such that the inequalities $n' > n_0$ and $n > n_0$ together imply the inequality

$$\mathsf{P}\{|\xi_{n'} - \xi_n| > \varepsilon\} < \delta .$$

This condition is called the *Cauchy condition* with respect to the probability of the sequence $\{\xi_n\}$ (cf. Theorem 5, Section 3, Chapter II).

9. If $\mathsf{P}\text{-lim}\, \xi_n^{(j)} = \eta_j$ for $j = 1, 2, \cdots, s$, then for an arbitrary continuous function $\varphi(t_1, t_2, \cdots, t_s)$ where $-\infty < t_j < \infty$ for $j = 1, 2, \cdots, s$, the sequence of random variables $\varphi(\xi_n^{(1)}, \xi_n^{(2)}, \cdots, \xi_n^{(s)})$ converges in probability to a random variable $\varphi(\eta_1, \eta_2, \cdots, \eta_s)$ (cf. Theorem 6, Section 3, Chapter II).

Definition 5. The integral

$$\mathsf{M}\xi = \int_U f(u)\mathsf{P}(du) , \qquad \xi = f(u) ,$$

if it is meaningful, is called the *mathematical expectation* of the random variable ξ and is denoted by $\mathsf{M}\xi$.

On the basis of Theorem 5, Section 4, Chapter II, the value of the mathematical expectation is independent of the function $f(u)$ representing the random variable ξ. The mathematical expectation is a functional defined on some subset of random variables and enjoying the following properties:

10. $\mathsf{M}(\alpha\xi + \beta\eta) = \alpha\mathsf{M}\xi + \beta\mathsf{M}\eta$, where α and β are constants and at least one of the quantities $\mathsf{M}\xi$, $\mathsf{M}\eta$ is finite.

11. If χ_A is the characteristic function of the event A, that is, if $\chi_A(u) = 1$ for $u \in A$ and $\chi_A(u) = 0$ for $u \notin A$, then $\mathsf{M}\chi_A = \mathsf{P}(A)$.

12. If $\xi \leqq \eta$ (mod P), then $\mathsf{M}\xi \leqq \mathsf{M}\eta$.

13. For an arbitrary sequence of nonnegative random variables $\{\xi_n\}$,

$$\mathsf{M}\sum_{k=1}^{\infty} \xi_k = \sum_{k=1}^{\infty} \mathsf{M}\xi_k$$

(cf. Theorem 1, Section 5, Chapter II).

14. If $|\xi_n| \leqq \eta$ for $n = 1, 2, \cdots$, if $\mathsf{M}\eta < \infty$, and if the sequence of random variables ξ_n converges in probability to ζ, that is, $\zeta = \mathsf{P}\text{-}\lim \xi_n$, then $\lim_{n\to\infty} \mathsf{M}\xi_n = \mathsf{M}\zeta$ (cf. Theorem 5, Section 5, Chapter II).

We shall now prove an important inequality that gives a bound for the probability of an event in terms of mathematical expectation.

15. For an arbitrary nonnegative Borel function $f(x)$ that is nondecreasing on $[0, \infty)$ and an arbitrary random variable ξ,

$$\mathsf{P}\{|\xi| \geqq a\} \leqq \frac{\mathsf{M}f(|\xi|)}{f(a)} .$$

This is known "Chebyshev's inequality." Proof follows from the inequality $f(|\xi|) \geqq f(a)\chi_{[a,\infty)}(|\xi|)$, where $\chi_{[a,\infty)}(x)$ is the characteristic function of the infinite left-closed interval $[a, \infty)$.

We now mention another inequality that is well known in analysis.

16. If $g(x)$ is a continuous convex (downward) function for all real x and if ξ is a random variable with finite mathematical expectation, then

$$\mathsf{M}g(\xi) \geqq g(\mathsf{M}\xi) .$$

This is known as "Jensen's inequality." The proof can be found in the book by Hardy, Littlewood, and Polya.

Let $\{U, \mathfrak{S}, \mathsf{P}\}$ denote a probability space and let X denote an arbitrary set with fixed σ-algebra of the sets $\mathfrak{B}$.

Definition 6. Let $\zeta = f(u)$ denote a function defined on U into the set X and suppose that for arbitrary $B \in \mathfrak{B}$, $\{u; f(u) \in B\} \in \mathfrak{S}$. Then the function $\zeta = f(u)$ is called a *random element* with range in X.

In other words, $\zeta = f(u)$ is a random element if under the mapping $u \to x = f(u)$, we have $f(\mathfrak{S}) \supset \mathfrak{B}$. In the case in which X is a metric space, we always understand by $\mathfrak{B}$ the σ-algebra of Borel subsets of X.

A random element with range in a finite-dimensional linear space is called a *random vector*.

2. CONSTRUCTION OF PROBABILITY SPACES

Let $\{F_{\theta_1,\theta_2,\cdots,\theta_n}(x_1, x_2, \cdots, x_n)\}$ where $n = 1, 2, \cdots$ and $\theta_k \in \Theta$ for $k = 1, 2, \cdots, n$, denote a family of distribution functions satisfying the compatibility conditions (cf. Section 1, Chapter I) and describing a random function in the broad sense. Is it possible to define a probability space $\{U, \mathfrak{S}, \mathsf{P}\}$ and a family of random variables $\xi_\theta = f_\theta(u)$, $u \in U$, $\theta \in \Theta$, on it in such a way that the joint distribution function of the sequence $\{\xi_{\theta_1}, \xi_{\theta_2}, \cdots, \xi_{\theta_n}\}$ coincides with the given function $F_{\theta_1,\theta_2,\cdots\theta_n}(x_1, \cdots, x_n)$ for arbitrary $n = 1, 2, \cdots$ and $\theta_k \in \Theta$, where $k = 1, 2, \cdots, n$?

The following problem is less difficult. Let $F(x_1, \cdots, x_n)$ denote a distribution function of n random variables $\xi_1, \xi_2, \cdots, \xi_n$ and let A denote an arbitrary n-dimensional Borel set. How should we define the probability of the event $\{\xi_1, \xi_2, \cdots, \xi_n\} \in A$?

A precise statement of this problem follows: Let $F(x_1, x_2, \cdots, x_n)$ denote a distribution function of n independent variables, let E_n denote n-dimensional space, and let $\mathfrak{S}$ denote the σ-algebra of all Borel subsets of E_n. How can one construct a probability space $\{E_n, \mathfrak{S}, \mathsf{P}\}$ in such a way that

$$F(a_1, a_2, \cdots, a_n) = \mathsf{P}(I_{a_1,a_2,\cdots a_n})$$

for arbitrary real a_k (for $k = 1, \cdots, n$)? (Here, $I_{a_1,a_2,\cdots,a_n}$ denotes the n-dimensional orthant of the form

$$I_{a_1,a_2,\cdots,a_n} = \{(x_1, \cdots, x_n); x_1 < a_1, \cdots, x_n < a_n\} .)$$

We begin by solving the latter problem.

Let $\boldsymbol{a}, \boldsymbol{b}, \boldsymbol{x}, \cdots$, where $\boldsymbol{a} = (a_1, a_2, \cdots, a_n)$ denote the points in the set E_n. Let us write $\boldsymbol{a} \leqq \boldsymbol{b}$ if $a_i \leqq b_i$ for $i = 1, 2, \cdots, n$. We shall refer to the set $I[\boldsymbol{a}, \boldsymbol{b}) = \{\boldsymbol{x}; a_i \leqq x_i < b_i, i = 1, \cdots, n\}$, where $\boldsymbol{a} \leqq \boldsymbol{b}$, as an *n-dimensional left-closed right-open interval* or, briefly, as a *left closed interval*.

Let $\mathfrak{J}$ denote the class of all left-closed intervals. This class constitutes a decomposable family:

$$I[\boldsymbol{a}, \boldsymbol{b}) \cap I[\boldsymbol{c}, \boldsymbol{d}) = \{\boldsymbol{x}; a_i \leqq x_i < b_i, c_i \leqq x_i < d_i, i = 1, \cdots, n\}$$
$$= \{\boldsymbol{x}; \max(a_i, c_i) \leqq x_i < \min(b_i, d_i)\} = I[\boldsymbol{a}', \boldsymbol{b}') ,$$

and

$$I[\boldsymbol{a}, \boldsymbol{b})\backslash I[\boldsymbol{c}, \boldsymbol{d}) = \{\boldsymbol{x};\ x_i \in [a_i, b_i)\backslash[c_i, d_i)\}$$

which is the sum of a finite number of left-closed intervals.

Let us define the probability that a random point $(\xi_1, \xi_2, \cdots, \xi_n)$ will fall in a particular left-closed interval. We introduce the notation

$$\begin{aligned}\Delta^{(k)}_{[a_k,b_k)}F(\boldsymbol{x}) &= F(x_1, \cdots, x_{k-1}, b_k, x_{k+1}, \cdots, x_n) \\ &\quad - F(x_1, \cdots, x_{k-1}, a_k, x_{k+1}, \cdots, x_n)\ .\end{aligned}$$

Because of the monotonicity of $F(\boldsymbol{x})$ with respect to each variable, the function $\Delta^{(k)}_{[a_k,b_k]}F(\boldsymbol{x})$ of the variables $x_1, \cdots, x_{k-1}, x_{k+1}, \cdots, x_n$ is nonnegative and nondecreasing with respect to each variable. The probability-theoretic meaning of the quantity $\Delta^{(k)}_{[a_k,b_k)}F(x)$ is as follows: This quantity is the probability that the inequalities $\xi_1 < x_1, \cdots, \xi_{k-1} < x_{k-1}, a_k \leqq \xi_k < b_k, \xi_{k+1} < x_{k+1}, \cdots, \xi_n < x_n$ will be satisfied. We obtain by induction

$$F(I[\boldsymbol{a}, \boldsymbol{b})) = \Delta^{(1)}_{[a_1,b_1)}\Delta^{(2)}_{[a_2,b_2)} \cdots \Delta^{(n)}_{[a_n,b_n)}F(x) \geqq 0\ , \tag{1}$$

where $F(I[\boldsymbol{a}, \boldsymbol{b}))$ is the probability of the event $(\xi_1, \xi_2, \cdots, \xi_n) \in I(\boldsymbol{a}, \boldsymbol{b})$. The quantity $F(I[\boldsymbol{a}, \boldsymbol{b}))$ can also be written

$$F(I[\boldsymbol{a}, \boldsymbol{b})) = \sum_{t_1,\cdots,t_n=0}^{1} (-1)^{\sum_{i=1}^{n} t_i} F[\boldsymbol{b} - \boldsymbol{t}(\boldsymbol{b} - \boldsymbol{a})]\ , \tag{2}$$

where

$$\boldsymbol{t} = (t_1, \cdots, t_n),\ \boldsymbol{t}(\boldsymbol{b} - \boldsymbol{a}) = [t_1(b_1 - a_1), \cdots, t_n(b_n - a_n)]\ .$$

The function $F(I[\boldsymbol{a}, \boldsymbol{b}))$ is an additive function on $\mathfrak{J}$, because if we partition $I[\boldsymbol{a}, \boldsymbol{b})$ into two subintervals $I[\boldsymbol{a}, \boldsymbol{c}_1)$ and $I[\boldsymbol{c}_1, \boldsymbol{b})$ where $\boldsymbol{c}_1 = (c_1, a_2, \cdots, a_n)$ with $a_1 < c_1 < b_1$, then

$$\begin{aligned}F(I[\boldsymbol{a}, \boldsymbol{b})) &= \Delta^{(1)}_{[a_1,b_1)} \cdots \Delta^{(n)}_{[a_n,b_n)}F(\boldsymbol{x}) \\ &= \Delta^{(1)}_{[a_1,c_1)}\Delta^{(2)}_{[a_2,b_2)} \cdots \Delta^{(n)}_{[a_n,b_n)}F(\boldsymbol{x}) \\ &\quad + \Delta^{(1)}_{[c_1,b_1)}\Delta^{(2)}_{[a_2,b_2)} \cdots \Delta^{(n)}_{[a_n,b_n)}F(\boldsymbol{x}) \\ &= F(I[\boldsymbol{a}, \boldsymbol{c}_1)) + F(I[\boldsymbol{c}_1, \boldsymbol{b}_1))\ .\end{aligned}$$

The same relation holds if the left-closed interval $I[\boldsymbol{a}, \boldsymbol{b})$ is partitioned into two left-closed intervals by partitioning any one of the sides $[a_k, b_k)$ into two parts. One can show by induction that the function F is additive for an arbitrary decomposition of $I[\boldsymbol{a}, \boldsymbol{b})$ into a union of left-closed intervals.

For the function F to be extendable as a measure on some σ-algebra of subsets of E_n, it is necessary and sufficient that it be subadditive (cf. Theorem 2, Section 7, Chapter II):

$$F(I) \leqq \sum_{k=1}^{\infty} F(I_k) \tag{3}$$

for an arbitrary system of left-closed intervals I_k (for $k = 1, 2, \cdots$) such that $\bigcup_{k=1}^{\infty} I_k \supset I$. Let us verify that this condition is satisfied in the present case. Let $I[\boldsymbol{a}_k, \boldsymbol{b}_k)$ be denoted by I_k and let $I[\boldsymbol{a}_0, \boldsymbol{b}_0)$ be denoted by I. Since the function $F(\boldsymbol{x})$ is continuous from the left, there exists an $\boldsymbol{\varepsilon}_k = (\varepsilon_k, \cdots, \varepsilon_k)$, where $\varepsilon_k > 0$, such that

$$0 \leqq F(I[\boldsymbol{a}_k - \boldsymbol{\varepsilon}_k, \boldsymbol{b}_k)) - F(I[\boldsymbol{a}_k, \boldsymbol{b}_k)) < 2^{-k}\eta ,$$

where $\eta > 0$ and $k = 1, 2, \cdots$. The open intervals $(\boldsymbol{a}_k - \boldsymbol{\varepsilon}_k, \boldsymbol{b}_k)$ cover the closed interval $[\boldsymbol{a}_0, \boldsymbol{b}_0 - \boldsymbol{\varepsilon}]$. From the Heine-Borel theorem, the collection of these open intervals contains a finite subcovering, for example, $\{(\boldsymbol{a}_k - \boldsymbol{\varepsilon}_k, \boldsymbol{b}_k)\}$ for $k = 1, \cdots, n$. Then the set of left-closed intervals $\{(\boldsymbol{a}_k - \boldsymbol{\varepsilon}_k, \boldsymbol{b}_k)\}$ for $k = 1, \cdots, n$ covers the left-closed interval $[\boldsymbol{a}_0, \boldsymbol{b}_0 - \boldsymbol{\varepsilon})$. Consider the collection of disjoint sets

$$[\boldsymbol{a}_0, \boldsymbol{b}_0 - \boldsymbol{\varepsilon}) \cap \left\{[\boldsymbol{a}_k - \boldsymbol{\varepsilon}_k, \boldsymbol{b}_k) \Big\backslash \bigcup_{i=1}^{k-1} [\boldsymbol{a}_i - \boldsymbol{\varepsilon}_i, \boldsymbol{b}_i)\right\} , \qquad k = 1, \cdots, n ,$$

each of which is the union of nonintersecting left-closed intervals $\Delta_j^{(k)}$ (for $j = 1, 2, \cdots, m_k$). Thus

$$[\boldsymbol{a}_0, \boldsymbol{b}_0 - \boldsymbol{\varepsilon}) = \bigcup_{k=1}^{n} \bigcup_{j=1}^{m_k} \Delta_j^{(k)}$$

and

$$F(I[\boldsymbol{a}_0, \boldsymbol{b}_0 - \boldsymbol{\varepsilon})) = \sum_{k=1}^{n} \sum_{j=1}^{m_k} F(\Delta_j^{(k)})$$
$$\leqq \sum_{k=1}^{n} F(I[\boldsymbol{a}_k - \boldsymbol{\varepsilon}_k, \boldsymbol{b}_k)) \leqq \sum_{k=1}^{\infty} F(I[\boldsymbol{a}_k - \boldsymbol{\varepsilon}_k, \boldsymbol{b}_k)) \leqq \sum_{k=1}^{\infty} F(I_k) + \eta .$$

Taking the limit as $\boldsymbol{\varepsilon} \longrightarrow 0$, we obtain

$$F(I[\boldsymbol{a}_0, \boldsymbol{b}_0)) \leqq \sum_{k=1}^{\infty} F(I_k) + \eta ,$$

and since η is arbitrary this proves inequality 3 and also:

Theorem 1. *For an arbitrary distribution function* $F(x_1, \cdots, x_n) = F(\boldsymbol{x})$, *it is possible to construct a probability space* $\{E_n, \mathfrak{S}, \mathsf{P}\}$ *such that* $\mathfrak{S}$ *contains all Borel sets in* E_n, *and* $F(x_1, \cdots, x_n)$ *is the joint distribution function of the random variables* $\xi_1 = f_1(\boldsymbol{x}) = x_1, \xi_2 = f_2(\boldsymbol{x}) = x_2, \cdots, \xi_n = f_n(\boldsymbol{x}) = x_n$.

Let us turn now to the general problem posed at the beginning of the section. Consider an arbitrary family of compatible distribution functions $F_{\theta_1,\theta_2,\cdots,\theta_n}(x_1, \cdots, x_n)$ where $\theta_k \in \Theta$ for $k = 1, \cdots, n$. As we have just shown, for arbitrary n in n-dimensional space (which is denoted by $E_{\theta_1,\cdots,\theta_n}$) we can define a σ-algebra $\mathfrak{S}_{\theta_1,\cdots,\theta_n}$ containing the Borel sets of the space and we can construct a probability space $\{E_{\theta_1,\cdots,\theta_n}, \mathfrak{S}_{\theta_1,\cdots,\theta_n}, \mathsf{P}_{\theta_1,\cdots,\theta_n}\}$ such that $\mathsf{P}_{\theta_1,\cdots,\theta_n}(I_x)$ will coincide with the given distribution function $F_{\theta_1,\cdots,\theta_n}(x)$. Now the problem is to construct a single probability space $\{U, \mathfrak{S}, \mathsf{P}\}$ and a family of random variables $\xi_\theta = f_\theta(u)$ for $u \in U$, such that the joint distribution of the variables $\xi_{\theta_1}, \xi_{\theta_2}, \cdots, \xi_{\theta_n}$ coincides with the function $\mathsf{P}_{\theta_1,\cdots,\theta_n}(I_x) = F_{\theta_1,\cdots,\theta_n}(x)$. Let us generalize the statement of the problem.

Let X denote some metric space and let $\mathfrak{B}$ denote the σ-algebra of Borel subsets of X. Let X^n denote the nth power of the space X and let $\mathfrak{B}^{(n)}$ denote the σ-algebra of Borel subsets of X^n. Furthermore, let us assume that for every positive integer n and for arbitrary $\theta_1, \theta_2, \cdots, \theta_n$ where the θ_k (for $k = 1, 2, \cdots, n$) are arbitrary points belonging to some set Θ, a probability measure

$$\mathsf{P}_{\theta_1,\cdots,\theta_n}(A^{(n)}),\ A^{(n)} \in \mathfrak{B}^{(n)} \tag{4}$$

is defined on $\mathfrak{B}^{(n)}$ and the family of measures (4) satisfies the following compatibility conditions (cf. Section 1, Chapter I):

a. $$\mathsf{P}_{\theta_1,\cdots,\theta_n,\theta_{n+1},\cdots,\theta_{n+m}}(A^{(n)} \times X^{(m)}) = \mathsf{P}_{\theta_1,\cdots,\theta_n}(A^{(n)}) . \tag{5}$$

b. Let S denote an arbitrary permutation of the elements $\theta_1, \theta_2, \cdots, \theta_n$. Let S also denote the transform in the space X^n that permutes the coordinates x_k of a point $(x_1, x_2, \cdots, x_n)$ by means of the permutation S. Then

$$\mathsf{P}_{\theta_1,\cdots,\theta_n}(A^{(n)}) = \mathsf{P}_{S(\theta_1,\cdots,\theta_n)}(SA^{(n)}) . \tag{6}$$

Definition 1. Let $\{U, \mathfrak{S}, \mathsf{P}\}$ denote a probability space and let $g(\theta, u)$ denote a function defined on $\Theta \times U$ with range in X. Suppose that $g(\theta, u)$ is $\mathfrak{S}$-measurable for every fixed $\theta \in \Theta$ and that the finite-dimensional distributions of the random function $g(\theta, u)$ coincide with the given family (4); that is, for every $A^{(n)} \in \mathfrak{B}^{(n)}$,

$$\mathsf{P}\{u; (g(\theta_1, u), g(\theta_2, u), \cdots, g(\theta_n, u)) \in A^{(n)}\} = \mathsf{P}_{\theta_1,\cdots,\theta_n}(A^{(n)}) . \tag{7}$$

Then the probability space $\{U, \mathfrak{S}, \mathsf{P}\}$ and the function $g(\theta, u)$ are called the *representation* of the family of distributions (4). We shall show that under rather broad assumptions a compatible family of distributions (4) admits a representation. The role of the space U will be played by the space Ω of all functions defined on Θ into X, and the elementary events will be the functions of θ (that is, $u = \omega(\theta)$), and the function $g(\theta, u) = \omega(\theta)$.

Definition 2. Let Ω denote the space of all functions $u = \omega(\theta)$ defined on a set Θ into a metric space X. Let $A^{(n)}$ denote a Borel subset of X^n. The set of functions $\omega(\theta) \in \Omega$ such that the point $\{\omega(\theta_1), \cdots, \omega(\theta_n)\}$ in X^n belongs to $A^{(n)}$; that is, the set

$$C_{\theta_1,\cdots,\theta_n}(A^{(n)}) = \{\omega(\theta); \big(\omega(\theta_1), \omega(\theta_2), \cdots, \omega(\theta_n)\big) \in A^{(n)}\}$$

is called a *cylindrical set in* Ω *with base* $A^{(n)}$ *over the coordinates* $\theta_1, \theta_2, \cdots, \theta_n$, or simply a *cylindrical set*.

Let us make some remarks concerning cylindrical sets and operations on them. If the number n and the points $\theta_1, \theta_2, \cdots, \theta_n$ are fixed, there exists an isomorphism between the cylindrical sets over the coordinates $\theta_1, \cdots, \theta_n$ and the Borel subsets of X^n: Every Borel set $A^{(n)} \in X^n$ defines a cylindrical set $C_{\theta_1,\cdots,\theta_n}(A^{(n)})$ for which it serves as base. Different cylindrical sets correspond to different bases. The sum, difference, and intersection of cylindrical sets correspond to the sum, difference, and intersection of bases. This follows immediately from the definition of a cylindrical set.

With regard to operations on cylindrical sets in the general case, we need to keep in mind that a fixed cylindrical set can be defined by means of different coordinate systems.

Thus, obviously, $C_{\theta_1,\cdots,\theta_n}(A^{(n)}) = C_{\theta_1,\cdots,\theta_n,\theta_{n+1},\cdots,\theta_{n+r}}(A^{(n)} \times X^r)$. It is easy to see that any two cylindrical sets $C = C_{\theta_1,\cdots,\theta_n}(A^{(n)})$ and $C' = C_{\theta'_1,\cdots,\theta'_m}(B^{(m)})$ can always be regarded as cylindrical sets over a single sequence of coordinates $\theta''_1 \cdots, \theta''_p$ containing both $\theta_1, \theta_2, \cdots, \theta_n$ and $\theta'_1, \theta'_2, \cdots, \theta'_m$. From this it follows that we may consider algebraic operations on a finite number of cylindrical sets as defined on a fixed coordinate sequence. Therefore we have

Theorem 2. *The class* $\mathfrak{C}$ *of all cylindrical sets constitutes an algebra of sets.*

To this we need to add that if Θ contains infinitely many points and X contains at least two points, then $\mathfrak{C}$ is not a σ-algebra. To see this, note that the set

$$\bigcup_{k=1}^{\infty} C_{\theta_k}(\{x_k\}) ,$$

where $\{x_k\}$ (for $k = 1, 2, \cdots$) is a sequence of points in X, is not a cylindrical set.

We shall now prove:

Theorem 3 (Kolmogorov). *Let X denote a complete separable metric space. The family of distributions* (4) *satisfying compatability conditions* (5) *and* (6) *admits a representation.*

Proof. Let us define a set function $\mathsf{P}'(C)$ on the algebra $\mathfrak{C}$ of cylindrical subsets of the space Ω by setting $\mathsf{P}'(C) = \mathsf{P}_{\theta_1,\cdots,\theta_n}(A^{(n)})$, if C is a cylindrical set with $\{A^{(n)}\}$ over the coordinates $\theta_1, \theta_2, \cdots, \theta_n$. The conditions of compatability ensure uniqueness of the definition of the function $\mathsf{P}'(C)$. Let $\{C_k\}$ for $k = 1, 2, \cdots, n$ denote an n-tuple of cylindrical sets. We may assume without loss of generality that these sets are defined by the bases $A_k^{(p)}$ over a single p-tuple of coordinates $\theta_1, \theta_2, \cdots, \theta_p$. Algebraic operations on the sets C_k correspond exactly to the same operations on the sets $A_k^{(p)}$. Since the measure $\mathsf{P}_{\theta_1,\cdots,\theta_n}(A^{(p)})$ is countably additive in X^p, it follows that the set function $\mathsf{P}'(C)$ is finitely additive on $\mathfrak{C}$. We now extend the function $\mathsf{P}'(C)$ from the algebra $\mathfrak{C}$ to a measure $\tilde{\mathsf{P}}$ on some σ-algebra $\tilde{\mathfrak{C}}$. For us to be able to apply Lebesgue's method, it will be sufficient (on the basis of Theorem 2, Section 7, Chapter II) to verify that for every $C \in \mathfrak{C}$ and an arbitrary covering $\{C_k\}$ for $k = 1, \cdots, n, \cdots$ (where $C_k \in \mathfrak{C}$) of the set C (thus we are assuming $C \subset \bigcup_{k=1}^{\infty} C_k$), the inequality

$$\mathsf{P}'(C) \leqq \sum_{k=1}^{\infty} \mathsf{P}'(C_k) \tag{8}$$

is valid.

Let us show that if

$$\sum_{k=1}^{\infty} C_k = C \quad (C \in \mathfrak{C},\ C_k \in \mathfrak{C},\ k = 1, 2, \cdots)$$

and $C_k \cap C_r = \varnothing$ for $k \neq r$, then

$$\mathsf{P}'(C) = \sum_{k=1}^{\infty} \mathsf{P}'(C_k) \ . \tag{9}$$

Inequality 8 will then follow for an arbitrary covering of the cylindrical set C by sets in $\mathfrak{C}$. Let us set

$$C \Big\backslash \bigcup_{k=1}^{n} C_k = D_n \ .$$

These sets D_n constitute a decreasing sequence of cylindrical sets and the intersection of all these sets is empty:

$$\bigcap_{n=1}^{\infty} D_n = C \Big\backslash \sum_{k=1}^{\infty} C_k = \varnothing \ . \tag{10}$$

It follows from the additivity of P' that

$$\mathsf{P}'(C) = \sum_{k=1}^{n} \mathsf{P}'(C_k) + \mathsf{P}'(D_n) \ .$$

To prove (6), it suffices to show that $\lim_{n\to\infty} \mathsf{P}'(D_n) = 0$.

Let us assume the opposite, that is, that

$$\lim_{n\to\infty} \mathsf{P}'(D_n) = L > 0 \ . \tag{11}$$

Let B_n denote the base of the cylindrical set D_n. Suppose that D_n is situated over the coordinates $\theta_1, \theta_2, \cdots, \theta_{m_n}$. Here we assume that with increasing n, the set of corresponding points $\theta_1, \theta_2, \cdots, \theta_{m_n}$ does not decrease. As shown above, this assumption does not restrict generality.

For every B_n, there exists a compact set $K_n \subset B_n$ such that

$$P_{\theta_1,\cdots,\theta_{m_n}}(B_n \backslash K_n) < \frac{L}{2^{n+1}}, \qquad n = 1, 2, \cdots$$

(cf. Theorem 6, Section 7, Chapter II). Let Q_n denote a cylindrical set over the coordinates $\theta_1, \theta_2, \cdots, \theta_{m_n}$ with base K_n. Define

$$G_n = \bigcap_{r=1}^{n} Q_r .$$

Let M_n denote the base of the set G_n. Obviously M_n is compact since it is the intersection of closed sets, at least one of which, K_n, is compact.

Since the sets G_n constitute a decreasing sequence, it follows that if $\omega(\theta) \in G_{n+p} (p > 0)$, then $\omega(\theta) \in G_n$. From this in turn it follows that if $\{x_1, x_2, \cdots, x_n, \cdots, x_{m_{n+p}}\} \in M_{n+p}$, $(p > 0)$, then $\{x_1, x_2, \cdots, x_{m_n}\} \in M_n$. The sets G_n are obviously nonempty. Furthermore, since

$$D_n \backslash G_n = \bigcup_{r=1}^{n} (D_n \backslash Q_r) \subset \bigcup_{r=1}^{n} (D_r \backslash Q_r) ,$$

we have

$$\mathsf{P}'(D_n \backslash G_n) \leqq \sum_{r=1}^{n} \mathsf{P}'(D_r \backslash Q_r) = \sum_{r=1}^{n} P_{\theta_1,\cdots,\theta_{m_r}}(B_r \backslash K_r) \leqq \frac{L}{2} ,$$

from which it follows that

$$\lim_{n\to\infty} \mathsf{P}'(G_n) = \lim_{n\to\infty} \mathsf{P}'(D_n) - \lim_{n\to\infty} \mathsf{P}'(D_n \backslash G_n) \geqq \frac{L}{2} .$$

For every set M_n, let us choose a point $\{x_1^{(n)}, \cdots, x_{m_n}^{(n)}\}$. On the basis of what was said above, for arbitrary k the sequence $\{x_k^{(n)}\}$, $n = 1, 2, \cdots$ belongs to a compact set in X and the sequence $(x_1^{(n+p)}, \cdots, x_{m_n}^{(n+p)}\}, p = 0, 1, 2, \cdots$, is contained in M_n. By means of a diagonal process let us find a sequence of indices n_j such that for each k the sequence $\{x_k^{(n_j)}\}$ converges to some limit $x_k^{(0)}$. Since the set M_n is closed, it follows that for arbitrary n, $\{x_1^{(0)}, \cdots, x_{m_n}^{(0)}\} \in M_n$.

Let us define a function $\omega(\theta)$ by setting $\omega(\theta_k) = x_k^{(0)}$ for $k = 1, \cdots, n$ and then extending its definition to the remaining points in an arbitrary manner. Then for arbitrary n we have $\omega(\theta) \in G_n \subset D_n$. Consequently, the set $\bigcap_{n=1}^{\infty} D_n$ is nonempty, which contradicts (10).

Thus, inequality (11) is impossible and

$$\lim_{n\to\infty} \mathsf{P}'(D_n) = 0 .$$

Consequently, the function P' satisfies inequality (8) and admits an extension as a complete measure $(\tilde{\mathsf{P}}, \tilde{\mathfrak{C}})$ such that $\tilde{\mathfrak{C}} \supset \mathfrak{C}$. Let us define the function $g(\theta, \omega)$, for $\omega \in \Omega$ and $\theta \in \Theta$, by $g(\theta, \omega) = \omega(\theta)$. For an arbitrary Borel set $A^{(n)} \subset X^n$ and arbitrary $n, \theta_1, \theta_2, \cdots, \theta_n$. we have

$$\tilde{\mathsf{P}}\{(g(\theta_1, \omega), g(\theta_2, \omega), \cdots, g(\theta_n, \omega)) \in A^{(n)}\}$$
$$= \tilde{\mathsf{P}}\{(\omega(\theta_1), \omega(\theta_2), \cdots, \omega(\theta_n)) \in A^{(n)}\} = \mathsf{P}_{\theta_1,\cdots,\theta_n}(A^{(n)}) .$$

Thus we have constructed a representation of the family of distributions (4). This completes the proof of theorem.

3. INDEPENDENCE

Let $\{U, \mathfrak{S}, \mathsf{P}\}$ denote a fixed probability space. We recall that two events A and B are said to be independent if $\mathsf{P}(A \cap B) = \mathsf{P}(A)\mathsf{P}(B)$. This definition has the following immediate consequences: The events U and A are independent for arbitrary A; if $\mathsf{P}(N) = 0$ and $A \in \mathfrak{S}$, then N and A are independent; A is independent of A if and only if $\mathsf{P}(A) = 0$ or $\mathsf{P}(A) = 1$; if A and B are independent, then the events $\bar{A}$ and B are also independent; if $\mathsf{P}(N) = 0$ and the events A and B are independent, then the events $A \cup N$ and B are independent, as are the events $A \backslash N$ and B.

Definition 1. Suppose that $\{A_\lambda\}$, for $\lambda \in \Lambda$, is an arbitrary class of events and Λ is an arbitrary set of indices. The events A_λ are said to be *independent* (or *mutually independent*) if for an arbitrary k-tuple of distinct indices,

$$\mathsf{P}(A_{i_1} \cap A_{i_2} \cap \cdots \cap A_{i_k}) = \mathsf{P}(A_{i_1})\mathsf{P}(A_{i_2}) \cdots \mathsf{P}(A_{i_k}) .$$

We can make the same remarks regarding a sequence of independent events that we did for two independent events. In particular, if we replace an arbitrary finite or infinite set of events A_n in a sequence of events $\{A_n\}$ with their complementary events, the new sequence of events will also be independent.

Let $\{A_n\}$ denote an infinite sequence of events. The event $\overline{\lim}\, A_n$ was defined above (see Section 1, Chapter II) as the set of those elementary events contained in infinitely many of the events A_n. What can we say about the probability of the event $\overline{\lim}\, A_n$?

Theorem 1. *If* $\sum_{n=1}^{\infty} \mathsf{P}(A_n) < \infty$, *then the event* $\overline{\lim}\, A_n$ *has prob-*

ability 0.

The proof follows from the formula

$$\overline{\lim}\, A_n = \bigcap_{n=1}^{\infty} \bigcup_{k=n}^{\infty} A_k \,,$$

on the basis of which

$$\mathsf{P}(\overline{\lim}\, A_n) = \lim_{n\to\infty} \mathsf{P}\Big(\bigcup_{k=n}^{\infty} A_k\Big) \leqq \lim_{n\to\infty} \sum_{k=n}^{\infty} \mathsf{P}(A_k) = 0 \,.$$

For a sequence of independent events, this theorem can be strengthened by the following remarkable fact:

Theorem 2 (Borel-Cantelli). *Let $\{A_n\}$ denote a sequence of independent events. Then the probability of the event $\overline{\lim}\, A_n$ is* 0 *or* 1 *according as the series*

$$\sum_{n=k}^{\infty} \mathsf{P}(A_n)$$

converges or diverges.

It suffices to show that the relation

$$\sum_{n=1}^{\infty} \mathsf{P}(A_n) = \infty$$

implies $\mathsf{P}(\overline{\lim}\, A_n) = 1$. If $A^* = \overline{\lim}\, A_n$, then

$$U \backslash A^* = \bigcup_{n=1}^{\infty} \bigcap_{k=n}^{\infty} (U \backslash A_k)$$

and

$$\begin{aligned} \mathsf{P}(U\backslash A^*) &= \lim_{n\to\infty} \mathsf{P}\Big(\bigcap_{k=n}^{\infty} (U\backslash A_k)\Big) \\ &= \lim_{n\to\infty} \prod_{k=n}^{\infty} \mathsf{P}(U\backslash A_k) = \lim_{n\to\infty} \prod_{k=n}^{\infty} (1 - \mathsf{P}(A_k)) = 0 \end{aligned}$$

by virtue of the divergence of the series $\sum_{k=1}^{\infty} \mathsf{P}(A_k)$.

Let us generalize Theorem 2 and the concepts we have presented to classes of events.

Definition 2. Let $\{\mathfrak{M}_\lambda\}$ (for $\lambda \in \Lambda$, where Λ is an arbitrary set) denote a family of sets of events. The sets $\mathfrak{M}_\lambda$ are said to be (mutually) independent if an arbitrary class of events $\{A_\lambda\}$, consisting, for each λ, of a single event $A_\lambda \in \mathfrak{M}_\lambda$ is a class of independent events.

Theorem 3. *Suppose that $\mathfrak{N}_\lambda$ are decomposable classes of events, that $U \in \mathfrak{N}_\lambda$* (cf. Section 6, Chapter II), *and that the $\mathfrak{N}_\lambda$ (for $\lambda \in \Lambda$) are independent. Then both the sets of minimal σ-algebras $\sigma\{\mathfrak{N}_\lambda\}$ (for*

$\lambda \in \Lambda$) *generated by the classes of events* $\mathfrak{N}_\lambda$ *and their completions* $\tilde{\sigma}\{\mathfrak{N}_\lambda\}$ (*for* $\lambda \in \Lambda$) *are independent.*

Proof. Let $\mathfrak{M}_{\lambda_0}$ denote the set of all events A such that the events $\{A, A_\lambda; \lambda \in \Lambda, \lambda \neq \lambda_0\}$ are mutually independent. Let us list certain simple properties of the set $\mathfrak{M}_{\lambda_0}$:

a. $\mathfrak{M}_{\lambda_0}$ is a monotonic class of events (cf. Definition 8, Section 1, Chapter II). This assertion is a simple consequence of the definition of independence of events.

b. If A_1 and A_2 belong to $\mathfrak{M}_{\lambda_0}$ and if $A_1 \cap A_2 = \emptyset$, then $A_1 \cup A_2 \in \mathfrak{M}_{\lambda_0}$. This is true because if

$$\mathsf{P}(A_i \cap A_{\lambda_1} \cap \cdots \cap A_{\lambda_n}) = \mathsf{P}(A_i)\mathsf{P}(A_{\lambda_1}) \cdots \mathsf{P}(A_{\lambda_n})$$

for $i = 1, 2$, then

$$\begin{aligned}\mathsf{P}(&[A_1 \cup A_2] \cap A_{\lambda_1} \cap \cdots \cap A_{\lambda_n}) \\ &= \mathsf{P}(A_1 \cap A_{\lambda_1} \cap \cdots \cap A_{\lambda_n}) + \mathsf{P}(A_2 \cap A_{\lambda_1} \cap \cdots \cap A_{\lambda_n}) \\ &= \mathsf{P}(A_1 \cup A_2)\mathsf{P}(A_{\lambda_1})\mathsf{P}(A_{\lambda_2}) \cdots \mathsf{P}(A_{\lambda_n}) \,.\end{aligned}$$

As we see, property b and property a are independent of the special structure assumed for the sets $\mathfrak{N}_\lambda$ (for $\lambda \in \Lambda, \lambda \neq \lambda_0$).

c. $\mathfrak{N}_{\lambda_0} \subset \mathfrak{M}$.

Let us now look at the set N of events that can be represented in the form $B = C_1 \cup \cdots \cup C_n$ for $C_i = 1, \cdots, n$, where n is arbitrary, and the C_i are incompatible events in $\mathfrak{N}_\lambda$. N is an algebra of sets. On the basis of property b we have $N \subset \mathfrak{M}_{\lambda_0}$. On the basis of property a and Theorem 3, Section 1, Chapter II, the set $\mathfrak{M}_{\lambda_0}$ contains the σ-algebra $\sigma\{\mathfrak{N}_{\lambda_0}\}$. Since addition of events of probability zero to an event $A \in \mathfrak{M}_{\lambda_0}$ or removal of such events yields an event in $\mathfrak{M}_{\lambda_0}$, the set $\tilde{\sigma}\{\mathfrak{N}_{\lambda_0}\}$ is also contained in $\mathfrak{M}_{\lambda_0}$. This completes the proof of the theorem.

Theorem 4. *Let* $\mathfrak{S}_\lambda$, *for* $\lambda \in \Lambda$, *denote a set of independent* σ-*algebras and suppose that* $\Lambda = \Lambda_1 \cup \Lambda_2$ ($\Lambda_1 \cap \Lambda_2 = \emptyset$). *Let* $\mathfrak{B}_i = \sigma\{\mathfrak{S}_\lambda, \lambda \in \Lambda_i\}$ *for* $i = 1, 2$, *denote the smallest* σ-*algebra containing all the* σ-*algebras* $\mathfrak{S}_\lambda$ *for* $\lambda \in \Lambda_i$. *Then the* σ-*algebras* $\mathfrak{B}_1$ *and* $\mathfrak{B}_2$ *are independent.*

Proof. We note that the events

$$C' = \bigcup_{k=1}^{n_1} B'_k \quad \text{and} \quad C'' = \bigcup_{k=1}^{n_2} B''_k \,,$$

where

$$B'_k = \bigcap_{r=1}^{m'_k} A'_{kr},\ B''_k = \bigcap_{r=1}^{m''_k} A''_{kr} \quad \text{and} \quad A'_{kr}\ (A''_{kr})$$

for some $\lambda = \lambda_{kr} \in \Lambda_1$ (resp. $\lambda = \lambda_{kr} \in \Lambda_2$) belongs to $\mathfrak{S}_{\lambda_{kr}}$, are mutually independent. The collection $\mathfrak{B}_1^{(0)}$ (resp. $\mathfrak{B}_2^{(0)}$) of all sets of the form C' (resp. C'') constitutes an algebra of sets. On the basis of Theorem 3, the smallest σ-algebras containing $\mathfrak{B}_1^{(0)}$ and $\mathfrak{B}_2^{(0)}$ are independent. On the other hand, $\sigma\{\mathfrak{B}_i^{(0)}\} = \mathfrak{B}_i$ for $i = 1, 2$. This completes the proof of the theorem.

REMARK 1. It follows from what has been shown that if Λ is partitioned into an arbitrary collection of disjoint subsets Λ_α (for $\alpha \in A$), the σ-algebras

$$\sigma\{\mathfrak{S}_\lambda, \lambda \in \Lambda_\alpha\} = \mathfrak{B}_\alpha \qquad (\text{for } \alpha \in A)$$

constitute a set of independent σ-algebras.

Let us now look at an arbitrary sequence of independent σ-algebras $\mathfrak{S}_n$ for $n = 1, 2, \cdots$. On the basis of the Borel-Cantelli theorem, every event A that is contained in infinitely many of the $A_n \in \mathfrak{S}_n$ has probability either 0 or 1. This result can be generalized to arbitrary events generated by the collection of all σ-algebras $\mathfrak{S}_n$, for $n = 1, 2, \cdots$, independently of the n-tuple of σ-algebras $\mathfrak{S}_1, \mathfrak{S}_2, \cdots, \mathfrak{S}_n$. Let us make this assertion more precise. Let

$$\sigma\{\mathfrak{S}_k, \mathfrak{S}_{k+1}, \cdots, \mathfrak{S}_n, \cdots\} = \mathfrak{B}_k$$

denote the smallest σ-algebra containing $\mathfrak{S}_n$ for $n = k, k+1, \cdots$. The $\mathfrak{B}_k$ constitute a decreasing sequence of σ-algebras. Their intersection

$$\mathfrak{B} = \bigcap_{k=1}^{\infty} \mathfrak{B}_k$$

is also a σ-algebra. Let us define

$$\overline{\lim}\, \mathfrak{S}_n = \mathfrak{B} = \bigcap_{k=1}^{\infty} \sigma\{\mathfrak{S}_k, \mathfrak{S}_{k+1}, \cdots\}\,.$$

Obviously, the σ-algebra $\overline{\lim}\, \mathfrak{S}_n$ is unchanged by replacing an arbitrary finite number of σ-algebras $\mathfrak{S}_1, \cdots, \mathfrak{S}_n$ with other σ-algebras.

Theorem 5 (Kolmogorov's general 0-or-1 law). *If $\mathfrak{S}_n$, for $n = 1, 2, \cdots$, are mutually independent σ-algebras, then every event in $\overline{\lim}\, \mathfrak{S}_n$ is of probability* 0 *or* 1.

Proof. Let A denote a member of $\overline{\lim}\, \mathfrak{S}_n$. Then $A \in \mathfrak{B}_k$ for arbitrary k. Consequently, A and $\sigma\{\mathfrak{S}_1, \cdots, \mathfrak{S}_{k-1}\}$ are independent. Hence A and $\sigma\{\mathfrak{S}_1, \cdots, \mathfrak{S}_n, \cdots\}$ are also independent. Since $A \in \sigma\{\mathfrak{S}_1, \cdots, \mathfrak{S}_n, \cdots\}$, A is independent of A. But this is possible only when $\mathsf{P}(A) = 0$ or 1.

Let us now turn to the question of independent random variables.

Definition 3. Random variables ζ_λ (for $\lambda \in \Lambda$) are said to be *(mutually) independent* if the sets $\mathfrak{M}_\lambda$ (for $\lambda \in \Lambda$) are mutually independent, where $\mathfrak{M}_\lambda$ consists of all events of the form $\{u; \zeta_\lambda < a\}$, where $-\infty < a < \infty$.

The definition of independence of a set of random variables is equivalent to the following: Random variables ζ_λ (for $\lambda \in \Lambda$) are independent if for arbitrary n and arbitrary $\lambda_j \in \Lambda$ (for $j = 1, \cdots, n$), the joint distribution function of the variables $\zeta_{\lambda_1}, \zeta_{\lambda_2}, \cdots, \zeta_{\lambda_n}$ is equal to the product of the distribution functions of the variables ζ_{λ_j}:

$$\mathsf{P}\{\zeta_{\lambda_1} < a_1, \zeta_{\lambda_2} < a_2, \cdots, \zeta_{\lambda_n} < a_n\} = \prod_{j=1}^{n} \mathsf{P}\{\zeta_{\lambda_j} < a_j\} .$$

The definition of independence of a set of classes of random variables is analogous. Let us consider a set of random variables $\{\zeta_\lambda^\mu, \lambda \in \Lambda_\mu\}$ where μ is a fixed index and λ ranges over a set Λ_μ depending on the index μ. For convenience, let us call this set a *class* and let us look at the set of such classes indexed by μ where μ ranges over a set M.

Definition 4. The classes of random variables $\{\zeta_\lambda^\mu, \lambda \in \Lambda_\mu\}$ for $\mu \in M$ are said to be *(mutually) independent* if the sets of events $\mathfrak{M}_\mu$ (for $\mu \in M$) are mutually independent, where $\mathfrak{M}_\mu$ consists of all events of the form

$$\{u; \zeta_{\lambda_1}^\mu < a_1, \cdots, \zeta_{\lambda_n}^\mu < a_n\} , \qquad n = 1, 2, \cdots, \lambda_j \in \Lambda_\mu, -\infty < a_j < \infty . \tag{1}$$

Definition 5. A σ-algebra of events $\sigma\{\zeta_\lambda, \lambda \in \Lambda\}$ generated by events of the form

$$\{u; \zeta_{\lambda_1} < a_1, \cdots, \zeta_{\lambda_n} < a_n\} , \qquad n = 1, 2, \cdots, \lambda_j \in \Lambda, -\infty < a_j < \infty ,$$

is called the *σ-algebra generated by* the class of random variables ζ_λ for $\lambda \in \Lambda$. Let $\tilde{\sigma}\{\zeta_\lambda, \lambda \in \Lambda\}$ denote the completion of the σ-algebra $\sigma\{\zeta_\lambda, \lambda \in \Lambda\}$.

Obviously, $\sigma\{\zeta_\lambda, \lambda \in \Lambda\}$ is the minimal σ-algebra of events with respect to which all the ζ_λ are random variables (that is, the minimal σ-algebra of sets with respect to which all the functions $\zeta_\lambda = f_\lambda(u)$ are measurable).

We note in particular that the σ-algebra of events generated by a single random variable ζ is a minimal σ-algebra containing events of the form $\{u; \zeta < a\}$ where $-\infty < a < \infty$.

Theorem 6. *If the classes of random variables* $\{\zeta_\lambda^\mu, \lambda \in \Lambda_\mu\}$, *for*

$\mu \in M$, are independent, then the sets of σ-algebras $\sigma\{\zeta_\lambda^\mu, \lambda \in \Lambda_\mu\}$, for $\mu \in M$, and their completions $\tilde{\sigma}\{\zeta_\lambda^\mu, \lambda \in \Lambda_\mu\}$ are also independent.

Proof. Let $\mathfrak{M}_\mu$ denote the set of events of the form

$$(\zeta_{\lambda_1}^\mu, \zeta_{\lambda_2}^\mu, \cdots, \zeta_{\lambda_n}^\mu) \in I_n[\boldsymbol{a}, \boldsymbol{b}) ,$$

where $I_n[\boldsymbol{a}, \boldsymbol{b})$ is an arbitrary n-dimensional left-closed interval (cf. Section 2), where the components a_j and b_j of the vectors $\boldsymbol{a}$ and $\boldsymbol{b}$ can assume the values $\pm\infty$, where n is a natural number, and where the λ_j are arbitrary members of Λ_μ. It follows from Definition 4 that

$$\mathsf{P}\left\{\bigcap_{k=1}^{m} [(\zeta_{\lambda_1}^{\mu_k}, \cdots, \zeta_{\lambda_{n_k}}^{\mu_k}) \in I_{n_k}^{(k)}[\boldsymbol{a}_k, \boldsymbol{b}_k)]\right\}$$

$$= \prod_{k=1}^{m} \mathsf{P}\{(\zeta_{\lambda_1}^{\mu_k}, \cdots, \zeta_{\lambda_{n_k}}^{\mu_k}) \in I_{n_k}^{(k)}[\boldsymbol{a}_k, \boldsymbol{b}_k)\} .$$

This means that the $\mathfrak{M}_\mu$ are independent sets of events. Since these are decomposable families of events, Theorem 3 can be applied, and the Theorem is proved.

Corollary. *Let $\{g_\mu(t_1, t_2, \cdots, t_s)\}$, for $\mu = 1, 2, \cdots$, denote a sequence of arbitrary finite Borel functions of s real variables. If the sequences of random vectors $\zeta_\mu = \{\zeta_1^\mu, \cdots, \zeta_s^\mu\}$, for $\mu = 1, 2, \cdots$, are mutually independent, then the random variables $\xi^\mu = g_\mu(\zeta_1^\mu, \cdots, \zeta_s^\mu)$ are also mutually independent.*

The concept of independence of random variables and the theorems proved are easily generalized to random elements in an arbitrary measurable space X. We shall not repeat all the preceding formulations and theorems. We note only that when the ζ_λ are random elements, we should consider in the preceding exposition not events of the form

$$\{u; \zeta_{\lambda_1} < a_1, \cdots, \zeta_{\lambda_n} < a_n\}$$

but events of the form

$$\{u; \zeta_{\lambda_1} \in A_1, \cdots, \zeta_{\lambda_n} \in A_n\} ,$$

where the A_k are sets in the σ-algebra $\mathfrak{B}$ of subsets of X (cf. Definition 6, Section 1). Thus, for example, the random elements $\zeta_k = f_k(u)$, for $k = 1, 2, \cdots, n$, with values in X_k are said to be *independent* if for arbitrary $A_k \in \mathfrak{B}_k$ (where the $\mathfrak{B}_k$ are the given σ-algebras of subsets of X_k) we have

$$\mathsf{P}\left\{\bigcap_{k=1}^{n} \{\zeta_k \in A_k\}\right\} = \prod_{k=1}^{n} \mathsf{P}\{\zeta_k \in A_k\} .$$

Each of the random elements ζ_k defines a measure on $\mathfrak{B}_k$:

$$\mu_k(A) = \mathsf{P}f_k^{-1}(A) = \mathsf{P}\{\zeta_k \in A\}\,, \qquad A \in \mathfrak{B}_k\,, \qquad k = 1, \cdots, n\,.$$

Let us now look at the sequence of random elements $(\xi_1, \xi_2, \cdots, \xi_n)$. It is easy to see that this sequence can be regarded as a random element with values in $\prod_{k=1}^n X_k$. Specifically, let $\mathfrak{B}^{(n)}$ denote the product of the σ-algebras $\mathfrak{B}_1, \cdots, \mathfrak{B}_n$. If $C = A_1 \times A_2 \times \cdots \times A_n$, for $A_i \in \mathfrak{B}_i$, where $i = 1, \cdots, n$, then

$$\{u;\, (f_1(u) \cdots f_n(u)) \in C\} = \bigcap_{i=1}^n \{u;\, f_i(u) \in A_i\}\,, \tag{2}$$

that is, the preimage of C is $\mathfrak{S}$-measurable. From this it follows that the preimage of an arbitrary set in the minimal σ-algebra containing all C (cf. Section 6, Chapter II), that is, the preimage of an arbitrary set in $\mathfrak{B}^{(n)}$, is $\mathfrak{S}$-measurable. Let $\mu_{1,2,\cdots,n}$ denote the measure in $\prod_{k=1}^n X_k$ induced by the n-tuple $(\xi_1, \cdots, \xi_n)$:

$$\mu_{1,2,\cdots,n}(C) = \mathsf{P}\{(\xi_1, \cdots, \xi_n) \in C\}\,.$$

Let us assume that the elements ξ_i, for $i = 1, \cdots, n$, are independent. Formula (2) shows that

$$\mu_{1,2,\cdots,n}(A_1 \times A_2 \times \cdots \times A_n) = \mu_1(A_1)\mu_2(A_2) \cdots \mu_n(A_n)\,.$$

Since the extension of the measure from a decomposable family of sets (cf. Theorem 3, Section 7, Chapter II) to the minimal σ-algebra is unique, the measure $\mu_{1,2,\cdots,n}$ is the product of measures $\mu_1, \mu_2, \cdots, \mu_n$. The converse is true: if the measure $\mu_{1,2,\cdots,n}$ coincides with the product of the measures $\mu_1, \mu_2, \cdots, \mu_n$, then the random elements $\xi_1, \xi_2, \cdots, \xi_n$ are independent. From this we get:

Theorem 7. *The random elements $\xi_1, \xi_2, \cdots, \xi_n$ are independent if and only if the measure $\mu_{1,2,\cdots,n}$ induced by the n-tuple $(\xi_1, \xi_2, \cdots, \xi_n)$ on the σ-algebra $\mathfrak{B}^{(n)} = \{\mathfrak{B}_1 \times \cdots \times \mathfrak{B}_n\}$ is the product of measures μ_i (for $i = 1, \cdots, n$) induced by the elements ξ_i on $\mathfrak{B}_i$.*

Theorem 8. *Let $g(x_1, x_2)$ denote a $\mathfrak{B}^{(2)}$-measurable finite function, and let ξ_1 and ξ_2 denote independent random elements. Suppose that $\mathsf{M}\,|\,g(\xi_1, \xi_2)\,| < \infty$. Then $\varphi(x) = \mathsf{M}g(x, \xi_2)$ is a $\mathfrak{B}_1$-measurable function of x and $\mathsf{M}g(\xi_1, \xi_2) = \mathsf{M}\varphi(\xi_1)$, or*

$$\mathsf{M}g(\xi_1, \xi_2) = \int_{X_1} \mu_1(dx_1) \int_{X_2} g(x_1, x_2)\mu_2(dx_2)\,.$$

The same formula remains valid for an arbitrary $\tilde{\mathfrak{B}}^{(2)}$-measurable function, where the tilde $\sim$ denotes the completion of the σ-algebra (measure) if the measures μ_1 and μ_2 are assumed complete.

Theorem 8 is a direct consequence of Theorem 3, Section 6, Chapter II. On the basis of Theorem 8 and Fubini's theorem,

$$\mathsf{M}g(\xi_1, \xi_2) = \int_{X_1 \times X_2} g(x_1, x_2)\mu_{1,2}(d(x_1, x_2)) .$$

Corollary. *If ξ_1 and ξ_2 are independent random variables with finite expectation, then $\mathsf{M}\xi_1\xi_2 = \mathsf{M}\xi_1\mathsf{M}\xi_2$.*

Let X denote an arbitrary space. Suppose that an infinite sequence of measures μ_n, where $\mu_n(X) = 1$, is defined on some σ-algebra $\mathfrak{B}$ of subsets of X. Is it possible to construct a probability space $\{U, \mathfrak{S}, \mathsf{P}\}$ and to define on it a sequence of independent random variables ζ_n with values in X such that the measure induced on $\mathfrak{B}$ by the element ζ_n coincides with μ_n? When X is a complete separable metric space, a positive answer to this question is contained in Kolmogorov's theorem (Theorem 3, Section 2). We note that in this case it is easy to prove the corresponding theorem with no restrictions on the set X.

We shall now give some consequences of the general 0-or-1 law (Theorem 5) for random variables.

Theorem 9. *Let $\{\zeta_n\}$ denote a sequence of independent random variables and let $\mathfrak{S}_n = \sigma\{\zeta_n\}$ denote the σ-algebra generated by ζ_n. Define $\mathfrak{S}_n^* = \sigma\{\mathfrak{S}_n, \mathfrak{S}_{n+1}, \cdots\}$. Then*

a. *the limit of the sequence $\{\zeta_n\}$ exists either with probability 1 or with probability 0;*

b. *the limit of the sequence $\{\zeta_n\}$ is constant with probability 1;*

c. *if $z = f(x_1, \cdots, x_n \cdots)$ is a function of an infinite number of variables $x_1, \cdots, x_n, \cdots$, and if $f(\zeta_1, \cdots, \zeta_n, \cdots)$ is $\mathfrak{S}_n^*$-measurable for arbitrary n, then it is constant with probability 1.*

Proof. (a) The set of points of convergence of the sequence $\{\zeta_n\}$ (cf. equation (1), Section 3, Chapter II) belongs to $\overline{\lim}\, \mathfrak{S}_n$. (b) Similarly, for arbitrary a, the set $\{u; \lim \zeta_n < a\}$ belongs to $\overline{\lim}\, \mathfrak{S}_n$; that is, for arbitrary a, we have $\mathsf{P}\{\lim \zeta_n < a\} = 0$ or 1. (c) This result is obtained from the same reasoning as in (b).

4. SERIES OF INDEPENDENT RANDOM VARIABLES

In this section we consider the conditions under which a series of independent random variables converges. We begin by giving some important inequalities that indicate bounds for the probability that the maximum partial sum of a series of independent random variables assumes a great value. The idea of obtaining these inequalities is Kolmogorov's. In what follows we always assume that $\{\xi_n\}$, for $n = 1, 2, \cdots$, is a sequence of independent random variables, we define

$$\zeta_n = \sum_{i=1}^{n} \xi_i ,$$

where $n = 1, 2, \cdots$, and we assume $\zeta_0 = 0$. We let $\chi(A)$ denote the characteristic function of the event A.

Theorem 1 (Kolmogorov's inequality). *Suppose that* $\mathsf{M}\xi_n = 0$ *for* $n = 1, 2, \cdots$. *Let* $g(x)$ *denote an arbitrary continuous nonnegative convex-downward function that increases for* $x \geqq 0$. *Let* t *denote a positive number. Then*

$$\mathsf{P}\left\{\max_{1 \leqq k \leqq n} |\zeta_k| \geqq t\right\} \leqq \frac{\mathsf{M}g(|\zeta_n|)}{g(t)} . \tag{1}$$

Proof. Let A_k, for $k = 1, 2, \cdots, n$, denote the event

$$A_k = \{|\zeta_1| < t, \cdots, |\zeta_{k-1}| < t, |\zeta_k| \geqq t\} .$$

The events A_k are pairwise incompatible, and $A = \bigcup_{k=1}^{n} A_k$ is the event $\{\max_{1 \leqq k \leqq n} |\zeta_k| \geqq t\}$. We have

$$\mathsf{M}g(|\zeta_n|) \geqq \mathsf{M}\sum_{k=1}^{n} \chi(A_k)g(|\zeta_n|) = \sum_{k=1}^{n} \mathsf{M}\chi(A_k)g(|\zeta_k + (\zeta_n - \zeta_k)|) .$$

Since the variables ξ_k are independent, we have (from Theorem 8, Section 3)

$$\mathsf{M}\chi(A_k)g(|\zeta_k + (\zeta_n - \zeta_k)|) = \mathsf{M}\chi(A_k)g_k(\zeta_k) ,$$

where $g_k(x) = \mathsf{M}g(|x + \zeta_n - \zeta_k|)$. Applying Jensen's inequality (inequality (17), Section 1) and remembering that the function $|x|$ is convex downward, we obtain.

$$\begin{aligned} g_k(x) &= \mathsf{M}g(|x + \zeta_n - \zeta_k|) \geqq g(\mathsf{M}|x + \zeta_n - \zeta_k|) \\ &\geqq g(|x + \mathsf{M}(\zeta_n - \zeta_k)|) = g(|x|) . \end{aligned}$$

Thus

$$\begin{aligned} \mathsf{M}g(|\zeta_n|) &\geqq \mathsf{M}\sum_{k=1}^{n} \chi(A_k)g_k(\zeta_k) \geqq \mathsf{M}\sum_{k=1}^{n} \chi(A_k)g(|\zeta_k|) \\ &\geqq \mathsf{M}\sum_{k=1}^{n} \chi(A_k)g(t) = g(t)\mathsf{P}\left\{\max_{1 \leqq k \leqq n} |\zeta_k| \geqq t\right\} \end{aligned}$$

(here we use the fact that if the event A_k occurs, then $|\zeta_k| > t$), which completes the proof.

Let us note a special case of Kolmogorov's inequality. Let us set $g(x) = x^2$. Then

$$\mathsf{P}\left\{\max_{1 \leqq k \leqq n} |\zeta_k| \geqq t\right\} \leqq \frac{1}{t^2} \sum_{k=1}^{n} \sigma_k^2 , \tag{2}$$

where $\sigma_k^2 = \mathsf{M}\xi_k^2 = \mathsf{D}\xi_k$.

In the case in which we do not assume the existence of any mathematical expectations instead of Kolmogorov's inequality we can use:

Theorem 2. *If* $\mathsf{P}\{|\zeta_n - \zeta_k| \leqq t\} \geqq \alpha$ *for* $k = 0, 1, \cdots, n$, *then*

$$\mathsf{P}\left\{\max_{1 \leqq k \leqq n} |\zeta_k| > 2t\right\} \leqq \frac{1-\alpha}{\alpha}. \tag{3}$$

Proof. We define

$$A_k = \{|\zeta_1| \leqq 2t, \cdots, |\zeta_{k-1}| \leqq 2t, |\zeta_k| > 2t\}$$

and

$$B_k = \{|\zeta_n - \zeta_k| \leqq t\}$$

for $k = 0, 1, \cdots, n$. Then

$$C = \{|\zeta_n| > t\} \supset \bigcup_{k=1}^{n} (A_k \cap B_k),$$

the events A_k, for $k = 0, 1, \cdots, n$, are pairwise incompatible, and the events A_k and B_k (for n fixed, $k = 0, 1, 2, \cdots, n$) are independent. Therefore,

$$1 - \alpha \geqq \mathsf{P}\{|\zeta_n| > t\} \geqq \mathsf{P}\left\{\bigcup_{k=1}^{n} (A_k \cap B_k)\right\}$$

$$= \sum_{k=1}^{n} \mathsf{P}(A_k)\mathsf{P}(B_k) \geqq \alpha \sum_{k=1}^{n} \mathsf{P}(A_k) = \alpha\mathsf{P}\left\{\max_{1 \leqq k \leqq n} |\zeta_k| > 2t\right\},$$

which implies inequality (3).

The following inequality sometimes enables us to find a lower bound for the probability of the event $\{\max_{1 \leqq k \leqq n} |\zeta_k| > t\}$:

Theorem 3. *If* $\mathsf{M}\xi_k = 0$ *and if* $|\zeta_k| \leqq c$ *with probability* 1, *then*

$$\mathsf{P}\left\{\max_{1 \leqq k \leqq n} |\zeta_k| \leqq t\right\} \leqq \frac{(c+t)^2}{\sum_{k=1}^{n} \sigma_k^2}, \tag{4}$$

where $\sigma_k^2 = \mathsf{M}\xi_k^2 = \mathsf{D}\xi_k$.

Proof. Let E_n denote the events $\{\max_{0 \leqq k \leqq n} |\zeta_k| \leqq t\}$ for $n = 0, 1, \cdots$. This is a decreasing sequence of events. We have

$$\mathsf{M}\chi(E_n)\zeta_n^2 = \sum_{k=1}^{n} \mathsf{M}\{\chi(E_k)\zeta_k^2 - \chi(E_{k-1})\zeta_{k-1}^2\}$$

$$= \sum_{k=1}^{n} \mathsf{M}\chi(E_{k-1})(\zeta_k^2 - \zeta_{k-1}^2) - \sum_{k=1}^{n} \mathsf{M}\{\chi(E_{k-1}\backslash E_k)\zeta_k^2\}. \tag{5}$$

Furthermore,

$$\mathsf{M}\chi(E_{k-1}\backslash E_k)\zeta_k^2 = \mathsf{M}\chi(E_{k-1}\backslash E_k)(\zeta_{k-1} + \xi_k)^2 \leqq (t+c)^2\mathsf{M}\chi(E_{k-1}\backslash E_k),$$

$$\sum_{k=1}^{n} \mathsf{M}\chi(E_{k-1}\backslash E_k)\zeta_k^2 \leqq (t+c)^2 \sum_{k=1}^{n} \mathsf{M}\chi(E_{k-1}\backslash E_k)$$

$$= (t+c)^2[1 - \mathsf{P}(E_n)]. \tag{6}$$

Also,

$$\begin{aligned} \mathsf{M}\chi(E_{k-1})(\zeta_k^2 - \zeta_{k-1}^2) &= \mathsf{M}\chi(E_{k-1})(2\zeta_{k-1}\xi_k + \xi_k^2) \\ &= 2\mathsf{M}\chi(E_{k-1})\zeta_{k-1}\mathsf{M}\xi_k + \mathsf{M}\chi(E_{k-1})\mathsf{M}\xi_k^2 = \sigma_k^2\mathsf{M}\chi(E_{k-1}) . \qquad (7) \end{aligned}$$

Relations (5) to (7) yield

$$\begin{aligned} t^2\mathsf{P}(E_n) \geqq \mathsf{M}\chi(E_n)\zeta_n^2 &\geqq \sum_{k=1}^{n} \sigma_k^2\mathsf{M}\chi(E_{k-1}) - (t+c)^2(1 - \mathsf{P}(E_n)) \\ &\geqq \mathsf{P}(E_n)\left\{\sum_{k=1}^{n} \sigma_k^2 + (t+c)^2\right\} - (t+c)^2 \end{aligned}$$

or

$$(t+c)^2 \geqq \mathsf{P}(E_n)\left\{\sum_{k=1}^{n} \sigma_k^2 + c^2 + 2ct\right\} ,$$

from which inequality (4) follows.

We turn now to the question of the convergence of a series of independent random variables. We know from the preceding section (Theorem 9) that such a series converges either with probability 1 or with probability 0.

Theorem 4. *If* $\mathsf{M}\xi_k = 0$, $D\xi_k = \sigma_k^2$, *and*

$$\sum_{k=1}^{\infty} \sigma_k^2 < \infty , \qquad (8)$$

then the series of independent random variables

$$\sum_{k=1}^{\infty} \xi_k \qquad (9)$$

converges with probability 1.

Proof. Define $\zeta_n = \sum_{k=1}^{n} \xi_k$. Let N denote the event that the series (9) diverges. We define

$$E_{m,r} = \left\{\sup_{k,k' \geqq m} |\zeta_k - \zeta_{k'}| > \frac{1}{r}\right\} .$$

Then

$$N = \bigcup_{r=1}^{\infty} \bigcap_{m} E_{m,r} .$$

From Kolmogorov's inequality we obtain

$$\mathsf{P}(N) = \lim_{r\to\infty} \lim_{m\to\infty} \mathsf{P}(E_{m,r}) \leqq \overline{\lim_{r\to\infty}}\ \overline{\lim_{m\to\infty}}\ r^2 \sum_{k=m}^{\infty} \sigma_k^2 = 0 .$$

In the general case, the question of convergence of the series (9) is completely answered by:

Theorem 5 (Kolmogorov's three-series theorem). *For the series* (9) *of independent random variables to converge, it is necessary (resp.*

sufficient) that for every (resp. some) $c > 0$, *the three series*

$$\sum_{n=1}^{\infty} \mathsf{P}\{|\xi_n| > c\} , \tag{10}$$

$$\sum_{n=1}^{\infty} \mathsf{M}\xi'_n , \tag{11}$$

$$\sum_{n=1}^{\infty} \mathsf{D}\xi'_n , \tag{12}$$

converge, where $\xi'_n = \xi_n$ *for* $|\xi_n| < c$ *and* $\xi'_n = 0$ *for* $|\xi_n| > c$.

Proof of the Sufficiency. According to Theorem 4, the series

$$\sum_{n=1}^{\infty} (\xi'_n - \mathsf{M}\xi'_n)$$

converges with probability 1. Since the series (11) converges, it follows that the series $\sum_{n=1}^{\infty} \xi'_n$ converges. The series

$$\sum_{n=1}^{\infty} \xi_n = \sum_{n=1}^{\infty} (\xi'_n + \xi''_n) ,$$

where $\xi''_n = \xi_n - \xi'_n$, has only finitely many nonzero terms ξ''_n on the basis of condition (10) and the Borel-Cantelli theorem (Theorem 2, Section 3). Therefore the series (9) converges with probability 1.

Proof of the Necessity. Suppose that the the series (9) converges. Then the sequence of its terms converges to zero with probability 1 since only finitely many terms in the series exceed the number c in absolute value. Therefore the series $\sum_{n=1}^{\infty} \xi'_n$ converges with probability 1. Let $\{\eta_n\}$, for $n = 1, 2, \cdots$, denote a sequence of independent random variables that do not depend on the sequence $\{\xi'_n\}$ and that have the same distributions as do the ξ'_n. Let us set $\tilde{\xi}_n = \xi'_n - \eta_n$. Then the series $\sum_{n=1}^{\infty} \tilde{\xi}_n$ converges with probability 1; also,

$$\mathsf{M}\tilde{\xi}_n = 0 , \qquad |\tilde{\xi}_k| \leq 2c , \qquad \mathsf{D}\tilde{\xi}_n = 2\mathsf{D}\xi'_n .$$

The convergence of the series $\sum_{n=1}^{\infty} \tilde{\xi}_n$ implies that

$$\mathsf{P}\left\{\max_{1 \leq n \leq \infty} \left|\sum_{k=1}^{n} \tilde{\xi}_k\right| < \infty\right\} = 1 .$$

Therefore, for some t,

$$\mathsf{P}\left\{\max_{1 \leq n \leq \infty} \left|\sum_{k=1}^{n} \tilde{\xi}_k\right| \leq t\right\} = a > 0 .$$

It follows from inequality (4) that for arbitrary n,

$$2\sum_{k=1}^{n} \mathsf{D}\xi'_k = \sum_{k=1}^{n} \mathsf{D}\tilde{\xi}_k \leq \frac{(2c + t)^2}{a} ,$$

which proves the convergence of the series (12). It then follows from Theorem 4 that the series $\sum_{n=1}^{\infty} \xi'_n - \mathsf{M}\xi'_n$ coverges with probability 1. This in turn implies convergence of the series (11). On the basis of Theorem 2 of Section 3, the series (10) must converge, since when the series (9) converges, $|\xi_n|$ can exceed c only for a finite number of values of n. This completes the proof of the theorem.

Corollary. *For the series* (9) *of independent nonnegative random variables to converge, it is necessary* (*resp. sufficient*) *that for every* (*resp. some*) $c > 0$ *the series*

$$\sum_{n=1}^{\infty} \mathsf{P}\{\xi_k > c\} , \qquad \sum_{n=1}^{\infty} \mathsf{M}\xi'_k$$

converge.

Proof. For nonnegative random variables ξ_n, we have $\mathsf{M}\xi'^2_n \leqq c\mathsf{M}\xi'_n$. Therefore convergence of the series (11) implies convergence of the series $\sum_{n=1}^{\infty} \mathsf{M}\xi'^2_n$ and hence of the series (12).

5. ERGODIC THEOREMS

Let $\xi(t)$ denote a stationary sequence, that is, a stationary random function that assumes real values and that is defined on the set Z of all integers ($t = 0, \pm 1, \pm 2, \cdots$). Let R^Z denote the space of all real sequences $\omega = \{a(t), t \in Z\}$ and let $\mathfrak{C}$ denote the σ-algebra generated by the cylindrical sets R^Z. The sequence $\xi(t)$ defines a measure $\mathsf{P}_\xi(C)$ on $\mathfrak{C}$:

$$\mathsf{P}_\xi(C) = \mathsf{P}(\{\xi(t); t \in Z\} \in C), C \in \mathfrak{C} .$$

Let $\{\mathsf{P}_\xi, \tilde{\mathfrak{C}}_\xi\}$ denote the completion of the measure P_ξ. In R^Z we define an operation S representing time displacement: $\omega' = S\omega$ if $a'(t) = a(t+1)$ for $t \in Z$, where $\omega = \{a(t); t \in Z\}$ and $\omega' = \{a'(t); t \in Z\}$. The operation S has an inverse S^{-1}:

$$S^{-1}\omega = \omega'' , \qquad \omega'' = \{a''(t); t \in Z\} , \qquad a''(t) = a(t-1) .$$

The condition for stationarity of the sequence $\xi(t)$ means (cf. Section 1, Chapter I) that for an arbitrary cylindrical set C,

$$\mathsf{P}_\xi(C) = \mathsf{P}_\xi(SC) . \tag{1}$$

Since a measure on cylindrical sets uniquely defines a measure on $\mathfrak{C}$ and on its completion $\tilde{\mathfrak{C}}_\xi$, equation (1) remains valid for arbitrary $A \in \tilde{\mathfrak{C}}_\xi$:

$$\mathsf{P}_\xi(A) = \mathsf{P}_\xi(SA) , \qquad A \in \tilde{\mathfrak{C}}_\xi . \tag{2}$$

Definition 1. Let $\{\Omega, \mathfrak{S}, \mu\}$ denote a space with a measure μ and let T denote a measurable mapping of $\{\Omega, \mathfrak{S}\}$ into itself. The mapping T is said to be *measure-preserving* if for arbitrary $A \in \mathfrak{S}$, $\mu(T^{-1}A) = \mu(A)$ where $T^{-1}A$ is the complete preimage of the set A.

A mapping T is said to be invertible if there exists a measurable transformation T^{-1} such that $TT^{-1} = T^{-1}T = I$, where I is the identity mapping. In this case the mapping T^{-1} is called the inverse of T.

The definition of a stationary sequence is equivalent to saying that a sequence $\{\xi(t); t \in Z\}$ is stationary if the time-displacement operator S preserves the measure P_ξ in R^Z.

Thus the problem of studying stationary sequences is included in the problem of studying measure-preserving invertible transformations (automorphisms) of some space with a measure.

Let us now look at the question of the asymptotic behavior of the mean

$$\frac{1}{n}\sum_{k=0}^{n-1} f(T^k\omega) , \qquad n \longrightarrow \infty , \tag{3}$$

where T^k is the kth power of the mapping T, where $f(\omega)$ is an arbitrary $\mathfrak{S}$-measurable function, and where $\{\Omega, \mathfrak{S}, \mu\}$ is a space with measure μ such that $\mu(\Omega) = 1$. To understand the meaning of this problem, let us look at the case in which $\{\Omega, \mathfrak{S}, \mu\}$ is our space $\{R^Z, \tilde{\mathfrak{C}}_\xi, \mathsf{P}_\xi\}$ and $T = S$. Suppose that $f(\omega)$ is $\chi_A(\xi(0))$, where $\chi_A(x)$ is the characteristic function of the Borel set A on the real line. Then $f(T^k\omega) = \chi_A(S^k\omega) = \chi_A(\xi(k))$ and

$$\frac{1}{n}\sum_{k=0}^{n-1} f(T^k\omega) = \frac{\nu_n(A, \omega)}{n} , \tag{4}$$

where $\nu_n(A, \omega)$ is the number of terms in the sequence $\xi(0), \xi(1), \cdots, \xi(n-1)$ with values in the set A; that is $\nu_n(A, \omega)/n$ is the frequency with which the first n terms in the sequence $\xi(t)$ (for $t = 0, 1, \cdots$) fall in the set A. Thus the question we must consider is the question regarding the frequency with which the random variable $\xi(t)$ assumes values in an arbitrary set A. Let us show first of all that the limit as $n \longrightarrow \infty$ of the mean (3) exists with probability 1. This assumption constitutes the substance of the famous Birkhoff-Khinchin theorem.

Lemma 1. *If T preserves the measure μ, if $D \in \mathfrak{S}$, and if $f(\omega)$ is a $\mathfrak{S}$-measurable nonnegative (μ-integrable) function, then*

$$\int_{T^{-1}D} f(T\omega)\mu(d\omega) = \int_D f(\omega)\mu(d\omega) . \tag{5}$$

Proof. If we set $f(\omega) = \chi_A(\omega)$, formula (5) becomes the equation

$$\mu(T^{-1}(A \cap D)) = \mu(A \cap D) ,$$

which is valid for arbitrary A and $D \in \mathfrak{S}$. From this it follows that it is valid for arbitrary $\mathfrak{S}$-measurable nonnegative and μ-integrable functions.

The following lemma is of an elementary arithmetic nature. Let $a_1, a_2, \cdots, a_n$ denote a sequence of real numbers and let p denote an integer. We shall say that a term a_k in a sequence $\{a_k\}$ is *p-distinguished* if at least one of the terms

$$a_k,\ a_k + a_{k+1},\ \cdots,\ a_k + a_{k+1} + \cdots + a_{k+p-1}$$

is nonnegative. Thus a_k is 1-distinguished if and only if it is nonnegative.

Lemma 2. *The sum of all p-distinguished elements is nonnegative.*

Proof. Let a_{k_1} denote the smallest p-distinguished element of a sequence, and let $a_{k_1} + a_{k_1+1} + \cdots + a_{k_1+r}$, where $r \leqq p - 1$, denote the nonnegative sum with the smallest number of terms. For $h < r$, we have

$$a_{k_1} + a_{k_1+1} + \cdots + a_{k_1+h} < 0 .$$

Consequently, $a_{k_1+h+1} + \cdots + a_{k_1+r} \geqq 0$, that is, all terms of the sequence $a_{k_1}, a_{k_1+1} + \cdots + a_{k_1+r}$ are p-distinguished, and their sum is nonnegative. We can extend this reasoning by considering the sequence beginning with the term a_{k_1+r+1}. Thus every sequence is broken into finite sums of p-distinguished terms, and each such sum is nonnegative. The set of p-distinguished elements in the entire sequence coincides with the union of the sets of p-distinguished elements in the parts chosen. This completes the proof of the lemma.

The following lemma constitutes a basic step in the proof of the Birkhoff-Khinchin theorem.

Lemma 3. *Let $f(\omega)$ denote a μ-integrable function and define*

$$E = \left\{\omega;\ \exists n \ \textit{for which} \ \sum_{k=0}^{n-1} f(T^k\omega) \geqq 0,\ n = 1, 2, \cdots\right\} .$$

Then

$$\int_E f(\omega)\mu(d\omega) \geqq 0 . \qquad (6)$$

Proof. Consider the sequence $f(\omega), f(T\omega), \cdots, f(T^{N+p-1}\omega)$. Let $s(\omega)$ denote the sum of all p-distinguished elements of this sequence. On the basis of Lemma 2 we have $s(\omega) \geqq 0$. Define

$D_k = \{\omega; f(T^k\omega)$ is a p-distinguished element$\}$, for $k = 0, 1, \cdots, N + p - 1$. Let $\chi_k(\omega)$ denote the characteristic function of the set D_k. We note that

$$D_0 = \left\{\omega; \sup_{n \leqq p} \sum_{k=0}^{n-1} f(T^k\omega) \geqq 0\right\} \text{ and } D_k = T^{-1}D_{k-1} \text{ for } k \leqq N .$$

Therefore $D_k = T^{-k}D_0$ for $k \leqq N$. Therefore

$$0 \leqq \int_\Omega s(\omega)\mu(d\omega) = \int_\Omega \sum_{k=0}^{N+p-1} f(T^k\omega)\chi_k(\omega)\mu(d\omega)$$
$$= \sum_{k=0}^{N+p-1} \int_{D_k} f(T^k\omega)\mu(d\omega) .$$

On the basis of Lemma 1,

$$\int_{D_k} f(T^k\omega)\mu(d\omega) = \int_{T^{-k}D_0} f(T^k\omega)\mu(d\omega) = \int_{D_0} f(\omega)\mu(d\omega) , \quad k \leqq N .$$

Consequently,

$$N\int_{D_0} f(\omega)\mu(d\omega) + \sum_{k=N+1}^{N+p-1} \int_{D_k} f(T^k\omega)\mu(d\omega) \geqq 0 . \tag{7}$$

Since

$$\left|\int_{D_k} f(T^k\omega)\mu(d\omega)\right| \leqq \int_\Omega |f(T^k\omega)| \mu(d\omega) = \int_\Omega |f(\omega)| \mu(d\omega) < \infty ,$$

by dividing inequality (7) by N and then letting N approach ∞, we obtain

$$\int_{D_0} f(\omega)\mu(d\omega) \geqq 0 . \tag{8}$$

The sets $D_0 = D_0(p)$, for $p = 1, 2, \cdots, n$, constitute an increasing sequence:

$$\lim_{p\to\infty} D_0(p) = \bigcup_{p=1}^{\infty} D_0(p) = E .$$

Taking the limit in (8) as $p \to \infty$, we obtain (6), which completes the proof of the lemma.

(Cf. also, A. Garsia, A simple proof of E. Hopf's maximal ergodic theorem, J. of Math. and Mech., 14, 381–382 (1965). Ed.)

Lemma 4 (The maximal ergodic theorem). *Suppose that $f(\omega)$ is a μ-integrable function, that λ is a real number, and that*

$$E_\lambda = \left\{\omega;\ \exists n \text{ such that } \frac{1}{n}\sum_{k=0}^{n-1} f(T^k\omega) \geqq \lambda\right\} .$$

Then

$$\int_{E_\lambda} f(\omega)\mu(d\omega) \geqq \lambda\mu(E_\lambda) . \tag{9}$$

This is proved by applying Lemma 3 to an appropriate choice of the function f.

Theorem 1 (Birkhoff-Khinchin). *Let $\{\Omega, \mathfrak{S}, \mu\}$ denote a space with a measure, let T denote a measurable μ-measure-preserving mapping of $\{\Omega, \mathfrak{S}\}$ into itself and let $f(\omega)$ denote an arbitrary μ-integrable function. Then the limit*

$$\lim_{n\to\infty} \frac{1}{n} \sum_{k=0}^{n-1} f(T^k\omega) = f^*(\omega) \pmod{\mu} \tag{10}$$

exists μ-almost-everywhere in Ω. The function $f^*(\omega)$ is T-invariant; that is,

$$f^*(T\omega) = f^*(\omega) \pmod{\mu} , \tag{11}$$

and it is integrable. Also, if $\mu(\Omega) < \infty$, then

$$\int_\Omega f^*(\omega)\mu(d\omega) = \int_\Omega f(\omega)\mu(d\omega) . \tag{12}$$

Proof. Without loss of generality we may assume that the function $f(\omega)$ is finite and nonnegative. Let us set

$$g^*(\omega) = \overline{\lim} \frac{1}{n} \sum_{k=0}^{n-1} f(T^k\omega) , \qquad g_*(\omega) = \underline{\lim} \frac{1}{n} \sum_{k=0}^{n-1} f(T^k\omega) .$$

We need to show that $g^*(\omega) = g_*(\omega) \pmod{\mu}$. Suppose that

$$K_{\alpha\beta} = \{\omega;\, g^*(\omega) \geqq \beta,\, g_*(\omega) \leqq \alpha\} , \qquad 0 \leqq \alpha \leqq \beta .$$

It will be sufficient to show that $\mu(K_{\alpha\beta}) = 0$ since

$$\{\omega;\, g^*(\omega) > g_*(\omega)\} = \bigcup_{\substack{\alpha<\beta \\ \alpha,\beta\in R}} K_{\alpha\beta} ,$$

where R is the set of nonnegative rational numbers. Note that

$$g^*(T\omega) = \overline{\lim} \left\{\frac{1}{n} \sum_{k=0}^{n-1} f(T^k\omega) + \frac{f(T^n\omega) - f(\omega)}{n}\right\} = g^*(\omega)$$

and, analogously, $g_*(T\omega) = g_*(\omega)$. This means that $T^{-1}K_{\alpha\beta} = K_{\alpha\beta}$. Therefore we can apply Lemma 4 to a space with measure

$$\{K_{\alpha\beta}, \mathfrak{S} \cap K_{\alpha\beta}, \mu\} .^*$$

From this it follows that

$$\int_{K_{\alpha\beta}} f(\omega)\mu(d\omega) \geqq \beta\mu(K_{\alpha\beta}) , \tag{13}$$

* Lemma 4 does not seem to be directly applicable. Instead (13) seems to follow most simply by choosing $0 < \varepsilon < \beta$ and noting that

$$\int_{K_{\alpha\beta}} f(\omega)\mu(d\omega) \geqq \int_{K_{\alpha\beta}} \{f(\omega) - (\beta - \varepsilon)\}\mu(d\omega) = \int_{K_{\alpha\beta}} \left\{\frac{1}{n}\sum_{k=0}^{n-1} f(T^k\omega) - (\beta - \varepsilon)\right\}\mu(d\omega) \geqq 0$$

if n is chosen appropriately. This implies that

$$\int_{K_{\alpha\beta}} f(\omega)\mu(d\omega) \geqq (\beta - \varepsilon)\mu(K_{\alpha\beta}) ;$$

(13) now follows by letting $\varepsilon \downarrow 0$.

and applying (9) to the function $-f(\omega)$, we obtain

$$\int_{K_{\alpha\beta}} (-f(\omega))\mu(d\omega) \geqq -\alpha\mu(K_{\alpha\beta}) , \quad \int_{K_{\alpha\beta}} f(\omega)\mu(d\omega) \leqq \alpha\mu(K_{\alpha\beta}) . \quad (14)$$

Since $\beta > 0$, it follows from (13) that $\mu(K_{\alpha\beta}) < \infty$, but then (14) is possible if and only if $\mu(K_{\alpha\beta}) = 0$. Thus the existence (mod μ) of the limit (10) is proven. Let us set $f^*(\omega) = g^*(\omega)$. Then (10) is satisfied and the function $f^*(\omega)$ is T-invariant everywhere in Ω.

To prove formula (12), we set

$$A_{kn} = \left\{\omega; \frac{k}{2^n} \leqq f^*(\omega) < \frac{k+1}{2^n}\right\} .$$

We have $\Omega = \bigcup_{k=-\infty}^{\infty} A_{kn}$, and

$$T^{-1}A_{kn} = \left\{\omega; \frac{k}{2^n} \leqq f^*(T\omega) < \frac{k+1}{2^n}\right\} = A_{kn} .$$

Let us apply Lemma 4 to the set A_{kn}. For arbitrary $\varepsilon > 0$ we have

$$\int_{A_{kn}} f(\omega)\mu(d\omega) > \left(\frac{k}{2^n} - \varepsilon\right)\mu(A_{kn}) .$$

By letting ε approach 0, we obtain the inequality

$$\int_{A_{kn}} f(\omega)\mu(d\omega) \geqq \frac{k}{2^n}\, \mu(A_{kn}) .$$

Analogously,

$$\int_{A_{kn}} f(\omega)\mu(d\omega) \leqq \frac{k+1}{2^n}\, \mu(A_{kn}) .$$

From this it follows that

$$\left|\int_{A_{kn}} f(\omega)\mu(d\omega) - \int_{A_{kn}} f^*(\omega)\mu(d\omega)\right| \leqq \frac{1}{2^n}\, \mu(A_{kn}) .$$

Summing these inequalities over all k, we obtain

$$\left|\int_{\Omega} f(\omega)\mu(d\omega) - \int_{\Omega} f^*(\omega)\mu(d\omega)\right| < \frac{1}{2^n}\, \mu(\Omega) .$$

Since n is arbitrary and $\mu(\Omega) < \infty$, we obtain formula (12). This completes the proof of the theorem.

Corollary 1. *Convergence of the means almost everywhere implies convergence with respect to the metric of L_p, if $\mu(\Omega) < \infty$.*

Suppose that $\mu(\Omega) < \infty$ and $f(\omega) \in L_p$ (for $p \geqq 1$). Then

$$\left\|\frac{1}{n}\sum_{k=0}^{n-1} f(T^k\omega) - f^*(\omega)\right\|_p \longrightarrow 0 \quad \text{as} \quad n \longrightarrow \infty , \quad (15)$$

where $\|f\|_p$ is the norm of the function f in L_p (cf. Section 5,

Chapter II). To prove this, let us take some bounded function $f_0(\omega)$ and let us set $\|f(\omega) - f_0(\omega)\|_p = \delta$. We have

$$\left\|\frac{1}{n}\sum_{k=0}^{n-1} f(T^k\omega) - f^*(\omega)\right\|_p \leqq \left\|\frac{1}{n}\sum_{k=0}^{n-1}[f(T^k\omega) - f_0(T^k\omega)]\right\|_p + \left\|\frac{1}{n}\sum_{k=0}^{n-1} f_0(T^k\omega) - f_0^*(\omega)\right\|_p + \|f_0^*(\omega) - f^*(\omega)\|_p .$$

From Jensen's inequality and Lemma 1,

$$\begin{aligned}\left\|\frac{1}{n}\sum_{k=0}^{n-1}[f(T^k\omega) - f_0(T^k\omega)]\right\|_p &= \left\{\int_\Omega\left[\frac{1}{n}\sum_{k=0}^{n-1}(f(T^k\omega) - f_0(T^k\omega))\right]^p \mu(d\omega)\right\}^{1/p} \\ &\leqq \left\{\int_\Omega \frac{1}{n}\sum_{k=0}^{n-1}|f(T^k\omega) - f_0(T^k\omega)|^p \mu(d\omega)\right\}^{1/p} \\ &= \left\{\frac{1}{n}\sum_{k=0}^{n-1}\int_\Omega |f(\omega) - f_0(\omega)|^p \mu(d\omega)\right\}^{1/p} = \delta .\end{aligned}$$

Using Fatou's lemma (Theorem 2, Section 5, Chapter II), we obtain

$$\begin{aligned}\|f_0^*(\omega) - f^*(\omega)\|_p &= \left\{\int_\Omega \underline{\lim}\left|\frac{1}{n}\sum_{k=0}^{n-1}[f(T^k\omega) - f_0^*(T^k\omega)]\right|^p \mu(d\omega)\right\}^{1/p} \\ &\leqq \underline{\lim}\left\|\frac{1}{n}\sum_{k=0}^{n-1}[f(T^k\omega) - f_0(T^k\omega)]\right\|_p \leqq \delta .\end{aligned}$$

Furthermore, since the function $f_0(\omega)$ is bounded, all its means are bounded by a single constant. Therefore if we let n approach ∞ in the expression

$$\left\|\frac{1}{n}\sum_{k=0}^{n-1} f_0(T^k\omega) - f_0^*(\omega)\right\|_p = \left\{\int_\Omega\left|\frac{1}{n}\sum_{k=0}^{n-1} f_0(T^k\omega) - f_0^*(\omega)\right|^p \mu(d\omega)\right\}^{1/p}$$

we can, by virtue of Lebesgue's theorem, take the limit under the integral sign. Consequently this expression approaches 0 and for sufficiently large n is less than δ. Hence

$$\left\|\frac{1}{n}\sum_{k=0}^{n-1} f(T^k\omega) - f^*(\omega)\right\|_p < 3\delta , \qquad n \geqq n_0 = n_0(\delta) ,$$

where the number $\delta > 0$ can be chosen arbitrarily small. Thus (15) is proved.

Let us return now to stationary sequences.

Corollary 2. *Let $\{\xi_n\}$, for $n = 0, \pm 1, \cdots$, denote a stationary sequence and suppose that $\mathsf{M}|\xi_0| < \infty$. Then with probability 1 the limit*

$$\lim_{n\to\infty}\frac{1}{n}\sum_{k=m}^{n+m}\xi_k = \xi^*$$

exists, and $\mathsf{M}\xi^* = \mathsf{M}\xi$.

Let us look at an arbitrary event $A \in \mathfrak{C}$ and the sequence of events obtained from A by a "time displacement": $A, S^{\pm 1}A, S^{\pm 2}A, \cdots$. Let χ_n denote the characteristic function of the event $S^n A$. Then the χ_n (for $n = 0, \pm 1, \cdots$) constitute a stationary sequence of random variables, and

$$\frac{1}{n} \sum_{k=0}^{n-1} \chi_k$$

is the frequency of the occurrence of the event A calculated from a single sample function of the sequence $\{\xi_n\}$, for $n = 0, 1, 2, \cdots$, and for n consecutive displacements of the origin of the time scale

$$\frac{1}{n} \sum_{k=0}^{n-1} \chi_k = \frac{\nu_n(A)}{n} .$$

On the basis of the Birkhoff-Khinchin theorem, the limit

$$\lim_{n \to \infty} \frac{\nu_n(A)}{n} = \pi(A)$$

exists with probability 1 and $\mathsf{M}\pi(A) = \mathsf{M}\chi_0 = \mathsf{P}(A)$. The quantity $\pi(A)$ can be called the *empirical probability* of the event A. It is a random variable and is determined by a single sample function of the infinite sequence $\{\xi_n\}$, for $n = 0, 1, 2, \cdots$. The question naturally arises: When is an empirical probability $\pi(A)$ independent of chance and when does it coincide with the probability $\mathsf{P}(A)$?

Definition 2. A stationary sequence $\{\xi_n\}$, for $n = 0, 1, \cdots$, is said to be *ergodic* if for arbitrary $A \in \tilde{\mathfrak{C}}_\xi$,

$$\lim_{n \to \infty} \frac{\nu_n(A)}{n} = \mathsf{P}(A) \pmod{\mathsf{P}} .$$

An event $A \in \tilde{\mathfrak{C}}_\xi$ is said to be *invariant* (mod P) if the events $S^{-1}A$ and A coincide with probability 1. A sequence $\{\xi_n\}$, for $n = 0, \pm 1, \cdots$, is said to be *transitive* if an arbitrary invariable event has probability 1 or 0.

Theorem 2. *For an ergodic stationary sequence, it is necessary and sufficient that one or the other of the following two conditions be satisfied*:

a. *The sequence is transitive*;

b. *for an arbitrary $\tilde{\mathfrak{C}}_\xi$-measurable function $f(\omega)$ such that $\mathsf{M}\,|f(\omega)| < \infty$, the function $f^*(\omega)$ is constant with probability* 1, *where*

$$f^*(\omega) = \lim \frac{1}{n} \sum_{k=0}^{n-1} f(S^k \omega) .$$

Proof. Let A denote an invariant event. Then, the events

$A, SA, S^2A, \cdots$ coincide with probability 1, and $\nu_n(A)/n = \chi_A(\omega)$. Consequently, the limit

$$\lim_{n\to\infty} \nu_n(A)/n$$

cannot be a constant (mod P) if $0 < \mathsf{P}(A) < 1$. Thus ergodicity implies transitivity. Suppose now that the sequence $\{\xi_n\}$ is transitive. Since the function $f^*(\omega)$ is S-invariant (mod P), the events

$$S^{-1}\{\omega; f^*(\omega) < x\} = \{\omega; f^*(S\omega) < x\} \quad \text{and} \quad \{\omega; f^*(\omega) < x\}$$

coincide with probability 1. It follows from the transitivity that for arbitrary real x the quantity $\mathsf{P}\{\omega; f^*(\omega) < x\}$ is either 0 or 1; that is, $f^*(\omega)$ is constant (mod P). Thus condition a implies condition b. Finally, the condition for ergodicity is a special case of condition b, namely the case in which $f(\omega)$ is the characteristic function of some event.

Let us indicate a few consequences of ergodicity.

Let $f(\omega)$ and $g(\omega)$ denote arbitrary functions belonging to the set.

$$L_2 = L_2(\Omega, \widetilde{\mathfrak{S}}_\xi, \mathsf{P}) .$$

It follows from Corollary 1 to Theorem 1 that

$$\lim_{n\to\infty} \int_\Omega \frac{1}{n} \sum_{k=0}^{n-1} f(S^k\omega)g(\omega)\mathsf{P}(d\omega) = \int_\Omega f^*(\omega)g(\omega)\mathsf{P}(d\omega) . \tag{16}$$

Let us set $g(\omega) = \eta$ and $f(S^k\omega) = \zeta_k$. Let us suppose that the original stationary sequence $\{\xi_n\}$, for $n = 0, \pm 1, \cdots$, is ergodic. Equation (16) takes the form

$$\lim_{n\to\infty} \mathsf{M}\frac{1}{n} \sum_{k=0}^{n-1} \zeta_k\eta = \mathsf{M}\zeta_0\mathsf{M}\eta . \tag{17}$$

Suppose that $\eta = \chi_B(\omega)$, $\zeta_0 = f(\omega) = \chi_A(\omega)$, where A and B belong to $\widetilde{\mathfrak{S}}_\xi$. It follows from (17) that

$$\lim_{n\to\infty} \frac{1}{n} \sum_{k=0}^{n-1} \mathsf{P}(S^{-k}A \cap B) = \mathsf{P}(A)\mathsf{P}(B) \tag{18}$$

or (if $\mathsf{P}(B) \neq 0$)

$$\lim_{n\to\infty} \frac{1}{n} \sum_{k=0}^{n-1} \mathsf{P}(S^{-k}A \mid B) = \mathsf{P}(A) , \tag{19}$$

where $\mathsf{P}(A \mid B)$ is the conditional probability of the event A, given B.

Corollary 1. *Equation* (18) (*or equation* (19)) *is for* $A, B \in \widetilde{\mathfrak{S}}_\xi$ *equivalent to ergodicity.*

To prove this, it suffices to show that (18) implies ergodicity. Let C denote an arbitrary S-invariant set. In (18) let us set $A = B = C$. Then equation (18) becomes $\mathsf{P}(C) = \mathsf{P}^2(C)$, so that $\mathsf{P}(C)$ is equal to either 0 or 1, and our assertion follows from Theorem 2.

Equation (19) has the following probability-theoretic meaning. Let A and B denote two events in $\widetilde{\mathfrak{C}}_\xi$. If the event A is displaced by an infinite amount in time, the events $S^{-1}A$ and B become independent in mean for every event B.

Condition (19) is a special case of the more stringent requirement

$$\lim_{n\to\infty} \mathsf{P}(S^{-n}A \mid B) = \mathsf{P}(A) , \tag{20}$$

which is called the displacement condition.

Let us look, for example, at a sequence $\{\xi_n\}$, for $n = 0, \pm 1, \cdots$, of independent identically distributed random variables and let us assume that $\mathsf{M}\,|\,\xi_n\,| < \infty$. This sequence is a stationary sequence. On the basis of the Birkhoff-Khinchin theorem,

$$\lim_{n\to\infty} \frac{1}{n} \sum_{k=0}^{n-1} \xi_k = \xi^* \pmod{\mathsf{P}} , \qquad \mathsf{M}\xi^* = \mathsf{M}\xi .$$

The random variable ξ^* is obviously independent of any finite number of variables $\xi_0, \xi_1, \cdots, \xi_p$. Therefore ξ^* is measurable with respect to $\overline{\lim}\, \tilde{\sigma}\{\xi_k\}$ and on the basis of the 0-or-1 law (Theorem 5, Section 3), it is constant: $\xi^* = c \pmod{\mathsf{P}}$, where $c = \mathsf{M}\xi$. Thus we obtain:

Corollary 2 (A strengthened form of the law of large numbers). *Let $\{\xi_n\}$, for $n = 0, \pm 1, \cdots$, denote a sequence of independent identically distributed random variables and suppose that $\mathsf{M}\,|\,\xi_k\,| < \infty$. Then with probability* 1,

$$\lim_{n\to\infty} \frac{1}{n} \sum_{k=0}^{n-1} \xi_k = \mathsf{M}\xi_0 . \tag{21}$$

In the present case we can say more than this. Let $f_\varepsilon(\omega)$ and $g_\varepsilon(\omega)$ denote arbitrary $\mathfrak{C}_\xi$-measurable functions that depend on a finite fixed number of coordinates of the point ω. For sufficiently large n, the functions $f_\varepsilon(S^n\omega)$ and $g_\varepsilon(\omega)$ will be functions of two pairs of finite sequences of independent random variables and hence will themselves be independent. Thus

$$\lim_{n\to\infty} \mathsf{M}f_\varepsilon(S^n\omega)g_\varepsilon(\omega) = \mathsf{M}f_\varepsilon(\omega)\mathsf{M}g_\varepsilon(\omega) . \tag{22}$$

Now let f and g denote arbitrary members of $L_2(\Omega, \widetilde{\mathfrak{C}}_\xi, \mathsf{P})$. On the basis of Theorem 7, Section 7, Chapter II, for arbitrary $\varepsilon > 0$ there exist functions $f_\varepsilon(\omega)$ and $g_\varepsilon(\omega)$ depending on finitely many coordinates of the points ω and satisfying the inequalities

$$\mathsf{M}\,|\,f(\omega) - f_\varepsilon(\omega)\,|^2 \leqq \varepsilon^2\,,$$
$$\mathsf{M}\,|\,g(\omega) - g_\varepsilon(\omega)\,|^2 \leqq \varepsilon^2\,.$$

Then

$$|\,\mathsf{M}f(S^n\omega)g(\omega) - \mathsf{M}f(\omega)g(\omega)\,| \leqq |\,\mathsf{M}f_\varepsilon(S^n\omega)g_\varepsilon(\omega) - \mathsf{M}f_\varepsilon(\omega)g_\varepsilon(\omega)\,| + 2\varepsilon(\|g\| + \|f_\varepsilon\|)$$

where $\|f\|^2 = \mathsf{M}\,|\,f(\omega)\,|^2$. Taking (22) into account, we then obtain

$$\lim_{n\to\infty} \mathsf{M}f(S^n\omega)g(\omega) = \mathsf{M}f(\omega)g(\omega) \tag{23}$$

for arbitrary $f, g \in L_2(\Omega, \tilde{\mathfrak{C}}_\xi, \mathsf{P})$. This proves:

Theorem 3. *A sequence of independent identically distributed random variables satisfies the displacement condition. In particular, it is ergodic.*

Another example of a process satisfying the displacement condition is a stationary Gaussian sequence whose covariance approaches 0. Thus let $\{\xi_n\}$, for $n = 0, \pm 1, \pm 2, \cdots$, denote a stationary Gaussian sequence and suppose that $\mathsf{M}\xi_n = m$ and $\mathsf{M}(\xi_n - m)(\xi_0 - m) = R_n$. Suppose that $f_\varepsilon(\omega) = f(a(0), a(1), \cdots, a(p))$ and $g_\varepsilon(\omega) = g(a(0), a(1), \cdots, a(p))$ are bounded sufficiently smooth functions of $p + 1$ variables and have absolutely integrable Fourier transforms $f_\varepsilon^*(\lambda_0, \cdots, \lambda_p)$ and $g_\varepsilon^*(\lambda_0, \cdots, \lambda_p)$. Then,

$$\mathsf{M}f_\varepsilon(S^n\omega)g_\varepsilon(\omega) = \mathsf{M}\int_{-\infty}^{+\infty} \cdots \int_{-\infty}^{+\infty} \exp\left\{i\left(\sum_{k=0}^{p} \lambda_k \xi(n+k) + \sum_{k=0}^{p} \mu_k \xi(k)\right)\right\}$$
$$\times f_\varepsilon^*(\lambda_0, \cdots, \lambda_p) g_\varepsilon^*(\mu_0, \cdots, \mu_p) d\lambda_0 \cdots d\lambda_p d\mu_0 \cdots d\mu_p$$
$$= \int_{-\infty}^{+\infty} \cdots \int_{-\infty}^{+\infty} \exp\left[-\frac{1}{2}\left\{\sum_{k,r=0}^{p} R_{k-r}(\lambda_k\lambda_r + \mu_k\mu_r)\right.\right.$$
$$\left.\left. + 2\sum_{k,r=0}^{p} R_{n+k-r}\lambda_k\mu_r\right\}\right]$$
$$\times f_\varepsilon^*(\lambda_0, \cdots, \lambda_p) g_\varepsilon^*(\mu_0, \cdots, \mu_p) d\lambda_0 \cdots d\lambda_p d\mu_0 \cdots d\mu_p\,.$$

If $\lim_{n\to\infty} R_n = 0$, then by taking the limit as $n \to \infty$ in the preceding equation, we obtain $\lim_{n\to\infty} \mathsf{M}f_\varepsilon(S^n\omega)g_\varepsilon(\omega) = \mathsf{M}f_\varepsilon(\omega)\mathsf{M}g_\varepsilon(\omega)$. By following the same reasoning we used in the proof of Theorem 2, that is, by approximating the arbitrary functions f_ε, we see that equation (23) is also valid for arbitrary functions $f, g \in L_2(\Omega, \tilde{\mathfrak{C}}_\xi, \mathsf{P})$.

Thus we have proved:

Theorem 4. *A stationary Gaussian sequence whose covariance* $R_n = \mathsf{M}(\xi_n - m)\ (\xi_0 - m)$ *(where* $m = \mathsf{M}\xi_n$*) approaches* 0 *as* $n \to \infty$ *satisfies the displacement condition* (23).

6. CONDITIONAL PROBABILITIES AND CONDITIONAL MATHEMATICAL EXPECTATIONS

Let us first recall the definition of conditional probability and conditional mathematical expectation in the elementary case. The conditional probability $\mathsf{P}(A \mid B)$ of an arbitrary event A conditioned by the event B is defined for $\mathsf{P}(B) \neq 0$, by

$$\mathsf{P}(A \mid B) = \frac{\mathsf{P}(A \cap B)}{\mathsf{P}(B)} .$$

For fixed B, the conditional probability $\mathsf{P}(A \mid B)$ is a normed measure defined on the same σ-algebra of sets as is the "unconditional" probability $\mathsf{P}(A)$. The conditional mathematical expectation of a random variable $\xi = f(u)$ conditioned by B is defined by

$$\mathsf{M}\{\xi \mid B\} = \int_U f(u)\mathsf{P}(du \mid B) .$$

Keeping in mind the definition of conditional probability, we can rewrite this equation in the form

$$\mathsf{P}(B)\mathsf{M}\{\xi \mid B\} = \int_B f(u)\mathsf{P}(du) . \tag{1}$$

If we are to be able to define the conditional mathematical expectations and conditional probabilities relative to events with probability 0, we need to modify these concepts somewhat. Let us note first of all that if ξ is the characteristic function of the event A, then $\mathsf{M}\{\xi \mid B\} = \mathsf{P}(A \mid B)$. Thus the conditional probabilities constitute a special case of conditional mathematical expectations, and we can for the moment confine ourselves to these. Let $\mathfrak{M}$ denote a countable class of incompatible events $\{B_i\}$ where $B_i \in \mathfrak{S}$ for $i = 1, 2, \cdots$ and $\bigcup_{i=1}^{\infty} B_i = U$. We define the random variable $\mathsf{M}\{\xi \mid \mathfrak{M}\}$ by setting it equal to $\mathsf{M}\{\xi \mid B_i\}$ if $u \in B_i$ and we call it the conditional mathematical expectation of the random variable ξ with respect to the given class $\mathfrak{M}$ of sets. We are unable to define it only for those values of u that belong to those B_i with zero probability; that is, it is defined with probability 1. On a set $B_i \in \mathfrak{M}$ for which $\mathsf{P}(B_i) \neq 0$, it is constant and equal to the conditional mathematical expectation ξ under the hypothesis B_i. Note that if we know the function $\mathsf{M}\{\xi \mid \mathfrak{M}\}$, we can define not only $\mathsf{M}\{\xi \mid B_i\}$ when $B_i \in \mathfrak{M}$ and $\mathsf{P}(B_i) \neq 0$ but also the conditional mathematical expectation of the random variable with respect to an arbitrary set B belonging to $\sigma\{\mathfrak{M}\}$ and such that $\mathsf{P}(B) \neq 0$. Specifically, if $B = \bigcup_{k=1}^{\infty} B_{j_k}$ then

$$\mathsf{P}(B)\mathsf{M}\{\xi \mid B\} = \sum_{k=1}^{\infty} \mathsf{P}(B_{j_k})\mathsf{M}\{\xi \mid B_{j_k}\} . \tag{2}$$

This formula shows that if we know the conditional mathematical expectations for given B_i (for $i = 1, 2, \cdots$), we can calculate the conditional mathematical expectations for given countable sums of these sets. Hence we can calculate the conditional probability with respect to an arbitrary set in the smallest σ-algebra containing all the B_i. Therefore it is meaningful to speak of the conditional mathematical expectations with regard to some σ-algebra of subsets. We note that equation (2) can be rewritten

$$\int_B \mathsf{M}\{\xi \mid \mathfrak{M}\}\mathsf{P}(du) = \int_B f(u)\mathsf{P}(du) , \qquad B \in \mathfrak{M} .$$

This formula is obtained on the basis of the definition of the conditional mathematical expectation with respect to an arbitrary σ-algebra of events.

Definition 1. Let $\mathfrak{B}$ denote an arbitrary σ-algebra of events that is contained in $\mathfrak{S}$ and let $\xi = f(u)$ denote an arbitrary random variable with finite mathematical expectation. Suppose that the random variable $\mathsf{M}(\xi \mid \mathfrak{B})$ is measurable with respect to $\mathfrak{B}$ and satisfies the equation

$$\int_B \mathsf{M}\{\xi \mid \mathfrak{B}\}\mathsf{P}(du) = \int_B f(u)\mathsf{P}(du) \tag{3}$$

for arbitrary $B \in \mathfrak{B}$. Then $\mathsf{M}\{\xi \mid \mathfrak{B}\}$ is called the *conditional mathematical expectation* of the random variable ξ with respect to the σ-algebra $\mathfrak{B}$.

The existence and uniqueness (up to equivalence) of the random variable $\mathsf{M}\{\xi \mid \mathfrak{B}\}$ follows immediately from the Radon-Nikodym theorem (Chapter II, Section 6). Specifically, the right-hand member of formula (3) is a finite countably additive set function defined on $\mathfrak{B}$ and absolutely continuous with respect to the measure P. Therefore there exists a $\mathfrak{B}$-measurable function $\psi(u)$ such that

$$\int_B f(u)\mathsf{P}(du) = \int_B \psi(u)\mathsf{P}(du) .$$

Here the function $\psi(u)$ is unique (up to equivalence). It is the conditional mathematical expectation of the random variable $\xi = f(u)$ with respect to the σ-algebra $\mathfrak{B}$, by definition.

REMARK 1. Let $\tilde{\mathfrak{B}}$ denote the completion of $\mathfrak{B}$ with respect to the probability P. One can easily show that

$$\mathsf{M}\{\xi \mid \mathfrak{B}\} = \mathsf{M}\{\xi \mid \tilde{\mathfrak{B}}\} \pmod{\mathsf{P}} .$$

Since the class of $\tilde{\mathfrak{B}}$-measurable functions is broader than the class of $\mathfrak{B}$-measurable functions, it is sometimes expedient to consider the conditional mathematical expectation with respect to completed σ-algebras.

It follows from the definition that

$$\text{if} \quad \xi \geqq 0, \quad \text{then} \quad \mathsf{M}\{\xi \mid \mathfrak{B}\} \geqq 0 \quad (\text{mod } \mathsf{P}) , \tag{4}$$

$$\mathsf{M}\mathsf{M}\{\xi \mid \mathfrak{B}\} = \mathsf{M}\xi , \tag{5}$$

if $\xi = \alpha\xi_1 + \beta\xi_2$ and $\mathsf{M}\xi_i (i = 1, 2)$ are finite, then

$$\mathsf{M}\{\xi \mid \mathfrak{B}\} = \alpha\mathsf{M}\{\xi_1 \mid \mathfrak{B}\} + \beta\mathsf{M}\{\xi_2 \mid \mathfrak{B}\} \; (\text{mod } \mathsf{P}) . \tag{6}$$

We define the conditional probabilities $\mathsf{P}\{A \mid \mathfrak{B}\}$ with respect to the σ-algebra $\mathfrak{B}$ as the special case of conditional mathematical expectations obtained by setting $\xi = \chi_A(u)$. In other words, for fixed A the conditional probability $\mathsf{P}\{A \mid \mathfrak{B}\}$ is a $\mathfrak{B}$-measurable random variable satisfying the equation

$$\int_B \mathsf{P}\{A \mid \mathfrak{B}\} \mathsf{P}(du) = \mathsf{P}(A \cap B) \tag{7}$$

for arbitrary $B \in \mathfrak{B}$. We now define the conditional mathematical expectations and the conditional probabilities with respect to a random variable. Let V denote a measurable space with σ-algebra $\mathfrak{B}$ and let $v = g(u)$ denote a measurable mapping of U into V. The conditional mathematical expectation (resp. conditional probability) with respect to the function $v = g(u)$ is defined as the conditional mathematical expectation (resp. conditional probability) with respect to the σ-algebra $g^{-1}\mathfrak{B}$. These are denoted by $\mathsf{M}\{\xi \mid v\}$ and $\mathsf{P}\{\xi < x \mid v\}$ respectively.

Theorem 1. *The conditional mathematical expectation with respect to the function $v = g(u)$ is a function of v:*

$$\mathsf{M}\{\xi \mid v) = \psi(g(u)) ,$$

where $\psi(v)$ is $\mathfrak{B}$-measurable.

Proof. Suppose that $\xi = f(u)$ has a finite mathematical expectation. Let B denote an arbitrary set belonging to $g^{-1}\mathfrak{B}$. Then

$$\int_B f(u) \mathsf{P}(du) = \varphi(B)$$

is a finite countably additive function on $g^{-1}\mathfrak{B}$. We have $\varphi(B) = \varphi g^{-1}(B')$, where $B' (\in \mathfrak{B})$ is the image of B. The function φg^{-1} is absolutely continuous with respect to the measure $\mathsf{P}' = \mathsf{P}g^{-1}$. On the basis of the Radon-Nikodym theorem there exists a $\mathfrak{B}$-measurable function $\psi(v)$ such that

$$\varphi(B) = \varphi g^{-1}(B') = \int_{B'} \psi(v)\mathsf{P}'(dv) .$$

If we now apply the rule for change of variables (Theorem 3, Section 6, Chapter II), we obtain

$$\int_B f(u)\mathsf{P}(du) = \int_B \psi[g(u)]\mathsf{P}(du) ,$$

from which it follows that $\mathsf{M}\{\xi \mid v\} = \psi[g(u)] = \psi(v)$.

This theorem shows that the conditional mathematical expectation with respect to a function $v = g(u)$ can be regarded as a function of the independent variable v in the space (with measure) $\{V, \mathfrak{B}, \mathsf{P}g^{-1}\}$, whereas in the original definition the conditional mathematical expectation is regarded as a function of the variable u. Accordingly, the conditional mathematical expectation can be defined by means of the formula

$$\int_{g^{-1}(B')} f(u)\mathsf{P}(du) = \int_{B'} \psi(v)\mathsf{P}g^{-1}(dv) \tag{8}$$

for arbitrary $B' \in \mathfrak{B}$, where $\psi(v) = \mathsf{M}\{\xi \mid v\}$.

Let us now look at certain properties of conditional mathematical expectations. It follows immediately from the definition of conditional mathematical expectation that if a random variable $\xi = f(u)$ is $\mathfrak{B}$-measurable, its conditional mathematical expectation with respect to $\mathfrak{B}$ coincides with ξ itself: $\mathsf{M}\{\xi \mid \mathfrak{B}\} = \xi$. From this it follows that

$$\text{if } \mathfrak{B} \subset \mathfrak{B}', \text{ then } \mathsf{M}\{\mathsf{M}(\xi \mid \mathfrak{B}) \mid \mathfrak{B}'\} = \mathsf{M}\{\xi \mid \mathfrak{B}\} . \tag{9}$$

Theorem 2. *Let ξ and η denote random variables having finite mathematical expectations. If ξ is $\mathfrak{B}$-measurable, then*

$$\mathsf{M}\{\xi\eta \mid \mathfrak{B}\} = \xi\mathsf{M}\{\eta \mid \mathfrak{B}\} . \tag{10}$$

Proof. Since, for every set $B \in \mathfrak{B}$,

$$\int_B \mathsf{M}\{\eta \mid \mathfrak{B}\}\mathsf{P}(du) = \int_B g(u)\mathsf{P}(du) , \qquad \eta = g(u) ,$$

it follows that the equation

$$\int_B f(u)\mathsf{M}\{\eta \mid \mathfrak{B}\}\mathsf{P}(du) = \int_B f(u)g(u)\mathsf{P}(du) \tag{11}$$

is satisfied for every function $f(u)$ that is simple with respect to $\mathfrak{B}$. Taking the limit in equation (11) with respect to a nondecreasing sequence of $\mathfrak{B}$-measurable functions, we see that (11) is valid for an arbitrary $\mathfrak{B}$-measurable nonnegative function. It follows from (6) that the theorem holds for arbitrary integrable and $\mathfrak{B}$-measurable functions.

Theorem 3. *If $\mathfrak{B} \subset \mathfrak{B}'$, and $M\xi$ is finite, then*

$$\mathsf{M}\{\mathsf{M}\{\xi \mid \mathfrak{B}'\} \mid \mathfrak{B}\} = \mathsf{M}\{\xi \mid \mathfrak{B}\} .$$

Proof. If $B \in \mathfrak{B}$, then $B \in \mathfrak{B}'$, and consequently

$$\int_B \mathsf{M}\{\mathsf{M}\{\xi \mid \mathfrak{B}') \mid \mathfrak{B}\}\mathsf{P}(du) = \int_B \mathsf{M}\{\xi \mid \mathfrak{B}'\}\mathsf{P}(du)$$

$$= \int_B f(u)\mathsf{P}(du) = \int_B \mathsf{M}\{\xi \mid \mathfrak{B}\}\mathsf{P}(du) .$$

In view of the extreme members of this chain of equalities, we see that

$$\mathsf{M}\{\mathsf{M}(\xi \mid \mathfrak{B}') \mid \mathfrak{B}\} = \mathsf{M}\{\xi \mid \mathfrak{B}\} .$$

Therefore, we have:

Corollary. *If $\mathfrak{B} \subset \mathfrak{B}'$ and η is a $\mathfrak{B}'$-measurable random variable, and ξ and η have finite expectations, then*

$$\mathsf{M}\{\xi\eta \mid \mathfrak{B}\} = \mathsf{M}\{\eta\mathsf{M}(\xi \mid \mathfrak{B}') \mid \mathfrak{B}\} . \tag{12}$$

The next theorem corresponds to the elementary fact that if A and B are independent events, the conditional probability of an event A with respect to B coincides with its unconditional probability. Let $\sigma\{\xi\}$ denote the σ-algebra of events that is generated by the random variable ξ.

Theorem 4. *If the σ-algebras $\mathfrak{B}$ and $\sigma\{\xi\}$ are independent and if $\mathsf{M}\xi$ is finite, then*

$$\mathsf{M}\{\xi \mid \mathfrak{B}\} = \mathsf{M}\xi \ (\text{mod } \mathsf{P}) . \tag{13}$$

Proof. For arbitrary $B \in \mathfrak{B}$, the random variables ξ and χ_B, where χ_B is the characteristic function of the set B, are independent. Therefore

$$\int_B f(u)\mathsf{P}du = \mathsf{M}\chi_B\xi = \mathsf{M}\chi_B\mathsf{M}\xi = \int_B (\mathsf{M}\xi)\mathsf{P}(du) ,$$

and since $\mathsf{M}\xi = \text{const.}$ is $\mathfrak{B}$-measurable, we may take $\mathsf{M}\xi$ as $\mathsf{M}\{\xi \mid \mathfrak{B}\}$.

Corollary. *If a random variable ξ and a random element η are independent and if $\mathsf{M}\xi$ is finite, then*

$$\mathsf{M}\{\xi \mid \eta\} = \mathsf{M}\xi . \tag{14}$$

Theorem 5. *If $\xi_n = f_n(u) \geqq 0$ and $\xi_n \leqq \xi_{n+1}$, then*

$$\mathsf{M}\{\eta \mid \mathfrak{B}\} = \lim_{n\to\infty} \mathsf{M}\{\xi_n \mid \mathfrak{B}\} \ (\text{mod } \mathsf{P}) \tag{15}$$

where $\eta = \lim_{n\to\infty} \xi_n$.

In the equations

$$\int_B \mathsf{M}\{\xi_n \mid \mathfrak{B}\}\mathsf{P}(du) = \int_B f_n(u)\mathsf{P}(du)$$

where $B \in \mathfrak{B}$, let us take the limit as $n \to \infty$, keeping in mind (4), (6) and Lebesgue's theorem (Chapter II). We obtain

$$\int_B \lim_{n\to\infty} \mathsf{M}\{\xi_n \mid \mathfrak{B}\}\mathsf{P}(du) = \int_B \lim_{n\to\infty} f_n(u)\mathsf{P}(du) .$$

Since the random variable $\lim_{n\to\infty} \mathsf{M}\{\xi_n \mid \mathfrak{B}\}$ is $\mathfrak{B}$-measurable, we can take it as the conditional mathematical expectation of the variable $\eta = \lim_{n\to\infty} f_n(u)$.

Let us now look at conditional probabilities. From the properties of conditional mathematical expectations that we have proved, we get:

Theorem 6. *With probability* 1,

$$\mathsf{P}\{A \mid \mathfrak{B}\} \geqq 0 , \qquad A \in \mathfrak{S} ; \tag{16}$$

$$\mathsf{P}\{U \mid \mathfrak{B}\} = 1 ; \tag{17}$$

if $A = \bigcup_{k=1}^{\infty} A_k$, *where* $A_k \cap A_r = \varnothing$ *for* $k \neq r$, $A_k \in \mathfrak{S}$, *then*

$$\mathsf{P}\{A \mid \mathfrak{B}\} = \sum_{k=1}^{\infty} \mathsf{P}\{A_k \mid \mathfrak{B}\} ; \tag{18}$$

if $\{A_n\}$, *for* $n = 1, 2, \cdots$, *is a decreasing or increasing sequence of events* $A_k \in \mathfrak{S}$ *and* $B = \bigcap_{k=1}^{\infty} A_k$ *(or* $B = \bigcup_{k=1}^{\infty} A_k$*), then*

$$\mathsf{P}\{B \mid \mathfrak{B}\} = \lim_{n\to\infty} \mathsf{P}(A_k \mid \mathfrak{B}) . \tag{19}$$

We recall that the conditional probabilities are defined only up to equivalence. Therefore relations (16) to (19) are only P-almost-certain. But can we regard the conditional probabilities $\mathsf{P}\{A \mid \mathfrak{B}\}$ as measures? In other words, the conditional probability $\mathsf{P}\{A \mid \mathfrak{B}\} = \varphi(A, u)$ is a function of $u \in U$ and $A \in \mathfrak{S}$ that is defined for every fixed A only up to a set of probability 0; we now ask whether there exists a function $\psi(A, u)$ (for $u \in U$ and $A \in \mathfrak{S}$) such that

a. $\varphi(A, u) = \psi(A, u)$ almost certainly for arbitrary fixed A and

b. on fixed u, the function $\psi(A, u)$ is a probability on the σ-algebra $\mathfrak{S}$.

We find that such a function $\psi(A, u)$ does not always exist.

Definition 2. Suppose that a function $\psi(A, u)$ satisfying requirements (a) and (b) does exist. The family of conditional probabilities $\mathsf{P}\{A \mid \mathfrak{B}\}$ is then said to be *regular*. In this case we identify $\mathsf{P}\{A \mid \mathfrak{B}\}$ with $\psi(A, u)$.

In the regular case the conditional mathematical expectations are, as we would expect, expressed by integrals with conditional probabilities used for measures.

Theorem 7. *If* $\mathsf{P}\{A \mid \mathfrak{B}\}$ *is a regular conditional probability and* $\xi = f(u)$ *has finite expectation, then*

$$\mathsf{M}\{\xi \mid \mathfrak{B}\} = \int f(u)\mathsf{P}(du \mid \mathfrak{B}) \pmod{\mathsf{P}} . \tag{20}$$

Proof of this assertion presents no difficulty. It is true by definition in the case in which ξ is the characteristic function of some event $A \in \mathfrak{S}$. It follows from the linearity of the two sides of equation (20) as functionals of f that the assertion holds for simple functions. If we take the limit with respect to an increasing sequence of simple functions (Theorem 5), we see that (20) holds for an arbitrary nonnegative random variable ξ. Repeated use of the linearity of the two sides of equation (20) completes the proof.

In certain cases we need to emphasize that the conditional probability is a function of an elementary event. In such a case we shall write $\mathsf{P}\{A \mid \mathfrak{B}\} = \mathsf{P}_{\mathfrak{B}}(u, A)$ or simply $\mathsf{P}(u, A)$ if the σ-algebra $\mathfrak{B}$ is fixed. Analogously, $\mathsf{P}_\xi(u, A)$ is another notation for $\mathsf{P}\{A \mid \xi\}$. If $\mathfrak{B}$ and $\mathfrak{B}'(\subset \mathfrak{B})$ are two σ-algebras contained in $\mathfrak{S}$ and if $\mathsf{P}\{A \mid \mathfrak{B}\}$ and $\mathsf{P}\{A \mid \mathfrak{B}'\}$ are regular, then equation (12) can be rewritten

$$\int f(v)g(v)\mathsf{P}_{\mathfrak{B}}(u, dv)$$
$$= \int\left(\int f(v')\mathsf{P}_{\mathfrak{B}'}(v, dv')\right)g(v)\mathsf{P}_{\mathfrak{B}}(u, dv) , \tag{21}$$

where $g(v)$ is $\mathfrak{B}'$-measurable, $f(v)$ is $\mathfrak{S}$-measurable, and both are P-integrable.

Since the property of regularity of conditional probabilities does not always obtain, it is convenient to extend this property.

Let us now introduce the concept of conditional distribution of a family of random variables. Let $\{\xi\}$ denote a family of random variables and let $\mathfrak{S}_{\{\xi\}}$ denote the σ-algebra of sets generated by the family $\{\xi\}$. Since $\mathfrak{S}_{\{\xi\}} \subset \mathfrak{S}$, it follows from the regularity of the conditional probability with respect to some σ-algebra $\mathfrak{B}$ in the probability space $\{U, \mathfrak{S}, \mathsf{P}\}$ that this conditional probability is regular in $\{U, \mathfrak{S}_{\{\xi\}}, \mathsf{P}\}$.

Definition 3. A function $\mathsf{P}_{\mathfrak{B}}\{u, A\}$ is called the *conditional distribution* of the family of random variables $\{\xi\}$ if

a. for fixed $A \in \mathfrak{S}_{\{\xi\}}$, it is $\mathfrak{B}$-measurable as a function of u and for fixed u it is a probability on $\mathfrak{S}_{\{\xi\}}$;

b. for arbitrary $A \in \mathfrak{S}_{\{\xi\}}$ and $B \in \mathfrak{B}$,

$$\int_B \mathsf{P}_{\mathfrak{B}}(u, A)\mathsf{P}(du) = \mathsf{P}(A \cap B) ;$$

Thus, the conditional distribution of a family of random variables is a regular representation (if it exists) of conditional probabilities of the probability space $\{U, \mathfrak{S}_{\{\xi\}}, \mathsf{P}\}$ with respect to the σ-algebra $\mathfrak{B}$.

Let $\xi_1, \xi_2, \cdots, \xi_n$ denote a sequence of random variables for which there exists a conditional distribution with respect to $\mathfrak{B}$ and let S denote an arbitrary Borel set in the n-dimensional space E_n. We define

$$Q_{\mathfrak{B}}(u, S) = \mathsf{P}_{\mathfrak{B}}\{u; [\xi_1, \xi_2, \cdots, \xi_n] \in S\} . \tag{22}$$

Then $Q_{\mathfrak{B}}(u, S)$ is a $\mathfrak{B}$-measurable function of u and is a measure on the σ-algebra of Borel sets S for fixed u. We note that if the function $Q_{\mathfrak{B}}(u, S)$ is given, it still does not define the conditional distribution $\mathsf{P}_{\mathfrak{B}}(u, A)$ since it is possible that two distinct subsets S_1 and S_2 of E_n correspond to a single set $A \in \mathfrak{S}_{\{\xi\}}$, that is,

$$A = \{u; [\xi_1, \xi_2, \cdots, \xi_n] \in S_i\} , \qquad i = 1, 2 .$$

Let $Q_{\mathfrak{B}}(u, S)$ denote a function that is $\mathfrak{B}$-measurable on U that is a probability on the σ-algebra of Borel subsets of the space E_n for fixed u, and that satisfies equation (22) for P-almost-all u with fixed S. We shall call this function the *conditional distribution* of a sequence of random variables $\xi_1, \xi_2, \cdots, \xi_n$ with respect to the σ-algebra $\mathfrak{B}$ *in the broad sense*. It is easy to prove the validity of:

Theorem 8. *Let $g(x_1, x_2, \cdots, x_n)$ denote a Borel function defined on a subset of E_n. Suppose that its mathematical expectation $\mathsf{M}g(\xi_1, \cdots, \xi_n)$ is finite and that the sequence $\{\xi_i\}$ of random variables ξ_i, for $i = 1, 2, \cdots, n$, has conditional distribution $Q_{\mathfrak{B}}(u, S)$ with respect to $\mathfrak{B}$ in the broad sense. Then*

$$\mathsf{M}\{g(\xi_1, \cdots, \xi_n) \mid \mathfrak{B}\} = \int_R g(x_1, x_2, \cdots, x_n) Q_{\mathfrak{B}}(u, dx)$$

P-*almost-everywhere in U.*

Theorem 9. *For an arbitrary n-tuple of random variables $\{\xi_i\}$, for $i = 1, 2, \cdots, n$, and an arbitrary σ-algebra $\mathfrak{B}$, there exists a conditional distribution $Q_{\mathfrak{B}}(u, S)$ with respect to $\mathfrak{B}$ in the broad sense.*

Let I_y, where $y = (y_1, y_2, \cdots, y_n)$, denote the set of points $x = (x_1, \cdots, x_n)$ contained in E_n and satisfying the inequalities $x_1 < y_1, x_2 < y_2, \cdots, x_n < y_n$. Let r denote a point in E_n with rational coordinates. We set

$$F(r) = Q_{\mathfrak{B}}(u, I_r) . \tag{23}$$

The set N of points u, such that $Q(u, I_r)$ does not satisfy, for arbitrary rational r, at least one of the relations $F(-\infty) = 0, F(+\infty) = 1$, $0 \leqq F(\boldsymbol{r}) \leqq F(\boldsymbol{r}') \leqq 1$ for $\boldsymbol{r} \leqq \boldsymbol{r}'$;

$$F(\boldsymbol{r}') \longrightarrow F(\boldsymbol{r}), \boldsymbol{r}' \leqq \boldsymbol{r}, \boldsymbol{r}' \longrightarrow \boldsymbol{r} ,$$

is a set of $\mathfrak{B}$-measure zero.

For arbitrary $\boldsymbol{x}$, we set

$$F(\boldsymbol{x}) = Q_{\mathfrak{B}}(u, I_x) = \lim_{r \to x, r < x} F(\boldsymbol{r}) , \qquad \text{if } u \notin N ,$$

$$F(\boldsymbol{x}) = Q_{\mathfrak{B}}(u, I_x) = F_0(\boldsymbol{x}) , \qquad \text{if } u \in N ,$$

where $F_0(\boldsymbol{x})$ is a fixed n-dimensional distribution function. The function $F(\boldsymbol{x})$ is for $u \notin N$ a distribution function, and it defines the probability measure $Q_{\mathfrak{B}}(u, S)$ unambiguously in E_n on the σ-algebra of Borel sets for $u \notin N$. Furthermore, the equation

$$Q_{\mathfrak{B}}(u, S) = \mathsf{P}_{\mathfrak{B}}\{u; [\xi_1, \cdots, \xi_n] \in S\}$$

which is valid for $S = I_r$ (equation (23)), remains valid for every set S in the minimal σ-algebra containing all the I_r (that is, for all Borel S) and for all u except possibly a set of points $N'(S)$ of probability 0. Furthermore, the function $Q_{\mathfrak{B}}(u, S)$ as a function of u is $\mathfrak{B}$-measurable by construction. This completes the proof of the theorem.

REMARK 2. The definition of conditional distribution in the broad sense can be applied to an arbitrary mapping $\xi = g(u)$ of a probability space into an arbitrary space X. Here Theorem 9 is generalized to the case of an arbitrary separable metric space. To prove the theorem, we take not the sets I_r but spheres with rational radii and centers belonging to a countable everywhere-dense set. Everything else remains unchanged.

Let $\xi_1, \xi_2, \cdots, \xi_n$ denote a sequence of random variables. Define $\mathfrak{B}_k = \sigma\{\xi_1, \xi_2, \cdots, \xi_k\}$. Just as in Theorem 1, we can show that the conditional probability (in the broad sense) $Q_{\mathfrak{B}_k}(u, S)$ can be regarded as a function of $\xi_1, \cdots, \xi_k$:

$$Q_{\mathfrak{B}_k}(u, S) = Q_k(\xi_1, \cdots, \xi_k; S) .$$

The function $Q_k(\xi_1, \cdots, \xi_k; S)$ is determined by the equation

$$\mathsf{M}Q_k(\xi_1, \cdots, \xi_k; S)\chi_{B_k}(\xi_1, \cdots, \xi_k) = \mathsf{P}\{S \cap B_k\} , \tag{24}$$

where B_k is the cylindrical set in E_n defined by $\{\xi_1, \cdots, \xi_k\} \in B_k^{(0)}$, $B_k^{(0)}$ being a Borel k-dimensional set.

It follows from equation (24) that

$$Q_k(\xi_1, \cdots, \xi_k; S) = Q_k(\xi_1, \cdots, \xi_k; S \cap B'_k)$$

almost certainly for arbitrary B'_k having the same structure as B_k. This means that

$$Q_k(\xi_1, \cdots, \xi_k; S) = Q_k(\xi_1, \cdots, \xi_k; R^k \times S_{\xi_1,\ldots,\xi_k}) \pmod{\mathsf{P}}$$

where $S_{\xi_1,\ldots,\xi_k}$ is the $(\xi_1, \cdots, \xi_k)$-section of the set S. Thus we may set

$$Q_k(\xi_1, \cdots, \xi_k; S) = \mathsf{P}(\xi_1, \cdots, \xi_k; S_{\xi_1,\ldots,\xi_k})$$

where $\mathsf{P}(x_1, x_2, \cdots, x_k; C)$ is a Borel function of $(x_1, x_2, \cdots, x_k)$ for fixed C and it is a probability in $(n-k)$-dimensional space for fixed $x_1, x_2, \cdots, x_k$.

By successive application of Theorem 8, we obtain the following equations for an arbitrary Borel function $g(x_1, x_2, \cdots, x_n)$ such that $\mathsf{M}g(\xi_1, \cdots, \xi_n)$ exists:

$$\mathsf{M}g(\xi_1, \cdots, \xi_n) = \int \mathsf{P}(dx_1) \int \mathsf{P}(x_1; dx_2) \cdots \int \mathsf{P}(x_1, \cdots, x_{n-1}; dx_n) g(x_1, \cdots, x_n) \,. \tag{25}$$

and almost certainly,

$$M\{g(\xi_1, \cdots, \xi_n) \mid \xi_1\} = \int \mathsf{P}(\xi_1; dx_2) \int \mathsf{P}(\xi_1, x_2; dx_3) \cdots \int \mathsf{P}(\xi_1, x_2, \cdots, x_{n-1}; dx_n) g(x_1, x_2, \cdots, x_n) \,. \tag{26}$$

IV

RANDOM FUNCTIONS

1. DEFINITION OF A RANDOM FUNCTION

In Chapter I a random function was defined as a family of random variables depending on a parameter. There we pointed out the difficulties associated with this concept in the broad sense, that is, as a set of finite-dimensional distribution functions satisfying compatibility conditions. The axiomatization of probability theory immediately suggests that it is natural to regard a random function as an arbitrary family of random variables, all defined on the same probability space.

Definition 1. Let $\{U, \mathfrak{S}, \mathsf{P}\}$ denote a probability space. Let $g(\theta, u) = \xi(\theta)$ denote a function of two variables defined on $\Theta \times U$ into a metric space X. If g is $\mathfrak{S}$-measurable as a function of u for every $\theta \in \Theta$, it is called a *random function.* The set Θ is called the *domain of definition* of the random function and X is called its *range*.

Let us look at a particular case of the general definition. Let U denote a functional space. Suppose that $u = u(\theta)$ for $\theta \in \Theta$ and suppose that the σ-algebra $\mathfrak{S}$ contains all sets of the space U of the form $\{u; u(\theta_0) \in A\}$ for every $\theta_0 \in \Theta$ and every Borel set $A \subset X$. Let P denote an arbitrary probability measure on $\mathfrak{S}$. It is natural to put the random variable $g(\theta, u) = u(\theta)$ in correspondence with such a probability space. In some problems it is convenient to identify the random function $g(\theta, u) = u(\theta)$ with the probability space $\{U, \mathfrak{S}, \mathsf{P}\}$ of the type described.

It is easy to see that the general definition of a random function can be reduced to the particular case just described. Specifically, if a random function $\xi(\theta)$ is defined as a function of two variables, that is, $\xi(\theta) = g(\theta, u)$, then by setting $u' = g(\theta, u)$ where u is a fixed member of U and letting U' denote the set of all functions $u' = g(\theta, u)$ obtained by letting u range over U, we obtain a mapping T of the set U onto U'. Here the σ-algebra $\mathfrak{S}$ of sets U

is mapped into some σ-algebra $\mathfrak{S}'$ of sets U', and the probability measure P on $\mathfrak{S}$ is mapped into a probability measure P' on $\mathfrak{S}'$ (cf. Section 6, Chapter II). For arbitrary fixed θ, the set $\{u'; u' = g(\theta, u) \in A$, where A is a Borel set in $x\}$ belongs to $\mathfrak{S}'$ since

$$T^{-1}\{u'; u' = g(\theta, u) \in A, \text{where } A \text{ is a Borel set in } x\}$$
$$= \{u; g(\theta, u) \in A, \text{ where } A \text{ is a Borel set in } x\} \in \mathfrak{S}\,.$$

Thus we have obtained a probability space $\{U', \mathfrak{S}', \mathsf{P}'\}$ where U' is a set of functions $u = u'(\theta)$. Here, for arbitrary $n, \theta_1, \theta_2, \cdots, \theta_n$ (where $\theta_k \in \Theta$ for $k = 1, \cdots, n$), the distribution of the sequence of random variables on $\{U, \mathfrak{S}, \mathsf{P}\}$

$$g(\theta_1, u), g(\theta_2, u), \cdots, g(\theta_n, u)$$

coincides with the distribution of the sequence

$$u'(\theta_1), u'(\theta_2), \cdots, u'(\theta_n)$$

of random variables defined on the probability space $\{U', \mathfrak{S}', \mathsf{P}'\}$.

It is important to note certain facts regarding the equivalence of random functions. In the solution of many problems we have no basis for distinguishing between random functions that are obtained from each other by transformations of probability space. We can go even farther in this direction. From a practical point of view, experiment enables us to distinguish only between hypotheses dealing with finite-dimensional distributions of a random function. Therefore, it is convenient to assume that experimental data do not enable us to distinguish between two random functions $\xi(\theta)$ and $\xi'(\theta)$ whose finite-dimensional distributions coincide, that is, the joint distributions of the sequences

$$\xi(\theta_1), \xi(\theta_2), \cdots, \xi(\theta_n) \tag{1}$$

and

$$\xi'(\theta_1), \xi'(\theta_2), \cdots, \xi'(\theta_n) \tag{2}$$

for arbitrary $n = 1, 2, \cdots$ and $\theta_k \in \Theta$ for $k = 1, \cdots, n$, coincide. Accordingly we take the following.

Definition 2. Two random functions $\xi(\theta)$ and $\xi'(\theta)$ defined on the same set Θ are said to be *stochastically equivalent* in the broad sense if for arbitrary $n = 1, 2, \cdots$ and $\theta_k \in \Theta$, for $k = 1, 2, \cdots, n$, the joint distributions of the sequences (1) and (2) of random variables coincide.

In what follows we shall often use the concept of stochastic equivalence of random functions in the narrow sense:

Definition 3. Two random functions $g_1(\theta, u)$ and $g_2(\theta, u)$, for $\theta \in \Theta$ and $u \in U$, defined on the same probability space $\{U, \mathfrak{S}, \mathsf{P}\}$

are said to be *stochastically equivalent* if for arbitrary $\theta \in \Theta$,

$$\mathsf{P}\{g_1(\theta, u) \neq g_2(\theta, u)\} = 0 .$$

Obviously if $g_1(\theta, u)$ and $g_2(\theta, u)$ are stochastically equivalent (in this narrow sense), they are stochastically equivalent in the broad sense.

Let us look at a few examples of random functions.

a. The random oscillation $\xi(t)$ for $-\infty < t < \infty$ (considered in Section 4 of Chapter I),

$$\xi(t) = \sum_{k=1}^{n} \gamma_k e^{i\lambda_k t}, \qquad \gamma_k = \alpha_k + i\beta_k ,$$

where α_k and β_k (for $k = 1, 2, \cdots, n$) are random variables, can be represented in the form $\xi(t) = g(t, \boldsymbol{u})$, where $\boldsymbol{u} = (\alpha_1, \alpha_2, \cdots, \alpha_n, \beta_1, \beta_2, \cdots, \beta_n)$ is a point in $2n$-dimensional Euclidean space U and $g(t, \boldsymbol{u})$ is for fixed t a linear function of $\boldsymbol{u}$ and for fixed $\boldsymbol{u}$ is the sum of trigonometric functions of t. The probability P in U is defined by the common distribution of the random variables $\alpha_1, \alpha_2, \cdots, \alpha_n, \beta_1, \beta_2, \cdots, \beta_n$. On the other hand, the process $\xi(t)$ can be regarded as a probability space $\{U', \mathfrak{S}', \mathsf{P}'\}$ where U' is the space of all complex-valued functions of the form $u = \sum_{k=1}^{n} \gamma_k e^{i\lambda_k t}$ with given bases of exponents $(\lambda_1, \lambda_2, \cdots, \lambda_n)$.

The measure P' in U' is induced by the measure P in U by means of the mapping $u \longrightarrow u' = g(t, \theta)$.

b. Consider a process $\xi(t)$ whose sample functions are constant on the time intervals $[k - 1, k)$ and assume on these intervals the values 0 and 1 with equal probabilities independent of the value of $\xi(t)$ on the preceding time intervals.

We can set up a one-to-one correspondence between the sample functions of the process $\xi(t)$ and the nonterminating binary fractions in accordance with the scheme

$$\xi(t) \longrightarrow u = 0.\, x_1 x_2 \cdots x_n \cdots , \tag{3}$$

where x_n is the value of $\xi(t)$ on the time interval $[n - 1, n)$.

To each nonterminating binary fraction u there corresponds some point in the interval $[0, 1]$. The one-to-one nature of this correspondence is violated only for fractions, all binary digits of which from some point on are the same. If a point in this interval has a representation with all 0's from some digit on, it also has a representation with all 1's and vice versa. (Exceptions are the end points 0 and 1 which have only the representations $0 = 0.00 \cdots 0 \cdots$ and $1 = 0.11 \cdots 1 \cdots$.) The sample functions corresponding to such nonterminating binary fractions become constant from some instant of time onward. Let $\mathfrak{A}$ denote this event or set of events and let $\mathfrak{A}_n$ denote the event that the random function $\xi(t)$ will be constant beginning at the instant t and thereafter. Then

$$\mathfrak{A} = \bigcup_{n=1}^{\infty} \mathfrak{A}_n .$$

Also, let $\mathfrak{A}_{nm}$ mean that the function $\xi(t)$ is constant on the time interval $[n, n + m)$. The events $\mathfrak{A}_{nm}$ (for $m = 1, 2, \cdots$) constitute a decreasing sequence $\mathfrak{A}_n = \bigcap_{m=1}^{\infty} \mathfrak{A}_{nm}$, and

$$\mathsf{P}(\mathfrak{A}_n) = \lim_{m\to\infty} \mathsf{P}(\mathfrak{A}_{nm}) .$$

Since $\mathsf{P}(\mathfrak{A}_{nm}) = 2^{1-m}$, it follows that $\mathsf{P}(\mathfrak{A}_n) = 0$ and $\mathsf{P}(\mathfrak{A}) = 0$. Consequently, if we neglect the random functions $\xi(t)$ in the set $\mathfrak{A}$, which is of probability 0, there exists a one-to-one correspondence between points of the interval $[0, 1]$ and the random functions $\xi(t)$.

Let Δ denote an interval of length 2^{-n} with dyadic endpoints. Since a dyadic fraction always has a terminating binary representation, Δ is either of the form

$$[0.j_1 \cdots j_{n-1}0, \qquad 0.j_1 \cdots j_{n-1}1)$$

or of the form

$$[0.j_1 \cdots 011 \cdots 1, \qquad 0.j_1 \cdots 100 \cdots 0) ,$$

where each j_k is either 0 or 1.

The set $\mathfrak{A}$ of random functions $\xi(t)$ corresponding to the interval Δ consists of functions satisfying the conditions

$$\xi(0) = j_1, \qquad \xi(\tau) = j_2, \cdots, \xi((n - 1)\tau) = j_n .$$

The probability that $\xi(t) \in \mathfrak{A}$ is equal to 2^{-n}; that is, $\mathsf{P}(\mathfrak{A}) = 2^{-n}$, which coincides with the length of the interval Δ. It follows from this that if A is an arbitrary Borel set of points belonging to the interval $[0, 1]$ and $\mathfrak{A}$ is a set of random functions $\xi(t)$ corresponding on the basis of (3) to the numerical set A, then $\mathsf{P}(\mathfrak{A})$ coincides with the Lebesgue measure of A. Thus, the choice of random function $\xi(t)$ is equivalent to the random choice of a point u belonging to the interval $[0, 1]$ with given Lebesgue measure on it. More precisely this means that the random process is completely described as follows: A point in the interval $[0, 1]$ is chosen at random. Here the probability that the point u belongs to A, where A is a Borel subset of the interval $[0, 1]$, is equal to the Lebesgue measure of A. The coordinate of the point u is described in the binary system by $u = 0.x_1x_2 \cdots x_n \cdots$. Then the value of the function $\xi(t)$ on the interval $[(n - 1), n)$ is equal to x_n. Accordingly, the random function $\xi(t)$ can be written in the form $\xi(t) = f(t, u)$ where $f(t, u)$

is a completely defined nonrandom function of two variables t and u (with $0 \leqq t < \infty$ and $0 \leqq u < 1$).

Let us consider an arbitrary random function in the broad sense with range contained in a complete separable metric space X and with domain of definition Θ. As we demonstrated in Section 2 of Chapter III, it is always possible to construct a probability space $\{U, \tilde{\mathfrak{S}}, \mathsf{P}\}$, where U is the set of all mappings $u = u(\theta)$ of the set Θ into X and such that the distribution of the sequence

$$\{u(\theta_1), u(\theta_2), \cdots, u(\theta_n)\}$$

in X^n for arbitrary natural n and $\theta_k \in \Theta$ for $k = 1, \cdots, n$, coincides with the corresponding distribution for the given random function in the broad sense. In other words, for an arbitrary random function in the broad sense it is possible to construct a stochastically equivalent (in the broad sense) random function (representation in the terminology of Section 3, Chapter III) in the sense of Definition 1. Unfortunately, the representation obtained does not always fully serve our purpose. Arbitrary functions $u = u(\theta)$ of a variable θ are elementary events in the resulting probability space, and we can no longer speak of such properties of the functions $u(\theta)$ as their continuity, integrability, differentiability, and so forth. (Of course we are assuming that Θ and X are such that the corresponding concepts are meaningful.)

Furthermore, in this probability space we cannot consider events of the form

$$\{u(\theta) \in A \text{ for all } \theta \in Q, A \in \mathfrak{B}\} \tag{4}$$

where Q is an unusual subset of Θ. Such events are important in solving many particular problems. For example, corresponding to the event

$$\{u; u = u(\theta) \in A \text{ for all } \theta \in Q\} = \bigcap_{\theta \in Q} \{u; u(\theta) \in A, \theta \text{ fixed}\}$$

in U is a set representing the intersection of an uncountable collection of sets belonging to $\tilde{\mathfrak{C}}$; hence, it does not necessarily belong to the σ-algebra $\tilde{\mathfrak{C}}$. These considerations suggest that in the construction of the representation of the family of distributions (1) it is desirable for U to be as narrow as possible, that is, for functions belonging to U to have the best possible analytic properties.

For example, to be able to solve the problem of calculating the probability of the event (4), we should like the random function $g(\theta, u)$, which is the representation of the family of distributions (1), to have the following property:

c. For sufficiently broad classes of sets $\mathfrak{A}$ in X and $\mathfrak{Q}$, in Θ, there exists a countable set Λ of points $\theta_j \in \Theta$ such that for arbitrary $A \in \mathfrak{A}$ and $Q \in \mathfrak{Q}$, the set of points

$$\{u; g(\theta, u) \in A \text{ for all } \theta \in Q\}, A \in \mathfrak{A}, Q \in \mathfrak{Q}\,, \tag{5}$$

differs from the set

$$\{u: g(\theta_j, u) \in A \text{ for all } \theta_j \in \Lambda \cap Q\} \tag{6}$$

only on a subset of some fixed set N of P-measure 0 that is independent of A and Q. Since the set (6) is the intersection of no more than countably many measurable sets, it is itself measurable and so is the set (5) (because of the completeness of the measure P). Here the probabilities of the events (5) and (6) are equal to each other.

A random variable that satisfies condition c is said to be *separable* (with respect to the class of sets $\mathfrak{A}$).

Suppose that Θ is a separable metric space, that $\mathfrak{Q}$, is a class of open sets, that $\mathfrak{A}$ is a class of closed subsets of the space X, and that the function $g(\theta, u)$ is a continuous function of the argument θ for almost all fixed u. Note that under these conditions the random function $g(\theta, u)$ is separable with respect to the class of closed sets $\mathfrak{A}$.

In problems dealing with integration of random variables with respect to a particular measure $\{\mu, \mathfrak{R}\}$ on Θ, it is desirable for the function $g(\theta, u)$ to be $\mathfrak{R}$-measurable as a function of θ for almost all fixed u.

In other cases, it is necessary to find a representation of a given random function in the broad sense under which almost all sample functions are continuous or have discontinuities of the first kind only, or are k times differentiable, and so forth. Finally, we should expect the possibility of obtaining a representation with special properties to be determined by finite-dimensional distributions of the random function.

Analogous problems can arise for functions other than random functions in the broad sense. A random function $g(\theta, u)$ can have "pathological" properties and yet there may exist a "smooth" function $g^*(\theta, u)$ that is equivalent to it,

$$\mathsf{P}\{g(\theta, u) \neq g^*(\theta, u)\} = 0\,,$$

but that is not "pathological." According to our definition of stochastically equivalent functions, in solving the problems with which we are concerned it is permissible to replace such a function with a stochastically equivalent regular function $g^*(\theta, u)$.

Let us consider an example. Suppose that A is the set of rational numbers on the real line $(-\infty, \infty)$, that $\chi(t)$ is the characteristic function of the set A, and that u is a random variable uniformly distributed on the interval $[0, 1]$.

Let us set $g(t, u) = \chi(t + u)$. For arbitrary fixed u, the function $g(t, u)$ is everywhere discontinuous. On the other hand, for fixed t the function $g(t, u)$ is equal to 0 with probability 1. Thus, with probability 1, the everywhere-discontinuous random function $g(t, u)$ is stochastically equivalent to the function $g^*(t, u) \equiv 0$.

In this chapter we shall consider different conditions under which there exists, for a given random function, a stochastically equivalent (or stochastically equivalent in the broad sense) function possessing specified regularity properties.

2. SEPARABLE RANDOM FUNCTIONS

We mentioned the concept of a separable random function in the preceding section. The property of separability, however, is not a stringent restriction on a random function. Under rather broad hypotheses having to do with the nature of the domain of definition Θ and the range X of the random function, there exists a separable random function stochastically equivalent to the given one. However it should be noted that in constructing the equivalent separable random function it is some times necessary to broaden the range of the function in order to make it a compact set.

Let us assume that Θ and X are metric spaces with metrics $r(\theta_1, \theta_2)$ and $\rho(x_1, x_2)$ respectively, and that Θ is separable. The role of the classes of sets $\mathfrak{A}$ and $\mathfrak{Q}$ in the definition of separability is played by the closed subsets of X and the open subsets of Θ. Thus separability of a random function will be understood in the sense of:

Definition 1. A random function $g(\theta, u)$ is said to be *separable* if Θ contains an everywhere dense countable set of points $\{\theta_j\}$, for $j = 1, 2, \cdots$, and if U contains a subset N of probability 0 such that for an arbitrary open set $G \subset \Theta$ and an arbitrary closed set $F \subset X$, the two sets

$$\{u;\, g(\theta_j, j) \in F,\ \theta_j \in G\}\ ,$$

$$\{u;\, g(\theta, u) \in F \text{ for all } \theta \in G\}$$

differ from each other only by a subset of N.

The countable set of points θ_j in this definition is called the *set of separability* of the random function.

Theorem 1. *Let X and Θ denote metric spaces. Suppose that X is compact and that Θ is separable. Then an arbitrary random function $g(\theta, u)$ defined on Θ into X is stochastically equivalent to some separable random function.*

Proof. Suppose that $\tilde{g}(\theta, u)$ is a separable random function, that I is its set of separability, and that N is the corresponding exceptional set of points u. Let V denote the class of all open spheres in the space Θ with rational radius and center at a point belonging to a fixed countable everywhere-dense subset of Θ. The class V is countable. On the other hand, an arbitrary open subset G of Θ can be represented as the union of countably many spheres contained in V.

Let $A(G, u)$ denote the closure of the range of the function $\tilde{g}(\theta, u)$ as θ ranges over the set $I \cap G$. Then

$$A(\theta, u) = \bigcap_{S \ni \theta} A(S, u)$$

is the intersection of all $A(S, u)$ as S ranges over the collection of spheres in V to which the point θ belongs. The family of closed sets $A(G, u)$, where $\theta \in G$, is centered; an arbitrary finite number of sets belonging to this family have common points, and because of the compactness of X their intersection $A(\theta, u)$ is nonempty. It follows from the separability of the function $\tilde{g}(\theta, u)$ that

$$\tilde{g}(\theta, u) \in A(\theta, u), \; u \notin N. \tag{1}$$

Conversely, if (1) is satisfied for every $u \notin N$ such that $\mathsf{P}(N) = 0$, then $\tilde{g}(\theta, u)$ is a separable random function. To see this, note that if $\tilde{g}(\theta, u) \in F$ for all $\theta \in I \cap S$ where F is a closed subset of X and $S \in V$, then $A(\theta, u) \subset A(S, u) \subset F$ for every $\theta \in S$ and consequently $\tilde{g}(\theta, u) \in F$ for every θ in S.

Let G denote an arbitrary open set in Θ. Then we may represent it as a sum $G = \bigcup_k S_k$ of subsets of V in order to see, on the basis of what has just been said, that the relation

$$\tilde{g}(\theta, u) \in F \text{ for all } \theta \in I \cap G, u \notin N,$$

implies

$$\tilde{g}(\theta, u) \in F \text{ for arbitrary } \theta \in G.$$

Let us formulate the result just obtained as:

Lemma 1. *For a random function $\tilde{g}(\theta, u)$ to be separable, it is necessary and sufficient that there exist a set N such that $\mathsf{P}(N) = 0$*

and (1) *is satisfied for every* $u \notin N$.

Thus, to construct a separable function stochastically equivalent to $g(\theta, u)$, it suffices to find a function $\tilde{g}(\theta, u)$ that satisfies (1) and coincides with the function $g(\theta, u)$ with probability 1,

$$\mathsf{P}\{\tilde{g}(\theta, u) \neq g(\theta, u)\} = 0 .$$

Lemma 2. *Let* B *denote an arbitrary Borel subset of* X. *Then there exists a finite or countably infinite set of points* $\theta_1, \theta_2, \cdots$, *such that the set*

$$N(\theta, B) = \{u; g(\theta_k, u) \in B, k = 1, 2, \cdots, g(\theta, u) \notin B\}$$

has probability 0 *for arbitrary* $\theta \in \Theta$.

Proof. Let θ_1 be arbitrary. If $\theta_1, \theta_2, \cdots, \theta_k$ are constructed, let us set

$$m_k = \sup_{\theta \in \Theta} \mathsf{P}\{g(\theta_1, u) \in B, \cdots, g(\theta_k, u) \in B; g(\theta, u) \notin B\} .$$

The sequence $\{m_k\}$ decreases monotonically. If $m_k = 0$, the corresponding sequence is constructed. If $m_k > 0$, let θ_{k+1} be a point such that

$$\mathsf{P}\{g(\theta_1, u) \in B, \cdots, g(\theta_k, u) \in B, g(\theta_{k+1}, u) \notin B) \geqq \frac{m_k}{2} .$$

Since the sets

$$L_k = \{u; g(\theta_i, u) \in B, i = 1, 2, \cdots, k, g(\theta_{k+1}, u) \notin B\}$$

have no points in common, we have

$$1 \geqq \sum \mathsf{P}(L_k) \geqq \frac{1}{2}\sum_{k=1}^{\infty} m_k .$$

Consequently, $m_k \longrightarrow 0$ as $k \longrightarrow \infty$. Thus for arbitrary θ,

$$\mathsf{P}\{g(\theta_k, u) \in B, k = 1, 2, \cdots, g(\theta, u) \notin B\} \leqq \lim m_k = 0$$

which completes the proof of Lemma 2.

From what has been shown, it is not difficult to prove

Lemma 3. *Let* $\mathfrak{M}_0$ *denote a countable class of sets and let* $\mathfrak{M}$ *denote the class consisting of the intersections of all possible subclasses of sets belonging to* $\mathfrak{M}_0$. *Then there exists a finite or countably infinite set of points* $\theta_1, \theta_2, \cdots, \theta_n, \cdots$ *and for every* θ *a set* $N(\theta)$ *such that* $\mathsf{P}\{N(\theta)\} = 0$ *and*

$$\{u; g(\theta_n, u) \in B, n = 1, 2, \cdots, g(\theta, u) \notin B\} \subset N(\theta)$$

for every $B \in \mathfrak{M}$.

Proof. Let I denote the countable set of points in Θ that is the union of the sets $\{\theta_n, n = 1, 2, \cdots\}$ constructed for each $B \in \mathfrak{M}_0$ in accordance with Lemma 2. Define

$$N(\theta) = \bigcup_{B \in \mathfrak{M}_0} N(\theta, B) .$$

If $B' \in \mathfrak{M}$ and if $B' \subset B \in \mathfrak{M}_0$, then

$$\{u; g(\theta_n, u) \in B', \theta_n \in I, g(\theta, u) \notin B\} \subset \{u; g(\theta, u) \in B, \theta_n \in I, g(\theta, u) \notin B\} \subset N(\theta, B) \subset N(\theta) .$$

Furthermore, if $B' = \bigcap_{k=1}^{\infty} B_k$ for $B_k \in \mathfrak{M}_0$, then

$$\{u; g(\theta_n, u) \in B', \theta_n \in I, g(\theta, u) \notin B'\}$$

$$\subset \bigcup_{k=1}^{\infty} \{u; g(\theta_n, u) \in B', \theta_n \in I, g(\theta, u) \notin B_k\} \subset \bigcup_{k=1}^{\infty} N(\theta, B_k) \subset N(\theta) ,$$

which completes the proof of the lemma.

Now we can easily conclude the proof of Theorem 1. Let us choose a countable everywhere-dense set of points $L \in X$. Let $\mathfrak{M}_0$ denote the class of complements of spheres of rational radius with centers at the points L. Then $\mathfrak{M}$ denotes the class of intersections of sets belonging to $\mathfrak{M}_0$, and this class contains all closed sets. Furthermore, for every $S \in V$, let us consider the random function $g(\theta, u)$ as defined only for $\theta \in S$ and let us construct a sequence $I = I(S)$ and sets $N(\theta) = N_S(\theta)$ in accordance with Lemma 3. We define

$$J = \bigcup_{S \in V} I(S), \qquad N_\theta = \bigcup_{S \in V} N_S(\theta) ,$$

we also define

$$\tilde{g}(\theta, u) = g(\theta, u) ,$$

if $\theta \in J$ or $u \notin N_\theta$; if $u \in N_\theta$ and $\theta \notin J$, we define $\tilde{g}(\theta, u)$ in an arbitrary manner so that $\tilde{g}(\theta, u) \in A(\theta, u)$. Since the values of the functions $\tilde{g}(\theta, u)$ and $g(\theta, u)$ coincide for $\theta \in J$, the sets $A(\theta,u)$ constructed for the functions $\tilde{g}(\theta, u)$ and $g(\theta, u)$ also coincide. It follows from the definition that $\tilde{g}(\theta, u) \in A(\theta, u)$ for arbitrary θ and u. Since $\{u; g(\theta, u) \neq \tilde{g}(\theta, u)\} \subset N_\theta$, we have

$$\mathsf{P}\{\tilde{g}(\theta, u) = g(\theta, u)\} = 1 ,$$

which completes the proof of the theorem.

Theorem 1 can be immediately generalized for random functions with range in a separable locally compact space.

Theorem 2. *Let X denote a separable locally compact space and let Θ denote an arbitrary separable metric space. For an arbitrary random function $g(\theta, u)$ defined on Θ with range in X, there exists a stochastically equivalent separable random function $\tilde{g}(\theta, u)$ with range* *in some compact extension* $\tilde{X}$ *of the space* $X(\tilde{X} \supset X)$.

Proof follows from the fact that every locally compact separable space X can be regarded as a subset of some compact set $\tilde{X}$. For example, if $g(\theta, u)$ is a random function with range in a finite-dimensional space X, then by completing X with an "infinitely distant" point ∞, we easily obtain the compact space $\tilde{X} = X \cup \{\infty\}$ with a new metric such that every closed set $F \subset X$ (in the topology of the space X) is also closed in $\tilde{X}$ (with respect to the new metric). In construction of the separable sample function of a random function, it may be convenient to assign to it the additional value "∞," but obviously for fixed θ the probability of this value is 0. In many questions it is important to know what set J can play the role of the set of separability. Before answering this question, we introduce the important concept of stochastic continuity of a random function, and present some simple theorems involving it. (A special case of this concept was encountered in Section 4, Chapter I).

Definition 2. A random function $g(\theta, u)$ with range in X is said to be *stochastically continuous* at a point $\theta_0 \in \Theta$ if

$$\mathsf{P}\{\rho(g(\theta_0, u), g(\theta, u)) > \varepsilon\} \longrightarrow 0 \text{ as } r(\theta, \theta_0) \longrightarrow 0 . \qquad (2)$$

If $g(\theta, u)$ is stochastically continuous at every point of a set $B \subset \Theta$, it is said to be stochastically continuous on B.

We note that the condition of stochastic continuity is the condition imposed on a "two-dimensional" distribution of a random variable, that is, on the joint distribution of random elements $g(\theta_1, u)$ and $g(\theta_2, u)$, where θ_1 and θ_2 belong to Θ. This concept can be applied to random functions in the broad sense.

Definition 3. If there exists a point $x \in X$ such that

$$\sup_{\theta \in B} \mathsf{P}\{\rho[g(\theta, u), x] > K\} \longrightarrow 0 \qquad (3)$$

as $K(>0) \longrightarrow \infty$, then the random function $g(\theta, u)$ is said to be *bounded in probability* on the set B.

Theorem 3. *If a random function* $g(\theta, u)$ *is stochastically continuous on a compact set* Θ, *it is bounded in probability on* Θ.

Proof. Let ε denote an arbitrary positive number. For every point θ let us construct a sphere S_θ with center at θ such that

$$\mathsf{P}\{\rho(g(\theta, u), g(\theta', u)) > 1\} < \frac{\varepsilon}{2}$$

for arbitrary $\theta' \in S_\theta$. From the set of spheres S_θ let us choose an n-tuple $S_{\theta_1}, S_{\theta_2}, \cdots, S_{\theta_n}$ constituting a finite covering of Θ. Let a denote their greatest radius. Then for arbitrary x,

$$\rho(g(\theta, u), x) \leqq \rho(g(\theta_1, u), x) + \max_{i=1,\cdots,n} \rho(g(\theta_1, u), g(\theta_i, u))$$
$$+ \rho(g(\theta_j, u), g(\theta, u))$$

where θ_j denotes the center of the sphere containing θ in its interior. The terms on the right side of this inequality are finite random variables. Therefore, for sufficiently large N,

$$\mathsf{P}\{\rho(g(\theta_1, u), x) + \max_{i=2,\cdots,n} \rho(g(\theta_1, u), g(\theta_i, u)) > N\} < \frac{\varepsilon}{2} .$$

If we assume that $N > 1$, then

$$\sup \mathsf{P}\{\rho(g(\theta, u), x) > 2N\} \leqq \sup \mathsf{P}\{\rho(g(\theta_1, u), x)$$
$$+ \max_{i=2,\cdots,n} \rho(g(\theta_1, u), g(\theta_i, u)) > N\}$$
$$+ \mathsf{P}\{\rho(g(\theta_j, u), g(\theta, u)) > 1\} < \varepsilon ,$$

which completes the proof.

Definition 4. A random function $g(\theta, u)$ is said to be *uniformly stochastically continuous* on Θ if for arbitrarily small positive numbers ε and ε_1 there exists a $\delta > 0$ such that

$$\mathsf{P}\{\rho(g(\theta, u), g(\theta', u)) > \varepsilon\} < \varepsilon_1 \tag{4}$$

whenever $r(\theta, \theta') < \delta$.

Theorem 4. *If $g(\theta, u)$ is stochastically continuous on a compact set Θ, then $g(\theta, u)$ is uniformly stochastically continuous.*

Proof. If this is not the case, there is a pair of positive numbers ε and ε_1, and for arbitrary $\delta_n > 0$ a pair of points θ_n and θ'_n, for which $r(\theta_n, \theta'_n) < \delta_n$ and

$$\mathsf{P}\{\rho(g(\theta_n, u), g(\theta'_n u)) > \varepsilon\} > \varepsilon_1 .$$

We may assume that $\delta_n \longrightarrow 0$ and $\theta_n \longrightarrow \theta_0$ as $n \longrightarrow \infty$. Then $\theta'_n \longrightarrow \theta_0$ as $n \longrightarrow \infty$, and

$$\varepsilon_1 < \mathsf{P}\{\rho(g(\theta_n, u), g(\theta'_n, u)) > \varepsilon\}$$
$$\leqq \mathsf{P}\left\{\rho(g(\theta_n, u), g(\theta_0, u)) > \frac{\varepsilon}{2}\right\}$$
$$+ \mathsf{P}\left\{\rho(g(\theta_0, u), g(\theta'_n, u)) > \frac{\varepsilon}{2}\right\} .$$

This inequality contradicts the hypothesis of stochastic continuity.

Theorem 5. *Let Θ denote a separable space and let $g(\theta, u)$ denote a separable stochastically continuous random function. Then an arbitrary countable everywhere-dense set of points in Θ can serve as set of separability for the random function $g(\theta, u)$.*

Proof. Let $V = \{S\}$ denote the countable set of spheres in Θ mentioned in the proof of Theorem 1, let $J = \{\theta_k, k = 1, 2, \cdots, n, \cdots\}$ denote a set of separability of the random function $g(\theta, u)$, let N denote the exceptional set of values of u that appears in the definition of separability, and let Λ denote an arbitrary countable everywhere-dense set of points in Θ. Let $B(S, u)$ denote the closure of the set of values $g(\gamma_k, u)$ as the point γ_k ranges over $\Lambda \cap S$ and let $N(S, k)$ denote the event that $g(\theta_k, u) \notin B(S, u)$ if $\theta_k \in S$. The events $N(S, k)$ have probability 0. To see this, let $\{\gamma_r\}$ for $r = 1, 2, \cdots, n, \cdots$ denote an arbitrary sequence of points in $\Lambda \cap S$ that converges to θ_k. Then

$$\mathsf{P}\{g(\theta_k, u) \notin B(S, u)\} \leqq \mathsf{P}\{\varliminf_{r\to\infty} \rho(g(\theta_k, u), g(\gamma_r, u)) > 0\}$$

$$\leqq \lim_{n\to\infty} \mathsf{P}\left\{\varliminf \rho(g(\theta_k, u), g(\gamma_r, u)) > \frac{1}{n}\right\}$$

$$\leqq \lim_{n\to\infty} \varliminf_{r\to\infty} \mathsf{P}\left\{\rho(g(\theta_k, u), g(\gamma_r, u)) > \frac{1}{n}\right\} = 0 .$$

Suppose that

$$N' = \bigcup_S \bigcup_{\theta_k \in S} N(S, k) .$$

Then $\mathsf{P}(N') = 0$. If $u \notin N \cup N'$ and $g(\gamma, u) \in F$ for all $\gamma \in \Lambda \cap G$, where G is some open set and $F \subset X$ is closed, then for every $\theta_k \in G$ and S such that $\theta_k \in S \subset G$, we have

$$g(\theta_k, u) \in B(S, u) \subset F .$$

From the definition of the set $\{\theta_k\}$, it then follows that $g(\theta, u) \in F$ for all $\theta \in G$ and $u \notin N \cup N'$. Thus the set Λ satisfies the condition in the definition of a set of separability of a random function.

3. MEASURABLE RANDOM FUNCTIONS

Let Θ and X denote metric spaces with distance $r(\theta_1, \theta_2)$ and $\rho(x_1, x_2)$ respectively, let $g(\theta, u)$ denote a random function with range in X and domain of definition Θ, and let u denote an elementary event of the probability space $\{U, \mathfrak{S}, \mathsf{P}\}$.

Let us suppose that a σ-algebra of sets $\mathfrak{K}$ containing Borel sets is defined on Θ and that a complete measure μ is defined on $\mathfrak{K}$. Let $\sigma\{\mathfrak{K} \times \mathfrak{S}\}$ denote the smallest σ-algebra generated in $\Theta \times U$ by the product of the σ-algebras $\mathfrak{K}$ and $\mathfrak{S}$ and let $\tilde{\sigma}\{\mathfrak{K} \times \mathfrak{S}\}$ denote its completion with respect to the measure $\mu \times \mathsf{P}$ (cf. Chapter II, Section 8).

Definition 1. A random function $g(\theta, u)$ is said to be *measurable* if it is measurable with respect to $\tilde{\sigma}\{\mathfrak{R} \times \mathfrak{S}\}$.

By definition a random function $g(\theta, u)$ is $\mathfrak{S}$-measurable for arbitrary $\theta \in \Theta$. On the other hand, if a random function is measurable, then on the basis of Fubini's theorem, $g(\theta, u)$ is $\mathfrak{R}$-measurable as a function of θ for P-almost-all u. In other words its sample functions are $\mathfrak{R}$-measurable with probability 1.

Let us now look at conditions that ensure the existence of a measurable separable function stochastically equivalent to a given random function.

Theorem 1. *Suppose that Θ and X are compact. If for μ-almost-all θ, a random function $g(\theta, u)$ is stochastically continuous, then there exists a measurable separable random function $g^*(\theta, u)$ that is stochastically equivalent to the function $g(\theta, u)$.*

On the basis of Theorem 1 of Section 2, corresponding to the function $g(\theta, u)$ there is a stochastically equivalent separable random function $\tilde{g}(\theta, u)$. Let I denote the set of separability of the function $\tilde{g}(\theta, u)$. As in Section 2, $A(G, u)$ denotes the closure of the range of $\tilde{g}(\theta, u)$ as θ ranges over the set $G \cap I$ and $A(\theta, u)$ denotes the intersection of all sets of the form $A(S, u)$ where S is an arbitrary open sphere in V that contains θ. By virtue of the separability, $\tilde{g}(\theta, u) \in A(\theta, u)$ almost certainly (that is, for $u \notin N$, where $\mathsf{P}(N) = 0$). On the other hand, if

$$\mathsf{P}\{g'(\theta, u) = \tilde{g}(\theta, u), \theta \in I\} = 1$$

and $g'(\theta, u) \in A(\theta, u)$ (where $u \notin N$), then $g'(\theta, u)$ is also a separable random function (cf. Lemma 1, Section 2). Let us construct a function $g^*(\theta, u)$ that is stochastically equivalent to the function $\tilde{g}(\theta, u)$ and measurable with respect to the σ-algebra $\tilde{\sigma}\{\mathfrak{R} \times \mathfrak{S}\}$. For arbitrary n, let us cover Θ with a finite number of spheres $S_k^{(n)} \in V$, for $k = 1, 2, \cdots, m_n$ of diameter not exceeding $1/n$. In each $S_k^{(n)}$, let us choose a point $\theta_k^{(n)} \in I$ and let us set

$$\bar{g}_n(\theta, u) = \tilde{g}(\theta_k^{(n)}, u)$$

for

$$\theta \in S_k^{(n)} \Big\backslash \bigcup_{j=1}^{k-1} S_j^{(n)}$$

where $k = 1, 2, \cdots, m_n$. Obviously the functions $\bar{g}_n(\theta, u)$ are $\sigma\{\mathfrak{R} \times \mathfrak{S}\}$-measurable. Furthermore,

$$\rho[\bar{g}_n(\theta, u), \tilde{g}(\theta, u)] = \rho[\tilde{g}(\theta_k^{(n)}, u), \tilde{g}(\theta, u)] \tag{1}$$

if

$$\theta \in S_k^{(n)} \Big\backslash \bigcup_{j=1}^{k-1} S_j^{(n)} .$$

Also, $r(\theta_k^{(n)}, \theta) \leqq 1/n$. If we set

$$G_{n,m}(\theta) = \mathsf{P}\{u; \rho[\bar{g}_n(\theta, u), \bar{g}_{n+m}(\theta, u)] > \varepsilon\} ,$$

then by virtue of the hypothesis of the theorem, the function $G_{n,m}(\theta)$ approaches 0 as $n \longrightarrow \infty$ for μ-almost-all θ. Therefore

$$\mu \times \mathsf{P}[\theta, u; \rho\{\bar{g}_n(\theta, u), \bar{g}_{n+m}(\theta, u)\} > \varepsilon] = \int_\Theta G_{n,m}(\theta)\mu(d\theta) \longrightarrow 0$$

as $n \longrightarrow \infty$; that is, the sequence $\{\bar{g}_n(\theta, u)\}$ is fundamental with respect to the measure $\mu \times \mathsf{P}$. It contains a subsequence $\{\bar{g}_{n_k}(\theta, u)\}$ that converges $(\mu \times \mathsf{P})$-almost-everywhere to some $\sigma\{\mathfrak{R} \times \mathfrak{S}\}$-measurable function $\bar{g}(\theta, u)$. Let M_1 denote the set of points (θ, u) at which this convergence does not take place. For $(\theta, u) \notin M_1$, we have $\bar{g}(\theta, u) \in A(\theta, u)$. Since the set M_1 has measure 0, it follows that μ-almost-all its θ-sections have P-measure 0 (cf. Section 8, Chapter II). Let K_1 denote the set of values of θ whose corresponding sections have nonzero P-measure. We set

$$g^*(\theta, u) = \begin{cases} \tilde{g}(\theta, u) & \text{for} \quad \theta \in I \cup K_1 \cup K , \\ \bar{g}(\theta, u) & \text{for} \quad \theta \notin I \cup K_1 \cup K , \end{cases}$$

where K is the set of all θ at which the limit relation (2) of Section 2 is not satisfied. Then $g^*(\theta, u) \in A(\theta, u)$ (where $u \notin N$); that is, $g^*(\theta, u)$ is separable. Furthermore it is $\tilde{\sigma}\{\mathfrak{R} \times \mathfrak{S}\}$-measurable since it coincides with a measurable function almost everywhere in $\Theta \times U$ (exclusive of points $\theta \in K_1 \cup K$ and $u \in N$). Furthermore, if $\theta \notin K_1 \cup K$, then by virtue of (1) and the condition of stochastic continuity,

$$\mathsf{P}\{\bar{g}(\theta, u) = \tilde{g}(\theta, u)\} = 1 ,$$

from which it follows that the random functions $g^*(\theta, u)$ and $\tilde{g}(\theta, u)$ are stochastically equivalent. This completes the proof of the theorem.

We can make a number of statements generalizing Theorem 1.

REMARK 1. In Theorem 1 the requirement that the spaces Θ and X be compact can be replaced with the requirement that they be locally compact and separable. The compactness of the space X was necessary only so that we might refer to Theorem 1 of Section 2. Now, however, we can refer to Theorem 2 of Section 2. Here the separable and measurable representation $g^*(\theta, u)$ of the function $g(\theta, u)$ assumes values, generally speaking, in some compact topological extension of the space X. Furthermore, if the space Θ

is locally compact and separable it can be represented as the sum of countably many compact sets. The reasoning can be applied to each such compact set in particular. The assertion of the theorem then follows for their union also. Furthermore, the measure μ need not be finite; it is sufficient that it be σ-finite. From this we get:

REMARK 2. The assertion of Theorem 1 holds for the case in which Θ and X are finite-dimensional Euclidean spaces and the measure $\{\mu, \mathfrak{K}\}$ is Lebesgue measure in Θ.

Now we note that the proof of Theorem 1 would be simplified if we did not require separability of the measurable representation of the given random function. The set I would not come into the picture and the points $\theta_k^{(n)}$ could be chosen arbitrarily from the corresponding sets. Of the properties of the space X, we should need to use only its completeness. Thus we have:

REMARK 3. If X is a complete metric space, if Θ is a locally compact separable space, and if μ is a σ-finite measure on a σ-algebra containing Borel subsets of Θ, then the random function $g(\theta, u)$, $\theta \in \Theta$, $u \in U$, with range in X, which is stochastically continuous for μ-almost-all θ, is stochastically equivalent to the measurable random function.

The following result, which has great significance, follows immediately from Fubini's theorem (Theorem 2, Section 8, Chapter II):

Theorem 2. *Let $\xi(\theta) = g(\theta, u)$ denote a measurable random function with real range. If*

$$\int_{\Theta} \mathsf{M} |\xi(\theta)| \mu(d\theta) < \infty ,$$

then for an arbitrary set $B \in \mathfrak{K}$,

$$\int_B \mathsf{M}\xi(\theta)\mu(d\theta) = \mathsf{M}\int_B \xi(\theta)\mu(d\theta) .$$

The last equation indicates the commutativity of the symbols representing the mathematical expectation and integration with respect to a parameter.

4. CONDITIONS FOR NONEXISTENCE OF DISCONTINUITIES OF THE SECOND KIND

Let $\xi(t)$ where $t \in \mathfrak{T} = [a, b]$ denote a random process with range in a complete metric space X.

Definition 1. If sample functions of a process have, with probability 1, left- and right-hand limits at every $t \in (a, b)$ and have right-hand (resp. left-hand) limits at the point a (resp. the point b), the process is said not to have discontinuities of the second kind.

Throughout the present section, we assume that the process $\xi(t)$ is separable. We let J denote the set of separability of the process.

Definition 2. Let ε denote a positive number. A function $y = f(t)$ with range in X is said to have no fewer than m ε-oscillations on a closed interval $[a, b]$ if there exist points $t_0, \cdots, t_m$, where $a \leqq t_0 < t_1 < \cdots < t_m \leqq b$, such that $\rho(f(t_{k-1}), f(t_k)) > \varepsilon$ for $k = 1, \cdots, m$.

Lemma 1. *For a function $y = f(t)$ not to have discontinuities of the second kind on a closed interval $[a, b]$, it is necessary and sufficient that for arbitrary $\varepsilon > 0$ it have only finitely many ε-oscillations on $[a, b]$.*

Proof of the Sufficiency. Let us prove the existence of the limit $f(t - 0)$ for arbitrary $t \in (a, b]$. Let $\{t_n\}$ denote an arbitrary sequence such that $t_n \uparrow t$. There exist only finitely many numbers t_{n_k} (where $n_k < n_{k+1}$) such that

$$\rho(f(t_{n_k}), f(t_{n_{k+1}})) > \varepsilon .$$

Consequently from some m onward, the inequality $\rho(f(t_n), f(t_{n+k})) \leqq \varepsilon$, where $k > 0$, holds for all $n \geqq m$, that is, the sequence $\{f(t_n)\}$ converges. This implies the existence of

$$f(t - 0) = \lim_{\tau \uparrow t} f(\tau) .$$

Proof of the existence of $f(t + 0)$ on $[a, b)$ is analogous.

Proof of the Necessity. Suppose that one of the one-sided limits (for example, the left-hand limit) does not exist at a point t_0. Then there exists a sequence $t_n \uparrow t_0$ such that for arbitrary n,

$$\sup_{m>n} \rho(f(t_m), f(t_n)) > \varepsilon ,$$

that is, there are infinitely many ε-oscillations.

REMARK 1. Definition 2 can be carried over in a trivial manner to random functions defined on an arbitrary set of real values of t. If the sample function of a separable process $\xi(t)$ has no fewer than m ε-oscillations on $[a, b]$, it also has no fewer than m ε-oscillations on the set of separability J except possibly for a set of sample functions N of probability 0.

Theorem 1. *Suppose that a separable random process $\xi(t)$, for*

$t \in [a, b]$, with range in X satisfies the following conditions:

a. *There exist numbers p, q, r, and C (where $p \geqq 0$, $q \geqq 0$, $r > 0$, and $C \geqq 0$) such that, for arbitrary t_1, t_2, and t_3 such that $a \leqq t_1 < t_2 < t_3 \leqq b$, we have*

$$\mathsf{M}\rho^p[\xi(t_1), \xi(t_2)]\rho^q[\xi(t_2), \xi(t_3)] \leqq C \mid t_3 - t_1 \mid^{1+r} ; \tag{1}$$

b. *the process $\xi(t)$ is stochastically continuous on $[a, b]$.*

Then $\xi(t)$ has no discontinuities of the second kind.

Proof. It follows from (1) and Chebyshev's inequality that

$$\begin{aligned} &\mathsf{P}[\{\rho[\xi(t_1), \xi(t_2)] \geqq \varepsilon_1\} \cap \{\rho[\xi(t_2), \xi(t_3)] \geqq \varepsilon_2\}] \\ &\leqq \mathsf{M}\frac{\rho^p[\xi(t_1), \xi(t_2)]}{\varepsilon_1^p}\frac{\rho^q[\xi(t_2), \xi(t_3)]}{\varepsilon_2^q} \leqq \frac{C \mid t_3 - t_1 \mid^{1+r}}{\varepsilon_1^p \varepsilon_2^q} . \end{aligned} \tag{2}$$

In proving the theorem, we actually use inequality (2) instead of condition (1). It follows from (b) that for the set of separability we can choose an arbitrary countable set that is everywhere-dense on $[a, b]$. For such a set, let us take the set J of all dyadic numbers belonging to $[a, b]$.

For the sake of convenience let us take $[a, b] = [0, 1]$. We shall break down the proof into several steps.

1. Let $A_{k,n}$ denote the event

$$\left\{\rho\left[\xi\left(\frac{k-1}{2^n}\right), \xi\left(\frac{k}{2^n}\right)\right] < \varepsilon_n\right\} ,$$

where

$$\varepsilon_n = C^{(1/p+q)}\left(\frac{1}{2}\right)r(n-1)/2(p+q) = L\alpha^n ,$$

$$L = (2^{r/2}C)^{(1/p+q)}, \qquad \alpha = 2^{-r/2(p+q)} < 1 ,$$

and

$$\begin{aligned} B_{kn} = A_{k,n} \cup A_{k+1,n} = &\left\{\rho\left[\xi\left(\frac{k-1}{2^n}\right), \xi\left(\frac{k}{2^n}\right)\right] < \varepsilon_n\right\} \\ &\cup\left\{\rho\left[\xi\left(\frac{k}{2^n}\right), \xi\left(\frac{k+1}{2^n}\right)\right] < \varepsilon_n\right\} . \end{aligned}$$

On the basis of inequality (2),

$$\mathsf{P}\{\bar{B}_{kn}\} \leqq 2^{-(1+r/2)(n-1)}, \qquad k = 1, 2, \cdots, 2^n - 1 \tag{3}$$

where $\bar{B}_{kn}$ is the event complementary to the event B_{kn}. Let us define

$$D_n = \bigcap_{m=n}^{\infty} \bigcap_{k=1}^{2^m-1} B_{km} .$$

Then for the complementary event $\bar{D}_n$ we have

$$\mathsf{P}(\bar{D}_n) \leqq \sum_{m=n}^{\infty} \sum_{k=1}^{2^m-1} \mathsf{P}(\bar{B}_{km}) \leqq \sum_{m=n}^{\infty} \sum_{k=1}^{2^m-1} 2^{-(1+r/2)(m-1)} ,$$

or

$$\mathsf{P}(\bar{D}_n) \leqq 2 \sum_{m=n}^{\infty} 2^{-(m-1)\frac{r}{2}} = \frac{2\beta^{n-1}}{1-\beta}, \qquad \beta = 2^{-r/2} < 1 \,. \tag{4}$$

From this it follows that for $D = \bigcup_{n=1}^{\infty} D_n$,

$$\mathsf{P}(D) = \lim_{n\to\infty} \mathsf{P}(D_n) = 1 \,.$$

2. Let us fix k, m, and n and let us set

$$\rho_{i,m} = \rho\left[\xi\left(\frac{k}{2^n}\right), \xi\left(\frac{k}{2^n} + \frac{i}{2^{m+n}}\right)\right], \qquad \omega_m = \max_{0\leqq i\leqq 2^m} \rho_{i,m} \,;$$

$$\rho'_{i,m} = \rho\left[\xi\left(\frac{k}{2^n} + \frac{i}{2^{m+n}}\right), \xi\left(\frac{k+1}{2^n}\right)\right], \omega'_m = \max_{0\leqq i\leqq 2^m} \rho'_{i,m}$$

$$(m = 0, 1, 2, \cdots) \,.$$

Let us choose a sample function $\xi(t)$ for which the event $A_{k+1,n} \cap D_{n+1}$ occurs. Then

$$\rho'_{0,0} = \rho_{1,0} = \rho\left[\xi\left(\frac{k}{2^n}\right), \xi\left(\frac{k+1}{2^n}\right)\right] < L\alpha^n \,.$$

Since the event D_{n+1} occurs, at least one of the two quantities

$$\rho\left[\xi\left(\frac{k}{2^n}\right), \xi\left(\frac{k}{2^n} + \frac{1}{2^{n+1}}\right)\right], \rho\left[\xi\left(\frac{k}{2^n} + \frac{1}{2^{n+1}}\right), \xi\left(\frac{k+1}{2^n}\right)\right] \tag{5}$$

is less than $L\alpha^{n+1}$. If this holds for the first of them, then $\rho_{1,1} \leqq L\alpha^{n+1}$ and *a fortiori* $\rho_{1,1} \leqq L(\alpha^n + \alpha^{n+1})$.

On the other hand,

$$\rho'_{1,1} = \rho\left[\xi\left(\frac{k}{2^n} + \frac{1}{2^{n+1}}\right), \xi\left(\frac{k+1}{2^n}\right)\right]$$
$$\leqq \rho'_{0,0} + \rho_{1,1} \leqq L(\alpha^n + \alpha^{n+1}) \,.$$

In the case in which the second of the quantities (5) is less than $L\alpha^{n+1}$, the same reasoning can be applied with the roles of the quantities $\rho_{1,1}$ and $\rho'_{1,1}$ reversed. In both cases,

$$\omega_1 \leqq L(\alpha^n + \alpha^{n+1}), \qquad \omega'_1 \leqq L(\alpha^n + \alpha^{n+1}) \,.$$

Analogous reasoning leads inductively to the inequalities

$$\omega_m \leqq L(\alpha^n + \alpha^{n+1} + \cdots + \alpha^{n+m}) \leqq \frac{L\alpha^n}{1-\alpha}, \tag{6}$$

$$\omega'_m \leqq L(\alpha^n + \alpha^{n+1} + \cdots + \alpha^{n+m}) \leqq \frac{L\alpha^n}{1-\alpha} \,. \tag{7}$$

Let us suppose that (6) and (7) are satisfied. Since

$$\rho_{2i+1,m+1} \leqq \rho_{i,m} + \rho\left[\xi\left(\frac{k}{2^n} + \frac{i}{2^{n+m}}\right), \xi\left(\frac{k}{2^n} + \frac{2i+1}{2^{n+m+1}}\right)\right],$$

$$\rho_{2i+1,m+1} \leqq \rho_{i+1,m} + \rho\left[\xi\left(\frac{k}{2^n} + \frac{2i+1}{2^{m+n+1}}\right), \xi\left(\frac{k}{2^n} + \frac{i+1}{2^{n+m}}\right)\right],$$

and since occurrence of the event D_{n+1} implies satisfaction of one of the inequalities

$$\rho\left[\xi\left(\frac{k}{2^n} + \frac{i}{2^{m+n}}\right), \xi\left(\frac{k}{2^n} + \frac{2i+1}{2^{n+m+1}}\right)\right] < L\alpha^{m+n+1},$$

$$\rho\left[\xi\left(\frac{k}{2^n} + \frac{i+1}{2^{n+m}}\right), \xi\left(\frac{k}{2^n} + \frac{2i+1}{2^{n+m+1}}\right)\right] < L\alpha^{n+m+1},$$

we have $\rho_{2i+1,m+1} \leqq \omega_m + L\alpha^{n+m+1}$. Furthermore, $\rho_{2i,m+1} = \rho_{i,m}$. These relationships prove inequality (6). Inequality (7) is proved in the same way.

3. Let us show that if the event D_n occurs for a sample function $\xi(t)$, then for arbitrary m there exists a number $j_{n,m}$ such that

$$\max_{0 \leqq j \leqq j_{n,m}} \rho\left[\xi\left(\frac{k-1}{2^n}\right), \xi\left(\frac{k-1}{2^n} + \frac{j}{2^{n+m}}\right)\right] \leqq \frac{L\alpha^n}{(1-\alpha)^2} \tag{8}$$

and

$$\max_{j_{n,m}+1 \leqq j \leqq 2^{m+1}} \rho\left[\xi\left(\frac{k-1}{2^n} + \frac{j}{2^{n+m}}\right), \xi\left(\frac{k+1}{2^n}\right)\right] \leqq \frac{L\alpha^n}{(1-\alpha)^2}. \tag{9}$$

Let us note that occurrence of D_n implies satisfaction of one of the inequalities

$$\rho\left[\xi\left(\frac{k-1}{2^n}\right), \xi\left(\frac{k}{2^n}\right)\right] < L\alpha^n, \quad \rho\left[\xi\left(\frac{k}{2^n}\right), \xi\left(\frac{k+1}{2^n}\right)\right] < L\alpha^n.$$

If both these inequalities hold, it follows from (6) and (7) that we can take the number 2^m for $j_{n,m}$. Suppose that only the first of these inequalities is satisfied. On the basis of (6), inequality (8) holds for $j \leqq 2^m$. Consider now the interval $[k/2^n, (k+1)/2^n]$. For this interval at least one of the inequalities

$$\begin{aligned} &\rho\left[\xi\left(\frac{k}{2^n}\right), \xi\left(\frac{2k+1}{2^{n+1}}\right)\right] < L\alpha^{n+1} \\ &\rho\left[\xi\left(\frac{2k+1}{2^{n+1}}\right), \xi\left(\frac{k+1}{2^n}\right)\right] < L\alpha^{n+1} \end{aligned} \tag{10}$$

is satisfied. If the first of these is satisfied, then on the basis of (6),

$$\max_{1 \leqq i \leqq 2^{m-1}} \rho\left[\xi\left(\frac{k}{2^n}\right), \xi\left(\frac{k}{2^n} + \frac{i}{2^{n+m}}\right)\right] \leqq \frac{L\alpha^{n+1}}{1-\alpha}.$$

Consequently

$$\max_{1 \leqq i \leqq 2^m + 2^{m-1}} \rho\left[\xi\left(\frac{k-1}{2^n}\right), \xi\left(\frac{k-1}{2^n} + \frac{i}{2^{n+m}}\right)\right] \leqq \frac{L}{1-\alpha}(\alpha^n + \alpha^{n+1}).$$

On the other hand, if the second of inequalities (10) is satisfied, it follows from (7) that

$$\max_{2^m+2^{m-1}\leqq i\leqq 2^{m+1}} \rho\left[\xi\left(\frac{k-1}{2^n}+\frac{i}{2^{n+m}}\right), \xi\left(\frac{k+1}{2^n}\right)\right] \leqq \frac{L\alpha^{n+1}}{1-\alpha} .$$

We can continue this process inductively, and after a finite number of steps we arrive at proof of inequalities (8) and (9).

4. Suppose that the event D occurs. Then beginning with some n_0, all the events D_n for $n \geqq n_0$ occur for the sample function of the process. For arbitrary $\varepsilon > 0$ let us find an $n \geqq n_0$ such that $2L\alpha^n/(1-\alpha)^2 < \varepsilon$. We note that on an arbitrary set of the form

$$J \cap \left[\frac{k-1}{2^n}, \frac{k+1}{2^n}\right],$$

there cannot be more than a single ε-oscillation. To see this, suppose that such a set contains three points t_0, t_1, and t_2 such that

$$2^n t_0 = k - 1 + \frac{j_0}{2^m} ,$$

$$2^n t_1 = k - 1 + \frac{j_1}{2^m} ,$$

$$2^n t_2 = k - 1 + \frac{j_2}{2^m} ,$$

where $0 \leqq j_0 < j_1 < j_2 \leqq 2^{m+1}$, and

$$\rho(\xi(t_1), \xi(t_0)) > \varepsilon ,$$
$$\rho(\xi(t_2), \xi(t_1)) > \varepsilon .$$

Then at least two of the three numbers j_0, j_1, j_2 lie on one side of the number $j_{m,n}$ (cf. paragraph 3 of the proof). Suppose for example that $j_1 \leqq j_{m,n}$. Then $|\xi(t_1) - \xi(t_0)| > \varepsilon$, which contradicts the inequality

$$\rho(\xi(t_1), \xi(t_0)) \leqq \rho\left(\xi(t_0), \xi\left(\frac{k-1}{2^n}\right)\right) + \rho\left(\xi\left(\frac{k-1}{2^n}\right), \xi(t_1)\right) \leqq \frac{2L\alpha^n}{(1-\alpha)^2} < \varepsilon .$$

Thus the set

$$J \cap \left[\frac{k-1}{2^n}, \frac{k+1}{2^n}\right]$$

cannot contain more than one ε-oscillation. From this it follows that the function $\xi(t)$ has no more than 2^n ε-oscillations on J.

Thus the sample functions of the process $\xi(t)$ have, with probability 1, finitely many ε-oscillations; that is, they do not have discontinuities of the second kind.

Corollary. *Suppose that a random process in the broad sense with range in a complete separable locally compact space is defined on the interval* $[a, b]$ *and that the "three-dimensional" distributions of this process satisfy condition* (1) *(or condition* (2)*). Then there exists a stochastically equivalent random process without discontinuities of the second kind.*

We now give a different test for nonexistence of discontinuities of the second kind. Suppose that $\mathfrak{F}_s = \sigma\{\xi(\tau); \tau \leqq s\}$ is the σ-algebra generated by random elements $\xi(\tau)$ where $\tau \leqq s$. We introduce the quantity

$$\alpha(\varepsilon, \delta) = \inf_{\Gamma'} \sup_{a \leqq s \leqq t \leqq s+\delta \leqq b} \mathsf{P}\{\rho(\xi(s); \xi(t)) \geqq \varepsilon \mid \mathfrak{F}_s\}, \tag{11}$$

where Γ' is an arbitrary set in $\mathfrak{F}_s$ such that $\mathsf{P}(\Gamma') = 1$. One can easily see that there exists a $\Gamma_0 \in \sigma\{\xi(\tau); \tau \leqq s\}$ such that $\mathsf{P}(\Gamma_0) = 1$ and the greatest lower bound in (11) is attained on the set Γ_0:

$$\alpha(\varepsilon, \delta) = \sup_{\substack{a \leqq s \leqq t \leqq s+\delta \leqq b \\ \xi(\tau) \in \Gamma_0}} \mathsf{P}\{\rho(\xi(s); \xi(t)) \geqq \varepsilon \mid \mathfrak{F}_s\}.$$

We shall show that the limit condition $\alpha(\varepsilon, \delta) \rightarrow 0$ as $\delta \rightarrow 0$ for arbitrary positive ε ensures nonexistence of discontinuities of the second kind for separable processes. Suppose that $[c, d]$ is a fixed interval contained in $[a, b]$ and that I is an arbitrary n-tuple of instants of time $t_1, t_2, \cdots, t_n$, where $s \leqq c \leqq t_1 < t_2 < \cdots < t_n \leqq d$. Let $A(\varepsilon, Z)$ denote the event that a sample function of the random process $\xi(t)$ on $[c, d] \cap Z$ has at least one ε-oscillation.

Lemma 2. *With probability* 1,

$$\mathsf{P}\{A(\varepsilon, I) \mid \mathfrak{F}_s\} \leqq 2\alpha\left(\frac{\varepsilon}{4}, d - c\right). \tag{12}$$

Proof. Let us note first of all that since $\mathfrak{F}_s \subset \mathfrak{F}_t$ for $s < t$, on the basis of the properties of conditional mathematical expectations (cf. Theorem 3, Section 6, Chapter III) it follows that for $s < t < u$,

$$\begin{aligned}\mathsf{P}\{\rho(\xi(t); \xi(u)) \geqq \varepsilon \mid \mathfrak{F}_s\} &= \mathsf{M}\{\mathsf{P}\{\rho(\xi(t); \xi(u)) \geqq \varepsilon \mid \mathfrak{F}_t\} \mid \mathfrak{F}_s\} \\ &\leqq \alpha(\varepsilon, u - t) (\mathrm{mod}\ \mathsf{P}).\end{aligned} \tag{13}$$

We now define

$$B_k = \left\{\rho(\xi(c), \xi(t_i)) < \frac{\varepsilon}{2}, \quad i = 1, 2, \cdots, k - 1, \right.$$
$$\left. \rho(\xi(c), \xi(t_k)) \geqq \frac{\varepsilon}{2}\right\},$$

$$C_k = \left\{\rho(\xi(t_k), \xi(d)) \geq \frac{\varepsilon}{4}\right\}, \quad D_k = B_k \cap C_k, \quad k = 1, \cdots, n\,,$$

$$C_0 = \left\{\rho(\xi(c), \xi(d)) \geq \frac{\varepsilon}{4}\right\}.$$

The events B_k are incompatible and $A(\varepsilon, I) \subset C_0 \cup D$ where $D = \bigcup_{k=1}^{n} D_k$. To see this, note that if $A(\varepsilon, I)$ holds, then there exists a k at which the quantity $\rho(\xi(c), \xi(t_k))$ first becomes equal to or greater than $\varepsilon/2$; that is, one of the events B_k (for $k = 1, \cdots, n$) occurs. If $\bar{C}_k$ occurs, that is, if $\rho(\xi(t_k), \xi(d)) < \varepsilon/4$, then

$$\rho(\xi(c), \xi(d)) \geq \rho(\xi(c), \xi(t_k)) - \rho(\xi(t_k), \xi(d)) > \frac{\varepsilon}{4}\,;$$

that is, the event C_0 occurs. Thus $A(\varepsilon, I) \subset C_0 \cup D$. Now with probability 1,

$$\begin{aligned}\mathsf{P}\{D_k \mid \mathfrak{F}_s\} &= \mathsf{M}\{\chi_{D_k} \mid \mathfrak{F}_s\} = \mathsf{M}\{\mathsf{M}\{\chi_{B_k}\chi_{C_k} \mid \mathfrak{F}_{t_k}\} \mid \mathfrak{F}_s\} \\ &= \mathsf{M}\{\chi_{B_k}\mathsf{P}\{C_k \mid \mathfrak{F}_{t_k}\} \mid \mathfrak{F}_s\} \leq \alpha\left(\frac{\varepsilon}{4}, d - c\right)\mathsf{M}\{\chi_{B_k} \mid \mathfrak{F}_s\}\,,\end{aligned}$$

where χ_A denotes, as usual, the characteristic function of the event A. From this we get

$$\begin{aligned}\mathsf{P}\{D \mid \mathfrak{F}_s\} &= \sum_{k=1}^{n} \mathsf{P}\{D_k \mid \mathfrak{F}_s\} \leq \alpha\left(\frac{\varepsilon}{4}, d - c\right)\mathsf{M}\left\{\sum_{k=1}^{n} \chi_{B_k} \mid \mathfrak{F}_s\right\} \\ &\leq \alpha\left(\frac{\varepsilon}{4}, d - c\right) (\mathrm{mod}\ \mathsf{P})\,.\end{aligned}$$

On the basis of (13), $\mathsf{P}\{C_0 \mid \mathfrak{F}_s\} \leq \alpha(\varepsilon/4, d - c)$. Thus

$$\mathsf{P}\{A(\varepsilon, I) \mid \mathfrak{F}_s\} \leq \mathsf{P}\{D \mid \mathfrak{F}_s\} + \mathsf{P}\{C_0 \mid \mathfrak{F}_s\} \leq 2\alpha\left(\frac{\varepsilon}{4}, d - c\right) (\mathrm{mod}\ \mathsf{P})\,,$$

which completes the proof of the lemma.

Lemma 3. *Suppose that $A^k(\varepsilon, I)$ denotes the event that $\xi(t)$ has at least k ε-oscillations on I. Then*

$$\mathsf{P}\{A^k(\varepsilon, I) \mid \mathfrak{F}_s\} \leq \left[2\alpha\left(\frac{\varepsilon}{4}, d - c\right)\right]^k (\mathrm{mod}\ \mathsf{P})\,. \tag{14}$$

Proof. Let $B_r(\varepsilon, I)$ denote the event that the sample function of the process $\xi(t)$ has at least $(k - 1)$ ε-oscillations on the set $(t_1, \cdots, t_r)$ but has fewer than $(k - 1)$ ε-oscillations on $(t_1, \cdots, t_{r-1})$. The events $B_r(\varepsilon, I)$ (for $r = 1, 2, \cdots, n$) are incompatible and

$$\bigcup_{r=1}^{n} B_r(\varepsilon, I) = A^{k-1}(\varepsilon, I) \supset A^k(\varepsilon, I)\,.$$

On the other hand, if $A^k(\varepsilon, I) \cap B_r(\varepsilon, I) \neq \phi$, there is at least one ε-oscillation on the set $(t_r, t_{r+1}, \cdots, t_n)$. Consequently

$$A^k(\varepsilon, I) \subset \bigcup_{r=1}^{n} (B_r(\varepsilon, I) \cap C_r(\varepsilon, I))$$

where $C_r(\varepsilon, I)$ means that $\xi(t)$ has at least one ε-oscillation on $(t_r, t_{r+1}, \cdots, t_n)$. Therefore

$$\mathsf{P}\{A^k(\varepsilon, I) \mid \mathfrak{F}_s\} \leqq \sum_{r=1}^{n} \mathsf{P}\{B_r(\varepsilon, I) \cap C_r(\varepsilon, I) \mid \mathfrak{F}_s\} \pmod{\mathsf{P}} . \tag{15}$$

Using Theorem 3, Section 6, Chapter III, we obtain

$$\begin{aligned} \mathsf{P}\{B_r(\varepsilon, I) \cap C_r(\varepsilon, I) \mid \mathfrak{F}_s\} &= \mathsf{M}\{\mathsf{M}\{\chi_{B_r(\varepsilon,I)}\chi_{C_r(\varepsilon,I)} \mid \mathfrak{F}_{t_r}\} \mid \mathfrak{F}_s\} \\ &\leqq \mathsf{M}\{\chi_{B_r(\varepsilon,I)}\mathsf{P}\{C_r(\varepsilon, I) \mid \mathfrak{F}_{t_r}\} \mid \mathfrak{F}_s\} \\ &\leqq 2\alpha\left(\frac{\varepsilon}{4}, d - c\right)\mathsf{P}\{B_r(\varepsilon, I) \mid \mathfrak{F}_s\} \pmod{\mathsf{P}} . \end{aligned}$$

Substituting this inequality into (15), we obtain

$$\begin{aligned} \mathsf{P}\{A^k(\varepsilon, I) \mid \mathfrak{F}_s\} &\leqq 2\alpha\left(\frac{\varepsilon}{4}, d - c\right)\sum_{r=1}^{n} \mathsf{P}\{B_r(\varepsilon, I) \mid \mathfrak{F}_s\} \\ &= 2\alpha\left(\frac{\varepsilon}{4}, d - c\right)\mathsf{P}\{A^{k-1}(\varepsilon, I) \mid \mathfrak{F}_s\} \pmod{\mathsf{P}} , \end{aligned}$$

from which the desired result follows.

Theorem 2. *If $\xi(t)$ is a separable process and*

$$\lim_{\delta \to 0} \alpha(\varepsilon, \delta) = 0 \tag{16}$$

for arbitrary $\varepsilon > 0$, the process $\xi(t)$ has no discontinuities of the second kind.

It will be sufficient to show that with probability 1 every sample function $\xi(t)$ has only finitely many ε-oscillations. Let J denote the set of separability of the process $\xi(t)$. Let us represent it in the form

$$J = \bigcup_{n=1}^{\infty} I_n$$

where I_n is an increasing sequence of sets each consisting of finitely many elements. Let ε denote any positive number. We partition the interval $[a, b]$ into m subintervals Δ_r for $r = 1, \cdots, m$, all of equal length, so that

$$2\alpha\left(\frac{\varepsilon}{4}, \frac{b - a}{m}\right) = \beta < 1 .$$

Then

$$\begin{aligned} \mathsf{P}\{A^\infty(\varepsilon, J \cap \Delta_r) \mid \mathfrak{F}_s\} &\leqq \mathsf{P}\{A^k(\varepsilon, J \cap \Delta_r) \mid \mathfrak{F}_s\} \\ &= \lim_{n\to\infty} \mathsf{P}\{A^k(\varepsilon, I_n \cap \Delta_r) \mid \mathfrak{F}_s\} \leqq \beta^k , \end{aligned}$$

so that

$$\mathsf{P}\{A^\infty(\varepsilon, I \cap \Delta_r) \mid \mathfrak{F}_s\} = 0 \pmod{\mathsf{P}} \quad \text{and} \quad \mathsf{P}\{A^\infty(\varepsilon, I \cap \Delta_r)\} = 0\,.$$

Consequently, $\mathsf{P}\{A^\infty(\varepsilon, J)\} = 0$. This completes the proof of the theorem.

REMARK 1. Under the hypotheses of Theorem 2 the process $\xi(t)$ is stochastically continuous. If $s < t$, then $\mathsf{P}\{\rho(\xi(s), \xi(t)) > \varepsilon\} = \mathsf{MP}\{\rho(\xi(s), \xi(t)) > \varepsilon \mid \mathfrak{F}_s\} \leqq \alpha(\varepsilon, \delta)$ for $t - s < \delta$.

Corollary. *A separable stochastically continuous process with independent increments has no discontinuities of the second kind.*

Proof. On the basis of the definition of processes with independent increments and uniform stochastic continuity (cf. Theorem 5, Section 2), we have

$$\mathsf{P}\{\rho(\xi(s), \xi(t)) \geqq \varepsilon \mid \mathfrak{F}_s\} = \mathsf{P}\{\rho(\xi(s), \xi(t)) \geqq \varepsilon\} \leqq \varphi(\varepsilon, \delta) \pmod{\mathsf{P}}$$

where $|t - s| \leqq \delta$, $\varphi(\varepsilon, \delta)$ is independent of t and s, and $\varphi(\varepsilon, \delta) \to 0$ as $\delta \to 0$. Thus the hypotheses of Theorem 2 are satisfied.

We recall that if a process is separable the values of the sample functions $\xi(t)$ are, with probability 1, the limiting values of sequences $\{\xi(t_i)\}$ as $t_i \to t$, where each t_i belongs to the set of separability. Here if a process has no discontinuities of the second kind, then $\xi(t)$ will, with probability 1, be equal to $\xi(t - 0)$ or $\xi(t + 0)$ for every t.

Theorem 3. *If $\xi(t)$ is a stochastically continuous process without discontinuities of the second kind, there exists a process $\xi'(t)$ equivalent to it whose sample functions are continuous from the right* (mod P).

Proof. Let A denote the event that the limit

$$\lim_{n\to\infty} \xi\left(t + \frac{1}{n}\right)$$

exists for each $t \in [a, b]$. The probability of this event is 1. Let us set $\xi'(t) = \lim_{n\to\infty} \xi(t + 1/n)$ for the outcome A and $\xi'(t) = \xi(t)$ for the outcome $\bar{A}$. We then have

$$\{\xi'(t) \neq \xi(t)\} = \bigcup_{m=1}^{\infty} \left\{\rho(\xi(t), \xi'(t)) > \frac{1}{m}\right\} \cap A\,,$$

$$\mathsf{P}\{\xi'(t) \neq \xi(t)\} = \lim_{m\to\infty} \mathsf{P}\left(\left\{\rho(\xi(t), \xi'(t)) > \frac{1}{m}\right\} \cap A\right).$$

On the other hand

$$\mathsf{P}\left\{\rho(\xi(t), \xi'(t)) > \frac{1}{m}\right\}$$
$$= \mathsf{P}\left\{\bigcup_{k=1}^{\infty} \bigcap_{n=k}^{\infty} \left\{\rho\left(\xi(t), \xi\left(t + \frac{1}{n}\right)\right) > \frac{1}{m}\right\}\right\}$$
$$= \lim_{k\to\infty} \mathsf{P}\left\{\bigcap_{n=k}^{\infty} \left\{\rho\left(\xi(t), \xi\left(t + \frac{1}{n}\right)\right) > \frac{1}{m}\right\}\right\}$$
$$\leqq \lim_{n\to\infty} \mathsf{P}\left\{\rho\left(\xi(t), \xi\left(t + \frac{1}{n}\right)\right) > \frac{1}{m}\right\}.$$

Thus $\mathsf{P}\{\xi'(t) \neq \xi(t)\} = 0$. We note that with outcome A the function $\xi'(t)$ is continuous from the right. This completes the proof of the theorem.

One can prove by an analogous method the existence of a stochastically equivalent process that is continuous from the left.

5. CONTINUOUS RANDOM FUNCTIONS

Let $\mathfrak{T}$ denote the interval $[a, b]$, let X denote a complete metric space, and let $\xi(t)$ denote a random process defined on $\mathfrak{T}$ into X.

Definition 1. A process $\xi(t)$ for $t \in \mathfrak{T}$ is said to be *continuous* if almost all sample functions of the process are continuous on $\mathfrak{T}$.

For processes without discontinuities of the second kind we can establish a rather simple sufficient condition for continuity.

Theorem 1. *Let $\{t_{n,k}\}$ for $n = 1, 2, \cdots$ and $k = 0, 1, \cdots, m_n$ denote a sequence of partitions of the interval $[a, b]$:*

$$a = t_{n,0} < t_{n,1} < \cdots < t_{n,m_n} = b\,,$$

define

$$\lambda_n = \max_{1\leqq k\leqq m_n} (t_{n,k} - t_{n,k-1}) \to 0 \quad as \quad n \to \infty\,.$$

If a separable process $\xi(t)$ has no discontinuities of the second kind, then the random process is continuous if for every $\varepsilon > 0$,

$$\sum_{k=1}^{m_n} \mathsf{P}\{\rho[\xi(t_{n,k}), \xi(t_{n,k-1})] > \varepsilon\} \to 0 \quad as \quad n \to \infty\,. \tag{1}$$

Proof. Let ν_ε (where $0 \leqq \nu_\varepsilon \leqq \infty$) denote the number of values of t at which $\rho[\xi(t + 0), \xi(t - 0)] > 2\varepsilon$ and let $\nu_\varepsilon^{(n)}$ denote the number of indices k for which $\rho[\xi(t_{n,k}), \xi(t_{n,k-1})] > \varepsilon$. Obviously $\nu_\varepsilon \leqq \underline{\lim}_{n\to\infty} \nu_\varepsilon^{(n)}$. On the other hand

$$\mathsf{M}\nu_\varepsilon^{(n)} = \sum_{k=1}^{m_n} \mathsf{P}\{\rho[\xi(t_{n,k}), \xi(t_{n,k-1})] > \varepsilon\}\,.$$

By virtue of Fatou's lemma (Theorem 2, Section 5, Chapter II),

$$\mathsf{M}\nu_\varepsilon \leqq \mathsf{M}\varliminf_{n\to\infty}\nu_\varepsilon \leqq \varliminf_{n\to\infty}\mathsf{M}\nu_\varepsilon^{(n)} .$$

Thus $\mathsf{M}\nu_\varepsilon = 0$; that is, $\nu_\varepsilon = 0$ with probability 1 for arbitrary $\varepsilon > 0$. Consequently, for arbitrary t we have $\xi(t - 0) = \xi(t + 0)$ with probability 1. By virtue of the separability of the process, $\xi(t) = \xi(t - 0) = \xi(t + 0)$; that is, the process is continuous.

Theorem 2. *Suppose that there exist three positive constants C, r, and β such that for arbitrary $\varepsilon > 0$,*

$$\mathsf{P}\{\rho[\xi(t_1), \xi(t_2)] > \varepsilon\} \leqq \frac{C\,|\,t_2 - t_1\,|^{r+1}}{\varepsilon^\beta} . \qquad (2)$$

If the process $\xi(t)$ is separable, it is continuous.

Proof. Condition (2) is a special case of condition (2) of Section 4 (with $q = 0$ and $\beta = p$). In addition, condition (2) ensures stochastic continuity of the process. Therefore the process $\xi(t)$ has no discontinuities of the second kind. If $\{t_{n,k}\}$ for $n = 1, 2, \cdots$ and $k = 0, 1, \cdots, m_n$ is a sequence of partitions of the interval $[a, b]$, then

$$\sum_{k=1}^{m_n} \mathsf{P}\{\rho[\xi(t_{n,k}), \xi(t_{n,k-1})] > \varepsilon\} \leqq \sum_{k=1}^{m_n} \frac{C\,|\,t_{n,k} - t_{n,k-1}\,|^{1+r}}{\varepsilon^\beta}$$

$$\leqq \frac{C(b - a)}{\varepsilon^\beta} \max_{1\leqq k\leqq m_n} |\,t_{n,k} - t_{n,k-1}\,|^r \longrightarrow 0$$

as

$$\lambda_n = \max_{1\leqq k\leqq m_n} |\,t_{n,k} - t_{n,k-1}\,| \longrightarrow 0 .$$

By virtue of Theorem 1, the process $\xi(t)$ is continuous.

REMARK 1. Condition (2) of Theorem 2 can be replaced with the somewhat more stringent but, from a practical point of view, more convenient inequality

$$\mathsf{M}\rho^\beta[\xi(t_1), \xi(t_2)] \leqq C\,|\,t_2 - t_1\,|^{1+r} . \qquad (3)$$

Applying Chebyshev's inequality to the left-hand member of inequality (2) and keeping (3) in mind, we obtain the right-hand member of (2).

These two theorems give only sufficient conditions for continuity of a random process. For the particular case of processes with independent increments, the conditions of Theorem 1 are also necessary.

Theorem 3. *If a process $\xi(t)$ for $t \in [a, b]$ with independent increments is continuous, then condition* (1) *is satisfied for an arbitrary* *sequence* $\{t_{n,k}\}$ *for* $n = 1, 2, \cdots$ *and* $k = 0, \cdots, m_n$ *of partitions of the interval* $[a, b]$ *such that*

$$\lambda_n = \max_{1 \leq k \leq m_n} (t_{n,k} - t_{n,k-1}) \longrightarrow 0 .$$

Proof. Let us set $\Delta_h = \sup_{|t_1 - t_2| \leq h} \rho[\xi(t_1), \xi(t_2)]$. Since the process $\xi(t)$ is continuous on $[a, b]$, it follows that $\Delta_h \longrightarrow 0$ as $h \longrightarrow 0$ with probability 1. Therefore $\lim_{h\to 0} \mathsf{P}\{\Delta_h > \varepsilon\} = 0$. On the other hand, if $\lambda_n < h$ we have

$$\begin{aligned}
\mathsf{P}\{\Delta_h > \varepsilon\} &\geqq \mathsf{P}\{\sup \rho[\xi(t_{n,k}), \xi(t_{n,k-1})] > \varepsilon\} \\
&= \mathsf{P}\{\rho[\xi(t_{n,1}), \xi(t_{n,0})] > \varepsilon\} \\
&\quad + \mathsf{P}\{\rho[\xi(t_{n,1}), \xi(t_{n,0})] \leqq \varepsilon\}\mathsf{P}\{\rho[\xi(t_{n,2}), \xi(t_{n,1})] > \varepsilon\} \\
&\quad \cdots + \prod_{k=1}^{m_n - 1} \mathsf{P}\{\rho[\xi(t_{n,k}), \xi(t_{n,k-1})] \leqq \varepsilon\} \\
&\quad \times \mathsf{P}\{\rho[\xi(t_{n,m_n}), \xi(t_{n,m_n - 1})] > \varepsilon\} \\
&\geqq \mathsf{P}\{\Delta_h \leqq \varepsilon\} \sum_{k=1}^{m_n} \mathsf{P}\{\rho[\xi(t_{n,k}), \xi(t_{n,k-1})] > \varepsilon\} ,
\end{aligned}$$

from which it follows that for arbitrary $\varepsilon > 0$,

$$\sum_{k=1}^{n} \mathsf{P}\{\rho[\xi(t_{n,k}), \xi(t_{n,k-1})] > \varepsilon\} \leqq \frac{\mathsf{P}\{\Delta_h > \varepsilon\}}{\mathsf{P}\{\Delta_h \leqq \varepsilon\}} \longrightarrow 0$$

as $h \longrightarrow 0$. This completes the proof of the theorem.

From Theorem 2 of Section 2 and Theorem 1 of the present section we get another test for continuity of a process.

Theorem 4. *If a process* $\xi(t)$ *is separable and*

$$\lim_{\delta \to 0} \frac{\alpha(\varepsilon, \delta)}{\delta} = 0 \tag{4}$$

for arbitrary $\varepsilon > 0$ *where* $\alpha(\varepsilon, \delta)$ *is determined by formula* (11) *of Section* 4, *then the process* $\xi(t)$ *is continuous.*

Since satisfaction of condition 4 implies that the process $\xi(t)$ has no discontinuities of the second kind, it will be sufficient to verify relation (1). Remembering that

$$\mathsf{P}\{\rho[\xi(t_{n,k}), \xi(t_{n,k-1})] > \varepsilon\} \leqq \alpha(\varepsilon, \Delta t_{n,k})$$

where $\Delta t_{n,k} = t_{n,k} - t_{n,k-1}$, we obtain the result that

$$\sum_{k=1}^{m_n} \mathsf{P}\{\rho[\xi(t_{n,k}), \xi(t_{n,k-1})] > \varepsilon\} \leqq (b - a) \max_{1 \leqq k \leqq n} \frac{\alpha(\varepsilon, \Delta t_{n,k})}{\Delta t_{n,k}} \longrightarrow 0$$

as $\lambda_n \longrightarrow 0$. This completes the proof of the theorem.

Let us look at the condition for continuity of a Gaussian process $\xi(t)$ that assumes real values. For the characteristic function of the random variable $\xi(t)$ we have the expression

$$\Psi(\lambda, t) = \mathsf{M}e^{i\lambda\xi(t)} = \exp\left(im(t)\lambda - \frac{\lambda^2}{2}\sigma^2(t)\right).$$

If a process $\xi(t)$ is continuous, it follows from Lebesgue's theorem on taking the limit under the integral sign that $\Psi(\lambda, t)$ is continuous with respect to t for arbitrary λ. This in turn implies continuity of the function

$$\ln \Psi(t, \lambda) = im(t)\lambda - \frac{\lambda^2}{2}\sigma^2(t)$$

with respect to t and hence continuity of the functions

$$\sigma^2(t) = \frac{\ln \Psi(\lambda, t) + \ln \Psi(-\lambda, t)}{-\lambda^2}$$

and

$$m(t) = \frac{1}{i\lambda}\left[\ln \Psi(\lambda, t) + \frac{\lambda^2}{2}\sigma^2(t)\right].$$

Thus continuity of the functions $m(t)$ and $\sigma^2(t)$ is a necessary condition for continuity of a Gaussian process.

Turning to sufficient conditions for continuity, let us suppose that $m(t) = \mathsf{M}\xi(t) = 0$. (If this is not the case, we may consider the process $\xi'(t) = \xi(t) - m(t)$ instead of $\xi(t)$.) Let $R(t_1, t_2)$ denote the correlation function of the process $\xi(t)$. Then

$$\mathsf{M}(\xi(t_2) - \xi(t_1))^2 = R(t_2, t_2) - 2R(t_1, t_2) + R(t_1, t_1) = \Delta R,$$
$$\mathsf{M}(\xi(t_2) - \xi(t_1))^{2m} = (2m - 1)!\,|\Delta R|^m.$$

Using Remark 1, we obtain the following result:

Theorem 5. *If there exist $C > 0$ and $\alpha > 0$ such that $m(t)$ is continuous and*

$$|R(t_2, t_2) - 2R(t_1, t_2) + R(t_1, t_1)| \leqq C|t_2 - t_1|^\alpha, \; t_i \in [a, b],$$

then a separable Gaussian process with mathematical expectation $m(t)$ and correlation function $R(t_1, t_2)$ is continuous.

For Gaussian processes with independent increments we can go even farther. In such a case, $R(t_1, t_2) = R(t_1, t_1) = \sigma^2(t_1)$ for $t_1 \leqq t_2$, $\Delta R = \sigma^2(t_2) - \sigma^2(t_1)$, and

$$\sum_{k=1}^{m} \mathsf{P}\{|\xi(t_k) - \xi(t_{k-1})| > \varepsilon\} \leqq \frac{1}{\varepsilon^4}\sum_{k=1}^{m} \mathsf{M}|\xi(t_k) - \xi(t_{k-1})|^4$$
$$\leqq \frac{3}{\varepsilon^4}\sum_{k=1}^{m} [\sigma^2(t_k) - \sigma^2(t_{k-1})]^2$$
$$\leqq \frac{3[\sigma^2(b) - \sigma^2(a)]}{\varepsilon^4} \max\,[\sigma^2(t_k) - \sigma^2(t_{k-1})].$$

Thus if the function $\sigma^2(t)$ is continuous, then by virtue of the corollary to Theorem 2 of Section 4 and Theorem 1 of the present section a separable Gaussian process with independent increments is continuous.

Theorem 6. *For a separable Gaussian process $\xi(t)$ with independent increments to be continuous for $(a \leqq t \leqq b)$, it is necessary and sufficient that the functions*

$$m(t) = \mathsf{M}\xi(t) \,,$$
$$\sigma^2(t) = \mathsf{M}[\xi(t) - m(t)]^2 \,,$$
$$a \leqq t \leqq b$$

be continuous.

V

LINEAR TRANSFORMATION OF RANDOM PROCESSES

In this chapter, we shall consider linear operations on random processes. Examples of such operations are differentiation and integration of processes, and transformations by means of differential and integral equations. The problem of prediction of the value of a random function or the problem of filtering a useful component from the observed values of a random function that is the sum of a transmitted signal and a distorting "noise" is actually solved, at the present time, only in the framework of the theory of linear transformations of random processes.

To study the problems that are presented here we shall consider a Hilbert space of random variables, which in a number of cases enables us to obtain final solutions that are suitable for applications. It is assumed that the reader is familiar with the elementary concepts of the theory of Hilbert spaces, a few facts of which are recalled in Section 1.

1. HILBERT SPACES

Let H denote a linear space over the field of complex numbers. Let (x, y) denote a complex-valued function defined for $x, y \in H$ with the following properties:

a. $(x, x) \geqq 0$ with equality holding only for the zero element of the space H; that is, $(x, x) = 0$ implies $x = 0$ and conversely;

b. $(x, y) = \overline{(y, x)}$;

c. $(\lambda x_1 + \mu x_2, y) = \lambda(x_1, y) + \mu(x_2, y)$.

We call (x, y) the *scalar product* of the elements x and y of the space H. We use this product to introduce the concepts of the length of a vector x, the distance between two vectors, and orthogonality of two vectors. Specifically, the *length* of a vector x is defined as $\sqrt{(x, x)}$ and is denoted by $\|x\|$. The distance between two vectors x and y is defined by $\rho(x, y) = \|x - y\| = \sqrt{(x - y, x - y)}$.

We have the following inequalities:

$$|(x, y)|^2 \leqq (x, x)(y, y) \quad \text{or} \quad |(x, y)| \leqq \|x\| \|y\|, \tag{1}$$

$$\|x + y\| \leqq \|x\| + \|y\|. \tag{2}$$

The first of these is called the "Cauchy-Schwarz inequality" and the second is called the "triangle inequality."

Two vectors x and y are said to be *orthogonal* if $(x, y) = 0$. We indicate their orthogonality by writing $x \perp y$. Two sets A and B contained in H are said to be orthogonal (and we write $A \perp B$) if every vector $x \in A$ is orthogonal to every vector $y \in B$. We note that for a pair of orthogonal vectors x and y,

$$\|x + y\|^2 = \|x\|^2 + \|y\|^2.$$

A linear space X with a scalar product defined on it is called a *Hilbert space* if it is complete, that is, if every sequence $\{x_n\}$ of elements of H such that $\|x_n - x_{n+m}\| \to 0$ as $n \to \infty$ uniformly with respect to $m \geqq 0$ converges to some limit in H.

An example of a Hilbert space is the space of functions L_2 that was introduced in Section 5 of Chapter II. The elements of this space are the complex-valued functions $f(u)$ defined on some space U with measure $\{U, \mathfrak{S}, \mu\}$ that are $\mathfrak{S}$-measurable and square-integrable:

$$\int_U |f(u)|^2 \mu(du) < \infty.$$

The scalar product of two functions f and g in L_2 is defined by

$$(f, g) = \int_U f(u)\overline{g(u)}\mu(du).$$

In particular, f and g are orthogonal in L_2 if $\int_U f(u)\overline{g(u)}\mu(du) = 0$.

Completeness of the space L_2 follows from Riesz's theorem (Theorem 7, Section 5, Chapter II).

Let $x(\theta)$ denote a function defined on an arbitrary set Θ into H and let $\psi(\theta)$ denote a nonnegative function defined on Θ that assumes arbitrarily small positive values. An element h of H is called the *limit* of the function $x(\theta)$ as $\psi(\theta) \to 0$ (and we write either $h = \lim_{\psi(\theta)\to 0} x(\theta)$ or $x(\theta) \to h$ as $\psi(\theta) \to 0$), if for arbitrary positive ε there exists a positive δ such that

$$\|x(\theta) - h\| < \varepsilon \quad \text{for} \quad 0 < \psi(\theta) < \delta.$$

The scalar product is a continuous function of both variables: If

$x_i(\theta) \longrightarrow h_i$ as $\psi(\theta) \longrightarrow 0$ (for $i = 1, 2$), then $(x_1(\theta), x_2(\theta)) \longrightarrow (h_1, h_2)$ as $\psi(\theta) \longrightarrow 0$. We have the following condition for the existence of a limit:

Lemma 1. *For the limit*

$$\lim_{\psi(\theta)\to 0} x(\theta)$$

to exist, it is necessary and sufficient that the limit

$$\lim_{\psi(\theta')+\psi(\theta'')\to 0} (x(\theta'), x(\theta'')) = A \tag{3}$$

exist.

Proof. The necessity follows from the continuity of the scalar product. Here if $h = \lim_{\psi(\theta)\to 0} x(\theta)$, then

$$\lim_{\psi(\theta')+\psi(\theta'')\to 0} (x(\theta'), x(\theta'')) = (h, h) \, . \tag{4}$$

To prove the sufficiency we note that

$$\begin{aligned} \| x(\theta') - x(\theta'') \|^2 &= (x(\theta') - x(\theta''), x(\theta') - x(\theta'')) \\ &= (x(\theta'), x(\theta')) - 2 \operatorname{Re} (x(\theta'), x(\theta'')) + (x(\theta''), x(\theta'')) \, . \end{aligned}$$

In equation (3) the number A is nonnegative. Therefore, for arbitrary positive ε there exists a positive δ such that $\| x(\theta') - x(\theta'') \| < \varepsilon^2$ whenever $0 < \psi(\theta') + \psi(\theta'') < \delta$. Now there exists a sequence $\{\theta_n\}$ such that $0 < \psi(\theta_n) < 1/n$ and $\{x(\theta_n)\}$ is a fundamental sequence. By virtue of the completeness of the space, there is a limit $\lim x(\theta_n) = h$. Then

$$\| x(\theta) - h \| \leqq \| h - x(\theta_n) \| + \| x(\theta_n) - x(\theta) \| \longrightarrow 0$$

as $n \longrightarrow \infty$ if $\psi(\theta_n) + \psi(\theta) \longrightarrow 0$ as $n \longrightarrow \infty$. This completes the proof of the lemma.

A subset H' of H is called a *subspace* if it is linear and closed in H. One of the basic elementary facts from the theory of Hilbert spaces is the following: For a fixed but arbitrary subspace F of a Hilbert space H and for an arbitrary vector $x \in H$, there exists a unique decomposition of the form

$$x = x_F + x_N$$

where $x_F \in F$ and $x_N \perp F$. The vector x_F is called the *projection* of the vector x onto the subspace F. The projection x_F of the vector x onto the subspace F enjoys the elementary property that the distance between x and a vector x' belonging to F attains a minimum value if and only if $x' = x_F$. One can easily see that the set of all vectors y that are orthogonal to F constitutes a subspace. It is called the *orthogonal complement* of F and is denoted by N. The fact that N is the orthogonal complement of F is indicated by $N = H \ominus F$ or $H = F \oplus N$. If this last relation is satisfied we say that the space H is the orthogonal sum of the subspaces F and N.

The equation

$$H_1 = F_1 \oplus F_2 \oplus \cdots \oplus F_n \oplus \cdots ,$$

where H_1 and F_k(for $k = 1, 2, \cdots$) are subspaces of H, means that the subspaces F_i and F_j are for arbitrary i and j (with $i \neq j$) orthogonal and that an arbitrary vector $x \in H$ can be represented in the form of a series $x = x_1 + x_2 + \cdots + x_n + \cdots$, where $x_n \in F_n$ for $n = 1, 2, \cdots$. In such a case, we say that H_1 is the orthogonal sum of the subspaces F_n.

Let G denote a subset of H. Let us consider all possible sums of the form $\sum_{k=1}^{n} \lambda_k g_k$ where n is an arbitrary integer, the λ_k are arbitrary complex numbers, and $g_k \in G$ (for $k = 1, \cdots, n$). The set of all such sums is called the *linear hull* of G. The closure of the linear hull of G is called the *closed linear hull* of G. This closed linear hull of the set G is the minimal subspace containing G.

Let $\{l_\alpha, \alpha \in \Lambda\}$ denote some set of mutually orthogonal normalized vectors, that is, such that $(l_\alpha, l_\beta) = 0$ for $\alpha \neq \beta$, $(l_\alpha, l_\alpha) = ||\, l_\alpha \,||^2 = 1$. Such a system of vectors is said to be *orthonormal*. An orthonormal system $\{l_\alpha, \alpha \in \Lambda\}$ is said to be a *basis* in H if the closed linear hull of the set $\{l_\alpha, \alpha \in \Lambda\}$ coincides with H.

For an arbitrary orthonormal system of vectors $\{l_\alpha, \alpha \in \Lambda\}$ and an arbitrary vector $x \in H$ we have the inequality

$$\sum_{\alpha \in \Lambda} |\,(x, l_\alpha)\,|^2 \leqq ||\, x \,||^2 .$$

The left-hand member of this inequality has only countably many nonzero terms. This inequality is called *Bessel's inequality*. For an orthonormal system to be a basis for a space, it is necessary and sufficient that equality hold in Bessel's inequality for an arbitrary vector x in that space, that is, it is necessary and sufficient that for arbitrary $x \in H$,

$$||\, x \,||^2 = \sum_{\alpha \in \Lambda} |\,(x, l_\alpha)\,|^2 . \qquad (5)$$

Equation (5) is called the "equation of closure."

If H has a basis with countably many vectors, every other basis is also countable. This will be the case if and only if H is a separable metric space.

Let A denote a linear operator mapping H into H. The quantity

$$||\, A \,|| = \sup_{||x|| < 1} ||\, A(x) \,|| .$$

is called the *norm* of the operator A. Finiteness of the norm of an operator is a necessary and sufficient condition for continuity of that operator. Operators having a finite norm are said to be bounded. For an arbitrary bounded operator A there exists a bounded operator A^* satisfying the relation

$$(Ax, y) = (x, A^* y)$$

for every pair of elements x, y in H. The operator A^* is called the *adjoint* of A. If $A = A^*$, we say that A is a *bounded self-adjoint operator*. An example of a bounded self-adjoint operator is the operator P_F indicating the projection of P onto a subspace F. This operator is defined by the equation

$$P_F(x) = x_F ,$$

where x_F is the projection of x onto F. The following relations are characteristic of the projection operator:

$$P_F^2 = P_F , \qquad || P_F || = 1 .$$

An operator S mapping H onto H is called a *unitary operator* if it preserves the scalar product, that is, if

$$(Sx, Sy) = (x, y) .$$

A unitary operator is always linear and always has an inverse that coincides with its adjoint: $S^{-1} = S^*$.

An example of a unitary operator is the Fourier transform Φ in the space L_2 of square-integrable functions defined on $(-\infty, \infty)$ with respect to Lebesgue measure: If $f \in L_2$ then

$$g(y) = \Phi(f) = \frac{1}{\sqrt{2\pi}} \int_{-\infty}^{\infty} e^{-ixy} f(x) dx \tag{6}$$

where the improper integral is to be understood to mean

$$\int_{-\infty}^{\infty} = \underset{A, B \to +\infty}{\text{l.i.m.}} \int_{-B}^{A}$$

(in L_2). The inverse (adjoint) operator to the Fourier transformation operator is given by the formula

$$\Phi^{-1}(g) = \Phi^*(g) = \frac{1}{\sqrt{2\pi}} \int_{-\infty}^{\infty} g(y) e^{ixy} dy .$$

The unitariness of the operator Φ means that

$$\int_{-\infty}^{\infty} f_1(x)\overline{f_2(x)} dx = \int_{\infty}^{\infty} g_1(y)\overline{g_2(y)} dy \tag{7}$$

where $g_k = \Phi(f_k)$ for $k = 1, 2$. The Fourier transform $g_t(y) = e^{-ity} g_0(y) = e^{-ity}\overline{g(y)}$ corresponds to the function $f_t(x) = \overline{f(t - x)}$;

hence it follows from formula 7 that

$$\int_{-\infty}^{\infty} f(t-x)f(x)dx = \int_{-\infty}^{\infty} e^{ity}g^2(y)dy . \tag{8}$$

Relationship (7) is called "Parseval's equality."

Let S denote a linear operator. A subspace H_1 of H is said to be S-invariant if $SH_1 \subset H_1$. If in addition the space $H_2 = H \ominus H_1$ is also S-invariant, we also say that H_1 reduces S.

Lemma 2. *If S is a unitary operator mapping H_1 onto H_1, then H_1 reduces S.*

Proof. We need to show that $(Sy, x) = 0$ for all $x \in H_1$ and $y \in H_2 = H \ominus H_1$. From the hypothesis of the lemma it follows that $S^{-1}H_1 \subset H_1$. Therefore $S^{-1}x = x' \in H_1$. Consequently $(Sy, x) = (y, S^{-1}x) = (y, x') = 0$ for all $x \in H_1$ and $y \in H_2$.

A generalization of the concept of a unitary operator is the concept of an isometric mapping. Let H_1 and H_2 denote two Hilbert spaces with scalar products $(x, y)_1$ and $(x, y)_2$. A mapping V of the space H_1 onto H_2 is said to be *isometric* if $(Vx, Vy)_2 = (x, y)_1$ for arbitrary $x, y \in H_1$. An isometric mapping is always linear and has an inverse isometric mapping. If there exists an isometric mapping of H_1 onto H_2, the spaces H_1 and H_2 are said to be isometrically *isomorphic*.

Let G_i denote a subset of H_i (for $i = 1, 2$). Let $L(G_i)$ denote the linear hull of the set G_i and let $H(G_i)$ denote its closure in H_i. Let us suppose that there exists a one-to-one mapping $y = \psi(x)$ (for $x \in G_1$ and $y \in G_2$) between G_1 and G_2 such that $(x_1, x_2)_1 = (y_1, y_2)_2$, $y_i = \psi(x_i) (i = 1, 2)$, $x_i \in G_1$, $y_i \in G_2$.

Theorem 1. *A one-to-one scalar-product-preserving mapping $y = \psi(x)$ on G_1 onto G_2 can be extended to an isometric mapping of $H(G_1)$ onto $H(G_2)$.*

Proof. Let us extend the correspondence $y = \psi(x)$ between G_1 and G_2 in two steps. First let us establish a one-to-one correspondence between $L(G_1)$ and $L(G_2)$ and then between $H(G_1)$ and $H(G_2)$. We define

$$\psi\left(\sum_{k=1}^{n} \lambda_k x_k\right) = \sum_{k=1}^{n} \lambda_k \psi(x_k) \tag{9}$$

for arbitrary n, $x_k \in G_1$, and complex numbers λ_k (for $k = 1, \cdots, n$). This is an unambiguous definition. Indeed, if

$$z = \sum_{k=1}^{n} \lambda_k x_k = \sum_{k=1}^{m} \mu_k x'_k , \quad x_k \in G_1 , \quad x'_k \in G_1 ,$$

then without loss of generality we can assume that $n = m$ and $x_k = x'_k$

(for $k = 1, \cdots, n$) (since some of the λ_k and μ_j may be equal to 0 we may arrange the sets as required). Then

$$\left|\left|\sum_{k=1}^{n} \lambda_k \psi(x_k) - \sum_{k=1}^{n} \mu_k \psi(x_k)\right|\right|_2^2$$

$$= \sum_{k,r=1}^{n} (\lambda_k - \mu_k)\overline{(\lambda_r - \mu_r)}(\psi(x_k), \psi(x_r))_2$$

$$= \sum_{k,r=1}^{n} (\lambda_k - \mu_k)\overline{(\lambda_r - \mu_r)}(x_k, x_r)_1$$

$$= \left|\left|\sum_{k=1}^{n} \lambda_k x_k - \sum_{k=1}^{n} \mu_k x_k\right|\right|_1^2 = 0\ ,$$

that is,

$$\psi\left(\sum_{k=1}^{n} \lambda_k x_k\right) = \psi\left(\sum_{k=1}^{n} \mu_k x_k\right) .$$

These same considerations also indicate that the converse is true, that is, that for every $y \in L(G_2)$, equation 9 puts exactly one element $x \in L(G_1)$ in correspondence with y. It also follows from (9) that this correspondence is linear and it follows from the formula

$$\left(\sum_{k=1}^{n} \alpha_k y_k, \sum_{k=1}^{n} \beta_k y_k\right)_2 = \sum_{k,r=1}^{n} \alpha_k \bar{\beta}_r (\psi(x_k), \psi(x_r))_2 = \sum_{k,r=1}^{n} \alpha_k \bar{\beta}_r (x_k\ x_r)_1$$

$$= \left(\sum_{k=1}^{n} \alpha_k x_k, \sum_{k=1}^{n} \beta_k x_k\right)_1, \ y_k = \psi(x_k)\ ,$$

that it preserves the scalar product. Thus we have established an isometric correspondence between $L(G_1)$ and $L(G_2)$. Suppose now that $x \in H(G_1)$. Then $x = \lim x_n$ (where each $x_n \in L(G_1)$) as $n \longrightarrow \infty$. If $y_n = \psi(x_n)$ for $y_n \in L(G_2)$, then $||y_n - y_{n+m}||^2 = ||x_n - x_{n+m}||^2 \longrightarrow 0$ where $m > 0$, as $n \longrightarrow \infty$; that is, l.i.m. $y_n = y_0$ exists. Let us set $y_0 = \psi(x)$. To prove the single-valuedness of this definition let us take an arbitrary sequence $\{x'_n\}$, where each $x'_n \in L(G_1)$, such that $\lim x'_n =$ l.i.m. x_n. We need to show that l.i.m. $y'_n = \lim \psi(x'_n) = y_0$. If $x''_{2n} = x'_n$ and $x''_{2n-1} = x_n$ (for $n = 1, 2, \cdots$), then l.i.m. $x''_n = x$ and on the basis of what we have shown, l.i.m. $\psi(x''_n) = y''$ exists. From this it follows that

$$\text{l.i.m. } \psi(x_{2n}) = \text{l.i.m. } \psi(x_n) = y''$$

and

$$y'' = \text{l.i.m. } \psi(x''_{2n-1}) = \text{l.i.m. } \psi(x_n) = y_0\ .$$

Now it is easy to show that the extended mapping $y = \psi(x)$ is a one-to-one mapping of $H(G_1)$ onto $H(G_2)$ and that it preserves the scalar product. This completes the proof of the theorem.

2. HILBERT RANDOM FUNCTIONS

Let $\{U, \mathfrak{S}, \mathsf{P}\}$ denote a probability space.

Definition 1. The set of complex-valued random variables $\zeta = f(u)$ for $u \in U$ such that $\mathsf{M} |\zeta|^2 < \infty$ is called the Hilbert space $L_2 = L_2(U, \mathfrak{S}, \mathsf{P})$ of random variables of the probability space $\{U, \mathfrak{S}, \mathsf{P}\}$.

The scalar product in L_2 is defined by

$$(\zeta, \eta) = \mathsf{M}\zeta\bar{\eta}, \qquad \zeta, \eta \in L_2 .$$

Corresponding to this definition, we have the norm $\|\zeta\|$ of the random variable ζ :

$$\|\zeta\| = \{\mathsf{M} |\zeta|^2\}^{1/2} .$$

Two random variables ζ and η are said to be *orthogonal* if

$$(\zeta, \eta) = \mathsf{M}\zeta\bar{\eta} = 0 .$$

The square of the norm $\|\zeta\|^2$ of a real random variable ζ coincides with the second-order moment $\|\xi\|^2 = \mathsf{M} |\zeta|^2$, and if $\mathsf{M}\zeta = 0$ it coincides with the variance. If ζ and η are both real and $\mathsf{M}\zeta = \mathsf{M}\eta = 0$, their orthogonality implies that they are uncorrelated.

Definition 2. A complex-valued random function $\zeta(\theta)$ for $\theta \in \Theta$ is called a *Hilbert random function* if

$$\mathsf{M} |\zeta(\theta)|^2 < \infty , \qquad \theta \in \Theta .$$

A Hilbert random function can be regarded as a function defined on Θ into the Hilbert space L_2:

$$\theta \longrightarrow \zeta(\theta) = f(\theta, u) \in L_2 .$$

In particular, if Θ is a real interval (a, b) the Hilbert random function should be regarded as some curve in the Hilbert space L_2. The representation $\zeta = \zeta(\theta)$ for $\theta \in (a, b)$ is a parametric equation for that curve. In the present chapter we consider only Hilbert random functions. For this reason we shall often drop the word "Hilbert."

Suppose that the nonnegative function $\psi(\theta)$ defined on Θ assumes arbitrarily small positive values.

Definition 3. A random variable $\eta \in L_2$ is called the *mean-square limit* (abbreviated m. s. limit) of a Hilbert random function $\zeta(\theta)$ as $\psi(\theta) \longrightarrow 0$ if $\zeta(\theta) \longrightarrow \eta$ as $\psi(\theta) \longrightarrow 0$ in the sense of convergence in the Hilbert space L_2, that is, if for arbitrary $\varepsilon > 0$ there exists a $\delta > 0$ such that

$$\mathsf{M} |\eta - \zeta(\theta)|^2 < \varepsilon^2$$

for all θ such that $0 < \psi(\theta) < \delta$.

In particular if Θ is a metric space with metric $r(\theta_1, \theta_2)$, the function $\zeta(\theta)$ is said to be *mean-square continuous* at a point $\theta_0 \in \Theta$ if

$$\mathsf{M} \mid \zeta(\theta_1) - \zeta(\theta_0) \mid^2 \longrightarrow 0 \quad \text{for} \quad r(\theta_1, \theta_0) \longrightarrow 0 . \tag{1}$$

Definition 4. The *covariance* $B(\theta_1, \theta_2)$, where $\theta_1, \theta_2 \in \Theta$, of a Hilbert random function $\zeta(\theta)$ is defined as the quantity

$$B(\theta_1, \theta_2) = \mathsf{M}\zeta(\theta_1)\overline{\zeta(\theta_2)} = (\zeta(\theta_1), \zeta(\theta_2)) . \tag{2}$$

Lemma 1. *For a random function $\zeta(\theta)$ to have a mean-square limit as $\psi(\theta) \longrightarrow 0$, it is necessary and sufficient that the limit* $\lim B(\theta_1, \theta_2)$ *as $\psi(\theta_1) + \psi(\theta_2) \longrightarrow 0$ exist. If this condition is satisfied, then*

$$\mathsf{M} \mid \eta \mid^2 = \lim_{\psi(\theta_1)+\psi(\theta_2)\to 0} B(\theta_1, \theta_2) \tag{3}$$

where

$$\eta = \underset{\psi(\theta)\to 0}{\text{l.i.m.}} \ \zeta(\theta) .$$

This lemma is a rephrasing of Lemma 1 of Section 1.

Corollary 1. *For $\zeta(\theta)$ to be mean-square continuous at a point θ_0 it is necessary and sufficient that the covariance $B(\theta_1, \theta_2)$ be continuous at the point (θ_0, θ_0).*

The necessity follows from Lemma 1 and the sufficiency can be proven by simple calculation.

REMARK 1. Mean-square continuity of $\zeta(\theta)$ at the point θ_0 implies stochastic continuity of $\zeta(\theta)$ at that point. This is true because, by virtue of Chebyshev's inequality,

$$\mathsf{P}\{\mid \zeta(\theta) - \zeta(\theta_0) \mid > \varepsilon\} \leqq \frac{\mathsf{M} \mid \xi(\theta) - \zeta(\theta_0) \mid^2}{\varepsilon^2} .$$

If $\zeta(\theta)$ is mean-square continuous on Θ (that is, at every point of Θ), the sample functions are not necessarily with probability 1 continuous on Θ. Indeed, for a Poisson process we have

$$\mathsf{M} \mid \zeta(t + h) - \zeta(t) \mid^2 = \lambda h + (\lambda h)^2 ,$$

but the sample functions $\zeta(t)$ are, with positive probability, discontinuous.

Let us examine successively differentiation, integration, and orthogonal-series expansion of Hilbert processes.

1. Differentiation of Hilbert Processes

a. A random process $\zeta(t)$ for $t \in \mathfrak{T}$ is said to be *mean-square differentiable* at a point t_0 if the limit

$$\zeta'(t_0) = \underset{h\to 0}{\text{l.i.m.}} \ \frac{\zeta(t_0 + h) - \zeta(t_0)}{h}, \ t_0, t_0 + h \in \mathfrak{T}$$

exists. The random variable $\zeta'(t_0)$ is called the *mean-square derivative* of the random process at the point t_0.

It is easy to find necessary and sufficient conditions for mean-square differentiability of a random process. Since

$$\mathsf{M}\frac{\zeta(t_0+h)-\zeta(t_0)}{h}\overline{\frac{\zeta(t_0+h_1)-\zeta(t_0)}{h_1}}$$
$$=\frac{1}{h_1 h}\{B(t_0+h,t_0+h_1)-B(t_0,t_0+h_1)$$
$$-B(t_0+h,t_0)+B(t_0,t_0)\}\ , \qquad (4)$$

it follows on the basis of Lemma 1 that a necessary and sufficient condition for mean-square differentiability of the process $\zeta(t)$ at the point t_0 is that the generalized mixed derivative

$$\left.\frac{\partial^2 B(t,t')}{\partial t\partial t'}\right|_{t=t'=t_0}$$
$$=\lim_{h,h_1\to 0}\frac{B(t_0+h,t_0+h_1)-B(t_0,t_0+h_1)-B(t_0+h,t_0)+B(t_0,t_0)}{hh_1}$$

exists. The mean-square differentiability of the process at the point t and the inequality

$$\left|\mathsf{M}\left(\zeta'(t)-\frac{\zeta(t+h)-\zeta(t)}{h}\right)\right|\leqq\left\{\mathsf{M}\left|\zeta'(t)-\frac{\zeta(t+h)-\zeta(t)}{h}\right|^2\right\}^{1/2}$$

imply that $d\mathsf{M}\zeta(t)/dt$ exists and that

$$\mathsf{M}\zeta'(t)=\frac{d}{dt}\mathsf{M}\zeta(t)\ . \qquad (5)$$

If a process $\zeta(t)$ is mean-square differentiable at every point of an interval $\mathfrak{T}$, then the derivative $\zeta'(t)$ constitutes a Hilbert random process on $\mathfrak{T}$.

Theorem 1. *Let $\zeta(t)$ for $t\in\mathfrak{T}$ denote a Hilbert random process and suppose that the generalized derivative*

$$\left.\frac{\partial^2 B(t,t')}{\partial t\partial t'}\right|_{t'=t}$$

exists at every $t\in\mathfrak{T}$. Then the process $\zeta(t)$ is mean-square differentiable on $\mathfrak{T}$ and

$$B_{\zeta'\zeta'}(t,t')=\frac{\partial^2 B(t,t')}{\partial t\partial t'}\ , \qquad (6)$$

$$B_{\zeta'\zeta}(t,t')=\frac{\partial B(t,t')}{\partial t}\ , \qquad (7)$$

where

$$B_{\zeta'\zeta'}(t, t') = \mathsf{M}\zeta'(t)\overline{\zeta'(t')}$$

is the covariance of the process $\zeta'(t)$ *and*

$$B_{\zeta'\zeta}(t, t') = \mathsf{M}\overline{\zeta(t')}\zeta'(t)$$

is the mutual covariance of the processes $\zeta'(t)$ *and* $\zeta(t)$.

We need to prove formulas (6) and (7). We have

$$\begin{aligned} B_{\zeta'\zeta}(t, t') &= \mathsf{M}\overline{\zeta(t')}\zeta'(t) \\ &= \lim \mathsf{M}\overline{\zeta(t')}\left(\frac{\zeta(t+h) - \zeta(t)}{h}\right) \\ &= \lim \frac{B(t+h, t') - B(t, t')}{h} . \end{aligned}$$

consequently the derivative $\dfrac{\partial B(t, t')}{\partial t}$ exists and mutual covariance of the processes $\zeta'(t)$ and $\zeta(t)$ is given by formula (7). Furthermore,

$$B_{\zeta'\zeta'}(t, t') = \lim_{h,h'\to 0} \mathsf{M}\overline{\frac{\zeta(t'+h') - \zeta(t')}{h'}}\,\frac{\zeta(t+h) - \zeta(t)}{h}$$

$$= \lim_{h,h'\to 0} \frac{B(t+h, t'+h') - B(t, t'+h') - B(t+h, t') + B(t, t')}{hh'} .$$

From this it follows that the generalized second derivative $\dfrac{\partial^2 B(t, t')}{\partial t \partial t'}$ exists (the hypothesis of the theorem assumes only that this derivative exists at $t = t'$). Formula (6) follows.

If the process $\zeta(t)$ is stationary in the broad sense, then $B(t, t') = B(t - t')$, and the results just obtained take the form of:

Corollary 2. *A necessary and sufficient condition for differentiability of a process* $\zeta(t)$ *for* $t \in \mathfrak{T}$ *that is stationary in the broad sense is that the generalized second derivative of the covariance* $B(t) = \mathsf{M}\zeta(t)\overline{\zeta(0)}$ *exist at* $t = 0$. *Then the generalized derivative* $d^2B(t)/dt^2$ *exists and*

$$B_{\zeta'\zeta'}(t_0, t_0 + t) = -\frac{d^2B(t)}{dt^2} ,$$

$$B_{\zeta'\zeta}(t_0 + t, t_0) = B_{\zeta'\zeta}(t) = \frac{dB(t)}{dt} .$$

Analogous results hold for mean-square derivatives of higher orders.

2. Integration of Hilbert Processes

Let $\zeta(t)$ for $t \in [a, b]$,denote a measurable process and suppose that

$$\int_a^b \mathsf{M} \mid \zeta(t) \mid^2 dt < \infty .$$

On the basis of Fubini's theorem,

$$\mathsf{M}\int_a^b \mid \zeta(t) \mid^2 dt = \int_a^b \mathsf{M} \mid \zeta(t) \mid^2 dt .$$

Consequently the integral $\int_a^b \mid \zeta(t) \mid^2 dt$, and hence the integral $\int_a^b \zeta(t)dt$, exists with probability 1. If the functions $f_i(t)$ are square-integrable in the sense of Lebesgue over the interval $[a, b]$, the integrals

$$\int_a^b f_i(t)\zeta(t)dt$$

also exist. By again using Fubini's theorem, we obtain

$$\mathsf{M}\int_a^b f_1(t)\zeta(t)dt\overline{\int_a^b f_2(t)\zeta(t)dt} = \mathsf{M}\int_a^b\int_a^b f_1(t)\overline{f_2(\tau)}\zeta(t)\overline{\zeta(\tau)}dtd\tau$$

$$= \int_a^b\int_a^b f_1(t)B(t, \tau)\overline{f_2(\tau)}dtd\tau . \qquad (8)$$

REMARK 2. The integral

$$\int_a^b \zeta(t)dt \qquad (9)$$

is defined for a measurable process as a quantity that exists with probability 1 for an arbitrary sample function of the process. But we can proceed in a somewhat different manner. In the first place, the integral (9) can be defined as the mean-square limit of integrals of simple functions that converge to $\zeta(t)$. It is easy to show that this definition coincides with the original one. To prove this, we may confine ourselves to nonnegative random variables. By definition the integral (9) is the limit as $n \longrightarrow \infty$ of

$$\int_a^b \zeta_n(t)dt$$

where $\{\zeta_n(t)\}$ is a nondecreasing sequence of random variables that assume finitely many values and such that $\{\zeta_n(t)\}$ converges with probability 1 to $\zeta(t)$. Since $\mid \zeta(t) - \zeta_n(t) \mid \leqq \mid \zeta(t) \mid$, it follows from Lebesgue's theorem that

$$\mathsf{M}\left|\int_a^b \zeta(t)dt - \int_a^b \zeta_n(t)dt\right|^2 \leqq \mathsf{M}\int_a^b \mid \zeta(t) - \zeta_n(t) \mid^2 dt \longrightarrow 0$$

as $n \longrightarrow \infty$, so that

$$\int_a^b \zeta(t)dt = \text{l.i.m.} \int_a^b \zeta_n(t)dt .$$

In the second place, the integral (9) is often defined as the mean-square limit of the Riemann approximating sums

$$\sum_{k=1}^{n} \zeta(t_{nk})\Delta t_{nk}, \ \Delta t_{nk} = t_{nk} - t_{n,k-1}, \ a = t_{n0} < t_{n1} < \cdots < t_{nn} = b .$$

For the mean-square limit of these sums to exist it is necessary and sufficient, by virtue of Lemma 1, that the limit

$$\mathsf{M} \sum_{k=1}^{n} \zeta(t_{nk})\Delta t_{nk} \sum_{k=1}^{m} \overline{\zeta(t_{mk})}\Delta t_{mk} = \sum_{k=1}^{n} \sum_{r=1}^{m} B(t_{nk}, t_{mr})\Delta t_{nk}\Delta t_{mr}$$

exist as $n, m \to \infty$, that is, that the function $B(t, \tau)$ (for $a \leqq t, \tau \leqq b$) be Riemann-integrable. As we can see, such a definition of the integral is more restricted than the original definition but it is not connected with the concept of measurability of the process. One can easily show that when the second definition of the integral is used, it leads (mod P) to the same result as the original one. Specifically,

$$\begin{aligned} \mathsf{M} \left| \int_a^b \zeta(t)dt - \sum_{k=1}^{n} \zeta(t_{nk})\Delta t_{nk} \right|^2 &= \sum_{k=1}^{n} \sum_{r=1}^{n} \int_{t_{n,k-1}}^{t_{nk}} \int_{t_{n,r-1}}^{t_{nr}} [B(t, \tau) - B(t, t_{nr}) - B(t_{nk}, \tau) \\ &\quad + B(t_{nk}, t_{nr})]dtd\tau \leqq 2 \sum_{k=1}^{n} \sum_{r=1}^{n} \Omega_{nkr} \longrightarrow 0 \end{aligned}$$

where Ω_{nkr} is the oscillation of the function $B(t, \tau)$ in the rectangle $t_{n,k-1} \leqq t \leqq t_{nk}, \ t_{n,r-1} \leqq \tau \leqq t_{nr}$.

Example 1. Let $\zeta(t)$ (for $a \leqq t \leqq b$) denote a Hilbert random process and suppose that the derivatives

$$\frac{\partial B(t, t')}{\partial t} \quad \text{and} \quad \frac{\partial^2 B(t, t')}{\partial t \partial t'}$$

exist for arbitrary t and t' and are finite and Lebesgue-integrable with respect to their arguments. Let us look at the process $\zeta'(t)$ where $\zeta'(t)$ is the mean-square derivative of the process $\zeta(t)$. Let us also assume that $\zeta'(t)$ is a measurable process. Then by virtue of equation (8) and Theorem 1 we obtain

$$\begin{aligned} \mathsf{M} \left| \int_a^t \zeta'(\tau)d\tau + \zeta(a) - \zeta(t) \right|^2 &= \int_a^t \int_a^t \frac{\partial^2 B(\tau, \tau')}{\partial \tau \partial \tau'} d\tau d\tau' - 2 \operatorname{Re} \int_a^t \left(\frac{\partial B(\tau, t)}{\partial \tau} - \frac{\partial B(\tau, a)}{\partial \tau} \right) d\tau \\ &\quad + \mathsf{M} \, | \zeta(t) - \zeta(a) |^2 . \end{aligned}$$

On the basis of the assumptions we have made, we can use the Newton-Leibnitz formula in evaluating the integrals on the right side of this equation. This leads to the equation

$$\mathsf{M}\left|\int_a^t \zeta'(\tau)d\tau + \zeta(a) - \zeta(t)\right|^2 = 0\;;$$

that is, for every t,

$$\zeta(t) = \zeta(a) + \int_a^t \zeta'(t)dt \qquad (\text{mod } \mathsf{P})\;. \tag{10}$$

Thus the process ζ is with probability 1 representable as an integral of its mean-square derivative, and $\zeta(t)$ is with probability 1 differentiable for almost all $t \in [a, b]$.

If we assume that $\zeta(t)$ is a separable process, equation (10) will with probability 1 be simultaneously satisfied for all t; that is, $\zeta(t)$ is, with probability 1, differentiable.

Example 2. Let $\zeta(t)$ (for $-\infty < t < \infty$) denote a measurable Hilbert random process that satisfies condition (8) for arbitrary finite a and b. Let us examine the question of the existence of

$$\underset{T\to\infty}{\text{l.i.m.}}\, \frac{1}{T}\int_a^{a+T} \zeta(t)dt\;. \tag{11}$$

For the mean-square limit (11) to exist it is, by virtue of Lemma 1, necessary and sufficient that the limit

$$\lim_{T,T'\to\infty} \mathsf{M}\frac{1}{T}\int_a^{a+T}\zeta(t)dt\frac{1}{T'}\int_a^{a+T'}\overline{\zeta(t)dt} = \lim_{T,T'\to\infty}\frac{1}{TT'}\int_a^{a+T}\int_a^{a+T'} B(t, \tau)dtd\tau$$

exist. Furthermore, for the equation

$$\text{l.i.m.}\left\{\frac{1}{T}\int_a^{a+T}\zeta(t)dt - \frac{1}{T}\int_a^{a+T}\mathsf{M}\zeta(t)dt\right\} = 0 \tag{12}$$

to hold, it is necessary and sufficient that

$$\lim \frac{1}{T^2}\int_a^{a+T}\int_a^{a+T} R(t, \tau)dtd\tau = 0 \tag{13}$$

where $R(t, \tau)$ is the correlation function of the process. For a process that is stationary in the broad sense, $R(t, \tau) = R(t - \tau)$ and $\mathsf{M}\zeta(t) = m$.

Since

$$\frac{1}{T^2}\int_a^{a+T}\int_a^{a+T} R(t - \tau)dtd\tau = \frac{1}{T^2}\int_0^T\int_0^T R(t - \tau)dtd\tau$$

$$= \frac{1}{T^2}\int_0^T\int_{t-T}^t R(u)dudt = \frac{1}{T}\int_{-T}^T R(u)\left(1 - \frac{|u|}{T}\right)du\;.$$

We obtain the following result: If $\zeta(t)$ is a process that is stationary in the broad sense, for the equation

$$\underset{T\to\infty}{\text{l.i.m.}}\, \frac{1}{T}\int_a^{a+T}\zeta(t)dt = \mathsf{M}\zeta(t) \tag{14}$$

to hold, it is necessary and sufficient that

$$\lim \frac{1}{T}\int_{-T}^{T} R(u)\left(1 - \frac{|u|}{T}\right)du = 0 . \tag{15}$$

In particular, this condition will be satisfied if the mean value of the correlation function is 0:

$$\lim_{T\to\infty} \frac{1}{2T}\int_{-T}^{T} R(u)du = 0 .$$

3. Expansion of a Random Process in a Series of Orthogonal Functions

Let $\zeta(t)$, for $t \in [a, b]$ denote a measurable mean-square continuous Hilbert process. Its covariance $B(t_1, t_2)$ is a continuous nonnegative-definite kernel in the square $[a, b] \times [a, b]$. According to integral-equation theory, the kernel $B(t_1, t_2)$ can be expanded in a uniformly convergent series of its eigenfunctions $\varphi_n(t)$:

$$B(t_1, t_2) = \sum_{n=1}^{\infty} \lambda_n \varphi_n(t_1)\overline{\varphi_n(t_2)} ,$$

where

$$\lambda_n \varphi_n(t) = \int_a^b B(t, \tau)\varphi_n(\tau)d\tau , \qquad \int_a^b \varphi_n(t)\overline{\varphi_m(t)}dt = \delta_{nm} ,$$

and the eigenvalues λ_n are positive.

We set

$$\xi_n = \int_a^b \zeta(t)\overline{\varphi_n(t)}dt .$$

This integral exists (cf. Part 2 of Section 2) and by virtue of (8),

$$\mathsf{M}\xi_n\bar{\xi}_m = \int_a^b\int_a^b B(t, \tau)\overline{\varphi_n(t)}\varphi_m(\tau)dtd\tau = \lambda_n\delta_{nm} ;$$

that is, the sequence of random variables $\{\xi_n\}$, for $n = 1, 2, \cdots$, is orthogonal. Furthermore,

$$\mathsf{M}\zeta(t)\bar{\xi}_n = \int_a^b B(t, \tau)\varphi_n(\tau)d\tau = \lambda_n\varphi_n(t) .$$

From this it follows that

$$\begin{aligned}\mathsf{M}\left|\zeta(t) - \sum_{k=1}^{n} \xi_k\varphi_k(t)\right|^2 &= B(t, t) - 2\sum_{k=1}^{n} \overline{\varphi_k(t)}\mathsf{M}\zeta(t)\bar{\xi}_k + \sum_{k=1}^{n} \lambda_k |\varphi_k(t)|^2 \\ &= B(t, t) - \sum_{k=1}^{n} \lambda_k |\varphi_k(t)|^2 ,\end{aligned}$$

which approaches 0 as $n \longrightarrow \infty$.

Theorem 2. *A measurable mean-square continuous Hilbert process $\zeta(t)$, for $t \in [a, b]$, can be expanded in a series*

$$\zeta(t) = \sum_{k=1}^{\infty} \xi_k \varphi_k(t) \tag{16}$$

that converges in L_2 for every $t \in [a, b]$. In this expansion, $\{\xi_k\}$ is an orthogonal sequence of random variables $\mathsf{M} \mid \xi_k \mid^2 = \lambda_k$ where the λ_k are the eigenvalues and the $\varphi_k(t)$ are the eigenfunctions of the covariance $B(t, \tau)$ of the process.

REMARK 3. If the process $\zeta(t)$ is a Gaussian process, its mean-square derivative and all integrals of the form $\int_a^b f(t)\zeta(t)dt$ are also Gaussian.

REMARK 4. If $\zeta(t)$ is a real Gaussian process and $\mathsf{M}\zeta(t) = 0$, then the coefficients ξ_k in the series (16) are independent Gaussian variables and the series (16) converges with probability 1 for every t. To see this, note that independence of the variables ξ_k implies that they are orthogonal and Gaussian. For the series (16) to converge with probability 1 it is sufficient that the series

$$\sum_{k=1}^{\infty} \mathsf{M}(\xi_k \varphi_k(t))^2 = \sum_{k=1}^{\infty} \lambda_k^2 \mid \varphi_k(t) \mid^2$$

converge. But we have already seen that this series converges (to $B(t, t)$).

Example 3. Let us look at the expansion of a process of Brownian motion $\zeta(t)$ (where $\zeta(0) = 0$ and $\mathsf{M}\zeta^2(t) = t$) on the interval $[0, 1]$ in an orthogonal series. Here $\mathsf{M}\zeta(t) = 0$, $B(t, s) = \mathsf{M}\zeta(t)\zeta(s) = \min(t, s)$. The eigenvalues and eigenfunctions of the kernel $B(t, s)$ are easily found. From the equation

$$\lambda_n \varphi_n(t) = \int_0^1 \min(t, s)\varphi_n(s)ds = \int_0^t s\varphi_n(s)ds + \int_t^1 t\varphi_n(s)ds$$

we see first of all that $\varphi_n(0) = 0$. Differentiating with respect to t, we obtain $\lambda_n \varphi_n'(t) = \int_t^1 \varphi_n(s)ds$ from which we get $\varphi_n'(1) = 0$. By successive differentiation we arrive at the equation $\lambda_n \varphi_n''(t) = -\varphi_n(t)$. The normalized solutions of this last equation that satisfy the boundary conditions are of the form

$$\varphi_n(t) = \sqrt{2}\sin\left(n + \frac{1}{2}\right)\pi t\,, \quad \lambda_n^{-1} = \left(n + \frac{1}{2}\right)^2 \pi^2\,, \quad n = 0, 1, \cdots .$$

Thus

$$\xi(t) = \sqrt{2} \sum_{n=0}^{\infty} \xi_n \frac{\sin\left(n + \frac{1}{2}\right)\pi t}{\left(n + \frac{1}{2}\right)\pi}, \tag{17}$$

where $\{\xi_n\}$ is a sequence of independent Gaussian random variables with parameters 0 and 1. For fixed t this series converges with probability 1. Another expansion of a process of Brownian motion can be obtained as follows: Let us set $\xi(t) = \zeta(t) - t\zeta(1)$. Then $\xi(t)$ is a Gaussian process with convariance $B_1(t, s) = \min(t, s) - ts$ and $\mathsf{M}\xi(t) = 0$. The eigenvalues and eigenfunctions of the kernel $B_1(t, s)$ are found in the same way as in the preceding case. We again arrive at the equation $\lambda_n \varphi_n''(t) = -\varphi_n(t)$ with boundary conditions $\varphi_n(0) = \varphi_n(1) = 0$. The solutions of this equation with these boundary conditions are of the form $\varphi_n(t) = \sqrt{2}\sin n\pi t$, $\lambda_n^{-1} = n^2\pi^2$, $n = 1, 2, \cdots$. Thus

$$\xi(t) = \zeta(t) - t\zeta(1) = \sqrt{2} \sum_{n=1}^{\infty} \xi_n \frac{\sin n\pi t}{n\pi}, \tag{18}$$

where $\{\xi_n\}$, for $n = 1, 2, \cdots$, is again a normalized sequence of independent Gaussian random variables. Moreover,

$$\xi_n = \sqrt{2}\int_0^1 \xi(t) \sin n\pi t dt .$$

Since $\mathsf{M}\zeta(1) = 0$, $\mathsf{M}\zeta^2(1) = 1$, and

$$\mathsf{M}\xi_n\zeta(1) = \sqrt{2}\int_0^1 \mathsf{M}(\zeta(t) - t\zeta(1))\zeta(1) \sin n\pi t dt = 0,$$

if we set $\xi_0 = \zeta(1)$ we obtain

$$\zeta(t) = t\xi_0 + \sqrt{2} \sum_{n=1}^{\infty} \xi_n \frac{\sin n\pi t}{n\pi}, \tag{19}$$

and the sequence $\{\xi_n\}$ for $n = 0, 1, 2, \cdots$ enjoys the same properties as the sequence $\{\xi_n\}$ where $n = 1, 2, \cdots$. The series (19) converges (for every t) with probability 1.

3. STOCHASTIC MEASURES AND INTEGRALS

In a number of processes an important role is played by integrals of the form

$$\int f(t) d\zeta(t), \tag{1}$$

where $f(t)$ is a given (nonrandom) function and $\zeta(t)$ is a random process. The sample functions of the process $\zeta(t)$ are, in general, functions of unbounded variation, and we cannot regard the integral

(1) as the Stieltjes or Lebesgue-Stieltjes integral that exists for almost all sample functions $\zeta(t)$. Nonetheless, the integral (1) can be defined in such a way that it enjoys the properties associated with an ordinary integral.

In this section we shall consider the definition and properties of the integral corresponding to integration with respect to a random measure. Such integrals are called *stochastic integrals.*

Let $\{U, \mathfrak{S}, \mathsf{P}\}$ denote a probability space, let L_2 denote $L_2\{U, \mathfrak{S}, \mathsf{P}\}$, let E denote some set, and let $\mathfrak{M}$ denote a decomposable family of subsets of E. Suppose that to each $\Delta \in \mathfrak{M}$ is assigned a complex-valued random variable $\mu(\Delta)$ satisfying the following conditions:

1. $\mu(\Delta) \in L_2$, $\mu(\phi) = 0$;
2. $\mu(\Delta_1 \cup \Delta_2) = \mu(\Delta_1) + \mu(\Delta_2)(\text{mod } \mathsf{P})$ if $\Delta_1 \cap \Delta_2 = \phi$;
3. $\mathsf{M}\mu(\Delta_1)\overline{\mu(\Delta_2)} = m(\Delta_1 \cap \Delta_2)$;

where $m(\Delta)$ is a set function defined on $\mathfrak{M}$.

We shall refer to the family of random variables $\{\mu(\Delta)\}$ for $\Delta \in \mathfrak{M}$ as an *elementary orthogonal stochastic measure* and we shall refer to $m(\Delta)$ as its *structural function.* The orthogonality of a stochastic measure is expressed by condition (3) if $\Delta_1 \cap \Delta_2 = \varphi$, the variables $\mu(\Delta_1)$ and $\mu(\Delta_1)$ and $\mu(\Delta_2)$ are orthogonal.

It follows from the definition of $m(\Delta)$ that $m(\Delta)$ is nonnegative:

$$m(\Delta) = \mathsf{M}\,|\,\mu(\Delta)\,|^2 \geqq 0\,, \qquad m(\phi) = 0$$

and additive:

$$\Delta_1 \cap \Delta_2 = \phi\,,$$
$$\begin{aligned} m(\Delta_1 \cap \Delta_2) &= \mathsf{M}\,|\,\mu(\Delta_1) + \mu(\Delta_2)\,|^2 \\ &= m(\Delta_1) + m(\Delta_2) + 2m(\Delta_1 \cap \Delta_2) = m(\Delta_1) + m(\Delta_2)\,. \end{aligned}$$

Thus $m(\Delta)$ is an elementary measure on $\mathfrak{M}$ (cf. Definition 5, Section 7, Chapter II).

Let $L(\mathfrak{M})$ denote the class of all simple functions $f(x)$:

$$f(x) = \sum_{k=1}^{n} c_k \chi_{\Delta_k}(x)\,, \qquad \Delta_k \in \mathfrak{M}\,, \qquad k = 1, 2, \cdots, n\,, \qquad (2)$$

where n is an arbitrary natural number and $\chi_A(x)$ is the characteristic function of the set A.

Let us define the stochastic integral of a function $f(x) \in L\{\mathfrak{M}\}$ with respect to an elementary stochastic measure μ by

$$\eta = \int f(x)\mu(dx) = \sum_{k=1}^{n} c_k\mu(\Delta_k)\,. \qquad (3)$$

Since $\mathfrak{M}$ is a decomposable family of sets, every pair of functions in $L\{\mathfrak{M}\}$ can be represented as linear combinations of the characteristic functions of the same sets in $\mathfrak{M}$. Therefore, if f and

g belong to $L\{\mathfrak{M}\}$ we assume that $f(x)$ is given by formula (2) and that

$$g(x) = \sum_{k=1}^{n} d_k \chi_{\Delta_k}(x)$$

where $\Delta_k \cap \Delta_r = \phi$ for $k \neq r$.

The orthogonality of μ implies that

$$\mathsf{M}\left(\int f(x)\mu(dx)\overline{\int g(x)\mu(dx)}\right) = \sum_{k=1}^{n} c_k \bar{d}_k m(\Delta_k) \; . \tag{4}$$

We assume that the elementary measure m is subadditive and hence can be extended as a complete measure $\{E, \mathfrak{L}, m\}$ (cf. Theorem 2, Section 7, Chapter II). Then $L\{\mathfrak{M}\}$ is a linear subset of the Hilbert space $L_2\{m\} = L_2\{E, \mathfrak{L}, m\}$, and $L_2\{m\}$ is the closure of $L\{\mathfrak{M}\}$ in the topology generated by the scalar product

$$(f, g) = \int f(x)\overline{g(x)}m(dx) \tag{5}$$

(cf. Theorem 7, Section 7, Chapter II).

Here equation (4) can be written in the form

$$\mathsf{M}\left(\int f(x)\mu(dx)\overline{\int g(x)\mu(dx)}\right) = \int f(x)\overline{g(x)}m(dx) \tag{6}$$

for an arbitrary pair of functions $f(x)$ and $g(x)$ in $L\{\mathfrak{M}\}$.

We now introduce the linear hull $L\{\mu\}$ of the family of random variables $\{\mu(\Delta)\}$ defined for $\Delta \in \mathfrak{M}$—that is, the set of random variables that can be represented in the form (3)—and the space $L_2\{\mu\}$, which is the closure of $L\{\mu\}$ in the Hilbert space of random variables $L_2\{U, \mathfrak{S}, \mathsf{P}\}$. We note that equation (3) sets up an isometric correspondence $\eta = \psi(f)$ between $L\{\mathfrak{M}\}$ and $L\{\mu\}$. On the basis of Theorem 1 of Section 1, this correspondence can be extended as an isometric correspondence between $L_2\{m\}$ and $L_2\{\mu\}$. If $\eta = \psi(f)$ for $f \in L_2\{\mathfrak{M}\}$, then we use $\eta = \psi(f)$ as the definition of

$$\int f(x)\mu(dx) \; , \tag{7}$$

and we call the random variable η the stochastic integral of the function $f(x)$ with respect to the measure μ. From this definition we get:

Theorem 1. a. *For an arbitrary function* (2) *the value of the stochastic integral is given by formula* (3);

b. *equation* (6) *holds for arbitrary* $f(x)$ *and* $g(x)$ *in* $L_2\{E, \mathfrak{L}, m\}$;

c. $\int [\alpha f(x) + \beta g(x)]\mu(dx) = \alpha \int f(x)\mu(dx) + \beta \int g(x)\mu(dx)$;

d. *for an arbitrary sequence of functions* $f^{(n)}(x) \in L_2\{E, \mathfrak{L}, m\}$ *such that*

$$\int |f(x) - f^{(n)}(x)|^2 m(dx) \longrightarrow 0 , \qquad n \longrightarrow \infty , \tag{8}$$

we have

$$\int f(x)\mu(dx) = \text{l.i.m.} \int f^{(n)}(x)\mu(dx) .$$

REMARK 1. In particular, if the $f^{(n)}(x)$ are simple functions,

$$f^{(n)}(x) = \sum_{k=1}^{m_n} c_k^{(n)} \chi_{\Delta_k^{(n)}}(x), \ \Delta_k^{(n)} \in \mathfrak{M}, \qquad n = 1, 2, \cdots ,$$

and (8) is satisfied, then

$$\int f(x)\mu(dx) = \text{l.i.m.} \sum_{k=1}^{m_n} c_k^{(n)} \mu(\Delta_k^{(n)}) .$$

It follows from Theorem 7, Section 7, Chapter II that an arbitrary function $f(x) \in L_2\{E, \mathfrak{L}, m\}$ can be approximated by some sequence of simple functions. Thus the stochastic integral can be regarded as the mean-square limit of suitable approximating sums.

Let $\mathfrak{L}_0$ denote the class of all sets $A \in \mathfrak{L}$ for which $m(\Delta) < \infty$. We define a random set function $\tilde{\mu}(A)$:

$$\tilde{\mu}(A) = \int \chi_A(x)\mu(dx) = \int_A \mu(dx) . \tag{9}$$

This function enjoys the following properties:

a. $\tilde{\mu}(A)$ is defined on the class of sets $\mathfrak{L}_0$ such that for arbitrary $A \in \mathfrak{L}_0$ the class of sets $\{B \cap A\}$ for $B \in \mathfrak{L}$ constitutes a σ-algebra and $\mathsf{M} \, | \tilde{\mu}(A) |^2 = m(A) < \infty$.

b. If $A_n \in \mathfrak{L}_0$ (for $n = 0, 1, 2, \cdots$), $A_0 = \bigcup_{n=1}^{\infty} A_n$, and $A_k \cap A_r = \phi$ for $k \neq r$ (where k and r are both positive), then

$$\tilde{\mu}(A_0) = \sum_{n=1}^{\infty} \tilde{\mu}(A_n) (\text{mod } \mathsf{P}) ;$$

c. $\mathsf{M}\tilde{\mu}(A)\tilde{\mu}(B) = m(A \cap B), \ A, B \in \mathfrak{L}_0$;

d. $\tilde{\mu}(\Delta) = \mu(\Delta)$ for $\Delta \in \mathfrak{M}$.

A random set function satisfying conditions a to c is called a *stochastic orthogonal measure*. Property (d) means that $\tilde{\mu}(A)$ is the extension of an elementary stochastic measure μ. Thus we have:

Corollary. *If the structural function of an elementary stochastic measure* μ *is subadditive, then* μ *can be extended as a stochastic measure.*

REMARK 2. Since $L_2\{\mu\} = L_2\{\tilde{\mu}\}$, we have

$$\int f(x)\mu(dx) = \int f(x)\tilde{\mu}(dx) .$$

In what follows we shall identify the stochastic integral with respect to an elementary orthogonal measure μ(whose structural function is subadditive) with the stochastic integral with respect to a measure $\tilde{\mu}$ defined by equation (9).

Let us make a few comments about the definition of a stochastic integral over an interval. Suppose that $\zeta(t)$ (for $a \leqq t < b$) is a process with orthogonal increments; that is,

$$\mathsf{M}(\zeta(t_2) - \zeta(t_1))\overline{(\zeta(t_4) - \zeta(t_3))} = 0$$

if $t_1 < t_2 < t_3 < t_4$, and suppose that $\zeta(t)$ is mean-square continuous from the left:

$$\mathsf{M} \mid \zeta(t) - \zeta(\tau) \mid^2 \longrightarrow 0 \quad \text{as} \quad \tau \uparrow t.$$

We define

$$F(t) = \mathsf{M} \mid \zeta(t) - \zeta(a) \mid^2 .$$

It follows from the orthogonality of the increments of the process $\zeta(t)$ that for $t_2 > t_1$,

$$\begin{aligned} F(t_2) &= \mathsf{M} \mid \zeta(t_2) - \zeta(t_1) + \zeta(t_1) - \zeta(a) \mid^2 \\ &= F(t_1) + \mathsf{M} \mid \zeta(t_2) - \zeta(t_1) \mid^2 , \end{aligned}$$

from which we get $F(t_2) \geqq F(t_1)$ and $F(t) = \lim_{\tau \uparrow t} F(\tau)$. Thus $F(t)$ is a nondecreasing function that is continuous from the left. Let $\mathfrak{M}$ denote the class of all left-closed right-open intervals $\Delta = [t_1, t_2)$ such that $a \leqq t_1 < t_2 < b$. Define

$$\zeta([t_1, t_2)) = \zeta(t_2) - \zeta(t_1)$$

and

$$m([t_1, t_2)) = F(t_2) - F(t_1) .$$

Then $\mathfrak{M}$ is a decomposable class of sets

$$\mathsf{M}\zeta(\Delta_1)\overline{\zeta(\Delta_2)} = m(\Delta_1 \cap \Delta_2) ,$$

and ζ is an elementary orthogonal stochastic measure whose structural function can be extended as a measure. Thus, we can define the Stieltjes stochastic integral

$$\int_a^b f(t) d\zeta(t) = \int f(t)\zeta(dt) ,$$

where $\zeta(t)$ is a process with orthogonal increments. A stochastic integral over an open interval (a, b) or over the entire real line $(-\infty, \infty)$ is defined analogously.

Let us prove some propositions regarding stochastic integrals. Let μ denote an orthogonal stochastic measure with structural function m that is a complete measure on $\{E, \mathfrak{L}\}$ and let $g(x)$ denote a member of $L_2\{m\}$. We set

$$\lambda(A) = \int \chi_A(x) g(x) \mu(dx) , \qquad A \in L .$$

Then

$$\mathsf{M}\lambda(A)\overline{\lambda(B)} = \int \chi_A(x)\chi_B(x) \mid g(x) \mid^2 m(dx) = \int_{A \cap B} \mid g(x) \mid^2 m(dx) ,$$

On $\mathfrak{L}$ we define a new measure

$$l(A) = \int_A \mid g(x) \mid^2 m(dx) .$$

We see that $\lambda(A)$ is an orthogonal stochastic measure with structural function $l(A)$ for $A \in \mathfrak{L}$.

Lemma 1. *If $f(x) \in L_2\{l\}$, then $f(x)g(x) \in L_2\{m\}$ and*

$$\int f(x)\lambda(dx) = \int f(x)g(x)\mu(dx) .$$

Proof. The assertion of the lemma is obvious for simple functions $f(x)$ of the form $f(x) = \sum_k c_k \chi_{A_k}(x)$, where $A_k \in \mathfrak{L}$. Furthermore, if $\{f_k(x)\}$ is a fundamental sequence of simple functions in $L_2\{l\}$, then

$$\left|\left| \int f_n(x)\lambda(dx) - \int f_{n+m}(x)\lambda(dx) \right|\right|^2$$

$$= \int \mid f_n(x) - f_{n+m}(x) \mid^2 l(dx)$$

$$= \int \mid f_n(x) - f_{n+m}(x) \mid^2 \mid g(x) \mid^2 m(dx) ,$$

that is, $\{f_n(x)g(x)\}$ is a fundamental sequence in $L_2\{m\}$. Taking the limit as $n \to \infty$ in the equation

$$\int f_n(x)\lambda(dx) = \int f_n(x)g(x)\mu(dx) ,$$

we obtain the assertion of the lemma in the general case.

Lemma 2. *If*

$$\lambda(A) = \int \chi_A(x)g(x)\mu(dx)$$

for $g \in L_2\{m\}$, then for every $A \in \mathfrak{L}_0$,

$$\mu(A) = \int \frac{1}{g(x)} \chi_A(x)\lambda(dx) .$$

Proof. We note that $g(x) = 0$ on a set of l-measure 0. Thus $1/g(x) \neq \infty$ (mod l). Furthermore,

$$\int \frac{1}{|g(x)|^2} \chi_A(x) l(dx) = \int_A \frac{1}{|g(x)|^2} |g(x)|^2 m(dx) = m(A) < \infty$$

if $A \in \mathfrak{L}$. Consequently we may use lemma 1:

$$\int \frac{1}{g(x)} \chi_A(x) l(dx) = \int \frac{1}{g(x)} \chi_A(x) g(x) \mu(dx) = \mu(A) .$$

This completes the proof of Lemma 2.

Let $\mathfrak{T}$ denote a finite or infinite interval, let $\mathfrak{B}$ denote the σ-algebra of Lebesgue-measurable subsets of $\mathfrak{T}$, and let l denote Lebesgue measure.

Let us suppose that $g(t, x)$ is $(\mathfrak{B} \times \mathfrak{L})$-measurable, that $g(t, x) \in L_2\{l \times m\}$, and that $g(t, x) \in L_2\{m\}$ for arbitrary $t \in \mathfrak{T}$. Consider the stochastic integral

$$\xi(t) = \int g(t, x) \mu(dx) . \tag{10}$$

For every t this integral is defined with probability 1.

Lemma 3. *The stochastic integral* (10) *can be defined as a function of t such that the process $\xi(t)$ is measurable.*

Proof. If

$$g(t, x) = \sum c_k \chi_{B_k}(t) \chi_{A_k}(x) , \tag{11}$$

for $B_k \in \mathfrak{B}$ and $A_k \in \mathfrak{L}$, then the function

$$\xi(t) = \sum c_k \chi_{B_k}(t) \mu(A_k)$$

is a $(\mathfrak{B} \times \mathfrak{S})$-measurable function of the variables t and u for $t \in \mathfrak{T}$ and $u \in U$. In the general case we can construct a sequence of simple functions $\{g_n(t, x)\}$ of the form (11) such that

$$\iint |g(t, x) - g_n(t, x)|^2 m(dx) dt \longrightarrow 0 \quad \text{as} \quad n \longrightarrow \infty .$$

Let $\{\xi_n(t)\}$ denote a sequence of processes constructed in accordance with formula (10) for $g = g_n$. Then there exists a process $\tilde{\xi}(t)$ such that

$$\int \mathsf{M} |\tilde{\xi}(t) - \xi_n(t)|^2 dt \longrightarrow 0 \quad \text{as} \quad n \longrightarrow \infty ,$$

and $\tilde{\xi}(t)$ is a $(\mathfrak{B} \times \mathfrak{S})$-measurable function of t and u. On the other hand,

$$\int \mathsf{M} |\xi(t) - \xi_n(t)|^2 dt = \iint |g(t, x) - g_n(t, x)|^2 m(dx) dt \longrightarrow 0 ,$$

so that the processes $\xi(t)$ and $\tilde{\xi}(t)$ coincide almost everywhere in $\mathfrak{T} \times U$. We define

$$\xi'(t) = \begin{cases} \tilde{\xi}(t), & \text{if } \mathsf{P}\{\xi(t) \neq \tilde{\xi}(t)\} = 0 , \\ \xi(t), & \text{if } \mathsf{P}\{\xi(t) \neq \tilde{\xi}(t)\} > 0 . \end{cases}$$

The process $\xi'(t)$ is measurable (since $\xi'(t)$ differs from the $(\mathfrak{B} \times \mathfrak{S})$-measurable function $\tilde{\xi}(t)$ on a set of measure 0), and it is stochastically equivalent to $\xi(t)$. This completes the proof of the lemma.

REMARK 3. Throughout what follows, when studying processes defined by stochastic integrals of the form (10), we shall assume that the processes are measurable if they satisfy the conditions of Lemma 3.

Lemma 4. *Let $g(t, \tau)$ and $h(t)$ denote Borel functions such that*

$$\int_a^b \int_{-\infty}^{\infty} | g(t, \tau) |^2 \, dt m(d\tau) < \infty , \int_a^b | h(t) |^2 \, dt < \infty , \tag{12}$$

and let μ denote an orthogonal stochastic measure on $\mathfrak{B}$. Then

$$\int_a^b h(t) \int_{-\infty}^{\infty} g(t, \tau) \mu(d\tau) dt = \int_{-\infty}^{\infty} g_1(\tau) \mu(d\tau) , \tag{13}$$

where

$$g_1(\tau) = \int_a^b h(t) g(t, \tau) dt .$$

Proof. The mathematical expectation of the square of the integral on the left side of equation (13) is equal to

$$\int_a^b \int_a^b h(t_1)\overline{h(t_2)} \int_{-\infty}^{\infty} g(t_1, \tau)\overline{g(t_2, \tau)} m(d\tau) dt_1 dt_2 = \int_{-\infty}^{\infty} \left| \int_a^b h(t) g(t, \tau) dt \right|^2 m(d\tau)$$

$$\leqq \int_a^b | h(t) |^2 \, dt \int_{-\infty}^{\infty} \int_a^b | g(t, \tau) |^2 \, dt m(d\tau) .$$

For the square of the mathematical expectation of the integral on the right side of equation (13), we have the inequality shown in the last two rows of the expression displayed above. Consequently the right and left sides of equation (13) are continuous with respect to passage to the limit on sequences of the form $\{g_n(t, \tau)\}$ that converge in $L_2\{\Phi\}$, where Φ is the direct product of Lebesgue measure and the measure m in the strip $[a, b] \times (-\infty, \infty)$. Furthermore, the set of functions $g(t, \tau)$ for which equation (13) holds is linear and it contains all functions of the form $\sum c_k \chi_{A_k}(t) \chi_{B_k}(\tau)$. Consequently it contains all functions belonging to $L_2\{\Phi\}$.

Lemma 5. *If the conditions of Lemma 4 are satisfied for every finite interval (a, b) and if the integral*

$$\int_{-\infty}^{\infty} h(t) g(t, \tau) dt = \lim_{\substack{a \to -\infty \\ b \to +\infty}} \int_a^b h(t) g(t, \tau) dt$$

exists in the sense of convergence in $L_2\{m\}$, *then*

$$\int_{-\infty}^{\infty} h(t) \int_{-\infty}^{\infty} g(t, \tau)\mu(d\tau) = \int_{-\infty}^{\infty} f_1(\tau)\mu(d\tau) , \tag{14}$$

where

$$f_1(\tau) = \int_{-\infty}^{\infty} h(t)g(t, \tau)dt .$$

The proof follows directly from the fact that the left-hand member of equation (14) is the mean-square limit of the left-hand member of equation (13), and from the possibility of taking the limit under the stochastic integral sign in the right-hand member of formula (13).

Let us now consider a generalization of the preceding results to vector-valued stochastic measures. We shall confine ourselves to the simplest case of integration of scalar-valued functions—which differs only slightly from integration with respect to numerical stochastic measures.

Let R_p denote a complex vector space of dimension p. For simplicity we assume that some basis in this space is fixed. Let us suppose that to every $\Delta \in \mathfrak{M}$ is assigned a vector-valued random variable $\boldsymbol{\mu}(\Delta)$ with values in R_p:

$$\boldsymbol{\mu}(\Delta) = \{\mu_1(\Delta), \mu_2(\Delta), \cdots, \mu_p(\Delta)\} .$$

Let $|\boldsymbol{\mu}(\Delta)|$ denote the norm of the vector $\boldsymbol{\mu}(\Delta)$:

$$|\boldsymbol{\mu}(\Delta)|^2 = \sum_{k=1}^{p} |\mu_k(\Delta)|^2 .$$

Let us assume that

1. $\mathsf{M} |\boldsymbol{\mu}(\Delta)|^2 < \infty$, $\boldsymbol{\mu}(\phi) = 0$,
2. $\boldsymbol{\mu}(\Delta_1 \cup \Delta_2) = \boldsymbol{\mu}(\Delta_1) + \boldsymbol{\mu}(\Delta_2) (\text{mod } \mathsf{P})$, if $\Delta_1 \cap \Delta_2 = \phi$,
3. $\mathsf{M}\mu_k(\Delta_1)\overline{\mu_j(\Delta_2)} = m_{kj}(\Delta_1 \cap \Delta_2), \Delta_i \in \mathfrak{M}, i = 1, 2; k, j = 1, \cdots, p.$

We shall call a family of random vectors $\{\boldsymbol{\mu}(\Delta)\}$, for $\Delta \in \mathfrak{M}$, an elementary vector-valued stochastic (orthogonal) measure and we shall call the matrix $m(\Delta) = (m_{kj}(\Delta)) = \mathsf{M}\boldsymbol{\mu}(\Delta)\boldsymbol{\mu}^*(\Delta)$ its structural matrix.

We note that the matrix $m(\Delta_1 \cap \Delta_2)$, as a function of Δ_1 and Δ_2, enjoys properties 14 to 17 (cf. Chapter I, Section 2) of the covariance matrix of a vector-valued random function. Furthermore, if $\Delta_1 \cap \Delta_2 = \Phi$, then

$$m(\Delta_1 \cup \Delta_2) = m(\Delta_1) + m(\Delta_2) .$$

From this it follows that the diagonal elements of the matrix $m(\Delta)$ are elementary measures. Furthermore it follows from the inequality

$$|m_{kj}(\Delta)| \leqq \sqrt{m_{kk}(\Delta)m_{jj}(\Delta)} \tag{15}$$

that

$$\sum_r |m_{kj}(\Delta_r)| \leqq \{\sum_r m_{kk}(\Delta_r) \sum_r m_{jj}(\Delta_r)\}^{1/2}, \tag{16}$$

so that the set functions $m_{kj}(\Delta)$ (for $k, j = 1, \cdots, p$) will be of bounded variation on Δ_0 if $m_{jj}(\Delta_0) < \infty$ for $j = 1, \cdots, p$. We set

$$m_0(\Delta) = \operatorname{tr} m(\Delta) = \sum_{k=1}^{p} m_{kk}(\Delta) .$$

From (16) it also follows that if $\sum_{r=1}^{m_N} m_0(\Delta_r^N)$ approaches 0 as $N \to \infty$, $\sum_{r=1}^{m_N} |m_{kj}(\Delta_r^N)|$ also approaches 0 as $N \to \infty$. From this it follows that the functions $m_{jk}(\Delta)$ can be extended as countably additive set functions on $\mathfrak{S}$ (in the sense established in Theorem 8, Section 7, Chapter II) if the function $m_0(\Delta)$ is countably additive (or subadditive) on $\mathfrak{M}$ (cf. proof of the theorem just cited).

In what follows we shall say that the matrix functions obtained by such an extension from a structural function of an elementary orthogonal stochastic measure are positive-definite matrix measures.

Above, $\mathfrak{S}$ denoted the completion of $s\{\mathfrak{M}\}$ with respect to the extended elementary measure $m_0(\Delta)$. We shall keep the original notations for the extensions of the functions m_{kj} and m_0 and of the matrix m on $\mathfrak{S}$. In addition we shall assume that $m_0(\Delta)$ is subadditive on $\mathfrak{M}$.

On $L\{\mathfrak{M}\}$ we define a stochastic integral by the formula

$$\eta = \int f(x)\mu(dx) = \sum_{k=1}^{n} c_k\mu(\Delta_k) , \tag{17}$$

where $f(x) = \sum c_k\chi_{\Delta_k}(x)$ and $\Delta_k \in \mathfrak{M}$ (for $k = 1, \cdots, n$). The value of this integral is a random vector (a column vector) with values in R_p. We let $L^p\{\mu\}$ denote the set of all random vectors η of the form (17). If

$$g(x) = \sum_{k=1}^{n} d_k\chi_{\Delta_k}(x) ,$$

then

$$\mathsf{M}\left(\int f(x)\mu(dx)\left(\int g(x)\mu(dx)\right)^*\right) = \sum_{k=1}^{n} c_k\bar{d}_k m(\Delta_k) ,$$

which can be written in the form

$$\mathsf{M}\left(\int f(x)\mu(dx)\left(\int g(x)\mu(dx)\right)^*\right) = \int f(x)\overline{g(x)}m(dx) . \tag{18}$$

From this it follows that

$$\mathsf{M}\left|\int f(x)\mu(dx)\right|^2 = \int |f(x)|^2 m_0(dx) . \tag{19}$$

In $L\{\mathfrak{M}\}$, we define a scalar product by

$$(f, g) = \operatorname{tr} \int f(x)\overline{g(x)}m(dx) \equiv \operatorname{trace} \int f(x)\overline{g(x)}m(dx) ,$$

Formula (18) establishes an isometric mapping $\eta = \psi(f)$ of the space $L(\mathfrak{M})$ onto $L^p\{\mu\}$ if the scalar product of two elements η_1 and η_2 of $L^p\{\mu\}$ is defined as $\mathsf{M}\eta_2^*\eta_1$. Let the closure of the space of random variables $L^p\{\mu\}$ be denoted by $L_2^p\{\mu\}$ and let the completion of $L\{\mathfrak{M}\}$ be denoted $L_2\{\mathfrak{M}\}$.

In a manner analogous to the derivation of inequality (16) we can derive (first for simple functions, and then by taking the limit, for arbitrary $\mathfrak{S}$-measurable functions) the inequality

$$\int |f(x)| \, |m_{kj}| \, (dx) \leqq \left\{\int |f(x)| \, m_{kk}(dx) \int |f(x)| \, m_{jj}(dx)\right\}^{1/2} , \quad (20)$$

where $|m_{kj}|\,(A)$ is the absolute variation of the function m_{kj}. Inequality (20) implies the existence and continuity of the integral $\int f(x)\overline{g(x)}m_{kj}(dx)$ as a functional of f and g in $L_2\{m_0\}$.

4. INTEGRAL REPRESENTATIONS OF RANDOM FUNCTIONS

By using the results of the preceding section, we can obtain various representations of random functions with the aid of stochastic integrals.

Let us first suppose that a p-measurable vector-valued random function $\xi(\theta)$ for $\theta \in \Theta$ can be represented in the form

$$\xi(\theta) = \int g(\theta, x)\mu(dx) , \quad (1)$$

where μ is a stochastic measure with values in R_p and with structural matrix $m(A)$, in the notation of the preceding section, and $g(\theta, x)$ is a scalar-valued function such that for every $\theta \in \Theta$ the function $g(\theta, x)$ belongs to $L_2\{m_0\}$, where $m_0(A)$ is the trace of $m(A)$. On the basis of formula (18) of Section 3, the covariance matrix of the random function $\xi(\theta)$ is of the form

$$R(\theta_1, \theta_2) = \mathsf{M}\xi(\theta_1)\xi^*(\theta_2) = \int g(\theta_1, x)\overline{g(\theta_2, x)}m(dx) , \quad (2)$$

and it follows from equation (19) of Section 3 that

$$\mathsf{M}\xi^*(\theta_2)\xi(\theta_1) = \int g(\theta_1, x)\overline{g(\theta_2, x)}m_0(dx) . \quad (3)$$

We recall that $L_2\{m_0\}$ is a Hilbert space of $\mathfrak{S}$-measurable complex-valued m_0-square-integrable functions. We let $\tilde{L}_2\{g\}$ denote the closure in $L_2\{m_0\}$ of the linear hull of the system of functions $g(\theta, x)$ for $\theta \in \Theta$. If $\tilde{L}_2\{g\} = L_2\{m_0\}$ the system of functions $\{g(\theta, x)\}$ for $\theta \in \Theta$ is said to be *complete* in $L_2\{m_0\}$.

Theorem 1. *Suppose that the covariance matrix of a vector-valued random function $\boldsymbol{\xi}(\theta)$, for $\theta \in \Theta$, has a representation of the form* (2), *where $\{m, \mathfrak{S}\}$ is an orthogonal positive-definite matrix measure, where $g(\theta, x) \in L_2\{m_0\}$, and where $m_0(A)$ is the trace of $m(A)$. Then there exists a stochastic measure $\boldsymbol{\mu}(A)$, for $A \in \mathfrak{S}$, with structural matrix $m(A)$ such that equation* (1) *holds with probability* 1 *for arbitrary $\theta \in \Theta$.*

Proof. To every linear combination of the form

$$f(x) = \sum_{k=1}^{n} \alpha_k g(\theta_k, x) , \qquad \theta_k \in \Theta , \tag{4}$$

let us assign the random vector $\eta = \psi(f)$ defined by

$$\eta = \sum_{k=1}^{n} \alpha_k \boldsymbol{\xi}(\theta_k) . \tag{5}$$

Let $L(\boldsymbol{\xi})$ and $L\{g\}$ respectively denote the sets of all random vectors (5) and functions (4). In $L\{g\}$ we define the scalar product by

$$(f_1, f_2) = \int f_1(x)\overline{f_2(x)} m_0(dx) . \tag{6}$$

The relation $\eta = \psi(f)$ is an isometric mapping of $L\{g\}$ into $L(\boldsymbol{\xi})$. Consequently (cf. Theorem 1 of Section 1), it can be extended as an isometric mapping of $\tilde{L}_2\{g\}$ onto $\tilde{L}_2\{\boldsymbol{\xi}\}$, where $\tilde{L}_2\{\boldsymbol{\xi}\}$ is the closure of $\tilde{L}\{\boldsymbol{\xi}\}$ in the sense of mean-square convergence of random vectors and $\tilde{L}_2\{g\}$ is the completion of the space $L\{g\}$ with respect to the metric generated by the scalar product (6).

Let us first suppose that the system of functions $\{g(\theta, x)\}$ for $\theta \in \Theta$ is complete in $L_2\{m_0\}$. If $A \in \mathfrak{S}$, then $\chi_A(x) \in L_2\{m_0\} = \tilde{L}_2\{g\}$. We set $\boldsymbol{\mu}(A) = \psi(\chi_A)$. Then $\boldsymbol{\mu}(A)$ is a vector-valued stochastic measure and its structural function coincides with m:

$$\mathsf{M}\boldsymbol{\mu}(A_1)\boldsymbol{\mu}^*(A_2) = \int \chi_{A_1}(x)\overline{\chi_{A_2}(x)} m(dx) = m(A_1 \cap A_2) .$$

Now let us define a random function $\boldsymbol{\zeta}(\theta)$ as the stochastic integral

$$\boldsymbol{\zeta}(\theta) = \int g(\theta, x)\boldsymbol{\mu}(dx) .$$

Since

$$\mathsf{M}\boldsymbol{\xi}(\theta)\boldsymbol{\mu}^*(A) = \int g(\theta, x)\chi_A(x) m(dx) ,$$

the isometry of the correspondence $\eta = \psi(f)$ implies that

$$\mathsf{M}\boldsymbol{\xi}(\theta)\boldsymbol{\zeta}^*(\theta) = \int g(\theta, x)\overline{g(\theta, x)} m(dx) .$$

From this we obtain

$$\begin{aligned}\mathsf{M}\,|\,\xi(\theta)-\zeta(\theta)\,|^2 &= \mathsf{M}([\xi(\theta)-\zeta(\theta)]^*[\xi(\theta)-\zeta(\theta)]) \\ &= \mathsf{M}\xi^*(\theta)\xi(\theta)-\mathsf{M}\zeta^*(\theta)\xi(\theta)-\mathsf{M}\xi^*(\theta)\zeta(\theta)+\mathsf{M}\zeta^*(\theta)\zeta(\theta)=0\,,\end{aligned}$$

which proves the theorem for the particular case that we are considering.

Let us now turn to the general case. Let Θ_1 denote an arbitrary set having no points in common with Θ. Let us define a system of functions $\{u(\theta, x)\}$, for $\theta \in \Theta_1$ and $x \in E$, that is complete in $L_2\{m_0\} \ominus \tilde{L}_2\{g\}$, and let us define a Gaussian vector-valued random process $\xi_1(\theta)$, for $\theta \in \Theta_1$, that is independent of $\xi(\theta)$ for $\theta \in \Theta$, with mean zero and with correlation matrix

$$R_1(\theta_1, \theta_2) = \int u(\theta_1, x)\overline{u(\theta_2, x)}m(dx)\ .$$

We set $\zeta(\theta) = \xi(\theta)$ if $\theta \in \Theta$ and $\zeta(\theta) = \xi_1(\theta)$ if $\theta \in \Theta_1$. We set $f(\theta, x) = g(\theta, x)$ for $\theta \in \Theta$ and $f(\theta, x) = u(\theta, x)$ for $\theta \in \Theta_1$. Then the covariance matrix of the process $\zeta(\theta)$ is equal to

$$\int f(\theta_1, x)\overline{f(\theta_2, x)}m(dx) \quad (\theta_i \in \Theta \cup \Theta_1,\ i = 1, 2)$$

and $\{f(\theta, x)\}$ for $\theta \in \Theta \cup \Theta_1$, is a complete system of functions in $L_2\{m_0\}$. According to what we have proved, there exists an orthogonal stochastic measure μ on $\mathfrak{S}$ such that

$$\zeta(\theta) = \int f(\theta, x)\mu(dx)$$

for $\theta \in \Theta \cup \Theta_1$. Therefore,

$$\xi(\theta) = \int g(\theta, x)\mu(dx)\ , \qquad \theta \in \Theta\ .$$

This completes the proof of the theorem.

REMARK 1. If $\{g(\theta, x)\}$, for $\theta \in \Theta$, is a complete system of functions in $L_2\{m_0\}$, then $\mu(A) \in \tilde{L}_2\{\xi\}$; that is, the value of the stochastic measure μ can be established from the values of the sample function of the process. In the general case, this is not true.

If $\mu(A) \in \tilde{L}_2\{\xi\}$ for arbitrary $A \in \mathfrak{S}$, we shall say that the stochastic measure μ is *subordinate to* the process ξ.

We now present a number of examples illustrating the use of Theorem 1. In what follows we shall consider processes $\xi(t)$ that are stationary in the broad sense and that satisfy the condition $\mathsf{M}\xi(t) = 0$. For the sake of brevity we shall use the word "stationary" instead of the phrase "stationary in the broad sense."

On the basis of Theorem 2, Section 5, Chapter I, the covariance matrix of a stationary mean-square continuous process can be represented in the form

$$R(t_1, t_2) = R(t_1 - t_2) = \int_{-\infty}^{\infty} e^{i\omega(t_1 - t_2)} dF(\omega) \tag{7}$$

where $\Delta F = F(u_2) - F(u_1)$ is a nonnegative-definite matrix interval function of bounded variation (the spectral matrix of the process). The function $F(\omega)$ defines a positive-definite matrix measure $\{\tilde{\mathfrak{B}}, F\}$, where $\tilde{\mathfrak{B}}$ is the σ-algebra of subsets consisting of all Borel sets of the real line. (This σ-algebra is complete with respect to $\operatorname{tr} F(A) = m_0(A)$.) Equation (7) is the particular case of equation (2) in which the functions $g(\theta, x)$ correspond to $e^{i\omega t}$, $\theta \leftrightarrow t$, $\omega \leftrightarrow x$. Here, the function set $\{e^{i\omega t}, -\infty < \omega < \infty\}$ is complete in $L_2\{m_0\}$ where m_0 is an arbitrary bounded measure on the real line. Thus Theorem 1 is applicable and we obtain the following result:

Theorem 2. *A vector-valued mean-square continuous stationary random process $\xi(t)$ (for $-\infty < t < \infty$) such that $\mathsf{M}\xi(t) = 0$ has a representation of the form*

$$\xi(t) = \int_{-\infty}^{\infty} e^{it\omega} \mu(d\omega) \tag{8}$$

where $\mu(A)$ is a vector-valued orthogonal stochastic measure on $\tilde{\mathfrak{B}}$ subordinate to $\xi(t)$. Between $L_2\{\xi\}$ and $L_2\{m_0\}$ there exists an isometric correspondence under which

a. $\xi(t) \leftrightarrow e^{i\omega t}$, $\mu(A) \leftrightarrow \chi_A(\omega)$;
b. if $\eta_i \leftrightarrow g_i(\omega)(i = 1, 2)$, then

$$\eta_i = \int g_i(\omega)\mu(d\omega)$$

and

$$\mathsf{M}\eta_1\eta_2^* = \int g_1(\omega)\overline{g_2(\omega)} dF(\omega) .$$

Formula (8) is called the *spectral decomposition* of the stationary process and the measure $\mu(A)$ is called the *stochastic spectral measure* of the process. It follows from Theorem 2 that

$$\mathsf{M}\mu(A_1)\mu^*(A_2) = \int_{A_1 \cap A_2} dF(\omega) . \tag{9}$$

REMARK 2. For arbitrary $\eta \in \tilde{L}_2\{\xi\}$ we have $\mathsf{M}\eta = 0$. In particular, for arbitrary $A \in \tilde{\mathfrak{B}}$ we have $\mathsf{M}\mu(A) = 0$.

REMARK 3. If $\mathsf{M}\xi(t) = \boldsymbol{a} \neq 0$, Theorem 2 can be applied to the process $\xi(t) - \boldsymbol{a}$. On the other hand, the representation (8) can be preserved in the general case if we add to $\mu(A)$ a measure concentrated at the point $\omega = 0$ where its value is $\boldsymbol{a}$.

As an example of the application of Theorem 2 we give the Kotel'nikov-Shannon formula for a random process whose spectral function is concentrated in a finite interval $[-B, B]$. Let us expand the function $e^{i\omega t}$ on the interval $[-B, B]$ in a Fourier series. We have

$$e^{i\omega t} = \sum_{n=-\infty}^{\infty} \frac{\sin(Bt - \pi n)}{Bt - \pi n} \exp\left(i\frac{\pi n}{B}\omega\right).$$

The series on the right converges uniformly in every interval of the form $[-B', B']$, where $B' < B$, and the sequence of its partial sums is bounded. Therefore it also converges in $L_2\{m_0\}$. By virtue of the isomorphism between the spaces $L_2\{m_0\}$ and $L_2\{\xi\}$, we have (in the sense of mean-square convergence)

$$\xi(t) = \sum_{n=-\infty}^{\infty} \frac{\sin(Bt - \pi n)}{Bt - \pi n} \xi\left(\frac{\pi n}{B}\right). \tag{10}$$

Thus the value of the random function $\xi(t)$ at an arbitrary instant t is uniquely determined from its values at equally spaced instants of time $\pi n/B$, for $n = 0, \pm 1, \pm 2, \cdots$.

Remark 4. For stationary vector-valued sequences $\{\xi_n\}$ for $n = 0, \pm 1, \pm 2, \cdots$, it is possible to formulate a theorem completely analogous to Theorem 2. The only difference is that the spectral measure of the sequence is concentrated on the half-closed half-open interval $[-\pi, \pi)$ and not on the entire real line as in the case of a process with continuous time (cf. Theorem 3, Section 8, Chapter I). We can treat processes with increments that are stationary in the broad sense in analogous fashion. Since the structural function of such a process has (cf. Theorem 4, Section 5, Chapter I) a representation of the form

$$D(t_1, t_2, t_3, t_4) = \int_{-\infty}^{\infty} \frac{e^{it_2\omega} - e^{it_1\omega}}{i\omega} \frac{e^{-it_4\omega} - e^{-it_3\omega}}{-i\omega} dF(\omega)$$

where $F(\omega)$ is the spectral matrix of the process, it follows, by virtue of Theorem 1 (with $\zeta(\theta) = \zeta(t_2) - \zeta(t_1)$, where $\theta = [t_1, t_2)$), that

$$\zeta(t_2) - \zeta(t_1) = \int_{-\infty}^{\infty} \frac{e^{it_2\omega} - e^{it_1\omega}}{i\omega} \mu(d\omega), \tag{11}$$

where $\mu(A)$ is an orthogonal stochastic measure with structural matrix $F(A)$.

Here, the system of functions $(e^{it\omega} - 1)/i\omega$ (for $-\infty < t < \infty$) is complete in $L_2\{m_0\}$. This is easily seen if we represent the characteristic function of the interval $\Delta = (a, b)$, at whose endpoints the spectral matrix $F(u)$ is continuous, in the form

$$\chi_\Delta(\omega) = \lim_{A\to\infty} \frac{1}{\pi} \int_a^b \frac{\sin A(t-\omega)}{t-\omega} dt ,$$

that is, as the limit of linear combinations of the functions $(e^{it\omega} - 1)/i\omega$.

Theorem 3. *A random mean-square continuous process with stationary increments* $\zeta(t)$ *has a representation of the form* (11), *where* $\mu(A)$ *is an orthogonal stochastic measure, where* $\mathsf{M}(\mu(A)\mu^*(A)) = F(A)$ *is the spectral matrix of the process, and where*

$$\int_{-\infty}^{\infty} K(\omega) d \operatorname{tr} F(\omega) < \infty , \qquad K(\omega) = \begin{cases} 1, \ |\omega| \leqq 1 , \\ \dfrac{1}{\omega^2}, \ |\omega| > 1 . \end{cases}$$

The measure $\mu(A)$ *is subordinate to* $\zeta(t)$.

Let us make a few comments regarding processes with stationary increments. We confine ourselves to the one-dimensional case. For a process $\zeta(t)$ to have a mean-square derivative at a point t, it is necessary and sufficient (cf. Theorem 1, Section 2) that the limit

$$\lim_{t_1,t_2,t_3,t_4\to t} \frac{D(t_1, t_2, t_3, t_4)}{(t_2 - t_1)(t_4 - t_3)}$$

$$= \lim_{t_1,t_2,t_3,t_4\to t} \int_{-\infty}^{\infty} \frac{e^{i\omega t_2} - e^{-i\omega t_1}}{i\omega(t_2 - t_1)} \, \frac{e^{i\omega t_4} - e^{i\omega t_3}}{-i\omega(t_4 - t_3)} dF(\omega)$$

exist. This condition will be satisfied if

$$\int_{-\infty}^{\infty} dF(\omega) < \infty .$$

In such a case the process is mean-square differentiable at every point and its mean-square derivative $\zeta'(t)$ has correlation function

$$B_1(t_1, t_2) = \lim_{\Delta t_1, \Delta t_2 \to 0} \frac{D(t_1, t_1 + \Delta t_1, t_2, t_2 + \Delta t_2)}{\Delta t_1 \Delta t_2} = \int_{-\infty}^{\infty} e^{i\omega(t_2 - t_1)} dF(\omega) .$$

On the other hand, we can take the limit under the integral sign in the relation

$$\frac{\zeta(t+\Delta) - \zeta(t)}{\Delta} = \int_{-\infty}^{\infty} \frac{e^{i(t+\Delta)\omega} - e^{it\omega}}{i\omega\Delta} \mu(d\omega)$$

as $\Delta \longrightarrow 0$. Thus

$$\zeta'(t) = \int_{-\infty}^{\infty} e^{it\omega} \mu(d\omega)$$

is a process that is stationary in the broad sense and that has spectral function $F(\omega)$. Consequently, for $F(+\infty) < \infty$ a process with stationary increments can be regarded as the mean-square integral of the stationary process and, in the general case, as the mean-square limit of such integrals. An interesting example of a process with stationary increments that is not an integral of a stationary process is a process for which $dF(\omega) = d\omega$. Such a process can be regarded as the mean-square limit as $A \to +\infty$ of integrals of stationary processes having spectral density equal to 1 on the interval $(-A, A)$, but equal to 0 outside that interval. Speaking crudely, we say that such a process is an integral of a fictitious stationary process having constant spectral density on the entire real line. This fictitious stationary process is called "white noise" and is often encountered in nonrigorous expositions of the theory of random processes and its applications. On the other hand, the theory of such processes is part of the theory of generalized random processes. The process $\zeta(t)$ with stationary increments, which was introduced above, may be interpreted as white noise integrated and will serve as a substitute for white noise itself in our discussion.

We note that this integrated white noise is a process with orthogonal increments. In fact, using Parseval's equality for the Fourier integral, and the fact that $(e^{-ib\omega} - e^{-ia\omega})/i\omega$ is the Fourier transform of the function $\chi_{(a,b)}(\omega)$, where l is Lebesgue measure on the real line, we have

$$\begin{aligned} &\mathsf{M}(\zeta(t_2) - \zeta(t_1))(\zeta(t_4) - \zeta(t_3)) \\ &\quad = \int_{-\infty}^{\infty} \frac{e^{it_2\omega} - e^{it_1\omega}}{i\omega} \frac{e^{-it_4\omega} - e^{-it_3\omega}}{-i\omega} d\omega \\ &\quad = 2\pi \int_{-\infty}^{\infty} \chi_{(t_1,t_2)}(\omega)\chi_{(t_4,t_3)}(\omega)d\omega = 2\pi l[(t_1, t_2) \cap (t_4, t_3)] \,. \end{aligned}$$

Let us now look at the matter of spectral decomposition of a mean-square continuous isotropic two-dimensional random field. On the basis of Theorem 5, Section 5, Chapter I the covariance function of the field takes the form

$$R(\boldsymbol{x}_1, \boldsymbol{x}_2) = R(\rho) = \int_0^\infty J_0(\omega\rho)dg(\omega) \,, \tag{12}$$

where $\boldsymbol{x}_1$ and $\boldsymbol{x}_2$ are points in the plane and ρ is the distance between them. If (r_i, φ_i) are the polar coordinates of the point $\boldsymbol{x}_i$ (for $i = 1, 2$), then

$$\rho = \sqrt{r_1^2 + r_2^2 - 2r_1r_2\cos(\varphi_1 - \varphi_2)} \,.$$

Using the addition formula for the Bessel function J_0 (see for example Gray and Mathews, p. 39), namely

$$J_0(\omega\rho) = \sum_{k=-\infty}^{\infty} J_k(\omega r_1)J_k(\omega r_2)e^{ik(\varphi_1-\varphi_2)} \,,$$

we rewrite formula (12) in the form

$$R(\rho) = \int_0^\infty \int_{-\infty}^\infty J_\nu(\omega r_1) e^{i\nu\varphi_1} J_\nu(\omega r_2) e^{-i\nu\varphi_2} dg(\omega)\varepsilon(d\nu) ,$$

where $\varepsilon(d\nu)$ is a measure concentrated at the points $k = 0, \pm 1, \pm 2, \cdots$ and $\varepsilon(\{k\}) = 1$. On the basis of Theorem 1, a plane isotropic homogeneous mean-square continuous field $\xi(x)$ (such that $M\xi(x) = 0$) where $x = re^{i\varphi}$, has a representation of the form

$$\xi(x) = \sum_{k=-\infty}^{\infty} e^{ik\varphi} \int_0^\infty J_k(\omega r) \mu_k(d\omega) \tag{13}$$

where $\{\mu_k\}$ is a sequence of mutually orthogonal stochastic measures on the ray $[0, \infty)$.

5. LINEAR TRANSFORMATIONS

Consider a system $\sum$ (a device or a procedure) designed to transform signals (functions) $x(t)$ depending on the time t. The function to be transformed is called the *input function* of the system and the transformed function is called the *output function* or the *response* to the input function. Mathematically, every system is defined by a class D of "admissible" input functions and a relation of the form

$$z(t) = T(x \mid t) ,$$

where $x = x(s)$ (for $-\infty < s < \infty$) is the input function (belonging to D) and $z(t)$ is the value of the output function at the instant t.

A system $\sum$ is said to be *linear* if (a) the class D of admissible functions is linear and (b) the operator T satisfies the superposition principle

$$T(\alpha x_1 + \beta x_2 \mid t) = \alpha T(x_1 \mid t) + \beta T(x_2 \mid t) .$$

Let us introduce the operation of displacement of time S_τ (for $-\infty < \tau < \infty$) with the aid of the equation

$$x_\tau(t) = S_\tau(x \mid t) = x(t + \tau) .$$

This operation is defined on the set of all functions of the variable t (for $-\infty < t < \infty$) and is a linear operation. The system $\sum$ is said to be *homogeneous with respect to time* (or simply *homogeneous*) if

$$T(x_\tau \mid t) = T(x \mid t + \tau) \quad \text{or} \quad T(S_\tau x \mid t) = S_\tau T(x \mid t) ;$$

that is, if the transformation T commutes with the time-displacement S_τ (for $-\infty < \tau < \infty$).

A simple example of a linear transformation is one of the form

$$z(t) = \int_{-\infty}^{\infty} h(t, \tau)x(\tau)d\tau \ , \tag{1}$$

for which the class of admissible functions D depends on the properties of the function $h(t, \tau)$. Consider the set of functions of the form

$$x_\varepsilon(t) = \begin{cases} 0 & \text{for } |t - a| \geqq \dfrac{\varepsilon}{2} \ , \\ \dfrac{1}{\varepsilon} & \text{for } |t - a| < \dfrac{\varepsilon}{2} \ , \end{cases}$$

where $\varepsilon > 0$. The function $x_\varepsilon(t)$ has the nature of an impulse: it acts for a brief instant of time (of duration ε) but its value is great, since $\int_{-\infty}^{\infty} x_\varepsilon(\tau)d\tau = 1$. It is convenient to regard the limit as $\varepsilon \rightarrow 0$ of the set of impulse functions $x_\varepsilon(t)$ as the generalized function $\delta_a(t)$, which is the instantaneous unit impulse at the instant a. This function is equal to 0 for $t \neq a$ but it satisfies the condition $\int_{-a-h}^{a+h} \delta(t, a)dt = 1$, where $h > 0$. There are no functions, in the usual sense of the word, with such properties. The theory of generalized functions including functions of the type described, is developed in the books by L. Schwartz, and I. M. Gel'fand and G. E. Shilov. The function $\delta_a(t)$ is called the *delta function acting at the point a.*

Corresponding to the output function $x_\varepsilon(t)$ of the system is the function

$$z_\varepsilon(t) = \frac{1}{\varepsilon}\int_{a-\varepsilon/2}^{a+\varepsilon/2} h(t, \tau)d\tau \ ,$$

which approaches $h(t, a)$ under certain hypotheses (for example, if $h(t, \tau)$ is a continuous function of τ at the point a).

Thus we should interpret $h(t, \tau)$ as the response of the system to the delta function at the instant τ. In accordance with this, $h(t, \tau)$ is called the *impulse transfer function* of the system. If the system $\sum$ is homogeneous with respect to time, then formally,

$$h(t, a - c) = T(\delta_{a-c} \mid t) = T(S_c\delta_a \mid t) = S_cT(\delta_a \mid t) = h(t + c, a) \ ,$$

or by setting $a \sim c$ and $t \sim t - c$,

$$h(t - c, 0) = h(t, c) \ .$$

The function $h(t) = h(t + c, c)$ is called the *impulse transfer function* of the homogeneous system.

Thus for a homogeneous system, equation (1) takes the form

$$z(t) = \int_{-\infty}^{\infty} h(t - \tau)x(\tau)d\tau \, . \tag{2}$$

The operation on the right side of equation (2) is called the *convolution* of the functions $h(t)$ and $x(t)$.

If the input function of the system differs from the output function only by a scalar factor (the transformation T does not change the form of the signal), that is, if $T(f \mid t) = \lambda f(t)$ for $(-\infty < t < \infty)$, then $f(t)$ is called an *eigenfunction* and λ an *eigenvalue* of the transformation T. For systems that are homogeneous with respect to time and have an integrable impulse transfer function, $\int_{-\infty}^{\infty} |h(t)| \, dt < \infty$, functions of the form $e^{it\omega}$ (where ω is an arbitrary real number) are eigenfunctions. This is true because all bounded measurable functions are admissible and

$$\int_{-\infty}^{\infty} h(t - \tau)e^{i\omega\tau}d\tau = \int_{-\infty}^{\infty} h(u)e^{i\omega(t-u)}du = H(i\omega)e^{i\omega t} \, ,$$

where

$$H(i\omega) = \int_{-\infty}^{\infty} h(u)e^{-i\omega u}du \, , \tag{3}$$

the Fourier transform of the impulse transfer function, is an eigenvalue of the transformation.

Thus the ratio of the response of the system to the simple harmonic input function $e^{i\omega t}$ to the latter function

$$H(i\omega) = \frac{T(e^{i\omega u} \mid t)}{e^{i\omega t}} \tag{3}$$

is independent of time. The function $H(i\omega)$ is called the *frequency characteristic* of the system or the *transfer coefficient*.

It is possible to interpret the frequency characteristic of the system (2) in a somewhat different way by considering another class of admissible functions. Suppose that $x(t)$ is integrable. On the basis of Fubini's theorem,

$$\int_{-\infty}^{\infty} | z(t) | \, dt \leqq \int_{-\infty}^{\infty}\int_{-\infty}^{\infty} | h(t - \tau) | \, | x(\tau) | \, d\tau dt$$

$$= \int_{-\infty}^{\infty} | x(\tau) | \, d\tau \int_{-\infty}^{\infty} | h(t) | \, dt < \infty \; ;$$

that is, the function $z(t)$ is also integrable. Now consider the Fourier transform of the function $z(t)$. Applying Fubini's theorem, we obtain

$$\tilde{z}(\omega) = \int_{-\infty}^{\infty} e^{-i\omega t} z(t) dt = \int_{-\infty}^{\infty}\int_{-\infty}^{\infty} e^{-i\omega(t-\tau)} h(t - \tau) e^{-i\omega\tau} x(\tau) d\tau dt$$
$$= H(i\omega)\tilde{x}(\omega), \ \tilde{x}(\omega) = \int_{-\infty}^{\infty} e^{-i\omega\tau} x(\tau) d\tau \ .$$

Consequently the ratio of the Fourier transform of the output function to the Fourier transform of the input function is independent of the input function of the system and is equal to its frequency characteristic:

$$H(i\omega) = \frac{\tilde{z}(\omega)}{\tilde{x}(\omega)} \ .$$

In formula (1) the response of the system at the instant t depends on the values of the input function both at instants $\tau < t$ and at instants $\tau > t$. In physical constructions, however, it is impossible to anticipate the future. Therefore, for physical constructions,

$$h(t, \tau) = 0 \quad \text{for} \quad t < \tau \ . \tag{4}$$

Equation (4) is called the *condition of physical realizability* of the system. For systems satisfying condition (4) formula (1) assumes the form

$$z(t) = \int_{-\infty}^{t} h(t, \tau) x(\tau) d\tau \ , \tag{5}$$

and if the system is homogeneous,

$$z(t) = \int_{-\infty}^{t} h(t - \tau) x(\tau) d\tau = \int_{0}^{\infty} h(u) x(t - u) du \ . \tag{6}$$

If a function is introduced at the input of the system beginning at the instant of time 0 (that is, $x(\tau) = 0$ for $\tau < 0$), then

$$z(t) = \int_{0}^{t} h(t - \tau) x(\tau) d\tau \ . \tag{7}$$

In studying such systems, it is convenient to use not the Fourier transform but the Laplace transform

$$\tilde{z}(p) = \int_{0}^{\infty} e^{-pt} z(t) dt \ . \tag{8}$$

It follows from formula (7) that

$$\tilde{z}(p) = H(p)\tilde{x}(p) \ , \qquad \tilde{x}(p) = \int_{0}^{\infty} e^{-pt} x(t) dt \tag{9}$$

for $\text{Re}\, p \geqq \alpha$ if the functions $e^{-\alpha t} h(t)$ and $e^{-\alpha t} x(t)$ are absolutely integrable.

Let us turn to the basic theme of the present section, namely, linear transformations of random processes. For the most part, we shall consider homogeneous (with respect to time) transformations of stationary processes. In regard to the more general case, we shall confine ourselves to some simple comments.

Let $\zeta(t)$ (for $-\infty < t < \infty$) denote a measurable Hilbert process with covariance $B(t, \tau)$. Suppose that the function $B(t, t)$ is integrable with respect to t over every interval and that for every fixed τ, the function $h(\tau, t)$ is also integrable over every interval. Then with probability 1, the integral $\int_a^b h(t, \tau)\xi(\tau)d\tau$ exists for arbitrary a and b. Let us define the improper integral from $-\infty$ to ∞ as the mean-square limit of the integral over finite intervals of integration as their endpoints approach $-\infty$ and $+\infty$:

$$\int_{-\infty}^{\infty} h(t, \tau)\xi(\tau)d\tau = \underset{\substack{a\to-\infty \\ b\to+\infty}}{\text{l.i.m.}} \int_a^b h(t, \tau)\xi(\tau)d\tau .$$

From part 2 of Section 2, we recall that for this limit to exist it is necessary and sufficient that the integral

$$\int_{-\infty}^{\infty}\int_{-\infty}^{\infty} h(t, \tau_1)B(\tau_1, \tau_2)\overline{h(t, \tau_2)}d\tau_1 d\tau_2$$

exist in the sense of a Cauchy improper integral over the plane. If it exists for $t \in \mathfrak{T}$, then $z(t)$ for $t \in \mathfrak{T}$ is a Hilbert random process with covariance

$$B_z(t_1, t_2) = \int_{-\infty}^{\infty}\int_{-\infty}^{\infty} h(t_1, \tau_1)B(\tau_1, \tau_2)\overline{h(t_2, \tau_2)}d\tau_1 d\tau_2 . \tag{10}$$

Let us suppose now that $\xi(t)$ is a stationary process in the broad sense with spectral function $F(\omega)$ and $\mathsf{M}\xi(t) = 0$. This assumption will be retained until the end of the present section. The integral

$$\eta(t) = \int_{-\infty}^{\infty} h(t - \tau)\xi(\tau)d\tau \tag{11}$$

exists (in the sense mentioned above) if and only if the integral

$$\int_{-\infty}^{\infty}\int_{-\infty}^{\infty} h(t - \tau_1)R(\tau_1 - \tau_2)\overline{h(t - \tau_2)}d\tau_1 d\tau_2$$
$$= \int_{-\infty}^{\infty}\int_{-\infty}^{\infty} h(\tau_1)R(\tau_2 - \tau_1)\overline{h(\tau_2)}d\tau_1 d\tau_2$$

exists, where $R(t)$ is the covariance function of the process. For this in turn it is sufficient that the function $h(t)$ be absolutely integrable over $(-\infty, \infty)$. In this case, by using the spectral representation of the covariance function $R(t)$ (cf. (1), Section 5, Chapter I), we obtain the following expression for the covariance function $R_\eta(t_1, t_2)$ of the process $\eta(t)$:

$$R_\eta(t_1, t_2) = \int_{-\infty}^{\infty}\int_{-\infty}^{\infty} h(t_1 - \tau_1)R(\tau_1 - \tau_2)\overline{h(t_2 - \tau_2)}d\tau_1 d\tau_2$$
$$= \int_{-\infty}^{\infty}\int_{-\infty}^{\infty}\int_{-\infty}^{\infty} h(t_1 - \tau_1)e^{i\omega(\tau_1-\tau_2)}\overline{h(t_2 - \tau_2)}d\tau_1 d\tau_2 dF(\omega)$$
$$= \int_{-\infty}^{\infty} e^{i(t_1-t_2)\omega} \mid H(i\omega) \mid^2 dF(\omega) = R_\eta(t_1 - t_2) .$$

Thus the process $\eta(t)$ is also stationary in the broad sense.

Definition 1. For a process $\xi(t)$, a transformation T is called an *admissible filter* (or, more briefly, a *filter*) if it is defined by formula (11) where $h(t)$ is an absolutely integrable function, or if it is the mean-square limit of a sequence of such transformations (in $L_2\{\xi\}$).

A condition for convergence of a sequence $\{\eta_n(t)\} = \{T_n(\xi \mid t)\}$ of transformations of the form (11), with impulse transfer functions $h_n(t)$ and frequency characteristics $H_n(i\omega)$, consists in the following:

$$\mathsf{M} \mid \eta_n(t) - \eta_{n+m}(t) \mid^2$$
$$= \int_{-\infty}^{\infty} \mid H_n(i\omega) - H_{n+m}(i\omega) \mid^2 dF(\omega) \longrightarrow 0 , \tag{12}$$

that is, in the requirement that the sequence $\{H_n(i\omega)\}$ be a fundamental sequence in $L_2\{F\}$. But then the limit $H(i\omega) = \text{l.i.m. } H_n(i\omega)$ exists in $L_2\{F\}$. This limit is called the *frequency characteristic* of the limiting filter. If $\eta(t) = \text{l.i.m. } \eta_n(t)$, then

$$R_\eta(t) = \int_{-\infty}^{\infty} e^{it\omega} \mid H(i\omega) \mid^2 dF(\omega) . \tag{13}$$

Conversely, every function $H(i\omega) \in L_2\{F\}$ can be approximated in the sense of convergence in $L_2\{F\}$ by functions that are the Fourier transforms of absolutely integrable functions. Thus it is convenient to define filters by their frequency characteristics.

Theorem 1. *For a function $H(i\omega)$ to be the frequency characteristic of an admissible filter, it is necessary and sufficient that $H(i\omega)$ belong to $L_2\{F\}$. The covariance function of the process at the output of the filter with frequency characteristic $H(i\omega)$ is given by formula* (13).

If we recall the energy interpretation of the spectral function, it follows from formula (13) that $\mid H(i\omega) \mid^2$ shows by how much the energy of simple harmonic components of a process with frequencies in the interval $(\omega, \omega + d\omega)$ is multiplied by passage through the filter.

Theorem 2. *If a process $\xi(t)$ at the input of a filter with frequency characteristic $H(i\omega)$ has the spectral representation*

$$\xi(t) = \int_{-\infty}^{\infty} e^{i\omega t}\mu(d\omega) , \tag{14}$$

then the process $\eta(t)$ at the output of the filter is of the form

$$\eta(t) = \int_{-\infty}^{\infty} e^{i\omega t}H(i\omega)\mu(d\omega) . \tag{15}$$

Proof. If the filter has an absolutely integrable impulse transfer function, then on the basis of Lemma 5 of Section 3,

$$\eta(t) = \int_{-\infty}^{\infty} h(t - \tau)\xi(\tau)d\tau = \int_{-\infty}^{\infty} e^{i\omega t}H(i\omega)\mu(d\omega) .$$

Proof in the general case is obtained by taking the limit with respect to sequences $\{H_n(i\omega)\}$ that converge in $L_2\{F\}$ to $H(i\omega)$.

Let $\eta_k(t)$ denote a process at the output of a filter with frequency characteristic $H_k(i\omega)$ (for $k = 1, 2$). Let us find the mutual covariance function of the processes $\eta_1(t)$ and $\eta_2(t)$. It follows immediately from the isomorphism of the spaces $L_2\{\mu\}$ and $L_2\{F\}$ that

$$R_{12}(t) = \mathsf{M}\eta_1(t + \tau)\overline{\eta_2(\tau)} = \int_{-\infty}^{\infty} e^{i\omega t}H_1(i\omega)\overline{H_2(i\omega)}dF(\omega) . \tag{16}$$

Let us give some examples of filters and their frequency characteristics.

1. A band filter admits (without modification) only harmonic components of a process with frequencies in a given interval (a, b). The frequency characteristic of the filter is equal to $H(i\omega) = \chi_{(a,b)}(\omega)$, and the filter is admissible for an arbitrary process. The impulse transfer function is found from Fourier's formula

$$h(t) = \frac{1}{2\pi}\int_a^b e^{it\omega}d\omega = \frac{e^{ibt} - e^{iat}}{2\pi it} .$$

2. A low-pass or a high-pass filter admits (without modification) only harmonic oscillations with frequencies not exceeding or not less than some value b. Such a filter is admissible for an arbitrary process. Its frequency characteristic is equal to

$$H(i\omega) = \chi_{(-\infty,b)}(\omega) \quad (\chi_{(b,+\infty)}(\omega)) ,$$

and the impulse transfer function does not exist.

3. Consider the operation of mean-square differentiation of a process that is stationary in the broad sense. A sufficient condition for a process $\xi(t)$ to have a mean-square derivative is the existence of $R''(0)$ (cf. Corollary 2, Section 2). This condition is equivalent to the requirement (cf. Theorem 2, Section 5, Chapter I) that

$$\int_{-\infty}^{\infty} \omega^2 dF(\omega) < \infty . \tag{17}$$

On the other hand if this condition is satisfied, then

$$\frac{e^{i\omega h} - 1}{h} \to i\omega \qquad (\text{in } L_2\{F\})$$

and in the relation

$$\frac{\xi(t + h) - \xi(t)}{h} = \int_{-\infty}^{\infty} e^{it\omega} \frac{e^{ih\omega} - 1}{h} \mu(d\omega)$$

we may take the limit as $h \to 0$ under the stochastic integral sign. Consequently,

$$\xi'(t) = \int_{-\infty}^{\infty} e^{it\omega} i\omega \mu(d\omega) . \tag{18}$$

Thus, corresponding to the operation of differentiation is a filter with frequency characteristic $i\omega$, which is admissible for all stationary processes satisfying condition (17). The impulse transfer function does not exist, but the filter can be regarded as limiting (as $\varepsilon \to 0$) for filters with impulse transfer functions of the form

$$h_\varepsilon(t) = \begin{cases} 0 & \text{for } |t| \geqq \varepsilon , \\ -\dfrac{\operatorname{sgn} t}{\varepsilon^2} & \text{for } |t| < \varepsilon , \end{cases}$$

to which the frequency characteristics

$$-\frac{4 \sin^2 \dfrac{\omega\varepsilon}{2}}{i\omega\varepsilon^2}$$

correspond.

4. The time-displacement operation. Since

$$\xi(t + \tau) = \int_{-\infty}^{\infty} e^{i\omega t} e^{i\omega\tau} \mu(d\omega) ,$$

corresponding to the time-displacement operation S_τ defined by $S_\tau(\xi \mid t) = \xi(t + \tau)$ we have the frequency characteristic $H(i\omega) = e^{i\omega\tau}$. There is no impulse transfer function.

5. Differential equations. Consider a filter defined by a linear differential equation with constant coefficients

$$L\eta = M\xi , \tag{19}$$

where

$$L = a_0 \frac{d^n}{dt^n} + a_1 \frac{d^{n-1}}{dt^{n-1}} + \cdots + a_n ,$$

$$M = b_0 \frac{d^m}{dt^m} + b_1 \frac{d^{m-1}}{dt^{m-1}} + \cdots + b_m .$$

Equation (19) is meaningful only when the process $\xi(t)$ is m times mean-square differentiable. Then we seek an n-times mean-square differentiable stationary process $\eta(t)$ satisfying equation (19). Let us suppose that (19) has a stationary solution. Then $\eta(t)$ can be represented in the form

$$\eta(t) = \int_{-\infty}^{\infty} e^{i\omega t} H(i\omega)\mu(d\omega) .$$

If we apply the operations M and L to the processes $\xi(t)$ and $\eta(t)$ respectively, we obtain

$$\int e^{i\omega t} L(i\omega)H(i\omega)\mu(d\omega) = \int e^{i\omega t} M(i\omega)\mu(d\omega) ,$$

so that if $L(i\omega)$ does not have real roots,

$$H(i\omega) = \frac{M(i\omega)}{L(i\omega)} . \tag{20}$$

Conversely, if the process $\xi(t)$ is m times mean-square differentiable, if $M(i\omega) \in L_2\{F\}$, and if $L(i\omega) \neq 0$ (for $-\infty < \omega < \infty$), then the process

$$\eta(t) = \int_{-\infty}^{\infty} e^{i\omega t} \frac{M(i\omega)}{L(i\omega)} \mu(d\omega)$$

is n times mean-square differentiable and satisfies condition (19). Thus under the conditions $M(i\omega) \in L_2\{F\}$ and $L(i\omega) \neq 0$, there exists a unique filter satisfying the differential equation (19). We note, however, that it is possible to determine the solution of equation (19) under more general conditions. Let us suppose that the polynomial $L(i\omega)$ has no real roots. A filter with frequency characteristic $M(i\omega)/L(i\omega)$ exists, and without requiring that $M(i\omega)$ belong to $L_2\{F\}$, it is sufficient to require that $M(i\omega)/L(i\omega)$ belong to $L_2\{F\}$. This will always be the case when the degree n of the polynomial L is not less than m. Thus for $n \geqq m$, the filter with frequency characteristic (20) whose denominator does not vanish for real ω is admissible for an arbitrary input process, and we identify the output process of the filter with the stationary solution of equation (19). Let us confine ourselves, as before, to differential equations for which the polynomial $L(x)$ does not have purely imaginary roots. Let $P(x)$ denote the polynomial portion of the rational function $M(x)/L(x)$ (which will be nonzero if $m \geqq n$) and let us decompose the remainer into partial fractions. Then

$$\frac{L(i\omega)}{M(i\omega)} = P(i\omega) + \sum_{k=1}^{n'} \sum_{s=1}^{l'_k} \frac{c'_{ks}}{(i\omega - p'_k)^s} + \sum_{k=1}^{n''} \sum_{s=1}^{l''_k} \frac{c''_{ks}}{(i\omega - p''_k)^s} ,$$

where

$$P(i\omega) = \sum_{k=0}^{m-n} a_k(i\omega)^k$$

for $m \geqq n$ and

$$P(i\omega) = 0$$

for $m < n$; $\operatorname{Re} p'_k < 0$ and $\operatorname{Re} p''_k > 0$, where p'_k and p''_k are the roots of the polynomial $L(x) = 0$. Since

$$\frac{1}{(i\omega - p)^s} = \frac{1}{(s-1)!}\frac{d^s}{dp^s}\int_0^\infty e^{pt}e^{-i\omega t}dt = \int_0^\infty \frac{t^s}{(s-1)!} e^{pt}e^{-i\omega t}dt$$
$$(\operatorname{Re} p < 0)$$

and

$$\frac{1}{(i\omega - p)^s} = \int_{-\infty}^0 - \frac{t^s}{(s-1)!} e^{pt}e^{-iwt}dt \qquad (\operatorname{Re} p > 0)\ ,$$

the output process $\eta(t)$ of the filter can be represented in the form

$$\eta(t) = \sum_{k=0}^{m-n} a_k\xi^{(k)}(t) + \int_0^\infty \xi(t-\tau)G_1(\tau)d\tau + \int_0^\infty \xi(t+\tau)G_2(-\tau)d\tau\ ,$$

where

$$G_1(t) = \sum_{k=1}^{n'}\Big(\sum_{s=1}^{l'_k}\frac{c'_{ks}t^s}{(s-1)!}\Big)\{\exp p'_k t\ (t > 0)\}\ ,$$

$$G_2(t) = -\sum_{k=1}^{n''}\Big(\sum_{s=1}^{l''_k}\frac{c''_{ks}t^s}{(s-1)!}\Big)\{\exp p''_k t\ (t < 0)\}\ .$$

We note that if the polynomial $L(x)$ has roots with positive real part, the corresponding filter is physically unrealizable.

6. PHYSICALLY REALIZABLE FILTERS

In the present section we shall consider the question: What spectral functions can be obtained at the output of a physically realizable filter? At the input of the filter we shall consider the random process that is simplest in a certain sense. The processes considered in the present section are invariably assumed to be homogeneous and stationary in the broad sense. Therefore, the word "stationary" will sometimes be omitted, and the words "in the broad sense" will always be omitted.

We shall begin by considering stationary sequences. We shall not carry over to sequences all the definitions and heuristic considerations that were given for processes with continuous time although we shall use the corresponding terminology. Consider a system whose states at the input and output are regulated only at integral-valued instants of time $t = 0, \pm 1, \pm 2, \cdots$. Suppose that a unit impulse takes place at the input of the system at the instant of time 0. We let a_t denote the response of the system to that impulse at the instant t. If the system does not anticipate the future, then $a_t = 0$ for $t < 0$. If the system is homogeneous with respect to time, the response of the system to the unit impulse applied to the system at the instant τ is equal to $a_{t-\tau}$. The response of a linear, homogeneous, physically realizable system at the instant t to the sequence of impulses $\xi(n)$ (for $-\infty < n < \infty$) will be

$$\eta(t) = \sum_{n=-\infty}^{t} a_{t-n}\xi(n) = \sum_{n=0}^{\infty} a_n \xi(t-n) . \tag{1}$$

In a certain sense, the simplest hypothesis is that $\xi(n)$ is a stationary sequence with mean value 0 and uncorrelated values

$$\mathsf{M}\xi(n) = 0, \ \mathsf{M}(\xi(n)\overline{\xi(m)}) = \delta_{nm}(-\infty < n, m < \infty) .$$

We shall call such a sequence uncorrelated. Its covariance function has the spectral representation

$$R(n) = \frac{1}{2\pi}\int_{-\pi}^{\pi} e^{in\omega}d\omega .$$

Consequently the spectral density of this sequence is constant.

For the series (1) to converge in mean-square in the case of an uncorrelated sequence $\xi(n)$, it is necessary and sufficient that

$$\sum_{n=0}^{\infty} |a_n|^2 < \infty . \tag{2}$$

If this condition is satisfied, the process $\eta(t)$ is also stationary in the broad sense and

$$\mathsf{M}\eta(t) = 0 , \qquad R_\eta(t) = \mathsf{M}(\eta(t+\tau)\overline{\eta(\tau)}) = \sum_{n=0}^{\infty} a_{n+t}\bar{a}_n . \tag{3}$$

What sequences can be obtained in this manner?

Lemma 1. *For a stationary sequence $\eta(n)$ to be the response of a physically realizable filter to an uncorrelated sequence, it is necessary and sufficient that the spectrum of $\eta(n)$ be absolutely continuous and that its spectral density $f(\omega)$ have a representation of the form*

$$f(\omega) = |g(e^{i\omega})|^2 , \quad g(e^{i\omega}) = \sum_{n=0}^{\infty} b_n e^{in\omega} , \quad \sum_{n=0}^{\infty} |b_n|^2 < \infty . \tag{4}$$

Proof of the Necessity. Suppose that the sequence has a representation of the form (1). We set

$$g(e^{i\omega}) = \frac{1}{\sqrt{2\pi}} \sum_{n=0}^{\infty} \bar{a}_n e^{in\omega} . \tag{5}$$

Then from Parseval's formula,

$$R_\eta(t) = \sum_{n=0}^{\infty} a_{n+t}\bar{a}_n = \int_{-\pi}^{\pi} e^{it\omega} \,|\, g(e^{i\omega}) \,|^2 \, d\omega \; ;$$

that is, the sequence $\eta(n)$ has an absolutely continuous spectrum with density $f(\omega) = |\, g(e^{i\omega}) \,|^2$, where $g(e^{i\omega})$ is defined by formula (5).

Proof of the Sufficiency. Suppose that $\eta(n)$ is a sequence with covariance function

$$R_\eta(t) = \int_{-\pi}^{\pi} e^{it\omega} f(\omega) d\omega \; ,$$

and suppose that this sequence admits a spectral representation

$$\eta(t) = \int_{-\pi}^{\pi} e^{it\omega} \mu(d\omega) \; .$$

Suppose that $f(\omega) = |\, g(e^{i\omega}) \,|^2$, where $g(e^{i\omega})$ is defined by relations (4).

On the σ-algebra of Borel subsets of the interval $(-\pi, \pi)$, let us construct the stochastic measure

$$\xi(A) = \int_{-\pi}^{\pi} \frac{1}{\sqrt{2\pi}\, g(e^{i\omega})} \chi_A(\omega) \mu(d\omega) \; .$$

Then,

$$\mathsf{M}(\xi(A)\overline{\xi(B)}) = \int_{-\pi}^{\pi} \chi_A(\omega)\chi_B(\omega) \frac{1}{2\pi \,|\, g(e^{i\omega}) \,|^2} f(\omega) d\omega$$

$$= \frac{1}{2\pi} \int_{A \cap B} d\omega \; ;$$

that is, $\xi(A)$ is an orthogonal measure with structural function $l(A \cap B)$, where l denotes Lebesgue measure. By using Lemmas 2 and 1 of Section 3, we obtain

$$\eta(t) = \int_{-\pi}^{\pi} e^{it\omega} \mu(d\omega) = \int_{-\pi}^{\pi} e^{it\omega} \sqrt{2\pi}\, \bar{g}(e^{i\omega}) \xi(d\omega)$$

$$= \sum_{n=0}^{\infty} \sqrt{2\pi}\bar{b}_n \int_{-\pi}^{\pi} e^{i(t-n)\omega} \xi(d\omega) = \sum_{n=0}^{\infty} a_n \xi(t-n) \; ,$$

where $a_n = \sqrt{2\pi}\bar{b}(n)$, $\xi_n = \int_{-\pi}^{\pi} e^{in\omega} \xi(d\omega)$, and

$$\mathsf{M}(\xi(n)\bar{\xi}(m)) = \frac{1}{2\pi} \int_{-\pi}^{\pi} e^{i(n-m)\omega} d\omega = \delta_{nm} \; .$$

Thus the sequence $\xi(n)$ is uncorrelated.

This lemma gives a simple answer to the question asked. But this answer is not sufficient for us in the general case since we still do not know when the spectral density can be represented by formula (4).

Let us find conditions under which $f(\omega)$ admits a representation of the form (4). Let H_2 denote the set of all functions $f(z)$ that are analytic in the disk $D = \{z, |z| < 1\}$ and that satisfy the relation $\| f(z) \|^2 = \lim_{r \uparrow 1} \int_{-\pi}^{\pi} |f(re^{i\theta})|^2 \, d\theta < \infty$. If $f(z) = \sum_{n=0}^{\infty} a_n z^n$, then

$$f(re^{i\theta}) = \sum_{n=0}^{\infty} a_n r^n e^{in\theta} ;$$

that is, the $a_n r^n$ are the Fourier coefficients of the function $f(re^{i\theta})$. On the basis of Parseval's equality,

$$\int_{-\pi}^{\pi} |f(re^{i\theta})|^2 \, d\theta = 2\pi \sum_{n=0}^{\infty} |a_n|^2 r^{2n} .$$

From this it is clear that $f(z) \in H_2$ if and only if $\sum_{n=0}^{\infty} |a_n|^2 < \infty$. Consequently for every function $f(z) \in H_2$ it is possible to define a series $f(e^{i\theta}) = \sum_{n=0}^{\infty} a_n e^{in\theta}$ that converges in $L_2(l)$, where l is Lebesgue measure on $(-\pi, \pi)$. The function $f(z)$ (for $|z| < 1$) is determined by the function $f(e^{i\theta})$ in accordance with Poisson's formula

$$f(re^{i\theta}) = \frac{1}{2\pi} \int_{-\pi}^{\pi} f(e^{i\omega}) P(r, \theta, \omega) d\omega , \tag{6}$$

where

$$P(r, \theta, \omega) = \frac{1 - r^2}{1 - 2r \cos(\theta - \omega) + r^2} = \sum_{n=-\infty}^{\infty} r^{|n|} e^{in(\theta - \omega)} .$$

Proof of this assertion follows immediately from Parseval's equality.

It is shown in the theory of functions (see Privalov, or Hoffman) that if the function $f(e^{i\omega})$ in formula (6) is Lebesgue-integrable, the limit $\lim_{r \uparrow 1} f(re^{i\theta}) = f(e^{i\theta})$ exists for almost all θ. The function $f(e^{i\theta})$ is called the limiting value of the function $f(z)$ (for $|z| < 1$).

Theorem 1. *Let $f(\omega)$ denote a nonnegative function that is Lebesgue-integrable on the interval $[-\pi, \pi]$. For the existence of a function $g(z) \in H_2$ such that*

$$f(\omega) = |g(e^{i\omega})|^2 , \tag{7}$$

it is necessary and sufficient that

$$\int_{-\pi}^{\pi} |\ln f(\omega)| \, d\omega < \infty . \tag{8}$$

Proof of the Necessity. Suppose that

$$g(z) = \sum_{n=0}^{\infty} a_n z^n \in H_2$$

and that (7) holds. We may assume that $g(0) \neq 0$ (because if

$g(0) = 0$ we can consider not $g(z)$ but $z^{-m}g(z)$, where m is the order of the zero of $g(z)$ at $z = 0$) and in fact we may assume $g(0) = 1$. Let r denote a number in the open interval $(0, 1)$ and define $A = \{\omega; |g(re^{i\omega})| \leqq 1\}$ and $B = \{\omega; |g(re^{i\omega})| > 1\}$. Then

$$\int_{-\pi}^{\pi} |\ln|g(re^{i\omega})||\,d\omega = \int_B \ln|g(re^{i\omega})|\,d\omega - \int_A \ln|g(re^{i\omega})|\,d\omega$$
$$= 2\int_B \ln|g(re^{i\omega})|\,d\omega - \int_{-\pi}^{\pi} \ln|g(re^{i\omega})|\,d\omega\,.$$

From Jensen's inequality it follows that

$$\frac{1}{2\pi}\int_{-\pi}^{\pi} \ln|f(re^{i\omega})|\,d\omega = \ln\prod_{k=1}^{n}\frac{r}{|z_k|} \geqq 0\,,$$

where the z_k are the zeros of the function $f(z)$ inside the disk $|z| < r$. Consequently,

$$\int_{-\pi}^{\pi} |\ln|g(re^{iw})||\,d\omega \leqq \int_B \ln|g(re^{i\omega})|\,d\omega \leqq \int_B |g(re^{i\omega})|^2\,d\omega$$
$$\leqq \int_{-\pi}^{\pi} |g(re^{i\omega})|^2\,d\omega = 2\pi\sum_{n=0}^{\infty}|a_n|^2\,.$$

Applying Fatou's lemma, we obtain

$$\overline{\lim_{r\uparrow 1}}\int_{-\pi}^{\pi} |\ln|g(re^{i\omega})||\,d\omega \leqq \int_{-\pi}^{\pi} \overline{\lim}\,|\ln|g(re^{i\omega})||\,d\omega$$
$$= \int_{-\pi}^{\pi} |\ln|g(e^{i\omega})||\,d\omega \leqq 2\pi\sum_{n=0}^{\infty}|a_n|^2\,,$$

which proves the necessity of condition (8).

Proof of the Sufficiency. Suppose that condition (8) is satisfied. The function

$$u(r, \theta) = \frac{1}{2\pi}\int_{-\pi}^{\pi} \ln f(\omega)P(r, \theta, \omega)d\omega$$

is harmonic in the disk D: $|z| < 1$. We note that it follows from Jensen's inequality that

$$u(r, \theta) \leqq \ln\left\{\frac{1}{2\pi}\int_{-\pi}^{\pi} f(\omega)P(r, \theta, d\omega)\right\}\,.$$

Let $\varphi(z)$ denote an analytic function defined in D with real part $u(r, \theta)$ that assumes a positive value at $z = 0$. We set

$$g(z) = \exp\left(\frac{1}{2}\varphi(z)\right)\,.$$

Then

$$|g(re^{i\theta})|^2 = e^{\mathrm{Re}\varphi(z)} = e^{u(r,\theta)} \leqq \frac{1}{2\pi}\int_{-\pi}^{\pi} f(\omega)P(r, \theta, \omega)d\omega$$

and

$$\int_{-\pi}^{\pi} | g(re^{i\theta}) |^2 \, d\theta \leqq \int_{-\pi}^{\pi} f(\omega) d\omega < \infty \, .$$

Thus $g(z) \in H_2$ and

$$\lim_{r \uparrow 1} | g(re^{i\omega}) |^2 = \exp\left(\lim_{r \uparrow 1} u(r, \theta)\right) = f(\theta)$$

almost everywhere. This completes the proof of the theorem.

REMARK 1. As we can see from the proof of the theorem, the function $g(z)$ can be chosen so that it has a positive value at $z = 0$ and has no zeros in D.

Combining Lemma 1 and Theorem 1, we obtain

Theorem 2. *For the sequence $\eta(t)$ to have a representation of the form*

$$\eta(t) = \sum_{n=0}^{\infty} a_n \xi(t - n) \, , \qquad \sum_{n=0}^{\infty} | a_n |^2 < \infty \, , \tag{1'}$$

where $\xi(n)$ is an uncorrelated sequence, it is necessary and sufficient that the spectrum of $\eta(t)$ be absolutely continuous and that the spectral density $f(\omega)$ satisfy the requirement

$$\int_{-\pi}^{\pi} \ln f(\omega) > -\infty \, .$$

REMARK 2. The function $\varphi(z)$ is an analytic function in D, and its real part has limiting values $\ln f(\omega)$. Consequently,

$$\varphi(z) = \frac{1}{2\pi} \int_{-\pi}^{\pi} \ln f(\omega) \frac{e^{i\omega} + z}{e^{i\omega} - z} d\omega \, . \tag{9}$$

Expanding the function $g(z) = \exp(1/2\varphi(z))$ in a power series $g(z) = \sum_{n=0}^{\infty} b_n z^n$, we obtain the following values for the coefficients a_n in formula (1'): $a_n = \sqrt{2\pi} \bar{b}_n$.

On the other hand the expression for $g(z)$ can be transformed as follows: Since

$$\frac{e^{i\omega} + z}{e^{i\omega} - z} = 1 + \frac{2ze^{-i\omega}}{1 - ze^{-iw}} = 1 + 2 \sum_{k=1}^{\infty} z^k e^{-ik\omega} \, ,$$

we have

$$\overline{g(z)} = \exp\left\{\frac{1}{2\pi} \int_{-\pi}^{\pi} \ln f(\omega) d\omega + \frac{1}{2\pi} \sum_{k=1}^{\infty} d_k \bar{z}^k\right\} \, ,$$

where

$$d_k = \int_{-\pi}^{\pi} e^{ik\omega} \ln f(\omega) d\omega \, .$$

Setting

$$P = \exp\left(\frac{1}{4\pi}\int_{-\pi}^{\pi} \ln f(\omega) d\omega\right), \quad \exp\left(\frac{1}{2\pi}\sum_{k=1}^{\infty} d_k z^k\right) = \sum_{k=0}^{\infty} C_k z^k$$

$$(C_0 = 1),$$

we obtain

$$\overline{g(z)} = P\sum_{k=0}^{\infty} C_k \bar{z}^k .$$

Thus,

$$a_n = \sqrt{2\pi} P C_n .$$

REMARK 3. The function $g(z)$, whose existence was established by Theorem 1, is not uniquely defined. However, if $g(z)$ satisfies the two conditions

a. $g(z) \neq 0$ for $z \in D$,

b. $g(0) > 0$,

it is unique and hence coincides with the function that we have found. To see this, let $g_1(z)$ and $g_2(z)$ denote two such functions. Then $\psi(z) = g_1(z)/g_2(z)$ is analytic in D, it does not vanish in D, and its absolute value is equal to 1 on the boundary of D. The function $\ln \psi(z)$ is analytic in D and its real part vanishes on the boundary of D. Therefore $\ln \psi(z) = ik$, where k is real. Since $\ln \varphi(0)$ is real, we have $\ln \psi(z) = 0$.

Let $\zeta_1(\theta)$ and $\zeta_2(\theta)$, for $\theta \in \Theta$, denote two Hilbert random functions. Let H_{ζ_i} denote the closed linear hull of the system of random variables $\{\zeta_i(\theta), \theta \in \Theta\}$ in L_2.

Definition 1. If $H_{\zeta_1} \subset H_{\zeta_2}$, then the random function $\zeta_1(\theta)$ is said to be *subordinate* to $\zeta_2(\theta)$. On the other hand, if $H_{\zeta_1} = H_{\zeta_2}$, then $\zeta_1(\theta)$ and $\zeta_2(\theta)$ are said to be *equivalent*.

REMARK 4. As one can see from the proof of Lemma 1, the sequences $\xi(n)$ and $\eta(n)$ are equivalent.

Let us turn to processes with continuous time. Consider a process $\eta(t)$ that has a representation of the form

$$\eta(t) = \int_0^{\infty} a(\tau) d\xi(t - \tau) , \tag{10}$$

where $\xi(t)$ is an integrated white noise, that is, a process with orthogonal increments, such that $\mathsf{M}\,\Delta\xi(t) = 0$, $\mathsf{M}\,|\,\Delta\xi(t)\,|^2 = \Delta t$. Corresponding to the process $\xi(t)$ is an orthogonal stochastic measure on the σ-algebra of Lebesgue-measurable sets (cf. Section 3). The integral (10) exists if and only if

$$\int_0^\infty |a(t)|^2 dt < \infty . \tag{11}$$

The process $\eta(t)$ can be regarded as a process at the output of a physically realizable filter at the input of which there is a white noise. Then formula (10) applies to all such admissible filters, since by definition every admissible physically realizable filter is either of the form (10) or is the limiting value of a sequence of filters of such a form. A condition for mean-square convergence of filters of the form (10) with impulse transfer functions $a_n(t)$ is

$$\int_0^\infty |a_n(\tau) - a_{n'}(\tau)|^2 d\tau \longrightarrow 0 \quad \text{as} \quad n, n' \longrightarrow \infty .$$

But if this condition is satisfied, the limit $\text{l.i.m.}\, a_n(t) = a(t)$ exists (with respect to Lebesgue measure on $(0, \infty)$) and

$$\text{l.i.m.}\, \eta_n(t) = \text{l.i.m.} \int_0^\infty a_n(\tau) d\xi(t - \tau) = \int_0^\infty a(\tau) d\xi(t - \tau) .$$

Thus passage to the limit in filters of the form (10) does not broaden the class of filters. Formula (10) can be rewritten

$$\eta(t) = \int_{-\infty}^\infty a(t - \tau) d\xi(\tau) , \quad a(t) = 0 \quad \text{for} \quad t < 0 .$$

Consequently the covariance function of the process $\eta(t)$ is equal to

$$R(t) = \int_{-\infty}^\infty a(t + \tau - u)\overline{a(\tau - u)} du$$

or

$$R(t) = \int_0^\infty a(t + u)\overline{a(u)} du . \tag{12}$$

Lemma 2. *For a process $\eta(t)$ that is stationary in the broad sense to be the response of a physically realizable filter to a white noise subordinate to the process, it is necessary and sufficient that the process $\eta(t)$ have an absolutely continuous spectrum and that the spectral density of the process $f(\omega)$ have a representation of the form*

$$f(\omega) = |h(i\omega)|^2 , \tag{13}$$

where

$$h(i\omega) = \int_0^\infty b(\tau) e^{-i\omega\tau} d\tau , \qquad \int_0^\infty |b(\tau)|^2 d\tau < \infty . \tag{14}$$

Proof of the Necessity. Suppose that the process $\eta(t)$ has a representation of the form (10). Define

$$h(i\omega) = \frac{1}{\sqrt{2\pi}} \int_0^\infty a(\tau) e^{-i\omega\tau} d\tau .$$

On the basis of Parseval's equality,

$$R(t) = \int_0^\infty a(t+u)\overline{a(u)}du = \int_{-\infty}^\infty e^{i\omega t} \mid h(i\omega) \mid^2 d\omega \ ;$$

that is, the spectrum of the process is absolutely continuous and the spectral density is of the form (13)–(14).

Proof of the Sufficiency. Suppose that the conditions of the lemma are satisfied. Consider a spectral representation of the process

$$\eta(t) = \int_{-\infty}^\infty e^{i\omega t}\mu(d\omega)$$

and a stochastic measure

$$\xi(A) = \int_{-\infty}^\infty \frac{\chi_A(\omega)}{h(i\omega)}\mu(d\omega) \ . \tag{15}$$

The stochastic integral (15) is meaningful for an arbitrary bounded Borel set A since $\chi_A(\omega)/h(i\omega) \in L_2\{F\}$, where F is the spectral function of the process and $F(A) = \int_A \mid h(i\omega) \mid^2 d\omega$.

We can easily see that $\zeta(A)$ is an orthogonal measure and that

$$\mathsf{M}(\zeta(A)\overline{\zeta(B)}) = \int_{A\cap B} d\omega \ .$$

We define the integrated white noise $\xi(t)$ by

$$\xi(t_2) - \xi(t_1) = \frac{1}{\sqrt{2\pi}}\int_{-\infty}^\infty \frac{e^{-i\omega t_2} - e^{i\omega t_1}}{i\omega}\zeta(d\omega) \ . \tag{16}$$

On the basis of Lemma 1 of Section 3 and formula (15) we obtain

$$\eta(t) = \int_{-\infty}^\infty e^{i\omega t}h(i\omega)\zeta(d\omega) \ . \tag{17}$$

We note now that for an arbitrary function

$$h(i\omega) = \frac{1}{2\pi}\int_{-\infty}^\infty a(\tau)e^{i\omega\tau}d\tau \ ,$$

where $\int_{-\infty}^\infty \mid a(\tau) \mid^2 d\tau < \infty$, we have

$$\int_{-\infty}^\infty h(i\omega)\zeta(d\omega) = \int_{-\infty}^\infty a(\tau)\xi(d\tau) \ . \tag{18}$$

This is true since the spaces $\tilde{L}_2\{\zeta\}$ and $\tilde{L}_2\{\xi\}$ are isomorphic to the $L_2\{l\}$, where l is Lebesgue measure on the real line $(-\infty, \infty)$ and the scalar product in $L_2\{l\}$ is invariant under the Fourier transform; hence, it will be sufficient to prove formula (18) for simple functions. Suppose that $a(t) = \sum_k C_k\chi_{\Delta_k}(t)$ where Δ_k is the interval (open or closed on the left, open on the right) (a_k, b_k). Then

$$\int_{-\infty}^{\infty} a(t)\xi(dt) = \sum c_k \xi(\Delta_k) = \int_{-\infty}^{\infty} \sum_k c_k \frac{e^{i\omega b_k} - e^{-i\omega a_k}}{\sqrt{2\pi} i\omega} \zeta(d\omega) ,$$

which is a special case of (18). Thus formula (18) is established. It follows from (18) that

$$\int_{-\infty}^{\infty} e^{i\omega t} h(i\omega)\zeta(d\omega) = \int_{-\infty}^{\infty} a(\tau) d\xi(t - \tau) , \tag{19}$$

since, on the basis of formula (16), multiplication of the measure ζ by $e^{i\omega t}$ leads to displacement of the argument of the function ξ by an amount t. (Here in formula (19) the sign of the argument of the function ξ is changed, which causes a change in the sign of the stochastic interval.) Applying formula (19) to the expression (17) for $\eta(t)$, we obtain

$$\eta(t) = \int_0^{\infty} a(\tau) d\xi(t - \tau), \quad \text{where} \quad a(t) = \frac{1}{\sqrt{2\pi}} \int_{-\infty}^{\infty} h(i\omega) e^{-i\omega t} d\omega .$$

This completes the proof of the lemma.

Suppose that we are given the spectral density $f(\omega)$ of a process $\eta(t)$. The following questions arise: When does the spectral density admit a representation of the form (13)–(14)? (When does it have a factorization?) How can we find the function $h(i\omega)$ (and hence the function $a(t)$) from the function $f(\omega)$? The answers to these function-theoretic questions can be obtained by reducing them to the questions that we have already answered for the case of functions on a circle.

We introduce the space H_2^{π} of functions $h(w)$, where $w = x + i\omega$, that are analytic in the right half-plane and that satisfy the relation $\int_{-\infty}^{\infty} | h(x + i\omega) |^2 d\omega \leqq c$ for all $x > 0$. The functions in H_2^{π} are just the functions that are necessary for realization of the factorization (13) and (14). This fact follows from the Paley-Wiener theorem:

For $h(w)$ to belong to H_2^{π}, it is necessary and sufficient that it have a representation of the form

$$h(w) = \int_0^{\infty} e^{-wt} b(t) dt, \quad \text{where} \quad \int_0^{\infty} | b(t) |^2 dt < \infty \tag{20}$$

(cf. Wiener and Paley, and Hoffman).

The sufficiency of the condition of the theorem follows immediately from Parseval's equality. If $h(x + i\omega)$ has a representation of the form (20), then

$$\int_{-\infty}^{\infty} | h(x + i\omega) |^2 d\omega = 2\pi \int_{-\infty}^{\infty} e^{-2xt} | b(t) |^2 dt .$$

We introduce the transformation $w = (1 + z)/(1 - z)$, which maps the disk D in the z-plane into the right half of the w-plane. On the boundary of the corresponding regions ($w = i\omega$, $z = e^{i\theta}$), this transformation takes the form $\omega = \cot(\theta/2)$. Let us suppose that $f(\omega)$ admits a factorization of the form (13)–(14). We set

$$g(z) = h(w) = h\left(\frac{1 + z}{1 - z}\right), \qquad \tilde{f}(\theta) = f\left(\cot\frac{\theta}{2}\right) = f(\omega) .$$

The function $\tilde{f}(\theta)$ is integrable over the circle and it admits a factorization $\tilde{f}(\theta) = |g(e^{i\theta})|^2$, where $g(z)$ is analytic in D and

$$\int_{-\infty}^{\infty} \frac{|f(\omega)|^2}{1 + \omega^2} d\omega = \int_{-\pi}^{\pi} |g(e^{i\theta})|^2 \, d\theta < \infty ,$$

that is, $g(z) \in H_2$. But then on the basis of Theorem 1,

$$\int_{-\infty}^{\infty} \frac{\ln f(\omega)}{1 + \omega^2} d\omega = \int_{-\pi}^{\pi} \ln \tilde{f}(\theta) d\theta > -\infty . \tag{21}$$

The converse follows from analytic considerations: If (21) is satisfied, the factorization (13)–(14) exists. With regard to this factorization, we might make remarks similar to those regarding the factorization of functions on the circle. The expression for $h(w)$ is obtained from formula (9) by replacing z with w and making the corresponding change of variables under the integral sign:

$$h(w) = \exp\left\{\frac{1}{2\pi} \int_{-\infty}^{\infty} \frac{\ln f(\omega)}{1 + \omega^2} \frac{i + \omega w}{\omega + iw} d\omega\right\} . \tag{22}$$

Theorem 3. *For a nonnegative integrable function $f(\omega)$ (for $-\infty < \omega < \infty$) to admit a representation of the form* (13)–(14), *it is necessary and sufficient that*

$$\int_{-\infty}^{\infty} \frac{\ln f(\omega)}{1 + \omega^2} d\omega > -\infty . \tag{23}$$

Under the additional conditions that $h(w) \neq 0$ (for $\operatorname{Re} w > 0$) *and $h(1) > 0$, the function $h(w)$ is unique and is given by formula* (21).

Theorem 4. *For a stationary process $\eta(t)$ (for $-\infty < t < \infty$) to have representation of the form* (10), *it is necessary and sufficient that it have an absolutely continuous spectrum and that its spectral density satisfy condition* (23).

7. PREDICTION AND FILTERING OF STATIONARY PROCESSES

An important problem in the theory of random processes, one that has numerous practical applications, is that of finding as close as possible an estimate of the value of a random variable ζ in terms of the values of the variables ξ_α(for $\alpha \in \mathfrak{A}$). It is a matter of finding a function $f(\xi_\alpha \mid \alpha \in \mathfrak{A})$ depending on ξ_α for $\alpha \in \mathfrak{A}$ with least possible error, and satisfying the approximate equation

$$\zeta \approx \hat{\zeta} = f(\xi_\alpha \mid \alpha \in \mathfrak{A}) . \tag{1}$$

An example of such a problem is the prediction (extrapolation) of a random process. In this case, the problem is that of estimating the value of the random process at the instant t^* from its values on some set of instants of time $\mathfrak{T}$ preceding t^*.

Another example is the problem of the filtering of a random process. This problem consists in the following: A process $\xi(t) = \eta(t) + \zeta(t)$, representing the sum of the "useful" signal $\zeta(t)$ and the "noise" $\eta(t)$, is observed at instants $t' \in \mathfrak{T}'$. The problem is to separate the noise from the signal; that is, for some $t^* \in \mathfrak{T} \supset \mathfrak{T}'$, we need to find the best approximation of $\zeta(t)$ of the form

$$\zeta(t^*) \approx \hat{\zeta} = f(\xi(t') \mid t' \in \mathfrak{T}') .$$

The statement of the problem is not yet complete, since it has not been shown what is meant by "best approximation." Of course the criterion of optimality depends on the practical nature of the problem in question. With regard to the mathematical theory, the methods of solving this problem are primarily based on the mean-square deviation as a measure of accuracy of the approximate equation (1).

The quantity

$$\delta = \{\mathsf{M}[\zeta - f(\xi_\alpha \mid \alpha \in \mathfrak{A})]^2\}^{1/2} \tag{2}$$

is called the *mean-square error* of the approximate formula (1). The problem consists in determining the function f so that (2) is minimized. In the case in which $\mathfrak{A}$ is a finite set, we mean by $f(\xi_\alpha \mid \alpha \in \mathfrak{A})$ a measurable Borel function of the arguments ξ_α for $\alpha \in \mathfrak{A}$. On the other hand, if $\mathfrak{A}$ is infinite this symbol denotes a random variable that is measurable with respect to the σ-algebra $\mathfrak{F} = \sigma(\xi_\alpha, \alpha \in \mathfrak{A})$.

In what follows we shall assume that both ζ and $f(\xi_\alpha \mid \alpha \in \mathfrak{A})$ have second-order moments.

We define

$$\gamma = g(\xi_\alpha \mid \alpha \in \mathfrak{A}) = \mathsf{M}\{\zeta \mid \mathfrak{F}\} \tag{3}$$

(cf. Section 8, Chapter III). Then

$$\begin{aligned} \delta^2 &= \mathsf{M}\{\zeta - f(\xi_\alpha \mid \alpha \in \mathfrak{A})\}^2 \\ &= \mathsf{M}(\zeta - \gamma)^2 + 2\mathsf{M}(\zeta - \gamma)(\gamma - f(\xi_\alpha \mid \alpha \in \mathfrak{A})) \\ &\quad + \mathsf{M}(\gamma - f(\xi_\alpha \mid \alpha \in \mathfrak{A}))^2 . \end{aligned}$$

Since $\gamma - f(\xi_\alpha \mid \alpha \in \mathfrak{A})$ is $\mathfrak{F}$-measurable, we have

$$\begin{aligned} \mathsf{M}(\zeta - \gamma)(\gamma - f(\xi_\alpha \mid \alpha \in \mathfrak{A})) \\ = \mathsf{M}\mathsf{M}\{(\zeta - \gamma)(\gamma - f(\xi_\alpha \mid \alpha \in \mathfrak{A})) \mid \mathfrak{F}\} \\ = \mathsf{M}(\gamma - f(\xi_\alpha \mid \alpha \in \mathfrak{A}))\, \mathsf{M}\{(\xi - \gamma) \mid \mathfrak{F}\} = 0 . \end{aligned}$$

Thus

$$\delta^2 = \mathsf{M}(\zeta - \gamma)^2 + \mathsf{M}(\gamma - f(\xi_\alpha \mid \alpha \in \mathfrak{A}))^2 ,$$

from which we get

Theorem 1. *Suppose that a random variable ζ has a finite second-order moment. An approximation of ζ with minimum mean-square error obtained with the aid of a $\sigma\{\xi_\alpha, \alpha \in \mathfrak{A}\}$-measurable random variable is unique* (mod P) *and is given by the formula*

$$\gamma = \mathsf{M}\{\zeta \mid \mathfrak{F}\} .$$

REMARK 1. The estimate $\hat{\zeta} = \gamma$ of a random variable ζ is unbiased; that is,

$$\mathsf{M}\gamma = \mathsf{M}\mathsf{M}\{\zeta \mid \mathfrak{F}\} = \mathsf{M}\zeta$$

and the variables $\zeta - \gamma$ and ξ_α are, for arbitrary $\alpha \subset \mathfrak{A}$, uncorrelated:

$$\mathsf{M}(\zeta - \gamma)\xi_\alpha = \mathsf{M}\mathsf{M}\{(\zeta - \gamma)\xi_\alpha \mid \mathfrak{F}\} = \mathsf{M}\xi_\alpha \mathsf{M}\{(\zeta - \gamma) \mid \mathfrak{F}\} = 0 .$$

Unfortunately, the use of Theorem 1 to obtain actual approximation formulas is extremely difficult in practice. In the case of Gaussian random variables, however, we can proceed further. We note first of all that the simplest statement of the problem leading in a number of cases to a final and analytically attainable solution is the problem of finding the best approximation $\hat{\zeta}$ not in the class of all measurable functions of given random variables but in the narrower class of linear functions. More precisely this means: Let $\{U, \mathfrak{S}, \mathsf{P}\}$ denote a basic probability space. Let us suppose that the variables ξ_α and ζ have finite second-order moments. We introduce the Hilbert space $H\{\xi_\alpha, \alpha \in \mathfrak{A}\}$, which is the closed linear hull of the variables ξ_α, for $\alpha \in \mathfrak{A}$, together with all constants. The subspace $H\{\xi_\alpha, \alpha \in \mathfrak{A}\}$ can be regarded as the set of all linear (nonhomogeneous) functions of ξ_α with finite variances. The best linear approximation $\tilde{\zeta}$ to the random variable ζ is that element $H\{\xi_\alpha, \alpha \in \mathfrak{A}\}$ that lies nearest ζ:

$$\delta^2 = \mathsf{M}\,|\,\tilde{\zeta} - \zeta\,|^2 \leq \mathsf{M}\,|\,\zeta' - \zeta\,|^2$$

for arbitrary $\zeta' \in H\{\xi_\alpha, \alpha \in \mathfrak{A}\}$. We know from the theory of Hilbert spaces (cf. Section 1 of this chapter) that the problem of finding the element $\tilde{\zeta}$ in the subspace H_0 that lies nearest the given element

ζ always has a unique solution. Specifically, $\tilde{\zeta}$ is the projection of ζ onto H_0. The element $\tilde{\zeta}$ can always be determined (uniquely) from the system of equations $(\zeta - \tilde{\zeta}, \zeta') = 0$ for arbitrary $\zeta' \in H_0$, where (x, y) denotes the scalar product of x and y. In the present case this system of equations can be written in the form

$$(\mathsf{M}\tilde{\zeta}\bar{\xi}_\alpha) = \mathsf{M}(\zeta\bar{\xi}_\alpha) \,. \tag{4}$$

Since (1) belongs to $H\{\xi_\alpha, \alpha \in \mathfrak{A}\}$, we have $\mathsf{M}\tilde{\zeta} = \mathsf{M}\zeta$, so that the best linear estimates of $\tilde{\zeta}$ are necessarily unbiased. Furthermore we may assume that $\mathsf{M}\xi_\alpha = 0$ for arbitrary α. Therefore in what follows we shall confine ourselves to a study of the subspace of random variables in $L_2\{U, \mathfrak{S}, \mathsf{P}\}$ with mathematical expection 0.

Of course, we do not always have grounds for assuming that a linear estimate of the quantity ζ is acceptable. For example, if $\xi(n) = e^{i(\nu n+\varphi)}$, where ν is uniformly distributed on $(-\pi, \pi)$, then $\mathsf{M}(\xi(n)\overline{\xi(m)}) = 0$ for $n \neq m$ and the best linear approximation of the variable $\xi(m)$ from the values of all $\xi(n)$ for $n \neq m$ is of the form $\tilde{\xi}(m) = 0$; that is, it does not use values of the variables $\xi(n)$, whereas any pair of observations $\xi(k)$ and $\xi(k + 1)$ are sufficient to determine the entire sequence $\xi(n)$ precisely,

$$\xi(n) = \left(\frac{\xi(k+1)}{\xi(k)}\right)^{n-k} \xi(k) \,.$$

Let us suppose now that all finite-dimensional distributions of the system $\{\zeta, \xi_\alpha, \alpha \in \mathfrak{A}\}$ are normal and $\mathsf{M}\xi_\alpha = \mathsf{M}\zeta = 0$. In this case it follows from the uncorrelatedness of the variables $\zeta - \tilde{\zeta}$ and ξ_α that they are independent. Therefore $\zeta - \tilde{\zeta}$ are independent of the σ-algebra $\mathfrak{F}$ and

$$\mathsf{M}\{\zeta \mid \mathfrak{F}\} = \mathsf{M}\{\zeta - \tilde{\zeta} + \tilde{\zeta} \mid \mathfrak{F}\} = \mathsf{M}\{\zeta - \tilde{\zeta}\} + \tilde{\zeta} = \tilde{\zeta} \,.$$

Theorem 2. *For a system of Gaussian random variables* $(\zeta, \xi_\alpha, \alpha \in \mathfrak{A})$, *the best approximation (from a standpoint of mean-square deviation) of the variable* ζ *with the aid of* $\sigma\{\xi_\alpha, \alpha \in \mathfrak{A}\}$ *of a measurable function coincides with the best linear approximation in* $H\{\xi_\alpha, \alpha \in \mathfrak{A}\}$.

We now consider a number of particular problems on the construction of best linear approximations.

1. The number of random variables ξ_a (for $\alpha = 1, 2, \cdots, n$) is finite. This problem has a simple solution as we know from linear algebra. Assuming that the ξ_α are linearly independent, we can construct the projection $\tilde{\zeta}$ of the variable ζ onto the finite-dimensional space H_n generated by the quantities ξ_α (for $\alpha = 1, \cdots, n$) by means of the formula

$$\tilde{\zeta} = -\frac{1}{\Gamma}\begin{vmatrix} (\xi_1, \xi_1)\cdots(\xi_n, \xi_1)\ \xi_1 \\ \cdots\cdots\cdots\cdots\cdots \\ (\xi_n, \xi_1)\cdots(\xi_n, \xi_n)\xi_n \\ (\zeta, \xi_1)\cdots(\zeta, \xi_n)\ 0 \end{vmatrix},$$

where $\Gamma = \Gamma(\xi_1, \xi_2, \cdots, \xi_n)$ is the Gram determinant of the system of vectors $\xi_1, \xi_2, \cdots, \xi_n$,

$$\Gamma(\xi_1, \xi_2, \cdots, \xi_n) = \begin{vmatrix} (\xi_1, \xi_1)\cdots(\xi_1, \xi_n) \\ \cdots\cdots\cdots\cdots\cdots \\ (\xi_n, \xi_1)\cdots(\xi_n, \xi_n) \end{vmatrix}$$

and where $(\xi, \eta) = \mathsf{M}(\xi\bar{\eta})$. The mean-square error δ of the approximate equation $\zeta \approx \tilde{\zeta}$ is equal to the length of the perpendicular dropped from the vector ζ onto the space H_n and is given by the formula $\delta^2 = \{\Gamma(\xi_1, \xi_2, \cdots, \xi_n, \zeta)/\Gamma(\xi_1, \xi_2, \cdots, \xi_n)\}$.

2. Consider the problem of approximating a random variable ζ from results of observation of a mean-square continuous random process $\xi(t)$ on a finite interval of time $T = [a, b]$. Let $R(t, \tau)$ denote the covariance function of the process $\xi(t)$. On the basis of Theorem 2 of Section 2, the process $\xi(t)$ can be expanded in a series,

$$\xi(t) = \sum_{k=1}^{\infty} \sqrt{\lambda_k}\varphi_k(t)\xi_k ,$$

where $\{\varphi_k(t)\}$ is an orthonormal sequence of eigenfunctions, where the λ_k are the eigenvalues of the correlation function on (a, b):

$$\lambda_k\varphi_k(t) = \int_a^b R(t, \tau)\varphi_k(\tau)d\tau ,$$

and where ξ_k is the normalized uncorrelated sequence $\mathsf{M}\xi_k\bar{\xi}_r = \delta_{kr}$. Obviously, $\{\xi_k\}$, for $k = 1, 2, \cdots$, constitutes a basis in $H\{\xi(t); t \in (a, b)\}$. Therefore $\tilde{\zeta} = \sum_{n=1}^{\infty} c_n\xi_n$, where $\xi_n = \int_a^b \xi(t)\overline{\varphi_n(t)}dt$, for $n = 1, 2, \cdots$, and

$$c_n = \mathsf{M}\zeta\bar{\xi}_n = \int_a^b R_{\zeta\xi}(t)\varphi_n(t)dt , \qquad R_{\zeta\xi}(t) = \mathsf{M}\zeta\bar{\xi}(t) .$$

The mean-square error δ of the approximation can be found from the formula

$$\delta^2 = \mathsf{M}\,|\,\zeta\,|^2 - \mathsf{M}\,|\,\tilde{\zeta}\,|^2 = \mathsf{M}\,|\,\zeta^2\,| - \sum_{n=0}^{\infty}\left|\int_a^b R_{\zeta\xi}(t)\varphi_n(t)dt\right|^2 .$$

In practice, the application of this method is made more difficult by the complication of calculating the eigenfunctions and eigenvalues of the kernel $R(t, \tau)$.

3. *The integral equation for filtration and prediction. Wiener's*

method. Let $\xi(t)$ and $\zeta(t)$ for $t \in \mathfrak{T}$ denote two Hilbert random functions. Let us suppose that the process $\xi(t)$ is observed over some set T of values of the argument t. Consider the problem of determining the best approximation of the value of $\zeta(t_0)$ for $t_0 \in \mathfrak{T}$ from the observed values of $\xi(t)$ where $t \in T$. If we assume that the desired approximation is of the form

$$\tilde{\zeta}(t_0) = \int_T c(\theta)\xi(\theta)m(d\theta) \tag{5}$$

where m is some measure on T, and if the conditions under which the above integral is meaningful are satisfied, then equation (4) takes the form

$$\int_T c(\theta)R_{\xi\xi}(\theta, t)m(d\theta) = R_{\zeta\xi}(t_0, t)\,, \qquad t \in T\,, \tag{6}$$

where $R_{\xi\xi}$ is the covariance function of $\xi(t)$ and $R_{\zeta\xi}$ is the mutual correlation function of $\zeta(t)$ and $\xi(t)$. Equation (6) is a Fredholm integral equation of the first kind with symmetric (Hermitian) kernel. By no means does it always have a solution. However, if

$$\int_T \mathsf{M} \mid \xi(t) \mid^2 m(dt) < \infty\,,$$

the integral equation (6) has a solution $c(\theta) \in L_2\{m\}$ if and only if the best linear approximation of $\tilde{\zeta}(t_0)$ of the quantity $\zeta(t)$ is of the form (5). Suppose that $\mathfrak{T}$ is the real axis and that $T = (a, b)$, that the processes ξ and ζ are stationary and stationarily connected (in the broad sense), and that the measure m is Lebesgue measure. Then equation (6) takes the form

$$\int_a^b c(\theta)R_{\xi\xi}(\theta - t)d\theta = R_{\zeta\xi}(t_0 - t)\,, \qquad t \in (a, b)\,. \tag{7}$$

If $\zeta(t) = \xi(t)$ (for $-\infty < t < \infty$) and $t_0 > b$, that is, if the problem consists in finding an approximation for the quantity $\xi(t_0)$ from the values of $\xi(t)$ in the past, we shall call the problem one of pure prediction.

Let us look in greater detail at the problem of prediction of the variable $\zeta(t + T)$ from the results of observation of the process $\xi(\theta)$ from the instant t, where $\theta \leqq t$. Let us treat the predicting variable $\tilde{\zeta}(t)$ as a function of t for fixed T. We can easily see that $\tilde{\zeta}(t)$ as defined by equation (5) is a stationary process. To see this, note that equation (7) is of the form

$$\int_{-\infty}^t c_t(\theta)R_{\xi\xi}(\theta - u)d\theta = R_{\zeta\xi}(t + T - u)\,, \qquad u \leqq t\,.$$

The change of variables $t - u = v$, $t - \theta = \tau$ transforms the preceding equation into the following:

$$\int_0^\infty c_t(t-\tau)R_{\xi\xi}(v-\tau)d\tau = R_{\zeta\xi}(T+v)\,, \qquad v \geqq 0\,. \tag{8}$$

From this we see that the function $c_t(t-\tau)$ is independent of t. Let us set $c(\tau) = c_t(t-\tau)$. Equation (8) can now be written

$$\int_0^\infty c(\tau)R_{\xi\xi}(v-\tau)d\tau = R_{\zeta\xi}(T+v)\,, \qquad v \geqq 0\,, \tag{9}$$

and formula (5) for the predicting function takes the form

$$\tilde{\zeta}(t) = \int_{-\infty}^{t} c(t-\tau)\xi(\tau)d\tau = \int_0^\infty c(\tau)\xi(t-\tau)d\tau\,. \tag{10}$$

Thus the process $\tilde{\zeta}(t) = \tilde{\zeta}_T(t)$ is stationary. It follows from formula (10) that $c(t)$ is the impulse transfer function of a physically realizable filter that transforms the observed process into the best approximation of the quantity $\zeta(t+T)$.

It is easy to exhibit an expression for the mean-square error δ of the predicting function $\tilde{\zeta}(t)$. Since δ^2 is the square of the length of the perpendicular dropped from the vector $\zeta(t+T)$ onto $H_2\{\xi(\tau), \tau \leqq t\}$, we have

$$\begin{aligned}\delta^2 &= \mathsf{M}\,|\,\zeta(t+T)\,|^2 - \mathsf{M}\,|\,\tilde{\zeta}(t)\,|^2 \\ &= R_{\zeta\zeta}(0) - \int_0^\infty\int_0^\infty \overline{c(t)}R_{\xi\xi}(t-\tau)c(\tau)dtd\tau\,.\end{aligned} \tag{11}$$

Setting $R_{\zeta\zeta}(0) = \sigma_\zeta^2$ and shifting to the spectral representation of the covariance function $R_{\xi\xi}(t)$, we obtain

$$\delta^2 = \sigma_\zeta^2 - \int_{-\infty}^\infty |\,C(i\omega)\,|^2\, dF_{\xi\xi}(\omega)\,, \tag{12}$$

where $F_{\xi\xi}(\omega)$ is the spectral function of the process $\xi(t)$ and

$$C(i\omega) = \int_0^\infty c(t)e^{-i\omega t}dt\,.$$

We shall explain briefly a method proposed by N. Wiener for solving equation (9). Suppose that the spectrum of the process $\xi(t)$ is absolutely continuous and that the spectral density $f_{\xi\xi}(\omega)$ admits a factorization (cf. Theorem 3, Section 6):

$$f_{\xi\xi}(\omega) = |\,h(i\omega)\,|^2,\; h(z) = \frac{1}{\sqrt{2\pi}}\int_0^\infty a(t)e^{-zt}dt, \qquad \operatorname{Re} z \geqq 0\,.$$

It follows from Parseval's equality for the Fourier transform that

$$R_{\xi\xi}(t) = \int_{-\infty}^\infty e^{-it\omega}\,|\,h(i\omega)\,|^2\, d\omega = \int_0^\infty a(t+u)\overline{a(u)}du\,.$$

Let us suppose also that the mutual spectral function of the processes

$\zeta(t)$ and $\xi(t)$ is absolutely continuous and that its density $f_{\zeta\xi}(\omega)$ satisfies the condition

$$\frac{f_{\zeta\xi}(\omega)}{h(i\omega)} = k(i\omega) \in L_2 \,. \tag{13}$$

Then

$$\begin{aligned} R_{\zeta\xi}(t) &= \int_{-\infty}^{\infty} e^{it\omega} f_{\zeta\xi}(\omega) d\omega = \int_{-\infty}^{\infty} e^{it\omega} k(i\omega) \bar{h}(i\omega) d\omega \\ &= \int_0^{\infty} b(t+u) \overline{a(u)} du \,, \end{aligned}$$

where

$$b(t) = \frac{1}{\sqrt{2\pi}} \int_{-\infty}^{\infty} k(i\omega) e^{it\omega} d\omega \,.$$

With the aid of the expressions obtained we can rewrite (9) as follows:

$$\int_0^{\infty} [b(T+v+u) - \int_0^{\infty} c(\tau) a(v-\tau+u) d\tau] \bar{a}(u) du = 0 \,, \quad v > 0 \,. \tag{14}$$

A sufficient condition for equation (14) to hold is that $c(t)$ satisfy the equation

$$b(T+x) = \int_0^{\infty} c(\tau) a(x-\tau) d\tau \,, \qquad x > 0 \,. \tag{15}$$

Equation (15) is of the same type as equation (9) except for the important fact that the function $a(t)$ vanishes for negative values of t. If we write equation (15) in the form

$$b(T+x) = \int_0^{x} c(\tau) a(x-\tau) d\tau \,, \qquad x > 0 \,, \tag{16}$$

we can immediately solve it with the aid of the Laplace transform. Multiplying equation (16) by e^{-zx} and integrating from 0 to ∞, we obtain $B_T(z) = C(z)h(z)$, where $B_T(z) = (1/\sqrt{2\pi})\int_0^{\infty} b(T+x)e^{-zx}dx$,

$$C(z) = \frac{1}{\sqrt{2\pi}} \int_0^{\infty} c(t) e^{-zt} dt \,.$$

Thus

$$C(z) = \frac{B_T(z)}{h(z)} \,, \qquad c(t) = \frac{1}{\sqrt{2\pi}} \int_{-\infty}^{\infty} e^{i\omega t} \frac{B_T(i\omega)}{h(i\omega)} d\omega \,, \tag{17}$$

where the expression for $B_T(z)$, for $\operatorname{Re} z > 0$, can be written in the form

$$B_T(z) = \frac{1}{2\pi} \int_{-\infty}^{\infty} e^{iT\omega} \frac{f_{\zeta\xi}(\omega)}{h(i\omega)} \frac{d\omega}{z - i\omega} \,. \tag{18}$$

Formulation of the assumptions under which formulas (17) and (18) are valid is extremely laborious. In solving specific problems, it is simpler to verify directly the validity of the proposed transformations that lead to the solution of the problem.

4. *Yaglom's method.* With this method, in contrast to that of Wiener, we seek not the impulse transfer function of the optimum filter, which may not even exist, but instead the frequency characteristic. We shall not give general formulas for solving the problem but shall present only a method of choosing the desired function by starting with those requirements that it must satisfy. In many important cases, this choice is easy to make.

Suppose that a two-dimensional stationary process $(\xi(t), \zeta(t))$ has a spectral representation of the form

$$\xi(t) = \int_{-\infty}^{\infty} e^{i\omega t}\mu_1(du) , \qquad \zeta(t) = \int_{-\infty}^{\infty} e^{i\omega t}\mu_2(du)$$

with spectral density matrix

$$\begin{pmatrix} f_{\xi\xi}(\omega), f_{\xi\zeta}(\omega) \\ f_{\zeta\xi}(\omega), f_{\zeta\zeta}(\omega) \end{pmatrix} .$$

As before, let us consider the problem of the best approximation of the quantity $\zeta(t + T)$ from the values of the process $\xi(\tau)$, for $\tau \leqq t$. The predicting process $\tilde{\zeta}(t)$ is subordinate to $\xi(t)$. Therefore

$$\tilde{\zeta}(t) = \int_{-\infty}^{\infty} e^{i\omega t}c(i\omega)\mu_1(d\omega) , \qquad \int_{-\infty}^{\infty} | c(i\omega) |^2 f_{\xi\xi}(\omega)d\omega < \infty . \tag{19}$$

The equation defining the process $\tilde{\zeta}(t)$, namely

$$\mathsf{M}\zeta(t + T)\overline{\xi(\tau)} = \mathsf{M}\tilde{\zeta}(t)\overline{\xi(\tau)} , \qquad \tau \leqq t ,$$

takes the form

$$\int_{-\infty}^{\infty} e^{i\omega v}\{e^{i\omega T}f_{\zeta\xi}(\omega) - c(i\omega)f_{\xi\xi}(\omega)\}d\omega = 0 , \qquad v > 0 . \tag{20}$$

In addition to conditions (19) and (20), we also have the requirement that $c(i\omega)$ be the frequency characteristic of a physically realizable filter. These conditions will be satisfied if

a. the function $f_{\xi\xi}(\omega)$ is bounded,

b. $c(i\omega)$ is the limiting value of the function $c(z) \in H_2^\pi$ (cf. Section 6),

c. $\psi(i\omega) = e^{i\omega T}f_{\zeta\xi}(\omega) - c(i\omega)f_{\xi\xi}(\omega)$ is the limiting value of the function $\psi(z) \in H_2^\lambda$, where H_2^λ, is defined analogously to H_2^π except that H_2^λ consists of functions that are analytic in the left half-plane.

To see the validity of this assertion, note that condition (b) implies that $\int_{-\infty}^{\infty} | c(i\omega) |^2 d\omega < \infty$, and this, together with condition a, ensures satisfaction of condition (19). Also, condition (b) implies that $c(i\omega)$ is the frequency characteristic of a physically realizable filter. It follows from condition (c) that $e^{i\omega T}f_{\zeta\xi}(\omega) - c(i\omega)f_{\xi\xi}(\omega)$ is the Fourier transform of a function that, by virtue of equation (20), vanishes for positive values of the argument.

If we confine ourselves to condition b we rule out filters whose frequency characteristics can increase at infinity. Such frequency characteristics correspond to operations connected with differentiation of the process $\xi(t)$ and are often encountered in the construction of optimum filters. Therefore it is desirable to replace condition b with a less restrictive one. Let us suppose that $c(z)$ is a function that is analytic in the right half-plane and that $|c(z)|$ approaches ∞ no faster than the rth power for some r as $|z| \longrightarrow \infty$. We define the functions $c_n(z)$ by

$$c_n(z) = \frac{c(z)}{\left(1 + \frac{z}{n}\right)^{r+1}} \in H_2^{\pi} .$$

Since $|c_n(z)| \leqq |c(z)|$, we have

$$\lim_{n\to\infty} \int_{-\infty}^{\infty} |c_n(i\omega) - c(i\omega)|^2 f_{\xi\xi}(\omega)d\omega = 0 ,$$

if condition (19) is satisfied. Thus $c(i\omega)$ is the limit in $L_2\{F_{\xi\xi}\}$ of frequency characteristics of admissible physically realizable filters. Therefore $c(i\omega)$ is also the frequency characteristic of such a filter. Thus we have proved:

Theorem 3. *If the spectral density $f_{\xi\xi}(\omega)$ of a process $\xi(t)$ is bounded, then the three conditions*

a. $\int_{-\infty}^{\infty} |c(i\omega)|^2 f_{\xi\xi}(\omega)d\omega < \infty$;

b. *$c(i\omega)$ is the limiting value of a function $c(z)$ that is analytic in the right half-plane and that increases no faster than some power of $|z|$ as $|z| \longrightarrow \infty$;*

c. *$\psi(i\omega) = e^{i\omega T}f_{\zeta\xi}(\omega) - c(i\omega)f_{\xi\xi}(\omega)$ is the limiting value of a function $\psi(z) \in H_2^{\lambda}$;*

determine uniquely the frequency characteristic $c(i\omega)$ of the optimum filter approximating the quantity $\xi(t + T)$.

The mean-square error δ of the best approximation is equal to

$$\delta = \{\mathsf{M}\,|\,\zeta(t + T)\,|^2 - \mathsf{M}\,|\,\tilde{\zeta}(t)\,|^2\}^{\frac{1}{2}}$$
$$= \left\{\sigma_\zeta^2 - \int_{-\infty}^{\infty} |c(i\omega)|^2 f_{\xi\xi}(\omega)d\omega\right\}^{\frac{1}{2}} . \quad (21)$$

***Example* 1.** Consider the problem of a simple prediction of a process $\xi(t)$ (where $\xi(t) = \zeta(t)$) with covariance function $R(t) = \sigma^2 e^{-\alpha|t|}$ for $\alpha > 0$. The spectral density is easily found;

$$f_{\xi\xi}(\omega) = \frac{\sigma^2 \alpha}{\pi} \frac{1}{\omega^2 + \alpha^2}.$$

The analytic continuation of the function $\psi(i\omega)$ is of the form

$$\psi(z) = \frac{c(z) - e^{zT}}{(z + \alpha)(z - \alpha)}.$$

This function $\psi(z)$ has only one pole in the left half-plane: $z = -\alpha$. To neutralize it with the aid of a function $c(z)$ that is analytic in the right half-plane, it suffices to set $c(z) = \text{const} = e^{-\alpha T}$. Here condition (a) of Theorem 3 is satisfied. Thus

$$c(i\omega) = e^{-\alpha T}, \qquad \tilde{\xi}(t) = \int_{-\infty}^{\infty} e^{i\omega t} e^{-\alpha T} \mu(d\omega);$$

that is, the best formula for the prediction of the quantity $\xi(t + T)$ is $\xi(t + T) \approx e^{-\alpha T}\xi(t)$, which depends only on the value of $\xi(\tau)$ at the last observed instant of time. The mean-square error of the extrapolation is equal to $\delta = \sigma\sqrt{1 - e^{-2\alpha T}}$.

Example 2. Let us consider the problem of pure prediction of a process $\xi(t)$, that is, of finding an estimate for $\xi(t + T)$ from observed values of $\xi(\tau)$ for $\tau < t$. If the spectrum of the process $\xi(t)$ is absolutely continuous and if condition (23) of Section 6 is satisfied, then the spectral density of the process admits a factorization $f_{\xi\xi}(\omega) = |h(i\omega)|^2$, where $h(z)$ belongs to H_2^π and has no zeros in the right half-plane.

Let us consider the case which is very important for practical applications and in which $h(z)$ is a rational function $h(z) = P(z)/Q(z)$, where $P(z)$ is a polynomial of degree m and $Q(z)$ is a polynomial of degree $n > m$. Let us suppose also that the spectral density $f_{\zeta\zeta}(\omega)$ is bounded and that it does not vanish. Then the zeros of the polynomials $P(z)$ and $Q(z)$ lie in the left half-plane. Let $P(z)$ and $Q(z)$ be represented in the forms

$$P(z) = A \prod_{j=1}^{p} (z - z_j)^{\alpha_j}, \qquad Q(z) = B \prod_{j=1}^{q} (z - \tilde{z}_j)^{\beta_j},$$

where

$$\sum_{j=1}^{p} \alpha_j = m, \qquad \sum_{j=1}^{q} \beta_j = n.$$

Let us define

$$P_1(z) = (-1)^m A \prod_{j=1}^{p} (z + \bar{z}_j)^{\alpha_j}, \qquad Q_1(z) = (-1)^n B \prod_{j=1}^{q} (z + \bar{\tilde{z}}_j)^{\beta_j}.$$

The analytic continuation of the function $\psi(i\omega)$ is of the form

$$\psi(z) = (e^{izT} - c(z))\frac{P(z)}{Q(z)}\frac{P_1(z)}{Q_1(z)} .$$

The function $c(z)$ must be analytic in the right half-plane and $\psi(z)$ must be analytic in the left half-plane. Therefore $c(z)$ must be analytic in the entire complex plane and it may have poles at the zeros of the polynomial $P(z)$. The order of such a pole cannot exceed the order of the corresponding zero of $P(z)$. Therefore, $c(z) = M(z)/P(z)$, where $M(z)$ is an analytic function in the z-plane that has no singularities for finite z. Since $c(z)$ has no more than a power order of growth, $M(z)$ is a polynomial. In view of the square-integrability of the absolute value of the function

$$c(i\omega)\frac{P(i\omega)}{Q(i\omega)} = \frac{M(i\omega)}{Q(i\omega)} ,$$

the degree m_1 of the polynomial $M(i\omega)$ does not exceed $n - 1$, that is, $m_1 \leqq n - 1$.

On the other hand, the indicated choice of the function $c(z)$ ensures satisfaction of conditions (a) and (b) of Theorem 3. We must choose the polynomial $M(z)$ in such a way that the function

$$\psi(z) = \frac{(e^{zT}P(z) - M(z))}{Q(z)}\frac{P_1(z)}{Q_1(z)}$$

or, what amounts to the same thing, the function

$$\psi_1(z) = \frac{e^{zT}P(z) - M(z)}{Q(z)}$$

has no poles in the left half-plane. For this it is necessary and sufficient that

$$\left.\frac{d^jM(z)}{dz^j}\right|_{z=\tilde{z}_k} = \left.\frac{d^j(e^{zT}P(z))}{dz^j}\right|_{z=\tilde{z}_k} , \qquad j = 0, 1, \cdots, \beta_k - 1 ,$$

$$k = 1, \cdots, q . \tag{21'}$$

The problem of constructing a polynomial $M(z)$ satisfying conditions (21′) is the usual problem in the theory of interpolation and it always has a unique solution in the class of polynomials of degree $n - 1$. In finding the polynomial $M(z)$, we automatically find the frequency characteristic of the optimum predicting filter $c(i\omega) = M(i\omega)/P(i\omega)$.

We may also use the following procedure for determining the function $c(z)$. Let us decompose the functions $P(z)Q^{-1}(z)$ and $M(z)Q^{-1}(z)$ into partial fractions:

$$\frac{P(z)}{Q(z)} = \sum_{k=1}^{q} \sum_{j=1}^{\beta_k} \frac{c_{kj}}{(z - z_k)^j}, \qquad \frac{M(z)}{Q(z)} = \sum_{k=1}^{q} \sum_{j=1}^{\beta_k} \frac{\gamma_{kj}}{(z - \tilde{z}_k)^j}.$$

For the function $\psi_1(z)$ not to have poles at the points $\tilde{z}_k$, for $k = 1, \cdots, q$, it is necessary and sufficient that

$$\frac{d^j}{dz^j}(z - \tilde{z}_k)^{\beta_k}\psi_1(z)\Big|_{z=\tilde{z}_k} = 0, \qquad j = 0, 1, \cdots, \beta_k - 1,$$

where $\psi_1(z) = \sum_{k=1}^{q} \sum_{j=1}^{\beta_k} (c_{kj}e^{zT} - \gamma_{kj})/(z - \tilde{z}_k)^j$. Simple calculation shows that

$$\gamma_{kj} = \left[c_{kj} + Tc_{k,j+1} + \frac{T^2}{2!}c_{k,j+2} + \cdots + \frac{T^{\beta_k - j}}{(\beta_k - j)!}c_{k\beta_k}\right]e^{\tilde{z}_k T},$$

$$k = 1, \cdots, q.$$

Knowing the coefficients γ_{kj}, we can write the expression for $c(i\omega)$:

$$c(i\omega) = \frac{1}{h(i\omega)} \sum_{k=1}^{q} \sum_{j=1}^{\beta_k} \frac{\gamma_{kj}}{(z - \tilde{z}_k)^j} = \frac{\displaystyle\sum_{k=1}^{q} \sum_{j=1}^{\beta_k} \frac{\gamma_{kj}}{(z - \tilde{z}_k)^j}}{\displaystyle\sum_{k=1}^{q} \sum_{j=1}^{\beta_k} \frac{c_{kj}}{(z - \tilde{z}_k)^j}}. \tag{22}$$

***Example* 3.** Suppose that we are observing a process $\zeta(\tau)$ (for $\tau \leqq t$) but that the results of our measurements of $\zeta(\tau)$ are distorted by various errors, so that the observed values yield a function $\xi(\tau)$, for $\tau < t$, different from $\zeta(\tau)$. Let us suppose that the magnitude of the error (or, as we say, the noise) $\eta(t) = \xi(t) - \zeta(t)$ is a stationary process with mean value 0. Suppose that we wish to find an estimate of the value of $\zeta(t + T)$ from the results of observation of the process $\xi(\tau) = \zeta(\tau) + \eta(\tau)$ for $\tau \leqq t$.

Such problems are called filtering or smoothing problems (we say that we need to filter the noise $\eta(t)$ out of the process $\xi(t)$ or that the process $\xi(t)$ must be "smoothed," that is, the irregular noise needs to be subtracted from it). Here for $T > 0$, we have a problem of filtering with prediction and, for $T < 0$, we have a problem of filtering with lag.

Let us suppose that the noise $\eta(t)$ and the process $\zeta(t)$ are uncorrelated and have spectral densities $f_{\zeta\zeta}(\omega)$ and $f_{\eta\eta}(\omega)$. Then

$$R_{\xi\xi}(t) = R_{\eta\eta}(t) + R_{\zeta\zeta}(t), \quad \text{and} \quad f_{\xi\xi}(\omega) = f_{\eta\eta}(\omega) + f_{\zeta\zeta}(\omega).$$

Since $R_{\zeta\xi}(t) = R_{\zeta\zeta}(t)$, there exists a mutual spectral density of the processes $\zeta(t)$ and $\xi(t)$ and we have $f_{\zeta\xi}(\omega) = f_{\zeta\zeta}(\omega)$. Suppose that

$$f_{\zeta\zeta}(\omega) = \frac{c_1}{\omega^2 + \alpha^2} \quad \text{and} \quad f_{\eta\eta}(\omega) = \frac{c_2}{\omega^2 + \beta^2}.$$

Then

$$f_{\xi\xi}(\omega) = \frac{c_3(\omega^2 + \gamma^2)}{(\omega^2 + \alpha^2)(\omega^2 + \beta^2)}, \qquad c_3 = c_1 + c_2, \qquad \gamma^2 = \frac{c_1\alpha^2 + c_2\beta^2}{c_1 + c_2}.$$

For the function $\psi(z)$ we obtain the expression

$$\psi(z) = \frac{-c_1 e^{zT}(z^2 - \beta^2) + c_3 c(z)(z^2 - \gamma^2)}{(z^2 - \alpha^2)(z^2 - \beta^2)}.$$

Suppose that $T > 0$. The function $\psi(z)$ must be analytic in the left half-plane and it must belong to H_2^λ. For this it is necessary that the numerator vanish at the points $z = -\alpha$ and $z = -\beta$. This leads us to the equations

$$c(-\beta) = 0, \qquad c(-\alpha) = \frac{c_1}{c_3}\frac{e^{-\alpha T}(\alpha^2 - \beta^2)}{\alpha^2 - \gamma^2}. \tag{23}$$

Furthermore $c(z)$ must be analytic in the left half-plane (and also in the right half-plane by virtue of condition b) except at the single point $z = -\gamma$ where it may have a simple pole.

Thus $c(z) = \varphi(z)/(z + \gamma)$, where $\varphi(z)$ is an entire function. From the condition of finiteness of the integral $\int_{-\infty}^{\infty} |c(i\omega)|^2 f_{\xi\xi}(\omega)d\omega$, it follows that $\varphi(z)$ is a linear function, $\varphi(z) = Az + B$.

From (23) we obtain

$$c(z) = A\frac{z + \beta}{z + \gamma}, \qquad A = \frac{c_1}{c_3}\frac{\beta + \alpha}{\gamma + \alpha}e^{-\alpha T}.$$

Therefore the formula for optimum smoothing with prediction is of the form

$$\tilde{\zeta}_T(t) = A\int_{-\infty}^{\infty} e^{i\omega t}\frac{i\omega + \beta}{i\omega + \gamma}\mu_1(d\omega).$$

Remembering that $(i\omega + \gamma)^{-1}$ is the frequency characteristic of a physically realizable filter with impulse transfer function $e^{-\gamma t}$, we obtain

$$\tilde{\zeta}_T(t) = \frac{c_1}{c_3}\frac{\beta + \alpha}{\gamma + \alpha}e^{-\alpha T}\left\{\xi(t) - (\beta - \gamma)\int_{-\infty}^{t} e^{-\gamma(t-\tau)}\xi(\tau)d\tau\right\}. \tag{24}$$

For $T < 0$, formula (24) is not valid. Formally, this is connected with the fact that the function $\psi(z)$ is not bounded in the left half-plane in this case. For $T < 0$ the function $\psi(z)$ may be determined from the following considerations. Suppose that

$$\psi_1(z) = -c_1 e^{zT}(z^2 - \beta^2) + c_3 c(z)(z^2 - \gamma^2).$$

Then $c(z)$ must be analytic in the left half-plane except at the point $z = -\gamma$ and we have $\psi_1(-\alpha) = \psi_1(-\beta) = 0$. Since

$$c(z) = \frac{\psi_1(z) + c_1 e^{zT}(z^2 - \beta^2)}{c_3(z^2 - \gamma^2)}$$

and $c(z)$ is analytic in the right half-plane, it follows that $\psi_1(z)$ is an entire function and we have

$$\psi_1(\gamma) = -c_1 e^{\gamma T}(\gamma^2 - \beta^2) . \tag{25}$$

If we set $\psi_1(z) = A(z)(z + \alpha)(z + \beta)$, the function $A(z)$ must be an entire function. It follows from condition (8) of Theorem 3 that $A(z) = \text{const} = A$. The value of A is determined from equation (25):

$$A = c_1 e^{\gamma T} \frac{\gamma + \beta}{\alpha - \gamma} .$$

From this we get

$$c(i\omega) = \frac{c_1}{c_3} \frac{(\alpha - \gamma)(\omega^2 + \beta^2)e^{i\omega T} - e^{\gamma T}(\gamma + \beta)(i\omega + \alpha)(i\omega + \beta)}{(\alpha - \gamma)(\omega^2 + \gamma^2)} . \tag{26}$$

For the prediction and filtering of stationary sequences, we apply methods analogous to those that were discussed for processes with continuous time. The general solution of the problem of prediction of stationary sequences is given in the next section. Here we shall confine ourselves to a single example.

Example 4. *A process of autogression* is defined as a stationary sequence $\xi(t)$ satisfying the finite-difference equation

$$a_0\xi(t) + a_1\xi(t-1) + \cdots + a_p\xi(t-p) = \eta(t) \tag{27}$$

and subordinate to $\eta(t)$, where $\eta(t)$ is an uncorrelated sequence such that $\mathsf{M}\eta(t) = 0$ and $\mathsf{D}\eta(t) = \sigma^2$. Suppose that

$$\eta(t) = \int_{-\pi}^{\pi} e^{it\omega} d\zeta(\omega)$$

is the spectral representation of the sequence $\eta(t)$ and that $\zeta(\omega)$ is a process with independent increments and with structural function $(1/2\pi)l(A \cap B)$, where l denotes Lebesgue measure. Then the spectral representation of the sequence $\xi(t)$ must have the form

$$\xi(t) = \int_{-\pi}^{\pi} e^{it\omega}\varphi(\omega)d\zeta(\omega) , \quad \text{where} \quad \int_{-\pi}^{\pi} |\varphi(\omega)|^2 d\omega < \infty . \tag{28}$$

Substituting (28) into (27), we obtain

$$\int_{-\pi}^{\pi} e^{it\omega}\bar{P}(e^{i\omega})\varphi(\omega)d\zeta(\omega) = \int_{-\pi}^{\pi} e^{it\omega}d\zeta(\omega) ,$$

where $P(z) = \sum_{k=0}^{n} \bar{a}_k z^k$. From this it follows that

$$\varphi(\omega) = \frac{1}{\overline{P(e^{i\omega})}} \quad (\text{mod } l) .$$

Let us suppose that the function $P(z)$ has no zeros in the closed disk $|z| \leqq 1$. Then $1/P(z) \in H_2$. If

$$\frac{1}{P(z)} = \sum_{k=0}^{\infty} \bar{b}_k z^k \left(b_0 = \frac{1}{a_0} \right),$$

then

$$\xi(t) = \sum_{n=0}^{\infty} b_n \eta(t - n)$$

and we have obtained a representation of the sequence $\xi(t)$ in the form of the response of a physically realizable filter to the uncorrelated sequence $\eta(t)$. Since

$$\xi(t) = -\frac{1}{a_0}[a_1\xi(t-1) + \cdots + a_p\xi(t-p) + \eta_t], \tag{29}$$

the optimum prediction $\xi(t)$ from given $\xi(t-n)$, where $n = 1, 2, \cdots$, is of the form

$$\tilde{\xi}(t) = -\frac{1}{a_0}(a_1\xi(t-1) + a_2\xi(t-2) + \cdots + a_p\xi(t-p)).$$

The minimum mean-square error of the prognosis is equal to

$$\delta(t) = \left\{ \mathsf{M}\frac{|\eta(t)|^2}{|a_0|^2} \right\}^{1/2} = \frac{\sigma}{|a_0|}.$$

Successive use of formula (29) enables us to obtain the optimum prediction several steps in advance.

8. GENERAL THEOREMS ON THE PREDICTION OF STATIONARY PROCESSES

In this section we shall consider certain general theorems on the prediction of stationary sequences and processes with respect to the infinite past. By a stationary process we mean a process that is stationary in the broad sense and that has mathematical expectation 0.

1. Prediction of stationary sequences

Let $\xi(t)$, for $t = 0, \pm 1, \pm 2, \cdots$, denote a stationary sequence. Let H_ξ denote the closed linear hull generated in L_2 by all the quantities $\xi(t)$, and let $H_\xi(t)$ denote the closed linear hull generated by the quantities $\xi(n)$ for $n \leq t$. Obviously $H_\xi(t) \subset H_\xi(t+1)$ and H_ξ is the closure of $\bigcup_{t=-\infty}^{\infty} H_\xi(t)$. Consider in H_ξ the operation S representing time displacement. For elements of H_ξ of the form $\eta = \sum c_k \xi(t_k)$ this operation is defined by

$$S\eta = \sum c_k \xi(t_k + 1).$$

The operation S has an inverse S^{-1}:

$$S^{-1}\eta = \sum c_k \xi(t_k - 1) ,$$

and it preserves the scalar product,

$$\mathsf{M}(S(\sum c_k \xi(t_k))\overline{S(\sum d_k \xi(\tau_k))}) = \sum_k \sum_r c_k \bar{d}_r \mathsf{M}(\xi(t_k + 1)\overline{\xi(\tau_r + 1)})$$

$$= \sum_k \sum_r c_k \bar{d}_r \mathsf{M}(\xi(t_k)\overline{\xi(\tau_r)}) = \mathsf{M}\Big(\sum_k c_k \xi(t_k) \overline{\sum_r d_r \xi(\tau_r)}\Big) .$$

Therefore S can be extended as a continuous operator to H_ξ. It then becomes a unitary operator in H_ξ.

We introduce the spectral representation of the sequence $\xi(t)$ (cf. Remark 4, Section 4),

$$\xi(t) = \int_{-\pi}^{\pi} e^{i\omega t} \mu(d\omega) ,$$

where μ is a spectral stochastic measure with structural function F. In what follows we shall not distinguish the measure $F(A)$ from the spectral function (generated by it) of the sequence $F(\omega) = F[-\pi, \omega)$.

We recall that a random variable η belongs to H_ξ if and only if $\eta = \int_{-\pi}^{\pi} \varphi(\omega)\mu(d\omega)$, where $\varphi \in L_2\{F\}$. Consider the sequence $\eta(t)$ of random variables $\eta(t) = S^t \eta$ $(t = 0, \pm 1, \pm 2, \cdots)$.

Lemma 1. *The sequence $\eta(t)$ is stationary and subordinate to $\xi(t)$ and it has the spectral representation*

$$\eta(t) = \int_{-\pi}^{\pi} e^{it\omega} \varphi(\omega)\mu(d\omega) . \tag{1}$$

That $\eta(t)$ is subordinate to the process $\xi(t)$ is obvious. That it is stationary follows from the unitariness of S:

$$\mathsf{M}(\eta(t + \tau)\overline{\eta(\tau)}) = (\eta(t + \tau), \eta(\tau)) = (S^{t+\tau}\eta, S^\tau \eta)$$
$$= (S^t \eta, \eta) = \mathsf{M}(\eta(t)\overline{\eta(0)}) .$$

Finally, the spectral representation (1) is easily verified for elements η of the form $\eta = \sum a_k \xi(t_k)$ (where $\varphi(\omega) = \sum a_k e^{i\omega t_k}$) and it is obtained for arbitrary η by passage to the limit.

We also note the following obvious properties of the operator S:

a. $SH_\xi(t) = H_\xi(t + 1)$;

b. if $\xi^{(p)}(t)$ is the projection of $\xi(t)$ onto $H_\xi(t - p)$, then

$$S\xi^{(1)}(t) = \xi^{(1)}(t + 1) , \qquad S^q \xi^{(p)}(t) = \xi^{(p)}(t + q) .$$

Since

$$\mathsf{M} \mid \xi^{(p)}(t + q) \mid^2 = \mathsf{M} \mid S^q \xi^{(p)}(t) \mid^2 = \mathsf{M} \mid \xi^{(p)}(t) \mid^2 ,$$

the quantity $\mathsf{M} \mid \xi^{(p)}(t) \mid^2$ is independent of t. Therefore the quantity

$$\delta^2(p) = \mathsf{M} \mid \xi(t) - \xi^{(p)}(t) \mid^2 = \mathsf{M} \mid \xi(t) \mid^2 - \mathsf{M} \mid \xi^{(p)}(t) \mid^2 ,$$

which is equal to the square of the minimum mean-square error of the prediction of the variable $\xi(t)$ with the aid of the quantity $\xi(n)$, for $n \leqq t - p$, is also independent of t. Obviously,

$$\delta^2(1) \leqq \delta^2(2) \leqq \cdots \leqq \sigma^2 = \mathsf{M} \mid \xi(t) \mid^2 .$$

The equation $\delta^2(n) = \sigma^2$ means that $\xi(t)$ is, for arbitrary t, uncorrelated with all the variables ξ_k, for $k \leqq t - n$, so that the value of these terms yields nothing for the prediction of the variable $\xi(t)$. If $\delta(1) = 0$, then $\xi(t) \in H_\xi(t-1)$, so that $H_\xi(t-1) = H_\xi(t)$. Let us set $H_\xi^S = \bigcap_t H_\xi(t)$. In the present case, $H_\xi^S = H_\xi$. This means that if we know the sequence of values of the process $\xi(k)$, for $k \leq t_0$, then all the subsequent terms of the sequence have with probability 1 a precise linear expression in terms of the observed values. In a certain sense we have a contradiction in the case in which $H_\xi^S = 0$, where 0 denotes the trivial subspace of H_ξ consisting only of the element 0. Here, knowing the terms of the sequence $\xi(k)$ (for $k \leq n$) yields little for the prediction of the variable $\xi(n+t)$ for large values of t since $\lim_{t\to\infty} \mathsf{M} \| \xi^{(t)}(n) \| = 0$ and $\lim_{t\to\infty} \delta^2(t) = \sigma^2$.

Definition 1. If $H_\xi^S = H_\xi$, the process $\xi(t)$ is called a *singular* (or *determined*) process. If $\delta(1) > 0$, the process $\xi(t)$ is called an *undetermined* process. If $H_\xi = 0$, the process is called a *regular* (or completely *undetermined*) process.

Theorem 1. *An arbitrary sequence has a unique representation of the form*

$$\xi(t) = \xi_S(t) + \eta(t) , \tag{2}$$

where $\xi_S(t)$ and $\eta(t)$ are mutually uncorrelated sequences subordinate to $\xi(t)$, where $\xi_S(t)$ is singular, and where $\eta(t)$ is regular.

Proof. Obviously, $SH_\xi^s = H_\xi^s$. Since S is a unitary operator, it follows that H_ξ^s reduces S; that is, S is a one-to-one mapping of the subspace $H_\xi^r = H_\xi \ominus H_\xi^S$ onto itself (cf. Lemma 2, Section 1).

Let $\xi_S(0)$ denote the projection of $\xi(0)$ onto H_ξ^S and let $\eta(0)$ denote the projection of $\xi(0)$ onto H_ξ^r. Let $\xi_S(t) = S^t\xi_S(0)$ and $\eta(t) = S^t\eta(0)$ for $t = 0, \pm 1, \pm 2, \cdots$. Since $\xi(0) = \xi_S(0) + \eta(0)$, we have $\xi(t) = S^t\xi(0) = \xi_s(t) + \eta(t)$. Here the sequences $\eta(t)$ and $\xi_s(t)$ are stationary, mutually uncorrelated, and subordinate to $\xi(t)$.

Furthermore, since, in the equation $\xi(t) = \xi_s(t) + \eta(t)$ it is true that $\xi_S(t) \in H_\xi^S$ and $\eta(t) \in H_\xi^r$, we have $H_\xi(t) \cap H_\xi^S \subset H_{\xi_S}(t)$. Therefore $H_\xi^S \subset H_{\xi_S}^S$. On the other hand, the inclusion relation $\xi_S(t) \in H_\xi^S$ implies that $H_{\xi_S}(t) \subset H_\xi^S$. Thus for arbitrary t, we have $H_{\xi_S}(t) =$

$H_\xi^S = H_{\xi_S}^S$; that is, the sequences $\xi_S(t)$ is singular. Furthermore, the equation $\eta(t) = \xi(t) - \xi_S(t)$ implies that $\eta(t) \in H_\xi(t)$. Therefore $H_\eta^S = \bigcap H_\eta(t) \subset H_\xi^S$. On the other hand, by definition $H_\eta(t)$ is orthogonal to H_ξ^S. Thus $H_\eta^S = 0$; that is, the process $\eta(t)$ is regular.

The uniqueness of the representation (2) follows from the fact that under the hypotheses of the theorem the projection of $\eta(t)$ onto H_ξ^S is equal to 0, $H_\xi^S = H_{\xi_S}^S = H_{\xi_S}$, and consequently $\xi_S(t)$ is the projection of $\eta(t)$ onto H_ξ^S. This completes the proof of the theorem.

The sequences $\eta(t)$ and $\xi_S(t)$ are called respectively the *regular* and *singular components* of the process $\xi(t)$.

Theorem 2. *The regular component $\eta(t)$ of a stationary sequence can be represented in the form*

$$\eta(t) = \sum_{n=0}^{\infty} a(n)\zeta(t-n) , \tag{3}$$

where $\zeta(t)$ (for $t = 0, \pm 1, \cdots$) is an uncorrelated sequence, where $H_\zeta(t) = H_\eta(t)$, and where $\sum_{n=0}^{\infty} | a(n) |^2 < \infty$.

Proof. We introduce the subspace $G(t) = H_\eta(t) \ominus H_\eta(t-1)$. This subspace is one-dimensional (if it were the zero space we would have $\delta^2(1) = 0$ and $\eta(t)$ would be a singular sequence). Let us choose in $G(0)$ the unit vector $\zeta(0)$. Then the sequence $\zeta(t) = S^t\zeta(0)$ is orthonormal $\big(\zeta(t) \in H_\eta(t) \ominus H_\eta(t-1)$, therefore $\zeta(t)$ is orthogonal to $H_\eta(t-1)$; also $\zeta(k) \in H_\eta(t-1)$ for $k < t\big)$, $H_\zeta(t) \subset H_\eta(t)$, and $\bigcap_t H_\zeta(t) \subset \bigcap_t H_\eta(t) = 0$. This means that the sequence $\zeta(t)$ constitutes a basis in H_ξ. Expanding $\eta(0)$ in elements of this basis, we obtain

$$\eta(0) = \sum_{n=0}^{\infty} a(n)\zeta(-n) ,$$

$$\sum_{k=0}^{\infty} | a(n) |^2 = || \eta(0) ||^2 = \mathsf{M} | \eta(0) |^2 < \infty .$$

Applying to $\eta(0)$ the operator S^t, we obtain

$$\eta(t) = \sum_{n=0}^{\infty} a(n)\zeta(t-n) .$$

The inclusion relation $H_\eta(t) \subset H_\zeta(t)$ follows immediately from (3), and the opposite inclusion relation follows from the definition of $\zeta(t)$. This completes the proof of the theorem.

REMARK 1. We may assume without loss of generality that $a(0)$ is positive.

Lemma 2. *Suppose that the spectral function $F(\omega)$ of a stationary process $\xi(t)$ is equal to $F_1(\omega) + F_2(\omega)$, where the $F_i(\omega)$ are nonnegative* *nondecreasing functions and the measures $F_i(A)$ corresponding to the functions $F_i(\omega)$ are singular. Then there exists a decomposition $\xi(t) = \xi_1(t) + \xi_2(t)$, where the processes $\xi_i(t)$ are subordinate to $\xi(t)$, are orthogonal, and have spectral functions $F_i(\omega)$ for $i = 1, 2$.*

To prove this, we represent the interval $[-\pi, \pi]$ as the union of two disjoint sets P_1 and P_2 such that $F_2(P_1) = F_1(P_2) = 0$. We set

$$\xi_1(t) = \int_{-\pi}^{\pi} e^{it\omega}\chi_{P1}(\omega)\mu(d\omega) \ ,\xi_2(t) = \int_{-\pi}^{\pi} e^{it\omega}\chi_{P2}(\omega)\mu(d\omega) \ ,$$

where μ is a stochastic spectral measure of the process $\xi(t)$ and $\chi_{P_i}(\omega)$ is the characteristic function of the set P_i. Then

$$\xi_1(t) + \xi_2(t) = \int_{-\pi}^{\pi} e^{it\omega}\mu(d\omega) = \xi(t) \ ,$$

$$\mathsf{M}(\xi_1(t_1)\overline{\xi_2(t_2)}) = \int_{-\pi}^{\pi} e^{i(t_1-t_2)\omega}\chi_{P_1}(\omega)\chi_{P_2}(\omega)dF(\omega) = 0 \ ,$$

$$\mathsf{M}(\xi_j(t_1)\overline{\xi_j(t_2)}) = \int_{-\pi}^{\pi} e^{i(t_1-t_2)\omega}\chi_{P_j}(\omega)dF(\omega) = \int_{-\pi}^{\pi} e^{i(t_1-t_2)\omega}dF_j(\omega) \ ,$$

which completes the proof of the lemma.

Theorem 3. *For a sequence $\xi(t)$ to be undetermined, it is necessary and sufficient that*

$$\int_{-\pi}^{\pi} \ln f(\omega)d\omega > -\infty \ , \tag{4}$$

where $f(u)$ is the derivative of an absolutely continuous component of the measure $F(A)$ (with respect to Lebesgue measure).

Proof of the Necessity. Let $F_r(\omega)$ and $F_s(\omega)$ denote the spectral functions of the sequences $\eta(t)$ and $\xi_S(t)$. It follows from the uncorrelatedness of $\eta(t)$ and $\xi_S(t)$ that

$$F(\omega) = F_r(\omega) + F_S(\omega) \ .$$

It follows from Theorem 2, Lemma 1, Section 6 and Theorem 1, Section 6 that $F_r(\omega)$ is absolutely continuous and that the condition

$$\int_{-\pi}^{\pi} \ln f_r(\omega)d\omega > -\infty$$

is satisfied for $f_r(\omega) = F'_w(\omega)$. Decomposing the measures $F(A)$ and $F_S(A)$ into absolutely continuous and singular components with respect to Lebesgue measure (cf. Theorem 1, Section 6, Chapter II), we obtain

$$F(A) = \int_A f(\omega)d\omega + F^*(A) \ , \qquad F_S(A) = \int_A f_S(\omega)d\omega + F_S^*(A) \ ,$$

from which it follows that $f(\omega) = f_r(\omega) + f_S(\omega)$ and

$$\int_{-\infty}^{\infty} \ln f(\omega)d\omega \geqq \int_{-\infty}^{\infty} \ln f_r(\omega)d\omega > -\infty .$$

Thus if the process is undetermined, (4) is satisfied.

Proof of the sufficiency. Let us assume the opposite. Then the decomposition

$$F(A) = F_S(A) = \int_A f_S(\omega)d\omega + F_S^*(A)$$

corresponds to the decomposition of $\xi(t)$ into uncorrelated components $\xi_1(t)$ and $\xi_2(t)$ that are subordinate to $\xi(t)$ (cf. Lemma 2). If it were true that

$$\int \ln f(\omega)d\omega = \int \ln f_S(\omega)d\omega > -\infty ,$$

then on the basis of Theorem 2 of section 6,

$$\xi_1(t) = \sum_{n=0}^{\infty} a'(n)\zeta'(t-n)$$

where $\zeta'(t)$ is an uncorrelated sequence. Since

$$H_\xi(t) \subset H_{\xi_1}(t) \oplus H_{\xi_2}(t) \quad \text{and} \quad \bigcap H_{\xi_1}(t) = 0 ,$$

we have $\bigcap_t H_\xi(t) \subset \bigcap H_{\xi_2}(t) \subset H_{\xi_2}$, which does not coincide with $H_\xi = H_{\xi_1} \oplus H_{\xi_2}$ so that the process $\xi(t)$ cannot be singular. Therefore

$$\int_{-\pi}^{\pi} \ln f(\omega)d\omega = -\infty .$$

This completes the proof of the theorem.

Let us consider the problem of prediction of undetermined processes. Using Theorems 1 and 2, let us write

$$\xi(t) = \xi_S(t) + \eta(t) , \qquad \eta(t) = \sum_{n=0}^{\infty} a_n\zeta(t-n) .$$

Since $\xi_S(t)$ is exactly determined by the past, it will be sufficient to consider the prediction of the regular component $\eta(t)$ of the process $\xi(t)$. It follows from Theorem 2 that the projection $\eta(t)$ onto $H_\eta(t-q)$ coincides with the projection onto $H_\zeta(t-q)$. Consequently,

$$\eta^{(q)}(t) = \sum_{n=q}^{\infty} a_n\zeta(t-n) . \tag{5}$$

The value of the mean-square error is determined from the equation

$$\delta^2(q) = \sum_{n=0}^{q-1} | a_n |^2 . \tag{6}$$

We shall now obtain a formula for the best prediction that does not contain the sequence $\zeta(n)$. Since $\zeta(0) \subset H_\eta$, we have

$$\zeta(0) = \int_{-\pi}^{\pi} \varphi(\omega)\nu(d\omega) , \qquad \int_{-\pi}^{\pi} |\varphi(\omega)|^2 dF(\omega) < \infty ,$$

where ν is a spectral stochastic measure of the process $\eta(t)$, $F(\omega) = |g(e^{i\omega})|^2 d\omega$, and $g(e^{i\omega}) = 1/\sqrt{2\pi} \sum_{n=0}^{\infty} \bar{a}_n e^{in\omega}$ (cf. Lemma 1, Section 6). Consequently (cf. Lemma 1),

$$\zeta(t) = S^t\zeta(0) = \int_{-\pi}^{\pi} e^{it\omega}\varphi(\omega)\nu(d\omega) .$$

To find the function $\varphi(\omega)$, let us use formula (3). We have

$$\eta(t) = \sum_{n=0}^{\infty} a_n\zeta(t-n) = \int_{-\pi}^{\pi} e^{it\omega}\varphi(\omega) \sum_{n=0}^{\infty} a_n e^{-in\omega}\nu(d\omega) .$$

Comparing this with the equation $\eta(t) = \int_{-\pi}^{\pi} e^{it\omega}\nu(d\omega)$, we obtain

$$\varphi(\omega) = \left(\sum_{n=0}^{\infty} a_n e^{-in\omega}\right)^{-1} = (\sqrt{2\pi}\, \overline{g(i\omega)})^{-1} .$$

We now have

$$\eta^{(q)}(t) = \int_{-\pi}^{\pi} \left(\sum_{n=q}^{\infty} a_n e^{-in\omega}\right)\varphi(\omega)e^{it\omega}\nu(d\omega) ,$$

so that

$$\eta^{(q)}(t) = \int_{-\pi}^{\pi} e^{it\omega}\left[1 - \frac{g_q(e^{i\omega})}{g(e^{i\omega})}\right]\nu(d\omega) , \tag{7}$$

where

$$g_q(e^{i\omega}) = \frac{1}{\sqrt{2\pi}} \sum_{n=0}^{q-1} \overline{a_n} e^{in\omega} . \tag{8}$$

We shall now demonstrate a method of determining the function $g(z) = \sum_{n=0}^{\infty} b_n z^n$, where $b_n = (1/\sqrt{2\pi})\overline{a^n}$. In doing so, we shall obtain both the general solution of the problem of the prediction of a stationary sequence and a formula for calculating the mean-square error of the prediction. The function $g(z) \in H_2$, $g(0) = a_0/\sqrt{2\pi}$ is real (cf. Remark 1). With the aid of this function, the spectral density of the sequence $\eta(t)$, $f_r(\omega) = |g(e^{i\omega})|^2$ can be factored. On the basis of Remark 3 of Section 6, if $g(z)$ has no zeros in the disk $|z| < 1$, it can be determined uniquely from $f_r(\omega)$. Therefore if the function $g(z)$, constructed in accordance with Theorem 1, does not vanish for $|z| < 1$, it is identical to the function $g(z)$ obtained in the proof of Theorem 1 of Section 6.

Lemma 3. *For $|z| < 1$, we have $g(z) \neq 0$.*

Proof. We note first of all that if $f_\eta(\omega) = |h(e^{i\omega})|^2$, where

$$h(z) = \sum_{n=0}^{\infty} c_n z^n$$

with $\sum_{n=0}^{\infty} |c_n|^2 < \infty$, then $\delta^2(1) > 2\pi |c_0|^2$. This is true because

$$\mathsf{M}\left|\eta(0) - \sum_{k=1}^{N} d_k \eta(-k)\right|^2$$
$$= \int_{-\pi}^{\pi} \left|\left(1 - \sum_{k=1}^{N} d_k e^{-ik\omega}\right)\left(\sum_{k=0}^{\infty} \bar{c}_k e^{-ik\omega}\right)\right|^2 d\omega \geqq 2\pi |c_0|^2 .$$

Since this is true for arbitrary d_k and N, we have

$$\delta^2(1) \geqq 2\pi |c_0|^2. \tag{9}$$

Let us suppose now that $g(z_0) = 0$ for $|z_0| < 1$. The function

$$g_1(z) = \frac{1}{\sqrt{2\pi}} \sum_{n=0}^{\infty} a_n z^n$$

vanishes at the point $\bar{z}_0$. Therefore $g_1(z) = (z - \bar{z}_0) \sum_{n=0}^{\infty} b'_n z^n$, where $b'_0 = -a_0/(\sqrt{2\pi}\, \bar{z}_0)$. Then,

$$|g(e^{i\omega})| = \left|\sum_{n=0}^{\infty} \frac{1}{\sqrt{2\pi}} a_n e^{-in\omega}\right|$$
$$= \left|\frac{1 - e^{-i\omega} z_0}{e^{-i\omega} - \bar{z}_0}\right| \left|e^{-i\omega} - \bar{z}_0\right| \left|\sum_{n=0}^{\infty} b'_n e^{-in\omega}\right|$$
$$= \left|\sum_{n=0}^{\infty} b''_n e^{-in\omega}\right| = \left|\sum_{n=0}^{\infty} \bar{b}''_n e^{in\omega}\right| ,$$

where $b''_0 = b'_0 = -a_0/(\sqrt{2\pi}\, \bar{z}_0)$. It follows from (9) that

$$\delta^2(1) \geqq 2\pi |\bar{b}''_0|^2 = \left|\frac{a_0}{z_0}\right|^2 ,$$

which by virtue of (6) is impossible for $|z_0| < 1$. This completes the proof of the lemma.

Corollary. *In formula* (7) *for the best prediction, the function* $g(z) \in H_2$ *is uniquely determined (under the hypothesis that* $g(0)$ *is positive) and it coincides with the function obtained in Theorem* 1 *of Section* 6.

We have solved the problem of prediction for the regular part of an undetermined sequence. Now we need to clarify the questions: How can we express the spectral density of a sequence $\eta(t)$ in terms of the spectral function of the process $\xi(t)$? What is the form of the formula for prediction for the sequence $\xi(t)$ expressed in terms of the characteristic function of $\xi(t)$?

Lemma 4. *Suppose that an undetermined process* $\xi(t)$ *is represented in the form* $\xi(t) = \eta(t) + \xi_S(t)$, *where* $\eta(t)$ *and* $\xi_S(t)$ *are uncorrelated,* $\xi_S(t)$ *is a singular process,* $\eta(t)$ *is regular, and* $F(\omega)$, $F_r(\omega)$, *and* $F_S(\omega)$ *are the spectral functions of the sequences* $\xi(t)$, $\eta(t)$, *and* $\xi_S(t)$. *Then* *the equation*

$$F(\omega) = F_r(\omega) + F_S(\omega) \tag{10}$$

is the decomposition of the function $F(\omega)$ *into an absolutely continuous component* $F_r(\omega)$ *and a singular component* $F_S(\omega)$ *with respect to Lebesgue measure.*

Proof. Formula (10) follows from the uncorrelatedness of the sequences $\eta(t)$ and $\xi_S(t)$. We introduce the spectral representation of the uncorrelated sequence $\zeta(t)$ that appears in the representation (3):

$$\zeta(t) = \int_{-\pi}^{\pi} e^{i\omega t}\zeta(d\omega) , \tag{11}$$

where $\zeta(A)$ is the stochastic measure with structural function $(1/2\pi)l(A)$ (l denotes Lebesgue measure). Substituting (11) into (3), we obtain

$$\eta(t) = \int_{-\pi}^{\pi} e^{it\omega}\sqrt{2\pi}\, \overline{g(e^{i\omega})}\zeta(d\omega) .$$

Suppose that

$$\xi_S(t) = \int_{-\pi}^{\pi} e^{it\omega}\mu_S(d\omega) \tag{12}$$

is the spectral representation of the sequence $\xi_S(t)$. Then

$$\xi(t) = \int_{-\pi}^{\pi} e^{it\omega}\mu(d\omega) = \int_{-\pi}^{\pi} e^{it\omega}(\sqrt{2\pi}\, \overline{g(e^{i\omega})}\zeta(d\omega) + \mu_S(d\omega)) . \tag{13}$$

It follows from equation (13) that

$$\int_{-\pi}^{\pi} \varphi(\omega)\mu(d\omega) = \int_{-\pi}^{\pi} \varphi(\omega)(\sqrt{2\pi}\, \overline{g(e^{iw})}\zeta(d\omega) + \mu_S(d\omega)) \tag{14}$$

for an arbitrary function $\varphi(\omega) \in L_2\{F\}$.

We can write yet another spectral representation for $\xi_S(t)$. Since $\xi_S(0) \in H_\xi$, we have

$$\xi_S(0) = \int_{-\pi}^{\pi} \varphi_S(\omega)\mu(d\omega) ,$$

so that

$$\xi_S(t) = S^t\xi_S(0) = \int_{-\pi}^{\pi} e^{i\omega t}\varphi_S(\omega)\mu(d\omega) .$$

Remembering (13) and (14), we obtain

$$\xi_S(t) = \int_{-\pi}^{\pi} e^{i\omega t}\varphi_S(\omega)[\sqrt{2\pi}\, \overline{g(e^{i\omega})}\zeta(d\omega) + \mu_S(d\omega)] .$$

Substituting the expression for $\xi_S(t)$ given by equation (12) into this equation and transposing the second term in the above integral to the left, we obtain

$$\int_{-\pi}^{\pi} e^{i\omega t}(1-\varphi_S(\omega))\mu_S(d\omega) = \int_{-\pi}^{\pi} e^{i\omega t}\varphi_S(\omega)\sqrt{2\pi}\,\overline{g(e^{i\omega})}\zeta(d\omega)\ .$$

The two sides of this equation contain elements of subspaces that are orthogonal to each other. Therefore they are both zero. Consequently,

$$\varphi_S(\omega) = 1 \quad (\mathrm{mod}\ F_S)\ ,$$
$$\varphi_S(\omega)g(e^{i\omega}) = 0 \quad (\mathrm{mod}\ l)\ .$$

Since $g(e^{i\omega})$ can be equal to zero only on a set of Lebesgue measure 0, it follows that $\varphi_S(\omega)$ is equal to 0 almost everywhere. Let S denote the set on which $\varphi_S(\omega) = 1$. Then $l(S) = 0$. Thus

$$F_S(A) = \int_A |\varphi_S(\omega)|^2\, dF(\omega) = F(A \cap S)\ ,$$

$$F_r(A) = \int_A 2\pi\, |g(e^{i\omega})|^2\, d\omega\ .$$

This completes the proof of the lemma.

Lemma 5. *Suppose that $\varphi_1(\omega)$, $\varphi_2(\omega)$, and $\varphi_3(\omega)$ are such that the three integrals*

$$\int_{-\pi}^{\pi} \varphi_1(\omega)\mu(d\omega)\ , \qquad \int_{-\pi}^{\pi} \varphi_2(\omega)\nu(d\omega)\ , \qquad \int_{-\pi}^{\pi} \varphi_3(\omega)\mu_S(d\omega)$$

are the projections of the quantities $\xi(t+q)$, $\eta(t+q)$, and $\xi_S(t+q)$ onto $H_\xi(t)$, $H_\eta(t)$, and $H_{\xi_S}(t)$ respectively. Then

$$\varphi_1(\omega) = \varphi_2(\omega) = \varphi_3(\omega) = e^{it\omega}\left(1 - \frac{g_q(e^{i\omega})}{g(e^{i\omega})}\right) \qquad (\mathrm{mod}\ F)\ .$$

Proof. In view of formula (7), it will be sufficient to prove that $\varphi_1(\omega) = \varphi_2(\omega) = \varphi_3(\omega)$. It follows from the equation

$$\xi(t+q) - \int_{-\pi}^{\pi} \varphi_1(\omega)\mu(d\omega) = \left[\eta(t+q) - \int_{-\pi}^{\pi} \varphi_1(\omega)\nu(d\omega)\right] + \left[\xi_S(t+q) - \int_{-\pi}^{\pi} \varphi_1(\omega)\mu_S(d\omega)\right] \tag{15}$$

and the orthogonality of the terms in the bracketed expressions on the right that

$$\delta_\xi^2(q) = \mathsf{M}\left|\eta(t+q) - \int_{-\pi}^{\pi} \varphi_1(\omega)\nu(d\omega)\right|^2 + \mathsf{M}\left|\xi_S(t+q) - \int_{-\pi}^{\pi} \varphi_1(\omega)\mu_S(d\omega)\right|^2 \geqq \delta_\eta^2(q)\ ,$$

with equality possible only when

$$\varphi_1(\omega) = \varphi_2(\omega) \qquad (\mathrm{mod}\ F_r)\ ,$$

$$\xi_S(t+q) = \int_{-\pi}^{\pi} \varphi_1(\omega)\mu_S(d\omega)\,, \qquad \varphi_1(\omega) = \varphi_3(\omega) \qquad (\text{mod } F_S)\,.$$

On the other hand, $\delta_\xi^2(q) = \delta_\xi^2(q)$ by virtue of the definition of $\xi_S(t)$. This completes the proof of the lemma.

The results obtained can be formulated as:

Theorem 4. *Let $\xi(t)$ denote an undetermined stationary sequence. Then the optimal prediction $\xi^{(q)}(t)$ of the quantity $\xi(t+q)$ from the results of observation of $\xi(\tau)$ for $\tau \leqq t$ is given by the formula*

$$\xi^{(q)}(t) = \int_{-\pi}^{\pi} e^{it\omega}\left[1 - \frac{\overline{g_q(e^{i\omega})}}{\overline{g(e^{i\omega})}}\right]\mu(d\omega)\,,$$

where μ is the spectral stochastic measure of the sequence $\xi(t)$,

$$g(z) = \sum_{n=0}^{\infty} b_n z^n\,, \qquad g_q(z) = \sum_{n=0}^{q-1} b_n z^n\,,$$

the function $g(z) \in H_2$ has no zeros in the disk $|z| < 1$, $g(0)$ is positive, and $|g(e^{i\omega})|^2 = f(\omega)$, where $f(\omega)$ is the derivative of the absolutely continuous component of the spectral function of the sequence $\xi(t)$. The square of the mean-square error of the prediction is equal to

$$\delta^2(q) = 2\pi \exp\left(\frac{1}{2\pi}\int_{-\pi}^{\pi} \ln f(\omega)d\omega\right)\sum_{n=0}^{q-1} |c_n|^2\,,$$

where the c_n are determined from the equation

$$\exp\left\{\frac{1}{2\pi}\sum_{n=1}^{\infty} z^n \int_{-\pi}^{\pi} e^{in\omega} \ln f(\omega)d\omega\right\} = \sum_{n=0}^{\infty} c_n z^n\,.$$

In particular,

$$\delta^2(1) = 2\pi \exp\left(\frac{1}{2\pi}\int_{-\pi}^{\pi} \ln f(\omega)d\omega\right)\,.$$

The theorem follows immediately from Lemmas 4 and 5 and formula (7) of the present section and from Theorem 1 and Remark 2 of Section 6.

2. Prediction of stationary processes with continuous time

Let $\xi(t)$ (for $-\infty < t < \infty$) denote a stationary process

$$\xi(t) = \int_{-\infty}^{\infty} e^{i\omega t}\mu(d\omega)\,,$$

where μ is an orthogonal stochastic measure on the real line $(-\infty < \omega < \infty)$ and

$$\mathsf{M}\xi(t) = 0\,, \qquad R(t) = \mathsf{M}(\xi(t+u)\overline{\xi(u)}) = \int_{-\infty}^{\infty} e^{it\omega}dF(\omega)\,,$$
$$F(+\infty) = \sigma^2\,.$$

We introduce the Hilbert space $H_\xi = H\{\xi(t), -\infty < t < \infty\}$ and its subspaces $H_\xi(t) = H\{\xi(\tau), -\infty < \tau \leqq t\}$. In H_ξ we define the group of operators representing time displacement S^τ (for $-\infty < \tau < \infty$) by setting

$$S^\tau\Big(\sum_k c_k\xi(t_k)\Big) = \sum_k c_k\xi(t_k + \tau)$$

and extending the definition of S^τ as a continuous operator to the entire space H_ξ. The S^τ constitute a group of unitary transformations of H_ξ. This group has the same properties, with obvious modifications, as the group of transformations S^n in the case of discrete time. The problem of optimal linear prediction for a process $\xi(t)$ consists in finding a random variable $\xi_T(t) \in H_\xi(t)$ such that

$$\mathsf{M}\,|\,\xi(t+T) - \xi_T(t)\,|^2 \leqq \mathsf{M}\,|\,\xi(t+T) - \eta\,|^2$$

for an arbitrary element η of $H_\xi(t)$. This problem has a unique solution: the variable $\xi_T(t)$ is the projection of $\xi(t+T)$ onto $H_\xi(t)$. We set

$$\delta_\xi(T) = \delta(T) = \sqrt{\mathsf{M}\,|\,\xi(t+T) - \xi_T(t)\,|^2}\,.$$

The quantity $\delta(T)$, the mean-square error of the prediction, is a nonincreasing function of T and $0 \leqq \delta(T) \leqq \sigma$. If $\lim_{T\to\infty}\delta(T) = \sigma$, the process is said to be *regular* (completely *undetermined*). If $\delta(T_0) = 0$ for some T_0, then $H_\xi(t) \subset H_\xi(t - T_0)$ for arbitrary t. Consequently,

$$H_\xi(t) \subset \bigcap_{k=1}^{\infty} H_\xi(t - kT_0)$$

for arbitrary t and $\delta(T) = 0$ for all $T > 0$. In this case the process is said to be *singular* (*determined*). We shall call nonsingular processes *undetermined processes.*

The proof of Theorem 1 can be carried over directly to processes with continuous time: an arbitrary stationary process admits a decomposition of the form

$$\xi(t) = \eta(t) + \xi_S(t)\,,$$

where $\eta(t)$ is a regular and $\xi_S(t)$ is a singular process and the two are uncorrelated and subordinate to $\xi(t)$.

The analogue of Theorem 2 is:

Theorem 5. *A regular stationary process $\eta(t)$ can be represented* *in the form*

$$\eta(t) = \int_{-\infty}^{t} a(t-\tau)d\zeta(\tau) , \tag{16}$$

where $\zeta(\tau)$ *is an integrated white noise,* $H_\zeta(t) = H_\xi(t)$, *and*

$$\int_{-\infty}^{\infty} | a(t) |^2 dt < \infty .$$

Here $H_\zeta(t)$ *denotes the closure in* L_2 *of the set of random variables of the form* $\sum c_k[\zeta(t_k) - \zeta(t_{k-1})]$, $t_k < t$.

Proof of this theorem can be obtained from the preceding results for stationary sequences by means of a passage to the limit applied to the sequence $\eta_n(k) = \eta(k/n)$, $k = 0, \pm 1, \pm 2, \cdots$.

The results obtained for the prediction of stationary sequences can now be carried over to processes with continuous time with certain modifications in the wording and the proofs. Here, we need to use the spectral representation of stationary processes with continuous time and to refer to the results of Lemma 2 and Theorem 3 of Section 6.

Thus we obtain the following theorem and formulas:

Theorem 6. *For a process* $\xi(t)$ *to be undetermined, it is necessary and sufficient that*

$$\int_{-\infty}^{\infty} \frac{\ln f(\omega)d\omega}{1+\omega^2} > -\infty ,$$

where $f(\omega)$ *is the derivative of the absolutely continuous component of the spectral measure* F *of the process* $\xi(t)$.

If $\xi(t) = \eta(t) + \xi_S(t)$ is the decomposition of the process $\xi(t)$ into regular and singular components and if in accordance with Theorem 5,

$$\eta(t) = \int_0^{\infty} a(\tau)d\zeta(t-\tau) ,$$

then

$$\xi_T(t) = \int_T^{\infty} a(\tau)d\zeta(t-\tau) + \xi_S(t+T) ,$$

and the optimum mean-square error of the prediction is determined from the relation

$$\delta^2(T) = \int_0^T | a(\tau) |^2 d\tau .$$

Another expression for the optimum prediction is

$$\xi_T(t) = \int_{-\infty}^{\infty} e^{it\omega}\left[1 - \frac{h_T(i\omega)}{h(i\omega)}\right]\mu(d\omega) ,$$

where μ is the stochastic spectral measure of the process $\xi(t)$,

$$h(i\omega) = \frac{1}{\sqrt{2\pi}} \int_0^\infty a(\tau) e^{-i\omega\tau} d\tau ,$$

and

$$h_T(i\omega) = \frac{1}{\sqrt{2\pi}} \int_0^T a(\tau) e^{-i\omega\tau} d\tau .$$

The function $h(i\omega)$ is determined from the spectral density of $f(\omega)$ in accordance with formula (22) of Section 6.

VI

PROCESSES WITH INDEPENDENT INCREMENTS

A process $\xi(t)$ defined on $\mathfrak{T}$ into $R^{(m)}$ is called a *process with independent increments* if for all $t_0 < t_1 < \cdots t_k$ in $\mathfrak{T}$, the quantities $\xi(t_0), \xi(t_1) - \xi(t_0), \cdots, \xi(t_k) - \xi(t_{k-1})$ are independent. The simplest familiar examples of random processes with independent increments are Wiener and Poisson processes. Every stochastically continuous process with independent increments can be represented as a sum of independent Poisson and Wiener processes. (In the general case, this sum contains a continuum of Poisson "addends," so that we should speak of the sum of a Wiener process and an integral of Poisson processes.) Such a representation of the process is the basic topic of the present chapter, and by means of it we shall study certain important properties of the sample functions.

1. MEASURES CONSTRUCTED FROM THE JUMPS OF A PROCESS

Let $\xi(t)$ denote a separable stochastically continuous process with independent increments that is defined for $t \in [0, T]$ with range in some finite-dimensional Euclidean space X. On the basis of Corollary 2, Section 4, Chapter IV, $\xi(t)$ fails with probability 1 to have discontinuities of the second kind.

For every $\varepsilon > 0$, there will be with probability 1 only finitely many points t for which $|\xi(t+0) - \xi(t-0)| > \varepsilon$. Let X_ε denote the set of all x for which $|x| > \varepsilon$ and let $\mathfrak{B}_\varepsilon$ denote the σ-algebra of Borel sets contained in X_ε. It follows that for every $A \in \mathfrak{B}_\varepsilon$, the number of points t in $[0, T]$ for which $\xi(t+0) - \xi(t-0) \in A$ is finite with probability 1. Let $\nu(t, A)$ denote the number of points $s \in [0, t)$ for which $\xi(s+0) - \xi(s-0) \in A$. The process $\nu(t, A)$ has independent increments, since $\nu(t_2, A) - \nu(t_1, A)$, where $t_1 \leqq t_2$, is completely determined by the increments $\xi(s) - \xi(t_1)$ for $s \in [t_1, t_2]$. Hence the increments of $\nu(t, A)$ on the disjoint intervals are expressed in terms of increments of $\xi(t)$ on disjoint intervals. Furthermore $\nu(t, A)$ is a stochastically continuous process. (If

$$\nu(t', A) - \nu(t, A)$$

does not approach 0 in probability as $t' - t$ approaches 0, we have $\mathsf{P}\{|\xi(t') - \xi(t)| > \varepsilon\} \nrightarrow 0$, and this is impossible by virtue of the stochastic continuity of $\xi(t)$.)

Theorem 1. *For every* $A \in \mathfrak{B}_\varepsilon$ *and every* $r > 0$, *we have*

$$\mathsf{M}\nu(t, A)^r < \infty .$$

Proof. Suppose that $0 = t_{0n} < t_{1n} < \cdots < t_{nn} = t$ and

$$\lim_{n\to\infty} \max_k (t_{k+1,n} - t_{kn}) = 0 .$$

If $\psi_1(t) = 1$ for $t \geqq 1$ and $\psi_1(t) = 0$ for $t < 1$, the monotonicity of $\nu(t, A)$ with respect to t and the fact that all the jumps of $\nu(t, A)$ are equal to 1 imply that with probability 1,

$$\nu(t, A) = \lim_{n\to\infty} \sum_{k=0}^{n-1} \psi_1(\nu(t_{k+1,n}, A) - \nu(t_{kn}, A)) .$$

Let us set $\eta_{nk} = \psi_1(\nu(t_{kn}, A) - \nu(t_{k-1,n}, A))$. Then

$$\begin{aligned}\mathsf{P}\left\{\sum_{k=1}^{n} \eta_{nk} > l\right\} &= \sum_{k=1}^{n} \mathsf{P}\left\{\sum_{j=1}^{k-1} \eta_{nj} = 0,\ \sum_{j=1}^{k} \eta_{nj} = 1\right\}\mathsf{P}\left\{\sum_{j=k+1}^{n} \eta_{nj} > l - 1\right\} \\ &\leqq \mathsf{P}\left\{\sum_{k=1}^{n} \eta_{nk} > l - 1\right\}\sum_{k=1}^{n} \mathsf{P}\left\{\sum_{j=1}^{k-1} \eta_{nj} = 0,\ \sum_{j=1}^{k} \eta_{nj} = 1\right\} \\ &= \mathsf{P}\left\{\sum_{j=1}^{n} \eta_{nj} > 0\right\}\mathsf{P}\left\{\sum_{j=1}^{n} \eta_{nj} > l - 1\right\} .\end{aligned}$$

Therefore

$$\mathsf{P}\left\{\sum_{k=1}^{n} \eta_{nk} > l\right\} \leqq \left(\mathsf{P}\left\{\sum_{j=1}^{n} \eta_{nj} > 0\right\}\right)^{l+1} .$$

Taking the limit as $n \longrightarrow \infty$, we see that

$$\mathsf{P}\{\nu(t, A) > l\} \leqq (\mathsf{P}\{\nu(t, A) > 0\})^{l+1} . \tag{1}$$

Observe that $\mathsf{P}\{\nu(t, A) > 0\} < 1$. If this were not true, then we would obtain

$$\mathsf{P}\{\nu(t, A) = 0\} = \prod_{k=1}^{n} \mathsf{P}\{\nu(t_k, A) - \nu(t_{k-1}, A) = 0\} = 0$$

for $0 = t_0 < t_1 < \cdots t_n = t$ and hence there would exist t' and t'' arbitrarily close to each other such that $\mathsf{P}\{\nu(t'', A) - \nu(t', A) \geqq 1\} = 1$, which would contradict the stochastic continuity of $\nu(t, A)$. The existence of $\mathsf{M}\nu(t, A)^r$ for $r > 0$ follows from the inequality

$$\begin{aligned}\mathsf{M}\nu(t, A)^r &\leqq \sum_{k=0}^{\infty} \mathsf{P}\{\nu(t, A) > k\}(k + 1)^r \\ &\leqq \sum_{k=0}^{\infty} (\mathsf{P}\{\nu(t, A) > 0\})^k(k + 1)^r < \infty .\end{aligned}$$

(The series $\sum_{k=0}^{\infty}(k+1)^r x^k$ converges for arbitrary $|x|<1$.) This completes the proof of the theorem.

Corollary. *Define* $\Pi(t, A) = \mathsf{M}\nu(t, A)$. *Then the set function* $\Pi(t, A)$ *is, for fixed* t, *a measure on* $\mathfrak{B}_\varepsilon$.

Proof. If $A = \bigcup_{k=1}^{\infty} A_k$ where the A_k are pairwise disjoint, then $\nu(t, A) = \sum_{k=1}^{\infty} \nu(t, A_k)$, and consequently $\mathsf{M}\nu(t, A) = \sum_{k=1}^{\infty} \mathsf{M}\nu(t, A_k)$, in view of the fact that $0 \leqq \sum_{k=1}^{n} \nu(t, A_k) \leqq \nu(t, A)$.

To study the properties of the quantity $\nu(t, A)$, we find it useful to consider the process $\xi(t, A)$ defined by

$$\xi(t, A) = \sum_{s<t} [\xi(s+0) - \xi(s-0)]\chi_A(\xi(s+0) - \xi(s-0)) , \qquad (2)$$

where $\chi_A(x)$ is the characteristic function of the set A. In other words $\xi(t, A)$ is the sum of the jumps (of the process $\xi(t)$) that occur up to the instant t and that fall in the set A. If $A \in \mathfrak{B}_\varepsilon$, then the number of such jumps is with probability 1 finite, so that $\xi(t, A)$ is meaningful. The stochastic continuity of $\nu(t, A)$ implies the stochastic continuity of $\xi(t, A)$. Furthermore $\xi(t, A)$ is a process with independent increments. The most important property of processes of the type $\xi(t, A)$, namely the independence of the processes $\xi(t, A_1)$, $\cdots$, $\xi(t, A_k)$ for pairwise disjoint sets $A_1, A_2, \cdots, A_k$, follows from:

Theorem 2. *For every* $A \in \mathfrak{B}_\varepsilon$, *the processes* $\xi(t, A)$ *and*

$$\xi(t) - \xi(t, A)$$

are independent processes with independent increments.

Proof. That $\xi(t) - \xi(t, A)$ is a stochastically continuous process follows from the stochastic continuity of the processes $\xi(t)$ and $\xi(t, A)$. The considerations that we took into account in connection with $\nu(t, A)$ enable us to assert that the process

$$[\xi(t, A);\ \xi(t) - \xi(t, A)]$$

in $X \times X$ is also a process with independent increments. To prove the independence of the processes $\xi(t, A)$ and $\xi(t) - \xi(t, A)$, it will therefore be sufficient to show that

$$\mathsf{M}e^{i(z_1, \xi(t,A) - \xi(s,A)) + i(z_2, \xi(t) - \xi(t,A) - \xi(s) + \xi(s,A))}$$
$$= \mathsf{M}e^{i(z_1, \xi(t,A) - \xi(s,A))}\mathsf{M}e^{i(z_2, \xi(t) - \xi(t,A) - \xi(s) + \xi(s,A))} . \qquad (3)$$

for z_1 and z_2 in X and $s < t$. Specifically, it follows from (3) and the independence of the increments of the process

$$[\xi(t, A);\ \xi(t) - \xi(t, A)]$$

that for arbitrary $0 < t_0 < \cdots < t_n = T$ and $z_1^{(j)}, z_2^{(k)} \in X$ we have

$$\mathsf{M} \exp\left\{i \sum_{k=1}^{n} [(z_1^{(k)}, \xi(t_k, A) - \xi(t_{k-1}, A)) + (z_2^{(k)}, \xi(t_k) - \xi(t_k, A) - \xi(t_{k-1}) + \xi(t_{k-1}, A))]\right\}$$
$$= \mathsf{M} \exp\left\{i \sum_{k=1}^{n} (z_1^{(k)}, \xi(t_k, A) - \xi(t_{k-1}, A))\right\} \times \mathsf{M} \exp\left\{i \sum_{k=1}^{n} (z_2^{(k)}, \xi(t_k) - \xi(t_k, A) - \xi(t_{k-1}) + \xi(t_{k-1}, A))\right\} .$$

and this equation means that the processes $\xi(t, A)$ and $\xi(t) - \xi(t, A)$ are independent. Let us prove equation (3) first for the case

$$\Pi(t, \Gamma_A) = 0 ,$$

where Γ_A is the boundary of the set A. We note that in this case the process $\xi(t)$ fails with probability 1 to have jumps on Γ_A. Suppose that $s = t_{n0} < t_{n1} < \cdots < t_{nn} = t$ and

$$\lim_{n\to\infty} \max_k (t_{n,k+1} - t_{nk}) = 0 .$$

Then if $\chi_A(x)$ is the characteristic function of the set A, the relationship

$$\xi(t, A) - \xi(s, A) = \lim_{n\to\infty} \sum_{k=1}^{\infty} \chi_A(\xi(t_{nk}) - \xi(t_{n,k-1}))[\xi(t_{nk}) - \xi(t_{n,k-1})] \tag{4}$$

holds with probability 1. We set

$$\xi_{nk} = \chi_A(\xi(t_{nk}) - \xi(t_{n,k-1}))[\xi(t_{nk}) - \xi(t_{n,k-1})] ,$$
$$\eta_{nk} = \xi(t_{nk}) - \xi(t_{n,k-1}) - \xi_{nk} .$$

To prove (3) it will be sufficient, in view of equation (4), to show that

$$\lim_{n\to\infty} \left| \mathsf{M} e^{i\sum_{k=1}^{n}(z_1,\xi_{nk}) + i\sum_{k=1}^{n}(z_2,\eta_{nk})} - \mathsf{M} e^{i\sum_{k=1}^{n}(z_1,\xi_{nk})} \mathsf{M} e^{i\sum_{k=1}^{n}(z_2,\eta_{nk})} \right| = 0 . \tag{5}$$

Using the independence of the pairs (ξ_{nk}, η_{nk}) and the inequality

$$\left| \prod_{k=1}^{n} a_k - \prod_{k=1}^{n} b_k \right| \leqq \sum_{k=1}^{n} \left| a_k - b_k \right| ,$$

for $|a_k| \leqq 1$ and $|b_k| \leqq 1$, we see that

$$\left| \mathsf{M} e^{i\left(\sum_{k=1}^{n}(z_1,\xi_{nk}) + \sum_{k=1}^{n}(z_2,\eta_{nk})\right)} - \mathsf{M} e^{i\sum_{k=1}^{n}(z_1,\xi_{nk})} \mathsf{M} e^{i\sum_{k=1}^{n}(z_2,\eta_{nk})} \right|$$
$$= \left| \prod_{k=1}^{n} \mathsf{M} e^{i(z_1,\xi_{nk}) + i(z_2,\eta_{nk})} - \prod_{k=1}^{n} \mathsf{M} e^{i(z_1,\xi_{nk})} \mathsf{M} e^{i(z_2,\eta_{nk})} \right|$$
$$\leqq \sum_{k=1}^{n} \left| \mathsf{M} e^{i(z_1,\xi_{nk}) + i(z_2,\eta_{nk})} - \mathsf{M} e^{i(z_1,\xi_{nk})} \mathsf{M} e^{i(z_2,\eta_{nk})} \right| .$$

Since $(z_1, \xi_{nk})(z_2, \eta_{nk}) = 0$ (at least one of these must be zero), it follows that

$$e^{i(z_1,\xi_{nk})+i(z_2,\eta_{nk})} = 1 + \sum_{m=1}^{\infty} \frac{(i(z_1, \xi_{nk}) + i(z_2, \eta_{nk}))^m}{m!}$$
$$= 1 + \sum_{m=1}^{\infty}\left[\frac{(i(z_1, \xi_{nk}))^m}{m!} + \frac{(i(z_2, \eta_{nk}))^m}{m!}\right]$$
$$= e^{i(z_1,\xi_{nk})} + e^{i(z_2,\eta_{nk})} - 1 .$$

Therefore,

$$|\mathsf{M}e^{i(z_1,\xi_{nk})+i(z_2,\eta_{nk})} - \mathsf{M}e^{i(z_1,\xi_{nk})}\mathsf{M}e^{i(z_2,\eta_{nk})}|$$
$$= |\mathsf{M}e^{i(z_1,\xi_{nk})} - 1|\,|\mathsf{M}e^{i(z_2,\eta_{nk})} - 1| .$$

But

$$|\mathsf{M}e^{i(z_1,\xi_{nk})} - 1| \leqq \mathsf{M}|e^{i(z_1,\xi_{nk})} - 1| \leqq 2\mathsf{P}\{\xi_{nk} > 0\}$$
$$= 2\mathsf{P}\{\chi_A(\xi(t_{nk}) - \xi(t_{n,k-1})) > 0\}$$

and for every $\rho > 0$,

$$|\mathsf{M}e^{i(z_2,\eta_{nk})} - 1| \leqq \sup_{|x|\leqq\rho}|1 - e^{i(z_2,x)}| + 2\mathsf{P}\{|\eta_{nk}| > \rho\} .$$

Consequently,

$$\overline{\lim_{n\to\infty}}\left|\mathsf{M}e^{i\left(z_1,\sum_{k=1}^n \xi_{nk}\right)+i\left(z_2,\sum_{k=1}^n \eta_{nk}\right)} - \mathsf{M}e^{i\left(z_1,\sum_{k=1}^n \xi_{nk}\right)}\mathsf{M}e^{i\left(z_2,\sum_{k=1}^n \eta_{nk}\right)}\right|$$
$$\leqq \overline{\lim_{n\to\infty}}\,(\sup_{|x|\leqq\rho}|e^{i(z_2,x)} - 1| + 2\sup_k \mathsf{P}\{|\eta_{nk}| > \rho\})$$
$$\times \overline{\lim_{n\to\infty}}\, 2\sum_{k=1}^n \mathsf{P}\{\chi_A(\xi(t_{nk}) - \xi(t_{n,k-1})) > 0\}$$
$$= 2\,\overline{\lim_{n\to\infty}}\,(\sup_{|x|\leqq\rho}|e^{i(z_2,x)} - 1| + 2\sup_k \mathsf{P}\{|\eta_{nk}| > \rho\})$$
$$\times \overline{\lim_{n\to\infty}}\,\mathsf{M}\sum_{k=1}^n \chi_A(\xi(t_{nk}) - \xi(t_{n,k-1})) .$$

Since $\mathsf{P}\{|\eta_{nk}| > \rho\} \leqq \mathsf{P}\{|\xi(t_{nk}) - \xi(t_{n,k-1})| > \rho\}$ for $\rho < \varepsilon$ (where $A \in \mathfrak{B}_\varepsilon$), the uniform stochastic continuity of $\xi(t)$ implies that

$$\overline{\lim_{n\to\infty}}\,\mathsf{P}\{|\eta_{nk}| > \rho\} = 0 .$$

Let us show that

$$\overline{\lim_{n\to\infty}}\,\mathsf{M}\sum_{k=1}^n \chi_A(\xi(t_{nk}) - \xi(t_{n,k-1})) \leqq C < \infty . \tag{6}$$

Just as in the proof of Theorem 1, we can show that

$$\mathsf{P}\{\sum_{k=1}^n \chi_A(\xi(t_{nk}) - \xi(t_{n,k-1})) > l\}$$
$$\leqq (\mathsf{P}\{\sum_{k=1}^n \chi_A(\xi(t_{nk}) - \xi(t_{n,k-1})) > 0\})^l . \tag{7}$$

Furthermore it is easy to see that in the present case,

$$\lim_{n\to\infty} \sum_{k=1}^{n} \chi_A(\xi(t_{nk}) - \xi(t_{n,k-1})) = \nu(t, A) - \nu((s, A)$$

with probability 1. Therefore

$$\overline{\lim_{n\to\infty}}\, \mathsf{P}\left\{\sum_{k=1}^{n} \chi_A(\xi(t_{nk}) - \xi(t_{n,k-1})) > 0\right\} < 1 \tag{8}$$

since $\mathsf{P}\{\nu(t, A) - \nu(s, A) > 0\} < 1$. (This inequality was established in the proof of Theorem 1.) From inequalities (7) and (8) we get (6). Thus,

$$\overline{\lim_{k=\infty}} \left| \mathsf{M}e^{i\left(z_1, \sum_{k=1}^{n} \xi_{nk}\right) + i\left(z_2, \sum_{k=1}^{n} \eta_{nk}\right)} - \mathsf{M}e^{i\left(z_1, \sum_{k=1}^{n} \xi_{nk}\right)} \mathsf{M}e^{i\left(z_2, \sum_{k=1}^{n} \eta_{nk}\right)} \right|$$
$$\leqq 2C \sup_{|x| \leqq \rho} | e^{i(z_2, x)} - 1 | . \tag{9}$$

Taking the limit as $\rho \to 0$, we obtain (5), which completes the proof of the theorem for the case $\Pi(t, \Gamma_A) = 0$.

Turning to the general case, let us note as a preliminary that the collection $\mathfrak{A}$ of sets for which the theorem is valid constitutes a monotonic class (cf. Section 1, Chapter II), since for an arbitrary sequence of sets $A_n \in \mathfrak{B}_\varepsilon$, the relations

$$\xi(t, \bigcup_n A_n) = \lim_{n\to\infty} \xi(t, \bigcup_{k=1}^{n} A_k) ,$$

$$\xi(t, \bigcap_n A_n) = \lim_{n\to\infty} \xi(t, \bigcap_{k=1}^{n} A_k)$$

are valid with probability 1, and the operation of passing to the limit does not destroy the independence of the random variables. One can easily show that in the case in which ε is a positive number such that $\Pi(t, \Gamma_{X_\varepsilon}) = 0$, the sets A in $\mathfrak{B}_\varepsilon$ for which $\Pi(t, \Gamma_A) = 0$ constitute an algebra of sets. But every monotonic algebra is a σ-algebra (cf. Theorem 3, Section 1, Chapter II), so that $\mathfrak{A}$ is a σ-algebra. Finally we note that at each point there are spheres of arbitrarily small radius that belong to $\mathfrak{A}$. (Since the boundaries of the spheres $S_\rho(x)$ with the same center x but different radii ρ have no common points, the inequality $\Pi(t, \Gamma_{S_\rho(x)}) > 0$ holds for only a countable set of values of ρ.) Therefore $\mathfrak{A}$ contains all open sets belonging to $\mathfrak{B}_\varepsilon$, so that $\mathfrak{A}$ contains $\mathfrak{B}_\varepsilon$. This completes the proof of the theorem.

Corollary 1. *If $A_1, A_2, \cdots, A_k$ are pairwise disjoint sets belonging to $\mathfrak{B}_\varepsilon$ for some $\varepsilon > 0$, then the processes $\xi(t, A_1), \xi(t, A_2), \cdots, \xi(t, A_k)$ and $\xi(t) - \sum_{j=1}^{k} \xi(t, A_j)$ are mutually independent.*

Proof. The process $\xi(t) - \sum_{j=1}^{k} \xi(t, A_j) = \xi(t) - \xi(t, \cup_{j=1}^{k} A_j)$ is independent of the process $\xi(t, \cup_{j=1}^{k} A_j)$. The processes $\xi(t, A_j)$ are completely determined by the process $\xi(t, \cup_{j=1}^{k} A_j)$. Therefore they are all independent of the process $\xi(t) - \sum_{j=1}^{k} \xi(t, A_j)$. Analogously, the collection of processes $\xi(t, A_j)$, for $j \neq i$, are completely determined by the process $\xi(t) - \xi(t, A_i)$, which is independent of $\xi(t, A_i)$, so that this collection is also independent of the process $\xi(t, A_i)$. Thus each of the processes

$$\xi(t, A_j),\ j = 1, 2, \cdots, k;\ \xi(t) - \sum_{j=1}^{k} \xi(t, A_j)$$

is independent of all the others. Our assertion now follows.

Corollary 2. *For pairwise disjoint sets $A_1, A_2, \cdots, A_k$ in $\mathfrak{B}_\varepsilon$, the processes $\nu(t, A_1), \cdots, \nu(t, A_k)$ are mutually independent.*

This follows from the preceding assertion since the process $\nu(t, A)$ is completely determined by the process $\xi(t, A)$.

Let us now study the distribution of the quantity $\nu(t, A)$.

Theorem 3. *The process $\nu(t, A)$ is a Poisson process, that is, a process with independent increments with Poisson distributions: For $s < t$,*

$$\mathsf{P}\{\nu(t, A) - \nu(s, A) = l\} = \frac{[\Pi(t, A) - \Pi(s, A)]^l}{l!} \exp\{-[\Pi(t, A) - \Pi(s, A)]\} .$$

Proof. Suppose that $s = t_{n0} < t_{n1} < \cdots < t_{nn} = t$ and

$$\lim_{n\to\infty} \max_k (t_{nk} - t_{n,k-1}) = 0 .$$

Then

$$\mathsf{P}\{\nu(t, A) - \nu(s, A) = l\} = \lim_{n\to\infty} \sum_{i=0}^{n-1} \mathsf{P}\{\nu(t_{ni}, A) - \nu(s, A) = 0\} \times P\{\nu(t_{n,i+1}, A) - \nu(t_{ni}, A) = 1\}\mathsf{P}\{\nu(t, A) - \nu(t_{n,i+1}, A) = l - 1\} .$$

We note that

$$\sum_{i=1}^{n} \mathsf{P}\{\nu(t_{n,i+1}, A) - \nu(t_{ni}, A) = 1\} \leqq \sum_{i=1}^{n} \mathsf{M}[\nu(t_{n,i+1}, A) - \nu(t_{ni}, A)] = \mathsf{M}[\nu(t, A) - \nu(s, A)] < \infty .$$

Furthermore it follows from the stochastic continuity of the process $\nu(t, A)$ that

$$\lim_{n\to\infty} \max_i \mathsf{P}\{\nu(t_{n,i+1}, A) - \nu(t_{ni}, A) \geqq 1\} = 0 .$$

It follows from inequality (1) that

$$\mathsf{P}\{\nu(t_{n,i+1}, A) - \nu(t_{ni}, A) > 1\} \leqq (\mathsf{P}\{\nu(t_{n,i+1}, A) - \nu(t_{ni}, A) > 0\})^2 .$$

Therefore

$$\sum_{i=0}^{n-1} (\mathsf{P}\{\nu(t_{n,i+1}, A) - \nu(t_{ni}, A) > 0\} - \mathsf{P}\{\nu(t_{n,i+1}, A) - \nu(t_{ni}, A) = 1\})$$
$$= \sum_{i=1}^{n-1} \mathsf{P}\{\nu(t_{n,i+1}, A) - \nu(t_{ni}, A) > 1\}$$
$$\leqq \max_i \mathsf{P}\{\nu(t_{n,i+1}, A) - \nu(t_{ni}, A) > 0\} \times \sum_{i=0}^{n-1} \mathsf{P}\{\nu(t_{n,i+1}, A) - \nu(t_{ni}, A) > 0\} ,$$

and the last expression approaches 0 as $n \to \infty$. This means that

$$\mathsf{P}\{\nu(t, A) - \nu(s, A) = l\}$$
$$= \lim_{n\to\infty} \sum_{i=0}^{n-1} \mathsf{P}\{\nu(t_{ni}, A) - \nu(s, A) = 0\}\mathsf{P}\{\nu(t_{n,i+1}, A) - \nu(t_{ni}, A) > 0\} \times \mathsf{P}\{\nu(t, A) - \nu(t_{n,i+1}, A) = l - 1\}$$
$$= \lim_{n\to\infty} \sum_{i=0}^{n-1} \mathsf{P}\{\nu(t, A) - \nu(t_{n,i+1}, A) = l - 1\} \times [\mathsf{P}\{\nu(t_{n,i+1}, A) - \nu(s, A) > 0\} - \mathsf{P}\{\nu(t_{ni}, A) - \nu(s, A) > 0\}] .$$

Since $\mathsf{P}\{\nu(t, A) - \nu(\tau, A) = l - 1\}$ is a continuous function of τ, by virtue of the stochastic continuity of $\nu(\tau, A)$, it follows that

$$\mathsf{P}\{\nu(t, A) - \nu(s, A) = l\} = \int_s^t \mathsf{P}\{\nu(t, A) - \nu(\tau, A) = l - 1\}d_\tau \mathsf{P}\{\nu(\tau, A) - \nu(s, A) > 0\} . \quad (10)$$

For $\tau_1 < \tau$, we have

$$\mathsf{P}\{\nu(\tau, A) = 0\} = \mathsf{P}\{\nu(\tau_1, A) = 0\}\mathsf{P}\{\nu(\tau, A) - \nu(\tau_1, A) = 0\} . \quad (11)$$

As shown in the proof of Theorem 1, $\mathsf{P}\{\nu(\tau, A) = 0\} > 0$, and we may set $\mathsf{P}\{\nu(\tau, A) = 0\} = e^{\lambda(\tau)}$. It follows from (11) that

$$\mathsf{P}\{\nu(\tau, A) - (\tau_1, A) = 0\} = e^{\lambda(\tau)-\lambda(\tau_1)} .$$

The function $\lambda(\tau)$ is nondecreasing and continuous, since the process $\nu(\tau, A)$ is stochastically continuous. Thus,

$$\mathsf{P}\{\nu(t, A) - \nu(s, A) = 1\} = \int_s^t e^{\lambda(t)-\lambda(\tau)}d(1 - e^{\lambda(\tau)-\lambda(s)})$$
$$= e^{\lambda(t)-\lambda(s)}\int_s^t e^{-\lambda(\tau)}d(- e^{\lambda(\tau)})$$
$$= e^{\lambda(t)-\lambda(s)}\int_s^t (- d\lambda(\tau)) = [\lambda(s) - \lambda(t)]e^{\lambda(t)-\lambda(s)} .$$

Substituting the value obtained for $\mathsf{P}\{\nu(t, A) - \nu(s, A) = 1\}$ into formula (10), we obtain

$$\mathsf{P}\{\nu(t, A) - \nu(s, A) = 2\} = \frac{[\lambda(s) - \lambda(t)]^2}{2} e^{\lambda(t)-\lambda(s)} .$$

By induction, it is easy to verify that for all $l > 0$,

$$\mathsf{P}\{\nu(t, A) - \nu(s, A) = l\} = \frac{[\lambda(s) - \lambda(t)]^l}{l!} e^{\lambda(t)-\lambda(s)} . \tag{12}$$

Thus $\nu(t, A)$ is a Poisson process whose finite-dimensional distributions are determined by formula (12) and the relation $\nu(0, A) = 0$. In particular, $\mathsf{M}\nu(t, A) = -\lambda(t)$. On the other hand, $\mathsf{M}\nu(t, A) = \Pi(t, A)$ by definition. If we replace $\lambda(s) - \lambda(t)$ with the expression

$$\Pi(t, A) - \Pi(s, A)$$

in equation (12), the proof is complete.

REMARK 1. The characteristic function of the process $\nu(t, A)$ is given by the formula

$$\mathsf{M}e^{i\lambda\nu(t,A)} = \exp\{(e^{i\lambda} - 1)\Pi(t, A)\} . \tag{13}$$

REMARK 2. Let us consider the set $[0, T) \times X_\varepsilon$. If B^* is a Borel set belonging to this Cartesian product, we can consider the random variable $\nu^*(B^*)$ that is equal to the number of points t for which the pair

$$(t; \xi(t + 0) - \xi(t - 0))$$

belongs to B^*. One can easily show that ν^* is a random measure on the σ-algebra $\mathfrak{B}_\varepsilon^*$ of all Borel subsets of $[0, T] \times X_\varepsilon$. Let us set

$$\Pi^*(B^*) = \mathsf{M}\nu^*(B^*) .$$

The quantity $\Pi^*(B^*)$ is a finite measure $\mathfrak{B}_\varepsilon^*$. The connection between the measures $\nu(t, A)$ and the measure ν^* is obvious:

$$\nu^*([t_1, t_2] \times A) = \nu(t_2, A) - \nu(t_1, A) . \tag{14}$$

Analogously,

$$\Pi^*([t_1, t_2] \times A) = \Pi(t_2, A) - \Pi(t_1, A) . \tag{15}$$

Theorem 4. *The measure ν^* is a Poisson random measure with independent values. The characteristic function of the quantity $\nu^*(B^*)$ is given by the formula*

$$\mathsf{M}e^{i\lambda\nu^*(B^*)} = \exp\{e^{i\lambda} - 1)\Pi^*(B^*)\} . \tag{16}$$

Proof. Let $\mathfrak{A}_0$ denote the algebra of sets that is generated by sets of the form $[t_1, t_2] \times A$, where $[t_1, t_2] \subset [0, T]$ and $A \in \mathfrak{B}_\varepsilon^*$. If

$A_1^*, \cdots, A_k^*$ are disjoint sets belonging to $\mathfrak{A}_0$, there exist disjoint sets $\Delta_i^* = [t_1^{(i)}, t_2^{(i)}] \times A_i$ for $i = 1, \cdots, N$ such that the A_j^* are the sums Δ_i^*. Here the sets Δ_i^* can be chosen so that for distinct i both the intervals $[t_1^{(i)}, t_2^{(i)}]$ and the sets A_i are either disjoint or coincident. In this case independence of the $\nu^*(\Delta_i^*)$ is a consequence of the independence of the $\nu(t, A_i)$ for distinct A_i and of the independence of the increments of $\nu(t, A_i)$. The independence of the $\nu^*(\Delta_i^*)$ implies independence of the $\nu^*(A_j^*)$. These variables have a Poisson distribution since they are the sums of independent Poisson variables. To complete the proof, we need only note that $\sigma(\mathfrak{A}_0) = \mathfrak{B}_\varepsilon^*$.

2. CONTINUOUS COMPONENTS OF A PROCESS WITH INDEPENDENT INCREMENTS

Let $\xi(t)$ denote a separable stochastically continuous random process with independent increments and range X. Take $x_1, x_2, \cdots, x_r$ to be an orthonormal basis for X. Denote by $\xi_\varepsilon(t)$ the process obtained from $\xi(t)$ after discarding the jumps exceeding ε in absolute value: $\xi_\varepsilon(t) = \xi(t) - \xi(t, X_\varepsilon)$, where X_ε is the set of all x such that $|x| > \varepsilon$. The process $\xi_\varepsilon(t)$ is stochastically continuous, has independent increments, and possesses no jumps exceeding ε. It is natural to expect that $\xi_\varepsilon(t)$ converges to some continuous process with independent increments as $\varepsilon \to 0$. This turns out to be the case if we subtract from $\xi_\varepsilon(t)$ specially chosen continuous nonrandom functions. To prove this fact, we need:

Lemma. *For every* $\varepsilon > 0$, $\mathsf{M}|\xi_\varepsilon(t)|^2 < \infty$.

Proof. Suppose that $0 = t_{n0} < \cdots < t_{nn} = t$ and

$$\lim_{n\to\infty} \max_k (t_{nk} - t_{n,k-1}) = 0 .$$

Let us set $\xi_{nk} = \psi_{2\varepsilon}(\xi_\varepsilon(t_{nk}) - \xi_\varepsilon(t_{n,k-1}))$, where $\psi_\alpha(x) = x$ for $|x| \leqq \alpha$ and $\psi_\alpha(x) = 0$ for $|x| > \alpha$. It is easy to see that with probability 1

$$\xi_\varepsilon(t) = \lim_{n\to\infty} \sum_{k=1}^n \xi_{nk} \cdot \tag{1}$$

We note that no term in this sum exceeds 2ε in absolute value. If the sum $\sum_{k=1}^n \mathsf{D}(\xi_{nk}, x_i)$ were unbounded for some i, it would be possible to choose a sequence of values of n such that

$$\sum_{k=1}^n \mathsf{D}(\xi_{nk}, x_i) \longrightarrow \infty .$$

In such a case, the quantities

$$\eta_{nk} = \frac{(\xi_{nk} - \mathsf{M}\xi_{nk}, x_i)}{\sqrt{\sum_{j=1}^{n} \mathsf{D}(\xi_{nj}, x_i)}}$$

would satisfy the conditions of the limit Theorem 5, Section 6, Chapter I since

$$\mathsf{P}\left\{|\eta_{nk}| > \frac{4\varepsilon}{\sqrt{\sum_{j=1}^{n} \mathsf{D}\xi_{nk}}}\right\} = 0, \ \mathsf{M}\eta_{nk} = 0, \ \sum_{k=0}^{n} \mathsf{D}\eta_{nk} = 1 .$$

Thus the sum $\sum_{k=1}^{n} \eta_{nk}$ would have a normal limiting distribution, so that for arbitrary α,

$$\lim_{n\to\infty} \mathsf{P}\left\{\sum_{k=1}^{n} (\xi_{nk}, x_i) > \alpha\sqrt{\sum_{k=1}^{n} \mathsf{D}(\xi_{nk}, x_i)} + \sum_{k=1}^{n} (\mathsf{M}\xi_{nk}, x_i)\right\}$$
$$= \frac{1}{\sqrt{2\pi}}\int_{\alpha}^{\infty} e^{-u^2/2}du ,$$

$$\lim_{n\to\infty} \mathsf{P}\left\{\sum_{k=1}^{n} (\xi_{nk}, x_i) < -\alpha\sqrt{\sum_{k=1}^{n} \mathsf{D}(\xi_{nk}, x_i)} + \sum_{k=1}^{n} (\mathsf{M}\xi_{nk}, x_i)\right\}$$
$$= \frac{1}{\sqrt{2\pi}}\int_{-\infty}^{-\alpha} e^{-u^2/2}du .$$

But these last relations contradict the boundedness $\sum_{k=1}^{n} (\xi_{nk}, x_i)$, which follows from relation (1). Thus $\sum_{k=1}^{n} \mathsf{D}(\xi_{nk}, x_i)$ is bounded for all i.

We also note that by Chebyshev's inequality,

$$\mathsf{P}\left\{\left|\sum_{k=1}^{n} (\xi_{nk}, x_i) - \sum_{k=1}^{n} \mathsf{M}(\xi_{nk}, x_i)\right| > L\right\} \leqq \frac{\mathsf{D}\sum_{k=1}^{n} (\xi_{nk}, x_i)}{L^2} .$$

From this and the boundedness in probability of the quantity

$$\sum_{k=1}^{n} (\xi_{nk}, x_i) ,$$

it follows that $\mathsf{M}\sum_{k=1}^{n} (\xi_{nk}, x_i)$ is bounded. Since

$$\mathsf{M}\left|\sum_{k=1}^{n} \xi_{nk}\right|^2 = \mathsf{M}\sum_{i=1}^{r}\left(\sum_{k=1}^{n} \xi_{nk}, x_i\right)^2$$
$$= \sum_{i=1}^{r}\left[\mathsf{D}\sum_{k=1}^{n} (\xi_{nk}, x_i) + \left(\mathsf{M}\sum_{k=1}^{n} (\xi_{nk}, x_i)\right)^2\right],$$

the quantity $\mathsf{M}|\sum_{k=1}^{n} \xi_{nk}|^2$ is also bounded, as is $\mathsf{M}|\xi_\varepsilon(t)|^2$, since $\mathsf{M}|\xi_\varepsilon(t)|^2 \leqq \overline{\lim_{n\to\infty}}\, \mathsf{M}|\sum_{k=1}^{n} \xi_{nk}|^2$. This completes the proof of the lemma.

Let $\{\varepsilon_n\}$ denote a sequence that decreases to 0. We let Δ_k denote the set of all x such that $\varepsilon_k < |x| \leqq \varepsilon_{k-1}$ for $k = 2, 3, \cdots$, and we let Δ_1 denote the set of all x such that $|x| > \varepsilon_1$. We note that

$$\xi_{\varepsilon_1}(t) = \sum_{k=2}^{m} \xi(t, \Delta_k) + \xi_{\varepsilon_{m+1}}(t) .$$

The terms on the right are independent by virtue of Corollary 1 to Theorem 2 of Section 1. Therefore, for arbitrary x,

$$\sum_{k=2}^{m} \mathsf{D}(\xi(t, \Delta_k), x) \leqq \mathsf{D}(\xi_{\varepsilon_1}(t), x) ,$$

and hence for arbitrary x the series $\sum_{k=2}^{\infty} \mathsf{D}(\xi(t, \Delta_k), x)$ converges. Let us choose a sequence $\{n_k\}$ (where $n_1 = 2$) such that

$$\sum_{j=n_k}^{\infty} \mathsf{D}(\xi(T, \Delta_j), x_i) \leqq \frac{1}{k^6} \quad \text{for} \quad i = 1, 2, \cdots, r .$$

Then the sequence

$$\sum_{j=2}^{n_k} [\xi(t, \Delta_j) - \mathsf{M}\xi(t, \Delta_j)] \tag{2}$$

will, with probability 1, converge uniformly to some limit as $k \longrightarrow \infty$. To see this, note that

$$\mathsf{P}\left\{\sup_{0 \leqq t \leqq T} \left| \sum_{j=n_k+1}^{n_{k+1}} [\xi(t, \Delta_j) - \mathsf{M}\xi(t, \Delta_j)] - \sum_{j=2}^{n_k} [\xi(t, \Delta_j) - \mathsf{M}\xi(t, \Delta_j)] \right| > \frac{1}{k^2}\right\}$$

$$\leqq \sum_{i=1}^{r} \mathsf{P}\left\{\sup_{0 \leqq t \leqq T} \left| \sum_{j=n_k+1}^{n_{k+1}} (\xi(t, \Delta_j) - \mathsf{M}\xi(t, \Delta_j), x_i) \right| \geqq \frac{1}{k^2\sqrt{r}}\right.$$

$$\leqq \sum_{i=1}^{r} \overline{\lim_{m\to\infty}}\, \mathsf{P}\left\{\sup_{l \leqq mT} \left| \sum_{j=n_k+1}^{n_{k+1}} \left(\xi\left(\frac{l}{m}, \Delta_j\right) - \mathsf{M}\xi\left(\frac{l}{m}, \Delta_j\right), x_i\right) \right| \geqq \frac{1}{k^2\sqrt{r}}\right\}$$

$$\leqq \sum_{i=1}^{r} \overline{\lim_{m\to\infty}}\, rk^4 \mathsf{M} \left| \sum_{j=n_k+1}^{n_{k+1}} (\xi(T, \Delta_j) - \mathsf{M}\xi(T, \Delta_j), x_i) \right|^2 \leqq \frac{r^2}{k^2} .$$

(Here we used Kolmogorov's inequality: Theorem 1, Section 4, Chapter III.)

Since the series $\sum_{k=1}^{\infty} r^2/k^2$ converges, it follows from the Borel-Cantelli theorem (Theorem 2, Section 3, Chapter III) that the terms of the series

$$\sum_{k=1}^{\infty} \sum_{j=n_k+1}^{n_{k+1}} (\xi(t, \Delta_j) - \mathsf{M}\xi(t, \Delta_j))$$

are eventually majorized, with probability 1, by the terms of the convergent series $\sum_{k=1}^{\infty} 1/k^2$. Hence, the sequence (2) converges uniformly with probability 1. Therefore, there exists a process $\xi_0(t)$ that is the uniform limit of the sequence

$$\xi_{\varepsilon_1}(t) - \sum_{j=2}^{n_k} [\xi(t, \Delta_j) - \mathsf{M}\xi(t, \Delta_j)] .$$

Since $\xi(t, \Delta_j)$ is a stochastically continuous process and

$$\sup_{0 \le t \le T} \mathsf{M} | \xi(t, \Delta_j) |^2 < \infty ,$$

it follows from Theorem 6, Section 5, Chapter II, that

$$\lim_{t \to s} \mathsf{M}\xi(t, \Delta_j) = \mathsf{M}\xi(s, \Delta_j) .$$

Consequently, the process $\xi_{\varepsilon_1}(t) - \sum_{j=2}^{n_k} [\xi(t, \Delta_j) - \mathsf{M}\xi(t, \Delta_j)]$ fails with probability 1 to have jumps exceeding ε_{n_k} in absolute value, and the process $\xi_0(t)$ (the uniform limit of such processes) is continuous with probability 1. We note that in accordance with Kolmogorov's theorem (Theorem 4, Section 4, Chapter III), the series

$$\sum_{j=2}^{\infty} [\xi(t, \Delta_j) - \mathsf{M}\xi(t, \Delta_j)]$$

converges by virtue of the convergence of the series,

$$\sum_{j=2}^{\infty} \mathsf{D}(\xi(t, \Delta_j), x) ,$$

for every x. For every t, the sum of the series $\sum_{j=2}^{\infty} [\xi(t, \Delta_j) - \mathsf{M}\xi(t, \Delta_j)]$ coincides (mod P) with the sum of the series

$$\sum_{k=1}^{\infty} \sum_{j=n_k+1}^{n_{k+1}} [\xi(t, \Delta_j) - \mathsf{M}\xi(t, \Delta_j)] .$$

Thus we have

Theorem 1. *For every separable stochastically continuous process with independent increments, there exists a continuous process $\xi_0(t)$ such that*

$$\xi(t) = \xi_0(t) + \xi(t, \Delta_1) + \sum_{j=2}^{\infty} [\xi(t, \Delta_j) - \mathsf{M}\xi(t, \Delta_j)] .$$

REMARK 1. The process $\xi_0(t)$, being the limit of the sequence of processes

$$\xi_{\varepsilon_1}(t) - \sum_{j=2}^{n} [\xi(t, \Delta_j) - \mathsf{M}\xi(t, \Delta_j)] ,$$

is independent of each of the processes $\xi(t, \Delta_j)$, where $j = 1, 2, \cdots, m$. Since

$$\xi_0(t) + \sum_{j=2}^{\infty} (\xi(t, \Delta_j) - \mathsf{M}\xi(t, \Delta_j)) = \xi_{\varepsilon_1}(t)$$

and $\mathsf{M} | \xi_{\varepsilon_1}(t) |^2 < \infty$, and since the terms on the left are independent, it follows that $\mathsf{M} | \xi_0(t) |^2 < \infty$.

The following theorem completely characterizes the process $\xi_0(t)$.

Theorem 2. *The process $\xi_0(t)$ has independent Gaussian increments; that is, for every $z \in X$,*

$$\mathsf{M} \exp i(z, \xi_0(t_2) - \xi_0(t_1))$$
$$= \exp i(z, \mathsf{M}[\xi_0(t_2) - \xi_0(t_1)]) - \frac{1}{2}\mathsf{D}(z, \xi_0(t_2) - \xi_0(t_1)) \,. \tag{3}$$

Proof. $\xi_0(t)$ is a process with independent increments since it is the limit of processes with independent increments. It follows from Theorem 3, Section 5, Chapter IV that if

$$t_1 = t_{n0} < t_{n1} < \cdots < t_{nn} = t_2$$

and $\lim_{n\to\infty} \max_k (t_{nk} - t_{n,k-1}) = 0$, then for every $\rho > 0$,

$$\lim_{n\to\infty} \sum_{k=1}^{n} \mathsf{P}\{| \xi_0(t_{nk}) - \xi_0(t_{n,k-1}) | > \rho\} = 0 \,. \tag{4}$$

On the basis of formula (4) we may assert that there exists a sequence $\{\rho_n\}$ that converges to 0 as $n \longrightarrow \infty$ and such that

$$\lim_{n\to\infty} \sum_{k=1}^{n} \mathsf{P}\{| \xi_0(t_{nk}) - \xi_0(t_{n,k-1}) | > \rho_n\} = 0 \,.$$

Set $\xi_{nk} = (z, \psi_{\rho_n}(\xi_0(t_{nk}) - \xi_0(t_{n,k-1})))$, where $\psi_\rho(x) = 0$ for $|x| > \rho$ and $\psi_\rho(x) = x$ for $|x| \leqq \rho$. The quantities ξ_{nk} are bounded by the numbers $|z| \rho_n$. Since

$$\mathsf{P}\left\{\sum_{k=1}^{n} \xi_{nk} \neq (z, \xi_0(t_2) - \xi_0(t_1))\right\}$$
$$\leqq \sum_{k=1}^{n} \mathsf{P}\{\xi_{nk} \neq (z, \xi_0(t_{nk}) - \xi_0(t_{n,k-1}))\}$$
$$\leqq \sum_{k=1}^{n} \mathsf{P}\{| \xi_0(t_{nk}) - \xi_0(t_{n,k-1}) | > \rho_n\} \,,$$

the sequence $\{\sum_{k=1}^{n} \xi_{nk}\}$ converges in probability to the quantity $\xi_0(t_2) - \xi_0(t_1)$. Let us suppose that $\mathsf{D}(z, \xi_0(t_2) - \xi_0(t_1)) > 0$. Then

$$\lim_{n\to\infty} \sum_{k=1}^{n} \mathsf{D}\xi_{nk} \geqq \mathsf{D}(z, \xi_0(t_2) - \xi_0(t_1)) > 0 \,.$$

Therefore Theorem 5, Section 3, Chapter I is applicable to the quantities

$$\frac{\xi_{nk} - \mathsf{M}\xi_{nk}}{\sqrt{\sum_{k=1}^{n} \mathsf{D}\xi_{nk}}}$$

and hence,

$$\lim_{n\to\infty} \mathsf{M} \exp\left\{ i\lambda \frac{\sum_{k=1}^{n} \xi_{nk} - \sum_{k=1}^{n} \mathsf{M}\xi_{nk}}{\sqrt{\sum_{k=1}^{n} \mathsf{D}\xi_{nk}}} \right\} = e^{-\frac{\lambda^2}{2}} \tag{5}$$

Furthermore,

$$\lim_{n\to\infty} \mathsf{M} \exp\left\{i\lambda \sum_{k=1}^{n} \xi_{nk}\right\} = \mathsf{M} \exp\{i\lambda(z, \xi_0(t_2) - \xi_0(t_1))\} .$$

From these two relations one easily shows that $(z, \xi_0(t_2) - \xi_0(t_1))$ has a normal distribution. On the other hand if $\mathsf{D}(z, \xi_0(t_2) - \xi_0(t_1)) = 0$, then $(z, \xi_0(t_2) - \xi_0(t_1)) = \mathsf{M}(z, \xi_0(t_2) - \xi_0(t_1))$, and formula (3) is obviously valid. This completes the proof of the theorem.

REMARK 1. The expressions

$$\mathsf{M}(z, \xi_0(t) - \xi_0(0)) \quad \text{and} \quad \mathsf{D}(z, \xi_0(t) - \xi_0(0))$$

are continuous functions of t. It follows from the theorem just proved that

$$\mathsf{P}\{|\,(z, \xi_0(t_2)) - (z, \xi_0(t_1))\,| > \varepsilon\} = \frac{1}{\sqrt{2\pi}}\int_U e^{-u^2/2}du , \tag{6}$$

where

$$U = \{u: |\,u + \mathsf{M}(z, \xi_0(t_2) - \xi_0(t_1))\,| > \varepsilon(\sqrt{\mathsf{D}(z, \xi_0(t_2) - \xi_0(t_1))})^{-1}\} .$$

If there were a t such that $\mathsf{D}(z, \xi_0(t_2) - \xi_0(t_1)) \not\to 0$ as $t_1 \to t$ and $t_2 \to t$, the right-hand member of (6) would also not approach 0 and this would contradict the stochastic continuity of the process $\xi_0(t)$. For $t_1 < t_2$,

$$\mathsf{D}(z, \xi_0(t_2) - \xi_0(0)) = \mathsf{D}(z, \xi_0(t_1) - \xi_0(0)) + \mathsf{D}(z, \xi_0(t_2) - \xi_0(t_1))$$

by virtue of the independence of the increments $\xi_0(t)$. Consequently $\mathsf{D}(z, \xi_0(t) - \xi_0(0))$ is a continuous function. Furthermore, if

$$\varlimsup_{\substack{t_1\to t\\ t_2\to t}} |\,\mathsf{M}(z, \xi_0(t_2) - \xi_0(t_1))\,| > \varepsilon ,$$

it follows from (6) that

$$\varlimsup_{\substack{t_1\to t\\ t_2\to t}} \mathsf{P}\{|\,(z, \xi_0(t_2) - \xi_0(t_1))\,| > \varepsilon\} \geqq \frac{1}{2} .$$

This inequality ensures the continuity of $\mathsf{M}(z, \xi_0(t) - \xi_0(0))$.

REMARK 2. If we let $a(t)$ denote the quantity $\mathsf{M}(\xi_0(t) - \xi_0(0))$ and if we let $A(t)$ denote a nonnegative symmetric linear operator in X such that

$$\mathsf{D}(z, \xi_0(t) - \xi_0(0)) = (A(t)z, z) ,$$

then the distribution of the process $\xi_0(t)$ is determined by the characteristic function

$$\mathsf{M} \exp\{i(z, \xi_0(t))\}$$
$$= \mathsf{M}e^{i(z,\xi_0(0))} \exp\left\{i(a(t), z) - \frac{1}{2}(A(t)z, z)\right\} . \tag{7}$$

3. REPRESENTATION OF STOCHASTICALLY CONTINUOUS PROCESSES WITH INDEPENDENT INCREMENTS

Let us consider stochastic integrals with respect to the measure $\nu(t, A)$. As we mentioned in Section 1, the measure $\nu(t, A)$ is a countably additive nonnegative function of the set $A \in \mathfrak{B}_\varepsilon$. Suppose that a measurable function $\varphi(x)$ is bounded on every compact subset of the space X and is equal to 0 for $|x| < \varepsilon$ (where ε is some positive number). Then the integral $\int \varphi(x)\nu(t, dx)$ can be defined in the usual way. This follows from the finiteness of the measure $\nu(t, A)$ on $\mathfrak{B}_\varepsilon$ and also from the fact that $\nu(t, X_\rho) = 0$, where X_ρ is the set of all x such that $|x| > \rho$ and

$$\rho = \max_{0 \leqq s \leqq t} |\xi(s + 0) - \xi(s - 0)| \, .$$

Thus we are actually considering the integral only over the set $\{\varepsilon < |x| \leqq \rho\}$ on which the function $\varphi(x)$ is bounded. Let us show that

$$\xi(t, A) = \int_A x\nu(t, dx) \, . \tag{1}$$

If $A = \bigcup_{k=1}^{\infty} B_k$ where the B_k are pairwise disjoint sets with diameters not exceeding δ and if $x_k \in B_k$, then

$$\begin{aligned} \Big|\xi(t, A) - \sum_k x_k\nu(t, B_k)\Big| &\leqq \sum_k \Big|\xi(t, B_k) - x_k\nu(t, B_k)\Big| \\ &\leqq \delta \sum_k \nu(t, B_k) \leqq \delta\nu(t, A) \, . \end{aligned}$$

(The process $\xi(t, B_k)$ represents the sum of the jumps $\nu(t, B_k)$ belonging to B_k and therefore differing from x_k by no more than δ.) Equation (1) follows from this.

Since

$$\mathsf{M}|\xi(t, A) - \sum_k x_k\nu(t, B_k)| \leqq \delta\mathsf{M}\nu(t, A)$$

$$\mathsf{M}|\xi(t, A) - \sum_k x_k\nu(t, B_k)|^2 \leqq \delta^2\mathsf{M}[\nu(t, A)]^2 \, ,$$

we have

$$\mathsf{M}\xi(t, A) = \lim_{\delta \to 0} \mathsf{M} \sum_k x_k\nu(t, B_k) = \int_A x\Pi(t, dx) \, ,$$

$$\mathsf{D}(\xi(t, A), z) = \int_A (z, x)^2\Pi(t, dx)$$

for every bounded set A lying at a positive distance from the zero point of the space X.

Let us now consider the random set function

$$\tilde{\nu}(t, A) = \nu(t, A) - \Pi(t, A) ,$$

which has the following properties:

$$\mathsf{M}\tilde{\nu}(t, A) = 0 ,$$

$$\mathsf{M}(\tilde{\nu}(t, A)\tilde{\nu}(t, B)) = \mathsf{M}\nu(t, A \cap B) = \Pi(t, A \cap B) . \tag{2}$$

(The last follows from the fact that $\nu(t, \cdot)$ is independent for disjoint sets and is Poisson distributed.) Equations (2) enable us to use the general construction of a stochastic integral with respect to an orthogonal measure to construct the integral $\int f(x)\tilde{\nu}(t, dx)$ for all measurable functions $f(x)$ such that

$$\int | f(x) |^2 \Pi(t, dx) < \infty .$$

In the preceding section it was shown that

$$\sum_{k=2}^{\infty} \mathsf{D}(\xi(t, \Delta_k), z) < \infty .$$

Since

$$\mathsf{D}(\xi(t, \Delta_k), z) = \int_{\Delta_k} (x, z)^2 \Pi(t, dx) ,$$

it follows that for arbitrary $z \in X$,

$$\lim_{\varepsilon \to 0} \int_{\varepsilon < |x| \leq \varepsilon_1} (x, z)^2 \Pi(t, dx) < \infty .$$

Consequently we have

$$\lim_{\varepsilon \to 0} \int_{\varepsilon < x \leq \varepsilon_1} | x |^2 \Pi(t, dx) < \infty .$$

Therefore the integral $\int_{0<|x|<\varepsilon_1} x\tilde{\nu}(t, dx)$ exists. We note also that

$$\sum_{k=2}^{n} (\xi(t,\Delta_k) - \mathsf{M}\xi(t, \Delta_k)) = \int_{\varepsilon_{n+1}<|x|\leq\varepsilon_1} x\tilde{\nu}(t, dx)$$

and hence the series $\sum_{k=2}^{\infty} (\xi(t, \Delta_k) - \mathsf{M}\xi(t, \Delta_k))$ converges in probability to $\int_{0<|x|\varepsilon\leq 1} x\tilde{\nu}(t, dx)$. Remembering that

$$\xi(t, \Delta_1) = \int_{|x|>\varepsilon_1} x\nu(t, dx) ,$$

we obtain from Theorem 1 and Remark 1 of Section 2 the following result (for definiteness we set $\varepsilon_1 = 1$).

Theorem 1. *Let $\xi(t)$ denote a separable stochastic continuous process with independent increments. Then there exists an independent increment Gaussian process $\xi_0(t)$ such that $\xi_0(t)$ is independent of all the processes $\nu(t, A)$ and has the representation*

$$\xi(t) = \xi_0(t) + \int_{|x|>1} x\nu(t, dx) + \int_{|x|\leq 1} x\tilde{\nu}(t, dx) . \tag{3}$$

Let us find the characteristic function of the process $\xi(t)$. Since the terms on the right side of (3) are mutually independent, we only need find the characteristic function of each term. The characteristic function of the process $\xi_0(t)$ is given by formula (3) or formula (7) of Section 2. Let A denote a bounded set belonging to $\mathfrak{B}_\varepsilon$. Then

$$\mathsf{M}e^{i(z,\xi(t,A))} = \lim_{\lambda\to 0} \mathsf{M}e^{i\left(z,\sum_k x_k\nu(t,B_k)\right)}, \tag{4}$$

where $A = \bigcup_{k=1}^{n} B_k$ and the B_k are pairwise disjoint sets with diameters not exceeding λ, and where each point x_k belongs to B_k. The quantities $\nu(t, B_k)$ are mutually independent, so that by using formula (13) of Section 1 we obtain

$$\begin{aligned}\mathsf{M}e^{i\left(z,\sum_k x_k\nu(t,B_k)\right)} &= \prod_{k=1}^{n} \mathsf{M}e^{i(z,x_k)\nu(t,B_k)} = \prod_{k=1}^{n} \exp\{(e^{i(z,x_k)} - 1)\Pi(t, B_k)\} \\ &= \exp\left\{\sum_{k=1}^{n} (e^{i(z,x_k)} - 1)\Pi(t, B_k)\right\}.\end{aligned}$$

Taking the limit as $\lambda \longrightarrow 0$ in this last equation, we see that

$$\mathsf{M}e^{i(z,\xi(t,A))} = \exp\left\{\int_A (e^{i(z,x)} - 1)\Pi(t, dx)\right\}. \tag{5}$$

Formula (5) is also valid for unbounded sets belonging to $\mathfrak{B}_\varepsilon$ since such sets can be represented as sums of increasing sequences of bounded sets A_n for which (5) is valid and we can take the limit as $n \longrightarrow \infty$. It follows from (5) that

$$\begin{aligned}&\mathsf{M}\exp i(z, \xi(t, A) - \mathsf{M}\xi(t, A)) \\ &\quad= \exp\left\{\int_A (e^{i(z,x)} - 1 - i(z, x))\Pi(t, dx)\right\},\end{aligned} \tag{6}$$

that is,

$$\begin{aligned}&\mathsf{M}\exp i\left(z, \int_A x\tilde{\nu}(t, dx)\right) \\ &\quad= \exp\left\{\int_A (e^{i(z,x)} - 1 - i(z, x))\Pi(t, dx)\right\}.\end{aligned} \tag{7}$$

By taking the limit with respect to A, we see that formula (7) is valid for all sets A for which the right-hand member of this equation is meaningful. Now by taking formula (7) of Section 2 and formulas (5) and (7) of the present section, we can write the characteristic function of the process $\xi(t)$:

$$\begin{aligned}\mathsf{M}e^{i(z,\xi(t))} = \mathsf{M}e^{i(z,\xi(0))} \exp&\left\{i(a(t), z) - \frac{1}{2}(A(t)z, z)\right. \\ &+ \int_{|x|>1} (e^{i(z,x)} - 1)\Pi(t, dx) \\ &\left.+ \int_{0<|x|\leq 1} (e^{i(z,x)} - 1 - i(z, x))\Pi(t, dx)\right\}.\end{aligned} \tag{8}$$

With the aid of this formula we can determine the distribution $\xi(t_2) - \xi(t_1)$ and hence all finite-dimensional distributions of the process $\xi(t)$.

Since stochastically equivalent processes have identical finite-dimensional distributions and since for every process there is a stochastically equivalent separable process (by Theorem 2, Section 2, Chapter IV), we have:

Theorem 2. *The characteristic function of a stochastically continuous process with independent increments is of the form* (8), *where*

a. *the function* $\Pi(t, A)$ *is continuous, is nondecreasing with respect to* t *for every* $A \in \bigcup_\varepsilon \mathfrak{B}_\varepsilon$, *and satisfies the condition*

$$\lim_{\varepsilon \to 0} \int_{\varepsilon < |x| \leqq 1} |x|^2 \Pi(t, dx) < \infty ;$$

b. $a(t)$ *is a continuous function with range in* X;

c. *the function* $A(t)$ *is continuous and its range consists of nonnegative symmetric linear operators in* X; *also,* $A(t_2) - A(t_1)$ *is nonnegative for* $t_1 < t_2$.

Sometimes instead of formula (8) we use the formula

$$\mathsf{M}e^{i(z,\xi(t))} = \mathsf{M}e^{i(z,\xi(0))} \exp\left\{i(a_1(t), z) - \frac{1}{2}(A(t)z, z)\right.$$

$$\left. + \int\left(e^{i(z,x)} - 1 - \frac{i(z, x)}{1 + |x|^2}\right)\frac{1 + |x|^2}{|x|^2} G(t, dx)\right\} . \qquad (9)$$

Here,

$$a_1(t) = a(t) - \int_{|x| \leqq 1} \frac{x|x|^2}{1 + |x|^2}\Pi(t, dx) + \int_{|x|>1} \frac{x}{1 + |x|^2}\Pi(t, dx) ,$$

and the measure $G(t, dx)$ is defined by the relation

$$G(t, A) = \int_A \frac{|x|^2}{1 + |x|^2}\Pi(t, dx) .$$

Formula (9) is convenient because the measure $G(t, A)$ is finite and is defined on the σ-algebra of all Borel subsets of the space X.

A random process $\xi(t)$ with independent increments is said to be *homogeneous* if it is defined for $t \geqq 0$, $\xi(0) = 0$, and the distribution of the quantity $\xi(t + h) - \xi(t)$ depends only on h.

Theorem 3. *For a stochastically continuous homogeneous process* $\xi(t)$ *with independent increments, there exist a vector* $a \in X$, *a symmetric linear nonnegative operator* A, *and a function* $\Pi(A)$ *that is for every positive* ε *a finite measure on* $\mathfrak{B}_\varepsilon$ *and satisfies the inequality*

$$\lim_{\varepsilon \to 0} \int_{\varepsilon < |x| \leqq 1} |x|^2 \Pi(dx) < \infty ,$$

and such that the characteristic function of $\xi(t)$ *is given by the formula*

$$\mathsf{M}e^{i(z,\xi(t))} = \exp\left\{t\left[i(z, a) - \frac{1}{2}(z, Az) + \int_{|x|>1}(e^{i(z,x)} - 1)\Pi(dx) + \int_{|x|\leq 1}(e^{i(z,x)} - 1 - i(z, x))\Pi(dx)\right]\right\}. \tag{10}$$

Proof. Since $\xi(0) = 0$ in the present case, it follows from (9) that the finite quantity $\varphi(t) = \ln \mathsf{M}e^{i(z,\xi(t))}$ exists for small t. The function $\varphi(t)$ is continuous by virtue of the stochastic continuity of $\xi(t)$. Since for $t_1 > 0$ and $t_2 > 0$,

$$\mathsf{M}e^{i(z,\xi(t_1+t_2))} = \mathsf{M}e^{i(z,\xi(t_1))+i(z,\xi(t_1+t_2)-\xi(t_1))}$$
$$= \mathsf{M}e^{i(z,\xi(t_1))}\mathsf{M}e^{i(z,\xi(t_1+t_2)-\xi(t_1))} = \mathsf{M}e^{i(z,\xi(t_1))}\mathsf{M}e^{i(z,\xi(t_2))},$$

we have $\varphi(t_1 + t_2) = \varphi(t_1) + \varphi(t_2)$. Therefore $\varphi(t) = t\varphi(1)$. Formula (10) follows from (8) if we set

$$a = a(1), \ A = A(1), \ \Pi(A) = \Pi(1, A).$$

This completes the proof of the theorem.

4. PROPERTIES OF THE SAMPLE FUNCTIONS OF A STOCHASTICALLY CONTINUOUS PROCESS WITH INDEPENDENT INCREMENTS

We noted in Section 1 that a separable stochastically continuous process with independent increments fails with probability 1 to have discontinuities of the second kind. In this section we shall establish certain relationships between the properties of the sample functions of a separable process $\xi(t)$ and the properties of the functions $a(t)$, $A(t)$, and $\Pi(t, A)$ which determine the finite-dimensional distributions of the process $\xi(t)$.

Definition. A process $\xi(t)$ is said to be *piecewise-constant* on $[0, T]$ if there exist points $0 = t_0 < t_1 < \cdots < t_n = T$ such that $\xi(t)$ is constant on each of the intervals (t_i, t_{i+1}). The points t_i will, of course, depend on the individual paths.

The following theorem gives conditions under which a process will be piecewise-constant with probability 1.

Theorem 1. *For a separable stochastically continuous process* $\xi(t)$ *with independent increments to be piecewise-constant, it is necessary and sufficient that the characteristic function of* $\xi(t)$ *be given by the formula*

$$\mathsf{M}e^{i(z,\xi(t))} = \mathsf{M}e^{i(z,\xi(0))} \exp\left\{\int [e^{i(z,x)} - 1]\Pi(t, dx)\right\} \tag{1}$$

where $\Pi(t, A)$ is for every $t \in T$ a finite measure defined on the σ-algebra of all Borel subsets of the space X, and for each Borel set A is a continuous nondecreasing function of t.

Proof of the Necessity. If $X_\varepsilon = \{x; |x| > \varepsilon\}$, then for the step process $\xi(t)$, the relation $\xi(t) - \xi(0) = \lim_{\varepsilon\to 0} \xi(t, X_\varepsilon)$ will with probability 1 be satisfied (since $\xi(t) - \xi(0)$ is the sum of its jumps).

The characteristic function of the process $\xi(t, X_\varepsilon)$ is given by formula (5) of Section 3. Therefore

$$\mathsf{M}e^{i(z,\xi(t))} = \mathsf{M}e^{i(z,\xi(0))} \lim_{\varepsilon\to 0} \exp\left\{\int_{|x|>\varepsilon} [e^{i(z,x)} - 1]\Pi(t, dx)\right\}$$

and hence formula (1) will be proved once we show that the extension of $\Pi(t, A)$ onto the σ-algebra of all Borel sets is finite. To do this, it will be sufficient to show that the finite limit $\lim_{\varepsilon\to 0} \Pi(t, X_\varepsilon)$ exists. But the limit $\lim_{\varepsilon\to 0} \nu(t, X_\varepsilon)$ exists for a step process with probability 1 and is equal to the number of jumps of the process. Being the limit of random variables with Poisson distribution,

$$\lim_{\varepsilon\to 0} \nu(t, X_\varepsilon)$$

will also have a Poisson distribution. It follows from the convergence of the Poisson distributions that their parameters also converge; that is,

$$\mathsf{M} \lim_{\varepsilon\to 0} \nu(t, X_\varepsilon) = \lim_{\varepsilon\to 0} \Pi(t, X_\varepsilon) < \infty .$$

This completes the proof of the necessity of the hypotheses of the theorem. (The properties of $\Pi(t, A)$ as a function of t follow from Theorem 2 of Section 3.)

Proof of the sufficiency. Let us set

$$\lambda(t) = \lim_{\varepsilon\to 0} \Pi(t, X_\varepsilon)$$

and

$$G(t_1, t_2, A) = [\Pi(t_2, A) - \Pi(t_1, A)][\lambda(t_2) - \lambda(t_1)]^{-1} .$$

Then $G(t_1, t_2, A)$ is a probability measure on X. Let $G_k(t_1, t_2, A)$ denote the k-fold convolution of the distribution $G(t_1, t_2, A)$ with itself, so that

$$\int e^{i(z,x)} G_k(t_1, t_2, dx) = \left[\int e^{i(z,x)} G(t_1, t_2, dx)\right]^k .$$

Then from formula (1) the characteristic function of the variable $\xi(t_2) - \xi(t_1)$ can be written in the form

$$\sum_{k=0}^{\infty} e^{-[\lambda(t_2)-\lambda(t_1)]} \frac{[\lambda(t_2) - \lambda(t_1)]^k}{k!} \int e^{i(z,x)} G_k(t_1, t_2, dx) , \qquad (2)$$

where $G_0(t_1, t_2, dx)$ is the distribution of a quantity that with probability 1 is equal to 0. Thus the distribution of the variable $\xi(t_2) - \xi(t_1)$ coincides with the distribution of the variable S_ν, where $S_0 = 0$ and $S_n = \xi_1 + \cdots + \xi_n$. Here $\xi_1, \cdots, \xi_n, \cdots$ are independent random vectors with distribution $G(t_1, t_2, A)$, and ν is a random variable that is independent of the sequence $\xi_1, \cdots, \xi_n, \cdots$ and has a Poisson distribution with parameter $\lambda(t_2) - \lambda(t_1)$. Indeed,

$$\mathsf{M}e^{i(z,S_\nu)} = \sum_{k=0}^{\infty} \mathsf{M}(e^{i(z,S_\nu)} \mid \nu = k)\mathsf{P}\{\nu = k\} = \sum_{k=0}^{\infty} \mathsf{M}e^{i(z,S_k)}\mathsf{P}\{\nu = k\} ,$$

which coincides with (2). Thus

$$\mathsf{P}\{\xi(t_2) - \xi(t_1) = 0\} = \mathsf{P}\{S_\nu = 0\} \geqq \mathsf{P}\{\nu = 0\} = e^{-[\lambda(t_2)-\lambda(t_1)]} .$$

Consequently

$$\begin{aligned}
&\mathsf{P}\{ \sup_{t_1 \leqq s \leqq t_2} \mid \xi(s) - \xi(t_1) \mid = 0\} \\
&= \lim_{n\to\infty} \mathsf{P}\left\{ \sup_{1\leqq k \leqq n} \left| \xi\left(t_1 + \frac{k}{n}(t_2 - t_1)\right) - \xi\left(t_1 + \frac{k-1}{n}(t_2 - t_1)\right) \right| = 0 \right\} \\
&= \lim_{n\to\infty} \prod_{k=1}^{n} \mathsf{P}\left\{ \xi\left(t_1 + \frac{k}{n}(t_2 - t_1)\right) - \xi\left(t_1 + \frac{k-1}{n}(t_2 - t_1)\right) = 0 \right\} \\
&\geqq \lim_{n\to\infty} \prod_{k=1}^{n} \exp\left\{ - \left[\lambda\left(t_1 + \frac{k}{n}(t_2 - t_1)\right) \right.\right. \\
&\qquad \left.\left. - \lambda\left(t_1 + \frac{k-1}{n}(t_2 - t_1)\right)\right]\right\} = e^{-[\lambda(t_2)-\lambda(t_1)]} .
\end{aligned}$$

Define $h_n = T/2^n$ and $t_k = kh_n$ for $k = 0, \cdots, 2^n$. Let us find a bound for the probability of the event that among the intervals $[t_k, t_{k+1}]$ there will be no fewer than l intervals for which

$$\sup_{t_k \leqq t \leqq t_{k+1}} \mid \xi(t) - \xi(t_k) \mid > 0 .$$

Let this probability be denoted by P_l. Then we may write

$$\begin{aligned}
\mathsf{P}_l &\leqq \sum_{i_1 < \cdots < i_l} \prod_{j=1} \mathsf{P}\{ \sup_{t_{i_j} \leqq s \leqq t_{i_j+1}} |\xi(s) - \xi(t_{i_j}) \mid > 0\} \\
&\leqq \sum_{i_1 < \cdots < i_l} \prod_{j=1}^{l} [1 - e^{-[\lambda(t_{i_j+1}) - \lambda(t_{i_j})]}] \\
&\leqq \frac{1}{l!} \left[\sum_{k=0}^{n-1} (1 - e^{-[\lambda(t_{k+1}) - \lambda(t_k)]}) \right]^l \\
&\leqq \frac{1}{l!} \left(\sum_{k=0}^{n-1} [\lambda(t_{k+1}) - \lambda(t_k)] \right)^l \leqq \frac{1}{l!} \lambda(T)^l .
\end{aligned}$$

(In this chain of inequalities we use the relation $1 - e^{-x} < x$ for $x > 0$.)

Let ν_n denote the number of intervals $[t_k, t_{k+1}]$ for which

$$\sup_{t_k \leqq s \leqq t_{k+1}} |\xi(s) - \xi(t_k)| > 0 .$$

Since $\nu_{n+1} \geqq \nu_n$ it follows that the limit $\nu = \lim_{n\to\infty} \nu_n$ exists with probability 1. For the quantity ν we also have

$$\mathsf{P}\{\nu \geqq l\} \leqq \frac{1}{l!}\lambda(T)^l .$$

The event $\{\nu = l\}$ means that $\xi(t)$ is constant on no more than $l + 1$ adjacent intervals. Since $\mathsf{P}\{\nu \geqq l\} \to 0$ as $l \to \infty$, the process $\xi(t)$ is constant on $\nu + 1$ adjacent intervals and the quantity ν is finite with probability 1. This completes the proof of the theorem.

Suppose now that $\xi(t)$ is a process with numerical values, that is, that X is the real line. Let us investigate the conditions under which the sample functions of the process $\xi(t)$ are, with probability 1, monotonic functions.

Theorem 2. *For the sample functions of a numerical separable stochastically continuous process $\xi(t)$ with independent increments to be nondecreasing with probability 1, it is necessary and sufficient that the characteristic function of the variadle $\xi(t)$ be given by the formula*

$$\mathsf{M}e^{i\lambda\xi(t)} = \mathsf{M}e^{i\lambda\xi(0)} \exp\left\{i\lambda\gamma(t) + \int_0^\infty (e^{i\lambda x} - 1)\Pi(t, dx)\right. \tag{3}$$

where the measure Π satisfies the condition $\int_0^1 x\Pi(t, dx) < \infty$ and $\gamma(t)$ is a nondecreasing function.

Proof of the Necessity. If $\xi(t)$ is a nondecreasing function, the process $\xi(t)$ has only positive jumps; hence $\Pi(t, A) = 0$ for every set A lying on the negative half-line. We note also that in the present case the process $\xi(t) - \xi(t, X_\varepsilon)$ is also nondecreasing (since removal of the jumps does not destroy monotonicity). Similarly, the process

$$\xi(t, X_\varepsilon) - \xi(t, X_1) \quad \text{for} \quad 0 < \varepsilon < 1 ,$$

is also a monotonic process; also

$$0 \leqq \xi(t, X_\varepsilon) - \xi(t, X_1) \leqq \xi(t) - \xi(0) - \xi(t, X_1) .$$

On the basis of the lemma of Section 2,

$$\mathsf{M}[\xi(t) - \xi(0) - \xi(t, X_1)] < \infty .$$

Therefore

$$\mathsf{M}[\xi(t, X_\varepsilon) - \xi(t, X_1)] = \int_\varepsilon^1 x\Pi(t, dx) \leqq \mathsf{M}[\xi(t) - \xi(0) - \xi(t, X_1)] .$$

Taking the limit as $\varepsilon \to 0$, we see that $\int_0^1 x\Pi(t, dx)$ has a finite value. We note also that the quantity $\xi(t) - \xi(t, X_\varepsilon)$ decreases as $\varepsilon \to 0$ (since more positive jumps are discarded with decreasing ε). Consequently the limit $\lim_{\varepsilon\to 0} [\xi(t) - \xi(t, X_\varepsilon)] = \xi_0(t)$ exists with probability 1 and the process $\xi_0(t)$ is, with probability 1, continuous. As shown in Section 2, the increments of the process $\xi_0(t)$ will have Gaussian distributions. But the process $\xi_0(t)$, being the limit of nondecreasing processes, will itself be nondecreasing so that

$$\mathsf{P}\{\xi_0(t) - \xi_0(0) \geqq 0\} = 1 .$$

It follows from this relation that $\mathsf{D}(\xi_0(t) - \xi_0(0)) = 0$ (since a normally distributed variable ξ can be nonnegative with probability 1 only when $\mathsf{D}\xi = 0$). Thus

$$\xi_0(t) = \xi_0(0) + \gamma(t), \quad \text{where} \quad \gamma(t) = \mathsf{M}[\xi_0(t) - \xi_0(0)]$$

and hence does not decrease. Formula (3) can be obtained from the relation

$$\mathsf{M}e^{i\lambda\xi(t)} = \lim_{\varepsilon\to 0} \mathsf{M}e^{i\lambda\xi_0(t)}\mathsf{M}e^{i\lambda\xi(t, X_\varepsilon)}$$

if we keep in mind the form of the process $\xi_0(t)$ and formula (5) of Section 3. This completes the proof of the necessity.

Proof of the Sufficiency. Let us show that $\mathsf{P}\{\xi(t_2) - \xi(t_1) \geqq 0\} = 1$. To do this, it will be sufficient to show that with probability 1 a nondecreasing random variable ξ whose characteristic function has the form

$$\mathsf{M}e^{i\lambda\xi} = \exp\left\{\int_0^\infty (e^{i\lambda x} - 1)dG(x)\right\} , \tag{4}$$

where $G(x)$ is a monotonic bounded function, ($\xi(t_2) - \xi(t_1)$) is the sum of $\gamma(t_2) - \gamma(t_1)$ and the limit of a quantity with characteristic function of the form (4)).

Let us set $F(x) = c[G(+\infty) - G(x)]$, $c = [G(+\infty) - G(+0)]^{-1}$. Then

$$\mathsf{M}e^{i\lambda\xi} = \sum_{k=0}^\infty \frac{c^k}{k!} e^{-c}\left(\int_0^\infty e^{i\lambda x}dF(x)\right)^k ,$$

so that the characteristic function of the variable ξ coincides with the characteristic function of the variable S_ν, where $S_0 = 0$ and $S_n = \xi_1 + \cdots + \xi_n$. Here $\xi_1, \xi_2, \cdots$ is a sequence of independent identically distributed nonnegative variables with distribution function

$F(x)$, and ν is a Poisson random variable independent of $\xi_1, \xi_2, \cdots$. Consequently ξ is nonnegative.

Thus

$$\mathsf{P}\{\xi(t_2) \geqq \xi(t_1)\} = 1 \quad \text{for} \quad t_1 < t_2 .$$

It follows from the last relation that the event that the inequality $\xi(t_1) \leqq \xi(t_2)$ will be satisfied for all pairs of rational t_1 and t_2, where $t_1 < t_2$, has probability 1. Using the fact that $\xi(t)$ is separable and has no discontinuities of the second kind, we conclude that

$$\mathsf{P}(\xi(t_1) \leqq \xi(t_2),\ t_1 < t_2) = 1 .$$

This completes the proof of the theorem.

Let us investigate the conditions under which the sample functions of the process $\xi(t)$ are with probability 1 of bounded variation.

We recall that the *variation* of a function $x(t)$ given on $[a, b]$ with range in X is defined as

$$\operatorname*{var}_{[a,b]} x(t) = \sup \sum_{i=0}^{n-1} |\, x(t_i) - x(t_{i+1}) \,| ,$$

with the supremum being taken over all possible partitions of the interval $[a, b]$: $a = t_0 < t_1 < \cdots < t_n = b$.

Theorem 3. *For the sample functions of a separable stochastically continuous process $\xi(t)$ with independent increments and defined on an interval $[0, T]$ to be of bounded variation on that interval with probability 1, it is necessary and sufficient that the characteristic function of the variable $\xi(t)$ be given by formula (8) of Section 3, with*

$$\operatorname*{var}_{[0,T]} a(t) < \infty ,$$

$A(t) = 0$, *and the measure $\Pi(t, A)$ such that*

$$\int_{0<|x|\leqq 1} |\, x \,| \Pi(t, dx) < \infty .$$

Proof of the Sufficiency. Since the process defined by

$$\xi(t, X_1) = \int_{|x|>1} x\nu(t, dx)$$

is with probability 1 piecewise-constant, its variation, which coincides with the sum of the absolute values of the jumps, is finite. The function $a(t)$ is, by hypothesis, of bounded variation. Therefore to prove the boundedness of the variation of $\xi(t)$, it will be sufficient to prove the boundedness of the variation of the integral

$$\int_{0<|x|\leqq 1} x\tilde{\nu}(t, dx)$$

as a function of t (cf. formula (3) of Section 3). Consider the process

$$\xi^{(\varepsilon)}(t) = \int_{\varepsilon<|x|\leq 1} x\tilde{\nu}(t, dx) .$$

It is easy to show that the variation of the process $\xi^{(\varepsilon)}(t)$ on the interval $[0, T]$ is equal to

$$\operatorname*{var}_{[0,T]} \xi^{(\varepsilon)}(t) = \operatorname*{var}_{[0,T]} \int_{\varepsilon<|x|\leq 1} x\nu(t, dx) + \operatorname*{var}_{[0,T]} \int_{\varepsilon<|x|\leq 1} x\Pi(t, dx) .$$

Therefore

$$\begin{aligned} \operatorname*{var}_{[0,T]} \xi^{(\varepsilon)}(t) &\leqq \int_{\varepsilon<|x|\leq 1} |x|\, \nu(T, dx) + \int_{\varepsilon<|x|\leq 1} |x|\, \Pi(T, dx) \\ &\leqq \lim_{\varepsilon\to 0} \int_{\varepsilon<|x|\leq 1} |x|\, \nu(T, dx) + \int_{0<|x|\leq 1} |x|\, \Pi(T, dx) . \end{aligned}$$

The existence and finiteness of the limit $\lim_{\varepsilon\to\infty} \int_{\varepsilon<|x|\leq 1} |x|\, \nu(T, dx)$ follows from the fact that the quantity $\int_{\varepsilon<|x|\leq 1} |x|\, \nu(T, dx)$ is a monotonic function of ε and

$$\mathsf{M}\int_{\varepsilon<|x|\leq 1} |x|\, \nu(T, dx) \leqq \int_{0<|x|\leq 1} |x|\, \Pi(T, dx) < \infty .$$

Since we can choose a sequence $\{\varepsilon_n\}$ that converges to zero as $n \longrightarrow \infty$ such that the sequence of the processes $\xi^{(\varepsilon_n)}(t)$ converges uniformly with probability 1 to the process $\int_{0<|x|\leq 1} x\tilde{\nu}(t, dx)$, it follows that with probability 1,

$$\operatorname*{var}_{[0,T]}\int_{0<|x|\leq 1} x\tilde{\nu}(t, dx) \leqq \int_{0<|x|\leq 1} |x|\, \nu(T, dx) + \int_{0<|x|\leq 1} |x|\Pi(T, dx) .$$

(It is easy to see that if $x_n(t) \longrightarrow x(t)$ as $n \longrightarrow \infty$ for every t, then

$$\sum_{k=0}^{m-1} |x(t_{k+1}) - x(t_k)| = \lim_{n\to\infty} \sum_{k=0}^{m-1} |x_n(t_{k+1}) - x_n(t_k)| \leqq \overline{\lim_{n\to\infty}} \operatorname*{var}_{[0,T]} x_n(t) ,$$

and hence $\operatorname{var}_{[0,T]} x(t) \leqq \overline{\lim}_{n\to\infty} \operatorname{var}_{[0,T]} x_n(t)$. From this it follows that the variation of $\int_{0<|x|\leq 1} x\tilde{\nu}(t, dx)$ is bounded.

Proof of the Necessity. Suppose that the variation of $\xi(t)$ on the interval $[0, T]$ is bounded. Then it will also be bounded on every interval contained in $[0, T]$. Consider the process

$$\zeta(t) = \operatorname*{var}_{[0,t]} \xi(s) .$$

This process is also a process with independent increments since the increment $\zeta(t_2) - \zeta(t_1)$, where $t_1 < t_2$, is the variation of a process on the interval $[t_1, t_2]$ and hence depends only on the increments of the process $\xi(t)$ on the interval $[t_1, t_2]$. Hence it is independent of the values of $\zeta(t)$ for $t \leqq t_1$. The process $\zeta(t)$ is a stochastically continuous process since the variation $\operatorname{var}_{[0,t]} x(s)$ of the function $x(s)$ can have discontinuities only at points of discontinuity of the function $x(s)$ itself and the process $\xi(t)$ has no fixed discontinuities (because of its stochastic continuity). Finally it is obvious that $\zeta(t)$ is a nondecreasing function of t. From what was said above it follows that

$$\zeta(t+0) - \zeta(t-0) = |\xi(t+0) - \xi(t-0)| .$$

Therefore, if $\zeta^{(\varepsilon)}(t)$ is the sum of all jumps of $\zeta(t)$ exceeding ε and occurring up to the instant t inclusive, then

$$\zeta^{(\varepsilon)}(t) = \int_{|x|>\varepsilon} |x|\, \nu(t, dx) .$$

Since $\zeta(0) = 0$, it follows from the monotonicity of the process $\zeta(t)$ that

$$\zeta(t) = \gamma(t) + \lim_{\varepsilon\to 0} \zeta^{(\varepsilon)}(t) = \gamma(t) + \lim_{\varepsilon\to 0} \int_{|x|>\varepsilon} |x|\,\nu(t, dx)$$

where $\gamma(t)$ is the variation of the continuous component of the process $\xi(t)$. Furthermore,

$$\int_{|x|\leq 1} |x|\, \Pi(t, dx) = \mathsf{M}[\zeta(t) - \zeta^{(1)}(t)] < \infty .$$

Finally we note that if $\xi_0(t)$ is the continuous component of the process $\xi(t)$, then

$$|\xi_0(t) - \xi_0(0)| \leqq \operatorname*{var}_{[0,T]} \xi_0(s) = \gamma(t)$$

and hence

$$|(\xi_0(t) - \xi_0(0), z)| \leqq \gamma(t)\, |z| .$$

But a normally distributed variable with positive variance cannot be bounded with probability 1 by any constant. Consequently for arbitrary z,

$$\mathsf{D}(z, \xi_0(t) - \xi_0(0)) = 0 ;$$

that is, $A(t) = 0$. Thus $\gamma(t)$ is the variation of the function

$$a(t) - \int_{0<|x|\leq 1} x\Pi(t, dx) .$$

Since the variation

$$\operatorname*{var}_{[0,T]} \int_{0<|x|\leq 1} x\Pi(t, dx) \leqq \int_{0<|x|\leq 1} |x|\, \Pi(T, dx)$$

is finite, the variation $\operatorname{var}_{[0,T]} a(t)$ is also finite. This completes the proof of the theorem.

REMARK. It follows from the proof given of the necessity that if a process $\xi(t)$ is a bounded variation, then we can write

$$\xi(t) = \xi(0) + a_1(t) + \int_{|x|>0} x\nu(t, dx) .$$

Also, we have seen that

$$\zeta(t) = \operatorname*{var}_{[0,t]} \xi(s) = \operatorname*{var}_{[0,t]} a_1(s) + \int_{|x|>0} |x| \nu(t, ds)$$

and the characteristic function of the variable $\zeta(t)$ is given by

$$\mathsf{M}e^{i\lambda\zeta(t)} = \exp\left\{i\lambda \operatorname*{var}_{[0,t]} a_1(s) + \int_{|x|>0} [e^{i\lambda|x|} - 1]\Pi(t, dx)\right\} .$$

5. PROCESSES OF BROWNIAN MOTION

A *process* $w(t)$ *of Brownian motion* is defined as a separable homogeneous Gaussian process with independent increments such that $\mathsf{M}w(t) = 0$ and $\mathsf{D}(z, w(t)) = ct(z, z)$. In this section we shall consider a one-dimensional process of Brownian motion (a process with numerical values) and for brevity in writing we shall assume that $c = 1$ (this can always be obtained by changing the units of the time scale). For such a process, the variable $w(t + h) - w(t)$ has a normal distribution with mean zero and variance h, so that the distribution density $p_h(x)$ of the variable $w(t + h) - w(t)$ is given by the formula

$$p_h(x) = \frac{1}{\sqrt{2\pi h}} e^{-(x^2/2h)} .$$

On the basis of Theorem 6, Section 5, Chapter IV a process of Brownian motion is continuous.

REMARK 1. The name "process of Brownian motion" can be explained by the following description of the corresponding physical phenomenon. Consider a sufficiently small particle suspended in a liquid and consider its motion under the influence of collisions with the molecules of the liquid, which are in chaotic thermal motion. In physics this phenomenon is known as "Brownian motion."

In the probabilistic study of this phenomenon it is natural to consider the velocities of the molecules with which the particle collides as random, and in the case of a homogeneous liquid we assume that the distribution of the velocity is independent of the position of a molecule. (It depends on the temperature, which is everywhere uniform.) If we also assume that the velocities of the different molecules are independent of each other and if we neglect the mass of the particle, then the displacement of the particle from an arbitrary position after a certain interval of time will not be dependent on the position of the particle or its previous motion. Consequently, a process $\boldsymbol{\zeta}(t)$ in $R^{(3)}$ that describes the position of a particle at an instant t is a process with independent increments. Furthermore it is obvious from physical considerations that it will be continuous and homogeneous with respect to time if the physical state of the liquid does not change with time. But every continuous homogeneous process with independent increments is a Gaussian process. We shall assume that at the initial instant of time the position of the particle coincides with the coordinate origin: $\boldsymbol{\xi}(0) = 0$. Suppose that $\mathsf{M}\boldsymbol{\xi}(t) = t\boldsymbol{a}$ and $\mathsf{D}(\boldsymbol{\xi}(t), z) = t(Az, z)$, where a is a vector in $R^{(3)}$ and A is a symmetric operator in $R^{(3)}$. In the case of a homogeneous liquid where there are no currents, the process must be isotropic (since the distribution of the projections of the velocity of the molecule of the liquid in an arbitrary direction is independent of that direction); that is, $(\boldsymbol{a}, z)$ and (Az, z) are independent of z for $|z| = 1$. This is possible when $(\boldsymbol{a}, z) = 0$ and $(Az, z) = c(z, z)$. Thus from the most general considerations of the physical phenomenon of Brownian motion, we have obtained our definition of a process of Brownian motion.

Let a denote some number other than 0. Let τ_a denote an instant of time at which $w(t)/a \leqq 1$ for $t \leqq \tau_a$ and for arbitrary $\delta > 0$,

$$\sup_{\tau_a \leqq t \leqq \tau_a + \delta} \frac{w(t)}{a} > 1 .$$

If $w(t)/a$ for all t, we take $\tau_a = +\infty$. We shall call τ_a the *instant of first crossing* of the level a by the process $w(t)$.

Now suppose that τ'_a is that instant of time at which $w(t)/a < 1$ for $t < \tau'_a$ and $w(\tau'_a) = a$. We shall call τ'_a the *instant of first attainment* of the level a by the process $w(t)$. Obviously, $\tau'_a \leqq \tau_a$.

Lemma 1. $\mathsf{P}\{\tau'_a = \tau_a\} = 1$.

Proof. By virtue of the symmetry of the process $w(t)$ (the process $-w(t)$ has the same distributions) we assume that $a > 0$. The event $\{\tau'_a < \tau_a\}$ implies at least one of the events

$$\{\max_{0 \leqq s \leqq r/m} w(s) = a\}, \quad r = 1, 2, \cdots; \; m = 1, 2, \cdots .$$

Therefore to prove the lemma, it will be sufficient to show that for $a > 0$,

$$\mathsf{P}\{\max_{0 \leqq s \leqq t} w(s) = a\} = 0$$

for every t. It is easy to see that for $t_1 < t$,

$$\mathsf{P}\{\max_{0 \leqq s \leqq t} w(s) = a\} \leqq \mathsf{P}\{\max_{0 \leqq s \leqq t_1} w(s) = a\} + \int_{-\infty}^{a} \mathsf{P}\{w(t_1) \in dx\} \mathsf{P}\{\sup_{t_1 \leqq s \leqq t} w(s) - w(t_1) = a - x\} .$$

But $\mathsf{P}\{\sup_{t_1 \leqq s \leqq t} w(s) - w(t_1) = z\}$ can be nonzero at only countably many distinct values of z; that is,

$$\mathsf{P}\{\sup_{t_1 \leqq s \leqq t} w(s) - w(t_1) = a - x\} = 0$$

for almost all x (in the sense of Lebesgue measure). This means that

$$\int_{-\infty}^{a} \mathsf{P}\{w(t_1) \in dx\} \mathsf{P}\{\sup_{t_1 \leqq s \leqq t} w(s) - w(t_1) = a - x\} = \frac{1}{\sqrt{2\pi t_1}} \int_{-\infty}^{a} \mathsf{P}\{\sup_{t_1 \leqq s \leqq t} w(s) - w(t_1) = a - x\} e^{-(x^2/2t_1)} dx = 0 ,$$

since the integrand vanishes almost everywhere. Thus

$$\mathsf{P}\{\max_{0 \leqq s \leqq t} w(s) = a\} \leqq \mathsf{P}\{\max_{0 \leqq s \leqq t_1} w(s) = a\} ;$$

that is, $\mathsf{P}\{\max_{0 \leqq s \leqq t} w(s) = a\}$ does not decrease as $t \downarrow 0$; whereas, by virtue of the continuity of $w(t)$, for $\varepsilon > 0$,

$$\mathsf{P}\{\max_{0 \leqq s \leqq t} w(s) > \varepsilon\} \longrightarrow 0 \quad \text{as} \quad t \longrightarrow 0 .$$

Therefore

$$\mathsf{P}\{\max_{0 \leqq s \leqq t} w(s) = a\} \leqq \lim_{t \downarrow 0} \mathsf{P}\left\{\max_{0 \leqq s \leqq t} w(s) > \frac{a}{2}\right\} = 0 .$$

This completes the proof of the lemma.

For our purposes in what follows, then, we shall not find it necessary to distinguish the instant of the first attainment from the instant of first crossing of the level a. We shall therefore let τ_a denote both.

To study certain characteristics of the process $w(t)$, we shall use:

Lemma 2. *Let $w(t)$ denote a process of Brownian motion, let a denote a number other than zero, and let τ denote the instant of first crossing of the level a by the process $w(t)$. We define a process $w_1(t)$*

by $w_1(t) = w(t)$ *for* $t < \tau_a$ *and* $w_1(t) = 2a - w(t)$ *for* $t \geqq \tau_a$. *Then the process* $w_1(t)$ *is also a process of Brownian motion.*

Proof. Let us set

$$w_{nk} = w\left(\frac{k}{n}\right) - w\left(\frac{k-1}{n}\right),$$

$w^{(n)}(t) = \sum_{k \leqq nt} w_{nk}$, and $w_1^{(n)}(t) = \sum_{k \leqq nt} (-1)^{\varepsilon_{nk}} w_{nk}$, where

$$\varepsilon_{nk} = \begin{cases} 0 \text{ if } \sup\limits_{j \leqq k-1} \dfrac{w_1^{(n)}\left(\dfrac{j}{n}\right)}{a} \leqq 1 \\ 1 \text{ if } \sup\limits_{j \leqq k-1} \dfrac{w^{(n)}\left(\dfrac{j}{n}\right)}{a} > 1 . \end{cases}$$

We note that the quantities $(-1)^{\varepsilon_{nk}} w_{nk}$ for $k = 1, 2, \cdots$ are mutually independent and identically distributed. Also, their distributions coincide with the distributions of the quantities w_{nk}. This follows from the fact that w_{nk} and $-w_{nk}$ are identically distributed and w_{nk} is independent of $\varepsilon_{nk}, w_{n1}, \cdots, w_{n,k-1}$. Consequently the $(-1)^{\varepsilon_{nk}} w_{nk}$ have a normal distribution with mean zero and variance $1/n$. Therefore the finite-dimensional distributions of the processes $w^{(n)}(t)$ and $w_1^{(n)}(t)$ coincide. Since a process of Brownian motion is continuous, and therefor $w^{(n)}(t) \to w(t)$ and $w_1^{(n)}(t) \to w_1(t)$ with probability 1 as $n \to \infty$, the lemma is proved.

We shall now use this lemma to find the distribution of such characteristics of the process as $\max_{0 \leqq t \leqq T} w(t)$, $\min_{0 \leqq t \leqq T} w(t)$, and $\sup_{0 \leqq t \leqq T} | w(t) |$.

Theorem 1. *For* $a > 0$,

$$\mathsf{P}\{\max_{0 \leqq t \leqq T} w(t) > a, w(T) \in [c, d]\}$$
$$\frac{1}{\sqrt{2\pi T}} \int_{\max[c,a]}^{\max[d,a]} e^{-(x^2/2T)} dx + \frac{1}{\sqrt{2\pi T}} \int_{\max[2a-d,a]}^{\max[2a-c,a]} e^{-(x^2/2T)} dx . \quad (1)$$

Proof. Let us use the relation

$$\mathsf{P}\{\max_{0 \leqq t \leqq T} w(t) > a, w(T) \in [c, d]\} = \mathsf{P}\{w(T) \in [c, d] \cap [a, \infty]\}$$
$$+ \mathsf{P}\{\max_{0 \leqq t \leqq T} w(t) > a, w(T) \in [c, d] \cap (-\infty, a]\} .$$

(The validity of this relation follows from the fact that the event $\{w(T) \in [c, d] \cap [a, \infty]\}$ implies the event $\{\max_{a \leqq t \leqq T} w(t) > a.\}$ Let us now find the probability

$$\mathsf{P}\{\max_{0 \leqq t \leqq T} w(t) > a, w(T) \in [c, d] \cap (-\infty, a]\} .$$

Let $w_1(t)$ denote the process defined in Lemma 2. Then the event

$$\{\max_{0 \leq t \leq T} w(t) \geqq a,\ w(T) \in [c, d] \cap (-\infty, a]\}$$

coincides with the event

$$\{\max_{0 \leq t \leq T} w_1(t) \geqq a,\ w_1(T) \in [2a - d, 2a - c] \cap [a, \infty)\} .$$

But the event

$$\{w_1(T) \in [2a - d, 2a - c] \cap [a, \infty)\}$$

implies the event $\{\max_{0 \leq t \leq T} w_1(t) \geqq a\}$. Therefore

$$\{\max_{0 \leq t \leq T} w_1(t) \geqq a,\ w_1(t) \in [2a - d, 2a - c] \cap [a, \infty)\}$$
$$= \{w_1(T) \in [2a - d, 2a - c] \cap [a, \infty)\} .$$

Consequently, on the basis of Lemma 2,

$$\mathsf{P}\{\max_{0 \leq t \leq T} w(t) > a,\ w(T) \in [c, d] \cap (-\infty, a]\}$$
$$= \frac{1}{\sqrt{2\pi T}} \int_{\max[2a-d,\, a]}^{\max[2a-c,\, a]} e^{-(x^2/2T)} dx . \tag{2}$$

Furthermore,

$$\mathsf{P}\{w(T) \in [c, d] \cap [a, \infty]\} = \int_{\max[c,a]}^{\max[d,a]} \frac{1}{\sqrt{2\pi T}} e^{-(x^2/2T)} dx . \tag{3}$$

Proof of the theorem follows from (2) and (3).

Corollary. *For* $a > 0$,

$$\mathsf{P}\{\max_{0 \leq t \leq T} w(t) > a\} = \frac{2}{\sqrt{2\pi T}} \int_a^\infty e^{-(x^2/2T)} dx .$$

This follows from Theorem 1 with $(c, d) = (-\infty, \infty)$.

Theorem 2. *Suppose that* $a_1 < 0 < a_2$ *and* $[c, d] \subset [a_1, a_2]$. *Then,*

$$\mathsf{P}\{\min_{0 \leq t \leq T} w(t) > a_1,\ \max_{0 \leq t \leq T} w(t) < a_2,\ w(T) \in [c, d]\}$$
$$= \frac{1}{\sqrt{2\pi T}} \sum_{k=-\infty}^{\infty} \int_c^d \Big[\exp\Big\{-\frac{1}{2T}(x + 2k(a_2 - a_1))^2\Big\}$$
$$- \exp\Big\{-\frac{1}{2T}(x - 2a_2 + 2k(a_2 - a_1))^2\Big\}\Big] dx . \tag{4}$$

Proof. Let $\mathfrak{A}_k^{(i)}$ denote the event that the process $w(t)$ defined on the interval $[0, T]$ crosses the level a_i earlier than it does the level a_j (where $j \neq i$ and $i, j = 1, 2$) and then crosses the interval $[a_1, a_2]$ no fewer than k times (we are assuming that the function $x(t)$ crosses the interval $[a_1, a_2]$ k times if the function

$$\operatorname{sgn}(x(t) - a_1) + \operatorname{sgn}(x(t) - a_2)$$

changes sign k times) and $w(T) \in [c, d]$. The desired probability can be expressed as follows:

$$\mathsf{P}\{w(T) \in [c, d]\} - \mathsf{P}\{\mathfrak{A}_0^{(1)}\} - \mathsf{P}\{\mathfrak{A}_0^{(2)}\} .$$

To calculate $\mathsf{P}\{\mathfrak{A}_0^{(i)}\}$, let us find the probabilities

$$\mathsf{P}\{\mathfrak{A}_k^{(i)}\} + \mathsf{P}\{\mathfrak{A}_{k+1}^{(j)}\} = \mathsf{P}\{\mathfrak{A}_k^{(i)} \cup \mathfrak{A}_{k+1}^{(j)}\} \qquad (i \neq j, i, j = 1, 2) .$$

As one can easily see, the $\mathfrak{A}_k^{(i)} \cup \mathfrak{A}_{k+1}^{(j)}$ is the event that the process $w(t)$ crosses the level a_i prior to the instant T (though not necessarily before it crosses the level a_j) and then crosses the interval $[a_1, a_2]$ no fewer than k times before, at time T falling into the interval $[c, d]$. Let τ_1 denote the instant of first crossing of the level a_i, let τ_2 denote the first crossing of a_j after the instant τ_1, let τ_3 denote the first instant of crossing of a_i after τ_2, etc. We set

$$w_1(t) = \begin{cases} w(t) & \text{for} \quad t < \tau_1 , \\ 2w(\tau_1) - w(t) & \text{for} \quad t \geqq \tau_1 \end{cases}$$

$$w_2(t) = \begin{cases} w_1(t) & \text{for} \quad t < \tau_2 , \\ 2w_1(\tau_2) - w_1(t) & \text{for} \quad t \geqq \tau_2 \end{cases}$$

$$w_3(t) = \begin{cases} w_2(t) & \text{for} \quad t < \tau_3 , \\ 2w_2(\tau_3) - w_2(t) & \text{for} \quad t \geqq \tau_3 , \text{ etc.} \end{cases}$$

We note that the processes $w_l(t)$ are processes of Brownian motion since τ_l is the instant of first crossing of the level

$$a_i + (l - 1)(a_i - a_j)$$

by the process $w_{l-1}(t)$. If the event $\mathfrak{A}_k^{(i)} \cup \mathfrak{A}_{k+1}^{(j)}$ occurs, the process $w_{k+1}(t)$ for $t < T$ crosses successively the levels

$$a_i, a_i + (a_i - a_j), \cdots, a_i + k(a_i - a_j)$$

and at the instant T falls in the interval $[c_k, d_k]$, where

$$\left.\begin{aligned} c_k &= c + (k + 1)(a_i - a_j), \\ d_k &= d + (k + 1)(a_i - a_j) \end{aligned}\right\} \quad \text{for odd} \quad k ,$$

$$\left.\begin{aligned} c_k &= 2a_i - d + k(a_i - a_j), \\ d_k &= 2a_i - c + k(a_i - a_j) \end{aligned}\right\} \quad \text{for even} \quad k .$$

Conversely, if $w_{k+1}(t)$ satisfies these conditions, the event $\mathfrak{A}_k^{(i)} \cup \mathfrak{A}_{k+1}^{(j)}$ occurs. Since $w_{k+1}(t)$ is a continuous process that vanishes at $t = 0$, for $w_{k+1}(T)$ to fall in the interval $[c_k, d_k]$ it must beforehand cross the levels $a_i + l(a_i - a_j)$ for $l = 0, \cdots, k$. Therefore

$$\mathsf{P}\{\mathfrak{A}_k^{(i)} \cup \mathfrak{A}_{k+1}^{(j)}\} = \mathsf{P}\{w_{k+1}(T) \in [c_k, d_k]\} = \mathsf{P}\{w(T) \in [c_k\, d_k]\} .$$

It follows from the continuity of the process $w(t)$ that $w(t)$ crosses, with probability 1, the interval $[a_1, a_2]$ finitely many times and hence,

$\mathsf{P}\{\mathfrak{A}_k^{(i)}\} \longrightarrow 0$ as $k \longrightarrow \infty$. Taking the limit in the equation

$$\mathsf{P}\{\mathfrak{A}_0^{(1)}\} + \mathsf{P}\{\mathfrak{A}_0^{(2)}\} = (-1)^{n+1}[\mathsf{P}\{\mathfrak{A}_{n+1}^{(1)}\} + \mathsf{P}\{\mathfrak{A}_{n+1}^{(2)}\}]$$
$$+ \sum_{k=0}^{n} (-1)^k (\mathsf{P}\{\mathfrak{A}_k^{(1)}\} + \mathsf{P}\{\mathfrak{A}_k^{(2)}\} + \mathsf{P}\{\mathfrak{A}_{k+1}^{(1)}\} + \mathsf{P}\{\mathfrak{A}_{k+1}^{(2)}\})$$

as $n \longrightarrow \infty$, we obtain

$$\mathsf{P}\{\mathfrak{A}_0^{(1)}\} + \mathsf{P}\{\mathfrak{A}_0^{(2)}\} = \sum_{k=0}^{\infty} (-1)^k \sum_{i=1}^{2} (\mathsf{P}\{\mathfrak{A}_k^{(i)}\} + \mathsf{P}\{\mathfrak{A}_{k+1}^{(i)}\})$$
$$= \frac{1}{\sqrt{2\pi T}} \sum_{k=0}^{\infty} \Bigg[\int_{2a_1-d+2k(a_2-a_1)}^{2a_1-c+2k(a_2-a_1)} e^{-(x^2/2T)} dx + \int_{2a_2-d+2k(a_2-a_1)}^{2a_2-c+2k(a_2-a_1)} e^{-(x^2/2T)} dx$$
$$- \int_{c+2(k+1)(a_2-a_1)}^{d+2(k+1)(a_2-a_1)} e^{-(x^2/2T)} dx - \int_{c+2(k+1)(a_2-a_1)}^{d+2(k+1)(a_2-a_1)} e^{-(x^2/2T)} dx \Bigg].$$

Therefore the desired probability is equal to

$$\frac{1}{\sqrt{2\pi T}} \sum_{k=-\infty}^{\infty} \Bigg[\int_{c+2k(a_2-a_1)}^{d+2k(a_2-a_1)} e^{-(x^2/2T)} dx - \int_{2a_2-c+2k(a_2-a_1)}^{2a_2-d+2k(a_2-a_1)} e^{-(x^2/2T)} dx \Bigg].$$

If we set $x - 2k(a_2 - a_1) = u$ in the first integral and

$$2k(a_2 - a_1) + 2a_2 - x = u$$

in the second, we obtain formula (4). This completes the proof of the theorem.

Corollary 1. *The joint distribution of the quantities*

$$\max_{0 \leq t \leq T} w(t), \quad \textit{and} \quad \min_{0 \leq t \leq T} w(t)$$

for $a_1 < 0$ *and* $a_2 > 0$ *is given by the formula*

$$\mathsf{P}\{\min_{0 \leq t \leq T} w(t) < a_1, \max_{0 \leq t \leq T} w(t) < a_2\}$$
$$= \frac{1}{\sqrt{2\pi T}} \sum_{k=-\infty}^{\infty} \int_{a_1}^{a_2} \Bigg[\exp\left\{-\frac{1}{2T}(x + 2k(a_2 - a_1))^2\right\}$$
$$- \exp\left\{-\frac{1}{2T}(x - 2a_2 + 2k(a_2 - a_1))^2\right\}\Bigg] dx. \qquad (5)$$

Corollary 2. *For* $a > 0$ *and* $[c, d] \subset [-a, a]$, *we have*

$$\mathsf{P}\{\sup_{0 \leq t \leq T} |w(t)| < a, w(T) \in [c, d)\}$$
$$= \frac{1}{\sqrt{2\pi T}} \int_c^d \sum_{k=-\infty}^{\infty} (-1)^k \exp\left(-\frac{(x - 2ka)^2}{2T}\right) dx. \qquad (6)$$

6. ON THE GROWTH OF HOMOGENEOUS PROCESSES WITH INDEPENDENT INCREMENTS

Let $\xi(t)$ denote a random process with independent increments. We are interested in the rate of growth of the quantities $\sup_{0 \leq t \leq T} \xi(t)$

and $\sup_{0 \le t \le T} |\xi(t)|$ as $T \to \infty$ and the rate at which the quantities $\sup_{0 \le t \le T} \xi(t) - \xi(0)$ and $\sup_{0 \le t \le T} |\xi(t) - \xi(0)|$ approach 0 as $T \to 0$. To get estimates for these rates, we shall study the quantities $\overline{\lim}_{t\to\infty} \xi(t)/g(t)$, $\overline{\lim}_{t\to\infty} |\xi(t)|/g(t)$, $\overline{\lim}_{t\to 0} \xi(t) - \xi(0)/\varphi(t)$, and $\overline{\lim}_{t\to 0} |\xi(t) - \xi(0)|/\varphi(t)$, where $g(t)$ and $\varphi(t)$ are nondecreasing functions satisfying the relations

$$\lim_{t\to\infty} g(t) = \infty \quad \text{and} \quad \lim_{t\to 0} \varphi(t) = 0 .$$

We shall call the function $g(t)$ an *upper function* for the process $\xi(t)$ if $\mathsf{P}\{\overline{\lim}_{t\to\infty} \xi(t)/g(t) < 1\} = 1$ and we shall call it a *lower function* if $\mathsf{P}\{\overline{\lim}_{t\to\infty} \xi(t)/g(t) > 1\} = 1$. If we replace $g(t)$ with $\varphi(t)$ and analogous relations are satisfied as $t \to 0$, $\varphi(t)$ is called a *locally upper function* and a *locally lower function* for the process $\xi(t)$ respectively. In what follows, we shall consider only regularly increasing upper and lower functions.

Suppose that $g(t)$ is an increasing continuous nonnegative function defined for $t > 0$. We shall say that $g(t)$ is a *function of regular growth* if there exist functions $k_1(\lambda)$ and $k_2(\lambda)$ such that

$$k_1(\lambda)g(t) \leqq g(\lambda t) \leqq k_2(\lambda)g(t) \quad \text{for} \quad \lambda \geqq 0,\ t > 0$$

and if $k_2(\lambda) \to 1$ as $\lambda \to 1$ and $k_1(\lambda) \to \infty$ as $\lambda \to \infty$.

We shall first look at upper and lower functions for symmetric homogeneous processes with independent increments. Then we shall consider upper and lower functions for $|\xi(t)|$, where $\xi(t)$ is not necessarily a symmetric process.

A process $\xi(t)$ is said to be *symmetric* if the processes $\xi(t)$ and $-\xi(t)$ have identical finite-dimensional distributions.

Lemma 1. *Let $\xi(t)$ denote a symmetric separable stochastically continuous process with independent increments. Then*

$$\mathsf{P}\{\sup_{0\le t\le T} \xi(t) > c\} \leqq 2\mathsf{P}\{\xi(T) > c\} . \tag{1}$$

Proof. Let $\xi_1, \xi_2, \cdots, \xi_n$ denote independent symmetrically distributed random variables and define $S_k = \xi_1 + \cdots + \xi_k$. Then

$$\mathsf{P}\{\sup_{1\le k\le n} S_k > c\} \leqq 2\mathsf{P}\{S_n > c\} . \tag{2}$$

To see this, note that since $2\mathsf{P}\{S_n - S_k \geqq 0\} \geqq 1$ we have

$$\begin{aligned}
\mathsf{P}\{\sup_k S_k > c\} &= \sum_{k=1}^n \mathsf{P}\{\sup_{i\le k-1} S_i \leqq c,\ S_k > c\} \\
&\leqq \sum_{k=1}^n \mathsf{P}\{\sup_{i\le k-1} S_i \leqq c,\ S_k > c\}2\mathsf{P}\{S_n - S_k \geqq 0\} \\
&= 2\sum_{k=1}^n \mathsf{P}\{\sup_{i\le k-1} S_i \leqq c,\ S_k > c,\ S_n - S_k \geqq 0\} .
\end{aligned}$$

The events $\{\sup_{i\leqq k-1} S_i \leqq c,\ S_k > c,\ S_n - S_k \geqq 0\}$ are incompatible and each of them implies the event $\{S_n > c\}$. Therefore

$$2\mathsf{P}\{S_n > c\}$$
$$\geqq 2\sum_{k=1}^{n} \mathsf{P}\{\sup_{i\leqq k-1} S_i \leqq c,\ S_k > c,\ S_n - S_k \geqq 0\} \geqq \mathsf{P}\{\sup_k S_k > c\}\ .$$

This establishes formula (2). Let us now apply formula (2) to the quantities $\xi_k = \xi(k/nT) - \xi(k - 1/nT)$ and $\xi_1 = \xi(T/n)$. Then

$$\mathsf{P}\left\{\sup_{1\leqq k\leqq n} \xi\left(\frac{kT}{n}\right) > c\right\} \leqq 2\mathsf{P}\{\xi(T) > c\}\ .$$

Taking the limit as $n \longrightarrow \infty$ and remembering that $\xi(t)$ fails with probability 1 to have discontinuities of the second kind, so that

$$\mathsf{P}\left\{\lim_{n\to\infty} \sup_{1\leqq k\leqq n} \xi\left(\frac{kT}{n}\right) = \sup_{0\leqq t\leqq T} \xi(t)\right\} = 1\ ,$$

we obtain (1). This completes the proof of the lemma.

Theorem 1. *Let $\xi(t)$ denote a symmetric homogeneous separable stochastically continuous process with independent increments and let $g(t)$ denote a function of regular growth such that*

$$\int_1^\infty \frac{1}{t}\mathsf{P}\{\xi(t) > g(t)\}dt < \infty\ .$$

Then for arbitrary $\lambda > 1$ the function $\lambda g(t)$ is an upper function for the process $\xi(t)$.

Proof. Let a denote a number greater than 1. Let $\mathfrak{A}_k$ denote the event that $\sup_{0\leqq t\leqq a^k} \xi(t) > g(a^{k+1})$. Then on the basis of Lemma 1, $\mathsf{P}\{\mathfrak{A}_k\} \leqq 2\mathsf{P}\{\xi(a^k) > g(a^{k+1})\}$. But for $t \in [a^k, a^{k+1}]$ we have

$$g(t) < g(a^{k+1}) \quad \text{and} \quad 2\mathsf{P}\{\xi(t) - \xi(a^k) \geqq 0\} \geqq 1\ .$$

Therefore

$$\mathsf{P}(\mathfrak{A}_k) \leqq 4\mathsf{P}\{\xi(a^k) > g(t)\}\mathsf{P}\{\xi(t) - \xi(a^k) \geqq 0\} \leqq 4\mathsf{P}\{\xi(t) > g(t)\}\ .$$

Consequently,

$$\int_{a^k}^{a^{k+1}} \frac{1}{t}\mathsf{P}\{\mathfrak{A}_k\}dt \leqq 4\int_{a^k}^{a^{k+1}} \frac{1}{t}\mathsf{P}\{\xi(t) > g(t)\}dt$$

and

$$\sum_{k=1}^{\infty} \mathsf{P}\{\mathfrak{A}_k\} \leqq \frac{1}{\ln a}\int_1^\infty \frac{1}{t}\mathsf{P}\{\xi(t) > g(t)\}dt\ .$$

It follows from the Borel-Cantelli lemma (Theorem 2, Section 3, Chapter III) that with probability 1, only finitely many of the events $\mathfrak{A}_k$ occur; that is, for some (generally speaking, random) number, k_0, the events $\mathfrak{A}_k$ do not occur if $k \geqq k_0$. This means that

$$\mathsf{P}\left\{\overline{\lim_{k\to\infty}} \frac{1}{g(a^{k+1})} \sup_{a^{k-1}\leqq t\leqq a^k} \xi(t) \leqq 1\right\} = 1 .$$

For $t \in [a^{k-1}, a^k]$ (where $k \geqq k_0$),

$$\frac{\xi(t)}{k_2(a^2)g(t)} \leqq \frac{1}{k_2(a^2)g(a^{k-1})} \sup_{a^{k-1}\leqq t\leqq a^k} \xi(t) \leqq \frac{1}{g(a^{k+1})} \sup_{a^{k-1}\leqq t\leqq a^k} \xi(t) .$$

Therefore, for arbitrary $a > 1$ and $\lambda > k_2(a^2)$ the function $\lambda g(t)$ is the upper function for $\xi(t)$. This completes the proof of the theorem.

Theorem 2. *Let $\xi(t)$ denote a symmetric homogeneous process with independent increments and let $g(t)$ denote a function of regular growth such that the series*

$$\sum_{k=1}^{\infty} \mathsf{P}\{\xi(a^k) > g(a^k)\}$$

diverges for all $a > 1$. Then for arbitrary $\lambda < 1$, $\lambda g(t)$ is the lower function for the process $\xi(t)$.

Proof. We first show that:

$$\mathsf{P}\left(\overline{\lim_{t\to\infty}} \frac{|\xi(t)|}{\lambda g(t)} > 1\right) \qquad (r < 1) . \tag{3}$$

1. Suppose that

$$\overline{\lim_{t\to\infty}} \mathsf{P}\{\xi(t) > g(t)\} \geqq \frac{1}{2} .$$

Then there exists a sequence $\{t_k\}$ such that

$$\mathsf{P}\{\xi(t_k) > g(t_k)\} \geqq \frac{1}{2} - \frac{1}{k^2} .$$

Hence, by symmetry, $\mathsf{P}(|\xi(t_k)| \leqq g(t_k)) \leqq 1/k^2$, and so the series

$$\sum_{k=1}^{\infty} \mathsf{P}\{|\xi(t_k)| \leqq g(t_k)\} \leqq \sum_{k=1}^{\infty} \frac{1}{k^2}$$

converges. It follows from the Borel-Cantelli lemma that with probability 1, $\xi(t_k) > g(t_k)$ from some k on. This means that for $\lambda < 1$,

$$\mathsf{P}\left\{\overline{\lim_{k\to\infty}} \frac{|\xi(t_k)|}{\lambda g(t_k)} > 1\right\} = 1 .$$

2. Suppose that there exists a $\delta > 0$ such that

$$\overline{\lim_{t\to\infty}} \mathsf{P}\{\xi(t) > g(t)\} < \frac{1}{2} - \delta .$$

Then for sufficiently large t,

$$\mathsf{P}\{|\,\xi(t)\,| \leqq g^{(t)}\} \geqq \delta\ .$$

Consider the independent events

$$\mathfrak{B}_k = \{\xi(a^{k+1}) - \xi(a^k) > g(a^{k+1}) - g(a^k)\}$$

where $a > 1$. Then

$$\mathsf{P}\{\mathfrak{B}_k\} \geqq \int_{-g(a^k)}^{g(a^k)} \mathsf{P}\{\xi(a^{k+1}) - z > g(a^{k+1}) - g(a^k)\}\mathsf{P}\{\xi(a^k) \in dz\}$$

$$\geqq \mathsf{P}\{\xi(a^{k+1}) > g(a^{k+1})\}\int_{-g(a^k)}^{g(a^k)} \mathsf{P}\{\xi(a^k) \in dz\}$$

$$= \mathsf{P}\{\xi(a^{k+1}) > g(a^{k+1})\}\mathsf{P}\{|\,\xi(a^k)\,| \leqq g(a^k)\} \geqq \mathsf{P}\{\xi(a^{k+1}) > g(a^{k+1})\}\delta$$

for sufficiently large k.

Consequently the series $\sum_{k=1}^{\infty} \mathsf{P}\{\mathfrak{B}_k\}$ diverges. Hence on the basis of the Borel-Cantelli lemma, infinitely many of the events $\mathfrak{B}_k$ occur with probability 1. We note that the event $\mathfrak{B}_k$ implies one of the events

$$\{-\xi(a^k) > g(a^k)\};\ \{\xi(a^{k+1}) > g(a^{k+1}) - 2g(a^k)\}\ .$$

Therefore, with probability 1, infinitely many of the events

$$\{|\,\xi(a^k)\,| > g(a^k) - 2g(a^{k-1})\}$$

occur. Given $r < 1$ choose a in such a way that

$$g(a^k) - 2g(a^{k-1}) = g(a^k)\left[1 - \frac{2g(a^{k-1})}{g(a^k)}\right]$$

$$\geqq g(a^k)\left[1 - \frac{2}{k_1(a)}\right] > \lambda g(a^k)$$

(the possibility of such a choice of a is ensured by the regularity of the growth of $g(t)$). We see that

$$\mathsf{P}\left\{\overline{\lim_{t\to\infty}} \frac{|\,\xi(t)\,|}{\lambda g(t)} > 1\right\} = 1\ .$$

We now show that (3) implies that $g(t)$ is a lower function for $\xi(t)$.

$$C = \left\{\overline{\lim_{t\to\infty}} \frac{\xi(t)}{\lambda g(t)} > 1\right\};\ D = \left\{\overline{\lim_{t\to\infty}} \frac{-\xi(t)}{\lambda g(t)} > 1\right\}.$$

Then it follows from (3) that $\mathsf{P}(C \cup D) = 1$. We conclude from the symmetry of the process $\xi(t)$ that $\mathsf{P}(C) = \mathsf{P}(D)$. Finally from the 0-or-1 law (cf. Theorem 5, Section 3, Chapter III) it follows that $\mathsf{P}(C)$ and $\mathsf{P}(D)$ can be only zero or one. This means that $\mathsf{P}(C) = \mathsf{P}(D) = 1$ since otherwise we would have $\mathsf{P}(C) = 0$ and hence $\mathsf{P}(C \cup D) = 0$ which contradicts equation (3). This completes the proof of the theorem.

REMARK 1. Suppose that in these two theorems we consider $a < 1$ and instead of the function $g(t)$ we consider the function $\varphi(t) = 1/g(1/t)$ where g is a function of regular growth. Without otherwise changing the proofs of the theorems, we see that the following assertions are true:

a. If $\xi(t)$ is a symmetric separable homogeneous stochastically continuous process with independent increments and if

$$\int_0^1 \frac{1}{t} \mathsf{P}\{\xi(t) > \varphi(t)\}dt < \infty ,$$

then for $\lambda > 1$ the function $\lambda\varphi(t)$ will be a locally upper function for the process $\xi(t)$; that is, $\mathsf{P}\{\overline{\lim}_{t\to 0} \xi(t)/\lambda\varphi(t) < 1\} = 1$.

b. If $\xi(t)$ is a homogeneous symmetric process with independent increments such that the series $\sum_{k=1}^{\infty} \mathsf{P}\{\xi(a^k) > \varphi(a^k)\}$ diverges for every $a < 1$, then for $\lambda < 1$ the function $\lambda\varphi(t)$ will be a locally lower function for $\xi(t)$; that is, $\mathsf{P}\{\overline{\lim}_{t\to 0} \xi(t)/\lambda\varphi(t) > 1\} = 1$.

Let us apply these results to a process of Brownian motion. Such a process is symmetric. By using the inequalities

$$\frac{1}{\sqrt{2\pi}}\int_z^\infty e^{-(u^2/2)}du \leqq \frac{1}{\sqrt{2\pi}}\int_z^\infty \frac{u}{z} e^{-(u^2/2)}du = \frac{1}{z\sqrt{2\pi}} e^{-(z^2/2)} (z > 0) ,$$

$$\frac{1}{\sqrt{2\pi}}\int_z^\infty e^{-(u^2/2)}du \geqq \frac{1}{\sqrt{2\pi}}\int_z^{z+1} e^{-(u^2/2)}du \geqq \frac{1}{\sqrt{2\pi}} e^{-(z+1)^2/2} (z > 0) ,$$

we see that

$$\frac{1}{\sqrt{2\pi}} e^{-(\frac{z}{\sqrt{t}}+1)^2/2} \leqq \mathsf{P}\{w(t) > z\} \leqq \frac{1}{z\sqrt{2\pi}} e^{-(z^2/2t)} .$$

Let us show that for arbitrary $\varepsilon \in (0, 1)$ $(1 + \varepsilon)\sqrt{2t \ln \ln t}$ and $(1 - \varepsilon)\sqrt{2t \ln \ln t}$ are upper and lower functions respectively. We have

$$\begin{aligned} &\mathsf{P}\{w(t) > (1 + \varepsilon)\sqrt{2t \ln \ln t}\} \\ &\leqq \frac{1}{\sqrt{2\pi(1 + \varepsilon)^2 2 \ln \ln t}} \exp\left\{-\frac{(1 + \varepsilon)^2 2 \ln \ln t}{2}\right\} \\ &= O((\ln t)^{-(1+\varepsilon)^2}) , \end{aligned}$$

and the integral $\int_c^\infty dt/t(\ln t)^{(1+\varepsilon)^2}$ converges for $c > 1$. On the other hand,

$$\begin{aligned} &\mathsf{P}\{w(a^k) > (1 - \varepsilon)\sqrt{2a^k \ln \ln a^k}\} \\ &\geqq \frac{e^{-(1/2)}}{\sqrt{2\pi}} \exp\{-(1 - \varepsilon^2)[\ln \ln a^k + \sqrt{2 \ln \ln a^k}\,]\} \\ &\geqq C \exp\{-\alpha \ln \ln a^k\} = C(\ln a^k)^{-\alpha} \end{aligned}$$

if $(1-\varepsilon^2)[1+(\sqrt{2^{-1}\ln\ln a^k})^{-1}] < \alpha < 1$ (as will be the case for sufficiently large k). Consequently the series

$$\sum \mathsf{P}\{w(a^k) > (1-\varepsilon)\sqrt{2a^k \ln\ln a^k}\}$$

diverges. Thus we have proved:

Theorem 3. *If $w(t)$ is a separable process of Brownian motion, then*

$$\mathsf{P}\left\{\overline{\lim_{t\to\infty}} \frac{w(t)}{\sqrt{2t\ln\ln t}} = 1\right\} = 1 .$$

By using Remark 1 we can prove:

Theorem 4. *If $w(t)$ is a separable process of Brownian motion, then*

$$\mathsf{P}\left\{\overline{\lim_{t\to 0}} \frac{w(t)}{\sqrt{2t\ln\ln \dfrac{1}{t}}} = 1\right\} = 1 .$$

Theorem 3 and 4 are called the "law of the iterated logarithm."

In studying upper and lower functions for $|\xi(t)|$ where $\xi(t)$ is a process with independent increments, we use:

Lemma 2. *Let $\xi(t)$ denote a separable stochastically continuous process with independent increments, for which there exists an $\alpha < 1$ such that*

$$\mathsf{P}\{|\xi(T) - \xi(s)| > C\} \leqq \alpha$$

for $0 < s \leqq T$. Then for every $x > 0$,

$$\mathsf{P}\{\sup_{0\leqq s\leqq T} |\xi(s)| > C + x\} \leqq \frac{1}{1-\alpha}\mathsf{P}\{|\xi(T)| > x\} . \tag{4}$$

Proof. It follows from Theorem 2, Section 4, Chapter III that

$$\mathsf{P}\left\{\sup_{1\leqq k\leqq n}\left|\xi\left(\frac{k}{n}T\right)\right| > x + C\right\} \leqq \frac{1}{1-\alpha}\mathsf{P}\{|\xi(T)| > x\} .$$

Taking the limit as $n\to\infty$, we obtain proof of the lemma.

Theorem 5. *Let $\xi(t)$ denote a separable homogeneous stochastically continuous process with independent increments and let $g(t)$ denote a function of regular growth such that for arbitrary $\varepsilon > 0$,*

$$\overline{\lim_{t\to\infty}}\,\mathsf{P}\{|\xi(t)| > \varepsilon g(t)\} < 1$$

and

$$\int_1^\infty \frac{1}{t}\mathsf{P}\{|\xi(t)| > g(t)\}dt < \infty .$$

Then for arbitrary $\lambda > 1$ *the function* $\lambda\, g(t)$ *is an upper function for* $|\,\xi(t)\,|$; *that is,*

$$\mathsf{P}\left\{\overline{\lim_{t\to\infty}}\frac{|\,\xi(t)\,|}{g(t)} \leqq 1\right\} = 1\ .$$

Proof. Let us choose $a \in (1, 2)$ and $\varepsilon > 0$. Let $\mathfrak{A}_k$ denote the event

$$\{\sup_{t\leqq a^k} |\,\xi(t)\,| > (1 + 2\varepsilon)g(a^{k+1})\}\ .$$

It follows from the hypotheses of the theorem that there exist a $c > 0$ and a T_0 such that

$$\mathsf{P}\{|\,\xi(t)\,| > \varepsilon g(t)\} < 1 - c$$

for $t \geqq T_0$. Since for $T_0 \leqq t < a^k$,

$$\mathsf{P}\{|\,\xi(t)\,| > \varepsilon g(a^{k+1})\} < 1 - c\ ,$$
$$\mathsf{P}\{|\,\xi(t)\,| > \varepsilon g(t)\} < 1 - c\ ,$$

and

$$\lim_{k\to\infty}\sup_{t\leqq T_0} \mathsf{P}\{|\,\xi(t)\,| > \varepsilon g(a^{k+1})\} = 0\ ,$$

it follows on the basis of Lemma 2 that for sufficiently large k,

$$\mathsf{P}\{\mathfrak{A}_k\} \leqq \frac{1}{c}\mathsf{P}\{|\,\xi(a^k)\,| > (1 + \varepsilon)g(a^{k+1})\}\ .$$

Remembering that $t - a^k \leqq (a - 1)a^k \leqq a^k$ for $t \in [a^k, a^{k+1}]$ and hence

$$\mathsf{P}\{|\,\xi(t) - \xi(a^k)\,| > \varepsilon g(a^k)\} < 1 - c$$

for sufficiently large k, we obtain

$$\mathsf{P}\{\mathfrak{A}_k\} \leqq \frac{1}{c^2}\mathsf{P}\{|\,\xi(a^k)\,| > (1 + \varepsilon)g(a^{k+1})\}\mathsf{P}\{|\,\xi(t) - \xi(a^k)\,|$$

$$\leqq \varepsilon g(a^{k+1})\} \leqq \frac{1}{c^2}\mathsf{P}\{|\,\xi(t)\,| > g(a^{k+1})\} \leqq \frac{1}{c^2}\mathsf{P}\{|\,\xi(t)\,| > g(t)\}$$

so that

$$\mathsf{P}\{\mathfrak{A}_k\} \leqq \frac{1}{c^2}\int_{a^k}^{a^{k+1}}\frac{1}{t}\mathsf{P}\{|\,\xi(t)\,| > g(t)\}dt\ ,$$

$$\sum_{k=l}^{\infty} \mathsf{P}\{\mathfrak{A}_k\} \leqq (\ln a)^{-1}c^{-2}\int_{a^l}^{\infty}\frac{1}{t}\mathsf{P}\{|\,\xi(t)\,| > g(t)\}dt\ .$$

This means that with probability 1, only finitely many events $\mathfrak{A}_k$ occur. By using the reasoning of Theorem 1, we see that the function $\lambda(1 + 2\varepsilon)k_2(a^2)g(t)$ is, for $\lambda > 1$, an upper function for $|\,\xi(t)\,|$. Since $k_2(a) \to 1$ $a \to 1$ and since $a > 1$ and $\varepsilon > 0$ are arbitrary, the assertion of the theorem follows.

Analyzing the proof of Theorem 2, one can easily prove:

Theorem 6. *Let $\xi(t)$ denote a homogeneous process with independent increments. If a function of regular growth is such that the series*

$$\sum_{k=1}^{\infty} \mathsf{P}\{|\,\xi(a^k)\,| > g(a^k)\}$$

diverges for every $a > 1$, then for every $0 < \lambda < 1$ the function $\lambda g(t)$ is a lower function for $|\,\xi(t)\,|$; that is,

$$\mathsf{P}\left\{\varlimsup_{t\to\infty} \frac{|\,\xi(t)\,|}{g(t)} \geqq 1\right\} = 1\,.$$

The results of Theorems 5 and 6 can be reformulated for the case in which $t \to 0$, in a manner analogous to that used in Remark 1.

VII

JUMP MARKOV PROCESSES

Let X denote an arbitrary space with fixed σ-algebra $\mathfrak{B}$. Let us interpret X as the phase space of some physical system $\sum$ and let us denote the state of $\sum$ at the instant t by $\xi(t)$ $(\in X)$. Let us suppose that the time t varies in discrete amounts ($t = 0, 1, 2, \cdots$).

Let us suppose that the change in the system $\sum$, from its state x at the instant t into another state at the next instant $t + 1$, is completely determined by the time t, the state x, and some random factor α_t that constitutes, for the different values of t, a sequence of independent random elements. Thus

$$\xi(t + 1) = f(t, \xi(t), \alpha_t) , \tag{1}$$

where $f(t, x, \alpha)$ is a function of the three variables t, x, and α, where $t = 0, 1, 2, \cdots$, $x \in X$, and $\alpha \in \Lambda$. Formula (1) enables us to express the state of the system $\sum$ at an arbitrary instant s by starting with the state $\xi(t)$ of the system $\sum$ at an instant $t < s$:

$$\xi(s) = g_{t,s}(\xi(t), \alpha_t, \alpha_{t+1}, \cdots, \alpha_{s-1}) . \tag{2}$$

We emphasize that $\xi(t)$ in this equation is independent of the set $\alpha_t, \alpha_{t+1}, \cdots, \alpha_{s-1}$.

Let $\{\Omega, \mathfrak{S}, \mathsf{P}\}$ denote the probability space on which the random elements α_t are defined. Let us suppose that for arbitrary fixed t and s (where $s > t$) the function $g_{t,s}(x, \alpha_t, \alpha_{t+1}, \cdots, \alpha_{s-1})$ is $(\mathfrak{B} \times \mathfrak{S})$-measurable. Then if the motion of the system $\sum$ begins at the instant t and its initial state $\xi(t) = x$ is known, formula (2) enables us to determine the probability that $\sum$ will fall in an arbitrary set $A \in \mathfrak{B}$ at the instant $s > t$. We shall call this probability the *transition probability* and we shall indicate it by $\mathsf{P}(t, x, s, A)$. If $\chi_A(x)$ denotes the characteristic function of the set A, then

$$\mathsf{P}(t, x, s, A) = \mathsf{M}\chi_A[g_{t,s}(x, \alpha_t, \cdots, \alpha_{s-1})] . \tag{3}$$

Let u and v denote two numbers such that $t < u < v$. It follows from formula (2) (cf. Theorem 7, Section 4, Chapter IV) and the independence of the random variables $\alpha_t, \cdots, \alpha_{v-1}$ that

$$\mathsf{P}(t, x, v, A) = \mathsf{M}\chi_A[g_{u,v}(\xi(u), \alpha_u, \cdots, \alpha_{v-1})]$$
$$= \mathsf{M}[\{\mathsf{M}\chi_A[g_{u,v}(y, \alpha_u, \cdots, \alpha_{v-1})]\}_{y=\xi(u)}] = \mathsf{M}\mathsf{P}(u, \xi(u), v, A) ,$$

which may be rewritten

$$\mathsf{P}(t, x, v, A) = \int \mathsf{P}(u, y, v, A)\mathsf{P}(t, x, u, dy), \; t < u < v . \quad (4)$$

Equation (4) is called the *Chapman-Kolmogorov equation.* It expresses an important property of the systems that we are considering, namely the absence of aftereffects: if we know the state of a system at a certain instant u, the probabilities of transition from that state do not depend on the motion of the system at previous instants of time. Systems enjoying this property are called *Markov systems.* They are frequently encountered in equations of science and technology.

Chapters 7 and 8 are devoted to a study of Markov processes. In this chapter, we shall consider systems whose motion can be characterized by the fact that the system $\sum$ is immobile in phase space for some period of time and at a random instant its position changes by a jump. In the next chapter we shall consider systems whose states change continuously with time.

1. TRANSITION PROBABILITIES

Let X denote an arbitrary space with fixed σ-algebra of sets $\mathfrak{B}$ and let $\mathfrak{T}$ denote a set of real numbers.

Definition 1. A family of functions $\mathsf{P}(t, x, u, A)$, where $t, u \in \mathfrak{T}$, $t < u$, $x \in X$, and $A \in \mathfrak{B}$, is called a *Markov process in the broad sense* in the phase space X if the functions P satisfy the following conditions:

a. $\mathsf{P}(t, x, u, A)$ is, for fixed t, x, and u, a probability measure on $\mathfrak{B}$,

b. for fixed t, u, and A, the function $\mathsf{P}(t, x, u, A)$ as a function of the variable x is $\mathfrak{B}$-measurable, and

c. for arbitrary t, u, v, x, and A (with $t < u < v$) the functions $\mathsf{P}(t, x, u, A)$ satisfy the Chapman-Kolmogorov equation

$$\mathsf{P}(t, x, v, A) = \int_X \mathsf{P}(t, x, u, dy)\mathsf{P}(u, y, v, A) . \quad (1)$$

The functions $\mathsf{P}(t, x, u, A)$ are called the *transition probabilities.*

According to our interpretation of transition probabilities we naturally assume that

$$P(t, x, t, A) = \chi_A(x) , \tag{2}$$

where $\chi_A(x)$ is the characteristic function of the set $A \in \mathfrak{B}$.

There are two families of operators connected with transition probabilities:

1. Suppose that the distribution of the position of the system $\sum$ in phase space is given at the instant t_0 and suppose that $\mu_0(A) = P\{\xi(t_0) \in A\}$ where $A \in \mathfrak{B}$. Let the distribution of the system $\sum$ at the instant $t > t_0$ be denoted by $\mu_t(A)$. Then

$$\mu_t(A) = \int_X \mu_0(dx)P(t_0, x, t, A) . \tag{3}$$

Formula (3) defines an operator $T^{[t_0,t]}$ (for $t_0 \leqq t$), which maps the probability measure $\mu_0(A)$ into a new probability measure $\mu_t(A)$. This is true because (obviously) $\mu_t(A)$ is nonnegative, $\mu_t(X) = \int_X \mu_0(dx)P(t_0, x, t, X) = \int_X \mu_0(dx) = 1$, and the countable additivity of $\mu_t(A)$ follows from the countable additivity of the integral and the transition probability $P(t_0, x, t, A)$. Furthermore, if instead of the probability measure μ_0 in formula (3) we substitute an arbitrary finite charge W_0, the right-hand side of this formula remains meaningful and we obtain a transformation $T^{[t_0,t]}$ in the space $\boldsymbol{W}$ of all finite charges (cf. Section 1, Chapter II):

$$W_t(A) = \int_X W_0(dx)P(t_0, x, t, A) . \tag{4}$$

If $W_0 = W_0^+ - W_0^-$, where W_0^+ and W_0^- are the positive and negative variations respectively of the function W_0, then

$$W_t(A) = \int W_0^+(dx)P(t_0, x, t, A) - \int W_0^-(dx)P(t_0, x, t, A) . \tag{5}$$

Consequently, for the positive and negative variations of the charge $W_t(A)$ we have

$$W_t^+(A) \leqq \int_X W_0^+(dx)P(t_0, x, t, A) ,$$

$$W_t^-(A) \leqq \int_X W_0^-(dx)P(t_0, x, t, A) ,$$

from which it follows that

$$\| W_t \| = W_t^+(X) + W_t^-(X) \leqq W_0^+(X) + W_0^-(X) = \| W_0 \| .$$

Thus the operators $T^{[t_0,t]}$ map $\boldsymbol{W}$ into itself, they are linear, and $\| T^{[t_0,t]} \| \leqq 1$.

The Chapman-Kolmogorov formula provides a simple law of

composition of the operators $T^{[t_0,t]}$. To establish this law, we need an auxiliary proposition.

Let X and Y denote two spaces, let $\mathfrak{S}_1$ and $\mathfrak{S}_2$ denote σ-algebras of subsets of X and Y respectively, and let μ denote a finite charge on $\mathfrak{S}_1$. Suppose that $\nu(B \mid x)$ is for fixed $B \in \mathfrak{S}_2$ a $\mathfrak{S}_1$-measurable function of the argument $x \in X$, and that for fixed x it is a finite measure on $\mathfrak{S}_2$. Suppose that

$$\sup_{x \in X} \nu(Y \mid x) \leqq c .$$

We define

$$\lambda(B) = \int_X \nu(B \mid x)\mu(dx) .$$

Then $\lambda(B)$ is a finite charge on $\mathfrak{S}_2$.

Lemma 1. *For an arbitrary bounded $\mathfrak{S}_2$-measurable function $f(y)$,*

$$\int_X \mu(dx)\int_Y f(y)\nu(dy \mid x) = \int_Y f(y)\lambda(dy) . \tag{6}$$

Equation (6) can be written more expressively in the form

$$\int_X \mu(dx)\int_Y f(y)\nu(dy \mid x) = \int_Y f(y)\int_X \nu(dy \mid x)\mu(dx) . \tag{7}$$

It will be sufficient to prove the assertion of the lemma for the case in which $\mu(A)$ is a measure. We then have $\lambda(B) \leqq c\mu(X)$ and, by virtue of Lebesgue's theorem on the integration of a series of nonnegative functions (cf. Theorem 1, Section 5, Chapter II), $\lambda(B)$ is a measure on $\mathfrak{S}_2$. Let K denote the class of functions for which (6) holds. Then K contains the characteristic functions of the sets $\mathfrak{S}_2$, it is linear, and therefore it contains all simple functions. Furthermore, K is closed under the operation of passing to the limit with respect to monotonic sequences. Consequently it contains all nonnegative and all bounded $\mathfrak{S}_2$-measurable functions. This completes the proof of the lemma.

Let us now turn to formula (4). Suppose that $t_0 < u < t$. By using equation (1) and the lemma just proved, we obtain

$$\begin{aligned} W_t(A) &= \int_X W_0(dx)\int_X \mathsf{P}(u, y, t, A)\mathsf{P}(t_0, x, u, dy) \\ &= \int_X \left(\int_X \mathsf{P}(t_0, x, u, dy)W_0(dx)\right)\mathsf{P}(u, y, t, A) \\ &= \int_X W_u(dy)\mathsf{P}(u, y, t, A) , \end{aligned}$$

This equation can be written

$$T^{[t_0,t]} = T^{[u,t]}T^{[t_0,u]} \qquad (t_0 < u < t) . \tag{8}$$

Consequently if we consider the set of operators $T^{[t_0,t]}$ as a function of the interval $[t_0, t]$, it is in a certain sense a directed noncommutative multiplicative family on the interval.*

2. If we know that the system is in a state x at an instant t, the mathematical expectation of a function $f(\xi(v))$ of the state of the system at the instant $v > t$ is given by the expression

$$f_t(x) = \int_X f(y)\mathsf{P}(t, x, v, dy), \qquad t < v . \tag{9}$$

where $f(y)$ is bounded and $\mathfrak{B}$-measurable. Here,

$$f_v(x) = f(x) . \tag{10}$$

Formula (9) defines an operator $S^{[t,v]}$ defined on the normed linear space B of all $\mathfrak{B}$-measurable bounded functions with norm $||f|| = \sup_{x \in X} |f(x)|$.

Obviously, $||f_t(x)|| \leqq ||f(x)||$, $S^{[t,v]}(1) = 1$, so that $||S^{[t,v]}|| = 1$. By using formula (1) and Lemma 1, we obtain for $t < u < v$,

$$\begin{aligned} f_t(x) &= \int_X f(y)\mathsf{P}(t, x, v, dy) \\ &= \int_X f(y)\int_X \mathsf{P}(u, z, v, dy)\mathsf{P}(t, x, u, dz) \\ &= \int_X \mathsf{P}(t, x, u, dz)\int_X f(y)\mathsf{P}(u, z, v, dy) \\ &= \int_X f_u(z)\mathsf{P}(t, x, u, dz) , \end{aligned}$$

or

$$S^{[t,v]} = S^{[t,u]}S^{[u,v]} \qquad (t < u < v) , \tag{11}$$

that is, the operators $S^{[t,v]}$ also constitute a multiplicative family on the interval, although it is directed in a different way from that of the family of operators $T^{[t,v]}$.

Definition 2. A Markov process is said to be *homogeneous in time* if the transition probabilities $\mathsf{P}(t, x, u, A)$ as functions of t and u depend only on $u - t$:

$$\mathsf{P}(t, x, u, A) = \mathsf{P}(u - t, x, A) .$$

For processes that are homogeneous in time, the operators $T^{[t,v]}$ and $S^{[t,v]}$ depend only on the single scalar parameter $v - t > 0$. Here the Chapman-Kolmogorov equation becomes

---

* For a noncommutative multiplicative family on an interval, the directedness characterizes one of the two possible orders of the factors corresponding to the given partition of the interval.

$$P(t+\tau, x, A) = \int_X P(t, x, dy)P(\tau, y, A) , \tag{12}$$

and the relations (8) and (11) become

$$T^{t+\tau} = T^t T^\tau , \tag{13}$$

$$S^{t+\tau} = S^\tau S^t, \qquad t > 0, \qquad \tau > 0 , \tag{14}$$

where

$$T^t = T^{[t_0, t_0+t]}, \qquad S^t = S^{[t_0, t_0+t]} .$$

Here $T^0 = S^0 = I$, where I is the identity operator.

It follows from equations (13) and (14) that the operators T^t (S^t) commute with each other. If $\mathfrak{T}$ is the infinite interval $[a, \infty)$, then T^t (S^t) constitute semigroups of operators.

Markov processes can be classified according to their phase spaces. In this sense the simplest of them are Markov processes with finitely or countably many states, that is, processes for which X consists of a finite or countable number of points. In such a case it is sufficient to give the transition probabilities for one-point sets. Specifically, if we let the points of X be denoted by the letters $i, j, \cdots$ and if $p_{ij}(t, u) = P(t, i, u, j)$ is the probability of transition into the set consisting of the single point j, then

$$P(t, i, u, A) = \sum_{j \in A} p_{ij}(t, u) \qquad (u > t) .$$

The Chapman-Kolmogorov equation for the probabilities $p_{ij}(t, \tau)$ is written

$$p_{ij}(t, v) = \sum p_{ik}(t, u) p_{kj}(u, v), \qquad t < u < v, i, j \in X . \tag{15}$$

In the homogeneous case we write $p_{ij}(t, v) = p_{ij}(v - t)$, and equations (15) take the form

$$p_{ij}(t+\tau) = \sum_{k \in X} p_{ik}(t) p_{kj}(\tau), \qquad t > 0, \tau > 0 . \tag{16}$$

In the following section the transition probabilities of a homogeneous process with countably many states will be considered in greater detail.

2. HOMOGENEOUS PROCESSES WITH COUNTABLY MANY STATES

Let us consider a homogeneous Markov process in the broad sense with a countable set of states J and with transition probabilities $\{p_{ij}(t); i, j \in J\}$. It follows from Definition 1 of Section 1 that the transition probabilities satisfy the following relations:

a. $p_{ij}(t) \geqq 0, \qquad t \geqq 0$;

b. $\sum_{j \in J} p_{ij}(t) = 1$;

c. $p_{ij}(t + s) = \sum_{k \in J} p_{ik}(t) p_{kj}(s), \qquad t \geqq 0, s \geqq 0$.

To these three conditions we add a fourth

$$\text{d.} \quad \lim_{t \to 0} p_{ij}(t) = p_{ij}(0) = \delta_{ij}, \qquad \delta_{ij} = \begin{cases} 1 & \text{if } i = j , \\ 0 & \text{if } i \neq j , \end{cases}$$

which we shall assume throughout to be satisfied.

From the probability point of view condition d indicates the stochastic continuity of the process at $t = 0$ no matter what the initial state of the system was.

Sometimes condition b is replaced with the weaker condition, $\sum_{j \in J} p_{ij}(t) \leqq 1$.

The case $\sum_{j \in J} p_{ij}(t) < 1$ can be interpreted as follows: A system finding itself in the ith state at some instant of time will, with positive probability equal to $1 - \sum_{i \in J} p_{ij}(t)$, be absent from the phase space after an interval of time t. In other words, there are too few points in the phase space to describe all possible states of the system. Let us agree to call processes of this type *improper Markov processes*. One can easily see that by adding a certain set of points to the phase space, we can extend the domain of definition of an improper Markov space and turn it into a Markov process in the proper sense, without changing the given transition probabilities. This can be done by simply adding to the phase space an "absorbing" "infinitely distant" state "∞". We define

$$J^* = J \cup \{\infty\}, \qquad p_{i\infty}(t) = 1 - \sum_{j \in J} p_{ij}(t) ,$$

$$p_{\infty i}(t) = 0, \qquad i \in J, \qquad p_{\infty\infty}(t) = 1 .$$

It is easy to see that the set of transition probabilities $\{p_{ij}(u)\}$; $i, j \in J^*$ constitutes a Markov process in the proper sense. To prove this it will be sufficient to show that condition c is satisfied. We have

$$p_{ij}(t + s) = \sum_{\alpha \in J} p_{i\alpha}(t) p_{\alpha j}(s) = \sum_{\alpha \in J} p_{i\alpha}(t) p_{\alpha j}(s)$$
$$+ p_{i\infty}(t) p_{\infty j}(s) = \sum_{\alpha \in J^*} p_{i\alpha}(t) p_{\alpha j}(s), \qquad i, j \in J ,$$

$$p_{\infty j}(t + s) = 0 = \sum_{\alpha \in J} p_{\infty\alpha}(t) p_{\alpha j}(s) + p_{\infty\infty}(t) p_{\infty j}(s)$$
$$= \sum_{\alpha \in J^*} p_{\infty\alpha}(t) p_{\alpha j}(s), \qquad j \in J ,$$

$$p_{\infty\infty}(t + s) = 1 = \sum_{\alpha \in J} p_{\infty\alpha}(t) p_{\alpha\infty}(s) + p_{\infty\infty}(t) p_{\infty\infty}(s)$$
$$= \sum_{\alpha \in J^*} p_{\infty\alpha}(t) p_{\alpha\infty}(s) ,$$

$$p_{i\infty}(t+s) = 1 - \sum_{\alpha \in J} p_{i\alpha}(t+s) = 1 - \sum_{\alpha \in J} \sum_{\beta \in J} p_{i\beta}(t) p_{\beta\alpha}(s)$$
$$= 1 - \sum_{\beta \in J} p_{i\beta}(t) \sum_{\alpha \in J} p_{\beta\alpha}(s) = 1 - \sum_{\beta \in J} p_{i\beta}(t)(1 - p_{\beta\infty}(s))$$
$$= p_{i\infty}(t) + \sum_{\beta \in J} p_{i\beta}(t) p_{\beta\infty}(s) = p_{i\infty}(t) p_{\infty\infty}(s)$$
$$+ \sum_{\beta \in J} p_{i\beta}(t) p_{\beta\infty}(s) = \sum_{\beta \in J^*} p_{i\beta}(t) p_{\beta\infty}(s) .$$

Thus improper Markov processes do not actually broaden the class of Markov processes. The general properties of transition probabilities of homogeneous Markov processes are also properties of transition probabilities in improper Markov processes. Furthermore, for improper processes the quantity $p_{i\infty}(t) = 1 - \sum_{\alpha \in J} p_{i\alpha}(t)$ enjoys the same properties as the transition probability $p_{i\alpha}(t)$ for $\alpha \in J$. It might be noted that the quantity $p_{i\infty}(t)$ is a nondecreasing function since as we have just seen,

$$p_{i\infty}(t+s) = p_{i\infty}(t) + \sum_{\beta \in J} p_{i\beta}(t) p_{\beta\infty}(s) \geqq p_{i\infty}(t), \qquad s > 0 .$$

We shall therefore confine ourselves to a consideration of Markov processes in the proper sense.

Theorem 1. *The transition probabilities of a homogeneous Markov process with countably many states are uniformly continuous for $t \geqq 0$ (with i, j fixed).*

Proof. Suppose that $h > 0$. Then

$$p_{ij}(t+h) - p_{ij}(t) = \sum_{\alpha \in J} p_{i\alpha}(h) p_{\alpha j}(t) - p_{ij}(t)$$
$$= -(1 - p_{ii}(h)) p_{ij}(t) + \sum_{\alpha \in J \setminus \{i\}} p_{i\alpha}(h) p_{\alpha j}(t) ,$$

from which it follows that

$$-(1 - p_{ii}(h)) \leqq p_{ij}(t+h) - p_{ij}(t) \leqq \sum_{\alpha \in J \setminus \{i\}} p_{i\alpha}(h) = 1 - p_{ii}(h) ;$$

that is, $| p_{ij}(t+h) - p_{ij}(t) | \leqq 1 - p_{ii}(h)$.

For negative h it follows from the preceding inequality that

$$| p_{ij}(t+h) - p_{ij}(t) | \leqq 1 - p_{ii}(|h|) , \tag{1}$$

from which the uniform continuity of the function $p_{ij}(t)$ for $t \geqq 0$ follows.

Let us look at the question of the differentiability of the transition probabilities $p_{ij}(t)$.

Theorem 2. *The limit $q_i = \lim_{h \to 0} [\{1 - p_{ii}(h)\}/h] \leqq \infty$ always exists, and for arbitrary $t > 0$, $\{1 - p_{ii}(t)\}/t \leqq q_i$.*

Proof. Let us set $\tilde{q}_i = \underline{\lim}_{h \to 0} [\{1 - p_{ii}(h)\}/h]$.

If $\tilde{q}_i = \infty$, then q_i exists and $q_i = \infty$. Suppose that $\tilde{q}_i < \infty$.

Let us show that

$$\frac{1 - p_{ii}(t)}{t} \leqq \tilde{q}_i \tag{2}$$

for arbitrary $t > 0$. Let ε and t denote arbitrary positive numbers. There exists an h such that

$$\frac{1 - p_{ii}(h)}{h} \leqq \tilde{q}_i + \frac{\varepsilon}{2}$$

and $1 - p_{ii}(s) < \varepsilon t/2$ for all $0 \leqq s \leqq h$. Let us set $t = nh + s$ for $0 \leqq s < h$. It follows from condition c that $p_{ii}(t + s) \geqq p_{ii}(t)p_{ii}(s)$. Therefore

$$\begin{aligned} p_{ii}(t) &\geqq [p_{ii}(h)]^n p_{ii}(s) \geqq \left(1 - h\left(\tilde{q}_i + \frac{\varepsilon}{2}\right)\right)^n \left(1 - \frac{\varepsilon t}{2}\right) \\ &\geqq \left[1 - nh\left(\tilde{q}_i + \frac{\varepsilon}{2}\right)\right]\left(1 - \frac{\varepsilon t}{2}\right) \\ &\geqq 1 - \frac{\varepsilon t}{2} - t\left(\tilde{q}_i + \frac{\varepsilon}{2}\right) = 1 - t(\tilde{q}_i + \varepsilon) . \end{aligned}$$

Taking the limit as $\varepsilon \to 0$ in this inequality, we obtain inequality (2). From that inequality it follows that

$$\overline{\lim_{t \to 0}} \frac{1 - p_{ii}(t)}{t} \leqq \tilde{q}_i ,$$

so that

$$\tilde{q}_i = \overline{\lim_{t \to 0}} \frac{1 - p_{ii}(t)}{t} = \underline{\lim_{t \to 0}} \frac{1 - p_{ii}(t)}{t} = \lim_{t \to 0} \frac{1 - p_{ii}(t)}{t} ,$$

which completes the proof.

Theorem 3. *If $q_i < \infty$, then for all $t > 0$ the derivatives $p'_{ij}(t)$ for $j \in J$ exist and are continuous. They satisfy the following relations:*

$$p'_{ij}(t + s) = \sum_{k \in J} p'_{ik}(t)p_{kj}(s) , \qquad t > 0 , \quad s > 0 , \tag{3}$$

$$\sum_{k \in J} p'_{ik}(t) = 0 , \tag{4}$$

$$\sum_{k \in J} | p'_{ik}(t) | \leqq 2q_i . \tag{5}$$

Proof. We set

$$\Delta_{ij}(t, t + s) = \frac{p_{ij}(t + s) - p_{ij}(t)}{s} .$$

On the basis of condition (c) and inequality (2) we have

$$\Delta_{ij}(t, t + s) \geqq p_{ij}(t)\frac{p_{ii}(s) - 1}{s} \geqq -q_i p_{ij}(t) . \tag{6}$$

Summing this inequality over an arbitrary set $M \subset J$ of indices, we obtain

$$\sum_{j \in M} \Delta_{ij}(t, t + s) \geqq -q_i .$$

Since $\sum_{j \in J} \Delta_{ij}(t, t + s) = 0$, we have

$$\sum_{j \in M} \Delta_{ij}(t, t + s) = -\sum_{j \notin M} \Delta_{ij}(t, t + s) \leqq q_i .$$

Thus

$$\left| \sum_{j \in M} \Delta_{ij}(t, t + s) \right| \leqq q_i . \tag{7}$$

Let P denote the set of all j such that $\Delta_{ij}(t, t + s) \geqq 0$. Then

$$\sum_{j \in J} | \Delta_{ij}(t, t + s) | = \sum_{j \in P} \Delta_{ij}(t, t + s) + \left| \sum_{j \notin P} \Delta_{ij}(t, t + s) \right| ;$$

that is

$$\sum_{j \in J} | \Delta_{ij}(t, t + s) | \leqq 2q_i . \tag{8}$$

In particular, $| \Delta_{ij}(t, t + s) | \leqq 2q_i$; that is, the functions $p_{ij}(t)$ satisfy a Lipschitz condition. Therefore $p_{ij}(t) = \delta_{ij} + \int_0^t p'_{ij}(u)du$, where $p'_{ij}(t)$ is a Lebesgue-integrable function and $p_{ij}(t)$ has a derivative $p'_{ij}(t)$ almost everywhere.

It follows from (8) that for $t \geqq 0$,

$$\sum_{j \in J} | p'_{ij}(t) | \leqq 2q_i \tag{9}$$

almost everywhere. Using inequality (9), we obtain

$$0 = \sum_{j \in J} \int_0^t p'_{ij}(u)du = \int_0^t \sum_{i \in J} p'_{ij}(u)du ,$$

so that $\sum_{i \in J} p'_{ij}(t) = 0$ for almost all $t \geqq 0$.

Let us now consider the case in which $t = \tau$, for which inequality (9) is satisfied and the derivatives $p'_{ij}(\tau)$ exist for arbitrary j.

Let us begin by setting for $t > \tau$,

$$A = \sum_{k \in J} p'_{ik}(\tau) p_{kj}(t - \tau)$$

and let us consider the difference $\Delta_{ij}(t, t + s) - A$. Since

$$\Delta_{ij}(t, t + s) = \sum_{k \in J} \Delta_{ik}(\tau, \tau + s) p_{kj}(t - \tau) ,$$

we have

$$p'_{ij}(t) \leqq \sum_{k \in J} p'_{ik}(\tau) p_{kj}(t - \tau)$$

for almost every t by Fatou's inequality and (6). Now

$$0 = \sum_{j \in J} p'_{ij}(t) = \sum_{j \in J} \sum_{k \in J} p'_{ik}(\tau) p_{kj}(t - \tau)$$
$$= 0$$

and hence

$$p'_{ij}(t) = \sum_{k \in J} p'_{ik}(\tau) p_{kj}(t - \tau) \tag{10}$$

for almost every t. However, by (9), the right-hand side of (10) is uniformly continuous in $t \geqq \tau$. Hence $p_{ij}(t)$ is absolutely continuous with a derivative which is uniformly continuous on the complement of a null set; i.e., $p'_{ij}(u)$ is essentially continuous. Since $p_{ij}(t) = p_{ij}(\tau) + \int_{\tau}^{t} p'_{ij}(u)du$ we conclude that $p_{ij}(t)$ is differentiable for all $t \geqq \tau$, and that (10) holds.

Since τ can be chosen arbitrarily small, it follows that for arbitrary $t > 0$, the derivative $p'_{ij}(t)$ exists and

$$p'_{ij}(t) = \sum_{k \in J} p'_{ik}(\tau) p_{kj}(t - \tau) \qquad \text{for arbitrary } \tau < t\,.$$

Furthermore, if relations (4) and (5) are satisfied for some $t = \tau$, it follows from (10) that they are satisfied for arbitrary $t > \tau$. To prove this we merely sum equation (10) over all $j \in J$. Furthermore, it follows from (10) that the function $p'_{ij}(t)$ is continuous for all $t > 0$. This completes the proof of the theorem.

Let us now show that finite derivatives $p'_{ij}(0)$ exist for $j \neq i$. (Here we do not use the restriction $q_i < \infty$.)

To prove this we need:

Lemma 1. *Suppose that for some $i, j, H > 0$ and ε,*

$$1 - p_{ii}(t) < \varepsilon\,, \quad 1 - p_{jj}(t) < \varepsilon\,, \quad 0 \leqq t \leqq H\,.$$

Then for $nh \leqq t \leqq H$, where n is an integer,

$$p_{ij}(h) \leqq \frac{p_{ij}(t)}{n} \frac{1}{1 - 3\varepsilon}\,.$$

Proof. Let us observe the states of the Markov system at the instants $0, h, 2h, \cdots, nh$, where $nh \leqq t$. We let p_k denote the probability that the system leaving the ith state at the initial instant $t = 0$ and observed at the instant rh, where $r = 1, 2, \cdots, k$, falls in the jth state for the first time at the instant kh. We let $p = \sum_{k=1}^{n} p_k$ denote the probability that the system leaving the state i

is in the state j at one of the instants kh, $k = 1, 2, \cdots, n$. We have

$$\varepsilon \geqq p_{ij}(t) \geqq \sum_{k=1}^{n} p_k p_{jj}(t - kh) \geqq (1 - \varepsilon)p$$

so that $p \leqq \{\varepsilon/(1 - \varepsilon)\}$.

Let Q_k denote the probability that the system leaving the ith state will return to it at the instant kh without entering the state j at the instants rh, for $r = 1, 2, \cdots, k - 1$. Then

$$Q_k \geqq p_{ii}(kh) - \sum_{r=1}^{k-1} p_i \geqq 1 - \varepsilon - \frac{\varepsilon}{1 - \varepsilon} .$$

We now have

$$\begin{aligned} p_{ij}(t) &\geqq \sum_{k=1}^{n} Q_{k-1} p_{ij}(h) p_{jj}(t - kh) \\ &\geqq n(1 - \varepsilon)\left(1 - \varepsilon - \frac{\varepsilon}{1 - \varepsilon}\right) p_{ij}(h) \geqq (1 - 3\varepsilon) n p_{ij}(h) , \end{aligned}$$

from which the assertion follows.

Theorem 4. *The finite limits*

$$\lim_{t \to 0} \frac{p_{ij}(t)}{t} = p'_{ij}(0) = q_{ij} < \infty , \qquad i \neq j$$

always exist.

Proof. Let us choose $H > 0$ so small that the conditions of Lemma 1 are satisfied. Let t and h denote an arbitrary pair of positive numbers not exceeding H. We set $nh \leqq t < (n + 1)h$. Then on the basis of Lemma 1,

$$\frac{p_{ij}(h)}{h} \leqq \frac{p_{ij}(t)}{t - h} \cdot \frac{1}{1 - 3\varepsilon} .$$

By letting h approach 0 we obtain

$$\overline{\lim_{h \to 0}} \frac{p_{ij}(h)}{h} \leqq \frac{p_{ij}(t)}{t} \cdot \frac{1}{1 - 3\varepsilon} < \infty .$$

If we then let t approach 0, we get

$$\overline{\lim_{h \to 0}} \frac{p_{ij}(h)}{h} \leqq \underline{\lim_{t \to 0}} \frac{p_{ij}(t)}{t} \cdot \frac{1}{1 - 3\varepsilon} < \infty .$$

Since ε can be chosen arbitrarily small (though positive), it follows that the limit superior and the limit inferior of the ratio $p_{ij}(t)/t$ for $t > 0$ are equal. This completes the proof of the theorem.

REMARK 1. In the case of a finite number of states it follows from the equation

$$1 - p_{ii}(t) = \sum_{j \in J \setminus \{i\}} p_{ij}(t) \tag{11}$$

and Theorem 4 that the derivative $p'_{ii}(0) = -q_i$ exists and is finite, and that

$$q_i = \sum_{j \in J \setminus \{i\}} q_{ij} \; . \tag{12}$$

If the number of states is infinite, Theorem 4 does not imply finiteness of q_i. Also, finiteness of q_i does not imply equation (12).

In the general case equation (12) is replaced with the inequality

$$q_i \geqq \sum_{j \in J \setminus \{i\}} q_{ij} \tag{13}$$

which we obtain by using only a finite number of terms on the right side of (11), dividing the resulting inequality by $t > 0$, taking the limit as $t \to 0$, and then letting the number of terms on the right side of the equation approach ∞ so that their sum ultimately coincides with the right-side of inequality (13).

Equation (3) is proven for $t > 0$. If we set $t = 0$ this equation becomes formally

$$p'_{ij}(s) = \sum_{k \in J} p'_{ik}(0) p_{kj}(s)$$

or

$$p'_{ij}(t) = -q_i p_{ij}(t) + \sum_{k \in J \setminus \{i\}} q_{ik} p_{kj}(t), \qquad i \in J, \qquad t > 0 \; . \tag{14}$$

Equations (14) are called the first Kolmogorov equations. Let us find conditions under which they hold.

Theorem 5. *For the transition probabilities $p_{ij}(s)$ to satisfy the first system of Kolmogorov equations for given i and arbitrary $j \in J$, it is necessary and sufficient that*

$$q_i < \infty, \qquad \sum_{j \in J \setminus \{i\}} q_{ij} = q_i \; . \tag{15}$$

Proof of the Necessity. If $q_i = \infty$, equation (14) has no meaning. Suppose that $q_i < \infty$. Let us sum (14) over all $j \in J$. On the basis of (4),

$$0 = -q_i + \sum_{k \in J \setminus \{i\}} q_{ik}$$

which proves the second of equations (15).

Proof of the Sufficiency. Suppose that $h > 0$. We have

$$\frac{p_{ij}(t+h) - p_{ij}(t)}{h} = -\frac{1 - p_{ii}(h)}{h} p_{ij}(t) + \sum_{k \in J \setminus \{i\}} \frac{p_{ik}(h)}{h} p_{kj}(t) \; .$$

Let us choose a finite set of indices $J' \subset J$ (where $i \notin J'$) so that

$$\sum_{k \in J \backslash J'} q_{ik} < \varepsilon .$$

Then

$$0 \leqq \sum_{k \in J \backslash J' \backslash \{i\}} \frac{p_{ik}(h)}{h} p_{kj}(t) \leqq \sum_{k \in J \backslash J' \backslash \{i\}} \frac{p_{ik}(h)}{h}$$

$$= \frac{1 - p_{ii}(h)}{h} - \sum_{k \in J'} \frac{p_{ik}(h)}{h} \rightarrow q_i - \sum_{k \in J'} q_{ik} < \varepsilon \quad \text{as} \quad h \rightarrow 0 .$$

Consequently

$$\overline{\lim_{h \to 0}} \left| \frac{p_{ij}(t+h) - p_{ij}(t)}{h} - \left\{ q_i p_{ij}(t) + \sum_{k \in J'} q_{ik} p_{kj}(t) \right\} \right| < \varepsilon .$$

Remembering that the derivatives $p'_{ij}(t)$ (for $t > 0$) exist (cf. Theorem 3) and taking the limit as $\varepsilon \rightarrow 0$ in this last inequality, we obtain equation (14).

We recall that in the case of a Markov process with finitely many states, equations (14) follow from the continuity of the transition probabilities. Furthermore, in this case the transition probabilities satisfy yet another system of differential equations, known as the second system of Kolmogorov differential equations.

We have (for $h > 0$)

$$\frac{p_{ij}(t+h) - p_{ij}(t)}{h} = p_{ij}(t) \frac{p_{jj}(h) - 1}{h} + \sum_{k \in J \backslash \{i\}} p_{ik}(t) \frac{p_{kj}(h)}{h} .$$

From this it follows that the derivative of the function $p_{ij}(t)$ satisfies the equation

$$\frac{dp_{ij}(t)}{dt} = -p_{ij}(t) q_j + \sum_{k \in J \backslash \{j\}} p_{ik}(t) q_{kj}, \; i, j \in J . \tag{16}$$

As we can see, the derivation of equation (16) remains valid for processes with a countable set of states if we assume that $q_j < \infty$ and that the relations

$$\lim_{h \to 0} \frac{p_{kj}(h)}{h} = q_{kj} \tag{17}$$

are satisfied uniformly with respect to k. This brings us to:

Theorem 6. *If the transition probabilities $p_{ij}(t)$ are continuous for $t = 0$, if $q_j < \infty$, and if relations* (17) *are satisfied uniformly with respect to k, then the system of equations* (16) *holds.*

The system (16) can be carried over to unconditional probabilities $p_i(t)$ for $i \in J$. If the $p_k(0)$ are the "initial" probabilities (the probabilities that at the initial instant $t = 0$ the Markov system is in the kth state), then at the instant t the system is in the state i with probability

$$p_i(t) = \sum_{k \in J} p_k(0) p_{ki}(t) .$$

Multiplying equations (16) by $p_i(0)$ and summing over all $i \in J$, we obtain

$$\frac{dp_j(t)}{dt} = -p_j(t)q_j + \sum_{k \in J \setminus \{i\}} p_k(t)q_{kj} . \tag{18}$$

The validity of our termwise differentiation of the series is easily shown. A homogeneous process is stationary if $p_i(t) = \text{const.}$

It follows from (18) that the unconditional probabilities $p_i(t) = p_i$ for a stationary process satisfy the system of equations

$$\sum_{k \in J \setminus \{i\}} p_k q_{kj} = p_j q_j . \tag{19}$$

3. JUMP PROCESSES

We might expect that under sufficiently regular conditions a Markov system with countably many states would behave as follows: In the course of some random interval of time the system $\sum$ is in the initial state, after which $\sum$ changes over to another state with definite probabilities, where it stays for some random interval of time, and so forth. But processes of such a kind can be considered in an arbitrary phase space X. In this section we introduce a class of Markov processes that can be regarded as a generalization of a process with a countable number of states. Here we do not require that the process be homogeneous in time, and the phase space X is arbitrary, but we impose on the transition probabilities restrictions more stringent than those discussed in Section 2. When we speak of a Markov process here, we mean a Markov process in the broad sense.

Let us consider a Markov process with phase space X, with fixed σ-algebra $\mathfrak{B}$. Let $\mathfrak{T}$ denote a finite interval of time, let $\mathsf{P}(t, x, s, A)$ denote the transition probabilities for $s > t$, where $s, t \in \mathfrak{T}$, $x \in X$, and $A \in \mathfrak{B}$. For $s = t$ we have $\mathsf{P}(t, x, t, A) = \chi_A(x)$, where $\chi_A(x)$ is the characteristic function of the set A. A generalization of condition d in Section 2 is the condition that the transition probabilities be continuous at $s = t$:

$$\lim_{s \downarrow t} \mathsf{P}(t, x, s, A) = \chi_A(x) .$$

But we impose additional requirements. Suppose that

a. $$\frac{\mathsf{P}(t, x, s, A) - \chi_A(x)}{s - t} \to q(t, x, A) \tag{1}$$

uniformly with respect to (t, x, A), where $t \in \mathfrak{T}$, $x \in X$, and $A \in \mathfrak{B}$, as s approaches t from above.

b. For fixed (x, A) (where $x \in X$ and $A \in \mathfrak{B}$), the function $q(t, x, A)$ is continuous with respect to $t \in \mathfrak{T}$ and uniformly continuous with respect to (x, A).

It follows from (a) that for some $k > 0$

c. $$| q(t, x, A) | \leqq k \tag{2}$$

for all

$$t \in \mathfrak{T}, \qquad x \in X, \qquad A \in \mathfrak{B} .$$

Definition 1. A Markov process in the broad sense that satisfies conditions a and b is called a jump process.

It follows from the definition of $q(t, x, A)$ that it is an additive function of the set $A \in \mathfrak{B}$. On the left side of the relation (1) is a charge, depending on the parameter s, that converges to a limit uniformly as s approaches t from above. Therefore $q(t, x, A)$ is a charge on $\mathfrak{B}$ (cf. Section 1, Chapter II). Furthermore, it is obvious that

$$q(t, x, X) = 0 ,$$

$$q(t, x, A) = \lim_{s \downarrow t} \frac{\mathsf{P}(t, x, s, A)}{s - t} \geqq 0 \quad \text{for} \quad x \notin A ,$$

$$q(t, x, \{x\}) = -q(t, x, X \backslash \{x\}) \leqq 0 ,$$

where $\{x\}$ is the set consisting only of the point x. Let us set

$$q(t, x) = -q(t, x, \{x\}) = q(t, x, X \backslash \{x\}) ,$$

$$\Pi(t, x, A) = \begin{cases} \dfrac{q(t, x, A \backslash \{x\})}{q(t, x)} & \text{for} \quad q(t, x) \neq 0 , \\ \chi_A(x) & \text{for} \quad q(t, x) = 0 . \end{cases} \tag{3}$$

Then $\Pi(t, x, A)$ is a probability on $\mathfrak{B}$ for fixed (t, x). It is easy to obtain a probabilistic interpretation of the quantities $q(t, x)$ and $\Pi(t, x, A)$.

It follows from (1) that

$$\mathsf{P}(t, x, t + \Delta t, \{x\}) = 1 - (q(t, x) + \varepsilon)\Delta t ,$$

where $\varepsilon \to 0$ as $\Delta t \to 0$. Thus, up to infinitesimals of order higher than Δt, the quantity $q(t, x)\Delta t$ is the probability that a system in state x at the instant t will not be in that state at the instant $t + \Delta t$ (where $\Delta t > 0$). Furthermore, for $q(t, x) \neq 0$,

$$\Pi(t, x, A) = \lim_{\Delta t \to 0} \frac{\mathsf{P}(t, x, t + \Delta t, A \backslash \{x\})}{\mathsf{P}(t, x, t + \Delta t, X \backslash \{x\})} \tag{4}$$

which can be regarded as the conditional probability that the system in the state x at the instant t and leaving that state at that instant will, as the result of a jump, be in the state A. This interpretation of the function $\Pi(t, x, A)$ will be proved in Section 7.

The relation (1) can be rewritten in the form

$$\mathsf{P}(t, x, s, A) = \chi_A(x) + q(t, x, A)(s - t) + r(t, x, s, A)(s - t) , \qquad (5)$$

where $r(t, x, s, A)$ is some charge whose absolute variation approaches 0 uniformly with respect to t and x as s approaches t from above (where $t \in \mathfrak{T}$).

In particular, it follows from (5) that

$$\| \mathsf{P}(t, x, s, A) - \chi_A(x) \| \leqq k_1 | s - t | , \qquad (6)$$

where $\| w \|$ is the norm of the charge in the space $\boldsymbol{W}$ (cf. Section 1, Chapter II) with convergence in variation and where k_1 is a constant not depending on $(t, x) \in \mathfrak{T} \times X$.

Let us now look at the smoothness properties of the operators $T^{[t_0,t]}$ and $S^{[t,v]}$ introduced in Section 1. Let us begin with the operator $T^{[t_0,t]}$. Suppose that t_0 is fixed and that w is an arbitrary finite charge defined on $\mathfrak{B}$. For $t > t' > t_0$ we have

$$T^{[t_0,t]}w - T^{[t_0,t']}w = \int_X w_{t'}(dx)[\mathsf{P}(t', x, t, A) - \chi_A(x)] \qquad (7)$$

where $w_t(A) = \int_X w(dx)\mathsf{P}(t_0, x, t, A)$. It follows from (6) and (7) that

$$\| (T^{[t_0,t]} - T^{[t_0,t']})w \| = \| w_t - w_{t'} \| \leqq k_1(t - t')\| w \| . \qquad (8)$$

Furthermore, it follows from (7) and (5) that

$$(T^{[t_0,t]} - T^{[t_0,t']})w(A) = (t - t')\left[\int_X q(t', x, A)w_{t'}(dx) + \int_X r(t', x, t, A)w_{t'}(dx)\right] .$$

Let s denote a member of the interval (t, t') and define

$$r(t', t) = \sup_{A,x} | r(t', x, t, A) | .$$

The following estimate obtains if $| t - t' |$ is small:

$$\left\|\frac{w_t(A) - w_{t'}(A)}{t' - t} - \int_X q(s, x, A)w_s(dx)\right\| \leqq \| w \|r(t', t) + \left\|\int_X [q(t', x, A) - q(s, x, A)]w_s(dx))\right\| + \left\|\int_X q(t', x, A)(w_t(dx) - w_s(dx))\right\| \leqq \| w \| \, \| r(t', t) \| + \varepsilon\| w \| + k\| w_t - w_s \| ,$$

where, by virtue of condition b in the definition of a jump process,

$$\varepsilon = \sup_{x,A} |q(t', x, A) - q(s, x, A)| \longrightarrow 0 \quad \text{as} \quad t' \longrightarrow s ,$$

$r(t', t) \longrightarrow 0$ as $t', t \longrightarrow s$ by virtue of condition a, and $\| w_t - w_s \| \longrightarrow 0$ in accordance with (8).

Theorem 1. *As a function of t, the operators* $T^{[t_0,t]}$ *of a jump process are uniformly differentiable and*

$$\frac{dT^{[t_0,t]}}{dt} = Q_t T^{[t_0,t]}, \; Q_t w = \int w(dx) q(t, x, A) . \tag{9}$$

Formula (9) means that the distribution $\mu_t(A)$ of the system $\sum$ in phase space at the instant t is differentiable with respect to t and satisfies the equation

$$\frac{d\mu_t(A)}{dt} = \int_X q(t, x, A)\mu_t(dx) \qquad (t > t_0) , \tag{10}$$

to which we still need to add the initial condition $\mu_{t_0}(A) = \mu(A)$. This is the so-called "second Kolmogorov equation."

With the aid of the functions $q(t, x)$ and $\Pi(t, x, A)$, equation (10) can be rewritten as follows (by setting $q(t, x, A) = q(t, x)[(\Pi(t, x, A) - \chi_A(x)]$):

$$\frac{d\mu_t(A)}{dt} = -\int_A q(t, x)\mu_t(dx) + \int_X q(t, x)\Pi(t, x, A)\mu_t(dx)(t > t_0) . \tag{11}$$

If we choose $\mu(A) = \chi_A(y_0)$, that is, if we assume that at the instant t_0 we know precisely the state of the system $\sum$, then

$$\mu_t(A) = \mathsf{P}(t_0, y_0, t, A)$$

and we obtain the following corollary to Theorem 1:

Corollary. *The transition probabilities* $\mathsf{P}(t_0, y, t, A)$ *of a jump Markov process are differentiable with respect to t, and*

$$\begin{aligned} \frac{\partial \mathsf{P}(t_0, y, t, A)}{\partial t} &= \int_X q(t, x, A)\mathsf{P}(t_0, y, t, dx) \\ &= -\int_A q(t, x)\mathsf{P}(t_0, y, t, dx) \\ &\quad + \int_X q(t, x)\Pi(t, x, A)\mathsf{P}(t_0, y, t, dx), \; t > t_0 . \end{aligned} \tag{12}$$

Let us now turn to the operators $S^{[t,v]}$ with v fixed and $t' \in (t, v)$. Using the notation of Section 1, we have

$$(S^{[t,v]} - S^{[t',v]})f = f_t(x) - f_{t'}(x) = \int_X [f_{t'}(z) - f_{t'}(x)]$$
$$\times \mathsf{P}(t, x, t', dz) = (t' - t)\int_X [f_{t'}(z) - f_{t'}(x)]q(t, x, dz)$$
$$+ \int_X [f_{t'}(z) - f_{t'}(x)]r(t, x, t', dz) , \tag{13}$$

from which we get

$$\| f_t(x) - f_{t'}(x) \| \leqq k_1 | t' - t | \, \| f_{t'}(x) \| \leqq k_1 | t' - t | \, \| f(x) \| . \tag{14}$$

Furthermore, if $| t' - t |$ is small with $t \leqq t_0 \leqq t'$, we obtain

$$\left\| \frac{S^{[t,v]}(f) - S^{[t',v]}(f)}{t' - t} - \int f_{t_0}(z) q(t_0, x, dz) \right\|$$
$$\leqq \left\| \int [f_{t'}(z) - f_{t_0}(z)] q(t_0, x, dz) \right\|$$
$$+ \left\| \int f_{t_0}(z)[q(t_0, x, dz) - q(t, x, dz)] \right\|$$
$$+ \left\| \int [f_{t'}(z) - f_{t'}(x)] r(t, x, t', dz) \right\|$$
$$\leqq k_1 | t' - t_0 | \, \| f \| k + \varepsilon \| f \| + 2 \| f \| r(t, t') ,$$

where just as before, $\varepsilon = \sup_{x,A} | q(t_0, x, A) - q(t', x, A) | \longrightarrow 0$ as $t' \longrightarrow t_0$.

Theorem 2. *As functions of t, the operators $S^{[t,v]}$ of a jump Markov process are uniformly differentiable and satisfy the condition*

$$-\frac{dS^{[t,v]}}{dt} = Q_t^* S, \qquad Q_t^* = \int_X f(z) q(t, x, dz) . \tag{15}$$

In particular, the function $f_t(x) = \int_X f(z)\mathsf{P}(t, x, v, dz)$ is differentiable with respect to t and

$$-\frac{\partial f_t(x)}{\partial t} = \int_X f_t(z) q(t, x, dz), \qquad t < v, \qquad f_v(x) = f(x) \tag{16}$$

or

$$\frac{\partial f_t(x)}{\partial t} = q(t, x)\left[f_t(x) - \int_X f_t(z)\Pi(t, x, dz)\right], \qquad t < v . \tag{17}$$

Equations (16) and (17) are called the "first Kolmogorov equations." They are integro-differential equations for determining the function $f_t(x)$. The equation

$$f_v(x) = f(x) , \tag{18}$$

together with the condition of continuity of the function $f_t(x)$ at $t = v$, plays the role of a boundary condition (that is, of a supplementary condition on the endpoint of the interval of time for which equation (17) is applicable).

Corollary. *The transition probabilities* $\mathsf{P}(t, y, v, A)$ *are differentiable with respect to* t *and they satisfy the condition*

$$\frac{\partial \mathsf{P}(t, y, v, A)}{\partial t} = -\int_X \mathsf{P}(t, z, v, A) q(t, y, dz)$$

$$= q(t, y)\left[\mathsf{P}(t, y, v, A) - \int_X \mathsf{P}(t, z, v, A)\Pi(t, y, dz)\right] \quad (19)$$

for $t < v$.

In the case in which the space X is countable ($X = J$), we introduce the probabilities $p_{ij}(t, v) = \mathsf{P}(t, i, v, \{j\})$ just as in Section 1. For a jump process, the limits

$$\lim_{v \downarrow t} \frac{p_{ij}(t, v)}{v - t} = q_{ij}(t), \qquad i \neq j ,$$

$$\lim_{v \downarrow t} \frac{1 - p_{ii}(t, v)}{v - t} = q_i(t)$$

exist uniformly with respect to (i, j, t), the functions $q_{ij}(t)$ and $q_i(t)$ are continuous, and $q_i(t) = \sum_{j \in J \setminus \{i\}} q_{ij}(t)$.

Equations (12) and (19) take the forms

$$\frac{\partial p_{ij}(t, v)}{\partial v} = \sum_{k \neq j} p_{ik}(t, v) q_{kj}(v) - p_{ij}(t, v) q_j(v) , \quad (20)$$

$$\frac{\partial p_{ij}(t, v)}{\partial t} = q_i(t) p_{ij}(t, v) - \sum_{k \neq i} q_{ik}(t) p_{kj}(t, v) . \quad (21)$$

For processes that are homogeneous in time, the systems of equations (20) and (21) coincide with equations (16) and (15) of the preceding section, but there they were obtained under broader assumptions.

Suppose that there exists a point x_0 such that $q(t, x_0) = 0$ for all $t \in [a, v]$. Then $\mathsf{P}(t, x, v, A) = \chi_A(x_0)$ for $t \in [a, v]$. This means that during the interval $[a, v]$, the point x_0 in phase space is an absorbing state: If the system $\sum$ is in the state x_0 at the instant $t \in [a, v]$, then it will be in that state up to the instant v.

Let us now investigate to what extent the function $q(t, x, A)$ defines a Markov process.

We shall find it convenient to deal with the two functions

$$q(t, x) = -q(t, x, \{x\}) \quad \text{and} \quad \tilde{q}(t, x, A) = q(t, x, A \setminus \{x\}) .$$

From our description of jump Markov processes, we note that these functions satisfy the following conditions:

a′. For fixed $t \in \mathfrak{T}$ and $x \in X$, the function $\tilde{q}(t, x, A)$ for $A \in \mathfrak{B}$ is a measure and $q(t, x) = \tilde{q}(t, x, X)$.

b′. For fixed $x \in X$ and $A \in \mathfrak{B}$, the function $\tilde{q}(t, x, A)$ is continuous with respect to t, uniformly continuous with respect to (x, A), and $\mathfrak{B}$-measurable for fixed $t \in \mathfrak{T}$ and $A \in \mathfrak{B}$.

c′. $\tilde{q}(t, x, A) \leqq k$ for $(t, x) \in \mathfrak{T} \times X$ and $A \in \mathfrak{B}$.

Suppose that we are given an arbitrary pair of functions $q(t, x)$ and $\tilde{q}(t, x, A)$ and suppose that

$$q(t, x, A) = -q(t, x)\chi_A(x) + \tilde{q}(t, x, A) .$$

Let us look at the question of the existence of solutions to equations (10) and (16).

Theorem 3. *If the functions $q(t, x)$ and $\tilde{q}(t, x, A)$ satisfy conditions* a′, b′, *and* c′, *then equations* (10) *and* (16) *have unique solutions. The first of these is in the class of all finite charges and the second is in the class of all bounded functions $f_t(x)$ that are $\mathfrak{B}$-measurable with respect to x.*

We introduce the space W of all finite charges on $\mathfrak{B}$ with absolute variation of the charge used as norm (cf. Section 1, Chapter II):

$$\| w \| = | w |(X) = w^+(X) + w^-(X) .$$

The operator Q_t defined by formula (9),

$$w_t(A) = Q_t[w](A) = -\int_A q(t, x)w(dx) + \int_X \tilde{q}(t, x, A)w(dx) , \tag{22}$$

maps W into W. Here

$$\| w_t \| \leqq k\| w \|, \qquad t \in \mathfrak{T} . \tag{23}$$

We also note that if $| \tilde{q}(t, x, A) - \tilde{q}(t', x, A) | \leqq \varepsilon$ for all $(x, A) \in X \times \mathfrak{B}$, then $\| w_t - w_{t'} \| \leqq 2\varepsilon\| w \|$; that is, the operator Q_t is continuous with respect to t. In particular, for arbitrary $A \in \mathfrak{B}$, the function $w_t(A)$ as a function of $t \in \mathfrak{T}$ is continuous.

We introduce the space C^W of continuous functions $w = w(t)$ defined on $[t_0, v]$ into W.

Equation (10) can be written

$$\hat{\mu} = \hat{\mu}_0 + D\hat{\mu} , \tag{24}$$

where

$$\hat{\mu} = \hat{\mu}(t) = \mu_t(A), \qquad \hat{\mu}_0 = \mu_0(A) ,$$

D is a linear transformation in C^W, and

$$D\hat{\mu} = \int_{t_0}^t Q_\tau \hat{\mu}(\tau)d\tau .$$

Let D^n denote the nth power of D. Then

$$\| D\hat{\mu}_1 - D\hat{\mu}_2 \| \leqq 2k \| \hat{\mu}_1 - \hat{\mu}_2 \| (t - t_0) ,$$

$$\| D^n\hat{\mu}_1 - D^n\hat{\mu}_2 \| \leqq (2k)^n \| \hat{\mu}_1 - \hat{\mu}_2 \| \frac{(t - t_0)^n}{n!} .$$

Thus, some power of the mapping D is a contraction mapping and on the basis of the principle of contraction mappings, equation (24) has a unique solution in C^W (cf. Kolmogorov and Fomin).

We may treat equation (16) analogously. The substitution

$$f_t(x) = \exp\left(- \int_t^v q(\tau, x) d\tau \right) g_t(x) \tag{25}$$

reduces equation (16) to the somewhat simpler form

$$\frac{\partial g_t(x)}{\partial t} = - \int_X g_t(z) \exp\left(\int_t^v [q(\tau, x) - q(\tau, z)] d\tau \right) \times \tilde{q}(t, x, dz), \; g_v(x) = f(x) , \tag{26}$$

or

$$g_t(x) = f(x) + \int_t^v \int_X g_\theta(z) \exp\left(\int_\theta^v [q(\tau, x) - q(\tau, z)] d\tau \right) \tilde{q}(\theta, x, dz) d\theta . \tag{27}$$

We introduce the space B of all $\mathfrak{B}$-measurable bounded functions $f(x)$ for $x \in X$ with norm $\| f \| = \sup_{x \in X} | f(x) |$ and we introduce the space C^B of continuous functions $f_t(x)$ of the argument $t \in [t_0, v]$ with range in B and with norm $\| f_t(x) \| = \sup_{t_0 \leqq t \leqq v} \| f_t(x) \|$.

The linear operator $\tilde{D}$ defined by

$$\tilde{D} f_t(x) = f_t(x) + \int_t^v \int_X g_\theta(z) \exp\left(\int_\theta^v [q(\tau, x) - q(\tau, z)] d\tau \right) \tilde{q}(\theta, x, dz) d\theta ,$$

where $g_t(x)$ is a fixed nonnegative function in C^B, maps the set of nonnegative functions in C^B into itself. Here

$$\| \tilde{D} f_t^{(1)}(x) - \tilde{D} f_t^{(2)}(x) \| \leqq k_1 \| f_t^{(1)}(x) - f_t^{(2)}(x) \| \, | v - t | ,$$

$$k_1 = e^{k(v - t_0)} ,$$

$$\| \tilde{D}^n f_t^{(1)}(x) - \tilde{D}^n f_1^{(2)}(x) \| \leqq k_1 \| f_t^{(1)}(x) - f_t^{(2)}(x) \| \frac{| v - t |^n}{n!} ;$$

that is, some power of the operator $\tilde{D}$ is a contraction operator. Thus equation (27) and equation (16) have a unique solution in C^B. This completes the proof of the theorem.

Corollary. *The transition probabilities of a jump Markov process are uniquely defined by the function* $q(t, x, A)$.

REMARK 1. The solutions of equations (10), (16), and (17) can be obtained by the method of successive approximations. Applying this method to equation (27), we obtain for the solution of equation (16) the formula

$$f_t(x) = \sum_{n=0}^{\infty} f_t^{(n)}(x) \tag{28}$$

where

$$f_t^{(0)}(x) = \exp\left(-\int_t^v q(\tau, x)d\tau\right)f(x),\ f_t^{(n+1)}(x)$$
$$= \int_t^v\int_X f_\theta^{(n)}(z) \exp\left(-\int_t^\theta q(\tau, x)d\tau\right)\tilde{q}(\theta, x, dz)d\theta\ . \tag{29}$$

For the transition probabilities $\mathsf{P}(t, x, v, A)$, we obtain the expressions

$$\mathsf{P}(t, x, v, A) = \sum_{n=0}^{\infty} \mathsf{P}^{(n)}(t, x, v, A) \tag{30}$$

where

$$\mathsf{P}^{(0)}(t, x, v, A) = \exp\left(-\int_t^v q(\tau, x)d\tau\right)\chi_A(x),$$

$$\mathsf{P}^{(n+1)}(t, x, v, A)$$
$$= \int_t^v\int_X \mathsf{P}^{(n)}(\theta, z, v, A) \exp\left(-\int_t^0 q(\tau, x)d\tau\right)\tilde{q}(\theta, x, dz)d\theta\ .$$
$$(n = 0, 1, \cdots)\ .$$

The functions $\mathsf{P}^{(n)}(t, x, v, A)$ have a simple probability-theoretic interpretation, which will be discussed in Section 7.

4. EXAMPLES

1. Random Signals

Imagine a channel that can be in one or the other of two states at an arbitrary instant of time: Either a signal is being transmitted or a signal is not being transmitted. We let these states of the channel be denoted by 1 and 0 respectively.

Suppose that transitions from one state to the other correspond to a homogeneous Markov process and suppose that

$$p_{01}(h) = \lambda h + o(h)\ ,$$
$$p_{10}(h) = \mu h + o(h)\ .$$

The quantities q_i and q_{ij} (for $i, j = 1, 2$), introduced in Section 2, have the values $q_0 = q_{01} = \lambda$, $q_1 = q_{10} = \mu$.

The second system of Kolmogorov's differential equations is written

$$p'_{00}(t) = -\lambda p_{00}(t) + \mu p_{01}(t); \qquad p'_{10}(t) = -\lambda p_{10}(t) + \mu p_{11}(t),$$
$$p'_{01}(t) = -\mu p_{01}(t) + \lambda p_{00}(t); \qquad p'_{11}(t) = -\mu p_{11}(t) + \lambda p_{10}(t) .$$

Solving these equations with initial conditions $p_{00}(0) = p_{11}(0) = 1$ and $p_{10}(0) = p_{01}(0) = 0$, we obtain

$$p_{00}(t) = \lambda_0 e^{-(\lambda+\mu)t} + \mu_0 ;$$
$$p_{01}(t) = \lambda_0[1 - e^{-(\lambda+\mu)t}] ,$$
$$p_{10}(t) = \mu_0(1 - e^{-(\lambda+\mu)t}), \qquad p_{11}(t) = \lambda_0 + \mu_0 e^{-(\lambda+\mu)t} ,$$

where $\lambda_0 = (\lambda)/(\lambda + \mu)$ and $\mu_0 = (\mu)/(\lambda + \mu)$.

As $t \longrightarrow \infty$, the probabilities $p_{ij}(t)$ approach the values

$$\lim_{t\to\infty} p_{j0}(t) = p_0 = \mu_0 \qquad \lim_{t\to\infty} p_{j1}(t) = p_1 = \lambda_0 \qquad (j = 1, 2) ,$$

which are independent of i. On the other hand, the probabilities $p_0 = \mu_0$ and $p_1 = \lambda_0$ coincide with the probabilities of the states 0 and 1 in a stationary mode. If $\nu(t)$ denotes the state of the channel in the stationary case, then

$$m = \mathsf{M}\nu(t) = 0p_0 + 1p_1 = \lambda_0$$

and the covariance function of the process $\nu(t)$ is calculated as follows:

$$R(t) = \mathsf{M}(\nu(t + \tau)\nu(\tau)) - m^2 ,$$
$$\mathsf{M}(\nu(t + \tau)\nu(\tau)) = p_1 p_{11}(t) ,$$

from which we get

$$R(t) = \lambda_0 \mu_0 e^{-(\lambda+\mu)t} .$$

2. Processes of Pure Growth

As an example of a Markov process with countably many states, let us consider a so-called *process of pure growth*. The possible states in this example are the numbers 0, 1, 2, ···, and the sample functions do not decrease with increasing t, but with each jump they increase by unity. A process of this kind can serve as a mathematical model of processes of registration of a phenomenon that takes place at random instants of time.

For example, when there is successive radioactive discharge from an original radioactive substance (the parent substance), another radioactive substance (the first daughter substance) is formed; from that is formed a second daughter substance, and so forth. Let us consider one particular atom of the original substance. In the course of a random instant of time this atom is in the original state. Then it disintegrates, becoming an atom of the first daughter substance, and so forth. Here the probability of disintegration of the atom in the instant (t, s) does not depend on the time the atom has "lived" up to the instant t, and each state of the atom has a definite mean life span $l_k = 1/\lambda_k$. Examples of such chains of successive radioactive discharge are the transformations of natural isotopes of uranium and thorium. These terminate with the formation of stable isotopes of lead.

Let us state mathematically our assumptions regarding the random process $\xi(t)$ in the form of conditions on the transition probabilities.

It is natural to assume that $p_{i,i+1}(t) = \lambda_i t + o(t), p_{ij}(t) = o(t)$ for $j - i \neq 0, 1, p_{ii}(t) = 1 - \lambda_i t + o(t)$.

The first system of Kolmogorov equations take the form

$$p'_{ij}(t) = -\lambda_i p_{ij}(t) + \lambda_i p_{i+1,j}(t), \qquad i \geqq 0 , \tag{1}$$

and the second system of Kolmogorov equations takes the form

$$\begin{aligned} p'_{ij}(t) &= -p_{ij}(t)\lambda_j + p_{i,j-1}(t)\lambda_{j-1}, \qquad j \geqq 1 , \\ p'_{i0}(t) &= -p_{i0}(t)\lambda_0 . \end{aligned} \tag{2}$$

To these equations we need to add the initial conditions $p_{ij}(0) = \delta_{ij}$.

Let us solve the second system. If $i > j$, then according to the definition of a process of pure growth we set $p_{ij}(t) = 0$ which of course is also a solution of the system (2) and satisfies the initial conditions. The system (2), for fixed i is then recursive.

First let us determine $p_{ii}(t)$ from the equation

$$p'_{ii}(t) = -\lambda_i p_{ii}(t), \qquad p_{ii}(0) = 1 ,$$

and then let us determine successively the functions $p_{i,i+1}(t)$ and $p_{i,i+2}(t)$, for each of which the system (2) is an ordinary linear differential equation. For the first step we have $p_{ii}(t) = e^{-\lambda_i t}$. Then

$$p_{ij}(t) = \lambda_{j-1}\int_0^t \exp\{-\lambda_j(t - s)\}p_{i,j-1}(s)ds .$$

One can easily obtain a solution in explicit form for the system (2) by the usual methods of operational calculus. To do this, consider the transforms of the functions $p_{ij}(t)$:

$$\varphi_{ij}(z) = z\int_0^\infty e^{-zt}p_{ij}(t)\,dt .$$

Then

$$z\int_0^\infty e^{-zt}p'_{ij}(t)dt = z(\varphi_{ij}(z) - \delta_{ij})$$

and when we shift to the transforms of the functions, the system (2) takes the form

$$z\varphi_{ij}(z) = -\lambda_j\varphi_{ij}(z) + \lambda_{j-1}\varphi_{i,j-1}(z), \qquad j > i ,$$
$$z(\varphi_{ii}(z) - 1) = -\lambda_i\varphi_{ii}(z) ,$$

from which we get

$$\varphi_{ii}(z) = \frac{z}{z + \lambda_i}, \qquad \varphi_{ij} = \frac{\lambda_{j-1}}{z + \lambda_j}\varphi_{i,j-1}, \qquad j > i ,$$

and $\varphi_{ij}(z) = (\prod_{k=1}^{j-i}\lambda_k)z/\psi(z)$, where $\psi(z) = \prod_{k=1}^{j}(z + \lambda_k)$. If all the numbers λ_k are distinct, then

$$\frac{1}{\psi(z)} = \sum_{k=i}^{j} \frac{1}{(z + \lambda_k)\psi'(-\lambda_k)} .$$

By means of this formula, the expression for $\varphi_{ij}(z)$ can be written in the form

$$\varphi_{ij}(z) = \left(\prod_{k=i}^{j-1}\lambda_k\right)\sum_{k=i}^{j} \frac{z}{(z + \lambda_k)\psi'(-\lambda_k)} .$$

Since $\varphi_{ii}(z) = z/(z + \lambda i)$ is the transform of the function $e^{-\lambda_i t}$, we have $p_{ij} = 0$ if $j < i$ and

$$p_{ij}(t) = \left(\prod_{k=i}^{j-1}\lambda_k\right)\sum_{k=i}^{j} \frac{e^{-\lambda_k t}}{\psi'(-\lambda_k)}, \qquad j > i , \tag{3}$$

$$p_{ii}(t) = e^{-\lambda_i t} , \tag{4}$$

where

$$\psi'(-\lambda_k) = \prod_{\substack{r=i \\ r \neq k}}^{j} (\lambda_r - \lambda_k) .$$

These expressions are solutions in explicit form of equations (2).

In the simplest case, in which $\lambda_k = k\lambda$, the corresponding process is called a *process of linear growth*, and

$$\begin{aligned} p_{ij}(t) &= i(i + 1) \cdots (j - 1)\lambda^{j-i} \\ &\quad \times \sum_{k=i}^{j} \frac{e^{-\lambda k t}}{\lambda^{j-i}(i - k)(i + 1 - k) \cdots (-1)\cdot 1 \cdots (j - k)} \\ &= e^{-i\lambda t}C_{j-1}^{j-i}(1 - e^{-\lambda t})^{j-i} . \end{aligned}$$

The corresponding distribution is called a Yule-Furry distribution. We note that for $\xi(0) = i$, $\mathsf{M}\xi(t) = ie^{\lambda t}$, $\mathsf{D}\xi(t) = ie^{\lambda t}(e^{\lambda t} - 1)$.

In this example of a process of pure growth it is easy to analyze the possibility that a particle executing a Markov process will go out "to infinity" or, in other words, the possibility that it will disappear from the phase space. Let us suppose that at the instant 0 the particle is in the state 0. It will be in that state for some random interval of time τ_0, after which it will jump to the state 1 where it will remain τ_1 units of time, and so forth. If the series

$$\xi = \sum_{k=0}^{\infty} \tau_k \tag{5}$$

converges with positive probability, after an interval of time ξ the particle will disappear from the space. In the sum (5) the quantities τ_k are positive, and by virtue of the definition of a Markov process they are independent. (This fact seems obvious but requires proof, which will be given in Section 7).

A condition for convergence of this series is given by:

Theorem 1. *The series* (5) *converges with probability* 1 *or* 0 *according as the series*

$$\sum_{k=0}^{\infty} \frac{1}{\lambda_k} \tag{6}$$

converges or diverges.

To prove Theorem 1 we use Corollary 1 of Section 4, Chapter III. The quantity τ_k, the duration of occupancy of the kth state, has the distribution function

$$\mathsf{P}\{\tau_k < x\} = 1 - p_{kk}(x) = 1 - e^{-\lambda_k x}$$

and $\mathsf{M}\tau_k = 1/\lambda_k$. Define $\tau'_k = \tau_k$ for $\tau_k \leqq 1$ and $\tau'_k = 0$ for $\tau_k > 1$. Then

$$\sum_{k=0}^{\infty} \mathsf{M}\tau'_k \leqq \sum_{k=1}^{\infty} \mathsf{M}\tau_k, \qquad \sum_{k=0}^{\infty} \mathsf{P}\{\tau_k > 1\} \leqq \sum_{k=1}^{\infty} \mathsf{M}\tau_k$$

and on the basis of the corollary mentioned, the series (5) converges with probability 1 if the series (6) converges. On the other hand,

$$\sum_{k=0}^{\infty} \mathsf{P}\{\tau_k > 1\} = \sum_{k=0}^{\infty} e^{-\lambda_k}, \tag{7}$$

$$\sum_{k=1}^{\infty} \mathsf{M}\tau'_k = \sum_{k=0}^{\infty} e^{-\lambda_k} + \sum_{k=0}^{\infty} \frac{1 - e^{-\lambda_k}}{\lambda_k} \tag{8}$$

and if the series (6) diverges but the series (7) converges, then the series (8) diverges.

Thus in the case of divergence of the series (6), one of the series (7) or (8) diverges and consequently the series (5) diverges with probability 1. This completes the proof of the theorem.

It follows from Theorem 1 that if the series (6) converges, then after a finite interval of time the system moves out to infinity with probability 1 (or disappears from the phase space).

With probability 1, a process of linear growth does not go out to infinity.

3. Birth and Death Processes

Birth and death processes are homogeneous Markov processes with possible states $0, 1, 2, \cdots, n, \cdots$ in which transitions from the state n into the states $n-1$ and $n+1$ are possible. Accordingly we set

$$p_{i,i+1}(t) = \lambda_i t + o(t), p_{ij}(t) = o(t), |i-j| > 1 ;$$
$$p_{i,i-1}(t) = \mu_i t + o(t), p_{ii}(t) = 1 - (\lambda_i + \mu_i)t + o(t) .$$

In the case of the first system of Kolmogorov equations, the differential equations for the transition probabilities take the form

$$p'_{ij}(t) = -(\lambda_i + \mu_i)p_{ij}(t) + \lambda_i p_{i+1,j}(t) + \mu_i p_{i-1,j}(t) ,$$
$$i = 0, 1, 2, \cdots (\mu_0 = 0) \tag{9}$$

and in the case of the second system of Kolmogorov equations, they take the form

$$p'_{ij}(t) = -p_{ij}(t)(\lambda_j + \mu_j) + p_{i,j-1}(t)\lambda_{j-1} + p_{i,j+1}(t)\mu_{j+1} . \tag{10}$$

For the unconditional probabilities $p_i(t)$ we have the system of equations

$$p'_i(t) = -p_i(t)(\lambda_i + \mu_i) + p_{i-1}(t)\lambda_{i-1} + p_{i+1}(t)\mu_{i+1} ,$$
$$i = 0, 1, 2, \cdots , \tag{11}$$
$$p_{-1}(t) = 0 .$$

Let us find a stationary distribution of the probabilities, that is, a distribution of the probabilities $p_i(t)$ for $i = 0, 1, 2, \cdots$ that satisfies the system (11) and does not change in time: $p_i(t) = \text{const.}$

For such a distribution the system of differential equations (11) degenerates into a homogeneous algebraic system:

$$-(\lambda_i + \mu_i)p_i + \lambda_{i-1}p_{i-1} + \mu_{i+1}p_{i+1} = 0, \qquad i = 1, 2, \cdots ,$$
$$-\lambda_0 p_0 + \mu_1 p_1 = 0 .$$

Suppose that $\mu_k > 0$ for $k = 1, 2, \cdots$. Then $p_1 = (\lambda_0/\mu_1)p_0$ and as we easily find by induction,

$$p_k = \frac{\lambda_0\lambda_1 \cdots \lambda_{k-1}}{\mu_1\mu_2 \cdots \mu_k} p_0 \tag{12}$$

and

$$\sum_{k=0}^{\infty} p_k = p_0\left(1 + \sum_{k=1}^{\infty} \frac{\lambda_0\lambda_1 \cdots \lambda_{k-1}}{\mu_1\mu_2 \cdots \mu_k}\right) . \tag{13}$$

Theorem 2. *For a stationary probability distribution to exist in a birth and death process it is necessary and sufficient that the series*

$$\sum_{k=1}^{\infty} \frac{\lambda_0\lambda_1 \cdots \lambda_{k-1}}{\mu_1\mu_2 \cdots \mu_k} \qquad (\mu_k > 0, k = 1, 2, \cdots) \tag{14}$$

converge.

It is interesting to note the connection between stationary distributions of a process and the so-called "final probabilities" $p_{ik}(\infty)$, where $p_{ik}(\infty) = \lim_{t\to\infty} p_{ik}(t)$.

Let us suppose that the final probabilities $p_{ik}(\infty)$ exist. When we integrate equation (11) from h to $h + T$, divide by T, and then take the limit as $T \to \infty$, we obtain

$$-p_{ij}(\infty)(\lambda_i + \mu_i) + p_{i,j-1}(\infty)\lambda_{j-1} + p_{i,j+1}(\infty)\mu_{j+1} = 0, \qquad j > 0\,;$$
$$-p_{i0}(\infty)\lambda_0 + p_{i1}(\infty)\mu_1 = 0\,;$$

that is, the final probabilities $p_{ij}(\infty)$ coincide with the stationary distribution (for fixed i) and they are independent of i.

In technology, physics, and natural science there are many problems that involve birth and death processes. Let us look at some of these.

1. *The servicing of lathes.* Suppose that m lathes are serviced by a crew of s repair men. When a lathe fails to function properly it is repaired immediately unless all of the repairmen are working on lathes that have already failed in which case the lathe must await repair. The lathes are repaired in the order in which they fail.

Let us make the following assumptions: For an individual functioning lathe, the probability of getting out of order during an interval of time $(t, t + \Delta t)$ is independent of t and is equal to $\lambda(\Delta t) = \lambda\Delta t + o(\Delta t)$, independently of the "history" of its operation (the length of time that it has been in use, the number of times that it has become out of order, and the length of service) up to the instant t. Analogously, if a lathe is being repaired, the probability of its being put back into operation during an interval of time $(t, t + \Delta t)$ is equal to $\mu(\Delta t) = \mu\Delta t + o(\Delta t)$ and is independent of the nature of its work and its length of service up to the instant t. The lathes are used, get out of order and are repaired independently of each other.

Let $\sum$ denote the state of the industrial process. Let us agree to say that $\sum$ is in the state E_k if at a given instant the number of lathes being repaired or awaiting repair (that is, the total number of lathes not in operation) is equal to k. Subsequent removal of a single lathe from service denotes transition to the state E_{k+1}, and completion of the repair of one of the lathes indicates transition into the state E_{k-1}. Thus we have a homogeneous Markov system with finitely many states $E_0, E_1, \cdots, E_m$. It follows from our assumptions that

$$p_{k,k+1}(\Delta t) = (m-k)\lambda\Delta t + o(\Delta t), \qquad k = 0, \cdots, m-1\,;$$
$$p_{k,k-1}(\Delta t) = k\mu\Delta t + o(\Delta t) \quad \text{for} \quad 1 \leqq k \leqq s\,,$$
$$p_{k,k-1}(\Delta t) = s\mu\Delta t + o(\Delta t) \quad \text{for} \quad m \geqq k \geqq s\,,$$
$$p_{k,k\pm r}(\Delta t) = o(\Delta t), \qquad r \geqq 2\,;$$

that is, we have a birth and death process with finitely many possible states. In the preceding notation,

$$\lambda_k = (m-k)\lambda, \qquad k = 0, 1, \cdots, m\,;$$
$$\mu_k = k\mu \quad \text{for} \quad 0 \leqq k \leqq s; \qquad \mu_k = s\mu \quad \text{for} \quad s \leqq k \leqq m\,.$$

A stationary distribution always exists. By virtue of equations (12) and (13) it is given by the formulas

$$p_k = C_m^k\left(\frac{\lambda}{\mu}\right)^k p_0, \qquad k \leqq s\,,$$

$$p_k = C_m^k \frac{k(k-1)\cdots(s+1)}{s^{k-s}}\left(\frac{\lambda}{\mu}\right)^k p_0, \qquad s < k \leqq m\,,$$

$$p_0 = \left[1 + \sum_{k=0}^{s} C_m^k\left(\frac{\lambda}{\mu}\right)^k + \sum_{k=s+1}^{m} C_m^k \frac{k(k-1)\cdots(s+1)}{s^{k-s}}\left(\frac{\lambda}{\mu}\right)^k\right]^{-1},$$

which can be used to calculate the most suitable state for the number of lathes and the number of repairmen under specific industrial conditions.

2. *Telephone networks.* In many cases we have the same type of set-up as in the preceding example, but with $m = \infty$. Such a situation exists in the case of a central interurban telephone station with s lines serving, to all intents and purposes, an infinite number of subscribers. The subscribers are the units that need to be served, and the connecting lines are the servicing units. The number of the state of the system $\sum$ indicates the number of subscribers requiring service at a given instant. If this number exceeds s, the subscribers must wait their turn for a free line. In contrast to the assumptions of the preceding example, we now assume that the probability that a single subscriber will ask for service in the course of an interval $(t, t + \Delta t)$ is equal to $\lambda(\Delta t) = \lambda\Delta t + o(\Delta t)$, and the probability that more than one subscriber will require service in that instant is equal to $o(\Delta t)$. These probabilities are independent of the number

of subscribers requiring service up to the given instant. We keep the assumptions of the preceding example regarding the mode of service for the subscribers. We then have

$$\lambda_k = \lambda,\ \mu_k = k\mu \quad \text{for} \quad k \leqq s, \qquad \mu_k = s\mu \quad \text{for} \quad k \geqq s\,.$$

For a stationary distribution to exist, it is necessary and sufficient that the series

$$S = 1 + \sum_{k=0}^{s} \frac{\lambda^k}{k!\,\mu^k} + \sum_{k=s+1}^{\infty} \frac{1}{s!\,s^{k-s}} \left(\frac{\lambda}{\mu}\right)^k < \infty$$

converge, that is, that λ be less than $s\mu$.

In this case, a stationary distribution is given by the formulas

$$p_k = \frac{1}{k!}\left(\frac{\lambda}{\mu}\right)^k \frac{1}{S}, \qquad 0 \leqq k \leqq s\,,$$

$$p_k = \frac{1}{s!\,s^{k-s}}\left(\frac{\lambda}{\mu}\right)^k \frac{1}{S}, \qquad k \geqq s\,.$$

3. *Textile threads.* A textile thread is a sheaf of fibers, the number of which varies from point to point along the length of the thread. If we assume that fiber length has a fixed negative exponential distribution independent both of the number of fibers at any point of the thread and of their lengths, and if we assume that the probability that the end of a new fiber will appear in a segment $(t, t + \Delta t)$ of the thread is equal to $\lambda \Delta t + o(\Delta t)$ and is independent of the number of fibers and their lengths, then the number of threads $\nu(t)$ at the point t on the thread is a birth and death process with parameters $\lambda_n = \lambda$, $\mu_n = n\mu$, where μ is the reciprocal of the mean length of a fiber.

5. BRANCHING PROCESSES

An important class of Markov processes with countably many states is the class of branching processes. Let us suppose that we are observing a physical system Σ consisting of a finite number of particles either of the same type or of several different types. With the passage of time each particle can disappear or turn into a group of new particles, independently of the other particles. Phenomena described by such a scheme are frequently encountered in natural science and technology, and include, for example, showers of cosmic rays, the passage of elementary particles through a substance, the development of biological populations, and the spread of epidemics.

A precise definition of processes of this type within the framework of the theory of Markov processes brings us to the concept of a branching process.

Let n denote the number of distinct possible types of particles. The state of the system $\sum$ at the instant t is characterized by an integer-valued vector

$$\nu(t) = \{\nu_1(t), \nu_2(t), \cdots, \nu_n(t)\} ,$$

where $\nu_i(t)$ is the number of particles of the ith type that exist at the instant t. Let us identify the state of the system $\sum$ at the instant t with the vector $\nu(t)$. With regard to the nature of the evolution of the system $\sum$ in time we make the following assumption: For every particle that exists at the instant t, its subsequent evolution is independent of when and how this particle first appeared and of the nature of the evolution of all the remaining particles in $\sum$ at the instant t.

Let $\boldsymbol{\alpha}, \boldsymbol{\beta}, \cdots$ denote n-dimensional vectors with integer-valued nonnegative coordinates $\boldsymbol{\alpha} = \{a_1, a_2, \cdots, a_n\}$, $\boldsymbol{\beta} = \{b_1, b_2, \cdots, b_n\}$. We introduce the transition probabilities $p_{\alpha\beta}(t_1, t_2)$ of the system $\sum$ from the state $\boldsymbol{\alpha}$ that exists at the instant t_1 to the state $\boldsymbol{\beta}$ at the instant t_2:

$$p_{\alpha\beta}(t_1, t_2) = \mathsf{P}\{\nu(t_2) = \boldsymbol{\beta} \mid \nu(t_1) = \boldsymbol{\alpha}\} .$$

Let $\{i\}$, for $i = 1, \cdots, n$, denote the state of the system $\sum$ that consists only of a single particle of the ith type.

The assumption regarding the nature of the evolution of the system $\sum$ in time can be described by the formula

$$p_{\alpha\beta}(t_1, t_2) = \sum_{\sum_{i=1}^{n} \sum_{j=1}^{a_i} \beta^{(ij)} = \beta} \prod_{i=1}^{n} \prod_{j=1}^{a_i} \mathsf{P}_{\{i\}\beta^{(ij)}}(t_1, t_2) , \qquad (1)$$

where the summation on the right is over all possible vectors $\boldsymbol{\beta}^{(ij)}$ with nonnegative integer-valued components $(i = 1, \cdots, n; j = 1, \cdots, a_i)$ that together constitute the vector $\boldsymbol{\beta}$. Here if $a_i = 0$ we assume

$$\prod_{j=1}^{a_i} \mathsf{P}_{\{i\}\beta^{(ij)}}(t_1, t_2) = 1 .$$

Thus a branching process is a Markov process whose space of all possible states N is the set of all integer-valued n-dimensional vectors with nonnegative coordinates, and whose transition probabilities satisfy equation (1).

In what follows, we consider only homogeneous branching processes, that is, processes for which $p_{\alpha\beta}(t_1, t_2) = p_{\alpha\beta}(t_2 - t_1)$.

Let us consider processes with a single type of particle ($n = 1$). Since in branching processes every particle evolves independently of the others, we may assume that at the initial instant only one particle exists. With the passage of time it either disappears or turns into k particles that are all of the same type and constitute the first generation.

Every particle of the first generation "lives" independently of the other particles, and the same probability-theoretic laws obtain for it as for the initial particle. At some instant of time the particle either disappears or turns into second-generation particles, and so forth. The entire process is described by a single integer-valued random function $\nu(t)$ that is equal to the number of particles existing at the instant t. By our assumption, $\nu(0) = 1$.

The set of possible states of the process is the sequence of natural numbers $0, 1, 2, \cdots$. Here 0 is an absorbing state: If $\nu(t_0) = 0$ then at all subsequent instants of time $\nu(t) = 0$. If, with probability 1, $\nu(t) = 0$ at some instant t, the process is said to be *degenerate*. In other cases the quantity $\nu(t)$ vanishes after a finite interval of time with probability less than one. This probability is called the *probability of degeneration* of the process. It is possible that the quantity $\nu(t)$ increases without bound in time. In the case of a nuclear reaction this situation can be interpreted as an explosion. Thus in the theory of branching processes we need to ask the following questions: What is the probability of degeneration of the branching process and what is the asymptotic behavior of the quantity $\nu(t)$?

Let $p_{ij}(t)$ denote the conditional probability that the system consists of j particles at the instant $t + \tau$ if there were i particles at the instant τ. To solve problems that arise in the theory of branching processes, we shall find it convenient to use the method of generating functions. We therefore introduce the generating functions $f_i(z, t)$ of the distributions $\{p_{ij}(t)\}$, for $j = 0, 1, \cdots$,

$$f_i(z, t) = \sum_{k=0}^{\infty} z^k p_{ik}(t), \qquad |z| \leqq 1 .$$

The probabilities $p_{ij}(t)$ with i fixed correspond to the distribution of i independent random variables all having the same distribution with generating function $f_1(z, t)$:

$$f_i(z, t) = [f_1(z, t)]^i . \tag{2}$$

To determine the function $f_1(z, t)$, we may use either of the systems of Kolmogorov equations (cf. Section 2).

According to the general theory, the derivatives

$$\lim_{t\to 0}\frac{p_{1j}(t)}{t} = b_j \quad j \neq 1, \quad \lim_{t\to 0}\frac{1 - p_{11}(t)}{t} = b_1$$

exist. Let us suppose that

$$b_1 = b_0 + \sum_{j=2}^{\infty} b_j < \infty .$$

Then we have the first system of Kolmogorov equations

$$\frac{dp_{1j}(t)}{dt} = -b_1 p_{1j}(t) + \sum_{\substack{k=0\\k\neq 1}}^{\infty} b_k p_{kj}(t) . \tag{3}$$

Multiplying both sides of equations (3) by z^j and summing with respect to j from 0 to ∞, we obtain

$$\frac{\partial f_1(z, t)}{\partial t} = -b_1 f_1(z, t) + \sum_{\substack{k=0\\k\neq 1}}^{\infty} b_k f_k(z, t)(|z| \leqq 1) ,$$

or on the basis of (2),

$$\frac{\partial f(z, t)}{\partial t} = -b_1 f(z, t) + \sum_{\substack{k=0\\k\neq 1}}^{\infty} b_k f^k(z, t) ,$$

where we set $f(z, t) = f_1(z, t)$. Finally we obtain the following nonlinear differential equation

$$\frac{\partial f}{\partial t} = u(f) \tag{4}$$

where

$$u(z) = b_0 - b_1 z + \sum_{k=2}^{\infty} b_k z^k \qquad (|z| \leqq 1) . \tag{5}$$

To equation (4) we must add the initial condition

$$f(z, 0) = z . \tag{6}$$

The solution of equation (4) with initial condition (6) can be written in the form

$$\varphi(f) - \varphi(z) = t,\ \varphi(z) = \int \frac{dz}{u(z)} . \tag{7}$$

Let us suppose that the conditions under which the second system of Kolmogorov differential equations may be applied are satisfied. From the definition of branching processes we get the formulas

$$p_{kk}(t) = (1 - b_1 t)^k + o(t) = 1 - kb_1 t + o(t) ,$$
$$p_{k,k-1}(t) = k(1 - b_1 t)^{k-1} b_0 t + o(t) = kb_0 t + o(t) ,$$
$$p_{k,k-j}(t) = o(t),\ j \geqq 2 ,$$
$$p_{k,k+j}(t) = k(1 - b_1 t) b_{j+1} t + o(t) = kb_{j+1} t + o(t),\ j \geqq 1 ,$$

from which it follows (in the notation of Section 2) that

$$q_{jj} = jb_1,\ q_{j,j-1} = jb_0,\ q_{j,j-k} = 0,\ k \geqq 2\ ;$$
$$q_{j,j+k} = jb_{k+1},\ k \geqq 1\ .$$

Thus for the functions $p_{1j}(t)$, the system (16) of Section 2 takes the form

$$\frac{dp_{1j}(t)}{dt} = (j+1)b_0 p_{1,j+1}(t) - jb_1 p_{1j}(t) + (j-1)b_2 p_{1,j-1}(t) + \cdots + b_j p_{11}(t)\ . \tag{8}$$

Multiplying (8) by z^j and summing with respect to j from 0 to ∞, we obtain a new equation for the generating function $f(z, t)$,

$$\frac{\partial f(z,t)}{\partial t} = u(z)\frac{\partial f(z,t)}{\partial z} \qquad (|z| < 1)\ , \tag{9}$$

where $u(z)$ is defined by equation (5). To this equation we add the initial condition (6). The solution of equations (9) and (6) takes the form

$$f(z,t) = \psi\left(t + \int \frac{dz}{u(z)}\right), \tag{10}$$

where $\psi(t)$ is the inverse of the function

$$t = \varphi(z) = \int \frac{du}{u(z)}\ .$$

This solution coincides with (7).

Example. Let us set

$$u(z) = p - (p+q)z + qz^2\ .$$

In this case a particle disappears with probability $pt + c(t)$ after an interval of time t, divides into two particles with probability $qt + o(t)$ at the end of the same interval, or keeps its original state with probability $1 - (p+q)t + o(t)$ for that interval. The probability of dividing into more than two particles is equal to $o(t)$. The process in question coincides with a linear birth and death process (cf. Section 4, Example 3). We have

$$\varphi(z) = \int \frac{dz}{u(z)} = \int \frac{dz}{p - (p+q)z + z^2}$$
$$= \frac{1}{q-p} \ln \frac{z-1}{z - \dfrac{p}{q}} \qquad (p \neq q)\ .$$

For the inverse function $z = \psi(t)$ we obtain the expression

$$z = \psi(t) = \frac{1 - \beta e^{qt(1-\beta)}}{1 - e^{q(1-\beta)t}}, \qquad \beta = \frac{p}{q},$$

from which we get

$$f(z, t) = \psi(t + \varphi(z)) = \frac{z - \beta - \beta(z - 1)e^{qt(1-\beta)}}{z - \beta - (z - 1)e^{qt(1-\beta)}}.$$

Expanding $f(z, t)$ in a series of powers of z, we get

$$f(z, t) = \frac{1 - e^{t(q-p)}}{1 - \frac{1}{\beta}e^{t(q-p)}} + \sum_{n=1}^{\infty} \frac{(1 - \beta)^2(1 - e^{t(q-p)})^{n-1}e^{t(q-p)}}{[\beta - e^{t(q-p)}]^{n+1}} z^n$$

which leads us to the formulas

$$p_{10}(t) = \frac{1 - e^{t(q-p)}}{1 - \frac{1}{\beta}e^{t(q-p)}}, \tag{11}$$

$$p_{1n}(t) = (1 - \beta)^2 \frac{(1 - e^{t(q-p)})^{n-1}e^{t(q-p)}}{[\beta - e^{t(q-p)}]^{n+1}}, \qquad n = 1, 2, \cdots. \tag{12}$$

Let us now consider the case in which $p = q$, which we have excluded up to this point. If $p = q$, then

$$\varphi(z) = \int_0^z \frac{dz}{p(1 - z)^2} = \frac{1}{p}\frac{z}{1 - z}, \qquad z = \psi(t) = 1 - \frac{1}{1 + pt},$$

from which we get

$$f(z, t) = \psi(t + \varphi(z)) = 1 - \frac{1 - z}{1 + pt(1 - z)}$$

$$= \frac{pt}{1 + pt} + \sum_{n=1}^{\infty} \frac{(pt)^{n-1}}{(1 + pt)^{n+1}} z^n.$$

Consequently

$$p_{10}(t) = \frac{pt}{1 + pt}, \tag{13}$$

$$p_{1n}(t) = \frac{(pt)^{n-1}}{(1 + pt)^{n+1}}, \qquad n = 1, 2, \cdots \tag{14}$$

The following asymptotic relationships as $t \to \infty$ are immediate consequences of formulas (11) to (14): If $q \leqq p$, then

$$p_{10}(t) \to 1, p_{1n}(t) \to 0 \quad \text{for} \quad n \geqq 1, t \to \infty;$$

if $q > p$, then

$$p_{10}(t) \to \beta(\beta < 1), p_{1n}(t) \to 0 \quad \text{for} \quad n \geqq 1, t \to \infty.$$

If $q \leqq p$, the branching process degenerates with probability 1; that is, all particles eventually disappear. If $q > p$, the probability of degeneration of the process is equal to $\beta = p/q < 1$. However if the particles do not disappear the number of them increases without bound in time, because

$$\mathsf{P}\{\nu(t) > N \mid \nu(t) > 0\} = 1 - \frac{1}{p_{10}(t)} \sum_{k=1}^{N} p_{1k}(t) \longrightarrow 1$$

for every N.

Let us look at the question of asymptotic behavior as $t \longrightarrow \infty$ of a branching process in the general case. In what follows we shall need the moments of the quantity $\nu(t)$. Since we are using generating functions we shall find it convenient to introduce factorial moments and use them instead of ordinary moments. We define

$$\mathsf{M}_k(t) = \mathsf{M}[\nu(t)(\nu(t) - 1) \cdots (\nu(t) - k + 1)] \, .$$

It is not difficult to set up linear differential equations that are satisfied by the factorial moments $\mathsf{M}_k(t)$. Let us suppose that $\sum_{k=1}^{\infty} kb_k < \infty$. Then for $|z| < 1$ the differential equation (4) can be differentiated with respect to z, and this gives us

$$\frac{\partial \mathsf{M}_1(z, t)}{\partial t} = u'(f)\mathsf{M}_1(z, t), \ \mathsf{M}_1(z, 0) = 1 \, , \tag{15}$$

where we set

$$\mathsf{M}_1(z, t) = \frac{\partial}{\partial z} f(z, t) = \mathsf{M}\nu(t) z^{\nu(t)-1}$$

$$u'(z) = -b_1 + \sum_{k=2}^{\infty} kb_k z^{k-1} \, .$$

It follows from (15) that

$$\mathsf{M}_1(z, t) = \exp\left(\int_0^t u'(f)dt\right)(|z| < 1) \, .$$

As $z \longrightarrow 1$, $u'(z) \longrightarrow -b_1 + \sum_{k=2}^{\infty} kb_k = m_1 < \infty$ and $f(z, t) \longrightarrow 1$, increasing monotonically and hence uniformly with respect to t. Consequently $\lim \mathsf{M}_1(z, t) = e^{m_1 t}$. On the other hand, by virtue of Lebesgue's theorem,

$$\lim_{z \uparrow 1} \mathsf{M}_1(z, t) = \lim_{z \uparrow 1} \mathsf{M}\nu(t) z^{\nu(t)-1} = \mathsf{M}\nu(t) = \sum_{k=1}^{\infty} kp_k(t) \, .$$

Thus

$$\mathsf{M}_1(t) = e^{m_1 t} \, . \tag{16}$$

This result is immediately generalized to factorial moments of higher orders.

Lemma 1. *Suppose that*

$$m_r = \sum_{n=1}^{\infty} n(n-1) \cdots (n-r+1) b_n < \infty, \qquad r = 1, \cdots k \, . \tag{17}$$

Then the factorial moments $\mathsf{M}_r(t)$, *for* $r = 1, 2, \cdots, k$, *of the variable* $\nu(t)$ *are finite and they satisfy first-order linear differential equations with constant coefficients.*

Proof. Let us differentiate equation (4) successively with respect to z (for $|z| < 1$). Setting

$$\mathsf{M}_k(z, t) = \frac{\partial^k f(z, t)}{\partial z^k} = \mathsf{M}(\nu(t)(\nu(t) - 1) \cdots (\nu(t) - k + 1)z^{\nu(t)-k}) ,$$

we obtain

$$\begin{aligned} \frac{\partial \mathsf{M}_2(z, t)}{\partial t} &= u'(f)\mathsf{M}_2(z, t) + u''(f)\mathsf{M}_1^2(z, t) , \\ &\cdots\cdots\cdots\cdots\cdots\cdots \\ \frac{\partial \mathsf{M}_k(z, t)}{\partial t} &= u'(f)\mathsf{M}_k(z, t) + F_k(z, t) , \end{aligned} \tag{18}$$

where $F_k(z, t)$ is a polynomial in $\mathsf{M}_1(z, t), \cdots, \mathsf{M}_{k-1}(z, t)$ with coefficients that depend linearly on $u''(f), \cdots, u^{(k)}(f)$. To these equations we add the initial conditions

$$\mathsf{M}_k(z, 0) = 0 \qquad (k = 2, 3, \cdots) .$$

The solution of equation (18) is of the form

$$\mathsf{M}_k(z, t) = \exp\left(\int_0^t u'(f)dt\right)\int_0^t F_k(z, \tau) \exp\left(-\int_0^\tau u'(f)d\theta\right)d\tau .$$

Using the same considerations as in the case of $k = 1$, we obtain by induction

$$\mathsf{M}_k(t) = \lim_{z\to 1} \mathsf{M}_k(z, t) = e^{m_1 t}\int_0^t F_k(1, \tau)e^{-m_1\tau}d\tau .$$

Here $\mathsf{M}_k(t)$ obviously satisfies equation (18), in which we have set $z = 1$. In particular,

$$\mathsf{M}_2(t) = \frac{m_2}{m_1}(e^{m_1 t} - 1)e^{m_1 t} \qquad (m_1 \neq 0) , \tag{19}$$

$$\mathsf{M}_2(t) = m_2 t \qquad (m_1 = 0) . \tag{20}$$

In the study of the asymptotic behavior of a branching process, an important role is played by the function $u(z)$ (cf. (5)), which we now consider for real values of z. We note that

$$u(0) = b_0 \geqq 0, u(1) = b_0 - b_1 + \sum_{k=2}^{\infty} b_k = 0 ,$$

$$u''(z) > 0 \quad \text{for} \quad z > 0 .$$

Here we assume that not all the b_k (for $k \geqq 2$) are equal to 0. Thus $u''(z)$ is convex downward for $z > 0$ and hence has no more than one zero in the interval (0, 1). Let us turn to the definition of the probability α of degeneration of the branching process $\nu(t)$. Since the events $\nu(t) = 0$ constitute an increasing class of events, we have

$$\alpha = \mathsf{P}\{\lim_{t\to\infty} \nu(t) = 0\} = \lim_{t\to\infty} \mathsf{P}\{\nu(t) = 0\} = \lim_{t\to\infty} p_{10}(t) .$$

Theorem 1. *The probability of degeneration of a branching process coincides with the smallest nonnegative root of the equation* $u(x) = 0$. *If*

$$u'(1) = m_1 = -b_1 + \sum_{k=2}^{\infty} kb_k < \infty ,$$

then for $u'(1) \leqq 0$ *the probability of degeneration is* $\alpha = 1$; *if* $u'(1) > 0$, *then* $\alpha < 1$.

Proof. If $p_{10}(t) = f(0, t)$, it follows from (4) that

$$\frac{dp_{10}(t)}{dt} = u(p_{10}(t)), p_{10}(0) = 0 . \tag{21}$$

If $b_0 = 0$, then $p_{10}(t) \equiv 0$ is a solution of equation (21) and the theorem is trivial. Suppose that $b_0 > 0$. We note that if x_0 is the smallest positive root of the equation $u(x) = 0$, then $p_{10}(t) < x_0$ for all $t > 0$, because if $p_{10}(t_0)$ were equal to x_0 for $t_0 > 0$, then by virtue of the uniqueness of the solution of equation (21) we would have $p_{10}(t) \equiv x_0$, which is impossible.

Furthermore since the limit $\alpha = \lim_{t\to\infty} p_{10}(t) \leqq 1$ exists, it follows from (21) that the limit $\lim_{t\to\infty} p'_{10}(t) = u(\alpha)$ also exists. But this implies that $u(\alpha) = 0$ because otherwise the quantity

$$p_0(t) = \int_{t_0}^{t} p'_0(t)dt + p_0(t_0)$$

would increase without bound. Thus we have shown that $\alpha = x_0$. If $x_0 < 1$, then the function $u(x)$ is an increasing function at the point $x = 1$ and the derivative $u'(1) > 0$ if it exists. On the other hand, if $x_0 = 1$, then $u'(1) \leqq 0$. This completes the proof of the theorem.

Let us now investigate the asymptotic behavior of the probability $p_{10}(t)$ as $t \to \infty$ for degenerating processes ($\alpha = 1$).

Theorem 2. *If* $m_1 = u'(1) \leqq 0$ *and* $m_2 = u''(1) < \infty$, *then*

$$1 - p_{10}(t) \approx ke^{m_1 t} \quad \text{for} \quad m_1 < 0$$

and

$$1 - p_{10}(t) \approx \frac{2}{m_2 t} \quad \text{for} \quad m_1 = 0 .$$

Proof. Let us define

$$q(t) = 1 - p_{10}(t) .$$

The function $q(t)$ satisfies the equation

$$\frac{dq}{dt} = -u(1 - q(t)), q(0) = 1 .$$

Using the formula for finite increments, we obtain

$$\frac{dq}{dt} = -u(1) + q(t)u'(\xi) = q(t)u'(\xi) ,$$

where ξ lies between $p_{10}(t)$ and 1. Since $u'(x)$ is an increasing function and $\xi \longrightarrow 1$ as $t \longrightarrow \infty$, it follows that $u'(\xi) = u'(1) - \varepsilon(t)$, where $\varepsilon(t) > 0$, and $\lim_{t\to\infty} \varepsilon(t) = 0$. Thus

$$\frac{dq}{dt} = q(t)(m_1 - \varepsilon(t)) ,$$

from which we get

$$q(t) = \exp\left(m_1 t - \int_0^t \varepsilon(\tau)d\tau\right) .$$

We note that for $\xi < \zeta < 1$,

$$0 < \varepsilon(t) = u'(1) - u'(\xi) = u''(\zeta)(1 - \xi)$$
$$\leqq u''(1)(1 - p_{10}(t)) \leqq m_2 e^{m_1 t} .$$

Therefore the integral $\int_0^\infty \varepsilon(t)dt$ is finite. From this it follows that for $m_1 < 0$,

$$q(t) \approx ke^{m_1 t} \quad \text{where} \quad k = \exp\left(-\int_0^\infty \varepsilon(t)dt\right) .$$

Consider the case $m_1 = 0$. We have

$$\frac{dq}{dt} = -u(1 - q(t)) = -u(1) + q(t)u'(1) - \frac{q^2(t)}{2}u''(\xi_1) ,$$

where ξ_1 is a number in the interval $(p_{10}(t), 1)$. Since $u''(\xi_1) \longrightarrow u''(1)$ as $t \longrightarrow \infty$, we have

$$\frac{dq}{dt} = -\frac{q^2(t)}{2}(m_2 + \varepsilon(t)) ,$$

where $\varepsilon(t) \longrightarrow 0$ as $t \longrightarrow \infty$. From this it follows that

$$q(t) = \frac{2}{m_2 t + \int_0^t \varepsilon(\tau)d\tau + 2} = \frac{2}{m_2 t} + o\left(\frac{1}{t}\right) .$$

This completes the proof of the theorem.

We shall supplement Theorem 2 with a result dealing with the asymptotic behavior of the probabilities $p_{1k}(t)$ for degenerating processes.

Since $\lim_{t\to\infty} p_{1n}(t) = 0$ (for $n > 0$), we have $\lim_{t\to\infty} f(z, t) = 1$. We define $q(z, t) = 1 - f(z, t)$. For $z = 0$ we have

$$q(0, t) = 1 - f(0, t) = 1 - p_{10}(t) = q(t) \approx ke^{m_1 t} .$$

We may assume that the same is true of the rate of decrease of the function $q(z, t)$ at $z \neq 0$ also. In connection with this we define

$$\varphi(z, t) = \frac{q(z, t)}{q(t)} = \frac{1 - f(z, t)}{q(t)} . \tag{22}$$

We note that the function

$$f^*(z, t) = 1 - \varphi(z, t) = \sum_{n=1}^{\infty} \frac{p_{1n}(t)}{q(t)} z^n \tag{23}$$

can be regarded as the generating function for the conditional distribution of the number $\nu(t)$ of particles under the hypothesis that it is nonzero up to the instant t.

Theorem 3. *If $m_1 = u'(1) < 0$ and $m_2 = u''(1) < \infty$ then as $t \longrightarrow \infty$ the conditional distribution of the number of particles $\nu(t)$, under the hypothesis that the process has not degenerated ($\nu(t) \neq 0$) up to the instant t, approaches a definite limit, the generating function $f^*(z)$ of which is equal to*

$$f^*(z) = 1 - \exp\left(m_1 \int_0^z \frac{dz}{u(z)}\right) . \tag{24}$$

Proof. Let us consider the function $\varphi(z, t)$. It follows from (4) that $\varphi(z, t)$ satisfies the equation

$$\frac{\partial \varphi}{\partial t} = -\frac{1}{q(t)} u(1 - q(t)\varphi) + \frac{\varphi}{q(t)} u(1 - q(t)) . \tag{25}$$

Expanding the right-hand member of the equation obtained in accordance with Taylor's formula, we get

$$\frac{\partial \varphi}{\partial t} = -\frac{1}{q(t)}\left[u(1) - q(t)\varphi u'(1) + \frac{(q(t)\varphi)^2}{2}(u''(1) + \varepsilon_1)\right]$$
$$+ \frac{\varphi}{q(t)}\left[u(1) - q(t)u'(1) + \frac{q^2(t)}{2}(u''(1) + \varepsilon_2)\right],$$

where $\varepsilon_1 = u''(\xi_1) - u''(1)$ and $\varepsilon_2 = u''(\xi_2) - u''(1)$, the number ξ_1 (resp. ξ_2) lying in the interval $(f(z, t), 1)$ (resp. $(f(0, t), 1)$). As $t \longrightarrow \infty$, the functions ε_1 (for $i = 1, 2$) approach 0 uniformly in an arbitrary region $|z| \leqq \rho < 1$. The preceding equation can be written in the form

$$\frac{\partial \varphi}{\partial t} = -\frac{q(t)\varphi^2}{2}(m_2 + \varepsilon_1) + \frac{q(t)\varphi}{2}(m_2 + \varepsilon_2) . \tag{25'}$$

Beginning with some sufficiently large t, we have

$$\frac{\partial \varphi}{\partial t} < \frac{q(t)\varphi}{2}\left(m_2 + \frac{m_2}{2}\right) = \frac{3}{4} m_2 q(t)\varphi$$

so that

$$\varphi(z, t) \leqq \varphi(z, t_0) \exp\left(\frac{3}{4} m_2 \int_0^t q(\tau)d\tau\right).$$

The convergence of the integral $\int_{t_0}^{\infty} q(\tau)d\tau$ implies that the function $\varphi(z, t)$ remains bounded as $t \to \infty$. Therefore equation (25) can be rewritten in the form

$$\frac{\partial \varphi}{\partial t} = \frac{q(t)\varphi}{2}[m_2(1 - \varphi) + \varepsilon], \qquad \varphi(z, 0) = 1 - z,$$

where $\varepsilon = \varepsilon_2 - \varphi\varepsilon_1 \to 0$ as $t \to \infty$. Representing the solution of the last equation in the form

$$\varphi(z, t) = (1 - z) \exp\left(\frac{1}{2}\int_0^t q(\tau)[m_2(1 - \varphi(z, \tau)) + \varepsilon]d\tau\right),$$

we see that the limit $\lim_{t\to\infty} \varphi(z, t) = K(z)$ exists. Furthermore, it follows from (25′) that $\lim_{t\to\infty} d\varphi/dt = 0$. Since $\varphi(z, t)$ is an analytic function inside the disk $|z| < 1$ and since all the limit relationships that we have used hold uniformly inside every disk $|z| \leqq \rho < 1$, it follows that the function $K(z)$ is also analytic inside the disk and that

$$\lim_{t\to\infty} \frac{\partial \varphi(z, t)}{\partial z} = \frac{dK(z)}{dz}$$

uniformly inside an arbitrary disk $|z| \leqq \rho < 1$. To determine the function $K(z)$, we may use equation (9). Setting $f(z, t) = 1 - q(t)\varphi(z, t)$ in that equation, we obtain

$$-q'(t)\varphi(z, t) - q(t)\frac{\partial \varphi(z, t)}{\partial t} = -u(z)q(t)\frac{\partial \varphi(z, t)}{\partial z}.$$

Dividing this equation by $q(t)$ and letting t approach ∞, and remembering that $q'(t)/q(t) \to m_1$ (cf. proof of Theorem 2), we obtain

$$m_1 K(z) = u(z)\frac{dK(z)}{dz}.$$

Here, $K(0) = \lim_{t\to\infty} \varphi(0, t) = 1$. Thus

$$K(z) = \exp\left(m_1 \int_0^z \frac{dz}{u(z)}\right),$$

$$1 - f(z, t) \approx q(t)K(z) = \exp\left(m_1\left(t + \int_0^z \frac{dz}{u(z)}\right)\right),$$

which completes the proof of the theorem.

Let us now derive an expression for the mean number of particles at the instant t under the hypothesis that the process has not degenerated up to that instant. We have

$$m^*(t) = \mathsf{M}\{\nu(t) \mid \nu(t) > 0\} = \frac{e^{m_1 t}}{q(t)}, \tag{26}$$

from which (keeping Theorem 2 in mind) we get the following asymptotic relations as $t \to \infty$:

$$m^*(t) \approx \begin{cases} \dfrac{1}{k} & \text{for } m_1 < 0\,, \\ \dfrac{m_2 t}{2} & \text{for } m_1 = 0\,, \\ \dfrac{e^{m_1 t}}{1-\alpha} & \text{for } m_1 > 0\,. \end{cases}$$

For $m_1 \geqq 0$, the number of particles $\nu(t)$, under the hypothesis that $\nu(t) > 0$, increases without bound. We define

$$\nu^*(t) = \frac{\nu(t)}{m^*(t)}\,.$$

Then $\mathsf{M}\{\nu^*(t) \mid \nu^*(t) > 0\} = 1$. Let us study the limiting behavior of the quantity $\nu^*(t)$ as $t \to \infty$.

As we might expect, the limiting distribution of the quantity $\nu^*(t)$ under the hypothesis $\nu(t) > 0$ will, if it exists, be a continuous distribution on the halfline $[0, \infty)$, and therefore it is convenient for us to shift over from the generating functions to the characteristic functions. For the characteristic function $g(\lambda, t)$ of the random variable $\nu^*(t)$, under the hypothesis that $\nu(t) > 0$, we have the value

$$g(\lambda, t) = \sum_{n=1}^{\infty} \exp\left(\frac{i\lambda n}{m^*(t)}\right) \frac{p_{1n}(t)}{q(t)} = \frac{f\left\{\exp\left(\frac{i\lambda}{m^*(t)}\right), t\right\} - f(0, t)}{q(t)}$$

or

$$g(\lambda, t) = 1 - \frac{1 - f\left\{\exp\left(\frac{i\lambda}{m^*(t)}\right), t\right\}}{q(t)}\,. \tag{27}$$

Consider the case $m_1 = 0$.

Theorem 4. *If $m_1 = 0$ and $m_2 < \infty$, then*

$$\lim_{t\to\infty} \mathsf{P}\left\{\frac{2\nu(t)}{m_2 t} < x \mid \nu(t) > 0\right\} = 1 - e^{-x}\,. \tag{28}$$

Proof. Defining $\psi(z, t) = 1 - f(z, t)$, we obtain from equation (4)

$$\frac{\partial\psi}{\partial t} = -u(1-\psi) = -\frac{\psi^2}{2}[u''(1) + \varepsilon(t)], \qquad \psi(z, 0) = 1 - z .$$

Since the process is degenerate for $m_1 = 0$, we see that $\psi(z, t) \longrightarrow 0$ uniformly in the region $|z| \leqq 1$ as $t \longrightarrow \infty$. From this it follows that $\varepsilon(t)$ also approaches 0 uniformly with respect to z, for $|z| \leqq 1$, as $t \longrightarrow \infty$. Integrating the last equation, we obtain

$$\frac{1}{\psi(z, t)} - \frac{1}{1-z} = \frac{1}{2}\left[m_2 t + \int_0^t \varepsilon(t)dt\right],$$

so that

$$\lim_{t\to\infty} \frac{q(t)}{\psi\left\{\exp\left(\frac{i\lambda}{m^*(t)}\right), t\right\}} = \lim_{t\to\infty} \left\{\frac{q(t)}{1 - \exp\left(\frac{i\lambda}{m^*(t)}\right)} + \frac{m_2 t q(t)}{2} + q(t)\int_0^t \varepsilon(t)dt\right\} = -\frac{1}{i\lambda} + 1 ,$$

and consequently

$$g(\lambda) = \lim_{t\to\infty} g(\lambda, t) = \frac{1}{1 - i\lambda} .$$

The function $g(\lambda)$ is the characteristic function of the distribution $F(x) = 1 - e^{-x}$ for $x > 0$, $F(x) = 0$ for $x < 0$. This completes the proof of the theorem.

In the case $m_1 > 0$, the quantity $q(t)$ approaches the nonzero limit $1 - \alpha$. Therefore normalization with the aid of the function $q(t)$ or shifting to the conditional mathematical expectations under the hypothesis that $\nu(t) > 0$ cannot play a significant role.

Theorem 5. *If $m_1 > 0$ and $m_2 < \infty$, then the quantity $\nu(t)e^{-m_1 t}$ converges in the sense of mean-square as $t \longrightarrow \infty$ to the random variable $\eta = \text{l.i.m. } \nu(t)e^{-m_1 t}$, whose characteristic function $g(\lambda)$ satisfies the functional equation*

$$(1 - g(\lambda)) \exp\left(-\int_1^{g(\lambda)} \frac{u(v) - m_1(v-1)}{u(v)(v-1)} dv\right) = -i\lambda . \tag{29}$$

To prove the convergence of $\tilde{\nu}(t) = \nu(t)e^{-m_1 t}$ in the sense of mean square, we use Cauchy's criterion. Suppose that $t < t'$. Then

$$\mathsf{M}(\tilde{\nu}(t) - \tilde{\nu}(t'))^2 = \mathsf{M}\tilde{\nu}(t)^2 + \mathsf{M}\tilde{\nu}(t')^2 - 2\mathsf{M}(\tilde{\nu}(t)\tilde{\nu}(t')) .$$

On the basis of formulas (16) and (19),

$$\mathsf{M}\tilde{\nu}(t)^2 \approx \frac{m_2}{m_1} \quad \text{as} \quad t \longrightarrow \infty .$$

Using the definition of a branching process and the fact that it is homogeneous, we obtain

$$\mathsf{M}(\nu(t)\nu(t')) = \mathsf{M}\nu(t)\mathsf{M}\{\nu(t') \mid \nu(t)\} = \mathsf{M}\nu^2(t)\mathsf{M}\nu(t'-t) \, ,$$

from which it follows (also on the basis of formulas (16) and (19)) that

$$\mathsf{M}(\tilde{\nu}(t)\tilde{\nu}(t')) \approx \frac{m_2}{m_1} \, .$$

Thus $\mathsf{M}(\tilde{\nu}(t) - \tilde{\nu}(t'))^2 \to 0$ as $t, t' \to \infty$ and the limit $\eta = \text{l.i.m.}_{t\to\infty} \tilde{\nu}(t)$ exists. Writing equation (4) in the form

$$\frac{df}{f-1} - \frac{u(f) - m_1(f-1)}{u(f)(f-1)} df = m_1 dt$$

and integrating with respect to t from 0 to t, we obtain

$$\ln(1-f) - \int_z^f \frac{u(v) - m_1(v-1)}{u(v)(v-1)} dv = m_1 t + \ln(1-z) \, .$$

Here if we set $z = e^{i\lambda/m^*(t)}$ and let t approach ∞, we obtain the equation

$$\ln(1-g) - \int_1^g \frac{u(v) - m_1(v-1)}{u(v)(v-1)} dv = \ln(-i\lambda) \, ,$$

from which (29) follows. This completes the proof of the theorem.

The theory of branching processes with particles of several types is analogous but more complicated. We pause only for the basic relationships in that theory. Just as in the case of particles of a single type, it is convenient to use the method of generating functions.

Let M denote the set of all possible states of a process, that is, the set of all vectors $\boldsymbol{\alpha} = (a_1, a_2, \cdots, a_n)$ with nonnegative integral components. Let us agree to denote n-dimensional vectors by Greek letters $\boldsymbol{\alpha}, \boldsymbol{\beta}, \boldsymbol{\sigma}, \cdots$ and their components with the corresponding Roman letters. We define the generating functions

$$F_i(t, \boldsymbol{\sigma}) = F_i(t, s_1, s_2, \cdots, s_n)$$

of the transition probabilities $p_{\{i\}\beta}(t)$ by the relations

$$F_i(t, \boldsymbol{\sigma}) = F_i(t, s_1, \cdots, s_n) = \sum_{\beta \in M} p_{\{i\}\beta}(t) s_1^{b_1} s_2^{b_2} \cdots s_n^{b_n},$$

$$(\boldsymbol{\sigma} = \{s_1, s_2 \cdots, s_n\}, \boldsymbol{\beta} = \{b_1, b_2, \cdots, b_n\}) \, . \tag{30}$$

We recall that $\{i\}$ denotes the vector $\{i\} = \{\delta_{i1}, \delta_{i2}, \cdots, \delta_{in}\}$. The functions $F_i(t, \boldsymbol{\sigma})$ are analytic functions of the variables $s_1, s_2, \cdots, s_n$ in the region $|s_i| < 1$ (for $i = 1, \cdots, n$). Also,

$$|F_i(t, \boldsymbol{\sigma})| \leqq 1 \quad \text{for} \quad |s_i| \leqq 1, F_i(t, 1, \cdots, 1) = 1, F_i(0, \boldsymbol{\sigma}) = s_i \, . \tag{31}$$

If we define the n-dimensional vector-valued function

$$\boldsymbol{\Phi}(t, \boldsymbol{\sigma}) = \{F_1(t, \boldsymbol{\sigma}), \cdots, F_n(t, \boldsymbol{\sigma})\} \, ,$$

it follows from (31) that

$$\Phi(0, \sigma) = \sigma . \tag{32}$$

Let us now find the equivalent of the Kolmogorov-Chapman formula for branching processes expressed in terms of the generating functions. We have

$$p_{\{i\}\beta}(t + \tau) = \sum_{\alpha \in M} p_{\{i\}\alpha}(t) p_{\alpha\beta}(\tau), \ t > 0, \tau > 0 .$$

If we substitute into this equation the value of $p_{\alpha\beta}(\tau)$ corresponding to formula (1), we obtain

$$p_{\{i\}\beta}(t + \tau) = \sum_{\alpha \in M} p_{\{i\}\alpha}(t) \sum_{\sum_k \sum_j \beta^{(k,j)} = \beta} \prod_{k=1}^{n} \prod_{j=1}^{a_i} p_{\{k\}\beta^{(k,j)}}(\tau) .$$

If we multiply both sides of this equation by $s_1^{b_1} \cdots s_n^{b_n}$ and sum over all β, we obtain the relations:

$$\begin{aligned} &F_i(t + \tau, \sigma) \\ &= \sum_{\alpha \in M} p_{\{i\}}(t) \sum_{\beta \in M} \sum_{\sum_k \sum_j \beta^{(k,j)} = \beta} \prod_{k=1}^{n} \prod_{j=1}^{a_k} p_{\{k\}\beta^{(k,j)}}(\tau) s_1^{b_1(k,j)} \cdots s_n^{b_n(k,j)} \\ &= \sum_{\alpha \in M} p_{\{i\}\alpha}(t) \prod_{k=1}^{n} \prod_{j=1}^{a_k} \left(\sum_{\beta^{(k,j)} \in M} p_{\{k\}\beta^{(k,j)}}(\tau) s_1^{b_1(k,j)} \cdots s_n^{b_n(k,j)} \right) \\ &= \sum_{\alpha \in M} p_{\{i\}\alpha}(t) \prod_{k=1}^{n} F_k^{a_k}(\tau, \sigma) , \end{aligned}$$

from which it follows that

$$F_i(t + \tau, \sigma) = F_i(t, F_1(\tau, \sigma), \cdots, F_n(\tau, \sigma)), \qquad i = 1, \cdots, s ,$$

or

$$\Phi(t + \tau, \sigma) = \Phi(t, \Phi(\tau, \sigma)) . \tag{33}$$

Theorem 6. *The system of generating functions of a branching process satisfies the system of functional equations* (33) *and the initial condition* (32).

Let us derive for the generating functions the differential equations corresponding to the first and second systems of Kolmogorov equations for the transition probabilities.

Suppose that

$$\lim_{t \downarrow 0} \frac{p_{\{i\}\beta}(t)}{t} = b_{i\beta} \ (\beta \neq \{i\}), \qquad \lim_{t \downarrow 0} \frac{1 - p_{\{i\}\{i\}}(t)}{t} = b_{ii}$$

and

$$b_{ii} = \sum_{\beta \in M, \beta \neq \{i\}} b_{i\beta} < \infty, \qquad i = 1, \cdots, n .$$

Then the transition probabilities $p_{\{i\}\beta}(t)$ satisfy the first system of Kolmogorov equations (cf. Section 2):

$$\frac{dp_{\{i\}\beta}(t)}{dt} = -b_{ii}p_{\{i\}\beta}(t) + \sum_{\alpha\in M,\alpha\neq\{i\}} b_{i\alpha}p_{\alpha\beta}(t) .$$

Multiplying this equation by $s_1^{b_1}, \cdots, s_n^{b_n}$, summing over all β, and noting that equation (1) implies

$$\sum_{\beta\in M} p_{\alpha\beta}(t)s_1^{b_1}s_2^{b_2}\cdots s_n^{b_n} = \prod_{i=1}^{n}[F_i(t, \sigma)]^{a_i}$$

(this equation expresses the independence of the evolution of the particles that exist at a given instant of time), we obtain

$$\frac{\partial F_i(t, \sigma)}{\partial t} = -b_{ii}F_i(t, \sigma) + \sum_{\alpha\in M,\alpha\neq\{i\}} b_{i\alpha}\prod_{i=1}^{n}[F_i(t, \sigma)]^{a_i}$$

or

$$\frac{\partial F_i(t, \sigma)}{\partial t} = u_i(F_1(t, \sigma), \cdots, F_n(t, \sigma)), \qquad i = 1, \cdots, n , \tag{34}$$

where

$$u_i(s_1, \cdots, s_n) = -b_{ii}s_i + \sum_{\alpha\in M,\alpha\neq\{i\}} b_{i\alpha}s_1^{a_1}\cdots s_n^{a_n}, i = 1, \cdots, n . \tag{35}$$

The functions $u_i(s_1, \cdots, s_n)$ are the generating functions of the systems of quantities $\{-b_{ii}, b_{i\alpha}, \alpha\in M, \alpha\neq\{i\}\}$.

To obtain the second equation, let us suppose that $|s_i| < 1$ (for $i = 1, \cdots, n$). Then $|F_i(t, \sigma)| < 1$ (for $i = 1, \cdots, n$) and let us differentiate equation (33) with respect to τ. Differentiating and then setting $\tau = 0$, we obtain

$$\frac{\partial\Phi(t, \sigma)}{\partial t} = \sum_{k=1}^{n} u_k(\sigma)\frac{\partial\Phi(t, \sigma)}{\partial s_k} . \tag{36}$$

Equation (36) is a system of the same type of equation for the generating functions $F_i(t, \sigma)$,

$$\frac{\partial F_i(t, \sigma)}{\partial t} = \sum_{k=1}^{n} u_k(\sigma)\frac{\partial F_i(t, \sigma)}{\partial s_k}, \qquad i = 1, \cdots, n ,$$

which must be solved under the initial conditions (31). Thus, we have obtained.

Theorem 7. *The system of generating functions* $F_i(t, \sigma)$ *for* $|s_i| < 1$, *where* $i = 1, \cdots, n$, *satisfies the system of ordinary differential equations* (34), *the partial differential equation* (36), *and the initial conditions* (31).

6. THE GENERAL DEFINITION OF A MARKOV PROCESS

At the basis of the concept of a Markov system (process) is the concept of a system whose future evolution depends only on the state of the system at the given instant of time (that is, its future evolution does not depend on the behavior of the system in the past). Let $\{U, \mathfrak{S}, \mathsf{P}\}$ denote a probability space on which a random process $\xi(t)$ is defined with range in a complete metric space X.

We shall call the space X the *phase space* of the system and we shall call $\xi(t)$ the *state* of the system at the instant $t \in \mathfrak{T}$, where $\mathfrak{T}$ is a finite or infinite interval of the real line. We let $\mathfrak{B}$ denote the algebra of Borel subsets of X. The hypothesis of absence of after-effect is most easily written with the aid of conditional probabilities:

$$\mathsf{P}\{\xi(t) \in A | \xi(t_1), \xi(t_2), \cdots, \xi(t_n)\} = \mathsf{P}\{\xi(t) \in A | \xi(t_n)\} \pmod{\mathsf{P}}, \quad (1)$$

for arbitrary $A \in \mathfrak{B}$ and $t_1 < t_2 < \cdots < t_n < t$. Since the conditional probability regarding a random variable can be regarded as a function of that variable, we set

$$\mathsf{P}\{\xi(t) \in A \mid \xi(s)\} = \mathsf{P}(s, \xi(s), t, A) \qquad (s < t). \qquad (2)$$

It follows from formula (26), Section 6, Chapter III that for $t_1 < t_2 < \cdots < t_n$ the equation

$$\begin{aligned}\mathsf{M}\{g(\xi(t_1), \xi(t_2), \cdots, \xi(t_n)) \mid \xi(t_1)\} \\ = \int \mathsf{P}(t_1, \xi(t_1), t_2, dx_2) \int \mathsf{P}(t_2, x_2, t_3, dx_3) \\ \cdots \int \mathsf{P}(t_{n-1}, x_{n-1}, t_n, dx_n) g(\xi(t_1), x_2, \cdots, x_n) \qquad \pmod{\mathsf{P}} \qquad (3)\end{aligned}$$

holds for an arbitrary bounded Borel function $g(x_1, x_2, \cdots, x_n)$ (where $x_k \in X$ for $k = 1, 2, \cdots, n$). In particular, if we set $g = \chi_A(x_3)$, where $\chi_A(x)$ is the characteristic function of the set $A \in \mathfrak{B}$, it follows from (3) that with probability 1,

$$\mathsf{P}(t_1, \xi(t_1), t_3, A) = \int \mathsf{P}(t_2, x_2, t_3, A) \mathsf{P}(t_1, \xi(t_1), t_2, dx_2). \qquad (4)$$

This equation appeared in Section 1 as the Chapman-Kolmogorov equation, and it served there as the basis for the definition of a Markov process in the broad sense. We now give the axiomatic definition of a Markov process within the framework of the general definitions of Chapter IV.

Definition 1. A random process $\xi(t)$ (for $t \in \mathfrak{T}$) with range in X is called a *Markov process* if

a. equation 1 is satisfied for arbitrary $t_1 < t_2 < \cdots < t_n < t$, where each t_k (for $k = 1, \cdots, n$) and t belong to $\mathfrak{T}$,

b. the conditional probabilities $\mathsf{P}(s, x, t, A)$ are $\mathfrak{B}$-measurable functions of x for fixed s, t, and A, and are probability measures on $\mathfrak{B}$ for fixed s, x, and t, and

c. the Chapman-Kolmogorov equations are satisfied for all $\xi(t_1) = x \in X$.

Thus, by definition, the family of conditional probabilities (1) is regular and the process $\xi(t)$ is independent of the "past". We shall call the property of the process expressed by equation (1) the *Markov property* or the *absence of after-effect.*

Let us show that the Markov property implies stronger assertions. Again using formula (26), Section 6, Chapter III and equation (1), we obtain for $t_1 < t_2 < \cdots < t_m < \cdots < t_{n+m}$, where $t_k \in \mathfrak{T}$ (for $k = 1, \cdots, n + m$), the equation

$$\mathsf{M}\{g(\xi(t_{m+1}), \xi(t_{m+2}) \cdots \xi(t_{n+m})) \mid \xi(t_1), \xi(t_2), \cdots, \xi(t_m)\}$$
$$= \int \mathsf{P}(t_m, \xi(t_m), t_{m+1}, dx_1) \cdots \int \mathsf{P}(t_{n+m-1}, x_{n-1}, t_{n+m}, dx_n)$$
$$\times\, g(x_1, \cdots, x_n) = \mathsf{M}\{g(\xi(t_{m+1}), \cdots, \xi(t_{n+m})) \mid \xi(t_m)\} \quad (\text{mod } \mathsf{P})\,.$$

If we set $g(x_1, \cdots, x_n) = \chi_{A^{(n)}}(x_1, \cdots, x_n)$, where $A^{(n)}$ is a Borel set in X^n, we get the following equation generalizing the Markov property of a process:

$$\mathsf{P}\{[\xi(t_{m+1}), \cdots, \xi(t_{m+n})] \in A^{(n)} \mid \xi(t_1), \cdots, \xi(t_m)\}$$
$$= \mathsf{P}\{[\xi(t_{m+1}), \cdots, \xi(t_{m+n})] \in A^{(n)} \mid \xi(t_m)\} \qquad (\text{mod } \mathsf{P})$$

for arbitrary $t_1 < t_2 < \cdots < t_{n+m} (\in \mathfrak{T})$, n and m. Let $\sigma\{\leqq t)$ denote the σ-algebra of events that is generated by the random variables $\xi(s)$, where $s \in \mathfrak{T}$ and $s \leqq t$, and let $\sigma\{>t\}$ denote the σ-algebra generated by the quantities $\xi(s)$, where $s \in \mathfrak{T}$ and $s > t$. Then with probability 1, for an arbitrary cylindrical set $C \in \sigma\{>t\}$,

$$\mathsf{P}\{C \mid \xi(t_1), \cdots, \xi(t_n)\} = \mathsf{P}\{C \mid \xi(t_n)\},$$
$$t_1 < t_2 < \cdots < t_n \leqq t \,(\text{mod } \mathsf{P})\,. \tag{5}$$

Let Λ denote the class of events for which equation (5) is valid. On the basis of the properties of conditional probabilities (cf. Theorem 6, Section 6, Chapter III), Λ is an algebra and a monotonic class of events. Therefore $\Lambda \supset \sigma\{>t\}$. On the other hand, let $\mathfrak{N}$ denote the class of events N such that for arbitrary $S \in \sigma\{>t\}$,

$$\int_N \mathsf{P}(S \mid \sigma\{\leqq t\}) d\mathsf{P} = \int_N \mathsf{P}(S \mid \xi(t)) d\mathsf{P}\,. \tag{6}$$

On the basis of (5), $\mathfrak{N}$ contains all cylindrical sets in $\sigma\{\leqq t\}$. Since the right and left-hand sides of equation (6) are countably additive functions on $\sigma\{\leqq t\}$, their coincidence on cylindrical sets contained in $\sigma\{\leqq t\}$ implies that they are identical on $\sigma\{\leqq t\}$. Thus we have

Theorem 1. *For arbitrary* $S \in \sigma\{>t\}$,

$$\mathsf{P}(S \mid \sigma\{\leqq t\}) = \mathsf{P}(S \mid \xi(t)) \qquad (\mathrm{mod}\ \mathsf{P}) . \tag{7}$$

Equation (7) shows that the conditional probability of an arbitrary event S, defined by the behavior of a Markov process in the "future" with completely given "past," depends only on the "present."

Let X denote a complete separable metric space and let $\mathfrak{T}$ denote some interval of the real line $[a, b]$ or $[0, \infty)$. Let us show how one can construct from a given Markov process in the broad sense, $\{\mathsf{P}(t, x, \tau, A), x \in X; t, \tau \in \mathfrak{T}\}$, a Markov process in the narrow sense whose conditional probabilities (2) coincide with the given transition probabilities $\mathsf{P}(t, x, \tau, A)$.

We introduce some initial distribution $\{X, \mathfrak{B}, \mu_0\}$, and for an arbitrary bounded Borel function $f(x_1, x_2, \cdots, x_n)$ of n variables for $x_k \in X$, and for arbitrary $t_k \in \mathfrak{T}$ (for $k = 1, \cdots, n$; $a < t_1 < t_2 \cdots < t_n$), we define

$$F_{t_1, t_2, \cdots, t_n}[f] = \int \mu_0(dx_0) \int \mathsf{P}(a, x_0, t_1, dx_1) \cdots \int f(x_1, x_2, \cdots, x_n) \mathsf{P}(t_{n-1}, x_{n-1}, t_n, dx_n) \tag{8}$$

and

$$\mathsf{P}_{t_1 \cdots t_n}(A^n) = F_{t_1, t_2, \cdots, t_n}[\chi_{A^{(n)}}] , \tag{9}$$

where $\chi_{A^{(n)}}$ is the characteristic function of the set $A^{(n)} \in \mathfrak{B}^{(n)}$, where $\mathfrak{B}^{(n)}$ is the nth power of the σ-algebra $\mathfrak{B}$. We note that for an arbitrary Borel function $f(x_1, x_2, \cdots, x_n)$ the function

$$f_1(x_1, x_2, \cdots, x_{n-1}) = \int f(x_1, x_2, \cdots, x_n) \mathsf{P}(t, x_{n-1}, s, dx_n)$$

is also a Borel function, since the integral is the limit of integrals of simple functions and these simple functions are Borel functions of the variables $x_1, x_2, \cdots, x_{n-1}$. By virtue of the properties of an integral, $\mathsf{P}_{t_1, \cdots, t_n}(A^{(n)})$ is a measure on $\mathfrak{B}^{(n)}$. Obviously the family of measures $\mathsf{P}_{t_1, \cdots, t_n}(B)$ satisfies the compatibility conditions and by virtue of Kolmogorov's theorem (Theorem 3, Section 2, Chapter III) has a representation of the form $\{\Omega, \tilde{\mathfrak{S}}, \mathsf{P}\}$, where Ω is the space of all functions $\omega(t)$ for $t \in \mathfrak{T}$, with range in X. Let $\xi(t)$ denote an arbitrary process stochastically equivalent to $\{\Omega, \tilde{\mathfrak{S}}, \mathsf{P}\}$. Let us show that

$$\mathsf{P}\{\xi(t) \in A \mid \xi(t_1), \xi(t_2) \cdots \xi(t_n)\} = \mathsf{P}(t_n, \xi(t_n), t, A) \qquad (\mathrm{mod}\ \mathsf{P}) ,$$

that is, that $\xi(t)$ is a Markov process with given transition probabilities. To do this, it will be sufficient to prove that

$$\int_{A^{(n)}} \mathsf{P}(t_n, x_n, t, B)\mathsf{P}_{t_1,t_2,\cdots,_n}(dx_1, dx_2, \cdots, dx_n)$$
$$= \mathsf{P}_{t_1,t_2,\cdots,t_n,t}(A^{(n)} \times B)$$

for arbitrary $A^{(n)} \in \mathfrak{B}^{(n)}$, $B \in \mathfrak{B}$, and $t_1 < t_2 < t_3 \cdots < t_n < t$. But this follows immediately from formula (8) and Lemma 1 of Section 1.

If X is separable and locally compact, the process $\{\Omega, \tilde{\mathfrak{S}}, \mathsf{P}\}$ is stochastically equivalent to the separable process $\xi(t)$ (by Theorem 2, Section 2, Chapter IV). In what follows we shall consider only separable Markov processes. In many cases, the following theorem characterizes the properties of regularity of Markov processes.

Let $\xi(t)$ denote a separable Markov process and suppose that

$$\alpha(\varepsilon, \delta) = \sup\{\mathsf{P}[t, x, s, \overline{S_\varepsilon(x)}]; t < s; t, s \in \mathfrak{T}; |t - s| < \delta; x \in X\},$$

where $\overline{S_\varepsilon(x)}$ is the complement of the sphere $S_\varepsilon(x)$ of radius ε with center at the point x.

Theorem 2. (a) *If $\alpha(\varepsilon, \delta) \to 0$ as $\delta \to 0$ for arbitrary $\varepsilon > 0$, then the process $\xi(t)$ does not have discontinuities of the second kind.*

(b) *if $\alpha(\varepsilon, \delta) \to 0$ as $\delta \to 0$ for arbitrary $\varepsilon > 0$ and if for an arbitrary sequence of partitions $\{t_{nk}, k = 0, 1, \cdots, m_n\}$ of the interval $[a, b]$,*

$$\sum_{k=1}^{m_n} \mathsf{P}\{\rho(\xi(t_{nk}), \xi(t_{n,k-1})) \geqq \varepsilon\} \to 0 \quad as \quad n \to \infty\ ,$$

then the process $\xi(t)$ is continuous.

(b′) *If $\alpha(\varepsilon, \delta) = o(\delta)$ for arbitrary $\varepsilon > 0$, then the process $\xi(t)$ is continuous.*

Proof. Assertion (b′) is a particular case of (b). Assertions (a) and (b) follow from Theorem 1 of the present section, Theorem 2 of Section 4, and Theorem 4 of Section 5, Chapter IV.

7. THE BASIC PROPERTIES OF JUMP PROCESSES

Jump Markov processes were introduced in Section 3. Let us now look at them in greater detail.

Let $\mu_t(A)$ denote the distribution of $\xi(t)$.

Lemma 1. *A jump Markov process is stochastically continuous.*

Proof. If $t' > t$, then on the basis of conditions a and c of Definition 1 in Section 3,

$$\mathsf{P}\{\xi(t) \neq \xi(t')\} = \int_X \mu_t(dx)\mathsf{P}(t, x, t', X - \{x\})$$
$$\leqq \int_X (q(t, x) + \varepsilon)(t' - t)\mu_t(dx) \leqq (k + \varepsilon)(t' - t)\ ,$$

from which the assertion follows.

Lemma 2. *If $\xi(t)$ is a separable jump process, then*

$$\mathsf{P}\{\xi(\tau) = x \text{ for all } \tau \in [t, s] \mid \xi(t) = x\} = \exp\left(-\int_t^s q(\tau, x)d\tau\right) \quad (1)$$

Proof. Let M denote the set of separability of the process $\xi(t)$ on the interval $[t, s]$. It follows from the stochastic continuity of the process and Theorem 5, Section 2, Chapter IV that we may take for M any countable set that is everywhere-dense on $[t, s]$. It follows from the separability of the process that $\mathsf{P} = \mathsf{P}\{\xi(\tau) = x$ for all $\tau \in [t, s] \mid \xi(t) = x\} = \mathsf{P}\{\xi(\tau) = x$ for all $\tau \in M \mid \xi(t) = x\}$. For the set M, we can take the set of points of the form $t_{nk} = t + kh/2^n$, for $k = 0, 1, \cdots, 2^n$ and $h = s - t$. We note that

$$\mathsf{P} = \lim_{n\to\infty} \mathsf{P}\{\xi(t_{nk}) = x, k = 1, \cdots, 2^n \mid \xi(t) = x\} ,$$

since the events $A_n = \{\xi(t_{nk}) = x, k = 1, \cdots, 2^n\}$ constitute a decreasing sequence and $\bigcap_{n=1}^{\infty} A_n = \{\xi(\tau) = x, \tau \in M\}$. Furthermore,

$$\begin{aligned} p_n &= \mathsf{P}\{\xi(t_{nk}) = x, k = 1, 2, \cdots, 2^n \mid \xi(t) = x\} \\ &= \prod_{k=1}^{2^n} \mathsf{P}\{t_{n,k-1}, x, t_{nk}, \{x\}\}, \ \ln p_n = \sum_{k=1}^{2^n} \ln \mathsf{P}\{t_{n,k-1}, x, t_{nk}, \{x\}\} . \end{aligned}$$

Let $f_n(t)$ denote a piecewise-constant function that on the interval $[t_{n,k-1}, t_{nk})$ is equal to $1/\Delta t_{nk} \ln \mathsf{P}(t_{n,k-1}, x, t_{nk}, \{x\})$, where $\Delta t_{nk} = t_{nk} - t_{n,k-1}$. It follows from the definition of a jump process (cf. Definition 1, Section 3) that $\mathsf{P}(t_{n,k-1}, x, t_{nk}, \{x\}) \to 1$ uniformly with respect to t. Setting $\mathsf{P}(t_{n,k-1}, x, t_{nk}, \{x\}) = 1 + \alpha_{nk}$, we see that

$$\frac{\ln(1 + \alpha_{nk})}{\Delta t_{nk}} + q(t_{nk}, x) = \frac{\ln(1 + \alpha_{nk})}{\alpha_{nk}} \frac{\alpha_{nk}}{\Delta t} + q(t_{nk}, x) \to 0$$

uniformly with respect to t_{nk}. Therefore $f_n(\tau) \to -q(\tau, x)$ uniformly on $[t, s]$ and

$$\ln p_n = \int_t^s f_n(\tau)d\tau \to -\int_t^s q(\tau, x)d\tau .$$

This completes the proof of the lemma.

Lemma 3. *A separable jump Markov process has no discontinuities of the second kind.*

This is true because

$$\begin{aligned} \alpha(\varepsilon, \delta) &\leq \sup_{\substack{t \leq \tau < t+\delta \\ x \in X}} \mathsf{P}(t, x, \tau, X - \{x\}) \\ &\leq \sup_{x,t} \left\{1 - \exp\left(-\int_t^{t+\delta} q(\tau, x)d\tau\right)\right\} \leq k\delta \end{aligned}$$

and the assertion of the lemma follows from Assertion (a) of Theorem 2, Section 6.

From this lemma and Theorem 3, Section 4, Chapter IV we get

Corollary 1. *For a separable jump Markov process* $\xi(t)$ *there exists an equivalent process* $\xi'(t)$ *whose sample functions are continuous from the right with probability* 1.

Corollary 2. *If* $\xi(t)$ *is a separable jump process and* $\xi(t) = x$, *then with probability* 1 *there exists an interval of time* $(t, t + h)$ *during which* $\xi(\tau) = x$ *for* $\tau \in (t, t + h)$.

Proof. The event A that such an interval does not exist can be represented in the form $A = \bigcap_{n=1}^{\infty} A_n$, where A_n is the event that $\xi(t) = x$ and that for at least one value of τ, $\xi(\tau) \neq x$ on the interval $(t, t + 1/n)$. Therefore,

$$\begin{aligned} \mathsf{P}(A \mid \xi(t) = x) &= \lim_{n\to\infty} \mathsf{P}(A_n \mid \xi(t) = x) \\ &= \lim_{n\to\infty} \left\{1 - \exp\left(\int_t^{t+1/n} q(\tau, x) d\tau\right)\right\} = 0 \,. \end{aligned}$$

Since an arbitrary separable jump process is equivalent to a process whose sample functions are continuous from the right (mod P), we can confine our study to these. In subsequent discussions in this section we shall always assume that $\xi(t)$ is a jump process that is continuous from the right (that is, a process whose sample functions are with probability 1 continuous from the right on $\mathfrak{T}$), and where $\mathfrak{T} = [0, \infty)$. The process is necessarily separable and we can take as the set of separability an arbitrary countable set that is everywhere-dense on $\mathfrak{T}$. Suppose that we know that $\xi(t_0) = x$. On the basis of Lemma 2 there exists a τ_1 such that either $\tau_1 = \infty$ and $\xi(t) = x$ for all $t \geqq t_0$ (as will be the case when $q(t, x_0) \equiv 0$ for all $t \geqq t_0$), or $\xi(t) = x$ for $t \in [t_0, t_0 + \tau_1)$ and $\xi(t_0 + \tau_1) = \xi_1 \neq \xi(t_0 + \tau_1 - 0)$. Let us find the probability of the event

$$B = \{\tau_1 \leqq t, \xi_1 \in A\}, \qquad x \notin A \,.$$

Define

$$B^{(n)}(r) = \left\{\xi\left(t_0 + \frac{j}{n}\right) = x, 1 \leqq j \leqq r, \xi\left(t_0 + \frac{r+1}{n}\right) \in A\right\}$$

and

$$B^{(n)} = \bigcup_{r=1}^{[nt]} B^{(n)}(r) \,.$$

If the sample function of the process $\xi(t)$ is continuous on the right, then

$$B = \bigcup_{n=1}^{\infty} \bigcap_{k=n}^{\infty} B^{(k)} \,.$$

For brevity in writing, we define

$$p_{nj}(x) = \mathsf{P}\Big(t_0 + \frac{j-1}{n}, x, t_0 + \frac{j}{n}, \{x\}\Big)$$

and

$$p_{nj}(x, A) = \mathsf{P}\Big(t_0 + \frac{j-1}{n}, x, t_0 + \frac{j}{n}, A\Big).$$

Then

$$\mathsf{P}(B^{(n)}(r) \mid \xi(t_0) = x) = \prod_{j=1}^{r} p_{nj}(x) p_{n,r+1}(x, A).$$

As shown in the proof of Lemma 2,

$$\prod_{j=1}^{r} p_{nj}(x) = \exp\Big(-\int_{t_0}^{t_0+(r/n)} q(\theta, x) d\theta\Big) + \varepsilon'_n,$$

where $\varepsilon'_n \to 0$ uniformly with respect to $r \leqq nT$ (where T is a fixed number) as $n \to \infty$. It follows from the definition of a jump process that

$$p_{n,r+1}(x, A) = \Big(q\Big(t_0 + \frac{r}{n}, x, A\Big) + \varepsilon''_n\Big)\frac{1}{n} \tag{2}$$

where ε''_n also approaches 0 uniformly with respect to $r \leqq nT$ as $n \to \infty$, Since

$$\mathsf{P}(B \mid \xi(t_0) = x) = \lim_{n\to\infty} \mathsf{P}(B^{(n)} \mid \xi(t_0) = x),$$

it follows from the preceding formulas that

$$\mathsf{P}(\tau_1 < t, \xi_1 \in A \mid \xi(t_0) = x) = \int_0^t \exp\Big(-\int_{t_0}^{s} q(\theta, x) d\theta\Big) q(s, x, A) ds. \tag{3}$$

Here, let us set $A = X\backslash\{x\}$. Remembering that $q(s, x, X\backslash\{x\}) = q(s, x)$, we obtain the distribution function of the quantity τ_1 under the hypothesis that $\xi(t_0) = x$:

$$\Psi(t_0, x, t) = 1 - \exp\Big(-\int_{t_0}^{t_0+t} q(\theta, x) d\theta\Big). \tag{4}$$

Equation (3) can be written in the form

$$\mathsf{P}(\tau_1 < t, \xi_1 \in A \mid \xi(t_0) = x) = \int_0^t \frac{q(t_0 + \theta, x, A)}{q(t_0 + \theta, x)} \Psi(t_0, x, d\theta). \tag{5}$$

From this it follows that

$$\mathsf{P}(\xi_1 \in A \mid \tau_1 = t, \xi(t_0) = x) = \frac{q(t_0 + t, x, A)}{q(t_0 + t, x)}, \quad (q(t, x) \neq 0). \tag{6}$$

Thus the interpretation of the quantity

$$\frac{q(t, x, A)}{q(t, x)} = \Pi\,(t, x, A), \qquad x \notin A \qquad (q(t, x) \neq 0),$$

of which we spoke in Section 3, is rigorously justified: $\prod(t, x, A)$ is the conditional distribution of the state $\xi(t+0)$ under the hypothesis that at the instant t the state of the process changes and $\xi(t-0) = x$.

Consider the event

$$B = \Big\{\xi(t) = \xi_{k-1} \quad \text{for} \quad t_0 + \sum_{j=1}^{k-1} \tau_j \leqq t < t_0 + \sum_{j=1}^{k} \tau_j ,$$
$$k = 1, 2, \cdots, m ,$$
$$\xi\Big(t_0 + \sum_{j=1}^{m} \tau_j\Big) = \xi_m, \xi_{k-1} \neq \xi_k, \xi_k \in A_k ,$$
$$\tau_1 < t_1, \cdots, \tau_m < t_m\Big\} .$$

Generalizing formula (5) let us find the conditional probability $F(t_1, t_2, \cdots, t_m; A_1, \cdots, A_m \mid t_0, x)$ of the event B under the hypothesis that $\xi(t_0) = x$. To do this, we introduce the events

$$B^{(n)}(r_1, r_2, \cdots, r_m) = \Big\{\xi\Big(t_0 + \frac{j}{n}\Big) = x, 1 \leqq j \leqq r_1 ;$$
$$\xi\Big(t_0 + \frac{j}{n}\Big) = \xi_1, r_1 + 1 \leqq j \leqq r_1 + r_2 ,$$
$$\xi_1 \in A_1, \cdots, \xi\Big(t_0 + \frac{j}{n}\Big) = \xi_{m-1}, r_1 + \cdots + r_{m-1}$$
$$+ 1 \leqq j \leqq r_1 + \cdots + r_m, \xi_{m-1} \in A_{m-1} ,$$
$$\xi\Big(t_0 + \frac{r_1 + \cdots + r_m + 1}{n}\Big) \in A_m\Big\}$$

and we define

$$B^{(n)} = \bigcup_{1 \leqq r_1 \leqq [nt_1]+1, 1 \leqq r_2 \leqq [nt_2]+1, \cdots, 1 \leqq r_m \leqq [nt_m]+1} B^{(n)}(r_1, r_2, \cdots, r_m) .$$

The continuity from the right of the sample functions of the process implies that with probability 1 the events B and $\lim_{n\to\infty} B^{(n)}$ coincide. Using the definition of a jump process, we obtain

$$P(B^{(n)}(r_1, \cdots, r_m) \mid \xi(t_0) = x)$$
$$= \exp\Big(-\int_{t_0}^{t_0+(r/n)} q(\theta, x)d\theta\Big)\int_{A_1} \frac{1}{n}\tilde{q}\Big(t_0 + \frac{r_1}{n}, x, dx_1\Big)$$
$$\times \exp\Big(-\int_{t_0+r_1/n}^{t_0+\{(r_1+r_2)/n\}} q(\theta, x_1)d\theta\Big)\int_{A_2} \frac{1}{n}\tilde{q}\Big(t_1 + \frac{r_1 + r_2}{n}, x_1, dx_2\Big) \cdots$$
$$\times \cdots \exp\Big(-\int_{t_0+\{(r_1+\cdots+r_{m-1})/n\}}^{t_0+\{(r_1+\cdots+r_m)/n\}} q(\theta, x_{m-1})d\theta\Big)$$
$$\times \frac{1}{n}\tilde{q}\Big(t_1 + \frac{r_1 + \cdots + r_m}{n}, A_m\Big) + \frac{1}{n^m}\varepsilon'_n , \tag{7}$$

where $\varepsilon'_n \to 0$ uniformly with respect to $r_1, \cdots, r_m$ as $n \to \infty$ (where $r_i/n \leqq T$ for $i = 1, \cdots, m$). From this it easily follows that

$$
\begin{aligned}
F(t_1, t_2, \cdots, t_m; A_1, A_2, \cdots, A_m \mid t_0, x) \\
= \int_{t_0}^{t_0+t_1} \exp\left(-\int_{t_0}^{s_1} q(\theta, x)d\theta\right)ds_1 \int_{A_1} \tilde{q}(s_1, x, dx_1) \\
\times \int_{s_1}^{s_1+t_2} \exp\left(-\int_{s_1}^{s_2} q(\theta, x)d\theta\right)\int_{A_2} \tilde{q}(s_2, x_1, dx_2) \cdots \\
\times \int_{s_{m-1}}^{s_{m-1}+t_m} \exp\left(-\int_{s_{m-1}}^{s_m} q(\theta, x_{m-1})d\theta\right)\tilde{q}(s_m, x_{m-1}, A_m)ds_m \,. \quad (8)
\end{aligned}
$$

The expression we have obtained shows that the joint conditional distribution of the variables $\tau_1, \tau_2, \cdots, \tau_m, \xi_1, \xi_2, \cdots, \xi_m$ can be expressed simply in terms of the separate conditional distributions of the variables τ_k and ξ_k. Specifically, formula (8) can be rewritten in the form

$$
\begin{aligned}
F(t_1, t_2, \cdots, t_m; A_1, A_2, \cdots, A_m \mid t_0, x) \\
= \int_0^{t_1} \Psi(t_0, x, ds_1)\int_{A_1} \Pi\,(t_0 + s_1, x, dx_1) \\
\times \int_0^{t_2} \Psi(t_0 + s_1, x_1, ds_2)\int_{A_2} \Pi\,(t_0 + s_1 + s_2, x_1, dx_2) \times \\
\cdots \times \int_0^{t_m} \Psi(t_0 + s_1 + \cdots + s_{m-1}, x_{m-1}, ds_m) \\
\times \Pi\,(t_0 + s_1 + \cdots + s_m, x_{m-1}, A_m) \,. \quad (9)
\end{aligned}
$$

It follows from this last relation that

$$
\begin{aligned}
\mathsf{P}\{\xi_m \in A_m \mid x_0, \xi_1, \cdots, \xi_{m-1}, \tau_1, \cdots, \tau_m\} \\
= \Pi\,(t_0 + \tau_1 + \cdots + \tau_m, \xi_{m-1}, A) \quad (10)
\end{aligned}
$$

and

$$
\begin{aligned}
\mathsf{P}(\tau_m \leqq t \mid x_0, \xi_1, \cdots, \xi_{m-1}, \tau_1, \tau_2, \cdots, \tau_{m-1}) \\
= \Psi(t_0 + \tau_1 + \cdots + \tau_{m-1}, \xi_{m-1}, t) \,. \quad (11)
\end{aligned}
$$

These formulas appear to follow immediately from the absence of after-effect in Markov processes. However, this is not quite the case since the absence of after-effect here is used at random instants of time depending on the progress of the process $\xi(t)$. Therefore formulas (10) and (11) do require proof. The intuitive simplicity of the relations obtained becomes especially clear in the case of a homogeneous Markov process. Then $\Psi(s, x, t)$ and $\Pi\,(s, x, A)$ are independent of s, and formulas (9) to (11) take the forms

$$F(t_1, t_2, \cdots, t_m; A_1, A_2, \cdots, A_m \mid t_0, x)$$
$$= \int_0^{t_1} \cdots \int_0^{t_m} \int_{A_1} \cdots \int_{A_m} \Psi(x, ds_1)\Psi(x_1, ds_2) \cdots \Psi(x_{m-1}, ds_m)$$
$$\times \prod (x, dx_1) \prod (x_1, dx_2) \cdots \prod (x_{m-1}, dx_m) , \tag{9'}$$
$$\mathsf{P}\{\xi_m \in A_m \mid x_0, \xi_1, \cdots, \xi_{m-1}, \tau_1, \cdots, \tau_m\} = \prod (\xi_{m-1}, A_m) , \tag{10'}$$
$$\mathsf{P}\{\tau_m < t \mid x_0, \xi_1, \cdots, \xi_{m-1}, \tau_1, \cdots, \tau_{m-1}\} = \Psi(\xi_{m-1}, t) . \tag{11'}$$

Theorem 1. *Let $\xi(t)$ denote a homogeneous jump process whose sample functions are continuous from the right with probability* 1. *Then*

a. *The sequence $\xi_1, \xi_2, \cdots, \xi_m, \cdots$ of distinct states of the system is connected in a simple homogeneous Markov chain with the transition probability $\prod (x, A) = q(x, A)/q(x)$,*

b. *the conditional distribution of the time of existence in the state ξ_{m+1} for given $x, \xi_1, \cdots, \xi_m, \tau_1, \cdots, \tau_m$ depends only on ξ_m.*

Let us return to the general formula (9).

Corollary 1. *In a jump Markov process that is continuous from the right, after a finite interval of time only finitely many transitions from one state to another are possible.*

Proof. Let us obtain a bound for the quantity

$$\mathsf{P}\{\tau_1 + \cdots + \tau_m < t \mid \xi(t_0) = x\} .$$

We have

$$\mathsf{P}\{\tau_1 + \cdots + \tau_m < t \mid \xi(t_0) = x\}$$
$$\leqq e^{Nt}\mathsf{M}\left\{\exp\left(-N \sum_{k=1}^{m} \tau_k\right) \middle| \xi(t_0) = x\right\} \tag{12}$$

for arbitrary $N > 0$. Suppose that $N > 2K$, where

$$K = \sup_{x \in X, \theta \in [t_0, t_0+t]} q(\theta, x) .$$

Since

$$\int_0^\infty e^{-Ns}\Psi(t, x, ds) = \int_0^\infty e^{-Ns}q(t + s, x) \exp\left(-\int_t^{t+s} q(t + s, x)\right)ds$$
$$\leqq K\int_0^\infty e^{-sN}ds = \frac{K}{N} < \frac{1}{2} ,$$

it follows from formulas (11) and (12) that

$$\mathsf{P}(\tau_1 + \cdots + \tau_m < t \mid \xi(t_0) = x) \leqq \left(\frac{1}{2}\right)^m e^{Nt} \longrightarrow 0 \quad \text{as} \quad m \longrightarrow \infty .$$

The last inequality proves the finiteness of the number of jumps of the function $\xi(t)$ on a finite interval of time $[t_0, t_0 + t]$ with probability 1.

Corollary 2. *The sample functions of a jump process that is continuous from the right have with probability* 1 *the following structure: There exists a sequence of positive random variables* $\tau_1, \tau_2, \cdots, \tau_m$ *such that*

$$\sum_{k=1}^{m} \tau_k \longrightarrow \infty$$

with probability 1 *as* $m \longrightarrow \infty$ *and* $\xi(t) = \xi_{k-1}$ *for*

$$\sum_{j=1}^{k-1} \tau_j \leqq t < \sum_{j=1}^{k} \tau_j \, .$$

We should also note that: If $\{\tau_k\}$ for $k = 1, 2, \cdots$ is a sequence of positive random variables, if $\{\xi_k\}$, for $k = 0, 1, \cdots$, is a sequence of random elements with values in X, if all the finite-dimensional distributions of the quantities τ_k and ξ_k are given, if

$$\sum_{k=1}^{\infty} \tau_k = \infty$$

with probability 1 and if

$$\eta(t) = \xi_k \quad \text{for} \quad \sum_{j=1}^{k} \tau_j \leqq t < \sum_{j=1}^{k+1} \tau_j, \, k = 0, 1, 2, \cdots ;$$

then $\eta(t)$ is a random process defined for all $t > 0$, the sample functions of the process $\eta(t)$ are continuous from the right, and all finite-dimensional distributions of the process are uniquely defined.

Let $\xi(t)$ for $t \geqq 0$ denote a jump process that is continuous from the right, and suppose that τ_k and ξ_k are the same as in the preceding paragraph. We can regard the process $\xi(t)$ as a function of the infinitely many variables $t, \xi_0, \tau_1, \xi_1, \tau_2, \cdots$. Then

$$\xi(t) = f(t, \xi_0, \tau_1, \xi_1, \tau_2, \cdots) = \xi_k, \quad \text{if} \quad \gamma_k \leqq t \leqq \gamma_{k+1} \, , \\ k = 0, 1, 2, \cdots \, ,$$

where $\gamma_0 = 0$ and $\gamma_k = \sum_{j=1}^{k} \tau_j$ Let Ω denote the space of all sequences $\omega = \{x_0, t_1, x_1, t_2, \cdots\}$ for $x_k \in X$ and $t_k \geqq 0$, and let $\mathfrak{F}$ denote the σ-algebra generated by cylindrical Borel sets in Ω.

It is easy to see that the function $f(t, \omega)$ for $\omega \in \Omega$ is a Borel function of the variables t and ω; that is, for arbitrary $A \in \mathfrak{B}$,

$$\{t, \omega; f(t, \omega) \in A\} \in \sigma\{\mathfrak{B}_1 \times \mathfrak{F}\} \, ,$$

where $\mathfrak{B}_1$ is the σ-algebra of subsets of the half-line $t \geqq 0$ and $\sigma\{\mathfrak{B}_1 \times \mathfrak{F}\}$ is the product of the σ-algebras $\mathfrak{B}_1$ and $\mathfrak{F}$ (cf. Definition 1, Section 8, Chapter II). Specifically, the set $\{t, \omega; f(t, \omega) \in A\}$ is the sum of countably many cylindrical Borel sets

$$\left\{x_k \in A, \sum_{j=1}^{k} t_j \leqq t < \sum_{j=1}^{k+1} t_{j+1}\right\}, \qquad k = 0, 1, \cdots$$

Let us now consider the random process

$$\xi'(t) = \xi(t + \gamma_m), \qquad t \geqq 0$$

Since $\xi'(t) = \xi_{k+m}$ for $\gamma_{k+m} - \gamma_m \leqq t < \gamma_{k+m-1} - \gamma_m$, we have $\xi'(t) = f(t, \xi_m, \tau_{m+1}, \xi_{m+1}, \tau_{m+2}, \cdots)$ and for arbitrary fixed $t \geqq 0$, $\xi'(t)$ is a Borel function of $\xi_m, \tau_{m+1}, \cdots$, that is, a random element of X.

Theorem 2. *Suppose that*

$$\mathsf{P}'(t, x, s, A) = \mathsf{P}\{\xi'(s) \in A \mid \xi'(t) = x, \gamma_m = T, \xi'(0) = z\} .$$

Then

$$\mathsf{P}'(t, x, s, A) = \mathsf{P}(t + T, x, s + T, A), \qquad t \leqq s .$$

In other words, for fixed $\gamma_m = T$ and $\xi'(0) = \xi_m = z$, the transition probabilities for the process $\xi'(t)$ coincide with the transition probabilities of the Markov process $\xi(t + T)$, where $t > 0$ and $\xi(T) = z$.

Proof. The theorem follows easily from formula (9), by virtue of which

$$\begin{aligned}
&\mathsf{P}\{\tau_{m+1} < t_1, \xi_{m+1} \in A_1', \cdots, \tau_r' < t_r' , \\
&\xi_r' \in A_r' \mid \xi_0, \xi_1, \cdots, \xi_m, \tau_1, \cdots, \tau_m\} \\
&= \int_0^{t_1} \Psi(t_0 + \gamma_m, \xi_m, ds) \int_{A_1'} \Pi\, (t_0 + \gamma_m + s_1, \xi_m, dx_1) \times \\
&\times \int_0^{t_r'} \Psi(t_0 + \gamma_m + s_1 + \cdots + s_{r-1}, x_{r-1}, ds_r) \\
&\times \Pi\, (t_0 + \gamma_m + s_1 + \cdots + s_r, x_{r-1}, A_r') ;
\end{aligned}$$

that is, the conditional distribution of the variables $\tau_{m+1}, \cdots, \tau_{m+r}, \xi_{m+1}, \cdots, \xi_{m+r}$ for given $\xi_0, \xi_1, \cdots, \xi_m, \tau_1, \cdots, \tau_m$ depends only on ξ_m and γ_m, and it coincides with the distribution that is obtained if we consider the Markov process $\xi''(t) = \xi(t + T)$, $T = \gamma_m$ with fixed initial state $\xi''(0) = \xi_m$. Since the joint distribution of the variables

$$\tau_{m+1}, \tau_{m+2}, \cdots, \tau_{m+r}, \xi_{m+1}, \cdots, \xi_{m+r} ,$$

where r is an arbitrary positive number, uniquely determines the finite-dimensional distributions of the process $\xi'(t)$, the theorem is proved.

Let us now give an important generalization of this theorem. Let $\alpha = \alpha(u)$ denote a nonnegative $\mathfrak{S}$-measurable function that is finite and defined on some $Q \in \mathfrak{S}$.

In the present section we shall call α a random variable, although this description is not exact since the function $\alpha(u)$ is not defined for all (mod P) elementary events.

Definition 1. The function $\alpha = \alpha(u)$ is said to be a *random variable independent of the future* (with respect to the process $\xi(t)$) if for arbitrary $t \geqq 0$,

$$\{u;\, \alpha(u) \leqq t\} \in \sigma\{\leqq t\} .$$

Thus $\alpha(u)$ is a quantity independent of the future if all we need to do to know whether the event $\{\alpha(u) \leqq t\}$ occurred or not is to observe the sample function of the process $\xi(t)$ up to the instant t. Let σ_α denote the class of all events $B \in \mathfrak{S}$ such that for arbitrary $t \geqq 0$,

$$B \cap \{u;\, \alpha(u) \leqq t\} \in \sigma\{\leqq t\} .$$

Obviously σ_α is a σ-algebra of events. If $\alpha(u) = t_0 = \text{const}$ for $u \in U$ (obviously this constant is a random variable independent of the future $\xi(t)$), then $\sigma_{t_0} = \sigma\{\leqq t_0\}$. If we interpret α as the time of occurrence of an event, then the σ-algebra σ_α consists of all those events that are determined by the progress of the process $\xi(t)$ up to a random instant of time α.

We are interested in the question of how the process $\xi(t + \alpha)$ where α is a random instant of time independent of the future, behaves. It is natural to expect that the conditional transition probabilities of the process $\xi(t + \alpha)$ with respect to α coincide with the transition probabilities of the process $\xi(t)$ (displaced by an appropriate amount along the time axis). Let us prove this for jump processes.

Let α denote a random variable that is independent of a future process $\xi(t)$. Suppose that $\xi(\alpha) = \xi'_0$ (which is meaningful only on a set $Q \subset U$). Let γ'_1 denote the earliest instant at which $\xi'_1 = \xi(\gamma'_1) \neq \xi'_0$ (where $\gamma'_1 > \alpha$). Define $\tau'_1 = \gamma'_1 - \alpha$ and define γ'_m inductively as the earliest instant at which $\xi'_m = \xi(\gamma'_m) \neq \xi'_{m-1}$ and $\tau'_m = \gamma'_m - \gamma'_{m-1}$ (for $m = 2, 3, \cdots$). We define the events

$$D = \bigcap_{k=1}^{m} \{\xi(t) = \xi'_{k-1} \quad \text{for} \quad \gamma'_{k-1} \leqq t < \gamma'_k,\ \xi'_{k-1} \in A'_{k-1}\, ,$$

$$\xi'_k \neq \xi'_{k-1},\, 0 < \gamma'_k - \gamma'_{k-1} < t'_k\} \cap \{\xi(\gamma'_m) = \xi'_m \in A'_m\}$$

and

$$\tilde{D}_n(j_0, j_1, \cdots, j_m) = B \cap \left\{\frac{j_0 - 1}{n} < \alpha \leqq \frac{j_0}{n}\right\} \cap \left(\bigcap_{k=1}^{m} \bigcap_{i=j_{k-1}}^{j_k} \xi\left(\frac{i}{n}\right)\right.$$

$$\left. = \xi'_{k-1} \in A'_{k-1},\, \xi'_{k-1} \neq \xi'_k\right) \cap \left\{\xi\left(\frac{j_m + 1}{n}\right) = \xi'_m \in A'_m\right\} .$$

Since the set of numbers of the form j/n is a set of separability, and the sample functions of a jump process on a finite interval of time assume a finite number of distinct values and are continuous from the right, it follows that for the event $\tilde{D} = B \cap \{\alpha \leqq s\} \cap D$, we have

$$\begin{aligned}\tilde{D} &= B \cap \{\alpha \leqq s\} \cap D \\ &= \lim_{n\to\infty} \bigcup_{0<j_0\leqq[sn]+1, 0<j_k-j_{k-1}\leqq[t'_k n]+1} \tilde{D}_n(j_0, j_1, \cdots, j_m)\end{aligned}$$

and

$$\mathsf{P}(\tilde{D}) = \lim_{n\to\infty} \sum_{\substack{0<j_0<[sn]+1 \\ 0<j_k-j_{k-1}<[t'_k n]+1, k=1,\cdots,m}} \mathsf{P}(\tilde{D}_n(j_0, j_1, \cdots, j_m)) .$$

Since

$$B \cap \left(\frac{j_0 - 1}{n} < \alpha \leqq \frac{j_0}{n}\right) \subset \sigma\left\{\leqq \frac{j_0}{n}\right\},$$

we have

$$\begin{aligned}&\mathsf{P}\{\tilde{D}_n(j_0, j_1, \cdots, j_m)\} \\ &= \int_{B\cap\{(j_0-1)/n<\alpha\leqq(j_0/n)\}} \mathsf{P}\Big\{\bigcap_{k=1}^{m} \bigcap_{j=j_{k-1}}^{j_k} \Big[\xi\Big(\frac{i}{n}\Big) \\ &= \xi'_{k-1} \in A'_{k-1}, \xi'_{k-1} \neq \xi'_k\Big] \cap \Big[\xi\Big(\frac{j_{m+1}}{n}\Big) = \xi'_m \in A'_m\Big] \Big| \xi\Big(\frac{j_0}{n}\Big)\Big\} d\mathsf{P} .\end{aligned} \quad (13)$$

The principal part of the integrand is found in accordance with formula (7). We now define the function

$$m(s, A) = \mathsf{P}\{B \cap (\alpha \leqq s) \cap (\xi(s) \in A)\} .$$

This function is a measure on $\mathfrak{B}$ for fixed s and it is a monotonic function with respect to s for fixed A. In accordance with the general theorems on the extension of set functions we can construct from the function $m(s, A)$ a measure on the Cartesian product of the σ-algebras $\{\mathfrak{B}_1 \times \mathfrak{B}\}$: $\mu([0, s] \times A) = m(s, A)$ for $\mu \leqq 1$. Let $F(t)$ denote the generalized distribution function of the variable α, $F(t) = \mathsf{P}\{\alpha \leqq t\}$. Since μ is absolutely continuous with respect to F, it follows from the Radon-Nikodym theorem that there exists a function $g(s, A)$ that is $\mathfrak{B}_1$-measurable for fixed A, that is a measure on $\mathfrak{B}$ for fixed s, and that satisfies the equation

$$\mu([0, s] \times A) = m(s, A) = \int_0^s g(t, A) dF(t) . \quad (14)$$

Applying formula (14) and the expression (7) to the principal part of the integrand in (13), we may rewrite this formula (13) in the form:

$$\mathsf{P}\{\tilde{D}(j_0, j_1, \cdots, j_m)\} = \int_{(j_0-1/n)}^{(j_0/n)} \int_X \exp\left(-\int_{(j_0/n)}^{(j_1/n)} q(\theta, x)d\theta\right)$$
$$\times \int_{A_1'} \frac{1}{n}\tilde{q}\left(\frac{j_1}{n}, x, dx_1\right) \exp\left(-\int_{(j_1/n)}^{(j_2/n)} q(\theta, x_1)d\theta\right)$$
$$\times \int_{A_2'} \frac{1}{n}\tilde{q}\left(\frac{j_2}{n}, x_1, dx_2\right) \cdots \exp\left(-\int_{(j_{m-1}/n)}^{(j_m/n)} q(\theta, x_{m-1})d\theta\right)$$
$$\times \frac{1}{n}\tilde{q}\left(\frac{j_m}{n}, x_{m-1}, A_m'\right)g(t, dx)dF(t) + \frac{1}{n^m}\varepsilon_n' , \tag{15}$$

where ε_n' approaches 0 uniformly with respect to $j_0, j_1, \cdots, j_m$ for $j_k \leqq Cn$, where $k = 1, \cdots, m$ and C is arbitrary. Here we shift from integration with respect to the variable u in the space $\{U, \mathfrak{S}, \mathsf{P}\}$ to integration with respect to the variables α and $\xi(j/n)$ (cf. Theorem 3, Section 6, Chapter II), and we use Lemma 1 of Section 1. It follows immediately from formula (15) that

$$\mathsf{P}\{\tilde{D}\} = \int_0^s dF(t)\int_X g(t, dx)\int_0^{t_1'} \Psi(t, x, dt_1)\int_{A1} \Pi\,(s + t_1, x, dx_1) \times$$
$$\cdots \times \int_0^{t_m'} \Psi(s + t_1 + \cdots + t_{m-1}, x_{m-1}, dx_m)$$
$$\times \Pi\,(s + t_1 + \cdots + t_m, x_{m-1}, A_m') , \tag{16}$$

From this it follows that

$$\mathsf{P}\{B \cap \{\alpha \leqq s\} \cap D \,|\, \alpha = s, \xi(\alpha) = x\}$$
$$= \int_0^{t_1'} \Psi(s, x, dt_1)\int_{A_1'} \Pi\,(s + t_1, x, dx_1) \times$$
$$\cdots \times \int_0^{t_m'} \Psi(s + t_1 + \cdots + t_{m-1}, x_{m-1}, dx_m)$$
$$\times \Pi\,(s + t_1 + \cdots + t_m, x_{m-1}, A_m') . \tag{17}$$

Formula (17) shows that the times at which the system remains in fixed states after a random instant α and the probabilities of transitions from certain states to others under the hypotheses that $\alpha = s$ and $\xi(\alpha) = x$ have the same conditional joint distributions as those for the process $\xi(t + s)$ for $t \geqq 0$, with the hypothesis $\xi(s) = x$. Since these joint distributions uniquely determine the conditional finite-dimensional distributions of the process, we have proven

Theorem 3. *Let $\xi(t)$ denote a jump Markov process and let α denote a random variable that is independent of the future of $\xi(t)$. Suppose that $B \in \sigma_\alpha$. Then if $0 < t_1 < \cdots < t_m$,*

$$\begin{aligned}\mathsf{P}\{B, \alpha \leqq s, \xi(t_1+\alpha) \in A_1, \cdots, \xi(t_m+\alpha) \in A_m\} \\ = \int_0^s \int_X \mathsf{P}\{\xi(t_1+t) \in A_1, \cdots, \xi(t_m+t) \in A_m \mid \xi(t) = x\} \\ \times g(t, dx)dF(t) ,\end{aligned}$$

where $\int_0^s g(t, A)dF(t) = \mathsf{P}\{B, \alpha \leqq s, \xi(\alpha) \in A\}$.

Corollary 1. *If the conditions of the preceding theorem are satisfied and the process $\xi(t)$ is homogeneous, then*

$$\begin{aligned}\mathsf{P}\{B, \alpha \leqq s, \xi(t_1+\alpha) \in A_1, \cdots, \xi(t_m+\alpha) \in A_m \mid \xi(\alpha) = x\} \\ = \mathsf{P}\{B, \alpha \leqq s\}\mathsf{P}\{\xi(t_1) \in A_1, \cdots, \xi(t_m) \in A_m \mid \xi(0) = x\} .\end{aligned}$$

Corollary 2. *Let $\xi(t)$ denote a homogeneous Markov jump process with countably many states, let α denote the time the system first falls in the sth state ($\alpha = \tau_1 + \cdots + \tau_k$ if $\xi_k = s$. Suppose that $\mathsf{P}(\alpha < \infty) = 1$. Then the process $\xi'(t) = \xi(\alpha + t)$ for $t \geqq 0$ is a Markov process with the same transition probabilities as the process $\xi(t)$, it satisfies the initial condition $\xi'(0) = s$, and it is independent of the σ-algebra of the events σ_α.*

For an example of the application of these results, consider the problem of determining the distribution function of the time of first transition from the sth state to the rth state in the case of a jump birth and death process. We recall (Example 3 of Section 4) that a birth and death process is a homogeneous Markov process with countably many states $(0, 1, 2, \cdots)$ for which $q(n) = \lambda_n + \mu_n$, $q(n, \{n+1\}) = \lambda_n$, and $q(n, \{n-1\}) = \mu_n$, for $n = 0, 1, 2, \cdots$, where $\mu_0 = 0$ and $\{n\}$ is the set consisting of the single element n. Thus in a jump birth and death process, transitions from the nth state are possible only into the adjacent $(n-1)st$ and $(n+1)st$ states with probabilities (cf. Theorem 1)

$$\pi'_n = \Pi(n, \{n+1\}) = \frac{\lambda_n}{\lambda_n + \mu_n} ,$$

$$\pi''_n = \Pi(n, \{n-1\}) = \frac{\mu_n}{\lambda_n + \mu_n} .$$

Suppose that $s > r$ and $\tau_{sr}(t)$ is the length of the interval of time to the first instant the system falls in the rth state if the system is in the sth state at the instant t. Then

$$\tau_{sr}(t) = \tau_{s,s-1}(t) + \tau_{s-1,r}(t + \tau_{s,s-1}(t)) . \tag{18}$$

It follows from Corollary 1 to Theorem 3 that the terms in the right-hand member of equation (18) are independent and that the second term has the same distribution as $\tau_{s-1,r}(t)$. Let the distribution function of the variables $\tau_{sr}(t)$ be denoted by $F_{sr}(x)$; that is, $F_{sr}(x) = \mathsf{P}\{\tau_{sr} < x\}$. Here we do not assume in advance that $F_{sr}(+\infty) = 1$. We introduce the Laplace-Stieltjes transform of the function $F_{sr}(x)$:

$$\varphi_{sr}(z) = \int_0^\infty e^{-zx} dF_{sr}(x) , \qquad \operatorname{Re} z \geqq 0 .$$

On the basis of what we have just said, it follows from (18) that

$$\varphi_{sr}(z) = \varphi_{s,s-1}(z)\varphi_{s-1,r}(z) . \tag{19}$$

Setting $\varphi_{s,s-1}(z) = \varphi_s(z)$, we obtain from (19)

$$\varphi_{sr}(z) = \varphi_s(z)\varphi_{s-1}(z) \cdots \varphi_{r+1}(z) . \tag{20}$$

Since the function $\varphi_{sr}(z)$ determines $F_{sr}(x)$ uniquely, the problem is reduced to determining $\varphi_s(z)$. Let $\tau_s(t)$ denote the length of the interval of time from t to the instant when the system leaves the sth state, which it is in at the instant t. Then with probability π_s'' we have $\tau_{s,s-1}(t) = \tau_s(t)$, and with probability π_s',

$$\begin{aligned} \tau_{s,s-1}(t) &= \tau_s(t) + \tau_{s+1,s}(t + \tau_s(t)) \\ &\quad + \tau_{s,s-1}[t + \tau_s(t) + \tau_{s+1,s}(t + \tau_s(t))] . \end{aligned} \tag{21}$$

The Laplace-Stieltjes transform $\varphi_s'(t)$ of the distribution function of the variable $\tau_s(t)$ is easily found. Using Lemma 2, we obtain

$$\varphi_s'(z) = \int_0^\infty e^{-zx} d(1 - e^{-q(s)x}) = \frac{q(s)}{q(s) + z} .$$

It again follows from Corollary 1 of Theorem 3 that

$$\varphi_s(z) = \pi_s''\varphi_s'(z) + \pi_s'\varphi_s'(z)\varphi_{s+1}(z)\varphi_s(z)$$

or

$$\varphi_{s+1}(z) = \frac{q(s) + z}{\lambda_s} - \frac{\mu_s/\lambda_s}{\varphi_s(z)} . \tag{22}$$

Successive applications of formula (10) lead to the representation of the function $\varphi_{s+1}(z)$ in the form of a continued fraction that is a rational function of $\rho_k = \mu_k/\lambda_k$ and $\varphi_1(z)$. Let us find the function $\varphi_1(z)$ from the following considerations. Define

$$p_{00}(t) = \mathsf{P}\{\xi(t) = 0 \mid \xi(0) = 0\} .$$

Let $\tau^{(k)}$ denote the length of the kth interval of time in the course of which the system finds itself in the state 0, and let $\tau_{10}^{(k)}$ denote the time spent by the system after leaving the state 0 for the kth time until the next return to 0. The probability $p_{00}(t)$ is the probability sum $\bigcup_{n=0}^\infty E_n$ of incompatible events E_n, where E_0 is the event $\tau^{(1)} > t$ and E_n is the event $(n \geqq 1)$

$$\sum_{k=1}^{n} (\tau^{(k)} + \tau_{10}^{(k)}) \leqq t < \sum_{k=1}^{n} (\tau^{(k)} + \tau_{10}^{(k)}) + \tau^{(n+1)} .$$

It follows from the independence of the variables $\tau^{(k)}$ and $\tau_{10}^{(k)}$ that (for $\mathsf{P}(\tau^{(1)} > t) = e^{-\lambda_0 t}$)

$$\mathsf{P}\left\{\sum_{k=1}^{n} (\tau^{(k)} + \tau_{10}^{(k)}) \leqq t < \sum_{k=1}^{n} (\tau^{(k)} + \tau_{10}^{(k)}) + \tau^{(n+1)}\right\}$$
$$= \int_0^t e^{-\lambda_0(t-x)} d\widetilde{F}^{(n)}(x) ,$$

where $\widetilde{F}^{(n)}(x)$ is the distribution function of the sum $\sum_{k=1}^{n} (\tau^{(k)} + \tau_{10}^{(k)})$,

$$\widetilde{F}^{(1)}(x) = \int_0^x e^{-\lambda_0(x-t)} dF_{10}(t) ,$$

and $\widetilde{F}^{(n)}(x)$ is the n-fold convolution of the function $\widetilde{F}^{(1)}(x)$ with itself. Taking the Laplace transform of both sides of the equation

$$p_{00}(t) = e^{-\lambda_0 t} + \sum_{n=1}^{\infty} \int_0^t e^{-\lambda_0(t-x)} d\widetilde{F}^{(n)}(x) ,$$

we obtain

$$p_0(z) = \frac{1}{\lambda_0 + z} + \sum_{n=1}^{\infty} \frac{1}{\lambda_0 + z}\left(\frac{\lambda_0}{\lambda_0 + z}\right)^n \varphi_1^n(z) = \frac{1}{z + \lambda_0(1 - \varphi_1(z))} ,$$

from which we get

$$\varphi_1(z) = \frac{\lambda_0 + z}{\lambda_0} - \frac{1}{\lambda_0 p_0(z)} , \qquad p_0(z) = \int_0^\infty p_{00}(t) e^{-zt} dt . \tag{23}$$

Formulas (20), (22), and (23) give the general solution of the problem.

We use analogous considerations to evaluate the distribution of the time necessary for a transition from a state S to a state R without entering the state N (where $R < S < N$). This problem is a particular case of the preceding one. Specifically, together with the given process $\xi(t)$ we introduce a new process $\xi'(t)$, where $\xi(t) = \xi'(t)$ for $t < \tau_{SN}(0)$ and $\xi'(t) = N$ for $t \geqq \tau_{SN}(0)$. The process $\xi'(t)$ is also a jump birth and death process with the same values of λ_n and μ_n as the process $\xi(t)$ for $n < N$. All the preceding formulas remain valid for this case. However, the function $p_{00}(t)$ now corresponds to the auxiliary process $\xi'(t)$ and we may not assume that it is given. In the present problem it is easy to find the distribution of the variable $\tau_{N-1,N-2}$:

$$\mathsf{P}(\tau_{N-1,N-2} < x) = \frac{\mu_{N-1}}{\lambda_{N-1} + \mu_{N-1}}(1 - e^{-(\lambda_{N-1}+\mu_{N-1})x}) ,$$

from which we get

$$\varphi_{N-1}(z) = \frac{\mu_{N-1}}{q(N-1) + z} \,.$$

Using (22), we obtain the representation of the function $\varphi_s(z)$ in the form of a continued fraction that is a rational function of z:

$$\varphi_s(z) = \cfrac{\rho_s}{1 + \rho_s + z - \cfrac{\rho_{s+1}}{1 + \rho_{s+1} + z - \cfrac{\rho_{s+2}}{\ddots - \cfrac{\rho_{N-1}}{1 + \rho_{N-1} + \cfrac{z}{\lambda_{N-1}}}}}} \tag{24}$$

Previously, we have always started with a given jump process. Let us now suppose that the function $q(t, x, A)$ is given. Can we construct a Markov process with transition probabilities $\mathsf{P}(t, x, s, A)$ connected with the function $q(t, x, A)$ by the relation

$$\lim_{s \to t} \frac{\mathsf{P}(t, x, s, A) - \chi_A(x)}{s - t} = q(t, x, A)? \tag{25}$$

If the the answer is affirmative and the process constructed is a jump process, then the sample functions of the process can be constructed by using the preceding results. This remark forms the basis of the solution of the problem posed. Let X denote a complete separable space, let $\mathfrak{B}$ denote the σ-algebra of Borel subsets of X. Suppose that the function

$$q(t, x, A) = -q(t, x)\chi_A(x) + \tilde{q}(t, x, A)$$

is defined for all $t \geqq 0$, $x \in X$, and $A \in \mathfrak{B}$ and that it satisfies the following conditions:

a′. For fixed $t \geqq 0$ and $x \in X$, the function $\tilde{q}(t, x, A)$ is a finite measure on $\mathfrak{B}$ and $q(t, x) = \tilde{q}(t, x, X)$.

b′. For fixed $x \in X$ and $A \in \mathfrak{B}$, the function $\tilde{q}(t, x, A)$ is continuous with respect to t and uniformly continuous with respect to A on every finite interval of variation of t.

We note that conditions a′ and b′ are more general than those that the function $q(t, x, A)$ of a jump Markov process must satisfy (conditions a-c of Section 3).

Consider the space Ω, introduced above, of the sequences $\omega = \{\xi_0, \tau_1, \xi_1, \cdots\}$, where $\xi_k \in X$ and $\tau_k \geqq 0$. On the algebra of cylindrical sets in Ω we introduce the measure $\mathsf{P}(C)$ as follows. If

$$C = \{\xi_k \in A_k, k = 0, 1, \cdots, m; \tau_j \leqq t_j, j = 1, \cdots, m\} \,,$$

then we set

$$\mathsf{P}(C) = \int_{A_0} F(t_1, \cdots, t_m; A_1, \cdots, A_m \mid 0, x)\mu_0(dx) ,$$

where $F(t_1, \cdots, t_m; A_1, \cdots, A_m \mid 0, x)$ is given by formula (8) and μ_0 is an arbitrary "initial" distribution on $\mathfrak{B}$. In accordance with Kolmogorov's theorem (Theorem 3, Section 2, Chapter III), the measure $\mathsf{P}(C)$ can be extended as a complete measure $\{\mathfrak{F}, \mathsf{P}\}$, where $\mathfrak{F}$ is the complete σ-algebra generated by the cylindrical sets in Ω. On Ω we define the function $\xi(t) = f(t, \omega)$ by

$$f(t, \omega) = x_k, \quad \text{if} \quad \gamma_k \leqq t < \gamma_{k+1}\left(\gamma_0 = 0, \gamma_k = \sum_{j=1}^{k} \tau_j\right) .$$

If $\gamma_\infty = \sum_{k=1}^{\infty} \tau_k = \infty$, then $f(t, \omega)$ is defined for all $t \geqq 0$ for given ω. Let N denote the set

$$N = \left\{\omega; \gamma_\infty = \sum_{k=1}^{\infty} \tau_k < \infty\right\}$$

which is $\mathfrak{F}$-measurable in Ω. If $\mathsf{P}(N) = 0$, then $\xi(t)$ is a random process defined with probability 1 for all $t \geqq 0$, and its sample functions are continuous on the right in the discrete topology on X. This will be the case, in particular, when the function $q(t, x)$ is bounded.

To see that this is so, note that the proof of Corollary 1 of Theorem 1 depends only on formula (8) and the boundedness of the function $q(t, x)$, so that $\mathsf{P}(N) = 0$ under this restriction upon the process in question.

To determine the random process $\xi(t)$ for all $t \geqq 0$ for P-almost-all ω in the case in which $\mathsf{P}(N) > 0$, we proceed in a different manner. The simplest method is as follows: We add to X a single point "∞." The extended space is denoted by X'; that is, $X' = X \cup \{\infty\}$. Let us assume that $\xi(t) = \infty$ for $t \geqq \gamma_\infty$. The process thus constructed is denoted by $\xi_0(t)$. Other extensions of the process $\xi(t)$ can be obtained as follows: Let Ω_k (for $k = 1, 2, \cdots$) denote a sequence of spaces that can be considered as distinct copies of the space Ω. On Ω_k let us consider the measure $\{\mathfrak{F}_k, \mathsf{P}_k\}$ defined in the same way that $\{\mathfrak{F}, \mathsf{P}\}$ is defined on Ω, but with measure $\{\mu_1, \mathfrak{B}\}$ as the initial distribution. Let N_k denote the set in Ω_k analogous to N in Ω:

$$N_k = \left\{\omega^{(k)}, \omega^{(k)} \in \Omega_k, \gamma_\infty^{(k)} = \sum_{j=1}^{\infty} \tau_j^{(k)} < \infty\right\} .$$

We shall assume that the σ-algebras of the events $\mathfrak{F}_k$ are independent. For $t > \gamma_\infty$ we obtain $\xi(t) = f(t - \gamma_\infty, \omega^{(1)})$. If $\mathsf{P}(\gamma_\infty^{(1)} < \infty) = 0$, then $\xi(t)$ is now defined for all t for almost all $(\omega, \omega^{(1)})$. On the other hand, if $\mathsf{P}(\gamma_\infty^{(1)} < \infty) > 0$ we set $\xi(t) = f(t - \gamma_\infty - \gamma_\infty^{(1)}, \omega^{(2)})$ for $t \geqq \gamma_\infty + \gamma_\infty^{(1)}$, and so forth. We note that if the function $q(t, x, A)$ is independent of t, then the variables $\gamma_\infty^{(k)}$ (for $k > 1$) are identically distributed and independent, so that the inductively defined process

$$\xi(t) = f(t - \gamma_\infty - \gamma_\infty^{(1)} - \cdots - \gamma_\infty^{(k)}, \omega^{(k+1)}) ,$$
$$\gamma_\infty^{(k)} \leqq t < \gamma_\infty^{(k+1)} \quad (\gamma_\infty^{(0)} = 0) ,$$

is defined for all t with probability 1.

Let us look at the process $\xi_0(t)$ in greater detail. It follows from (11) that

$$\mathsf{P}\left(\tau_m > t + x \mid \xi_0, \xi_1, \cdots, \xi_{m-1}, \tau_1, \cdots, \tau_{m-1}, \tau_m > t - \sum_{k=1}^{m-1} \tau_k\right)$$
$$= 1 - \Psi(t, \xi_{m-1}, x) ,$$

from which we get the important conclusion

$$\mathsf{P}\Big\{\tau_m - T < t_m, \tau_{m+1} < t_{m+1}, \cdots, \tau_{m+r} < t_{m+r}, \xi_m \in A_m,$$
$$\cdots, \xi_{m+r} \in A_{m+r} \mid \xi_0, \cdots, \xi_{m-1}, \tau_1, \cdots, \tau_{m-1}, \sum_{k=1}^{m-1} \tau_k < T < \sum_{k=1}^{m} \tau_k\Big\}$$
$$= F(t_m, t_{m+1}, \cdots, A_m, \cdots, A_{m+r} \mid T, \xi_{m-1}) . \tag{26}$$

But this means that the conditional probabilities of the events in $\sigma\{\xi_0(t), t > T\}$ with respect to the σ-algebra $\sigma\{\xi(t), t \leqq T\}$ depend only on $\xi_0(T) = \xi_{m-1}$ (or on $\xi_0(T) = \infty$, which is obvious); that is, $\xi_0(t)$ is a Markov process. One can easily show that the same conclusion holds for the process $\xi(t)$.

Theorem 4. *Let $q(t, x, A)$ satisfy the conditions* a′ *and* b′ (page 362). *Then*

a. *$\xi_0(t)$ is a Markov process and its sample functions are, with probability* 1, *continuous from the right*;

b. *the transition probabilities of the process $\xi_0(t)$ are defined by the relations*:

$$\bar{\mathsf{P}}(t, x, v, A) = \sum_{n=0}^{\infty} \bar{\mathsf{P}}^{(n)}(t, x, v, A) , \qquad v \geqq t, A \in \mathfrak{B} , \tag{27}$$

where

$$\bar{\mathsf{P}}^{(0)}(t, x, v, A) = \Psi(t, x, v)\chi_A(x) ,$$
$$\bar{\mathsf{P}}^{(n+1)}(t, x, v, A) = \int_t^v \int_X \bar{\mathsf{P}}^{(n)}(\theta, y, v, A)\Psi(t, x, \theta)\tilde{q}(\theta, x, dy)d\theta ;$$
$$\Psi(t, x, v) = \exp\left(-\int_t^v q(\theta, x)d\theta\right) , \qquad n = 0, 1, \cdots \tag{28}$$

and

$$\bar{\mathsf{P}}(t, x, v, \{\infty\}) = 1 - \bar{\mathsf{P}}(t, x, v, X) ;$$

c. *the function* $\bar{\mathsf{P}}(t, x, v, A)$ *satisfies the first Kolmogorov equation*:

$$\frac{\partial \bar{\mathsf{P}}(t, x, v, A)}{\partial t} = -\int_X \bar{\mathsf{P}}(t, z, v, A) q(t, x, dz) \qquad (29)$$

and the boundary condition

$$\lim_{t \uparrow v} \bar{\mathsf{P}}(t, x, v, A) = \chi_A(x) ;$$

d. *equation* (25) *is satisfied uniformly with respect to* t (*for* $0 \leqq t \leqq T$ *where* T *is an arbitrary number*) *for fixed* x *and* A.

Proof. a. This was proved before the statement of the theorem.

b. Note that Theorem 2 holds for the process $\xi_0(t)$ since its proof was based only on formula (8) and the fact that the first state ξ_0 changes by a jump. Let $\bar{\mathsf{P}}^{(n)}(t, x, v, A)$ denote the conditional probability that with the hypothesis $\xi(t) = x \neq \infty$, the function $\xi_0(v) \in A$, and the function $\xi_0(t)$ has exactly n jumps on the interval $[t, s]$. Then it follows from (26) that $\mathsf{P}^{(0)}(t, x, v, A) = \chi_A(x)$. Furthermore, if $\xi(\theta) = x$ for $t \leqq \theta < t + \tau_1$ and $\xi(t + \tau_1) = \xi_1$, then on the basis of Theorem 2,

$$\begin{aligned} \bar{\mathsf{P}}^{(n+1)}(t, x, v, A) &= \mathsf{M}\{\bar{\mathsf{P}}^{(n)}(t + \tau_1, \xi_1, v, A) \mid \xi(t) = x\} \\ &= \int_t^v \int_X \bar{\mathsf{P}}^{(n)}(\theta, y, v, A) \Psi(t, x, \theta) \tilde{q}(\theta, x, dy) d\theta . \end{aligned}$$

Therefore the function $\bar{\mathsf{P}}(t, x, v, A)$ defined by equation (27) is the probability of falling at the instant v into the set A after a finite number of jumps, after leaving the point x at the instant t. This completes the proof of (b).

c. It follows from (27) and (28) that

$$\begin{aligned} \bar{\mathsf{P}}(t, x, v, A) &= \Psi(t, x, v)\chi_A(x) \\ &\quad + \int_t^v \int_X \bar{\mathsf{P}}(\theta, y, v, A) \Psi(t, x, \theta) \tilde{q}(\theta, x, dy) d\theta . \end{aligned} \qquad (30)$$

From this it follows that

$$\lim_{t \uparrow v} \bar{\mathsf{P}}(t, x, v, A) = \chi_A(x)$$

and the function $\bar{\mathsf{P}}(t, x, v, A)$ is continuous with respect to t. It follows from the boundedness and continuity of the function $\bar{\mathsf{P}}(t, x, v, A)$ with respect to t and the continuity in t of $\tilde{q}(t, x, A)$ uniformly with respect to A (that is, the continuity of the variation of the measure $\tilde{q}(t, x, A)$ as a function of t) that the integral

$$\int_X \bar{\mathsf{P}}(\theta, z, v, A) \tilde{q}(\theta, x, dz)$$

is continuous with respect to θ. Consequently equation (30) can be differentiated with respect to t. Thus

$$\frac{\partial \bar{P}(t, x, v, A)}{\partial t} = q(t, x)\Psi(t, v, x)\chi_A(x) - \int_X \bar{P}(t, y, v, A)\Psi(t, x, t)\tilde{q}(t, x, dy) + \int_t^v \int_X \bar{P}(\theta, y, v, A)q(t, x)\tilde{q}(\theta, x, dz)d\theta$$

$$= q(t, x)\bar{P}(t, x, v, A) - \int_X \bar{P}(t, y, v, A)\tilde{q}(t, x, dy),$$

which completes the proof of (c).

d. It follows from the continuity of the function $q(t, x)$ with respect to t that

$$\lim_{t_1 \uparrow t, t_2 \downarrow t} \frac{\Psi(t_1, x, t_2)}{t_2 - t_1} = q(t, x)$$

uniformly with respect to t (for $0 \leqq t \leqq T$). In view of (30), it then follows that

$$\lim_{\substack{t_2 > t > t_1 \\ |t_1 - t_2| \to 0}} \frac{1 - \bar{P}(t_1, x, t_2, \{x\})}{t_2 - t_1} = q(t, x) .$$

It follows from (30) that $P(t_1, x, t_2, A)$ is a continuous function of t_1 and t_2. Keeping the continuity (mentioned above) of the inner integral on the right-hand side of formula (30), we obtain, for $x \notin A$,

$$\frac{P(t_1, x, t_2, A)}{t_2 - t_1} = \int_X \bar{P}(\theta, z, t_2, A)\Psi(t_1, x, \theta)\tilde{q}(\theta, x, dz), \quad (t_1 < \theta < t_2) .$$

Therefore the limit

$$\lim_{\substack{t_1 < t < t_2 \\ t_2 - t_1 \to 0}} \frac{P(t_1, x, t_2, A)}{t_2 - t_1} = \tilde{q}(t, x, A) \qquad (x \notin A)$$

exists and is uniform with respect to t (for $0 \leqq t \leqq T$). This completes the proof of the theorem.

REMARK 1. One can show in an analogous manner that the transition probabilities of the process $\xi(t)$ also satisfy the first Kolmogorov equation and equation (25). From this it follows that if $\bar{P}(t, x, v, X) < 1$ for some $v > t > 0$ and $x \in X$, then the first Kolmogorov equation and the problem of constructing a Markov process from the given function $q(t, x, A)$ have a nonunique solution.

Corollary. *The sample functions of the process $\xi_0(t)$ have with probability 1 a finite number of jumps in the course of a finite interval of time if and only if*

$$\mathsf{P}(t, x, v, X) = 1 , \qquad v > t > 0 , \qquad x \in X . \tag{31}$$

The question of whether or not a Markov process has infinitely many jumps in a finite interval of time is of great interest. However Equation (31) is not convenient for determining this. We can obtain a more suitable condition by confining ourselves to the homogeneous case, that is, by assuming that the function $q(t, x, A) = q(x, A)$ is independent of t. In the homogeneous case,

$$\bar{\mathsf{P}}^{(n)}(t, x, v, A) = \bar{\mathsf{P}}^{(n)}(v - t, x, A) ,$$
$$\bar{\mathsf{P}}^{(0)}(t, x, A) = e^{-tq(x)}\chi_A(x) ,$$
$$\bar{\mathsf{P}}^{(n)}(t, x, A) = \int_0^t\int_X \exp\{-\theta q(x)\}\bar{\mathsf{P}}^{(n-1)}(t - \theta, y, A)\tilde{q}(x, dy)d\theta ,$$

where $q(x) = \tilde{q}(x, X)$ and $\bar{\mathsf{P}}(t, x, v, A) = \bar{\mathsf{P}}(v - t, x, A)$. If we set $K(t, x) = 1 - \bar{\mathsf{P}}(t, x, X)$, then $K(0, x) = 0$ and it follows from equation (29) that

$$-\frac{\partial K(t, x)}{\partial t} = q(x)K(t, x) - \int_X K(t, y)q(x, dy) . \tag{32}$$

In this equation let us shift from the function $K(t, x)$ to its Laplace transform:

$$z(\lambda, x) = \lambda\int_0^\infty e^{-\lambda t}K(t, x)dt .$$

We obtain

$$(\lambda + q(x))z(\lambda, x) = \int_X z(\lambda, y)\tilde{q}(x, dy) .$$

Let $f(x)$ denote an arbitrary solution of the integral equation

$$(\lambda + q(x))f(x) = \int_X f(y)\tilde{q}(x, dy) , \tag{33}$$

and let $f(x)$ satisfy the condition $\sup_{x\in X} |f(x)| \leqq 1$.

Lemma 4. *For* $\lambda > 0$,

$$-z(\lambda, x) \leqq f(x) \leqq z(\lambda, x) .$$

Proof. We have $z(\lambda, x) = \lim_{n\to\infty} z^{(n)}(\lambda, x)$, where

$$z^{(n)}(\lambda, x) = \lambda\int_0^\infty e^{-\lambda t}\Big(1 - \sum_{k=0}^{n} \bar{\mathsf{P}}^{(k)}(t, x, X)\Big)dt .$$

Define

$$Q^{(n)}(\lambda, x) = \lambda\int_0^\infty e^{-\lambda t}\bar{\mathsf{P}}^{(n)}(t, x, X)dt .$$

Then

$$Q^{(0)}(\lambda, x) = \frac{\lambda}{\lambda + q(x)},$$

$$Q^{(k)}(\lambda, x) = \int_X Q^{(k-1)}(\lambda, y)\frac{\tilde{q}(x, dy)}{\lambda + q(x)},$$

$$z^{(n)}(\lambda, x) = 1 - \sum_{k=0}^{n} Q^{(k)}(\lambda, x)$$

$$= 1 - \frac{\lambda}{\lambda + q(x)} - \int_X (1 - z^{(n-1)}(\lambda, y))\frac{\tilde{q}(x, dy)}{\lambda + q(x)},$$

from which we get

$$(\lambda + q(x))z^{(n)}(\lambda, x) = \int_X z^{(n-1)}(\lambda, y)\tilde{q}(x, dy) . \tag{34}$$

Since $-z^{(0)}(\lambda, x) = -1 \leqq f(x) \leqq 1 = z^{(0)}(\lambda, x)$, by combining (33) and (34) we obtain by induction

$$-z^{(n)}(\lambda, x) \leqq f(x) \leqq z^{(n)}(\lambda, x) ,$$

and taking the limit as $n \longrightarrow \infty$, we obtain the desired assertion.

Theorem 5. *For $\bar{\mathsf{P}}(t, x, X)$ to be equal to 1 for all $t > 0$ and $x \in X$, it is necessary that equation (33) have no nontrivial bounded solutions for arbitrary $\lambda > 0$, and it is sufficient that this condition be satisfied for some $\lambda > 0$.*

Proof. If $\bar{\mathsf{P}}(t, x, X) = 1$, then $z(\lambda, x) = 0$ for arbitrary $\lambda > 0$, and on the basis of Lemma 4 every bounded solution of equation (33) is identically equal to 0. On the other hand, if equation (33) has no nontrivial bounded solutions for some λ, then $z(\lambda, x)$ which is a bounded solution of equation (33), is identically equal to 0 and $\bar{\mathsf{P}}(t, x, X) = 1$. This completes the proof of the theorem.

Example. Suppose that $\xi_0(t)$ for $t > 0$ is a birth and death process that is defined by the sequences $\{\lambda_n\}$ and $\{\mu_n\}$ (for $n = 0, 1, 2, \cdots$), where $\mu_0 = 0, \lambda_0 \geqq 0$, and $\lambda_n > 0$ for $n \geqq 1$ (cf. p. 324). Under what conditions do the sample functions of the process remain bounded with probability 1 in the course of a finite interval of time? Obviously, this will occur only when these sample functions have finitely many jumps in a finite interval of time. Let us use the preceding theorem. Equations (33) in this case become an infinite system of linear algebraic equations

$$(\lambda + \lambda_0)f(0) = \lambda_0 f(1) ,$$

$$\lambda_n(f(n + 1) - f(n)) = \mu_n(f(n) - f(n - 1)) + \lambda f(n) , \quad n \geqq 1 . \tag{35}$$

If $\lambda_0 \neq 0$, then equations (35) determine all the $f(n)$ for $n \geqq 1$ up to an arbitrary factor equal to $f(0)$. On the other hand, if $\lambda_0 = 0$

we can set $f(0) = 0$. Then all the $f(n)$ for $n \geqq 2$ are uniquely determined up to the factor $f(1)$. Since $f(1) > f(0) > 0$ (for $\lambda > 0$), we obtain by induction the result that $f(n+1) > f(n)$. Let us rewrite the system (35) in the form

$$f(n+1) - f(n) = \gamma_n(f(n) - f(n-1)) + \delta_n f(n) ,$$
$$n > 1 , \qquad \gamma_n = \frac{\mu_n}{\lambda_n} , \qquad \delta_n = \frac{\lambda}{\lambda_n} ,$$

and let us show that the sequence $f(n)$, where $f(0) = 1$, is bounded if and only if

$$\sum_{n=1}^{\infty} (\delta_n + \gamma_n \delta_{n-1} + \gamma_n \gamma_{n-1} \delta_{n-2} + \cdots + \gamma_n \gamma_{n-1} \cdots \gamma_2 \delta_1 + \gamma_n \gamma_{n-1} \cdots \gamma_1) < \infty . \tag{36}$$

We have

$$\begin{aligned} f(n+1) - f(n) &= \delta_n f(n) + \gamma_n \delta_{n-1} f(n-1) + \\ &\cdots + \gamma_n \gamma_{n-1} \cdots \gamma_2 \delta_1 f(1) + \gamma_n \gamma_{n-1} \cdots \gamma_1 (f(1) - 1) \\ &\leqq f(n)(\delta_n + \gamma_n \delta_{n-1} + \gamma_n \gamma_{n-1} \delta_{n-2} + \cdots + \gamma_n \gamma_{n-1} \cdots \gamma_1) \\ &= \rho(n) f(n) \end{aligned}$$

where

$$\rho(n) = \delta_n + \gamma_n \delta_{n-1} + \cdots + \gamma_n \gamma_{n-1} \cdots \gamma_1 .$$

On the other hand,

$$f(n+1) - f(n) \geqq \rho(n)(f(1) - 1) .$$

Thus

$$f(n) + \rho_n(f(1) - 1) \leqq f(n+1) \leqq f(n)(1 + \rho_n) ,$$

from which it follows that

$$f(1) + (f(1) - 1) \sum_{k=1}^{n} \rho_k \leqq f(n+1) \leqq f(1) \prod_{k=1}^{n} (1 + \rho_k) .$$

Since the series $\sum_{k=1}^{\infty} \rho_k$ and the infinite product $\prod_{k=1}^{\infty} (1 + \rho_k)$ converge simultaneously, we see that condition (36) is necessary and sufficient for the sequence $f(n)$ to be bounded.

Theorem 6. *For the sample functions of the birth and death process $\xi_0(t)$ to have with probability 1 finitely many jumps on an arbitrary finite interval of time, it is necessary and sufficient that*

$$\sum_{n=1}^{\infty} \left(\frac{1}{\lambda_n} + \frac{\mu_n}{\lambda_n \lambda_{n-1}} + \frac{\mu_n \mu_{n-1}}{\lambda_n \lambda_{n-1} \lambda_{n-2}} + \frac{\mu_n \mu_{n-1} \cdots \mu_2}{\lambda_n \lambda_{n-1} \cdots \lambda_2 \lambda_1} + \frac{\mu_n \cdots \mu_2 \mu_1}{\lambda_n \lambda_{n-1} \cdots \lambda_2 \lambda_1} \right) = \infty .$$

VIII

DIFFUSION PROCESSES

In this chapter we shall consider continuous Markov processes with range in m-dimensional Euclidean space $R^{(m)}$; Up to this point we have not completely described such processes. We shall now study an important class of these processes, the class of so-called diffusion processes which, as the name suggests, can serve as a probabilistic model of the physical process of diffusion. In Section 5, Chapter VI, we considered a process of Brownian motion as a probabilistic model of diffusion in a homogeneous medium. Using a similar construction in the case of a nonhomogeneous medium, we arrive at the concept of a general diffusion process. Let us clarify the basic concepts of diffusion processes by giving an example of a one-dimensional process.

Let x_t denote the coordinate of a sufficiently small particle suspended in a liquid at an instant t. Neglecting the inertia of the particle, we may assume that the displacement of the particle has two components: the "average" displacement caused by the macroscopic velocity of the motion of the liquid, and the fluctuation of the displacement caused by the chaotic nature of the thermal motion of the molecules of the liquid.

Suppose that the velocity of the macroscopic motion of the liquid at the point x and the instant t is equal to $a(t, x)$. We assume that the fluctuational component of the displacement is a random variable whose distribution depends on the position x of the particle, the instant t at which the displacement is observed, and the quantity Δt, which is the length of the interval of time during which the displacement is observed. We assume that the average value of this displacement is equal to 0 independently of t, x_t and Δt. Thus the displacement of the particle can be written approximately in the form

$$x_{t+\Delta t} - x_t = a(t, x_t)\Delta t + \xi_{t,x_t,\Delta t}\,; \tag{1}$$

here, $\mathsf{M}\xi_{t,x_t,\Delta t} = 0$. If $a(t, x)$ is equal to 0 and the distribution of

$\xi_{t,x_t,\Delta t}$ is independent of x and t, as we assumed when we were considering Brownian motion (cf. Remark 1, Section 5, Chapter VI), then $\mathsf{M}\xi_{t,\Delta t}^2 = \lambda\Delta t$. Since the properties of the medium are naturally assumed to change only slightly for small changes in t and x, the process is homogeneous in the small. Therefore, we may assume that

$$\xi_{t,x_t,\Delta t} = \sigma(t, x_t)\xi_{t,\Delta t}$$

where $\sigma(t, x)$ characterizes the properties of the medium at the point x at the instant t, and $\xi_{t,\Delta t}$ is the value of the increment that is obtained in the homogeneous case under the condition that $\sigma(t, x) = 1$. Thus $\xi_{t,\Delta t}$ must be distributed like the increment of a process of Brownian motion: $w(t + \Delta t) - w(t)$.

Consequently, for the increment $x_{t+\Delta t} - x_t$, we can write the approximate formula

$$x_{t+\Delta t} - x_t \approx a(t, x_t)\Delta t + \sigma(t, x_t)[w(t + \Delta t) - w(t)] . \qquad (2)$$

To make this formula precise, we replace the increments, as one frequently does in mathematical analysis, with differentials. When we do this we obtain the differential equation for x_t,

$$dx_t = a(t, x_t)dt + \sigma(t, x_t)dw(t) , \qquad (3)$$

which we may take as our starting point in determining the diffusion process.

Let $\boldsymbol{x}_t$ denote a multidimensional process with range in $R^{(m)}$. Then equation (1) remains meaningful if $\boldsymbol{a}(t, \boldsymbol{x}_t)$ is a function with range in $R^{(m)}$ and $\boldsymbol{\xi}_{t,x_t,\Delta t}$ is a random vector in $R^{(m)}$. In this case we assume that $\boldsymbol{\xi}_{t,x_t,\Delta t}$ can be represented in the form

$$\boldsymbol{\xi}_{t,x_t,\Delta t} = \sum_{k=1}^{m} \boldsymbol{b}_k(t, \boldsymbol{x}_t)[w_k(t + \Delta t) - w_k(t)] ,$$

where the $\boldsymbol{b}_k(t_t, \boldsymbol{x}_t)$ are functions with ranges in $R^{(m)}$ and the $w_k(t)$ are independent one-dimensional processes of Brownian motion. Such a representation corresponds to a nonisotropic medium: the displacements in the different directions have, in general, different distributions. The equation for the variable $\boldsymbol{x}_t$ in this case takes the form

$$d\boldsymbol{x}_t = \boldsymbol{a}(t, \boldsymbol{x}_t)dt + \sum_{k=1}^{m} \boldsymbol{b}_k(t, \boldsymbol{x}_t)dw_k(t) . \qquad (4)$$

We note that we cannot as yet give a precise meaning to equations (3) and (4). The difficulty lies in the fact that the quantity

$$\frac{w(t + \Delta t) - w(t)}{\Delta t} ,$$

where $w(t)$ is a process of Brownian motion, has a normal distribution with mean zero and variance $1/\Delta t$, and hence this quantity does not have a limit in any probabilistic sense. Since $w(t)$ does not have a derivative, the usual definition of the differential $dw(t)$ has no meaning.

We shall give a precise meaning to equations (3) and (4) when we introduce the concepts of a stochastic integral and stochastic differential in Section 2.

In Section 1 we shall define a diffusion process, beginning with the properties of transition probabilities.

In Sections 3 to 6, we shall study the solutions of equations (3) and (4) from the point of view of their existence and uniqueness, and of the properties that will enable us to determine the distributions of the basic characteristics of the process.

1. DIFFUSION PROCESSES IN THE BROAD SENSE

Let us first consider the one-dimensional case of diffusion processes in the broad sense. Let $\xi(t)$ denote a Markov process in the broad sense and defined on $[0, T]$ into $R^{(1)}$. This means that the transition probabilities $\mathsf{P}(t, x, s, A)$, of a process are given and satisfy conditions a-c of Section 1, Chapter VII.

A process $\xi(t)$ is called a *diffusion process* if the following conditions are satisfied.

a. For every x and every $\varepsilon > 0$,

$$\int_{|x-y|>\varepsilon} \mathsf{P}(t, x, s, dy) = o(s-t) \tag{1}$$

uniformly over $t < s$;

b. there exist functions $a(t, x)$ and $b(t, x)$ such that for every x and every $\varepsilon > 0$,

$$\int_{|x-y|\leq\varepsilon} (y-x)\mathsf{P}(t, x, s, dy) = a(t, x)(s-t) + o(s-t)\,, \tag{2}$$

$$\int_{|y-x|\leq\varepsilon} (y-x)^2\mathsf{P}(t, x, s, dy) = b(t, x)(s-t) + o(s-t) \tag{3}$$

uniformly over $t < s$.

The function $a(t, x)$ is called the *coefficient of transfer* and it is an interpretation of the function $a(t, x)$ defined above; the function $b(t, x)$ is called the *coefficient of diffusion*. A connection between a and σ on the one hand and the coefficients of diffusion and transfer on the other will be established in Section 3.

One can show that conditions (1) to (3) are insufficient for a unique determination of the transition probabilities of the process.

Let us show that under certain assumptions regarding the transition probability $\mathsf{P}(t, x, s, A)$ of the diffusion process, it is completely determined by the coefficients $a(t, x)$ and $b(t, x)$ if these coefficients are such that the Cauchy problem for the equation

$$-\frac{\partial u}{\partial t} = a(t, x)\frac{\partial u}{\partial x} + \frac{1}{2}b(t, x)\frac{\partial^2 u}{\partial x^2}\ ; \qquad u(s, x) = \varphi(x)$$

in the region $x \in R^{(1)}$, $t \in (0, s)$ has, for every $s \in [0, T]$, a unique solution for all initial functions $\varphi(x)$ belonging to some class of functions that is everywhere-dense with respect to the metric of uniform convergence in the space of all continuous functions.

To do this, let us prove:

Theorem 1 (Kolmogorov). *Let $\varphi(x)$ denote a continuous bounded function such that the function $u(t, x) = \int \varphi(y)\mathsf{P}(t, x, s, dy)$ has bounded continuous first and second derivatives with respect to x and let the functions $a(t, x)$ and $b(t, x)$ be continuous. Then $u(t, x)$ has a derivative $\partial u/\partial t$ and in the region $t \in (0, s)$, $x \in R^{(1)}$, which satisfies the equation*

$$-\frac{\partial u}{\partial t} = a(t, x)\frac{\partial u}{\partial x} + \frac{1}{2}b(t, x)\frac{\partial^2 u}{\partial x^2} \tag{4}$$

and $u(t, x)$ satisfies the boundary condition $\lim_{t \uparrow s} u(t, x) = \varphi(x)$.

Equation (4) is called the first (converse) equation of Kolmogorov.

Proof. Validity of the boundary condition follows from the equation

$$\begin{aligned} u(t, x) = \int \varphi(y)\mathsf{P}(t, s, x, dy) &= \varphi(x) \\ + \int [\varphi(y) - \varphi(x)]\mathsf{P}(t, s, x, dy) &= \varphi(x) \\ + \int_{|y-x| \leq \varepsilon} [\varphi(y) - \varphi(x)]\mathsf{P}(t, s, x, dy) &+ o(s - t)\ , \end{aligned}$$

the continuity of the function φ, and the arbitrariness of $\varepsilon > 0$.

To derive equation (4), we note that it follows from condition 3, Section 1, Chapter VII, that

$$u(t_1, x) = \int \mathsf{P}(t_1, x, t_2, dz)u(t_2, z) \qquad (t_1 < t_2 < s)\ .$$

Using the Taylor expansion of the function $u(t, z)$ to get

$$\begin{aligned} u(t, z) - u(t, x) &= \frac{\partial u(t, x)}{\partial x}(z - x) \\ &+ \frac{1}{2}\frac{\partial^2 u(t, x)}{\partial x^2}(z - x)^2(1 + \alpha_\varepsilon)\ , \qquad |z - x| \leq \varepsilon \end{aligned}$$

where $|\alpha_\varepsilon| \leqq \sup_{t,|z-x|\leqq\varepsilon} |(\partial^2 u/\partial x^2)(t, x) - (\partial^2 u/\partial x^2)(t, z)|$, we may write

$$u(t_1, x) - u(t_2, x) = \int [u(t_2, z) - u(t_2, x)]\mathsf{P}(t_1, x, t_2, dz)$$

$$= \int_{|z-x|\leqq\varepsilon} [u(t_2, z) - u(t_2, x)]\mathsf{P}(t_1, x, t_2, dz) + o(t_2 - t_1)$$

$$= \frac{\partial u}{\partial x}(t_2, x)\int_{|x-z|\leqq\varepsilon} (z - x)\mathsf{P}(t_1, x, t_2, dz)$$

$$+ \frac{1}{2}\frac{\partial^2 u(t_2, x)}{\partial x^2}\int_{|x-z|\leqq\varepsilon} (z - x)^2\mathsf{P}(t_1, x, t_2, dz)(1 + \alpha_\varepsilon) + o(t_2 - t_1) .$$

By using formulas (2) and (3), we obtain

$$\frac{u(t_1, x) - u(t_2, x)}{t_2 - t_1}$$

$$= a(t_1, x)\frac{\partial u}{\partial x}(t, x) + \frac{b(t_1, x)}{2}\frac{\partial^2}{\partial x^2}u(t, x)(1 + \alpha_\varepsilon) + \frac{o(t_2 - t_1)}{t_2 - t_1} ,$$

from which we get equation (4) by taking the limit as $t_2 - t_1 \to 0$ and $\varepsilon \to 0$; This completes the proof of the theorem.

If the solution of equation (4) is unique with initial function $\varphi(x)$, by solving it we find the unique function $u(t, x)$, and the integrals

$$\int \varphi(y)\mathsf{P}(t, x, s, dy) ,$$

as φ ranges over some class of functions that is everywhere-dense in the space of continuous functions, uniquely determine the measure $\mathsf{P}(t, x, s, A)$.

In the case in which the transition probability $p(t, x, s, y)$ has a density, that is, when there exists the function for which

$$\mathsf{P}(t, x, s, A) = \int_A p(t, x, s, y)dy ,$$

the Kolmogorov-Chapman equation takes the form

$$p(t_1, x, t_3, y) = \int p(t_1, x, t_2, z)p(t_2, z, t_3, y)dz \qquad (t_1 < t_2 < t_3) . \quad (5)$$

If $p(t, x, s, y)$ is sufficiently smooth as a function of s and y, we can derive from (5) an equation for this density, known as the Fokker-Planck equation or the second (direct) equation of Kolmogorov.

Theorem 2. *Suppose that relations* (1) *to* (3) *are satisfied uniformly with respect to* x *for the process* $\xi(t)$ *and that there exist continuous partial derivatives* $\partial p(t, x, s, y)/\partial s$, $(\partial/\partial y)(a(s, y)p(t, x, s, y))$,

$$\frac{\partial^2}{\partial y^2}(b(s, y)p(t, x, s, y)) ,$$

Then, $p(t, x, s, y)$ satisfies the equation

$$\frac{\partial p(t, x, s, y)}{\partial s} = -\frac{\partial}{\partial y}(a(s, y)p(t, x, s, y)) + \frac{1}{2}\frac{\partial^2}{\partial y^2}(b(s, y)p(t, x, s, y)) \qquad (6)$$

in the region $s \in (t, T)$, $-\infty < y < \infty$.

Proof. Let $g(y)$ denote an arbitrary function that is twice continuously differentiable and that vanishes outside some finite interval.

Just as in the proof of Theorem 1, we can show that

$$\lim_{h \to 0} \frac{1}{h}\left[\int g(y)p(s, x, s + h, y)dy - g(x)\right] = a(s, x)g'(x) + \frac{1}{2}b(s, x)g''(x) \qquad (7)$$

uniformly with respect to x.

Using (5) with $t_1 = t$, $t_2 = s$, $t_3 = s + h$ and (7), we obtain

$$\frac{\partial}{\partial s}\int p(t, x, s, y)g(y)dy = \int \frac{\partial}{\partial s}p(t, x, s, y)g(y)dy$$
$$= \lim_{h \to 0} \frac{1}{h}\left[\int p(t, x, s + h, y)g(y)dy - \int p(t, x, s, z)g(z)dz\right]$$
$$= \lim_{h \to 0} \frac{1}{h}\int p(t, x, s, z)\left[\int p(s, z, s + h, y)g(y)dy - g(z)\right]dz$$
$$= \int p(t, x, s, z)\left[a(s, z)g'(z)dz + \frac{1}{2}b(s, z)g''(z)dz\right] .$$

Integrating the last integral by parts, we obtain

$$\int \frac{\partial}{\partial s}p(t, x, s, y)g(y)dy = \int\left[-\frac{\partial}{\partial z}(a(s, z)p(t, x, s, z)) + \frac{1}{2}\frac{\partial^2}{\partial z^2}(b(s, z)p(t, x, s, z))\right]g(z)dz .$$

If instead of $g(z)$ we take the functions $ne^{-n|z-u|}$ and take the limit as $n \longrightarrow \infty$, we obtain proof of the theorem (by using the continuity of the derivatives, referred to in the hypothesis).

REMARK 1. The reader who is familiar with the concept of a generalized function will easily see that equation (6) is always satisfied (under the assumption that the diffusion process has a density) if the right and left members of this equation are regarded as generalized functions on the space of infinitely differentiable functions having compact support.

REMARK 2. Let $F_0(x)$ denote the distribution function of $\xi(0)$. Writing equation (6) for $p(0, x, t, y)$ and integrating it with respect to x, we obtain the following equation for $p(t, y)$, the distribution density of $\xi(t)$:

$$\frac{\partial}{\partial t}p(t, y) = -\frac{\partial}{\partial y}(a(t, y)p(t, y)) + \frac{1}{2}\frac{\partial^2}{\partial y^2}(b(t, y)p(t, y)) . \qquad (8)$$

If instead of the density $p(t, y)$ we consider the distribution function $F(t, y)$ of the variable $\xi(t)$, we obtain the following expression for it from equation (8):

$$\frac{\partial}{\partial t}F(t, y) = -a(t, y)\frac{\partial F(t, y)}{\partial y} + \frac{1}{2}\frac{\partial}{\partial y}\left(b(t, y)\frac{\partial}{\partial y}F(t, y)\right) . \qquad (9)$$

To equation (9) we must add the initial condition $F(0, y) = F_0(y)$.

In the multidimensional case a process $\boldsymbol{\xi}(t)$ is called a diffusion process if relations (1) and (2) are satisfied for the process ($\boldsymbol{a}(t, \boldsymbol{x})$, $\boldsymbol{x}$, and $\boldsymbol{y}$ are now vectors in the space $R^{(m)}$ containing the range of $\xi(t)$), and if there exists a symmetric linear nonnegative operator $B(t, \boldsymbol{x})$ such that for all $\boldsymbol{x}, \boldsymbol{z} \in R^{(m)}$ and all $\varepsilon > 0$,

$$\int_{|x-y|\leqq\varepsilon}(\boldsymbol{z}, \boldsymbol{y} - \boldsymbol{x})^2\mathsf{P}(t, \boldsymbol{x}, s, d\boldsymbol{y}) = (B(t, \boldsymbol{x})\boldsymbol{z}, \boldsymbol{z})(s - t) + o(s - t) \qquad (10)$$

uniformly over $s > t$.

The operator $B(t, \boldsymbol{x})$ is called the *diffusion operator* of the process $\boldsymbol{\xi}(t)$. Let $\boldsymbol{e}_1, \boldsymbol{e}_2, \cdots, \boldsymbol{e}_m$ denote an orthonormal basis in $R^{(m)}$, let $x^1, \cdots, x^m; y^1, \cdots, y^m$ denote the coordinates of the vectors $\boldsymbol{x}$ and $\boldsymbol{y}$ in $R^{(m)}$, and let $b_{ij}(t, \boldsymbol{x})$ denote the elements of the matrix of the operator $B(t, \boldsymbol{x})$ in that basis. Then condition (10) is equivalent to the condition

$$\int_{|x-y|\leqq\varepsilon}(y^i - x^i)(y^j - x^j)\mathsf{P}(t, \boldsymbol{x}, s, d\boldsymbol{y}) = b_{ij}(t, \boldsymbol{x})(s - t) + o(s - t) , \qquad i, j = 1, \cdots, m . \qquad (11)$$

For multidimensional diffusion processes, Theorems 1 and 2 can be reformulated as follows:

Theorem 3. *Let $f(\boldsymbol{x})$ denote a continuous function such that the function $u(t, \boldsymbol{x}) = \int f(\boldsymbol{y})\mathsf{P}(t, \boldsymbol{x}, s, d\boldsymbol{y})$ has continuous partial derivatives*

$$\frac{\partial}{\partial x^i} u(t, \boldsymbol{x}) , \qquad \frac{\partial^2}{\partial x^i \partial x^j} u(t, \boldsymbol{x})$$

for $i, (j = 1, \cdots, m)$, *and let the coefficients* $a_i(t, \boldsymbol{x})$ *and* $b_{ij}(t, \boldsymbol{x})$, *for* $i, j = 1, \cdots, m$, *be continuous. Then for* $\boldsymbol{x} \in R^{(m)}$ *and* $t \in (0, s)$, *the function* $u(t, \boldsymbol{x})$ *has a derivative* $\partial u / \partial t$ *which satisfies the equation*

$$-\frac{\partial u(t, x)}{\partial t} = \sum_{i=1}^{m} a_i(t, \boldsymbol{x}) \frac{\partial u(t, x)}{\partial x^i} + \frac{1}{2} \sum_{i,j=1}^{m} b_{ij}(t, \boldsymbol{x}) \frac{\partial^2 u(t, x)}{\partial x^i \partial x^j} \tag{12}$$

and $u(t, x)$ *satisfies the boundary condition* $\lim_{t \to s} u(t, \boldsymbol{x}) = f(\boldsymbol{x})$.

If the transition probability $p(t, \boldsymbol{x}, s, \boldsymbol{y})$ has a density

$$\mathsf{P}(t, \boldsymbol{x}, s, A) = \int_A p(t, \boldsymbol{x}, s, \boldsymbol{y}) d\boldsymbol{y} ,$$

we have:

Theorem 4. *If relations* (1), (2), *and* (10) *are satisfied uniformly with respect to* $\boldsymbol{x}$ *and if there exist continuous derivatives*

$$\frac{\partial}{\partial s} p(t, \boldsymbol{x}, s, \boldsymbol{y}) , \qquad \frac{\partial}{\partial x^i} (a_i(s, \boldsymbol{y}) p(t, \boldsymbol{x}, s, \boldsymbol{y})) ,$$

$$\frac{\partial^2}{\partial x^i \partial x^j} (b_{ij}(s, \boldsymbol{y}) p(t, \boldsymbol{x}, s, \boldsymbol{y})) ,$$

then $p(t, \boldsymbol{x}, s, \boldsymbol{y})$ *for* $\boldsymbol{y} \in R^{(m)}$ *and* $s \in (t, T)$ *satisfies the equation*

$$\frac{\partial p(t, \boldsymbol{x}, s, \boldsymbol{y})}{\partial s} = -\sum_{i=1}^{m} \frac{\partial}{\partial x^i} (a_i(s, \boldsymbol{y}) p(t, \boldsymbol{x}, s, \boldsymbol{y}))$$

$$+ \frac{1}{2} \sum_{i,j=1}^{m} \frac{\partial^2}{\partial x^i \partial x^j} (b_{ij}(s, \boldsymbol{y}) p(t, \boldsymbol{x}, s, \boldsymbol{y})) . \tag{13}$$

The proofs of Theorems 3 and 4 are analogous to the proofs of Theorems 1 and 2. If we integrate equation (13) with respect to $x^1, \cdots, x^m$, we obtain (as in Remark 2) an equation for the unconditional density of the distribution of the variable $\boldsymbol{\xi}(t)$.

In what follows, when verifying whether or not conditions (1) to (3) are satisfied, we shall find it convenient to use:

REMARK 3. For the process $\boldsymbol{\xi}(t)$ to be a diffusion process, it is sufficient that for some $\delta > 0$,

$$\int |\boldsymbol{y} - \boldsymbol{x}|^{2+\delta} \mathsf{P}(t, \boldsymbol{x}, s, d\boldsymbol{y}) = o(s - t) , \tag{14}$$

and that relations (2), (3), and (10) are satisfied for $\varepsilon = +\infty$. This is true because if equation (14) is satisfied, then for $k = 0, 1, 2$,

$$\int_{|y-x|>\varepsilon} |\boldsymbol{y} - \boldsymbol{x}|^k \mathsf{P}(t, \boldsymbol{x}, s, d\boldsymbol{y})$$

$$\leq \frac{1}{\varepsilon^2 + \delta - k} \int |\boldsymbol{y} - \boldsymbol{x}|^{2+\delta} \mathsf{P}(t, \boldsymbol{x}, s, d\boldsymbol{y}) = o(s - t) .$$

2. ITO'S STOCHASTIC INTEGRAL

Let $w(t)$ denote a process of Brownian motion. Suppose that a set of σ-algebras $\mathfrak{F}_t$ defined for $t \geqq 0$ on the basic probability space $\{U, \mathfrak{S}, \mathsf{P}\}$ is related to $w(t)$ as follows: (1) $\mathfrak{F}_{t_1} \subset \mathfrak{F}_{t_2}$ for $t_1 < t_2$; (2) $w(t)$ is measurable with respect to $\mathfrak{F}_t$ for every t; (3) for every h, the process $w(t + h) - w(h)$ is independent of any of the events of the σ-algebra $\mathfrak{F}_h$. Let $\mathfrak{M}_2[a, b]$, where $0 \leqq a < b$, denote the set of functions $f(t) = f(t, u)$ that are measurable with respect to the set of variables (t, u), that are defined for $t \in [a, b]$ and $u \in U$, and that satisfy the two conditions: (a) $f(t)$ is measurable with respect to the σ-algebra $\mathfrak{F}_t$ for every t in $[a, b]$ and (b) the integral $\int_a^b | f(t) |^2 dt$ is finite with probability 1.

For all functions in $\mathfrak{M}_2[a, b]$, the integral $\int_a^b f(t)dw(t)$ will be defined below.

A function $f(t)$ is called a *step function* if there exists a partition of the interval $[a, b]$, $a = t_0 < t_1 < \cdots < t_r = b$, such that $f(t) = f(t_i)$ for $t \in [t_i, t_{i+1})$ for $i = 0, \cdots, r - 1$; In what follows we shall often use:

Lemma 1. *For every function $f(t) \in \mathfrak{M}_2[a, b]$ there exists a sequence of step functions $f_n(t) \in \mathfrak{M}_2[a, b]$ such that with probability* 1,

$$\lim_{n \to \infty} \int_a^b | f(t) - f_n(t) |^2 dt = 0 .$$

Proof. If $f(t)$ is a continuous function with probability 1, the assertion of the lemma is obvious (since we can set $f_n(t) = f(k/n)$ for $k/n \leqq t < (k + 1)/n$). If $f(t)$ is a measurable function bounded by a constant independent of u, then there exists a sequence $\{\varphi_n(t)\}$, where all the $\varphi_n(t)$ are bounded by a single constant, such that $\{\varphi_n(t)\}$ converges to $f(t)$ for almost all t with probability 1. This means that with probability 1,

$$\lim_{n \to \infty} \int_a^b | f(t) - \varphi_n(t) |^2 dt = 0 .$$

$\left(\text{For example, we may set } \varphi_n(t) = n\int_a^t (e^{n(s-t)} f(s)ds.\right)$

Finally, an arbitrary function in $\mathfrak{M}_2[a, b]$ can be approximated in the mean-square by a bounded function with any desired degree of accuracy. Thus the step functions are dense in the sense of the convergence that we are considering in $\mathfrak{M}_2[a, b]$. This completes the proof of the lemma.

Let us define the integral $\int_a^b f(t)dw(t)$ for the case in which the function $f(t)$ is a step function. Suppose that $f(t) = f(t_i)$ for $t \in [t_i, t_{i+1})$, where $a = t_0 < t_1 < \cdots < t_r = b$ is a partition of the interval $[a, b]$. We then define

$$\int_a^b f(t)dw(t) = \sum_{k=0}^{r-1} f(t_k)[w(t_{k+1}) - w(t_k)] .$$

We mention a few properties of this integral.

1. If $\mathsf{M}(| f(t) | \mid \mathfrak{F}_a) < \infty$ for $t \in [a, b]$, then with probability 1,

$$\mathsf{M}\left(\int_a^b f(t)dw(t) \mid \mathfrak{F}_a\right) = 0 . \tag{1}$$

This is true because

$$\begin{aligned}
&\mathsf{M}(f(t_k)[w(t_{k+1}) - w(t_k)] \mid \mathfrak{F}_a) \\
&\quad = \mathsf{M}[\mathsf{M}(f(t_k)[w(t_{k+1}) - w(t_k)] \mid \mathfrak{F}_{t_k}) \mid \mathfrak{F}_a] \\
&\quad = \mathsf{M}[f(t_k)\mathsf{M}[w(t_{k+1}) - w(t_k) \mid \mathfrak{F}_{t_k}] \mid \mathfrak{F}_a] = 0
\end{aligned}$$

by virtue of the fact that $w(t_{k+1}) - w(t_k)$ is independent of $\mathfrak{F}_{t_k}$ and has mean zero.

2. If $\mathsf{M}(| f(t) |^2 \mid \mathfrak{F}_a) < \infty$ with probability 1 for $t \in [a, b]$, then

$$\mathsf{M}\left(\left[\int_a^b f(t)dw(t)\right]^2 \middle| \mathfrak{F}_a\right) = \int_a^b \mathsf{M}(f^2(t) \mid \mathfrak{F}_a)dt \qquad (\text{mod } \mathsf{P}) . \tag{2}$$

This is true because

$$\begin{aligned}
\mathsf{M}\left(\left[\int_a^b f(t)dw(t)\right]^2 \middle| \mathfrak{F}_a\right) &= \sum_{k=0}^{r-1} \mathsf{M}[f(t_k)^2(w(t_{k+1}) - w(t_k))^2 \mid \mathfrak{F}_a] \\
&\quad + 2\sum_{j<k} \mathsf{M}[f(t_j)f(t_k)(w(t_{j+1}) - w(t_j))(w(t_{k+1}) - w(t_k)) \mid \mathfrak{F}_a] \\
&= \sum_{k=0}^{r-1} \mathsf{M}\{f(t_k)^2\mathsf{M}[w(t_{k+1}) - w(t_k))^2 \mid \mathfrak{F}_{t_k}] \mid \mathfrak{F}_a\} \\
&\quad + 2\sum_{j<k} \mathsf{M}\{f(t_j)f(t_k)[w(t_{j+1}) \\
&\quad - w(t_j)]\mathsf{M}(w(t_{k+1}) - w(t_k) \mid \mathfrak{F}_{t_k}) \mid \mathfrak{F}_a\} \\
&= \sum_{k=0}^{r-1} \mathsf{M}(f(t_k)^2 \mid \mathfrak{F}_a)[t_{k+1} - t_k] = \int_a^b \mathsf{M}(f(t)^2 \mid \mathfrak{F}_a)dt .
\end{aligned}$$

Let us show that for a sequence of step functions $f_n(t)$ such that $\int_a^b [f(t) - f_n(t)]^2 dt \longrightarrow 0$ in probability, the sequence $\int_a^b f_n(t)dw(t)$ converges in probability to some limit. To do this, we shall need the following property of integrals of step functions:

3. If $\varphi(t)$ is a step function, then for every $N > 0$ and $c > 0$,

$$\mathsf{P}\left\{\left|\int_a^b \varphi(t)dw(t)\right| > c\right\} \leqq \frac{N}{c^2} + \mathsf{P}\left\{\int_a^b | \varphi(t) |^2 dt > N\right\} . \tag{3}$$

To see that this is true, suppose that $\varphi(t) = \varphi(t_i)$ for $t_i \leqq t <$

t_{i+1}, where $a = t_0 < t_1 < \cdots < t_r = b$. Since $\varphi(t)$ is measurable with respect to $\mathfrak{F}_{t_i}$ for $t \in [t_i, t_{i+1})$, it follows that the quantity $\int_a^{t_{i+1}} |\varphi(t)|^2 dt$ is also measurable with respect to $\mathfrak{F}_{t_i}$. Let us define $\varphi_N(t) = \varphi(t)$ if $\int_a^{t_{i+1}} |\varphi(t)|^2 dt \leqq N$, where t_{i+1} is such that $t_i \leqq t < t_{i+1}$, and let us define $\varphi_N(t) = 0$ for $t \in [t_i, t_{i+1}]$ if $\int_a^{t_{i+1}} |\varphi(t)|^2 dt > N$. The function $\varphi_N(t)$ is a step function belonging to $\mathfrak{M}_2[a, b]$.

Since

$$\int_a^b |\varphi_N(t)|^2 dt = \sum |\varphi_N(t_i)|^2 (t_{i+1} - t_i) \leqq N ,$$

it follows that $|\varphi_N(t_i)|^2 \leqq N/(t_{i+1} - t_i)$ and this means that $\mathsf{M} |\varphi_N(t)|^2 < \infty$. Consequently

$$\mathsf{M} \left| \int_a^b \varphi_N(t) dw(t) \right|^2 = \int_a^b \mathsf{M} |\varphi_N(t)|^2 dt \leqq N .$$

Finally,

$$\mathsf{P}\left\{ \sup_t |\varphi_N(t) - \varphi(t)| > 0 \right\} = \mathsf{P}\left\{ \int_a^b |\varphi(t)|^2 dt > N \right\} .$$

Using the last two relations, we obtain

$$\begin{aligned} &\mathsf{P}\left\{ \left| \int_a^b \varphi(t) dw(t) \right| > c \right\} \\ &\quad = \mathsf{P}\left\{ \left| \int_a^b \varphi_N(t) dw(t) + \int_a^b (\varphi(t) - \varphi_N(t)) dw(t) \right| > c \right\} \\ &\quad \leqq \mathsf{P}\left\{ \left| \int_a^b \varphi_N(t) dw(t) \right| > c \right\} + \mathsf{P}\left\{ \left| \int_a^b [\varphi(t) - \varphi_N(t)] dw(t) \right| > 0 \right\} \\ &\quad \leqq \frac{\mathsf{M} \left| \int_a^b \varphi_N(t) dw(t) \right|^2}{c^2} + \mathsf{P}\left\{ \int_a^b |\varphi(t)|^2 dt > N \right\} , \end{aligned}$$

which completes the proof.

Now let $\{f_n\}$ denote a sequence of step functions such that $\int_a^b |f(t) - f_n(t)|^2 dt \to 0$ in probability. Then $\int_a^b |f_n(t) - f_m(t)|^2 dt$ also approaches 0 in probability as n and m approach ∞. Consequently, for every $\varepsilon > 0$,

$$\lim_{n, m \to \infty} \mathsf{P}\left\{ \int_a^b |f_n(t) - f_m(t)|^2 dt > \varepsilon \right\} = 0 .$$

Using property 3, we can write, for arbitrary $\varepsilon > 0$ and $\delta > 0$,

$$\begin{aligned} &\overline{\lim_{n, m \to \infty}} \mathsf{P}\left\{ \left| \int_a^b f_n(t) dw(t) - \int_a^b f_m(t) dw(t) \right| > \delta \right\} \\ &\quad \leqq \frac{\varepsilon}{\delta^2} + \overline{\lim_{n, m \to \infty}} \mathsf{P}\left\{ \int_a^b |f_n(t) - f_m(t)|^2 dt > \varepsilon \right\} = \frac{\varepsilon}{\delta^2} , \end{aligned}$$

so that because of the arbitrariness of $\varepsilon > 0$ we have

$$\lim_{n,m\to\infty} \mathsf{P}\left\{\left|\int_a^b f_n(t)dw(t) - \int_a^b f_m(t)dw(t)\right|^2 > \delta\right\} = 0$$

for every $\delta > 0$. It follows from this equation that the sequence of the random variables $\int_a^b f_n(t)dw(t)$ converges in probability to some limit. This limit is independent of the choice of the sequence $\{f_n(t)\}$ for which $\int_a^b |f_n(t) - f(t)|^2 dt \to 0$. (If there are two such sequences $\{f_n(t)\}$ and $\{\bar{f}_n(t)\}$, by combining them into a single sequence we see that with probability 1, the two sequences have the same limit.) Let us define

$$\int_a^b f(t)dw(t) = \mathsf{P}\text{-}\lim \int_a^b f_n(t)dw(t) .$$

We shall call this limit *Ito's stochastic integral* of the function $f(t)$.

The definite integral defined in this manner is a homogeneous additive functional of the function $f(t)$ on $\mathfrak{M}_2[a, b]$. Furthermore, for $a < c < b$,

$$\int_a^c f(t)dw(t) + \int_c^b f(t)dw(t) = \int_a^b f(t)dw(t) . \tag{4}$$

Proof of these properties is obvious for step functions and it carries over to the general case by means of a trivial limiting operation.

Applying the limit for step functions obtained in (3) to arbitrary functions in $\mathfrak{M}_2[a, b]$, we see that property 3 is valid for all $f(t) \in \mathfrak{M}_2[a, b]$.

With the aid of this property we can show that:

4. If $f(t) \in \mathfrak{M}_2[a, b], f_n(t) \in \mathfrak{M}_2[a, b]$, and

$$\int_a^b |f_n(t) - f(t)|^2 dt \to 0$$

in probability, then

$$\mathsf{P}\text{-}\lim \int_a^b f_n(t)dw(t) = \int_a^b f(t)dw(t) .$$

Finally let us prove a property that generalizes properties 1 and 2.

2′. If the function f is such that

$$\int_a^b \mathsf{M}(|f(t)|^2 \mid \mathfrak{F}_a)dt < \infty$$

with probability 1, then

$$\mathsf{M}\left(\int_a^b f(t)dw(t) \mid \mathfrak{F}_a\right) = 0 \quad (\text{mod } \mathsf{P}) , \tag{5}$$

and

$$\mathsf{M}\left(\left[\int_a^b f(t)dw(t)\right]^2 \middle| \mathfrak{F}_a\right) = \int_a^b \mathsf{M}(|f(t)|^2 \mid \mathfrak{F}_a)dt \quad (\text{mod } \mathsf{P}) . \qquad (6)$$

To prove these properties, let us show that in the case in which $\int_a^b \mathsf{M}(|f(t)|^2 \mid \mathfrak{F}_a)dt < \infty$, there exists a sequence of step functions such that $\int_a^b \mathsf{M}(|f(t) - f_n(t)|^2 \mid \mathfrak{F}_a) \longrightarrow 0$ in probability. Let $\{\bar{f}_n(t)\}$ denote the sequence of step functions in $\mathfrak{M}_2[a, b]$ such that

$$\mathsf{P}\text{-lim} \int_a^b |f(t) - \bar{f}_n(t)|^2 dt = 0 .$$

Let us set $g_N(x) = x$ for $|x| \leqq N$ and $g_N(x) = Nx/|x|$ for $|x| > N$. Since $|g_N(x) - g_N(y)| \leqq |x - y|$, we have

$$\int_a^b |g_N(f(t)) - g_N(\bar{f}_n(t))|^2 dt \leqq \int_a^b |f(t) - \bar{f}_n(t)|^2 dt \longrightarrow 0$$

in probability. Since the quantities

$$\int_a^b |g_N(f(t)) - g_N(\bar{f}_n(t))|^2 dt$$

are bounded by the number $4N^2(b - a)$, and since their sequence converges to 0 in probability, it follows on the basis of Lebesgue's theorem that

$$\mathsf{M}\left(\int_a^b |g_N(f(t)) - g_N(\bar{f}_n(t))|^2 dt \mid \mathfrak{F}_a\right)$$
$$= \int_a^b \mathsf{M}(|g_N(f(t)) - g_N(\bar{f}_n(t))|^2 \mid \mathfrak{F}_a)dt \longrightarrow 0$$

in probability.

On the other hand,

$$\int_a^b \mathsf{M}(|g_N(f(t)) - f(t)|^2 \mid \mathfrak{F}_a)dt \longrightarrow 0$$

in probability (again by Lebesgue's theorem), since

$$|f(t) - g_N(f(t))|^2 \leqq |f(t)|^2 \quad \text{and} \quad \int_a^b \mathsf{M}(|f(t)|^2 \mid \mathfrak{F}_a)dt < \infty .$$

Therefore we can choose a sequence $\{N_n\}$ such that

$$\int_a^b \mathsf{M}(|f(t) - g_{N_n}(\bar{f}_n(t))|^2 \mid \mathfrak{F}_a)dt$$
$$\leqq 2\int_a^b \mathsf{M}(|f(t) - g_{N_n}(f(t))|^2 \mid \mathfrak{F}_a)$$
$$+ 2\int_a^b \mathsf{M}(|g_{N_n}(f(t)) - g_{N_n}(\bar{f}_n(t))|^2 \mid \mathfrak{F}_a)dt \longrightarrow 0$$

in probability.

The functions $g_{N_n}(\bar{f}_n(t))$ are step functions and can be chosen for the functions $f_n(t)$, whose existence we are proving.

Substituting into (1) the functions $f_n(t)$ that we have constructed and taking the limit as $n \longrightarrow \infty$, we obtain (5). To obtain (6), we note that if the sequence $\xi_n \longrightarrow \xi$ in probability and if $\mathsf{M} \mid \xi_n - \xi_m \mid^2 \longrightarrow 0$ as $n, m \longrightarrow \infty$, then $\mathsf{M} \mid \xi - \xi_n \mid^2 \longrightarrow 0$ and hence $\mathsf{M}\xi_n^2 \longrightarrow \mathsf{M}\xi^2$. Therefore

$$\mathsf{M}\left(\left[\int_a^b f(t)dw(t)\right]^2 \middle| \mathfrak{F}_a\right) = \lim_{n\to\infty} \mathsf{M}\left(\left[\int_a^b f_n(t)dw(t)\right]^2 \middle| \mathfrak{F}_a\right)$$
$$= \lim_{n\to\infty} \int_a^b \mathsf{M}(\mid f_n(t) \mid^2 \mid \mathfrak{F}_a)dt = \int_a^b \mathsf{M}(\mid f(t) \mid^2 \mid \mathfrak{F}_a)dt .$$

Let us now consider the stochastic integral as a function of the upper limit. Let $\psi_t(s)$ denote the function that is equal to 1 for $s < t$ and equal to 0 for $s > t$. If $f(s) \in \mathfrak{M}_2[a, b]$, then $f(s)\psi_t(s) \in \mathfrak{M}_2[a, b]$ for every $t \in [a, b]$. We define the integral $\int_a^t f(s)dw(s)$ for all t by

$$\int_a^t f(s)dw(s) = \int_a^b f(s)\psi_t(s)dw(s) .$$

It follows from the definition of a stochastic integral that this integral is defined probabilistically. Therefore, as a function of the upper limit, the integral is defined up to stochastic equivalence (cf. Section 1, Chapter IV). In what follows we shall always assume that the values of the integral as a function of the upper limit, for different values of t, are compatible in such a way that $\zeta(t) = \int_a^t f(s)dw(s)$ is a separable process. The possibility of doing this follows from Theorem 2, Section 2, Chapter IV.

Let us note the basic properties of the function

$$\zeta(t) = \int_a^t f(s)dw(s) .$$

5. If $\int_a^b \mathsf{M}(\mid f(s) \mid^2 \mid \mathfrak{F}_a)ds < \infty$, then

$$\mathsf{P}\left\{\sup_{a \leq t \leq b} \left|\int_a^t f(s)dw(s)\right| > c \middle| \mathfrak{F}_a\right\} \leq \frac{1}{c^2}\int_a^b \mathsf{M}(\mid f(s) \mid^2 \mid \mathfrak{F}_a)ds \tag{7}$$

and

$$\mathsf{P}\left\{\sup_{a \leq t \leq b} \left|\int_a^t f(s)dw(s)\right| > c\right\} \leq \frac{1}{c^2}\int_a^b \mathsf{M}(\mid f(s) \mid)^2 ds. \tag{8}$$

It will be sufficient to prove inequality (7). Let us choose a partition of the interval $[a, b]$: $a = t_0 < t_1 < \cdots < t_n = b$. We define $\zeta_k = \int_a^{t_k} f(s)dw(s)$, and we define $\chi_k = 1$ if $\mid \zeta_i \mid \leq c$ for $i < k$

and $|\zeta_k| \geqq c$, and $\chi_k = 0$ otherwise. Obviously $\sum_{k=1}^{n} \chi_k \leqq 1$ and χ_k is measurable with respect to $\mathfrak{F}_{t_k}$, and since $|\zeta_k|^2\chi_k \geqq c^2\chi_k$, we have

$$\zeta_n^2 \geqq \zeta_n^2 \sum_{k=0}^{n} \chi_k = \sum_{k=0}^{n} \zeta_k^2\chi_k + 2\sum_{k=0}^{n} \zeta_k(\zeta_n - \zeta_k)\chi_k$$
$$+ \sum_{k=0}^{n} (\zeta_n - \zeta_k)^2\chi_k \geqq c^2 \sum_{k=0}^{n} \chi_k + 2\sum_{k=0}^{n} \zeta_k(\zeta_n - \zeta_k)\chi_k .$$

While taking the conditional mathematical expectation of both sides with respect to $\mathfrak{F}_a$, we note that on the basis of (5),

$$\mathsf{M}(\zeta_n - \zeta_k \mid \mathfrak{F}_{t_k}) = \mathsf{M}\left(\int_{t_k}^{b} f(s)dw(s) \middle| \mathfrak{F}_{t_k}\right) = 0 .$$

Therefore

$$\mathsf{M}(\zeta_k(\zeta_n - \zeta_k)\chi_k \mid \mathfrak{F}_a) = \mathsf{M}(\zeta_k\chi_k\mathsf{M}(\zeta_n - \zeta_k \mid \mathfrak{F}_{t_k}) \mid \mathfrak{F}_a) = 0$$

and then using equation (6), we obtain

$$\mathsf{M}(\zeta_n^2 \mid \mathfrak{F}_a) \geqq c^2\mathsf{M}\left(\sum_{k=0}^{n} \chi_k \middle| \mathfrak{F}_a\right) .$$

We note that $\sum_{k=0}^{n} \chi_k$ is equal to 1 if $\sup |\zeta_k| > c$ and equal to 0 otherwise. Therefore

$$\mathsf{M}\left(\sum_{k=0}^{n} \chi_k \middle| \mathfrak{F}_a\right) = \mathsf{P}\left\{\sup_{1 \leqq k \leqq n} |\zeta_k| > c \middle| \mathfrak{F}_a\right\} .$$

Thus we have proved the inequality

$$\mathsf{P}\left\{\sup_{0 \leqq k \leqq n} \left[\int_0^{t_k} f(s)dw(s)\right] > c \middle| \mathfrak{F}_a\right\} \leqq \frac{1}{c^2}\int_a^b \mathsf{M}\left(|f(s)|^2 \middle| \mathfrak{F}_a\right)ds ,$$

from which we easily obtain property 5 by using the separability of the process $\int_a^t f(s)dw(s)$.

6. A separable process $\zeta(t) = \int_a^t f(s)dw(s)$ is continuous.

If $f(t)$ is a step function, the continuity of $\zeta(t)$ follows from the continuity of $w(t)$ and the formula defining $\zeta(t)$. Suppose that the function $f(t) \in \mathfrak{M}_2[a, b]$ is such that $\int_a^b \mathsf{M}|f(s)|^2ds < \infty$. Let $\{f_n(t)\}$ denote a sequence of step functions such that

$$\lim_{n\to\infty} \int_a^b \mathsf{M}|f(s) - f_n(s)|^2ds = 0 .$$

By virtue of property 5,

$$\mathsf{P}\left\{\sup_t \left|\int_a^t f(s)dw(s) - \int_a^t f_n(s)dw(s)\right| > \varepsilon\right\}$$
$$\leqq \frac{1}{\varepsilon^2}\int_a^b \mathsf{M}|f(s) - f_n(s)|^2ds .$$

Choosing a sequence $\{\varepsilon_k\}$ $(\to 0)$ and $\{n_k\}$ such that

$$\sum_{k=1}^{\infty} \frac{1}{\varepsilon_k^2} \int_a^b \mathsf{M} \, | f(t) - f_{n_k}(t) |^2 dt < \infty ,$$

we see that

$$\sum_{k=1}^{\infty} \mathsf{P}\left\{ \sup_{a \leq t \leq b} \left| \int_a^t f(s)dw(s) - \int_a^t f_{n_k}(s)dw(s) \right| > \varepsilon_k \right\} < \infty ,$$

and on the basis of the Borel-Cantelli lemma (Theorem 2, Section 3, Chapter III) we have with probability 1,

$$\sup_{a \leq t \leq b} \left| \int_a^t f(s)dw(s) - \int_a^t f_{n_k}(s)dw(s) \right| \leq \varepsilon_k$$

beginning with some number k.

Thus $\int_a^t f(s)dw(s)$ is with probability 1 the uniform limit of a sequence of continuous functions. Therefore, with probability 1 this limit function is also continuous. Suppose finally that $f(t)$ is an arbitrary function in $\mathfrak{M}_2[a, b]$. Let us define $f_N(s) = f(s)$ if $\int_a^s | f(u) |^2 du \leq N$, and $f_N(s) = 0$ if $\int_a^s | f(u) |^2 du > N$. Then

$$\mathsf{P}\left\{ \sup_{a \leq t \leq b} \left| \int_a^t f(s)dw(s) - \int_a^t f_N(s)dw(s) \right| > 0 \right\} \leq \mathsf{P}\left\{ \int_a^b | f(s) |^2 ds > N \right\} .$$

Since $\int_a^t f_N(s)dw(s)$ is continuous and since the probability in the right-hand member of the last inequality can be made arbitrarily small, the process $\int_a^t f(s)dw(s)$ is continuous in the general case.

We shall need the following inequality regarding the fourth moment of a stochastic integral.

7. If $f(t)$ is a member of $\mathfrak{M}_2[a, b]$ such that $\int_a^b \mathsf{M} | f(t) |^4 dt) < \infty$, then

$$\mathsf{M} \left| \int_a^b f(t)dw(t) \right|^4 \leq 36(b - a) \int_a^b \mathsf{M} \, | f(t) |^4 dt . \tag{9}$$

Suppose first that $f(t)$ is a step function for which $f(t) = f(t_i)$ for $t_i \leq t < t_{i+1}$, where $a = t_0 < t_1 < \cdots < t_r = b$ is a partition of the interval $[a, b]$. Then

$$\begin{aligned}
\mathsf{M} \left| \int_a^b f(t)dw(t) \right|^4 &= \mathsf{M} \left| \sum_{k=0}^{r-1} f(t_i)[w(t_{i+1}) - w(t_i)] \right|^4 \\
&= \mathsf{M} \sum_{k=0}^{r-1} | f(t_k) |^4 [w(t_{k+1}) - w(t_k)]^4 \\
&\quad + 6 \sum_{k=0}^{r-1} \mathsf{M} \left| \sum_{i=0}^{k-1} f(t_i)[w(t_{i+1}) - w(t_i)] \right|^2 | f(t_k) |^2 [w(t_{k+1}) - w(t_k)]^2 \\
&= 3 \sum_{k=0}^{r-1} \mathsf{M} \, | f(t_k) |^4 (t_{k+1} - t_k)^2 \\
&\quad + 6 \sum_{k=1}^{r-1} \mathsf{M} \left| \sum_{i=0}^{k-1} f(t_i)[w(t_{i+1}) - w(t_i)] \right|^2 | f(t_k) |^2 (t_{k+1} - t_k) ,
\end{aligned}$$

since the mathematical expectations of those terms for which the increment $w(t_{k+1}) - w(t_k)$ with highest index is raised to an odd power are equal to 0, and for $m = 1, 2, \cdots$, we have*

$$\mathsf{M}([w(t_{k+1}) - w(t_k)]^{2m} \mid \mathfrak{F}_{t_k}) = (2m - 1)!!(t_{k+1} - t_k)^m .$$

For an arbitrary step function $f(t)$ we may assume that the intervals $[t_k, t_{k+1}]$ are chosen in such a way that $\max_k [t_{k+1} - t_k]$ is arbitrarily small. Therefore, by taking the limit as $\max_k [t_{k+1} - t_k] \to 0$ in the last relation, we obtain

$$\mathsf{M}\left|\int_a^b f(t)dw(t)\right|^4 = 6\int_a^b \mathsf{M}\left|\int_a^t f(s)dw(s)\right|^2 |f(t)|^2 dt . \tag{10}$$

By applying the Cauchy-Schwarz inequality, we may write

$$\int_a^b \mathsf{M}\left|\int_a^t f(s)dw(s)\right|^2 |f(t)|^2 dt$$

$$\leqq \sqrt{\int_a^b \mathsf{M}\,|f(t)|^4 dt}\ \sqrt{\int_a^b \mathsf{M}\left|\int_a^t f(s)dw(s)\right|^4 dt} .$$

It follows from formula (10) that

$$\mathsf{M}\left|\int_a^t f(v)dw(v)\right|^4 = 6\int_a^t \mathsf{M}\left|\int_a^v f(s)dw(s)\right|^2 |f(v)|^2 dv$$

increases with increasing t. Hence,

$$\int_a^b \mathsf{M}\left|\int_a^t f(s)dw(s)\right|^4 dt \leqq (b - a)\mathsf{M}\left|\int_a^b f(s)dw(s)\right|^4 .$$

Thus,

$$\mathsf{M}\left|\int_a^b f(t)dw(t)\right|^4 \leqq 6\sqrt{(b - a)\left[\int_a^b \mathsf{M}\,|f(s)|^4 ds\right]\mathsf{M}\left|\int_a^b f(s)dw(s)\right|^4} .$$

Formula (9) then follows for step functions $f(t)$. Proof of this formula in the general case can be obtained by constructing for $f(t)$ a sequence of step functions $f_n(t)$ so that

$$\lim_{n\to\infty} \int_a^b \mathsf{M}\,|f(t) - f_n(t)|^4 dt = 0 .$$

(Such a sequence can be constructed by means of a slight modification of the device used in proving property 2.)

We now introduce the concept of a stochastic differential. Let $\zeta(t)$ denote a process that is measurable for every t with respect to $\mathfrak{F}_t$. Supposed that there exist $b(t) \in \mathfrak{M}_2[a, b]$ and $a(t)$, which is measurable for every t with respect to $\mathfrak{F}_t$ and has a finite integral $\int_a^b |a(t)|\, dt$ with probability 1, such that, for all $a \leqq t_1 \leqq t_2 \leqq b$,

Translator's note: The symbol $n!!$, common in Russian mathematical literature, indicates the product of all natural numbers $\leqq n$ that are congruent to n mod 2; thus $8!! = 8\cdot 6\cdot 4\cdot 2$ and $7!! = 7\cdot 5\cdot 3\cdot 1$.

$$\zeta(t_2) - \zeta(t_1) = \int_{t_1}^{t_2} a(t)dt + \int_{t_1}^{t_2} b(t)dw(t) .$$

Then, we say that $a(t)dt + b(t)dw(t)$ is the *stochastic differential* of the process $\zeta(t)$ and we write

$$d\zeta(t) = a(t)dt + b(t)dw(t) .$$

We shall now establish an important property of a stochastic differential, namely, Ito's formula for differentiating a composite function.

8. Suppose that the process $\zeta(t)$ has a stochastic differential $a(t)dt + b(t)dw(t)$ and the function $u(t, x)$ (nonrandom) is defined for $t \in [a, b]$ and $x \in R^{(1)}$; suppose that this function is continuous, and that it has continuous derivatives $u'_t(t, x)$, $u'_x(t, x)$, and $u''_{xx}(t, x)$. Then the process $\eta(t) = u(t, \zeta(t))$ also has a stochastic differential and

$$d\eta(t) = \Big[u'_t(t, \zeta(t)) + u'_x(t, \zeta(t))a(t) + \frac{1}{2}u''_{xx}(t, \zeta(t))b^2(t)\Big]dt + u'_x(t, \zeta(t))b(t)dw(t) . \tag{11}$$

Let us first prove formula (11) for the case in which $a(t)$ and $b(t)$ are independent of t. Suppose that $t_1 = t^{(0)} < \cdots < t^{(n)} = t_2$. Then

$$\eta(t_2) - \eta(t_1) = u(t_2, \zeta(t_2)) - u(t_1, \zeta(t_1)) = \sum_{k=0}^{n-1} [u(t^{(k+1)}, \zeta(t^{(k+1)})) - u(t^{(k)}, \zeta(t^{(k)}))] .$$

Using Taylor's formula, we can write

$$u(t^{(k+1)}, \zeta(t^{(k+1)})) - u(t^{(k)}, \zeta(t^{(k)})) = u'_t(t^{(k)} + \theta_k(t^{(k+1)} - t^{(k)}), \zeta(t^{(k+1)}))(t^{(k+1)} - t^{(k)}) + u'_x(t^{(k)}, \zeta(t^{(k)}))[\zeta(t^{(k+1)}) - \zeta(t^{(k)})] + \frac{1}{2}u''_{xx}(t^{(k)}, \zeta(t^{(k)}) + \theta'_k[\zeta(t^{(k+1)}) - \zeta(t^{(k)})])[\zeta(t^{(k+1)}) - \zeta(t^{(k)})]^2 ,$$

where $0 < \theta_k < 1$ and $0 < \theta'_k < 1$. Taking into account the continuity of $\zeta(t)$, u'_t, and u''_{xx} and the boundedness (mod P) of $\zeta(t)$, we see that there exist random variables α and β that approach 0 with probability 1 as $\max_k (t^{(k+1)} - t^{(k)}) \to 0$ and that satisfy the inequalities

$$| u'_t(t^{(k)} + \theta_k(t^{(k+1)} - t^{(k)}), \zeta(t^{(k)})) - u'_t(t^{(k)}, \zeta(t^{(k)})) | \leqq \alpha ,$$

and

$$u''_{xx}(t^{(k)}, \zeta(t^{(k)}) + \theta'_k[\zeta(t^{(k+1)}) - \zeta(t^{(k)})]) - u''_{xx}(t^{(k)}, \zeta(t^{(k)})) | \leqq \beta .$$

Since

$$\sum [\zeta(t^{(k+1)}) - \zeta(t^{(k)})]^2 \leqq 2a^2 \sum (t^{(k+1)} - t^{(k)})^2 + 2b^2 \sum [w(t^{k+1}) - w(t^{(k)})]^2$$

and

$$\mathsf{M} \sum [w(t^{(k+1)}) - w(t^{(k)})]^2 = t_2 - t_1 ,$$

we see that

$$| \sum u'_t(t^{(k)} + \theta_k(t^{(k+1)} - t^{(k)}), \zeta(t^{(k)}))(t^{(k+1)} - t^{(k)}) - \sum u'_t(t^{(k)}, \zeta(t^{(k)}))\Delta t_k | = O(\alpha) ,$$

$$| \sum u''_{xx}(t^{(k)}, \zeta(t^{(k)}) + \theta'_k[\zeta(t^{(k+1)}) - \zeta(t^{(k)})])[\zeta(t^{(k+1)}) - \zeta(t^{(k)})]^2 - \sum u''_{xx}(t^{(k)}, \zeta(t^{(k)}))[\zeta(t^{(k+1)}) - \zeta(t^{(k)})]^2 |$$
$$\leqq \beta \sum_{k=1}^{n} [\zeta(t^{(k+1)}) - \zeta(t^{(k)})]^2 \longrightarrow 0$$

in probability as $\max (t^{(k+1)} - t^{(k)}) \longrightarrow 0$.

If we replace $\zeta(t^{(k+1)}) - \zeta(t^{(k)})$ with the expression $a(t^{(k+1)} - t^{(k)}) + b(w(t^{(k+1)}) - w(t^{(k)}))$ and remember that

$$a^2 \sum_{k=0}^{n-1} (t^{(k+1)} - t^{(k)})^2 \longrightarrow 0$$

and

$$b \sum_{k=0}^{n-1} (w(t^{(k+1)}) - w(t^{(k)}))(t^{(k+1)} - t^{(k)}) \longrightarrow 0$$

in probability as

$$\max_k (t^{(k+1)} - t^{(k)}) \longrightarrow 0$$

(the latter in view of the continuity of the process $w(t)$), we obtain

$$\begin{aligned} \eta(t_2) - \eta(t_1) = \underset{\max(t^{(k+1)} - t^{(k)}) \to 0}{\mathsf{P}\text{-lim}} \Bigg[& \sum_{k=0}^{n-1} u'_t(t^{(k)}, \zeta(t^{(k)}))(t^{(k+1)} - t^{(k)}) \\ & + \sum_{k=0}^{n-1} u'_x(t^{(k)}, \zeta(t^{(k)}))a(t^{(k+1)} - t^{(k)}) \\ & + \sum_{k=0}^{n-1} u'_x(t^{(k)}, \zeta(t^{(k)}))b[w(t^{(k+1)}) - w(t^{(k)})] \\ & + \frac{b^2}{2} \sum_{k=0}^{n-1} u''_{xx}(t^{(k)}, \zeta(t^{(k)}))(t^{(k+1)} - t^{(k)}) \\ & + \frac{b^2}{2} \sum_{k=0}^{n-1} u''_{xx}(t^{(k)}, \zeta(t^{(k)}))[(w(t^{(k+1)}) - w(t^{(k)}))^2 - (t^{(k+1)} - t^{(k)})] \Bigg] . \end{aligned}$$

Obviously, the limits of all summations except the last are equal to the corresponding integrals. To prove formula (11) in the present case, it suffices to show that

$$\sum_{k=0}^{n-1} u''_{xx}(t^{(k)}, \zeta(t^{(k)}))[[w(t^{(k+1)}) - w(t^{(k)})]^2 - (t^{(k+1)} - t^{(k)})]$$

converges to 0 in probability. We define

$$\varepsilon_k = [w(t^{(k+1)}) - w(t^{(k)})]^2 - (t^{(k+1)} - t^{(k)})$$

and we let $\chi_k^{(N)}$ denote the characteristic function of the event

$$\{|\xi(t^{(i)})| \leqq N \text{ for } i \leqq k\} .$$

Then

$$\mathsf{M}\Big(\sum_{k=0}^{n-1} u''_{xx}(t^{(k)}, \zeta(t^{(k)}))\chi_k^{(N)}\varepsilon_k\Big)^2$$

$$\leqq \sum_{k=0}^{n-1} \mathsf{M}u''_{xx}(t^{(k)}, \zeta(t^{(k)}))^2\chi_k^{(N)}\varepsilon_k^2 \leqq \sup_{t,|x|\leqq N} |u''_{xx}(t, x)|^2 \sum \mathsf{M}\varepsilon_k^2$$

$$= 3 \sup_{t,|x|\leqq N} |u''_{xx}(t, x)|^2 \sum (t^{(k+1)} - t^{(k)})^2 \longrightarrow 0 ,$$

and

$$\mathsf{P}\Big\{\sum_{k=0}^{n-1} u''_{xx}(t^{(k)}, \zeta(t^{(k)}))^2(1 - \chi_k^{(N)})\varepsilon_k^2 \neq 0\Big\} \leqq \mathsf{P}\Big\{\sup_{a\leqq t\leqq b} |\zeta(t)| > N\Big\} \longrightarrow 0$$

as $N \longrightarrow \infty$. This completes the proof of formula (11) for constant a and b. It is easy to show that this formula remains valid for step functions since the domain of definition of a step function $\zeta(t)$ can be partitioned into finitely many intervals, throughout each of which a and b are constant. In the general case, we may choose sequences $\{a_n(t)\}$ and $\{b_n(t)\}$ of step functions such that with probability 1,

$$\int_a^b | a(t) - a_n(t) | \, dt \longrightarrow 0 ,$$

$$\int_a^b | b_n(t) - b(t) |^2 dt \longrightarrow 0$$

and the sequence of the processes

$$\zeta_n(t) = \zeta(a) + \int_a^t a_n(s)ds + \int_a^t b_n(s)dw(s)$$

converges uniformly, with probability 1, to $\zeta(t)$. Then the sequence of processes $\eta_n(t) = u(t, \zeta_n(t))$ also converges uniformly with probability 1 to $\eta(t)$. Taking the limit as $n \longrightarrow \infty$ in the formula

$$\eta_n(t_2) - \eta_n(t_1) = \int_{t_1}^{t_2}\Big[u'_t(t, \zeta_n(t)) + u'_x(t, \zeta_n(t))a_n(t)$$

$$+ \frac{1}{2}u''_{xx}(t, \zeta_n(t))b_n^2(t)\Big]dt + \int_{t_1}^{t_2} u'_x(t, \zeta_n(t))b_n(t)dw(t) ,$$

we obtain proof of formula (11) in the general case.

Let us also consider the integrals

$$\int_a^b F(t)dw(t)$$

of the functions $F(t)$ with range in $R^{(m)}$. If $f^1, \cdots, f^m$ are the coordinates of the function F relative to some basis, then

$$\int_a^b F(t)dw(t)$$

is a random variable with range in $R^{(m)}$ with coordinates

$$\int_a^b f^i(t)dw(t)$$

for $i = 1, \cdots, m$.

Properties 1 to 8 remain valid for stochastic integrals of vector-valued functions if by $|F|$ we understand

$$\sqrt{|f^1|^2 + \cdots + |f^m|^2} .$$

(Formulas characterizing these properties were deliberately written in such a form that they remain meaningful for integrals of functions with ranges in $R^{(m)}$.)

Let us suppose that k mutually independent processes of Brownian motion $w_1(t), \cdots, w_k(t)$ are such that condition (2) regarding the connection with the σ-algebra $\mathfrak{F}_t$ is satisfied for each of them and, for each h, the processes $w_i(t+h) - w_i(h)$, for $i = 1, \cdots, k$, are all independent of $\mathfrak{F}_h$. Then we can define the integral

$$\int_\alpha^b \boldsymbol{f}(t)dw_i(t)$$

for every $i = 1, \cdots, k$ and every function $\boldsymbol{f}(t) \in \mathfrak{M}_2[a, b]$ with range in $R^{(m)}$. Analogously to what we did in the one-dimensional case, we can define the differential:

$$d\boldsymbol{\zeta}(t) = \boldsymbol{a}(t)dt + \sum_{i=1}^{k} \boldsymbol{f}_i(t)dw_i(t) .$$

Let $\boldsymbol{u}(t, \boldsymbol{x})$ denote a function defined for $t \in [a, b]$ and $\boldsymbol{x} \in R^{(m)}$, with range in $R^{(m)}$. Let $x^1, \cdots, x^m$ denote the coordinates of the point $\boldsymbol{x}$ relative to some basis, and let $f_j^1, \cdots, f_j^m$ denote the coordinates of the function f_j.

8′. If the function $\boldsymbol{u}(t, \boldsymbol{x})$ is continuous and has continuous partial derivatives

$$\frac{\partial}{\partial t}\boldsymbol{u}(t, \boldsymbol{x}) , \qquad \frac{\partial}{\partial x^i}\boldsymbol{u}(t, \boldsymbol{x}) , \qquad \frac{\partial^2}{\partial x^i \partial x^j}\boldsymbol{u}(t, x)$$

for $i, j = 1, \cdots, m$, and if the process $\boldsymbol{\zeta}(t)$ has a differential

$$d\boldsymbol{\zeta}(t) = \boldsymbol{a}(t)dt + \sum_{i=1}^{k} \boldsymbol{f}_i(t)dw_i(t) ,$$

then the process $\boldsymbol{\eta}(t) = \boldsymbol{u}(t, \boldsymbol{\zeta}(t))$ also has a differential and

$$d\eta(t) = \left[\frac{\partial u(t, \zeta(t))}{\partial t} + \sum_i \frac{\partial}{\partial x^i} u(t, \zeta(t)) a^i(t) \right.$$
$$\left. + \frac{1}{2} \sum_{i,j,l} \frac{\partial^2}{\partial x^i \partial x^j} u(t, \zeta(t)) f_l^i(t) f_l^j(t) \right] dt$$
$$+ \sum_{i=1}^{k} \left(\sum_{j=1}^{m} \frac{\partial u}{\partial x^j} f_i^j(t) \right) dw_i(t) . \qquad (12)$$

The derivation of formula (12) is analogous to that of formula (11).

3. EXISTENCE AND UNIQUENESS OF SOLUTIONS OF STOCHASTIC DIFFERENTIAL EQUATIONS

Consider the stochastic differential equation

$$d\xi(t) = a(t, \xi(t))dt + \sigma(t, \xi(t))dw(t) , \qquad (1)$$

whose solution, it is natural for us to expect, is a diffusion process with coefficient of diffusion $\sigma^2(t, x)$ and coefficient of transfer $a(t, x)$. Let us assume that $a(t, x)$ and $\sigma(t, x)$ are Borel functions defined for $x \in R^{(1)}$ and $t \in [t_0, T]$.

Equation (1) is equivalent to the equation

$$\xi(t) = \xi(t_0) + \int_{t_0}^{t} a(s, \xi(s))ds + \int_{t_0}^{t} \sigma(s, \xi(s))dw(s) \qquad (2)$$

and it is solved under the condition that $\xi(t_0)$ is given. For the integrals in (2) and hence the differentials in (1) to be meaningful, we need to introduce the σ-algebras of events $\mathfrak{F}_t$.

In what follows, the quantity $\xi(t_0)$ will always be assumed to be independent of the process $w(t) - w(t_0)$ and by the σ-algebra $\mathfrak{F}_t$ we shall understand the minimal σ-algebra with respect to which the variables $\xi(t_0)$ and $w(s) - w(t_0)$ for $t_0 < s \leqq t$ are measurable. We shall consider $\xi(t)$ to be a solution of equation (2) if $\xi(t)$ is $\mathfrak{F}_t$-measurable, if the integrals in (2) exist, and if (2) holds for every $t \in [t_0, T]$ with probability 1.

We note that property 3 of the preceding section implies that for stochastically equivalent processes $f_1(s)$ and $f_2(s)$, the stochastic integrals

$$\int_{t_0}^{t} f_1(s)dw(s) , \qquad \int_{t_0}^{t} f_2(s)dw(s)$$

coincide with probability 1, since $f_1(s) = f_2(s)$ with probability 1 for every s and hence

$$\mathsf{P}\left\{\int_{t_0}^{T} |f_1(s) - f_2(s)|^2 ds > 0\right\} = 0 .$$

From this it follows that every process that is stochastically equivalent to a solution of equation (2) is itself a solution of the same equation. Since the right-hand member of equation (2) is stochastically equivalent to the left-hand member and with probability 1 is continuous, it follows that for every solution of (2), there exists a continuous solution stochastically equivalent to it. In what follows, we shall consider only continuous solutions of equation (2).

Theorem 1. *Let $a(t, x)$ and $\sigma(t, x)$ for $t \in [t_0, T]$ and $x \in R^{(1)}$ denote two Borel functions satisfying the following conditions for some K:*

a. *For all x and $y \in R^{(1)}$,*

$$|a(t, x) - a(t, y)| + |\sigma(t, x) - \sigma(t, y)| \leqq K|x - y| ,$$

b. *For all x,*

$$|a(t, x)|^2 + |\sigma(t, x)|^2 \leqq K^2(1 + |x|^2) .$$

Then, equation (2) has a solution. If $\xi_1(t)$ and $\xi_2(t)$ are two continuous solutions (for fixed $\xi_0(t)$) of equation (2), then

$$\mathsf{P}\left\{\sup_{t_0 \leqq t \leqq T} |\xi_1(t) - \xi_2(t)| > 0\right\} = 1 .$$

Proof. Let us first prove that a continuous solution is unique. Let $\xi_1(t)$ and $\xi_2(t)$ denote two continuous solutions of equation (2). Let $\chi_N(t)$ denote the random variable that is equal to 1 if $|\xi_1(s)| \leqq N$ and $|\xi_2(s)| \leqq N$ for all $s \in [t_0, t]$, but equal to 0 otherwise.

Since $\chi_N(t)\chi_N(s) = \chi_N(t)$ for $s < t$, we have

$$\chi_N(t)[\xi_1(t) - \xi_2(t)] = \chi_N(t)\left[\int_{t_0}^{t} \chi_N(s)[a(s, \xi_1(s)) - a(s, \xi_2(s))]ds \right.$$
$$\left. + \int_{t_0}^{t} \chi_N(s)[\sigma(s, \xi_1(s)) - \sigma(s, \xi_2(s)]dw(s)\right] .$$

Since

$$\chi_N(s)[|a(s, \xi_1(s)) - a(s, \xi_2(s))| + |\sigma(s, \xi_1(s)) - \sigma(s, \xi_2(s))|]$$
$$\leqq K\chi_N(s)|\xi_1(s) - \xi_2(s)| \leqq 2KN ,$$

the squares of the integrals on the right-hand side of the last equation have mathematical expectations. Applying the inequality $(a + b)^2 \leqq 2(a^2 + b^2)$, Cauchy's inequality, and property 2 of the preceding section, we obtain the inequality

$$\begin{aligned}
&\mathsf{M}\chi_N(t)[\xi_1(t) - \xi_2(t)]^2 \\
&\quad \leqq 2\mathsf{M}\chi_N(t)\left(\int_{t_0}^t \chi_N(s)[a(s, \xi_1(s)) - a(s, \xi_2(s))]ds\right)^2 \\
&\qquad + 2\mathsf{M}\chi_N(t)\left(\int_{t_0}^t [\sigma(s, \xi_1(s)) - \sigma(s, \xi_2(s))]dw(s)\right)^2 \\
&\quad \leqq 2(T - t_0)\int_{t_0}^t \mathsf{M}\chi_N(s)[a(s, \xi_1(s)) - a(s, \xi_2(s))]^2 ds \\
&\qquad + 2\int_{t_0}^t \mathsf{M}\chi_N(s)[\sigma(s, \xi_1(s)) - \sigma(s, \xi_2(s))]^2 dw(s) .
\end{aligned}$$

Taking into consideration condition a, we see that there exists a constant L such that

$$\mathsf{M}\chi_N(t)[\xi_1(t) - \xi_2(t)]^2 \leqq L\int_{t_0}^t \mathsf{M}\chi_N(s) \mid \xi_1(s) - \xi_2(s) \mid^2 ds . \tag{3}$$

We now prove an auxiliary proposition that will often be useful when we are seeking inequalities that provide bounds:

Lemma 1. *Let $\alpha(t)$ denote a nonnegative integrable function that is defined for $t \in [t_0, T]$ and that satisfies the inequality*

$$\alpha(t) \leqq H\int_{t_0}^t \alpha(s)ds + \beta(t) , \tag{4}$$

where H is a nonnegative constant and $\beta(t)$ is an integrable function. Then

$$\alpha(t) \leqq \beta(t) + H\int_{t_0}^t e^{H(t-s)}\beta(s)ds .$$

Proof. It follows from (4) that

$$\begin{aligned}
\alpha(t) &\leqq \beta(t) + H\int_{t_0}^t \left[\beta(s) + H\int_{t_0}^s \alpha(s_1)ds_1\right]ds \\
&\leqq \beta(t) + H\int_{t_0}^t \left[\beta(s_1) + H\int_{t_0}^{s_1}\left[\beta(s_2) + H\int_{t_0}^{s_2}\alpha(s_3)ds_3\right]ds_2\right]ds_1 \\
&\leqq \beta(t) + H\int_{t_0}^t \beta(s_1)ds_1 + H^2\int_{t_0}^t\int_{t_0}^{s_1}\beta(s_2)ds_2 ds_1 + \\
&\quad \cdots + H^n\int_{t_0}^t\int_{t_0}^{s_1}\cdots\int_{t_0}^{s_{n-1}}\beta(s_n)ds_1 \cdots ds_n \\
&\quad + H^{n+1}\int_{t_0}^t \cdots \int_{t_0}^{s_n}\alpha(s_{n+1})ds_1 \cdots ds_{n+1} .
\end{aligned}$$

Since

$$\int_{t_0}^t\int_{t_0}^{s_1}\cdots\int_{t_0}^{s_k}\alpha(s_{k+1})ds_{k+1}ds_k \cdots ds_1 = \int_{t_0}^t \frac{(t - s_{k+1})^k}{k!}\alpha(s_{k+1})ds_{k+1} ,$$

we have

$$\lim_{n\to\infty} H^{n+1}\int_{t_0}^{t} \cdots \int_{t_0}^{s_n} \alpha(s_{n+1})ds_{n+1} \cdots ds_1 = 0$$

Hence

$$\alpha(t) \leqq \beta(t) + H\int_{t_0}^{t} \beta(s)ds + H^2\int_{t_0}^{t}\int_{t_0}^{s_1} \beta(s_2)ds_2 ds_1 + \cdots$$
$$= \beta(t) + H\int_{t_0}^{t} \sum_{k=0}^{\infty} H^k \frac{(t-s)^k}{k!}\beta(s)ds .$$

This completes the proof of the lemma.

Let us set $\alpha(t) = \mathsf{M}\chi_N(t)[\xi_1(t) - \xi_2(t)]^2$ and $\beta(t) = 0$. Then

$$\mathsf{M}\chi_N(t)[\xi_1(t) - \xi_2(t)]^2 = 0 ;$$

that is,

$$\mathsf{P}\{\xi_1(t) \neq \xi_2(t)\} \leqq \mathsf{P}\left\{\sup_t |\xi_1(t)| > N\right\} + \mathsf{P}\left\{\sup_t |\xi_2(t)| > N\right\} .$$

The probabilities on the right approach 0 because of the continuity (and hence boundedness) with probability 1 of the processes $\xi_1(t)$ and $\xi_2(t)$. This means that $\xi_1(t)$ and $\xi_2(t)$ are stochastically equivalent. Since both processes are continuous with probability 1, we have $\mathsf{P}\{\sup_t |\xi_1(t) - \xi_2(t)| > 0\} = 1$. This completes the proof of the uniqueness of the solution to equation (2).

Let us now show that equation (2) has a solution. First, let us suppose that $\mathsf{M}|\xi(t_0)|^2 < \infty$. Consider the Banach space B of measurable random functions $\zeta(t)$ that for every t are measurable with respect to the σ-algebra $\mathfrak{F}_t$ and satisfy the relation

$$\sup_{t_0 \leqq t \leqq T} \mathsf{M}|\zeta(t)|^2 < \infty$$

with norm

$$||\zeta(t)|| = \sqrt{\sup_{t_0 \leqq t \leqq T} \mathsf{M}|\zeta(t)|^2} .$$

Let us define an operator S in the space B by

$$S\zeta(t) = \xi(t_0) + \int_{t_0}^{t} a(s, \zeta(s))ds + \int_{t_0}^{t} \sigma(s, \zeta(s))dw(s) .$$

The existence of both integrals follows from the relation

$$|a(s, \zeta(s))|^2 + |\sigma(s, \zeta(s))|^2 \leqq K^2(1 + |\zeta(s)|^2) .$$

Obviously, $S\zeta(t)$ is measurable with respect to $\mathfrak{F}_t$.

Using the inequality

$$(a + b + c)^2 \leqq 3(a^2 + b^2 + c^2)$$

and hypothesis b of Theorem 1, we obtain

$$\mathsf{M}\,|\,S\zeta(t)\,|^2 \leqq 3\mathsf{M}\,|\,\zeta(t_0)\,|^2 + 3(T-t_0)\mathsf{M}\int_{t_0}^{t} K^2(1+|\,\zeta(s)\,|^2)ds + 3\int_{t_0}^{t} \mathsf{M}K^2(1+|\,\zeta(s)\,|^2)ds \leqq 3\mathsf{M}\,|\,\zeta(t_0)\,|^2 + 3K^2(T-t_0)^2 + (3K^2 + 6K^2\,\|\,\zeta\,\|^2)(T-t_0)\,.$$

Thus the operator S maps B into B. Furthermore,

$$\mathsf{M}\,|\,S\zeta_1(t) - S\zeta_2(t)\,|^2 \leqq 2[T-t_0]\int_{t_0}^{t} \mathsf{M}[a(s,\zeta_1(s)) - a(s,\zeta_2(s))]^2 ds + 2\mathsf{M}\left(\int_{t_0}^{t} (\sigma(s,\zeta_1(s)) - \sigma(s,\zeta_2(s))dw(s)\right)^2 \leqq L\int_{t_0}^{t} \mathsf{M}\,|\,\zeta_1(s) - \zeta_2(s)\,|^2 ds$$

$$\leqq L(t-t_0)\,\|\,\zeta_1 - \zeta_2\,\|^2, \text{ if } L = 2K^2(T-t_0+1)\,.$$

This inequality shows that the operator S is continuous on B. Furthermore,

$$\mathsf{M}\,|\,S^n\zeta_1(t) - S^n\zeta_2(t)\,|^2 \leqq L\int_{t_0}^{t} \mathsf{M}\,|\,S^{n-1}\zeta_1(u) - S^{n-1}\zeta_2(u)\,|^2 du$$

$$\leqq L^n\int_{t_0}^{t}\int_{t_0}^{t_1}\cdots\int_{t_0}^{t_{n-1}} \mathsf{M}\,\|\,\zeta_1 - \zeta_2\,\|^2\,du\,dt_n\cdots dt_1$$

$$\leqq \frac{L^n(t-t_0)^n}{n!}\,\|\,\zeta_1 - \zeta_2\,\|^2\,.$$

Therefore, for every $\zeta(t)$ in B,

$$\|\,S^{n+1}\zeta - S^n\zeta\,\| \leqq \frac{L^n(T-t_0)^n}{n!}\,\|\,S\zeta - \zeta\,\|^2\,.$$

The convergence of the series

$$\sum_{n-1}^{\infty} \|\,S^{n+1}\zeta - S^n\zeta\,\|$$

implies that the process $S^n\zeta(t)$ has a limit as $n \to \infty$. If we let $\xi(t)$ denote this limit, it follows from the continuity of S that $S[S^n\zeta(t)] \to S\xi(t)$. But

$$S[S^n\zeta(t)] = S^{n+1}\zeta(t) \to \xi(t)\,.$$

Thus $\|\,S\xi(t) - \xi(t)\,\| = 0$. It then follows from the definition of the norm that $\xi(t) = S\xi(t)$ with probability 1 for every $t \in [t_0, T]$; that is, $\xi(t)$ is a solution of equation (2).

Let us now prove that equation (2) has a solution in the general case. Let $\xi^N(t_0)$ denote the quantity that is equal to $\xi(t_0)$ when $|\,\xi(t_0)\,| \leqq N$ and equal to 0 when $|\,\xi(t_0)\,| > N$. We let $\xi^N(t)$ denote the solution of the equation

$$\xi^N(t) = \xi^N(t_0) + \int_{t_0}^{t} a(s, \xi^N(s))ds + \int_{t_0}^{t} \sigma(s, \xi^N(s))dw(s) . \qquad (5)$$

Since $|\xi^N(t_0)| \leqq N$ and $\mathsf{M}|\xi^N(t_0)|^2 < \infty$, equation (5) has a solution, and it follows from what was shown above that

$$\sup_t \mathsf{M}|\xi^N(t)|^2 < \infty .$$

Let us show that as $N \to \infty$ the sequence $\{\xi^N(t)\}$ converges in probability to some process $\xi(t)$ that is a solution of equation (2). Suppose that $N' > N$. Let η denote the random variable that is equal to 1 for $|\xi(t_0)| \leqq N$ and equal to 0 for $|\xi(t_0)| > N$. Then $[\xi^N(t_0) - \xi^{N'}(t_0)]\eta = 0$. The quantity η is measurable with respect to the σ-algebra $\mathfrak{F}_{t_0}$. Starting with the inequality

$$|\xi^N(t) - \xi^{N'}(t)|^2\eta \leqq 2\left[\int_{t_0}^{t} [a(s, \xi^N(s)) - a(s, \xi^{N'}(s))]\eta ds\right]^2$$

$$+ 2\left(\int_{t_0}^{t} [\sigma(s, \xi^N(s)) - \sigma(s, \xi^{N'}(s))]\eta dw(s)\right)^2$$

and using condition a and the inequalities involving the integrals (which we have already applied), we see that the constant L is such that

$$\mathsf{M}|\xi^N(t) - \xi^{N'}(t)|^2\eta \leqq L\int_{t_0}^{t} \mathsf{M}|\xi^N(s) - \xi^{N'}(s)|^2\eta ds .$$

Consequently,

$$\mathsf{M}|\xi^N(t) - \xi^{N'}(t)|^2\eta = 0 ,$$

so that

$$\mathsf{P}\{|\xi^N(t) - \xi^{N'}(t)| > 0\} < \mathsf{P}\{|\xi(t_0)| > N\} .$$

It follows from the last inequality that $\xi^N(t)$ converges in probability to some limit $\xi(t)$ as $N \to \infty$, and that

$$\int_{t_0}^{T} (\xi^N(t) - \xi(t))^2 dt$$

converges in probability to 0 as $N \to \infty$. Using condition a of Theorem 1 and property 3 of Section 2, we obtain the inequalities

$$\left|\int_{t_0}^{t} a(s, \xi^N(s))ds - \int_{t_0}^{t} a(s, \xi(s))ds\right|$$

$$\leqq K\sqrt{(t - t_0)\int_{t_0}^{t} (\xi^N(s) - \xi(s))^2 ds} ,$$

$$\mathsf{P}\left\{\left|\int_{t_0}^{t} \sigma(s, \xi^N(s))dw(s) - \int_{t_0}^{t} \sigma(s, \xi(s))dw(s)\right| < \varepsilon\right\}$$

$$\leqq \frac{\delta^2}{\varepsilon^2} + \mathsf{P}\left\{K^2\int_{t_0}^{t} (\xi^N(s) - \xi(s))^2 ds > \delta^2\right\} ,$$

from which it follows that we can take the limit under the integral signs in the right-hand member of (5). Thus $\xi(t)$ is a solution of equation (2). This completes the proof of the theorem.

Let us show that under the hypotheses of Theorem 1, the solution of equation (2) is a Markov process. To do this, let us prove:

Theorem 2. *Let $a(t, x)$ and $\sigma(t, x)$ denote two functions satisfying the hypotheses of Theorem 1 and let $\xi_{t,x}(s)$ denote a process defined for $s \in [t, T]$, where $t > t_0$, that is a solution of the equation*

$$\xi_{t,x}(s) = x + \int_t^s a(u, \xi_{t,x}(u))du + \int_t^s \sigma(u, \xi_{t,x}(u))dw(u) . \qquad (6)$$

Then the process $\xi(t)$, which is a solution of equation (2), is a Markov process whose transition probabilities are given by the relation

$$\mathsf{P}(t, x, s, A) = \mathsf{P}\{\xi(s) \in A \mid \xi(t) = x\} = \mathsf{P}\{\xi_{t,x}(s) \in A\} .$$

Proof. Since $\xi(t)$ is measurable with respect to $\mathfrak{F}_t$ and since $\xi_{t,x}(s)$ is completely determined by the process $w(s) - w(t)$ for $s \in [t, T]$ (independently of $\mathfrak{F}_t$), it follows that $\xi_{t,x}(s)$ is independent of $\xi(t)$ and the events in $\mathfrak{F}_t$. For $s \in [t, T]$, the function $\xi(s)$ is (by virtue of Theorem 1) the unique solution of the equation

$$\xi(s) = \xi(t) + \int_t^s a(u, \xi(u))du + \int_t^s \sigma(u, \xi(u))dw(u) .$$

The process $\xi_{t,\xi(t)}(s)$ is also a solution of this equation. Therefore with probability 1 $\xi(s) = \xi_{t,\xi(t)}(s)$. Let us now show that

$$\mathsf{P}\{\xi(s) \in A \mid \xi(t)\} = \mathsf{P}\{\xi(s) \in A \mid \mathfrak{F}_t\} .$$

To do this, it will be sufficient to show that for an arbitrary bounded variable ζ that is measurable with respect to $\mathfrak{F}_t$ and an arbitrary bounded continuous function $\lambda(x)$,

$$\mathsf{M}\zeta\lambda(\xi(s)) = \mathsf{M}\zeta\mathsf{M}(\lambda(\xi(s)) \mid \xi(t)) . \qquad (7)$$

Let us suppose that we have a function $\varphi(x, u)$ of the form

$$\varphi(x, u) = \sum_{k=1}^n \varphi_k(x)\psi_k(u) , \qquad (8)$$

where $\psi_k(u)$ are independent of $\mathfrak{F}_t$. We have

$$\mathsf{M}\zeta \sum_{k=1}^n \varphi_k(\xi(t))\psi_k(u) = \sum_{k=1}^n [\mathsf{M}\zeta\varphi_k(\xi(t))]\mathsf{M}\psi_k(u)$$

$$= \mathsf{M} \sum_{k=1}^n \zeta\varphi_k(\xi(t))\mathsf{M}\psi_k(u) .$$

$$\mathsf{M}\Big(\sum_{k=1}^n \varphi_k(\xi(t))\psi_k(u) \mid \xi(t)\Big) = \sum_{k=1}^n \varphi_k(\xi(t))\mathsf{M}\psi_k(u) .$$

Since functions of the form (8) are dense we obtain for any function $\varphi(x, u)$ depending only on the future increments $w(s) - w(t)$, $s \geqq t$

$$\mathsf{M}(\varphi(\xi(t), u) \mid \mathfrak{F}_t) = \mathsf{M}(\varphi(\xi(t), u) \mid \xi(t)) = \mathsf{M}(\varphi(x, u))\big|_{x=\xi(t)} .$$

Let us now take $\varphi(x, u) = \lambda(\xi_{t,x}(s))$. Then $\varphi(\xi(t), u) = \lambda(\xi(s))$ and we obtain

$$\mathsf{M}(\lambda(\xi(s)) \mid \mathfrak{F}_t) = g(\xi(t)) , \tag{9}$$

where $g(x) = \mathsf{M}\lambda(\xi_{t,x}(s))$. Consequently,

$$\mathsf{P}\{\xi(t) \in A \mid \mathfrak{F}_t\} = \mathsf{P}_{t,\xi(t)}(s, A) ,$$

where $\mathsf{P}_{t,x}(s, A) = \mathsf{P}\{\xi_{t,x}(s) \in A\}$. This completes the proof of the theorem.

Let us show that under certain supplementary assumptions, the process $\xi(t)$ is a diffusion process with coefficient of diffusion $\sigma^2(t, x)$ and coefficient of transfer $a(t, x)$. To do this, we prove as a preliminary the auxiliary proposition:

Lemma 2. *Let $\xi_{t,x}(s)$ denote a solution of equation* (6) *in which the functions $a(t, x)$ and $\sigma(t, x)$ satisfy the conditions of Theorem* 1. *Then there exists a constant H such that*

$$\mathsf{M} \mid \xi_{t,x}(s) - x \mid^4 \leqq H(s - t)^2(1 + \mid x \mid^4) .$$

Proof. Define $\chi_N(s) = 1$ when

$$\sup_{t \leqq u \leqq s} \mid \xi_{t,x}(u) - x \mid \leqq N ,$$

and $\chi_N(s) = 0$ otherwise. Then

$$(\xi_{t,x}(s) - x)\chi_N(s) = \chi_N(s)\Big[\int_t^s \chi_N(u)a(u, \xi_{t,x}(u))du + \int_t^s \chi_N(u)\sigma(u, \xi_{t,x}(u))dw(u)\Big] .$$

Using the inequality $(a + b)^4 \leqq 8a^4 + 8b^4$, Cauchy's inequality, and property 7 of Section 2, we obtain

$$\begin{aligned}
\mathsf{M}(\xi_{t,x}(s) - x)^4\chi_N(s) &\leqq 8\Big\{\mathsf{M}\Big[\int_t^s \chi_N(u)a(u, \xi_{t,x}(u))du\Big]^4 + \mathsf{M}\Big[\int_t^s \chi_N(u)\sigma(u, \xi_{t,x}(u))dw(u)\Big]^4\Big\} \\
&\leqq 8(s - t)^3\int_t^s \mathsf{M}\chi_N(u)[a(u, \xi_{t,x}(u))]^4du \\
&\quad + 8\cdot 36(s - t)\int_t^s \mathsf{M}\sigma(u, \xi_{t,x}(u))^4\chi_N(u)du \\
&\leqq 64(s - t)^3\int_t^s \mathsf{M}\chi_N(u)a(u, x)^4du \\
&\quad + 64(s - t)^3\int_t^s \mathsf{M}\chi_N(u)[a(u, \xi_{t,x}(u)) - a(u, x)]^4du
\end{aligned}$$

$$+ 8\cdot 8\cdot 36(s - t)\int_t^s \mathsf{M}\chi_N(u)\sigma(u, x)^4 du$$

$$+ 8\cdot 8\cdot 36(s - t)\int_t^s \mathsf{M}\chi_N(u)[\sigma(u, \xi_{t,x}(u)) - \sigma(u, x)]^4 du$$

$$\leqq L(s - t)\int_t^s \mathsf{M}\chi_N(u) \mid \xi_{t,x}(u) - x \mid^4 du + R(s - t)^2 ,$$

where $L = K^4[(T - t_0)^2 \cdot 64 + 64 \cdot 38]$ and

$$R = L\Big[\max_t \mid \sigma(t, x) \mid^4 + \max_t \mid a(t, x) \mid^4\Big] \leqq h_1(1 + \mid x \mid^4) .$$

Thus if $\varphi(s) = \mathsf{M} \mid \xi_{t,x}(s) - x \mid^4 \chi_N(s)$, then

$$\varphi(s) \leqq L(s - t)\int_t^s \varphi(u)du + R(s - t)^2 \qquad (t < s) .$$

It follows from this inequality that $\varphi(s)/(s - t)$ is bounded. Therefore, for $t \leqq s \leqq t + 1$,

$$\frac{\varphi(s)}{s - t} \leqq L\int_t^s (u - t)\frac{\varphi(u)}{u - t}du + R(s - t)$$

$$\leqq L\int_t^s \frac{\varphi(u)}{u - t}du + R(s - t) .$$

Let us set $\alpha(z) = \varphi(t + z)/z$. Then

$$\alpha(z) \leqq L\int_0^z \alpha(u)du + Rz .$$

It follows from Lemma 1 that

$$\alpha(z) \leqq R\frac{e^{Lz} - 1}{L} .$$

Thus

$$\mathsf{M}\chi_N(s) \mid \xi_{t,x}(s) - x \mid^4 \leqq R(s - t)\frac{e^{L(s-t)} - 1}{L}$$

$$\leqq R_1 e^{L(s-t)}(s - t)^2(1 + \mid x \mid^4) .$$

Since the expression on the right is independent of N, and since

$$\mathsf{M}\chi_N(s) \mid \xi_{t,x}(s) - x \mid^4 \longrightarrow \mathsf{M} \mid \xi_{t,x}(s) - x \mid^4$$

as $N \longrightarrow \infty$, the assertion of the lemma follows from the last relation.

Corollary 1. *Let $\xi(t)$ denote a solution of equation* (2). *Suppose that $a(t, x)$ and $\sigma(t, x)$ satisfy the conditions of Theorem* 1 *and that $\mathsf{M} \mid \xi(t_0) \mid^4 < \infty$. Then*

$$\sup_{t_0 \leqq t \leqq T} \mathsf{M} \mid \xi(t) \mid^4 < \infty .$$

Proof. It follows from Lemma 2 that

$$\mathsf{M}(|\xi(t) - \xi(t_0)|^4 \,|\, \xi(t_0)) \leqq H(T - t_0)^2(1 + |\xi(t_0)|^4) ,$$
$$\mathsf{M}\,|\xi(t)|^4 \leqq 8\mathsf{M}\,|\xi(t_0)|^4 + 8\mathsf{M}\mathsf{M}(|\xi(t) - \xi(t_0)|^4 \,|\, \xi(t_0)) .$$

Corollary 2. *Under the hypotheses of the preceding corollary, there exists a constant* H_1 *such that*

$$\mathsf{M}\,|\xi(t) - \xi(s)|^4 \leqq H_1(t - s)^2 .$$

Proof. For $t < s$,

$$\mathsf{M}\,|\xi(t) - \xi(s)|^4 \leqq H(s - t)^2(\mathsf{M}\,|\xi(t)|^4 + 1) .$$

Theorem 3. *Suppose that the conditions of Theorem* 1 *are satisfied and that* $a(t, x)$ *and* $\sigma(t, x)$ *are continuous with respect to* t *for* $t \in [t_0, T]$. *Then* $\sigma^2(t, x)$ *and* $a(t, x)$ *are the coefficients of diffusion and transfer respectively for the process* $\xi(t)$, *which is a solution of equation* (2).

Proof. Since the transition probability $\mathsf{P}(t, x, s, A)$ for the process $\xi(t)$ coincides with the distribution of the variable $\xi_{t,x}(s)$ on the basis of Theorem 2, it follows from Lemma 2 that

$$\int (y - x)^4 \mathsf{P}(t, x, s, dy) = \mathsf{M}\,|\xi_{t,x}(s) - x|^4 = o(s - t)^2 .$$

Using Remark 1 of Section 1, we see that to prove the theorem it will be sufficient to show that

$$\int (y - x)\mathsf{P}(t, x, s, dy) = \mathsf{M}\xi_{t,x}(s) - x = a(t, x)(s - t) + o(s - t) \tag{10}$$

and

$$\int (y - x^2)\mathsf{P}(t, x, s, dy) = \mathsf{M}(\xi_{t,x}(s) - x)^2$$
$$= \sigma^2(t, x)(s - t) + o(s - t) . \tag{11}$$

Starting with equation (6), we may write

$$\mathsf{M}\xi_{t,x}(s) - x = \int_t^s \mathsf{M}a(u, \xi_{t,x}(u))du$$
$$= \int_t^s a(u, x)du + \int_t^s \mathsf{M}[a(u, \xi_{t,x}(u)) - a(u, x)]du .$$

But

$$\left|\mathsf{M}\int_t^s (a(u, \xi_{t,x}(u)) - a(u, x))du\right| \leqq \int_t^s \mathsf{M}\,|a(u, \xi_{t,x}(u)) - a(u, x)|\, du$$
$$\leqq (s - t)^{3/4}\left[\int_t^s \mathsf{M}\,|\xi_{t,x}(u) - x|^4 du\right]^{1/4}$$
$$\leqq (s - t)^{3/4}\left[\int_t^s H(u - t)^2 du\right]^{1/4} = O(s - t)^{3/4+3/4} = o(s - t) .$$

It follows from the continuity of $a(u, x)$ that

$$a(t, x)(t - s) - \int_t^s a(u, x)du = \int_t^s [a(t, x) - a(u, x)]du = o(s - t) ,$$

which proves formula (10).

Furthermore

$$\mathsf{M}(\xi_{t,x}(s) - x)^2 = \mathsf{M}\Big(\int_t^s a(u, \xi_{t,x}(u))du + \int_t^s \sigma(u, \xi_{t,x}(u))dw(u)\Big)^2 .$$

We note that

$$\mathsf{M}\Big[\int_t^s a(u, \xi_{t,x}(u))du\Big]^2 \leqq (s - t)\int_t^s \mathsf{M}a^2(u, \xi_{t,x}(u))du = o(s - t) ,$$

and

$$\mathsf{M}\Big(\int_t^s \sigma(u, \xi_{t,x}(u))dw(u)\Big)^2 = \int_t^s \mathsf{M}[\sigma(u, \xi_{t,x}(u))]^2 du$$

$$\leqq K^2\int_t^s \mathsf{M}(1 + \xi_{t,x}(u)^2)du = O(s - t) .$$

Therefore, using Cauchy's inequality, we see that

$$\Big|\mathsf{M}\Big[\int_t^s a(u, \xi_{t,x}(u))du\Big]\Big[\int_t^s \sigma(u, \xi_{t,x}(u))dw(u)\Big]\Big|$$

$$\leqq \sqrt{\mathsf{M}\Big(\int_t^s a(u, \xi_{t,x}(u))du\Big)^2 \mathsf{M}\Big(\int_t^s \sigma(u, \xi_{t,x}(u))dw(u)\Big)^2} = o(s - t) .$$

Thus

$$\mathsf{M}(\xi_{t,x}(s) - x)^2 = \mathsf{M}\Big(\int_t^s \sigma(u, \xi_{t,x}(u))dw(u)\Big)^2 + o(s - t)$$

$$= \int_t^s \mathsf{M}\sigma^2(u, \xi_{t,x}(u))du + o(s - t)$$

$$= \sigma^2(t, x)(s - t) + \int_t^s [\sigma^2(u, x) - \sigma^2(t, x)]du$$

$$+ \int_t^s \mathsf{M}[\sigma^2(u, \xi_{t,x}(u)) - \sigma^2(u, x)]du + o(s - t) .$$

It follows from the continuity of $\sigma(u, x)$ with respect to u that

$$\int_t^s [\sigma^2(u, x) - \sigma^2(u, x)]du = o(s - t) .$$

Furthermore,

$$\Big|\mathsf{M}\int_t^s [\sigma^2(u, \xi_{t,x}(u)) - \sigma^2(u, x)]du\Big|$$

$$\leqq \sqrt{\int_t^s \mathsf{M}[\sigma(u, \xi_{t,x}(u)) - \sigma(u, x)]^2 du}$$

$$\times \sqrt{\int_t^s \mathsf{M}[\sigma(u, \xi_{t,x}(u)) + \sigma(u, x)]^2 du}$$

$$\leqq \sqrt{O(s - t)(s - t)^{1/2}\Big(\int_t^s \mathsf{M}\,|\,\sigma(u, \xi_{t,x}(u)) - \sigma(u, x)\,|^4 du\Big)^{1/2}}$$

$$= O((s - t)^{1/2+1/4+1/2}) = o(s - t) .$$

Thus we have proved formula (11) and with it the theorem.

Let us now construct a multidimensional diffusion process $\boldsymbol{\xi}(t)$ with transfer vector $\boldsymbol{a}(t, \boldsymbol{x})$ and diffusion operator $B(t, \boldsymbol{x})$. We define

$$\boldsymbol{b}_k(t, \boldsymbol{x}) = \sqrt{\lambda_k(t, \boldsymbol{x})}\,\boldsymbol{e}_k(t, \boldsymbol{x}) ,$$

where the $\boldsymbol{e}_k(t, \boldsymbol{x})$ are the eigenvectors of the operator $B(t, \boldsymbol{x})$ and the $\lambda_k(t, \boldsymbol{x})$ are the eigenvalues corresponding to them.

We seek the process $\boldsymbol{\xi}(t)$ as a solution of the equation

$$d\boldsymbol{\xi}(t) = \boldsymbol{a}(t, \boldsymbol{\xi}(t))dt + \sum_{k=1}^{m} \boldsymbol{b}_k(t, \boldsymbol{\xi}(t))dw_k(t) , \tag{12}$$

where the $w_k(t)$ for $k = 1, 2, \cdots, m$ are independent processes of Brownian motion.

The functions $\boldsymbol{a}(t, \boldsymbol{x})$ and $\boldsymbol{b}_k(t, \boldsymbol{x})$ are defined for $t \in [t_0, T]$ and $\boldsymbol{x} \in R^{(m)}$, and their ranges are contained in $R^{(m)}$. Equation (12) is equivalent to the equation

$$\boldsymbol{\xi}(t) = \boldsymbol{\xi}(t_0) + \int_{t_0}^{t} \boldsymbol{a}(s, \boldsymbol{\xi}(s))ds + \sum_{k=1}^{m} \int_{t_0}^{t} \boldsymbol{b}_k(s, \boldsymbol{\xi}(s))dw_k(s) . \tag{13}$$

This equation is solved for given $\boldsymbol{\xi}(t_0)$, which we shall always assume to be independent of the processes $w_k(t)$.

Let $\mathfrak{F}_t$ denote the minimal σ-algebra generated by the quantities $\boldsymbol{\xi}(t_0)$ and $w_k(s) - w_k(t_0)$ for $k = 1, 2, \cdots, m$ and $s \in [t_0, t]$. We shall assume that the solution of equation (13) is a process $\boldsymbol{\xi}(t)$ such that the integrals on the right side of the equation exist, and such that for every $t \in [t_0, T]$ equation (13) is satisfied with probability 1.

Let us formulate the basic properties of the solutions of equation (13).

Theorem 4. *Let*

$$\boldsymbol{a}(t, \boldsymbol{x}), \boldsymbol{b}_1(t, \boldsymbol{x}), \cdots , \boldsymbol{b}_m(t, \boldsymbol{x})$$

denote Borel functions defined for $t \in [t_0, T]$ and $\boldsymbol{x} \in R^{(m)}$ with ranges in $R^{(m)}$. If there exists a K such that

$$| \boldsymbol{a}(t, \boldsymbol{x}) |^2 + \sum_{k=1}^{m} | \boldsymbol{b}_k(t, \boldsymbol{x}) |^2 \leq K^2(1 + | \boldsymbol{x} |^2) ,$$

$$| \boldsymbol{a}(t, \boldsymbol{x}) - \boldsymbol{a}(t, \boldsymbol{y}) | + \sum_{k=1}^{m} | \boldsymbol{b}_k(t, \boldsymbol{x}) - \boldsymbol{b}_k(t, \boldsymbol{y}) | \leq K | \boldsymbol{x} - \boldsymbol{y} |$$

for every $\boldsymbol{x}$ and $\boldsymbol{y}$ in $R^{(m)}$, then equation (13) *has a solution $\boldsymbol{\xi}(t)$ that is unique up to stochastic equivalence and continuous with probability* 1. *This solution $\boldsymbol{\xi}(t)$ is a Markov process whose transition probabilities $\mathsf{P}(t, \boldsymbol{x}, s, A)$ for $t < s$ are given by the relation*

$$\mathsf{P}(t, \boldsymbol{x}, s, A) = \mathsf{P}\{\boldsymbol{\xi}_{t,\boldsymbol{x}}(s) \in A\} ,$$

where $\xi_{t,x}(s)$ is a solution of the equation

$$\xi_{t,x}(s) = x + \int_t^s a(u, \xi_{t,x}(u))du + \sum_{k=1}^m \int_t^s b_k(u, \xi_{t,x}(u))dw(u) . \qquad (14)$$

If the functions $a(t, x)$ and $b_k(t, x)$ are continuous with respect to t, then the process $\xi(t)$ is a diffusion process with transfer vector $a(t, x)$ and diffusion operator $B(t, x)$ that satisfies the equation

$$(B(t, x)z, z) = \sum_{k=1}^m (b_k(t, x), z)^2 .$$

The proof follows essentially the proof given for the one-dimensional processes treated in Theorems 1 to 3.

Remark 1. If $\xi_{t,x}(s)$ is a solution of equation (14), in which the functions $a(s, x)$ and $b_k(s, x)$ satisfy the conditions of Theorem 4, there exists a constant H such that for $s > t$,

$$\mathsf{M} \mid \xi_{t,x}(s) - x \mid^4 \leq H(s - t)^2 .$$

This assertion is analogous to the one proved in Lemma 1 for a one-dimensional process.

Remark 2. If the coefficients $a(t, x)$ and $b_k(t, x)$ in equation (12) are independent of t, that is, if the equation is of the form

$$d\xi(t) = a(\xi(t))dt + \sum_{k=1}^m b_k(\xi(t))dw_k(t)$$

and $a(x)$ and the $b_k(x)$ for $k = 1, 2, \cdots, m$ satisfy the hypotheses of Theorem 4, then the solution $\xi(t)$ of the equation is a homogeneous Markov process; that is, the transition probability $\mathsf{P}(t, x, t + h, A)$ is independent of t. To see this, note that $\mathsf{P}(t, x, t + h, A)$ coincides with the distribution of $\zeta_{t,x}(h) = \xi_{t,x}(t + h)$ but according to Theorem 4, $\zeta_{t,x}(h)$ is a solution of the equation

$$d\zeta_{t,x}(h) = a(\zeta_{t,x}(h))dt + \sum_{k=1}^m b_k(\zeta_{t,x}(h))d[w_k(t + h) - w_k(t)]$$

with initial condition $\zeta_{t,x}(0) = x$.

Since the common distribution of $[w_k(t + h) - w_k(t)]$ for $k = 1, 2, \cdots, m$ is independent of t, the distribution $\zeta_{t,x}(h)$ is also independent of t.

4. DIFFERENTIABILITY OF SOLUTIONS OF STOCHASTIC EQUATIONS WITH RESPECT TO INITIAL CONDITIONS

The purpose of this section is to show that the function $\xi_{t,x}(s)$, defined in Section 3 as a solution of equation (14) of Section 3, is a differentiable function of $\boldsymbol{x}$ when the coefficients $\boldsymbol{a}(t, \boldsymbol{x})$ and $\boldsymbol{b}_k(t, \boldsymbol{x})$ are sufficiently smooth. Since $\boldsymbol{\xi}_{t,\boldsymbol{x}}(s)$ is a random function of $\boldsymbol{x}$, we need to show just what we mean by a derivative of $\boldsymbol{\xi}_{t,\boldsymbol{x}}(s)$. For our purposes it is convenient to consider mean-square differentiability of random functions. If $\varphi(\boldsymbol{x}, u)$ is a random function depending on the point $\boldsymbol{x}$ in the space $R^{(m)}$ and $x^1, \cdots, x^m$ are the coordinates of $\boldsymbol{x}$, then by $\partial\varphi/\partial x^i$ we mean the random variable for which

$$\lim_{\Delta x^i \to 0} \mathsf{M}\left| \frac{1}{\Delta x^i}(\varphi(x^i, \cdots, x^i + \Delta x^i, \cdots, x^m, u) - \varphi(x^1, \cdots, x^i, \cdots, x^m, u)) - \frac{\partial\varphi(\boldsymbol{x}, u)}{\partial x^i}\right|^2 = 0 .$$

Just as in the preceding section, we shall give complete proofs only for one-dimensional processes.

Theorem 1. *Suppose that two functions $a(t, x)$ and $\sigma(t, x)$ are defined and continuous for $x \in (-\infty, \infty)$ and $t \in [t_0, T]$ and suppose that they have the continuous bounded partial derivatives*

$$a'_x(t, x), a''_{xx}(t, x), \sigma'_x(t, x), \sigma''_{xx}(t, x) .$$

Then the solution $\xi_{t,x}(s)$ of equation (6), Section 3 is twice differentiable with respect to x and the derivatives are mean-square continuous functions of x.

Proof of this theorem rests on:

Lemma 1. *Suppose that processes $\zeta_n(t)$ where $n = 0, 1, \cdots$ are solutions of the stochastic equations*

$$\zeta_n(t) = \varphi_n(t) + \int_{t_0}^{t} \psi_n(s)\zeta_n(s)ds + \int_{t_0}^{t} \chi_n(s)\zeta_n(s)dw(s) . \tag{1}$$

We assume that the functions $\varphi_n(t)$, $\psi_n(t)$, and $\chi_n(t)$ are for every t measurable with respect to $\mathfrak{F}_t$, and that there exists a K such that with probability 1, $|\psi_n(s)| \leqq K$ and $|\chi_n(s)| \leqq K$. If $\sup_t \mathsf{M}|\varphi_n(t) - \varphi_0(t)|^2 \to 0$ as $n \to \infty$ and if $\psi_n(t) \to \psi_0(t)$ and $\chi_n(t) \to \chi_0(t)$ in probability for every t as $n \to \infty$, then $\sup_t \mathsf{M}|\zeta_n(t) - \zeta_0(t)|^2 \to 0$ as $n \to \infty$.

Proof. We note first of all that the existence and uniqueness of the solution to equation (1) is proved just as in the proof of Theorem 1 of Section 3, and that in the present case,

$$\sup_t \mathsf{M}|\zeta_n(t)|^2 < \infty .$$

Therefore, we may write

$$\begin{aligned}
\mathsf{M}\,|\,\zeta_n(t) - \zeta_0(t)\,|^2 &\leqq 3\mathsf{M}\,|\,\varphi_n(t) - \varphi_0(t)\,|^2 \\
&\quad + 3\mathsf{M}\Big(\int_{t_0}^t (\psi_n(s)\zeta_n(s) - \psi_0(s)\zeta_0(s))ds\Big)^2 \\
&\quad + 3\mathsf{M}\Big(\int_{t_0}^t (\chi_n(s)\zeta_n(s) - \chi_0(s)\zeta_0(s))dw(s)\Big)^2 \\
&= 3\mathsf{M}\,|\,\varphi_n(t) - \varphi_0(t)\,|^2 + 3\mathsf{M}\Big(\int_{t_0}^t \psi_n(s)[\zeta_n(s) - \zeta_0(s)]ds \\
&\quad + \int_{t_0}^t \zeta_0(s)[\psi_n(s) - \psi_0(s)]ds\Big)^2 + 3\mathsf{M}\Big(\int_{t_0}^t \chi_n(s)[\zeta_n(s) \\
&\quad - \zeta_0(s)]dw(s) + \int_{t_0}^t \zeta_0(s)[\chi_n(s) - \chi_0(s)]dw(s)\Big)^2 \\
&\leqq \delta_n(t) + 6\mathsf{M}\Big(\int_{t_0}^t K\,|\,\zeta_n(s) - \zeta_0(s)\,|\,ds\Big)^2 \\
&\quad + 6K^2\int_{t_0}^t \mathsf{M}\,|\,\zeta_n(s) - \zeta_0(s)\,|^2 ds\,,
\end{aligned}$$

where

$$\begin{aligned}
\delta_n(t) &= 3\mathsf{M}\,|\,\varphi_n(t) - \varphi_0(t)\,|^2 + 6(t - t_0)\int_{t_0}^t \mathsf{M}\zeta_0(s)^2\,|\,\psi_n(s) \\
&\quad - \psi_0(s)\,|^2 ds + 6\int_{t_0}^t \mathsf{M}\,|\chi_n(s) - \chi_0(s)\,|^2\zeta_0(s)^2 ds\,.
\end{aligned}$$

Since

$$\begin{aligned}
\sup_t \delta_n(t) &\leqq 3\sup_t \mathsf{M}\,|\,\varphi_n(t) - \varphi_0(t)\,|^2 \\
&\quad + 6(T - t_0 + 1)\int_{t_0}^T \mathsf{M}\zeta_0(s)^2(|\,\psi_n(s) - \psi_0(s)\,|^2 \\
&\quad + |\,\chi_n(s) - \chi_0(s)\,|^2)ds\,,
\end{aligned}$$

the sequence $\{\delta_n(t)\}$ converges to 0 uniformly with respect to t. (The integral approaches 0 by virtue of Lebesgue's theorem since the integrand is bounded by $4K^2\zeta_0(s)^2$, and approaches 0 in probability.)

If we set $H = 6K^2(T - t_0 + 1)$, then

$$\mathsf{M}\,|\,\zeta_n(t) - \zeta_0(t)\,|^2 \leqq \sup_t \delta_n(t) + H\int_{t_0}^t \mathsf{M}\,|\,\zeta_n(s) - \zeta_0(s)\,|^2 ds\,.$$

Then on the basis of Lemma 1, Section 3,

$$\mathsf{M}\,|\,\zeta_n(t) - \zeta_0(t)\,|^2 \leqq \sup_t \delta_n(t)e^{H(T-t_0)}\,.$$

This inequality proves the lemma.

REMARK 1. The lemma remains valid if the processes depend on a continuous parameter α and the corresponding limits exist as $\alpha \to 0$.

Let us now prove Theorem 1. Define

$$\zeta_{x,\Delta x}(s) = \frac{1}{\Delta x}[\xi_{x+\Delta x,t}(s) - \xi_{x,t}(s)] .$$

Then the process $\zeta_{x,\Delta x}(s)$ is a solution of the equation

$$\zeta_{x,\Delta x}(s) = 1 + \int_t^s \psi_{x,\Delta x}(u)\zeta_{x,\Delta x}(u)du + \int_t^s \chi_{x,\Delta x}(u)\zeta_{x,\Delta x}(u)dw(u) ,$$

where

$$\psi_{x,\Delta x}(s) = \frac{a(s, \xi_{t,x+\Delta x}(s)) - a(s, \xi_{t,x}(s))}{\xi_{t,x+\Delta x}(s) - \xi_{t,x}(s)} ,$$

$$\chi_{x,\Delta x}(s) = \frac{\sigma(s, \xi_{t,x+\Delta x}(s)) - \sigma(s, \xi_{t,x}(s))}{\xi_{t,x+\Delta x}(s) - \xi_{t,x}(s)} .$$

Since $a(s, x)$ and $\sigma(s, x)$ have bounded derivatives with respect to x, there exists a constant K such that with probability 1, $|\psi_{x,\Delta x}(s)| \leqq K$ and $|\chi_{x,\Delta x}(s)| \leqq K$. Therefore

$$\mathsf{M}\zeta_{x,\Delta x}(s)^2 \leqq 3 + 3(T - t_0)\int_{t_0} K^2\mathsf{M}\zeta_{x,\Delta x}(u)^2du$$

$$+ 3K^2\int_t^s \mathsf{M}\,|\zeta_{x,\Delta x}(u)|^2du \leqq 3 + H\int_t^s \mathsf{M}\,|\zeta_{x,\Delta x}(u)|^2du ,$$

where $H = 3(T - t_0 + 1)K^2$. It follows from Lemma 1, Section 3 that

$$\mathsf{M}\zeta_{x,\Delta x}(s)^2 \leqq 3e^{H(s-t)}$$

and consequently, for some H_1,

$$\mathsf{M}\,|\xi_{t,x+\Delta x}(s) - \xi_{t,x}(s)|^2 \leqq H_1(\Delta x)^2 . \tag{2}$$

This inequality means that $\xi_{t,+\Delta x}(s) - \xi_{t,x}(s) \longrightarrow 0$ in probability as $\Delta x \longrightarrow 0$. Therefore

$$\psi_{x,\Delta x}(s) \longrightarrow a'_x(s, \xi_{t,x}(s)),\ \chi_{x,\Delta x}(s) \longrightarrow \sigma'_x(s, \xi_{t,x}(s))$$

in probability as $\Delta x \longrightarrow 0$. We let $\zeta_x(s)$ denote the solution of the equation

$$\zeta_x(s) = 1 + \int_t^s a'_x(s, \xi_{t,x}(s))\zeta_x(s)ds + \int_t^s \sigma'_x(s, \xi_{t,x}(s))\zeta_x(s)dw(s) . \tag{3}$$

It follows from Lemma 1 that $\mathsf{M}\,|\zeta_{x,\Delta x}(s) - \zeta_x(s)|^2 \longrightarrow 0$ as $\Delta x \longrightarrow 0$, that is, that

$$\zeta_x(s) = \frac{\partial}{\partial x}\xi_{t,x}(s) .$$

The mean-square continuity of $\zeta_x(s)$ with respect to x again follows from Lemma 1 since the derivatives

$$a'_x(s, \xi_{t,x}(s)) \quad \text{and} \quad \sigma'_x(s, \xi_{t,x}(s))$$

are continuous with respect to x in the sense of convergence in probability, because of the continuity of $a'_x(s, x)$ and $\sigma'_x(s, x)$ with respect to x and the mean-square continuity of $\xi_{t,x}(s)$ with respect to x.

To prove the existence of $\partial^2\xi_{t,x}(s)/\partial x^2$, we introduce the processes $\eta_{x,\Delta x}(s) = (1/\Delta x)[\zeta_{x,\Delta x}(s) - \zeta_x(s)]$. These processes satisfy the equations

$$\eta_{x,\Delta x}(s) = \varphi^{(1)}_{x,\Delta x}(s) + \varphi^{(2)}_{x,\Delta x}(s) + \int_t^s a'_x(u, \xi_{t,x}(u))\eta_{x,\Delta x}(u)du + \int_t^s \sigma'_x(u, \xi_{t,x}(u))\eta_{x,\Delta x}(u)dw(u) ,$$

where

$$\varphi^{(1)}_{x,\Delta x}(s) = \int_t^s \frac{a'_x(u, \xi_{t,x+\Delta x}(u)) - a'_x(u, \xi_{t,x}(u))}{\xi_{t,x+\Delta x}(u) - \xi_{t,x}(u)}\zeta_{x,\Delta x}(u)\zeta_x(u)du ,$$

$$\varphi^{(2)}_{x,\Delta x}(s) = \int_t^s \frac{\sigma'_x(u, \xi_{t,x+\Delta x}(u)) - \sigma'_x(u, \xi_{t,x}(u))}{\xi_{t,x+\Delta x}(u) - \xi_{t,x}(u)}\zeta_{x,\Delta x}(u)\zeta_x(u)dw(u) .$$

Define

$$\varphi^{(1)}_x(s) = \int_t^s a''_{xx}(u, \xi_{t,x}(u))\zeta_x(u)^2du ,$$

$$\varphi^{(2)}_x(s) = \int_t^s \sigma''_{xx}(u, \xi_{t,x}(u))\zeta_x(u)^2dw(u) .$$

Let us show that

$$\lim_{\Delta x \to 0} \mathsf{M}\,|\,\varphi^{(i)}_{x,\Delta x}(s) - \varphi^{(i)}_x(s)\,|^2 = 0 , \qquad i = 1, 2$$

uniformly with respect to t. To do this, we use

Lemma 2. *If $\zeta_x(s)$ is a solution of equation* (3), *then*

$$\sup_s \mathsf{M}\,|\,\zeta_x(s)\,|^4 < \infty \quad \text{and} \quad \lim_{\Delta x \to 0} |\,\zeta_{x+\Delta x}(s) - \zeta_x(s)\,|^4 = 0 .$$

Proof. We note that the solution $\zeta_x(s)$ of equation (3) can be obtained by the method of successive approximations, in which we set $\zeta^{(0)}_x(s) = 1$:

$$\zeta^{(n)}_x(s) = 1 + \int_t^s a'_x(u, \xi_{t,x}(u))\zeta^{(n-1)}_x(u)du + \int_t^s \sigma'_x(u, \xi_{t,x}(u))\zeta^{(n-1)}_x(u)dw(u) .$$

Then by using the inequalities $(a + b + c)^4 \leqq 27(a^4 + b^4 + c^4)$, $|\,a'_x\,| \leqq K$, $|\,\sigma'_x\,| \leqq K$ and property 7 of Section 2, we obtain

$$\mathsf{M}\zeta_x^{(n)}(s)^4 \leqq 27\Big[1 + \mathsf{M}\Big(\int_t^s K \mid \zeta_x^{(n-1)}(u) \mid du\Big)^4 + 36(s-t)\int_t^s K^4 \mathsf{M}\zeta_x^{(n-1)}(u)^4 du\Big]$$

$$\leqq R_1 + R_2\int_t^s \mathsf{M} \mid \zeta_x^{(n-1)}(u) \mid^4 du \,,$$

where $R_1 = 27$ and $R_2 = 27K^4[(T-t_0)^3 + 36(T-t_0)]$. Remembering that $\mathsf{M}\zeta^{(0)}(s)^4 = 1$, we may write

$$\mathsf{M} \mid \zeta_x^{(n)}(s) \mid^4 \leqq R_1 + R_2\int_t^s R_1 ds + R_2^2\int_t^s\int_t^u \mathsf{M} \mid \zeta_x^{(n-2)}(v) \mid^4 dv\, du$$

$$\leqq R_1\Big(1 + R_2(s-t) + \frac{R_2^2(s-t)^2}{2} + \cdots + \frac{R_2^{n-1}(s-t)^{n-1}}{(n-1)!}\Big)$$

$$+ R_2^n \int_t^s\int_t^{u_1}\int_t^{u_2} \cdots \int_t^{u_{n-1}} \mathsf{M} \mid \zeta_x^{(0)}(v) \mid^4 dv\, du_{n-1} \cdots du_1$$

$$\leqq R_1 e^{R_2(s-t)} + R_2^n \frac{(s-t)^n}{n!} \,.$$

This inequality implies boundedness of $\mathsf{M} \mid \zeta_x(s) \mid^4$. Furthermore

$$\zeta_x(s) - \zeta_{x+\Delta x}(s) = \int_t^s \tilde{\psi}_{x,\Delta x}(u)\zeta_x(u)du + \int_t^s \tilde{\chi}_{x,\Delta x}(u)\zeta_x(u)dw(u)$$

$$+ \int_t^s a'_x(u, \xi_{t,x}(u))[\zeta_x(u) - \zeta_{x+\Delta x}(u)]du$$

$$+ \int_t^s \sigma'_x(u, \xi_{t,x}(u))[\zeta_x(u) - \zeta_{x+\Delta x}(u)]dw(u) \,,$$

where

$$\tilde{\psi}_{x,\Delta x}(s) = a'_x(s, \xi_{t,x}(s)) - \frac{a(s, \xi_{t,x+\Delta x}(s)) - a(s, \xi_{t,x}(s))}{\xi_{t,x+\Delta x}(s) - \xi_{t,x}(s)} \,,$$

$$\tilde{\chi}_{x,\Delta x}(s) = \sigma'_x(s, \xi_{t,x}(s)) - \frac{\sigma(s, \xi_{t,x+\Delta x}(s)) - \sigma(s, \xi_{t,x}(s))}{\xi_{t,x+\Delta x}(s) - \xi_{t,x}(s)} .$$

Taking the fourth power of both sides of these equations and using Cauchy's inequality and property 7 of Section 2, just as we did above, we see that there exist constants A and B such that

$$\mathsf{M} \mid \zeta_x(s) - \zeta_{x+\Delta x}(s) \mid^4 \leqq A\int_t^s \mathsf{M}(\mid \tilde{\psi}_{x,\Delta x}(u) \mid^4 + \mid \tilde{\chi}_{x,\Delta x}(u) \mid^4) \mid \zeta_x(u) \mid^4 du + B\int_t^s \mathsf{M} \mid \zeta_x(u) - \zeta_{x+\Delta}(u) \mid^4 du \,.$$

Consequently,

$$\mathsf{M} \mid \zeta_x(s) - \zeta_{x+\Delta x}(s) \mid^4 \leqq A\int_t^T \mathsf{M}(\mid \tilde{\psi}_{x,\Delta x}(u) \mid^4 + \mid \tilde{\chi}_{x+\Delta x}(u) \mid^4) \mid \zeta_x(u) \mid^4 du e^{B(s-t)} \,.$$

Since $|\tilde{\psi}_{x,\Delta x}(s)|^4 + |\chi_{x,\Delta x}(s)|^4 \to 0$ in probability as $\Delta x \to 0$, and since the expression on the left does not exceed $32K^4$, it follows from Lebesgue's theorem that

$$\lim_{\Delta x \to 0} \int_{t_0}^{T} \mathsf{M}(|\tilde{\psi}_{x,\Delta x}(s)|^4 + |\tilde{\chi}_{x,\Delta x}(s)|^4)\,|\zeta_x(s)|^4 ds = 0 .$$

This completes the proof of the lemma.

We note that from the boundedness of a''_{xx} it follows that there exists a constant C such that

$$\left| \frac{a'_x(s, x) - a'_x(s, y)}{x - y} \right| \leq C .$$

Therefore:

$$\begin{aligned} &\mathsf{M}\,|\varphi^{(1)}_{x,\Delta x}(s) - \varphi^{(1)}_x(s)|^2 \\ &= \mathsf{M}\Big\{\int_t^s \frac{a'_x(u, \xi_{t,x+\Delta x}(u)) - a'_x(u, \xi_{t,x}(u))}{\xi_{t,x+\Delta x}(u) - \xi_{t,x}(u)} \zeta_x(u)[\zeta_{x,\Delta x}(u) - \zeta_x(u)]du \\ &\quad + \int_t^s \Big[\frac{a'_x(u, \xi_{t,x+\Delta x}(u)) - a'_x(u, \xi_{t,x}(u))}{\xi_{t,x+\Delta x}(u) - \xi_{t,x}(u)} - a''_{xx}(u, \xi_{t,x}(u))\Big]\zeta_x(u)^2 du\Big\}^2 \\ &\leq 2(s-t)\Big[\int_t^s C^2 \mathsf{M}\zeta_x(u)^2[\zeta_{x,\Delta x}(u) - \zeta_x(u)]^2 du \\ &\quad + \int_t^s \mathsf{M}\Big[\frac{a'_x(u, \xi_{t,x+\Delta x}(u)) - a'_x(u, \xi_{t,x}(u))}{\xi_{t,x+\Delta x}(u) - \xi_{t,x}(u)} - a''_{xx}(u, \xi_{t,x}(u))\Big]^2 \zeta_x(u)^4 du\Big] . \end{aligned}$$

The second integral approaches 0 by virtue of Lebesgue's theorem, since the first factor in the integrand is bounded and approaches 0 in probability as $\Delta x \to 0$, and $\sup_u \mathsf{M}\,|\zeta_x(u)|^4 < \infty$. We have the following bound for the first integral:

$$\begin{aligned} &\int_t^s \mathsf{M}\zeta_x(u)^2 \mathsf{M}\,|\zeta_{x,\Delta x}(u) - \zeta_x(u)|^2 du \\ &\qquad \leq \sqrt{\mathsf{M}\int_t^s \zeta_x(u)^4 du}\ \sqrt{\int_t^s \mathsf{M}\,|\zeta_{x,\Delta x}(u) - \zeta_x(u)|^4 du} , \end{aligned}$$

so that it too approaches 0 in accordance with Lemma 2. This means that

$$\lim_{\Delta x \to 0} \mathsf{M}\,|\varphi^{(1)}_{x,\Delta x}(s) - \varphi^{(1)}_x(s)|^2 = 0 .$$

Analogously, we can show that

$$\lim_{\Delta x \to 0} \mathsf{M}\,|\varphi^{(2)}_{x,\Delta x}(s) - \varphi^{(2)}_x(s)|^2 = 0 .$$

On the basis of Lemma 1, the quantity $\eta_{x,\Delta x}(s)$ converges in the sense of mean square to the process $\eta_x(s)$, which is a solution of the equation

$$\eta_x(s) = \int_t^s a''_{xx}(u, \xi_{t,x}(u))\zeta_x(u)^2 du + \int_t^s \sigma''_{xx}(u, \xi_{t,x}(u))\zeta_x(u)^2 dw(u)$$
$$+ \int_t^s a'_x(u, \xi_{t,x}(u))\eta_x(u) du + \int_t^s \sigma'_x(u, \xi_{t,x}(u))\eta_x(u) dw(u) ;$$

that is, $\partial^2\xi_{t,x}(s)/\partial x^2 = \eta_x(s)$. The mean-square continuity of $\eta_x(s)$ with respect to x follows from the mean-square continuity of the functions $\varphi_x^{(1)}(s)$ and $\varphi_x^{(2)}(s)$, and the continuity in probability of $a''_{xx}(u, \xi_{t,x}(u))$ and $\sigma''_{xx}(u, \xi_{t,x}(u))$ as functions of x.

For example, for $\varphi_x^{(1)}(s)$ we have the inequality

$$\mathsf{M} \mid \varphi_x^{(1)}(s) - \varphi_y^{(1)}(s) \mid^2$$
$$\leqq 2(s-t)\int_t^s \mathsf{M} a''_{xx}(u, \xi_{t,x}(u))^2[\zeta_x(u) - \zeta_y(u)]^2 du.$$
$$+ 2(s-t)\int_t^s \mathsf{M}[a''_{xx}(u, \xi_{t,x}(u)) - a''_{xx}(u, \xi_{t,y}(u))]^2\zeta_x(u)^4 du .$$

According to Lebesgue's theorem, the second integral approaches 0 as $y \to x$, since

$$a''_{xx}(u, \xi_{t,x}(u)) - a''_{xx}(u, \xi_{t,y}(u))$$

is bounded and approaches 0 in probability as $y \to x$ and

$$\sup_u \mathsf{M}\zeta_x(u)^4 < \infty .$$

For the first integral we have the inequality

$$\int_t^s \mathsf{M} a''_x(u, \xi_{t,x}(u))^2[\zeta_x(u) - \zeta_y(u)]^2 du$$
$$\leqq C^2\sqrt{\int_s^t \mathsf{M} \mid \zeta_x(u) + \zeta_y(u) \mid^4 du \int_s^t \mathsf{M} \mid \zeta_x(u) - \zeta_y(u) \mid^4 du} ,$$

so that it approaches 0 in accordance with Lemma 2. The mean-square convergence of the function $\varphi_x^2(s)$ is proved in an analogous manner. This completes the proof of the theorem.

In the case of multidimensional processes, we have:

Theorem 2. *Let $\xi_{t,x}(s)$ denote a solution of equation* (14) *of Section* 3. *Suppose that the functions $\boldsymbol{a}(t, \boldsymbol{x})$, $\boldsymbol{b}_1(t, \boldsymbol{x})$, $\cdots$, $\boldsymbol{b}_k(t, \boldsymbol{x})$ are defined and continuous for $t \in [t_0, T]$ and $\boldsymbol{x} \in R^{(m)}$ and that they have bounded continuous first and second derivatives with respect to the variables $x^1, x^2, \cdots, x^m$. Then the function $\boldsymbol{\xi}_{t,x}(s)$ as a function of $\boldsymbol{x}$ is twice mean-square differentiable with respect to $\boldsymbol{x}$, and the derivatives*

$$\frac{\partial}{\partial x^i}\boldsymbol{\xi}_{t,x}(s) , \qquad \frac{\partial^2}{\partial x^i \partial x^j}\boldsymbol{\xi}_{t,x}(s)$$

as functions of x are mean-square continuous.

The proof of this theorem is similar to that of Theorem 1, and for this reason we omit it.

REMARK 2. If the hypotheses of Theorem 2 are satisfied and $f(x)$ is a bounded continuous function with continuous bounded derivatives of first and second orders, then $u(\boldsymbol{x}) = \mathsf{M}f(\boldsymbol{\xi}_{t,\boldsymbol{x}}(s))$ is twice continuously differentiable with respect to $\boldsymbol{x}$. Again, we shall prove the validity of this assertion only for the case of one-dimensional processes. Let us show that

$$u'_x(x) = \mathsf{M}f'_x(\xi_{t,x}(s))\zeta_x(s) \,. \tag{4}$$

By virtue of the boundedness of $[f(x) - f(y)]/(x - y)$, the convergence of

$$\frac{f(\xi_{t,x+\Delta x}(s)) - f(\xi_{t,x}(s))}{\xi_{t,x+\Delta x}(s) - \xi_{t,x}(s)} - f'_x(\xi_{t,x}(s))$$

to 0 in probability, and the relations

$$\lim_{\Delta x \to 0} \mathsf{M} \mid \zeta_{x,\Delta x}(s) - \zeta_x(s) \mid^2 = 0 \,, \qquad \mathsf{M}\zeta_x(s)^2 < \infty \,,$$

we have

$$\mathsf{M}\left[\frac{f(\xi_{t,x+\Delta x}(s)) - f(\xi_{t,x}(s))}{\Delta x} - f'_x(\xi_{t,x}(s))\zeta_x(s)\right]^2$$

$$\leqq 2\mathsf{M}\left[\frac{f(\xi_{t,x+\Delta x}(s)) - f(\xi_{t,x}(s))}{\xi_{t,x+\Delta x}(s) - \xi_{t,x}(s)}[\zeta_{x,\Delta x}(s) - \zeta_x(s)]\right]^2$$

$$+ 2\mathsf{M}\left[\frac{f(\xi_{t,x+\Delta x}(s)) - f(\xi_{t,x}(s))}{\xi_{t,x+\Delta x}(s) - \xi_{t,x}(s)} - f'_x(\xi_{t,x}(s))\right]^2\zeta_x(s)^2 \to 0 \,.$$

But

$$\left|\frac{u(x + \Delta x) - u(x)}{\Delta x} - \mathsf{M}f'_x(\xi_{t,x}(s))\zeta_x(s)\right|$$

$$\leqq \sqrt{\mathsf{M}\left[\frac{1}{\Delta x}(f(\xi_{t,x+\Delta x}(s)) - f(\xi_{t,x}(s))) - f'_x(\xi_{t,x}(s))\zeta_x(s)\right]^2} \,,$$

from which equation (4) follows. Analogously, we can show that

$$u''_{xx} = \mathsf{M}f''_{xx}(\xi_{t,x}(s))\xi_x(s)^2 + \mathsf{M}f'_x(\xi_{t,x}(s))\eta_x(s) \,. \tag{5}$$

The continuity of u'_x and u''_{xx} follows from the mean-square continuity of the processes $\zeta_x(s)$ and $\eta_x(s)$.

REMARK 3. The processes

$$\xi_{t,x}(s) \,, \quad \frac{\partial}{\partial x}\xi_{t,x}(s) \,, \quad \frac{\partial^2}{\partial x^2}\xi_{t,x}(s)$$

are stochastically continuous functions of t for fixed s such that $t_0 \leqq s \leqq T$, and this continuity is uniform with respect to x on every compact set. Suppose that $t < t' < s$. Then for $\xi_{t,x}(s)$ we have

$$\xi_{t,x}(s) - \xi_{t',x}(s) = \xi_{t,x}(t') - x + \int_{t'}^{s} (a(u, \xi_{t,x}(u)) - a(u, \xi_{t',x}(u)))du$$

$$+ \int_{t'}^{s} [\sigma(u, \xi_{t,x}(u)) - \sigma(u, \xi_{t',x}(u))]dw(u) .$$

Consequently, for some H,

$$\mathsf{M}[\xi_{t,x}(s) - \xi_{t',x}(s)]^2 \leqq 3\mathsf{M} \mid \xi_{t,x}(t') - x \mid^2$$

$$+ H\int_{t'}^{s} \mathsf{M} \mid \xi_{t,x}(u) - \xi_{t',x}(u) \mid^2 du ,$$

from which it follows that

$$\mathsf{M} \mid \xi_{t,x}(s) - \xi_{t',x}(s) \mid^2 \leqq 3\mathsf{M} \mid \xi_{t,x}(t') - x \mid^2 e^{H(s-t')} .$$

But $\mathsf{M} \mid \xi_{t,x}(t') - x \mid^2 = O(t' - t)$ uniformly on every compact set (by virtue of Lemma 2, Section 3). Consequently

$$\mathsf{M} \mid \xi_{t,x}(s) - \xi_{t',x}(s) \mid^2 = O(t' - t) .$$

Using the equations $\partial\xi_{t,x}(s)/\partial x = \zeta_x(s)$ and $\partial^2\xi_{t,x}(s)/\partial x^2 = \eta_x(s)$, we can obtain analogous inequalities for these functions.

Combining the assertions in Remarks 1 and 2, we obtain:

Theorem 3. *Let $\xi_{t,x}(s)$ denote a solution of equation* (14) *of Section* 3 *in which the functions $\boldsymbol{a}(t, \boldsymbol{x}), \boldsymbol{b}_1(t, \boldsymbol{x}), \cdots, \boldsymbol{b}_m(t, \boldsymbol{x})$ satisfy the conditions of Theorem* 2. *Let $f(\boldsymbol{x})$ denote a bounded continuous function defined on $R^{(m)}$ with bounded continuous first and second order derivatives. Then the function*

$$u(t, \boldsymbol{x}) = \mathsf{M}f(\xi_{t,\boldsymbol{x}}(s))$$

defined for $t_0 \leqq t \leqq s$ and $\boldsymbol{x} \in R^{(m)}$ is continuous and has first and second derivatives that are continuous with respect to the set of variables $t, \boldsymbol{x}$.

5. THE METHOD OF DIFFERENTIAL EQUATIONS

In this section we shall derive differential equations that will enable us to determine the distributions of certain functionals of diffusion processes. In so doing, we shall present a new derivation of Kolmogorov's first equation for diffusion processes.

Let $\boldsymbol{\xi}(t)$ denote a solution of equation (13) of Section 3. Let $\mathsf{P}(t, \boldsymbol{x}, s, A)$ denote the transition probabilities for the process $\boldsymbol{\xi}(t)$. As we have noted above, $\mathsf{P}(t, \boldsymbol{x}, s, A)$ coincides with the distribution of the process $\boldsymbol{\xi}_{t,\boldsymbol{x}}(s)$, which is a solution of equation (14) of Section

3. Let $M_{t,x}$ denote the conditional mathematical expectation corresponding to the distribution $P(t, x, s, A)$. To determine the transition probabilities, it is sufficient for us to determine $M_{t,x}\varphi[\xi(s)]$ for all sufficiently smooth functions $\varphi(x)$. To do this we can use the following theorem (cf. Theorem 3, Section 1):

Theorem 1. *Suppose that the hypotheses of Theorem 2 of Section 4 are satisfied for the process $\xi(t)$. Suppose that the function $\varphi(x)$ is bounded and continuous and that it has bounded and continuous derivatives of the first and second orders. Then the function*

$$u(t, x) = M_{t,x}\varphi[\xi(s)] \qquad (t \in [t_0, s)])$$

has continuous first and second order derivatives with respect to x^i, it is differentiable with respect to t, and it satisfies the equation

$$\frac{\partial}{\partial t} u(t, x) + \sum_{i=1}^{m} a^i(t, x)\frac{\partial}{\partial x^i} u(t, x) + \frac{1}{2}\sum_{k,i,j=1}^{m} b_k^i(t, x)b_k^j(t, x)\frac{\partial^2}{\partial x^i \partial x^j} u(t, x) = 0 \qquad (1)$$

and the condition $\lim_{t \uparrow s} u(t, x) = \varphi(x)$, where a^i, b_k^i, and x^i are components of the vectors a, b_k and x respectively.

Proof. The differentiability of the function $u(t, x)$ with respect to x, and also the continuity and boundedness of the partial derivatives, follow from Theorem 3 of Section 4. We note also that

$$u(t, x) = M_{t,x}\varphi(\xi(s)) = M_{t,x}M_{t+\Delta t, \xi(t+\Delta t)}\varphi(\xi(s)) = M_{t,x}u(t + \Delta t, \xi(t + \Delta t)) .$$

Using formula (10) of Section 2, we may write

$$u(t + \Delta t, \xi(t + \Delta t)) - u(t + \Delta t, \xi(t)) = \int_t^{t+\Delta t}\left[\sum_{j=1}^{m} \frac{\partial u}{\partial x^j}(t + \Delta t, \xi(s))a^j(s, \xi(s)) + \frac{1}{2}\sum_{k,i,j=1}^{m} b_k^i(s, \xi(s))b_k^j(s, \xi(s))\frac{\partial^2 u(t + \Delta t, \xi(s))}{\partial x^i \partial x^j}\right]ds + \int_t^{t+\Delta t}\sum_{k,i=1}^{m} \frac{\partial u}{\partial x^i}(t + \Delta t, \xi(s))b_k^i(s, \xi(s))dw_k(s) .$$

Consequently,

$$M_{t,x}u(t + \Delta t, \xi(t + \Delta t)) - M_{t,x}u(t + \Delta t, \xi(t)) = \int_t^{t+\Delta t} M_{t,x}\left[\sum_{i=1}^{m} \frac{\partial u}{\partial x^i}(t + \Delta t, \xi(s))a^i(s, \xi(s)) + \frac{1}{2}\sum_{k,i,j=1}^{m} b_k^i(s, \xi(s))b_k^j(s, \xi(s))\frac{\partial^2}{\partial x^i \partial x^j} u(t + \Delta t, \xi(s))\right]ds .$$

Since

$$\mathsf{M}_{t,x}u(t+\Delta t, \boldsymbol{\xi}(t)) = u(t+\Delta t, \boldsymbol{x}) ,$$

we have

$$u(t, \boldsymbol{x}) - u(t+\Delta t, \boldsymbol{x}) = \mathsf{M}_{t,x}\Big[\sum_{i=1}^{m} \frac{\partial u}{\partial x^i}(t+\Delta t, \boldsymbol{\xi}(s'))a^i(s', \boldsymbol{\xi}(s'))$$

$$+ \frac{1}{2}\sum_{k,i,j=1}^{m} b_k^i(s', \boldsymbol{\xi}(s'))b_k^j(s', \boldsymbol{\xi}(s'))\frac{\partial^2}{\partial x^i \partial x^j}u(t+\Delta t, \boldsymbol{\xi}(s'))\Big]\Delta t ,$$

where s' is some number in the interval $(t, t+\Delta t)$.

Remembering that $\partial u/\partial x^i$ and $\partial^2 u/\partial x^i \partial x^j$ are bounded, and that $s' \to t$ as $\Delta t \to 0$, using Lebesgue's theorem and taking the limit under the integral sign, we obtain

$$\frac{u(t, \boldsymbol{x}) - u(t+\Delta t, \boldsymbol{x})}{\Delta t} \longrightarrow \mathsf{M}_{t,x}\Big[\sum_{i=1}^{m} \frac{\partial}{\partial x^i}u(t, \boldsymbol{\xi}(t))a^i(t, \boldsymbol{\xi}(t))$$

$$+ \frac{1}{2}\sum_{k,i,j=1}^{m} \frac{\partial^2}{\partial x^i \partial x^j}u(t, \boldsymbol{\xi}(t))b_k^i(t, \boldsymbol{\xi}(t))b_k^j(t, \boldsymbol{\xi}(t))\Big] .$$

Equation (1) is proved. The fact that $\mathsf{M}_{t,x}\varphi(\boldsymbol{\xi}(s)) \to \varphi(\boldsymbol{x})$ follows from the relation $\mathsf{M}_{t,x}\varphi(\boldsymbol{\xi}(s)) = \mathsf{M}\varphi(\boldsymbol{\xi}_{t,x}(s))$ and the continuity of $\boldsymbol{\xi}_{t,x}(s)$:

$$\mathsf{P}\Big\{\lim_{t \uparrow s}(\boldsymbol{\xi}_{t,x}(s) - \boldsymbol{\xi}_{t,x}(t)) = \lim_{s \downarrow t}\boldsymbol{\xi}_{t,x}(s) - \boldsymbol{x} = 0\Big\} = 1 .$$

This completes the proof of the theorem.

Let us now derive equations that will enable us to determine the distribution of the random variable

$$I = \int_{t_0}^{T} f(s, \boldsymbol{\xi}(s))ds$$

where $f(s, \boldsymbol{x})$ is a sufficiently smooth function and $\boldsymbol{\xi}(s)$ is a process that satisfies equation (13) of Section 3. We introduce the function

$$v_\lambda(t, \boldsymbol{x}) = \mathsf{M}_{t,x}\exp\Big\{\lambda\int_t^T f(s, \boldsymbol{\xi}(s))ds\Big\} . \qquad (2)$$

To determine the distribution of the variable I, it will be sufficient for us to determine the function $v_\lambda(t, \boldsymbol{x})$ for $t \in [t_0, T]$ and $\boldsymbol{x} \in R^{(m)}$ and all imaginary values of λ, since $v_\lambda(t_0, \boldsymbol{x})$ will then give us the conditional characteristic function of the random variable I under the hypothesis $\boldsymbol{\xi}(t_0) = \boldsymbol{x}$. If we integrate $v_\lambda(t_0, \boldsymbol{x})$ with respect to the initial distribution, we obtain the unconditional characteristic function of the variable I.

Theorem 2. *If $\boldsymbol{\xi}(t)$ satisfies the hypotheses of Theorem 2 of Section 4, and if the functions $f(t, \boldsymbol{x})$, $f'_{x^i}(t, \boldsymbol{x})$ and $f''_{x^i x^j}(t, \boldsymbol{x})$ for $i, j =$*

$1, \cdots, m$ *are continuous and bounded, then for* $t \in [t_0, T]$, *the function* $v_\lambda(t, \boldsymbol{x})$ *satisfies the equation*

$$\begin{aligned}\frac{\partial}{\partial t} v_\lambda(t, \boldsymbol{x}) &+ \sum_{i=1}^{m} a^i(t, \boldsymbol{x}) \frac{\partial}{\partial x^i} v_\lambda(t, \boldsymbol{x}) \\ &+ \frac{1}{2} \sum_{i,j,k=1}^{m} b_k^i(t, \boldsymbol{x}) b_k^j(t, \boldsymbol{x}) \frac{\partial^2}{\partial x^i \partial x^j} v_\lambda(t, \boldsymbol{x}) \\ &+ \lambda f(t, \boldsymbol{x}) v_\lambda(t, \boldsymbol{x}) = 0 \end{aligned} \tag{3}$$

and the condition $\lim_{t \uparrow T} v_\lambda(t, \boldsymbol{x}) = 1$.

Proof. Validity of the last condition is obvious. The continuity and differentiability of $v_\lambda(t, \boldsymbol{x})$ and the continuity and boundedness of the derivatives

$$\frac{\partial}{\partial x^i} v_\lambda(t, \boldsymbol{x}) \quad \text{and} \quad \frac{\partial^2}{\partial x^i \partial x^j} v_\lambda(t, \boldsymbol{x})$$

follow from formula (2) and the differentiability of $\boldsymbol{\xi}_{t,x}(s)$ and $f(s, \boldsymbol{x})$ with respect to $\boldsymbol{x}$, just as in the proof of Theorem 3 of Section 4. From the equation

$$\begin{aligned}\exp\left(\lambda \int_{t'}^{T} f(s, \boldsymbol{\xi}(s)) ds\right) &- \exp\left(\lambda \int_{t}^{T} f(s, \boldsymbol{\xi}(s)) ds\right) \\ &= \lambda \int_{t'}^{t} \exp\left(\lambda \int_{s}^{T} f(u, \boldsymbol{\xi}(u)) du\right) f(s, \boldsymbol{\xi}(s)) ds\end{aligned}$$

we obtain for $t' < t''$,

$$\begin{aligned}\mathsf{M}_{t',x} \exp\left(\lambda \int_{t'}^{T} f(s, \boldsymbol{\xi}(s)) ds\right) &- \mathsf{M}_{t',x} \exp\left(\lambda \int_{t''}^{T} f(s, \boldsymbol{\xi}(s)) ds\right) \\ &= \lambda \int_{t'}^{t''} \mathsf{M}_{t',x} f(s, \boldsymbol{\xi}(s)) \mathsf{M}_{s,\boldsymbol{\xi}(s)} \exp\left(\lambda \int_{s}^{T} f(u, \boldsymbol{\xi}(u)) du\right).\end{aligned}$$

But

$$\begin{aligned}\mathsf{M}_{t',x} &\exp\left\{\lambda \int_{t''}^{T} f(s, \boldsymbol{\xi}(s)) ds\right\} \\ &= \mathsf{M}_{t',x} \mathsf{M}_{t'',\boldsymbol{\xi}(t'')} \exp\left\{\lambda \int_{t''}^{T} f(s, \boldsymbol{\xi}(s)) ds\right\} = \mathsf{M}_{t',x} v_\lambda(t'', \boldsymbol{\xi}(t'')) .\end{aligned}$$

Therefore,

$$v_\lambda(t', \boldsymbol{x}) - \mathsf{M}_{t',x} v_\lambda(t'', \boldsymbol{\xi}(t'')) = \lambda \int_{t'}^{t''} \mathsf{M}_{t',x} f(s, \boldsymbol{\xi}(s)) v_\lambda(s, \boldsymbol{\xi}(s)) .$$

Since

$$f(s, \boldsymbol{\xi}_{t',x}(s)) v_\lambda(s, \boldsymbol{\xi}_{t',x}(s)) - f(t', \boldsymbol{x}) v_\lambda(t', \boldsymbol{x}) \longrightarrow 0$$

in probability as $s - t' \longrightarrow 0$, the limit

$$\lim_{\substack{t''-t' \downarrow 0 \\ t' \to t}} \frac{1}{t'' - t'} \int_{t'}^{t''} \mathsf{M}_{t',x} f(s, \boldsymbol{\xi}(s)) v_\lambda(s, \boldsymbol{\xi}(s)) ds = f(t, \boldsymbol{x}) v_\lambda(t, \boldsymbol{x})$$

exists. Therefore the limit

$$\lim_{\substack{t''-t'\downarrow 0\\ t'\to t}} \frac{v_\lambda(t', \mathbf{x}) - \mathsf{M}_{t',\mathbf{x}} v_\lambda(t'', \boldsymbol{\xi}(t''))}{t'' - t'}$$

$$= \lim_{\substack{t''-t'\downarrow 0\\ t'\to t}} \left[\frac{v_\lambda(t', \mathbf{x}) - v_\lambda(t'', \mathbf{x})}{t'' - t'} + \frac{v_\lambda(t'', \mathbf{x}) - \mathsf{M}_{t',\mathbf{x}} v_\lambda(t'', \boldsymbol{\xi}(t''))}{t'' - t'}\right]$$

also exists. But as shown in the proof of Theorem 1,

$$\lim_{\substack{t''-t'\downarrow 0\\ t'\to t}} \frac{\mathsf{M}_{t',\mathbf{x}} v_\lambda(t'', \boldsymbol{\xi}(t'')) - v_\lambda(t'', \mathbf{x})}{t'' - t'} = \sum_{i=1}^{m} a^i(t, \mathbf{x}) \frac{\partial}{\partial x^i} v_\lambda(t, \mathbf{x})$$

$$+ \frac{1}{2} \sum_{i,j,k=1}^{m} b_k^i(t, \mathbf{x}) b_k^j(t, \mathbf{x}) \frac{\partial^2 v_\lambda(t, \mathbf{x})}{\partial x^i \partial x^j}.$$

Consequently, the limit

$$\lim_{\substack{t''-t'\downarrow 0\\ t'\to t}} \frac{v_\lambda(t'', \mathbf{x}) - v_\lambda(t', \mathbf{x})}{t'' - t'} = \frac{\partial}{\partial t} v_\lambda(t, \mathbf{x})$$

exists and equation (3) is satisfied. This completes the proof of the theorem.

Let us look in greater detail at the case of a one-dimensional process of Brownian motion. If the function f is independent of t, to find the distribution of the random variable I, we can in this case obtain equations of simpler form and with less rigid restrictions. For a process of Brownian motion, the coefficients in equation (2) of Section 3 are: $a = 0$, $\sigma = 1$. It is easy to see that the solution of equation (6) of Section 3 can then be written in the explicit form $\xi_{t,x}(s) = x + w(s) - w(t)$, where $w(t)$ is a process of Brownian motion.

It follows from Theorem 2 that the function

$$v_\lambda(t, x) = \mathsf{M} \exp\left\{\lambda \int_t^T f(x + w(s) - w(t)) ds\right\}$$

satisfies the equation

$$\frac{\partial v_\lambda(t, x)}{\partial t} + \frac{1}{2} \frac{\partial^2 v_\lambda(t, x)}{\partial x^2} + \lambda f(x) v_\lambda(t, x) = 0$$

and the condition $\lim_{s\uparrow T} v_\lambda(s, x) = 1$. We note that $v_\lambda(t, x) = u_\lambda(T - t, x)$, where

$$u_\lambda(t, x) = \mathsf{M} \exp\left(\lambda \int_0^t f(x + w(s)) ds\right).$$

The function $u_\lambda(t, x)$ satisfies the equation

$$\frac{\partial}{\partial t} u_\lambda(t, x) = \frac{1}{2} \frac{\partial^2}{\partial x^2} u_\lambda(t, x) + \lambda f(x) u_\lambda(t, x) \tag{4}$$

and the initial condition $\lim_{t\downarrow 0} u_\lambda(t, x) = 1$.

Equation (4) can be solved by using the Laplace transform with respect to t. We define

$$z_{\mu,\lambda}(x) = \int_0^\infty e^{-\mu t} u_\lambda(t, x) dt .$$

Here and below, λ denotes a purely imaginary number and μ denotes a real nonnegative number. Under these conditions $z_{\mu,\lambda}(x)$ is meaningful. Multiplying (4) by $e^{-\mu t}$ and integrating with respect to t from 0 to ∞, we obtain

$$\mu z_{\mu,\lambda}(x) - 1 = \frac{1}{2} \frac{\partial^2}{\partial x^2} z_{\mu,\lambda}(x) + \lambda f(x) z_{\mu,\lambda}(x) . \qquad (5)$$

Equation (5) is valid when the function f and its derivatives are twice continuously differentiable and bounded. Now let $f(x)$ denote a piecewise-continuous bounded function. Let us choose a sequence of functions $f_n(x)$ each of which satisfies equation (5) as applied to $z_{\mu,\lambda}^{(n)}(x)$ where

$$z_{\mu,\lambda}^{(n)}(x) = \int_0^\infty e^{-\mu t} \mathsf{M} \exp\left(\lambda \int_0^t f_n(x + w(s)) ds\right) dt .$$

In addition, let us suppose that the $f_n(x)$ are uniformly bounded and that they converge to $f(x)$ for each x. One can easily see that the $z_{\mu,\lambda}^{(n)}(x)$ are bounded by the number $1/\mu$ and that for every μ, λ, and x, the sequence $\{z_{\mu,\lambda}^{(n)}(x)\}$ converges to

$$z_{\mu,\lambda}(x) = \int_0^\infty e^{-\mu t} \mathsf{M} \exp\left(\lambda \int_0^t f(x + w(s)) ds\right) dt . \qquad (6)$$

It follows from (5) that the $\partial^2 z_{\mu,\lambda}^{(n)}(x)/\partial x^2$ are bounded by the number $4 + 2 |\lambda| C/\mu$ where C is a constant bounding the $f_n(x)$, and that $\partial^2 z_{\mu,\lambda}^{(n)}(x)/\partial x^2$ converges to $2[\mu - \lambda f(x)] z_{\mu,\lambda}(x) - 2$ as $n \longrightarrow \infty$. Therefore the derivative $\partial^2 z_{\mu,\lambda}(x)/\partial x^2$ exists and the sequence $\{\partial^2 z_{\mu,\lambda}^{(n)}(x)/\partial x^2\}$ converges to $\partial^2 z_{\mu,\lambda}(x)/\partial x^2$ (at every point of continuity of the function $f(x)$). Consequently at points of continuity of the function $f(x)$, the function $z_{\mu,\lambda}(x)$ satisfies equation (5).

Thus we have:

Theorem 3. *If $w(t)$ is a process of Brownian motion and $z_{\mu,\lambda}(x)$ for $\mu > 0$ and* $\operatorname{Re} \lambda = 0$ *is defined by formula* (6), *where $f(x)$ is a bounded piecewise-continuous function, then $z_{\mu,\lambda}(x)$ is continuously differentiable, it has a second derivative at all points of continuity of $f(x)$, and it satisfies equation* (5).

Example. Let us find the distribution of the quantity

$$I_t = \int_0^t \operatorname{sgn} w(s) ds .$$

In this case, equation (5) takes the form

$$z''_{\mu,\lambda}(x) + 2(\lambda \operatorname{sgn} x - \mu)z_{\mu,\lambda}(x) = -2 .$$

Solving this equation separately for the cases $x > 0$ and $x < 0$, we obtain

$$z_{\mu,\lambda}(x) = \frac{1}{\mu - \lambda} + C_1 \exp(\sqrt{2\mu - 2\lambda}x) + C_2 \exp(-\sqrt{2\mu - 2\lambda}x) \quad \text{for } x > 0 ,$$

$$z_{\mu,\lambda}(x) = \frac{1}{\mu + \lambda} + C_3 \exp(\sqrt{2\mu + 2\lambda}x) + C_4 \exp(-\sqrt{2\mu + 2\lambda}x) \quad \text{for } x < 0 .$$

From the assumption that $z_{\mu,\lambda}(x)$ is bounded as $x \longrightarrow \pm\infty$, we obtain $C_1 = C_4 = 0$. Furthermore, using the continuity of the functions $z_{\mu,\lambda}(x)$ and $\partial z_{\mu,\lambda}(x)/\partial x$ at the point $x = 0$, we may write

$$\frac{1}{\mu - \lambda} + C_2 = \frac{1}{\mu + \lambda} + C_3; \; -C_2\sqrt{2\mu - 2\lambda} = C_3\sqrt{2\mu + 2\lambda} .$$

From this we get

$$C_3 = \frac{-1 + \sqrt{\dfrac{\mu + \lambda}{\mu - \lambda}}}{\mu + \lambda} .$$

To determine the distribution of random variable I_t, it suffices to know $z_{\mu,\lambda}(0)$. Substituting the value found for C_3, we obtain for $|\lambda| < \mu$,

$$z_{\mu,\lambda}(0) = \frac{1}{\sqrt{\mu^2 - \lambda^2}} = \frac{1}{\mu}\left(1 - \frac{\lambda^2}{\mu^2}\right)^{-1/2} = \frac{1}{\mu}\sum_{n=0}^{\infty} \frac{(2n-1)!!}{2n!!}\frac{\lambda^{2n}}{\mu^{2n}} .$$

Since

$$\int_0^\infty t^n e^{-\mu t} dt = \frac{n!}{\mu^{n+1}}$$

and

$$\int_{-\pi/2}^{\pi/2} \sin^k \varphi d\varphi = \begin{cases} 0 & \text{for } k \text{ odd} , \\ \dfrac{(2n-1)!!}{(2n)!!}\pi & \text{for } k = 2n , \end{cases}$$

we have

$$\mathsf{M} \exp\left(\lambda\int_0^t \operatorname{sgn} w(s)ds\right) = \sum_{n=0}^{\infty} \frac{(2n-1)!!}{(2n)!!}\frac{\lambda^{2n}t^{2n}}{(2n)!}$$

$$= \sum_{k=0}^{\infty} \frac{1}{\pi}\int_{-\pi/2}^{\pi/2} \sin^k \varphi d\varphi \frac{\lambda^k t^k}{k!} = \frac{1}{\pi}\int_{-\pi/2}^{\pi/2} e^{\lambda t \sin\varphi} d\varphi .$$

Consequently,

$$\mathsf{M}\exp\left(\lambda\frac{1}{t}\int_0^t \operatorname{sgn} w(s)ds\right) = \frac{1}{\pi}\int_{-\pi/2}^{\pi/2} e^{\lambda\sin\varphi}d\varphi = \int_{-\infty}^{\infty} e^{\lambda u}dF(u) ,$$

where

$$F(x) = \begin{cases} 0 & \text{for } x < -1 , \\ \dfrac{1}{\pi}\left(\arcsin x + \dfrac{\pi}{2}\right) & \text{for } |x| \leqq 1 , \\ 1 & \text{for } |x| > 1 . \end{cases}$$

Thus

$$\mathsf{P}\left\{\int_0^t \operatorname{sgn} w(s)ds < x\right\} = \frac{1}{\pi}\left(\arcsin\frac{x}{t} + \frac{\pi}{2}\right) .$$

REMARK 1. Suppose that the function $\varphi_t(x)$ is defined by

$$\varphi_t(x) = \frac{1 + \operatorname{sgn} x}{2} .$$

Then the quantity

$$\tau_t^+ = \int_0^t \varphi^+(w(s))ds$$

represents the time passed on the positive half-axis by the process $w(s)$ during the time t. Using the result obtained, we can find the distribution τ_t^+. Specifically, since

$$\tau_t^+ = \frac{t}{2} + \frac{1}{2}\int_0^t \operatorname{sgn} w(s)ds ,$$

we have

$$\mathsf{P}\left\{\frac{1}{t}\tau_t^+ < x\right\} = \mathsf{P}\left\{\frac{1}{t}\int_0^t \operatorname{sgn} w(s)ds < 2x - 1\right\}$$

$$= \frac{1}{\pi}\left(\arcsin(2x-1) + \frac{\pi}{2}\right) \quad \text{for} \quad 0 \leqq x \leqq 1 .$$

Let us use some elementary transformations to obtain the distribution of the variable τ_t^+ in a simpler (and more commonly used) form. If

$$\frac{1}{2}\arccos(1 - 2x) = z ,$$

then on the one hand,

$$\cos 2z = 1 - 2x , \quad x = \frac{1 - \cos 2z}{2} = \sin^2 z , \quad z = \arcsin\sqrt{x} ,$$

and on the other hand,

$$\arcsin(2x - 1) + \frac{\pi}{2} \arccos(1 - 2x) = 2 \cdot \frac{z}{2} = 2 \arcsin\sqrt{x} .$$

Thus we have obtained what is known as the "arc sine law."

Theorem 4. *If τ_t^+ is the time a process of Brownian motion has spent, up to the instant t, on the half-line $x > 0$, then*

$$\mathsf{P}\{\tau_t^+ < x\} = \begin{cases} 0 & \text{for } x < 0 , \\ \dfrac{2}{\pi} \arcsin\sqrt{\dfrac{x}{t}} & \text{for } 0 \leqq x \leqq t , \\ 1 & \text{for } x > t . \end{cases}$$

6. ONE-DIMENSIONAL DIFFUSION PROCESSES WITH ABSORPTION

Let $g_1(t)$ and $g_2(t)$ denote two continuous functions defined on $[0, T]$ and satisfying the condition $g_1(t) < g_2(t)$. Let G denote the region in the tx-plane bounded by these curves and the straight lines $t = 0$ and $t = T$. We shall call the process $\xi^*(t)$ a *process with absorption* on the boundary G if it is $\mathfrak{F}_t$-measurable and if it has the property: The equation $\xi^*(t') = g_i(t')$ imples $\xi^*(t) = g_i(t)$ for all $t \in [t', T]$. In this section we shall consider diffusion processes with absorption on the boundary of the region G, and we shall assume that the functions $g_i(t)$ are continuously differentiable. The conditions for absorption on the boundary of G make obvious restrictions on the diffusion coefficients $a^*(t, x)$ and $\sigma^*(t, x)$ at points of the boundary of G: Since $d\xi^*(g_i(t)) = dg_i(t) = g_i'(t)dt$, we have

$$a^*(t, g_i(t)) = g_i'(t) , \qquad \sigma^*(t, g_i(t)) = 0 . \tag{1}$$

Within the region G, the coefficients $a^*(t, x)$ and $\sigma^*(t, x)$ can in principle be arbitrary. Suppose that $a^*(t, x)$ and $\sigma^*(t, x)$ are defined in the region G and satisfy conditions (1). Consider the stochastic differential equation

$$d\xi^*(t) = a^*(t, \xi^*(t))dt + \sigma^*(t, \xi^*(t))dw(t) . \tag{2}$$

Let us examine the question of the existence and uniqueness of a solution to this equation, a solution which is a process with absorption on the boundary of G.

Theorem 1. *If there exists a number K such that*

$$| a^*(t, x_1) - a^*(t, x_2) | + | \sigma^*(t, x_1) - \sigma^*(t, x_2) | \leqq K | x_1 - x_2 | ,$$
$$| a^*(t, x) | + | \sigma^*(t, x) | \leqq K$$

for $x_1, x_2 \in (g_1(t), g_2(t))$, and if ξ_0^* is independent of the process $w(t)$

and satisfies the equation $\mathsf{P}\{\xi_0^* \in [g_1(0), g_2(0)]\} = 1$, *then equation* (2) *has a unique solution satisfying the initial condition* $\xi_0^* = \xi^*(0)$ *in the form of a process with absorption on the boundary of* G.

Proof. Let us first show that equation (2) has a solution that is a process with absorption on the boundary of G. Suppose that $a(t, x)$ and $\sigma(t, x)$ are defined for $t \in [0, T]$ and $x \in (-\infty, \infty)$ in such a way that $a^*(t, x) = a(t, x)$ and $\sigma^*(t, x) = \sigma(t, x)$ for $x \in (g_1(t), g_2(t))$, and the functions $a(t, x)$ and $\sigma(t, x)$ satisfy a Lipschitz condition for all x. Then the equation

$$d\xi(t) = a(t, \xi(t))dt + \sigma(t, \xi(t))dw(t) \tag{3}$$

with initial condition $\xi(0) = \xi_0^*$ has a solution. Let τ denote the smallest root of the equation $(\xi(t) - g_1(t))(\xi(t) - g_2(t)) = 0$ on the interval $[0, T]$. If the equation has no root in $[0, T]$, we shall take $\tau = T$.

Let us set $\xi^*(t) = \xi(t)$ for $t < \tau$ but $\xi^*(t) = g_i(t)$ if $t \geqq \tau$, and $\xi(\tau) = g_i(\tau)$. By construction, this will be a process with absorption on the boundary of the region G. Let us show that $\xi^*(t)$ is a solution to equation (2). If $\chi(t) = 1$ for $t < \tau$ but $\chi(t) = 0$ for $t \geqq \tau$, and if $\xi(\tau) = g_i(\tau)$, then

$$\begin{aligned}
&\xi^*(t) - \xi_0^* - \int_0^t a^*(s, \xi^*(s))ds - \int_0^t \sigma^*(s, \xi^*(s))dw(s) \\
&\quad = \left[\xi(t) - \xi_0^* - \int_0^t a(s, \xi(s))ds - \int_0^t \sigma(s, \xi(s))dw(s)\right]\chi(t) \\
&\quad\quad + \left[\xi^*(t) - \xi_0^* - \int_0^\tau a^*(s, \xi^*(s))ds - \int_0^\tau \sigma^*(s, \xi^*(s))dw(s)\right. \\
&\quad\quad \left. - \int_\tau^t g_i'(s)ds\right](1 - \chi(t)) = 0 \,.
\end{aligned}$$

This is true because if $\chi(t) = 1$, then for $s < t$,

$$a^*(s, \xi^*(s)) = a(s, \xi(s)) \,, \qquad \sigma^*(s, \xi^*(s)) = \sigma(s, \xi(s)) \,,$$
$$\xi^*(s) = \xi(s) \,,$$

and if $\chi(t) = 0$,

$$\int_0^t \sigma^*(s, \xi^*(s))dw(s) = \int_0^\tau \sigma(s, \xi(s))dw(s) \,,$$
$$\int_0^t a^*(s, \xi(s))ds = \int_0^\tau a(s, \xi(s))ds + \int_\tau^t g_i'(s)ds$$

and

$$\begin{aligned}
&\xi^*(t) - \xi_0^* - \int_0^\tau a(s, \xi(s))ds - \int_0^\tau \sigma(s, \xi(s))dw(s) \\
&\quad = \xi^*(t) - \xi^*(\tau) = g_i(t) - g_i(\tau) = \int_\tau^t g_i'(s)ds \,.
\end{aligned}$$

Let us now prove that a solution of equation (2) that is a process with absorption on the boundary is unique. (As usual, by uniqueness we mean uniqueness up to stochastic equivalence.)

Let $\xi_1^*(t)$ and $\xi_2^*(t)$ denote two such solutions. We note that a sufficient condition for them to coincide is that they coincide inside the reigion G, since such solutions reach a given point of the boundary at the same instant and hence coincide. Furthermore, they are processes with absorption on the boundary of G. Let $\chi(t)$ denote the function that is equal to 1 if $\xi_i^*(s) \in (g_1(s), g_2(s))$ for $s \in [0, t]$ and $i = 1, 2$, and equal to 0 otherwise. Then

$$[\xi_1^*(t) - \xi_2^*(t)]\chi(t) = \chi(t)\int_0^t \chi(s)[a^*(s, \xi_1^*(s)) - a^*(s, \xi_2^*(s))]ds$$
$$+ \chi(t)\int_0^t \chi(s)[\sigma^*(s, \xi_1^*(s)) - \sigma^*(s, \xi_2^*(s))]dw(s) ,$$

from which we get

$$\mathsf{M}(\xi_1^*(t) - \xi_2^*(t))^2\chi(t) \leqq 2\mathsf{M}\left[\int_0^t \chi(s)(a^*(s, \xi_1^*(s)) - a^*(s, \xi_2^*(s)))ds\right]^2$$
$$+ 2\mathsf{M}\left[\int_0^t \chi(s)[\sigma^*(s, \xi_1^*(s)) - \sigma^*(s, \xi_2^*(s))]dw(s)\right]^2$$
$$\leqq L\int_0^t \mathsf{M}(\xi_1^*(s) - \xi_2^*(s))^2\chi(s)ds ,$$

where L is some constant. (Here we used the fact that $a^*(s, x)$ and $\sigma^*(s, x)$ satisfy a Lipschitz condition for $x \in (g_1(s), g_2(s))$.) Just as in the proof of the uniqueness in Theorem 1 of Section 3, it now follows from the last relation that $\mathsf{M}(\xi_1^*(t) - \xi_2^*(t))^2\chi(t) = 0$; that is, keeping in mind the continuity of the processes $\xi_i(t)$, we conclude that $\mathsf{P}\{\xi_1(t) = \xi_2(t), t \leqq \tau\} = 1$. This completes the proof of the theorem.

REMARK 2. A solution of equation (2) that is a process with absorption on the boundary of the region G, is a Markov process whose transition probabilities $\mathsf{P}^*(t, x, s, dy)$ coincide with the distribution of the process $\xi_{t,x}^*(s)$, with absorption on the boundary of G, that is a solution of the equation

$$\xi_{t,x}^*(s) = x + \int_t^s a^*(u, \xi_{t,x}^*(u))du$$
$$+ \int_t^s \sigma^*(u, \xi_{t,x}^*(u))dw(u) \qquad (s > t) . \tag{4}$$

The process $\xi_{t,x}^*(s)$ can be obtained from the process $\xi_{t,x}(s)$, which is a solution of equation (6) of Section 3, if in that equation we substitute those $a(t, x)$ and $\sigma(t, x)$ that were used in the proof of Theorem 1. Let $\tau_{t,x}$ denote the smallest root of the equation $(\xi_{t,x}(s) - g_1(s)) \times (\xi_{t,x}(s) - g_2(s)) = 0$ on the interval $s \in [t, T]$, and set $\tau_{t,x} = T$ when this equation has no root. Then $\xi^*_{t,x}(s) = \xi_{t,x}(s)$ for $s < \tau_{t,x}$ but $\xi^*_{t,x}(s) = g_i(s)$ for $s \geqq \tau_{t,x}$ if $\xi_{t,x}(\tau_{t,x}) = g_i(\tau_{t,x})$. The first assertion is proved in a manner analogous to that of the proof of Theorem 2, Section 3, and the second assertion follows from Theorem 1.

Let us look at certain transformations that enable us to simplify equation (2) in the region G. In addition to the process $\xi^*(t)$, let us consider the process $\eta^*(t) = f(t, \xi^*(t))$, where $f(t, x)$ is for each t an increasing twice continuously differentiable function of x for $x \in [g_1(t), g_2(t)]$ and differentiable with respect to t.

The process $\eta^*(t)$ is also a Markov process since $\xi^*(t)$ is uniquely determined from $\eta^*(t)$. This is a process with absorption on the boundary of the region $\bar{G}$ that is bounded by the curves

$$\bar{g}_1(t) = f(t, g_1(t)) , \qquad g_2(t) = f(t, g_2(t)) .$$

Using property 8 of Section 2, we see that $\eta^*(t)$ satisfies the equation

$$d\eta^*(t) = \Big[f'_t(t, \xi^*(t)) + f'_x(t, \xi^*(t)) a^*(t, \xi^*(t)) + \frac{1}{2} f''_{xx}(t, \xi^*(t)) \sigma^*(t, \xi^*(t))^2 \Big] dt + f'_x(t, \xi^*(t)) \sigma^*(t, \xi^*(t)) dw(t) .$$

Thus $\eta^*(t)$ is a solution of a stochastic equation of the form (2):

$$d\eta^*(t) = \bar{a}^*(t, \eta^*(t))dt + \bar{\sigma}^*(t, \eta^*(t))dw(t) ,$$

where

$$\bar{a}^*(t, y) = f'(t, \varphi(t, y)) + f'_x(t, \varphi(t, y)) a^*(t, \varphi(t, y)) + \frac{1}{2} f''_{xx}(t, \varphi(t, y)) \sigma^*(t, \varphi(t, y))^2 ,$$

$$\bar{\sigma}^*(t, y) = f'_x(t, \varphi(t, y)) \sigma^*(t, \varphi(t, y)) , \tag{5}$$

and the function $\varphi(t, y)$ is the inverse of $f(t, x)$ with respect to x; that is, $f(t, \varphi(t, y)) = y$, and $\varphi(t, f(t, x)) = x$. If we set $f'_x(t, x) = 1/\sigma^*(t, x)$ for $x \in (g_1(t), g_2(t))$, then $\bar{\sigma}^*(t, y) = 1$ for $y \in (\bar{g}_1(t), \bar{g}_2(t))$. Suppose that

$$f(t, z) = \int_{g_1(t)}^{z} \frac{du}{\sigma^*(t, u)} .$$

Then

$$\bar{g}_1(t) = 0 \,, \qquad \bar{g}_2(t) = \int_{g_1(t)}^{g_2(t)} \frac{du}{\sigma^*(t, u)} = C(t) \,.$$

Let us introduce the process $\eta_1^*(t) = \eta^*(t)/C(t)$. This is also a Markov process, with absorption on the boundary of the rectangular region Δ: $\{0 \leqq t \leqq T, 0 \leqq x \leqq 1\}$, and it satisfies the equation

$$d\eta_1^*(t) = \left[-\frac{C'(t)}{C(t)}\eta_1^*(t) + \bar{a}^*(t, C(t)\eta_1^*(t))\right]dt + \frac{\bar{\sigma}^*(t, C(t)\eta_1^*(t))}{C(t)}dw(t) \,. \tag{6}$$

Now let us define the process $\zeta(t) = \int_0^t (dw(s)/C(s))$. This process is, with probability 1, a continuous process with independent increments and satisfies the relations

$$\mathsf{M}\zeta(t) = 0 \,, \qquad \mathsf{D}\zeta(t) = \int_0^t \frac{ds}{C^2(s)} = \lambda(t) \,.$$

Therefore $\zeta(t) = w_1(\lambda(t))$, where $w_1(t)$ is a process of Brownian motion. Setting $\lambda(t) = s$ (so that $t = \lambda^{-1}(s)$) and

$$\eta_1^*(\lambda^{-1}(s)) = \xi_1^*(s) \,,$$

$$-\frac{C'(\lambda^{-1}(s))}{C(\lambda^{-1}(s))}x + \bar{a}^*(\lambda^{-1}(s), C(\lambda^{-1}(s))x) = a_1^*(s, x)$$

and letting $\chi_{(0,1)}(x)$ denote the characteristic function of the interval (0, 1), we obtain from (6) the following equation for $\xi_1^*(t)$:

$$d\xi_1^*(t) = a_1^*(t, \xi_1^*(t))dt + \chi_{(0,1)}(\xi_1^*(t))dw_1(t) \,. \tag{7}$$

Thus an arbitrary problem associated with finding the distribution of any characteristic of the process $\xi^*(t)$ that is a solution of equation (2), can be reduced to finding the distribution of some other characteristic of the process $\xi_1^*(t)$ that is a solution of equation (7). The latter is somewhat simpler than equation (2). The transition probabilities of the process $\xi^*(t)$ can easily be obtained in terms of the transition probabilities for the process $\xi_1^*(t)$.

Consider the question of determining the transition probabilities for the process $\xi_1^*(t)$. From the remarks made above we note that the transition probabilities $\mathsf{P}_1^*(t, x, s, dy)$ of the process $\xi_1^*(t)$ coincide with the distribution of the process $\xi_{t,x}^{(1)}(s)$ constructed as follows: Let $\xi_{t,x}(s)$ denote a solution of the equation

$$\xi_{t,x}(s) = x + \int_t^s a_1(u, \xi_{t,x}(u))du + w_1(s) - w(t) \,, \qquad 0 < x < 1 \,. \tag{8}$$

Suppose that $a_1(u, x)$ coincides with $a_1^*(u, x)$ for $0 < x < 1$. Suppose that this function is bounded and satisfies a Lipschitz condition, and that $\tau_{t,x}$ is the smallest root of the equation $\xi_{t,x}(s)(1 - \xi_{t,x}(s)) = 0$ on the interval $s \in [t, T]$. If this equation has no root, let $\tau_{t,x} = T$. Then $\xi_{t,x}^{(1)}(s) = \xi_{t,x}(s)$ for $s < \tau_{t,x}$ and $\xi_{t,x}^{(1)}(s) = \xi_{t,x}^{(1)}(\tau_{t,x})$ for $s \geqq \tau_{t,x}$. Let us show that $\tau_{t,x}$ is with probability 1 a continuous function of t and x. Since $\tau_{t,x} = \min[\tau_{t,x}^{(0)}, \tau_{t,x}^{(1)}]$, where $\tau_{t,x}^{(i)}$ is the smallest root of the equation $\xi_{t,x}(s) - i = 0$ on $[t, T]$ ($\tau_{t,x}^{(i)} = T$ if that equation has no root), it follows that to prove the continuity of $\tau_{t,x}$ it suffices to prove the continuity of $\tau_{t,x}^{(0)}$ and $\tau_{t,x}^{(1)}$. For this we use:

Lemma 1. *For every $h > 0$,*

$$\mathsf{P}\Big\{\sup_{\tau_{t,x}^{(1)} < s < \tau_{t,x}^{(1)} + h} \xi_{t,x}(s) > 1,\ \tau_{t,x}^{(1)} < T\Big\} = \mathsf{P}\{\tau_{t,x}^{(1)} < T\}\,.$$

Proof. Let us choose $\varepsilon > 0$ and $\delta > 0$. Let ν_δ denote the smallest natural number such that $\xi_{t,x}(\delta\nu_\delta) > 1 - \varepsilon$. Then

$$\begin{aligned}
&\mathsf{P}\Big\{\sup_{\delta\nu_\delta < s < \delta\nu_\delta + h} \xi_{t,x}(s) > 1,\ \delta\nu_\delta < T\Big\} \\
&\quad = \sum_{k\delta < T} \mathsf{P}\Big\{\sup_{k\delta < s < k\delta + h} \xi_{t,x}(s) > 1,\ \nu_\delta = k\Big\} \\
&\quad \geqq \sum_{k\delta < T} \mathsf{P}\Big\{\sup_{k\delta < s < k\delta + h} (\xi_{t,x}(s) - \xi_{t,x}(k\delta)) > \varepsilon,\ \nu_\delta = k\Big\} \\
&\quad \geqq \sum_{k\delta < T} \mathsf{P}\Big\{\sup_{k\delta < s < k\delta + h} [w(s) - w(k\delta)] > \varepsilon + Ch,\ \nu_\delta = k\Big\}\,,
\end{aligned}$$

where C is a constant bounding $a_1(t, x)$. (Here we used the inequality $\sup_{k\delta < s < k\delta + h} (\xi_{t,x}(s) - \xi_{t,x}(k\delta)) \leqq hC + \sup_{k\delta < s < k\delta + h} (w(s) - w(k\delta))$.) The process $w(s) - w(k\delta)$ for $s > k\delta$ is independent of the event $\{\nu_\delta = k\}$ since this event is completely determined by the behavior of the process $w(u)$ on the interval $[t, k\delta]$. Therefore, on the basis of Corollary 1, Section 5, Chapter VI,

$$\begin{aligned}
&\mathsf{P}\Big\{\sup_{\delta\nu_\delta < s < \delta\nu_\delta + h} \xi_{t,x}(s) > 1,\ \delta\nu_\delta < T\Big\} \\
&\quad \geqq \sum_{k\delta < T} \mathsf{P}\Big\{\sup_{k\delta < s < k\delta + h} (w(s) - w(k\delta)) > \varepsilon + Ch\Big\}\mathsf{P}\{\nu_\delta = k\} \\
&\quad = \frac{2}{\sqrt{2\pi}} \int_{\varepsilon/\sqrt{h} + C\sqrt{h}}^{\infty} \exp\Big(-\frac{u^2}{2}\Big) du\, \mathsf{P}\{\delta\nu_\delta < T\}\,.
\end{aligned}$$

Since

$$\mathsf{P}\{\tau_{t,x}^{(1)} < \delta\nu_\delta\} \leqq \mathsf{P}\Big\{\sup_{|s' - s''| \leqq \delta} |\xi(s') - \xi(s'')| > \varepsilon\Big\}\,,$$

we have

$$\mathsf{P}\Big\{\sup_{\tau_{(x)}^{(1)}<s<\tau_x^{(1)}+h} \xi_{t,x}(s) > 1,\ \tau_{t,x}^{(1)} < T\Big\}$$

$$\geqq \mathsf{P}\Big\{\sup_{\delta\nu_\delta<s<\delta\nu_\delta+h} \xi_{t,x}(s) > 1,\ \tau_{t,x}^{(1)} \geqq \delta\nu_\delta,\ \delta\nu_\delta < T\Big\}$$

$$\geqq \mathsf{P}\Big\{\sup_{\delta\nu_\delta<s<\delta\nu_\delta+h} \xi_{t,x}(s) > 1,\ \delta\nu_\delta < T\Big\} - \mathsf{P}\,\{\tau_{t,x}^{(1)} < \delta\nu_\delta\}$$

$$\geqq \frac{2}{\sqrt{2\pi}}\int_{\varepsilon/\sqrt{h}+C\sqrt{h}}^{\infty} \exp\Big(-\frac{u^2}{2}\Big)du\,\mathsf{P}\,\{\delta\nu_\delta < T\}$$

$$- \mathsf{P}\Big\{\sup_{|s'-s''|\leqq\delta} |\,\xi_{t,x}(s') - \xi_{t,x}(s'')\,| > \varepsilon\Big\}\,.$$

We also note that $\delta\nu_\delta < \tau_x^{(1)}$ for sufficiently small δ and hence

$$\lim_{\delta\to 0} \mathsf{P}\{\delta\nu_\delta < T\} \geqq \mathsf{P}\{\tau_{t,x}^{(1)} < T\}\,.$$

Therefore, taking the limit as $\delta \geqq 0$ and remembering that $\xi_{t,x}(s)$ is continuous with probability 1, we obtain

$$\mathsf{P}\Big\{\sup_{\tau_{t,x}^{(1)}<s<\tau_{t,x}^{(1)}+h} \xi_{t,x}(s) > 1,\ \tau_{t,x}^{(1)} < T\Big\}$$

$$\geqq \frac{2}{\sqrt{2\pi}}\int_{\varepsilon/\sqrt{h}+C\sqrt{h}}^{\infty} \exp\Big(-\frac{u^2}{2}\Big)du\,\mathsf{P}\,\{\tau_{t,x}^{(1)} < T\}\,.$$

Since for $\bar{h} < h$,

$$\mathsf{P}\Big\{\sup_{\tau_{t,x}^{(1)}<s<\tau_{t,x}^{(1)}+h} \xi_{t,x}(s) > 1,\ \tau_{t,x}^{(1)} < T\Big\}$$

$$\geqq \mathsf{P}\Big\{\sup_{\tau_{t,x}^{(1)}<s<\tau_{t,x}^{(1)}+\bar{h}} \xi_{t,x}(s) > 1,\ \tau_{t,x}^{(1)} < T\Big\}$$

$$\geqq \frac{2}{\sqrt{2\pi}}\int_{\varepsilon/\sqrt{\bar{h}}+C\sqrt{\bar{h}}}^{\infty} \exp\Big(-\frac{u^2}{2}\Big)du\,\mathsf{P}\,\{\tau_{t,x}^{(1)} < T\}\,,$$

by taking the limit first as $\varepsilon \to 0$ and then as $\bar{h} \to 0$, we obtain the assertion of the lemma.

Let us now prove:

Theorem 2. *The random variable $\tau_{t,x}$ is, with probability* 1, *continuous as a function of t and x.*

Proof. As already noted, to prove the assertion it will be sufficient to show that the random variables $\tau_{t,x}^{(1)}$ and $\tau_{t,x}^{(0)}$ are continuous with probability 1. Let us show, for example, that $\tau_{t,x}^{(1)}$ is continuous. We note that for $t_1 < t_2$ and $x_1, x_2 \in (0, 1)$,

$$| \xi_{t_1,x_1}(s) - \xi_{t_2,x_2}(s) | \leqq | x_1 - x_2 | + \left| \int_{t_1}^{t_2} a_1(u, \xi_{t,x}(u))du \right|$$
$$+ \int_{t_2}^{s} | a_1(u, \xi_{t_1,x_1}(u)) - a_1(u, \xi_{t_2,x_2}(u)) | \, du + | w(t_2) - w(t_1) |$$
$$\leqq | x_1 - x_2 | + C | t_1 - t_2 | + | w(t_2) - w(t_1) |$$
$$+ K \int_{t_2}^{s} | \xi_{t_1,x_1}(u) - \xi_{t_2,x_2}(u) | \, du \, .$$

Therefore

$$| \xi_{t_1,x_1}(s) - \xi_{t_2,x_2}(s) | \leqq (| x_1 - x_2 | + C | t_1 - t_2 | + | w(t_2) - w(t_1) |) \exp \{K(s - t_2)\} \, ,$$

so that for some L,

$$| \xi_{t_1,x_1}(s) - \xi_{t_2,x_2}(s) | \leqq L(| x_1 - x_2 | + | t_1 - t_2 | + | w(t_1) - w(t_2) |) \, . \tag{9}$$

The event $\tau^{(1)}_{t_1,x_1} > \tau^{(1)}_{t_2,x_2} + \delta$ means that

$$\xi_{t_1,x_1}(s) < 1 \quad \text{for} \quad s \in (\tau^{(1)}_{t_2,x_2}, \tau^{(1)}_{t_2,x_2} + \delta) \, .$$

Since

$$| \xi_{t_1,x_1}(s) - \xi_{t_2,x_2}(s) | \leqq L(| x_1 - x_2 | + | t_1 - t_2 | + | w(t_1) - w(t_2) |) \, ,$$

it follows that

$$\xi_{t_2,x_2}(s) < 1 + L(| x_1 - x_2 | + | t_1 - t_2 | + | w(t_1) - w(t_2) |)$$

for $s \in (\tau^{(1)}_{t_2,x_2}, \tau^{(1)}_{t_2,x_2} + \delta)$. We define

$$\eta_\delta = \sup_{\tau^{(1)}_{t_2,x_2} < s < \tau^{(1)}_{t_2,x_2} + \delta} \xi_{t_2,x_2}(s) - 1 \, .$$

According to Lemma 1, $\mathsf{P}\{\eta_\delta > 0\} = 1$ for every $\delta > 0$. As soon as

$$L(| x_1 - x_2 | + | t_1 - t_2 | + | w(t_1) - w(t_2) |) \leqq \eta_\delta \, ,$$

the event $\tau^{(1)}_{t_1,x_1} > \tau^{(1)}_{t_2,x_2} + \delta$ becomes impossible. Thus $\tau^{(1)}_{t_1,x_1} \leqq \tau^{(1)}_{t_2,x_2} + \delta$, if

$$L(| x_1 - x_2 | + | t_1 - t_2 | + | w(t_1) - w(t_2) |) \leqq \eta_\delta \, .$$

(This inequality holds for sufficiently small $| x_1 - x_2 |$ and $| t_1 - t_2 |$ by virtue of the continuity of $w(t)$.)

Let us also define

$$\zeta_\delta = 1 - \sup_{s < \tau^{(1)}_{t_2,x_2} - \delta} \xi_{t_2,x_2}(s) \, .$$

Obviously $\mathsf{P}\{\zeta_\delta > 0\} = 1$ for all $\delta > 0$. If

$$L(| x_1 - x_2 | + | t_1 - t_2 | + | w(t_1) - w(t_2) |) < \zeta_\delta \, ,$$

then

$$\sup_{s<\tau^{(1)}_{t_2,x_2}-\delta} \xi_{t_1,x_1}(s) < \sup_{s<\tau^{(1)}_{t_2,x_2}-\delta} \xi_{t_2,x_2}(s) + \zeta_\delta = 1$$

and hence $\tau^{(1)}_{t_1,x_1} > \tau^{(1)}_{t_2,x_2} - \delta$. This proves the continuity of $\tau^{(1)}_{t,x}$.

The continuity of $\tau^{(0)}_{t,x}$ is proved in an analogous manner, and this completes the proof of the theorem.

We shall now prove a theorem that will enable us to determine the transition probabilities of the process $\xi_1^*(t)$ that is a solution of (7).

Theorem 3. *Suppose that $a_1^*(t, x)$ for $x \in (0, 1)$ has uniformly continuous partial derivatives $\partial a_1^*(t, x)/\partial x$ and $\partial^2 a_1^*(t, x)/\partial x^2$ in the region $x \in (0, 1)$, $t \in [0, T]$. Suppose also that the function $f(x)$ is twice continuously differentiable and*

$$f(0) = f(1) = f'(0) = f'(1) = f''(0) = f''(1) = 0 .$$

Then the function

$$u^*(t, x) = \mathsf{M}(f(\xi(s)) \mid \xi(t) = x) = \int f(y)\mathsf{P}_1^*(t, x, s, dy) ,$$

where $\mathsf{P}_1^*(t, x, s, dy)$ is the transition probability of the process $\xi_1^*(t)$ in the region $0 < t < s$, $x \in (0, 1)$, satisfies the differential equation

$$\frac{\partial}{\partial t} u^*(t, x) + a_1^*(t, x)\frac{\partial}{\partial x} u^*(t, x) + \frac{1}{2}\frac{\partial^2}{\partial x^2} u^*(t, x) = 0 \qquad (10)$$

and the boundary conditions

$$\lim_{t\to s} u^*(t, x) = f(x) , \qquad \lim_{x\to 0} u^*(t, x) = 0 , \qquad \lim_{x\to 1} u^*(t, x) = 0 .$$

Proof. Let us first show that the process $\xi_{t,x}(s)$ is, with probability 1, twice differentiable with respect to x, and that the derivatives $\partial\xi_{t,x}(s)/\partial x$ and $\partial^2\xi_{t,x}(s)/\partial x^2$ with probability 1 are continuous with respect to the set of variables t and x and bounded by constants independent of chance. We shall assume that $a_1(t, x)$ in (8) is chosen in such a way that $a_1(t, x)$, $\partial a_1(t, x)/\partial x$, and $\partial^2 a_1(t, x)/\partial x^2$ exist for all x and are bounded and uniformly continuous for $t \in [0, T]$, $x \in (-\infty, \infty)$. (By virtue of the uniform continuity of

$$a_1^*(t, x), \qquad \frac{\partial}{\partial x} a_1^*(t, x) , \qquad \frac{\partial^2}{\partial x^2} a_1^*(t, x)$$

it is always possible to make such a choice.)

We note that continuity of $\xi_{t,x}(s)$ follows from inequality (9); specifically, $|\xi_{t,x_1}(s) - \xi_{t,x_2}(s)| \leq L|x_1 - x_2|$. Let $\eta_{x,t}(s)$ denote a solution of the equation

$$\eta_{x,t}(s) = 1 + \int_t^s a'_x(u, \xi_{t,x}(u))\eta_{x,t}(u)du \,. \tag{11}$$

Then

$$\begin{aligned}
&\frac{\xi_{t,x+\Delta x}(s) - \xi_{t,x}(s)}{\Delta x} - \eta_{x,t}(s) \\
&= \int_t^s \Bigg[\frac{a(u, \xi_{t,x+\Delta x}(u)) - a(u, \xi_{t,x}(u))}{\xi_{t,x+\Delta x}(u) - \xi_{t,x}(u)} \frac{\xi_{t,x+\Delta x}(u) - \xi_{t,x}(u)}{\Delta x} \\
&\quad - a'_x(u, \xi_{t,x}(u))\eta_{x,t}(u)\Bigg]du \\
&= \int_t^s \Bigg[\frac{a(u, \xi_{t,x+\Delta x}(u)) - a(u, \xi_{t,x}(u))}{\xi_{t,x+\Delta x}(u) - \xi_{t,x}(u)} - a'_x(u, \xi_{t,x}(u))\Bigg] \\
&\quad \times \frac{\xi_{t,x+\Delta x}(u) - \xi_{t,x}(u)}{\Delta x} du + \int_t^s a'_x(u, \xi_{t,x}(u)) \\
&\quad \times \left[\eta_{t,x} - \frac{\xi_{t,x+\Delta x}(u) - \xi_{t,x}(u)}{\Delta x}\right]du \,.
\end{aligned}$$

Since

$$\left[\frac{a(u, \xi_{t,x+\Delta x}(u)) - a(u, \xi_{t,x}(u))}{\xi_{t,x+\Delta x}(u) - \xi_{t,x}(u)} - a'_x(u, \xi_{t,x}(u))\right] \longrightarrow 0$$

uniformly with respect to u and x as $\Delta x \longrightarrow 0$, and since

$$\left|\frac{\xi_{t,x+\Delta x}(u) - \xi_{t,x}(u)}{\Delta x}\right| \leqq L \quad \text{and} \quad |a'_x(u, \xi_{t,x}(u))| \leqq K \,,$$

we conclude from the inequality

$$\begin{aligned}
&\left|\frac{\xi_{t,x+\Delta x}(s) - \xi_{t,x}(s)}{\Delta x} - \eta_{x,t}(s)\right| \\
&\leqq L\int_t^s \left|\frac{a(u, \xi_{t,x+\Delta x}(u)) - a(u, \xi_{t,x}(u))}{\xi_{t,x+\Delta x}(u) - \xi_{t,x}(u)} - a'_x(u, \xi_{t,x}(u))\right|du \\
&\quad + K\int_t^s \left|\frac{\xi_{t,x+\Delta x}(u) - \xi_{t,x}(u)}{\Delta x} - \eta_{t,x}(u)\right|du
\end{aligned}$$

that

$$\begin{aligned}
&\left|\frac{\xi_{t,x+\Delta x}(s) - \xi_{t,x}(s)}{\Delta x} - \eta_{x,t}(s)\right| \\
&\leqq L\int_t^s \left[\frac{a(u, \xi_{t,x+\Delta x}(u)) - a(u, \xi_{t,x}(u))}{\xi_{t,x+\Delta x}(u) - \xi_{t,x}(u)} - a'_x(u, \xi_{t,x}(u))\right]du e^{kT} \,;
\end{aligned}$$

that is, $\eta_{t,x}(s) = \partial\xi_{t,x}(s)/\partial x$. The continuity of $\eta_{t,x}(s)$ follows from the continuity of $a'_x(u, \xi_{t,x}(u))$ as a function of t and x since equation (11) is meaningful for every sample function and is an ordinary linear integral equation of the Volterra type.

Analogously we can show that

$$\frac{\partial}{\partial x}\eta_{t,x}(s) = \frac{\partial^2}{\partial x^2}\xi_{t,x}(s)$$

satisfies the equation

$$\frac{\partial \eta_{t,x}(s)}{\partial x} = \int_t^s a''_{xx}(u, \xi_{t,x}(u))\eta_{t,x}(u)^2 du + \int_t^s a'_x(u, \xi_{t,x}(u))\frac{\partial}{\partial x}\eta_{t,x}(u) du \,. \tag{12}$$

This too is an ordinary Volterra equation and its solution depends continuously on the parameters t and x. It follows from the inequality

$$\left| \frac{\xi_{t,x_1}(s) - \xi_{t,x_2}(s)}{x_1 - x_2} \right| \leqq L$$

that $\partial\xi_{t,x}(s)/\partial x \leqq L$ with probability 1.

If C is a constant bounding $a'_x(t, u)$ and $a''_{xx}(t, u)$, it follows from equation (12) that

$$\left| \frac{\partial \eta_{t,x}(s)}{\partial x} \right| \leqq L^2 C(s - t) + C\int_t^s \left| \frac{\partial}{\partial x}\eta_{t,x}(u) \right| du \,,$$

and hence

$$\left| \frac{\partial^2}{\partial x^2}\xi_{t,x}(s) \right| \leqq e^{CT} L^2 CT$$

with probability 1.

Suppose that $\xi^{(1)}_{t,x}(s)$ is defined as indicated above. Let us show that $f(\xi^{(1)}_{t,x}(s))$ is then, with probability 1, a twice continuously differentiable function of x. If $s < \tau_{t,x}$, then for sufficiently small δ, we also have $s < \tau_{t,x'}$ if $|x' - x| < \delta$. Therefore $\xi^{(1)}_{t,x'}(s) = \xi_{t,x'}(s)$ for $|x' - x| < \delta$. Hence $\xi^{(1)}_{t,x'}(s)$ is twice continuously differentiable with respect to x' in a neighborhood of the point x.

Furthermore, if $s > \tau_{t,x'}$, we can easily show that in some sufficiently small neighborhood of the point x', the function $\xi^{(1)}_{t,x}(s)$ is constant with respect to x, so that its derivatives of all orders are equal to 0. In both cases $f(\xi^{(1)}_{t,x}(s))$ has derivatives with respect to x since f is differentiable.

Suppose now that $s = \tau_{t,x}$. Then

$$f(\xi^{(1)}_{t,x+\Delta x}(\tau_{t,x})) - f(\xi^{(1)}_{t,x}(\tau_{t,x})) = f'(\xi^{(1)}_{t,x}(\tau_{t,x}) + \theta(\xi^{(1)}_{t,x+\Delta x}(\tau_{t,x}) - (\xi^{(1)}_{t,x}(\tau_{t,x})))[\xi^{(1)}_{t,x+\Delta x}(\tau_{t,x}) - \xi^{(1)}_{t,x}(\tau_{t,x})]$$

where $0 < \theta < 1$. Since

$$|\xi_{t,x+\Delta x}^{(1)}(\tau_{t,x}) - \xi_{t,x}^{(1)}(\tau_{t,x})| \leqq |\xi_{t,x+\Delta x}(\tau_{t,x}) - \xi_{t,x}(\tau_{t,x})| \leqq L\Delta x$$

and $f'(\xi_{t,x}^{(1)}(\tau_{t,x})) = 0$, because $\xi_{t,x}^{(1)}(\tau_{t,x})$ is equal to either 0 or 1, it follows that $\partial f(\xi_{t,x}^{(1)}(s))/\partial x = 0$ for $s = \tau_{x,t}$.

In an analogous manner we can show that $(\partial^2/\partial x^2)f(\xi_{t,x}^{(1)}(s)) = 0$ for $s = \tau_{x,t}$. Continuity of the derivatives of $f(\xi_{t,x}^{(1)}(s))$ follows from the relations

$$\frac{\partial}{\partial x}f(\xi_{t,x}^{(1)}(s)) = f'_x(\xi_{t,x}^{(1)}(s))\frac{\partial}{\partial x}\xi_{t,x}(s) ,$$

$$\frac{\partial^2}{\partial x^2}f(\xi_{t,x}^{(1)}(s)) = f''(\xi_{t,x}^{(1)}(s))\left(\frac{\partial}{\partial x}\xi_{t,x}(s)\right)^2 + f'_x(\xi_{t,x}^{(1)}(s))\frac{\partial^2}{\partial x^2}\xi_{t,x}(s) ,$$

which are valid for all s. These derivatives are bounded by constants independent of chance by virtue of the boundedness of $\partial\xi_{t,x}(s)/\partial x$ and $\partial^2\xi_{t,x}(s)/\partial x^2$. Therefore the function

$$u^*(t, x) = \int f(y)\mathsf{P}_1^*(t, x, s, dy) = \mathsf{M}f(\xi_{t,x}^{(1)}(s))$$

is twice continuously differentiable with respect to x. (We can differentiate under the mathematical expectation sign.)

After we have established the existence and continuity of the partial derivatives $\partial u^*(t, x)/\partial x$ and $\partial^2 u^*(t, x)/\partial x^2$, equation (10) can be derived just as equation (1) of Section 5 was derived (cf. proof of Theorem 1, Section 5). From the continuity of $\xi_{t,x}^{(1)}(s)$ with respect to t and the relation $\xi_{x,s}^{(1)}(s) = x$ it follows that $u^*(t, x) \to f(x)$ as $t \to s$; the boundary conditions $u^*(t, 0) = u^*(t, 1) = 0$ follow from the continuity of $\xi_{t,x}^{(1)}(s)$ for $x = 0$ and $x = 1$ and from the relation $\xi_{t,x}^{(1)}(s) = x$ for $x = 0$ and $x = 1$. This completes the proof of the theorem.

As an example of the application of this theorem, let us determine the probability that the process $\xi(t) = x + \gamma t + w(t)$ for $t \in [0, T]$ will not get outside the interval $[0, 1]$.

Let $\xi_1^*(t)$ denote a process that satisfies equation (7) if $a_1^*(t, x) = \gamma\chi_{(0,1)}(x)$, where $\chi_{(0,1)}(x)$ is the characteristic function of the interval $(0, 1)$. Then the probability in question is equal to

$$\mathsf{M}(\chi_{(0,1)}(\xi_1^*(T)) \mid \xi(0) = x) .$$

Let $f(x)$ denote an arbitrary function that satisfies the conditions of Theorem 3. Then if $u^*(t, x)$ is defined as in Theorem 3 (with $s = T$), we have

$$-\frac{\partial u^*(t, x)}{\partial t} = \gamma\frac{\partial u^*(t, x)}{\partial x} + \frac{1}{2}\frac{\partial^2 u^*(t, x)}{\partial x^2} .$$

In view of the homogeneity of the process $w(t)$, we see that

$u^*(t, x) = v(T - t, x)$, where $v(t, x) = \mathsf{M}f(\xi_1^*(t) \mid \xi(0) = x)$. For the function $v(t, x)$ we obtain the equation

$$\frac{\partial v(t, x)}{\partial t} = \gamma \frac{\partial}{\partial x} v(t, x) + \frac{1}{2} \frac{\partial^2 v(t, x)}{\partial x^2}$$

with initial condition $v(0, x) = f(x)$ and boundary conditions $v(0) = v(1) = 0$. Solving this problem by Fourier's method, we obtain

$$v(t, x) = \sum_{k=1}^{\infty} \alpha_k e^{-\gamma x} \exp\left(-\frac{[\gamma^2 + k^2\pi^2]}{2} t\right) \sin k\pi x$$

where $\alpha_k = 2\int_0^1 e^{\gamma x} f(x) \sin k\pi x dx$.

Taking the limiting case, that is, with $f(x) = \chi_{(0,1)}(x)$, we obtain the probability that we are seeking. Then

$$\alpha_k = \frac{k\pi}{\gamma^2 + k^2\pi^2}[e^{\gamma}(-1)^{k-1} + 1] ,$$

so that

$$\mathsf{P}\{\xi_1^*(t) \in (0, 1), 0 \leqq t \leqq T\}$$
$$= \sum_{k=1}^{\infty} \frac{k\pi}{\gamma^2 + k^2\pi^2}[(-1)^{k-1}e^{-\gamma(x-1)} + e^{-\gamma x}] \exp\left(-\frac{[\gamma^2 + k^2\pi^2]}{2} t\right) \sin k\pi x .$$

This formula is convenient to use when t is not very close to 0. Hence when $t \geqq 1$, to find the desired probability with an accuracy of 10^{-7}, it will be sufficient to take only the first term of the series.

When $\gamma = 0$, we can use another expression for the probability $\mathsf{P}\{0 < w(t) < 1, 0 \leqq t \leqq T\}$, which we found in Corollary 2, Section 5, Chapter VI:

$$\mathsf{P}\{w(t) \in (-x, 1 - x), 0 \leqq t \leqq T\}$$
$$= \sum_{k=1}^{\infty} \frac{2}{(2k - 1)\pi} \exp\left(-\frac{(2k - 1)^2\pi^2}{2} T\right) \sin (2k - 1)\pi x ,$$

from which we obtain the more general formula

$$\mathsf{P}\{w(t) \in [a_1, a_2], 0 \leqq t \leqq T\}$$
$$= \mathsf{P}\left\{w(t) \in \left(\frac{a_1}{a_2 - a_1}, \frac{a_2}{a_2 - a_1}\right), 0 \leqq t \leqq \frac{T}{(a_2 - a_1)^2}\right\}$$
$$= \sum_{k=1}^{\infty} \frac{1}{(2k - 1)\pi} \exp\left(-\frac{(2k - 1)^2\pi^2}{2(a_2 - a_1)^2} T\right) \sin (2k - 1)\pi \frac{a_1}{a_1 - a_2} .$$

This formula can be used when $T/(a_2 - a_1)^2$ is not very small, whereas formula (5), Section 5, Chapter VI is more convenint to use when $T/(a_2 - a_1)^2$ is small.

Using the differentiability of $f(\xi_{t,x}^{(1)}(s))$ with respect to x, we can prove the following theorem just as we proved Theorem 2 of Section 5.

Theorem 4. *Let $f(x)$ denote a twice continuously differentiable function on the interval $[0, 1]$. Suppose that $f(x)$ and its first and second order derivatives vanish at the points 0 and 1 and that $\xi_1^*(t)$ is a solution of equation (7). We define*

$$v_\lambda(t, x) = \mathsf{M}\left(\exp\left\{\lambda\int_t^T f(\xi_1^*(s))ds\right\}\middle|\, \xi_1^*(t) = x\right). \tag{13}$$

Then, the function $v_\lambda(t, x)$ is continuous for $t \in [t_0, T]$ and $x \in [0, 1]$, and it satisfies the equation

$$\frac{\partial v_\lambda}{\partial t} + a(t, x)\frac{\partial v_\lambda}{\partial x} + \frac{1}{2}\frac{\partial^2 v_\lambda}{\partial x^2} + \lambda f(x)v_\lambda = 0 \tag{14}$$

with boundary conditions $v_\lambda(T, x) = v_\lambda(t, 0) = v_\lambda(t, 1) = 1$.

Let us apply Theorem 4 to the solution of the "second diffusion problem" (in the terminology of A. Ya. Khinchin, 1936, p. 43). The problem is to determine the probability $\mathsf{P}_x(\alpha, \beta)$ that the diffusion process beginning at a point x will attain the point α before it attains the point β. Let us assume that $\alpha < x < \beta$. Consider a diffusion process $\xi(t)$ with coefficient of transfer $a(x)$ and coefficient of diffusion 1. Since the diffusion coefficients are independent of t, the process is homogeneous (cf. Remark 2, Section 3). Let us choose an arbitrary point $\beta_1 > \beta$. We let $\xi_{t,x}^{(1)}(s)$ denote a process with absorption on the interval $[\alpha, \beta_1]$ and satisfying the equation

$$d\xi_{t,x}^{(1)}(s) = a(\xi_{t,x}^{(1)}(s))\chi_{(\alpha,\beta_1)}(\xi_{t,x}^{(1)}(s))ds + \chi_{(\alpha,\beta_1)}(\xi_{t,x}^{(1)}(s))dw(s) \tag{15}$$

with initial condition $\xi_{t,x}^{(1)}(t) = x$. Let $\chi_{(\alpha,\beta)}$ denote the characteristic function of the interval (α, β_1). It follows from Theorem 4 that the function

$$v_\lambda(t, x) = \mathsf{M}\exp\left\{\lambda\int_t^T f(\xi_{t,x}^{(1)}(s))ds\right\}$$

satisfies the equation

$$\frac{\partial}{\partial t}v_\lambda(t, x) + a(x)\frac{\partial}{\partial x}v_\lambda(t, x) + \frac{1}{2}\frac{\partial^2}{\partial x^2}v_\lambda(t, x) + \lambda v_\lambda(t, x)f(x) = 0$$

and the boundary conditions $v_\lambda(T, x) = v_\lambda(t, \alpha) = v_\lambda(t, \beta) = 1$ if f is twice continuously differentiable on $[\alpha, \beta_1]$ and satisfies the conditions $f(\alpha) = f'(\alpha) = f''(\alpha) = f(\beta_1) = f'(\beta_1) = f''(\beta_1) = 0$.

We note that

$$v_\lambda(t, x) = \mathsf{M} \exp\left\{\lambda\int_0^t f(\xi_{0,x}^{(1)}(s))ds\right\}$$

because of the homogeneity of the process $\xi_{t,x}^{(1)}(s)$. Consequently, the function

$$u_\lambda(t, x) = \mathsf{M} \exp\left\{\lambda\int_0^t f(\xi_{0,x}^{(1)}(s))ds\right\},$$

for $t > 0$ and $x \in (\alpha, \beta_1)$, satisfies the equation

$$\frac{\partial u_\lambda}{\partial t} = a(x)\frac{\partial u_\lambda}{\partial x} + \frac{1}{2}\frac{\partial^2 u_\lambda}{\partial x^2} + \lambda f(x)u_\lambda$$

and the boundary conditions $u_\lambda(0, x) = u_\lambda(t, \alpha) = u_\lambda(t, \beta_1) = 1$. Suppose now that $f \geqq 0$ and $\lambda < 0$. Then the limit

$$u_\lambda(x) = \lim_{t\to\infty} u_\lambda(t, x) = \mathsf{M} \exp\left\{\lambda\int_0^\infty f(\xi_{t,x}^{(1)}(s))ds\right\}$$

exists (if we take $e^{-\infty} = 0$).

To establish certain properties of the function $u_\lambda(x)$, we need:

Lemma 2. *Let τ_x denote the instant at which the process $\xi_{0,x}^{(1)}(s)$ attains the boundary of the interval $[\alpha, \beta_1]$. This quantity is, with probability* 1, *finite and for every $k > 0$, $\sup_{\alpha \leqq x \leqq \beta_1} \mathsf{M}\tau_x^k < \infty$.*

Proof. Suppose that $\sup_{\alpha \leqq x \leqq \beta_1} | a(x) | = a_0$. Then

$$\mathsf{P}\{\tau_x < t_0\} > \mathsf{P}\left\{\sup_{0\leqq t\leqq t_0}\left[x + \int_0^t a(\xi_{0,x}^{(1)}(s))ds + w(t)\right] > \beta_1\right\}$$

$$\geqq \mathsf{P}\left\{\sup_{0\leqq t\leqq t_0} w(t) > \beta_1 - \alpha + t_0 a_0\right\} = \delta_0 > 0\,.$$

Define

$$\lambda_n = \max_x \mathsf{P}\{\tau_x > nt_0\}\,.$$

Then

$$\mathsf{P}\{\tau_x > nt_0\} = \mathsf{P}\{\alpha < \xi_{0,x}(t_0) < \beta_1, \tau_x > nt_0\}$$
$$= \int_\alpha^{\beta_1} \mathsf{P}\{\tau_x > nt_0 \mid \xi_{0,x}(t_0) = y\}\mathsf{P}\{\xi_{0,x}(t_0) \in dy\}$$
$$= \int_\alpha^{\beta_1} \mathsf{P}\{\tau_y > (n-1)t_0\}\mathsf{P}\{\xi_x(t_0) \in dy\}$$
$$\leqq \lambda_{n-1}\mathsf{P}\{\alpha < \xi_x(t_0) < \beta_1\} \leqq \lambda_{n-1}\lambda_1\,.$$

This means that $\lambda_n \leqq \lambda_{n-1}\lambda_1 \leqq \lambda_1^n$. Since $0 < \lambda_1 \leqq 1 - \delta_0 < 1$, it follows that $\lambda_1^n \to 0$ as $n \to \infty$ and that τ_x is finite with probability 1. Furthermore, the series

$$\sum_n n^k \lambda_1^n$$

converges for arbitrary $k > 0$, and

$$\mathsf{M}\tau_x^k \leqq \sum_n n^k \mathsf{P}\{\tau_x > n - 1\} \leqq \sum_n n^k \lambda_1^{n-1} < \infty .$$

This completes the proof of the lemma.

Since

$$\frac{\partial}{\partial t} u_\lambda(t, x) = \mathsf{M} \exp\left(\lambda \int_0^t f(\xi_{0,x}^{(1)}(s))ds\right) \lambda f(\xi_{0,x}^{(1)}(s))$$

and $f(\xi_{0,x}^{(1)}(t)) \to 0$ with probability 1 as $t \to \infty$ (because when $t > \tau_x$, the quantity $\xi_{0,x}^{(1)}$ is equal either to α or to β_1, and $f(\alpha) = f(\beta_1) = 0$ for $\lambda < 0$), it follows that

$$\lim_{t\to\infty} \frac{\partial u_\lambda(t, x)}{\partial t} = 0 .$$

Furthermore,

$$\frac{\partial}{\partial x} u_\lambda(t, x) = \mathsf{M} \exp\left(\lambda \int_0^t f(\xi_{0,x}^{(1)}(s))ds\right) \lambda \int_0^t \frac{\partial}{\partial x} f(\xi_{0,x}(s))ds$$

so that the limit

$$\lim_{t\to\infty} \frac{\partial}{\partial x} u_\lambda(t, x) = \lambda \mathsf{M} \exp\left(\lambda \int_0^\infty f(\xi_{0,x}^{(1)}(s))ds\right) \int_0^\infty \frac{\partial}{\partial x} f(\xi_{0,x}(s))ds$$

exists since

$$\mathsf{M}\left|\int_0^\infty \frac{\partial}{\partial x} f(\xi_{0,x}^{(1)}(s))ds\right| \leqq L\mathsf{M}\tau_x ,$$

where L is a constant bounding $\partial f(\xi_{0,x}^{(1)}(s))/\partial x$. Obviously

$$\lim_{t\to\infty} \frac{\partial u_\lambda(t, x)}{\partial x} = \frac{d}{dx} u_\lambda(x) .$$

Analogously we can show that

$$\lim_{t\to\infty} \frac{\partial^2 u_\lambda(t, x)}{\partial x^2} = \frac{\partial^2}{\partial x^2} u_\lambda(x) .$$

This means that the function $u_\lambda(x)$ satisfies the equation

$$a(x)\frac{du_\lambda}{dx} + \frac{1}{2}\frac{d^2 u_\lambda}{dx^2} + \lambda f(x) u_\lambda = 0 . \tag{16}$$

Suppose now that $f(x)$ is positive for $x \in (\beta, \beta_1)$ but equal to 0 on $[\alpha, \beta]$. Then on the interval $[\alpha, \beta]$, $u_\lambda(x)$ satisfies the equation

$$a(x)\frac{du_\lambda}{dx} + \frac{1}{2}\frac{d^2 u_\lambda}{\partial x^2} = 0 \tag{17}$$

and the condition $u_\lambda(\alpha) = 1$. Consider the limit

$$\lim_{\lambda\to-\infty} u_\lambda(x) = \lim_{\lambda\to\infty} \mathsf{M} \exp\left(-\lambda \int_0^\infty f(\xi_{0,x}^{(1)}(s))ds\right)$$

$$= \mathsf{M}\left[\lim_{\lambda\to\infty} \exp\left(-\lambda \int_0^\infty f(\xi_{0,x}^{(1)}(s))ds\right)\right] = \mathsf{M}\chi_0\left(\int_0^\infty f(\xi_{0,x}^{(1)}(s))ds\right) ,$$

where $\chi_0(x) = 1$ for $x = 0$ and $\chi_0(x) = 0$ for $x \neq 0$. The quantity

$$\chi_0\left(\int_0^\infty f(\xi_{0,x}^{(1)}(s))ds\right)$$

is equal to 0 if and only if the process $\xi_{0,x}^{(1)}(s)$ reaches the point α before it reaches the point β, since in the case in which it reaches β, the quantity $\xi_{0,x}^{(1)}(s)$ falls in (β, β_1) with probability 1 (on the basis of Lemma 1), so that

$$\int_0^\infty f(\xi_{0,x}^{(1)}(s))ds$$

is nonzero. Thus the probability in question is determined from the relation

$$\mathsf{P}_x(\alpha, \beta) = \lim_{\lambda \to -\infty} u_\lambda(x) .$$

We note that $u_\lambda(x)$ decreases monotonically as $\lambda \longrightarrow -\infty$, that $u_\lambda(x) \geqq 0$, and that $u_\lambda(\alpha) = 1$. If we define

$$C = \lim_{\lambda \to -\infty} u_\lambda(\beta) , \qquad C > 0 ,$$

we easily see that $u(x) = \lim u_\lambda(x)$ is a solution of the equation

$$a(x)\frac{d}{dx}u(x) + \frac{1}{2}\frac{d^2}{dx^2}u(x) = 0$$

with boundary conditions $u(\alpha) = 1$ and $u(\beta) = C$. To determine C, we note that

$$u(x) = \mathsf{P}_x(\alpha, \beta) = 1 - \mathsf{P}_x(\beta, \alpha)$$

so that $u(\beta) = 0$, since $\mathsf{P}_\beta(\beta, \alpha) = 1$. Solving this equation we find

$$\mathsf{P}_x(\alpha, \beta) = \frac{\int_x^\beta B(t)dt}{\int_\alpha^\beta B(t)dt} , \tag{18}$$

where

$$B(t) = \exp\left\{\int_\alpha^t 2a(u)du\right\} .$$

REMARK 2. We have seen that under certain conditions on the smoothness of $a(x)$ and $\sigma(x)$, we can, by a transformation of the unknown function, reduce a stochastic equation to the case in which $\sigma(x) = 1$. By means of such a transformation, we can easily show that formula (18) remains valid for arbitrary $\sigma(x)$ if we set

$$B(t) = \exp\left\{\int_0^t \frac{2a(u)}{\sigma^2(u)}du\right\} . \tag{19}$$

Sometimes it is useful to know the mathematical expectation

of the instant at which the process goes beyond the boundary of interval $[\alpha, \beta_1]$. To determine this quantity, let us differentiate equation (16) with respect to λ and then set $\lambda = 0$. We see that the function

$$\left.\frac{\partial u_\lambda}{\partial \lambda}\right|_{\lambda=0} = \varphi(x) = \mathsf{M}\int_0^\infty f(\xi_{0,x}^{(1)}(s))ds$$

satisfies the equation

$$a(x)\frac{\partial \varphi}{\partial x} + \frac{1}{2}\frac{d^2\varphi}{dx^2} + f = 0 \tag{20}$$

and the boundary conditions $\varphi(\alpha) = \varphi(\beta_1) = 0$. Solving equation (20), we obtain

$$\varphi(x) = \int_\alpha^x B(t)\left[C - \int_\alpha^t \frac{2f(u)}{B(u)}du\right]dt ,$$

where

$$C = \frac{\int_\alpha^{\beta_1} B(t)\int_\alpha^t \frac{2f(u)}{B(u)}du}{\int_\alpha^{\beta_1} B(t)dt} .$$

By means of a limiting operation wherein $f(x) \to 1$ for $x \in (\alpha, \beta_1)$ while remaining bounded, we obtain

$$\mathsf{M}\tau_x = \frac{2\int_\alpha^x dt\int_x^{\beta_1} dz\int_t^z \frac{B(t)B(z)}{B(u)}du}{\int_\alpha^{\beta_1} B(z)dz} . \tag{21}$$

IX

LIMIT THEOREMS FOR RANDOM PROCESSES

Throughout the remainder of this book we shall frequently encounter processes that are obtained from simpler random processes by a limiting operation. These limiting operations yield either sample functions of the original process (for example, in the solution of stochastic differential equations in Chapter VIII, Section 3, or in the expansion of a Gaussian process in a series of eigenfunctions of the kernel corresponding to the correlation function in Chapter V, Section 2) or simply finite-dimensional distributions of the process (for example, in the construction of stationary Gaussian processes by taking the limit of trigonometric sums, Chapter I, Section 4).

In the study of random processes considerable attention is given to methods of finding the distributions of different functionals of a random process, for example

$$\int_{t_1}^{t_2} f(\xi(s))ds \,, \quad \sup_{t_1 \leqq t \leqq t_2} \xi(t) \,, \quad \inf_{t_1 \leqq t \leqq t_2} \xi(t) \,.$$

Finding the distribution of functionals of random processes is one of the fundamental problems of the theory of random processes, and it is a method that is used in the solution of many practical problems. Therefore we ask the question: If a process $\xi(t)$ is obtained from a sequence of processes $\{\xi_n(t)\}$ by a particular limiting operation, is it possible to obtain the distributions of the functionals of the process $\xi(t)$ if we know the distributions of the corresponding functionals of the processes $\xi_n(t)$?

Of course if we show that the sequence $\{\xi_n(t)\}$ converges uniformly with probability 1 to $\xi(t)$, that is, if

$$\mathsf{P}\left\{\sup_t |\, \xi_n(t) - \xi(t) \,| \rightarrow 0\right\} = 1 \,,$$

then for every functional f that is continuous in the topology of uniform convergence, $f(\xi_n(t)) \rightarrow f(\xi(t))$ with probability 1, and hence the distribution of $f(\xi_n(t))$ converges to the distribution of $f(\xi(t))$. However there are two points here that we should consider. First, the processes $\xi_n(t)$ and $\xi(t)$ are not always defined on the same probability space; second, even when they are defined on the same probability space, the hypothesis of convergence for the distributions of functionals that are continuous in the topology of uniform convergence is too narrow, and restriction to the cases in which it is satisfied considerably decreases the usefulness of the method.

In what follows we shall assume that the sequence of processes $\xi_n(t)$ at least converges weakly to some process $\xi(t)$; that is, the sequences of finite-dimensional distributions of $\xi_n(t)$ converge to finite-dimensional distributions of $\xi(t)$ (cf. Definition 4, Section 1, Chapter I). These requirements are too weak to imply convergence of the distributions for a sufficiently broad class of functions (for example, for the functionals listed above). Therefore, we must seek supplementary conditions under which the sequence of distributions of a functional (belonging to some class F) of the processes $\xi_n(t)$ converges to the distribution of the corresponding functional of the process $\xi(t)$. The class F of functionals must be such that $f(\xi_n(t))$ and $f(\xi(t))$ are random variables for $f \in F$. Consequently, choice of the class F must depend on the properties of the processes $\xi_n(t)$ and $\xi(t)$.

We shall consider the cases in which (1) the processes are continuous with probability 1, (2) they have, with probability 1, no discontinuities of the second kind, and (3) they are measurable and q-integrable for some q. In each of these cases, we shall consider a different class of functions.

Limit theorems for random processes are important for the determination of the distributions of functionals of the final limit process from simpler processes by passage to the limit. However, it is also natural to use such limiting processes to describe the limiting behavior of discrete processes: processes with independent increments (to describe a sequence of sums of independent random variables) and continuous Markov processes (to describe Markov chains with discrete time). In this case we shall examine the procedure of obtaining processes that vary continuously in time by taking the limit, as the distances between the instants at which the changes take place approach 0, of processes for which changes take place only at certain fixed instants of time. For such processes we shall investigate the conditions for weak convergence.

As we have already seen, for processes with independent increments and Markov processes we have an analytic apparatus that enables us to find the distributions of certain important classes of functionals. Using the known distributions of functionals and applying the limit theorems to random processes, we can obtain limit theorems for sums of independent random variables that are different from the classical limit theorems.

1. WEAK CONVERGENCE OF DISTRIBUTIONS IN A METRIC SPACE

Suppose that the sample functions of the processes $\xi_n(t)$ and $\xi(t)$ (for $t \in [a, b]$) belong, with probability 1, to some metric space X of functions with metric $\rho_X(x, y)$ for $x, y \in X$. For example, if $\xi_n(t)$ and $\xi(t)$ are continuous with probability 1, they belong to the space C of continuous functions with metric

$$\rho_C(x, y) = \sup_t | x(t) - y(t) | .$$

For the class of functionals F, for which we are seeking conditions under which the distribution of $f(\xi_n(t))$ converges to the distribution of $f(\xi(t))$, we take the set of functions on X that are continuous with respect to the metric ρ_X. For $f(\xi(t))$ to be a random variable, it will be sufficient to require that the space X be separable and that for each open sphere S of the space X, the set $\{u: \xi(t) \in S\}$ be measurable with respect to the original probability space (since in this case the set $\{u: \xi(t) \in A\}$ will also be measurable for every Borel subset A of X).

Let us assume that X is a separable space and that the processes $\xi_k(t)$ and $\xi(t)$ satisfy the requirement stated above. The measure μ (resp. μ_n) corresponding to the process $\xi(t)$ (resp. $\xi_n(t)$) is defined on the σ-algebra $\mathfrak{B}$ of all Borel subsets:

$$\mu(A) = \mathsf{P}\{\xi(t) \in A\} \text{ (resp. } \mu_n(A) = \mathsf{P}\{\xi_n(t) \in A\}) .$$

For every bounded $\mathfrak{B}$-measurable functional f,

$$\mathsf{M}\, f(\xi(t)) = \int f(x)\mu(dx)$$

(cf. Theorem 3, Section 6, Chapter II).

A necessary and sufficient condition for convergence of the sequence of distributions of $f(\xi_n(t))$ to the distribution of $f(\xi(t))$ for all continuous functionals f is that

$$\lim_{n\to\infty} \int f(x)\mu_n(dx) = \int f(x)\mu(dx) \qquad (1)$$

for all continuous bounded functionals f. To see that this is true, note that convergence of the sequence of distributions of $f(\xi_n(t))$ to the distribution of $f(\xi(t))$ and the boundedness of f imply convergence of the sequence $\{\mathsf{M}f(\xi_n(t))\}$ to $\mathsf{M}f(\xi(t))$ and hence the validity of (1). On the other hand, equation (1) implies that for every continuous functional f, the sequence of characteristic functions of the variables $f(\xi(t))$ converges to the characteristic function of the variable $f(\xi(t))$:

$$\lim_{n\to\infty} \mathsf{M}e^{i\lambda f(\xi_n(t))} = \lim_{n\to\infty}\int e^{i\lambda f(x)}\mu_n(dx) = \int e^{i\lambda f(x)}\mu(dx) = \mathsf{M}e^{i\lambda f(\xi(t))} .$$

Definition 1. Suppose that equation (1) is satisfied for all continuous bounded functions $f(x)$. We then say that the sequence of the measures μ_n *converges weakly* to the measure μ and we write $\mu_n \Rightarrow \mu$.

Definition 2. The sequence of measures μ_n is said to be *weakly compact* if every subsequence of it contains a weakly convergent subsequence of measures.

Theorem 1. *Let X denote a complete separable space and let $\mathfrak{B}$ denote a σ-algebra of Borel sets. For the sequence of measures μ_n on $\mathfrak{B}$ to be weakly compact, it is necessary and sufficient that*

a. $\sup_n \mu_n(x) < \infty$,

b. *for every $\varepsilon > 0$, there exists a compact set K such that* $\sup_n \mu_n(X\backslash K) < \varepsilon$.

To prove this theorem, we need:

Lemma 1. *Let X denote a separable compact space and suppose that $\sup_n \mu_n(X) = H < \infty$. Then the sequence of measures μ_n is weakly compact.*

Proof. Let C_X denote the space of continuous functions f on X and define $\| f \| = \sup_{x\in X} | f(x) |$. Then C_X is a complete separable normed linear space. Let $\{f_k\}$, for $k = 1, 2, \cdots$, denote a sequence that is everywhere-dense in C_X. Using the diagonal method, we can choose a subsequence of measures μ_{n_k} such that for all i the limit

$$\lim_{k\to\infty}\int f_i(x)\mu_{n_k}(dx) = l[f_i]$$

exists. Since

$$| l[f_i] - l[f_j] | \leqq H \| f_i - f_j \| ,$$

$l[f_i]$ is a uniformly continuous functional on the set $\{f_k\}$ for $k = 1, 2, \cdots$, and consequently it can be extended as a continuous functional to the entire space C_X: $l[f] = \lim_{f_{n_i}\to f} l[f_{n_i}]$. Here the relation

$$l[f] = \lim_{k\to\infty}\int f(x)\mu_{n_k}(dx)$$

remains valid for all f. Being the limit of a sequence of linear nonnegative functionals, $l[f]$ is itself a linear nonnegative functional on C_X.

We introduce the set function $\lambda(A)$ defined on $\mathfrak{B}$ by

$$\lambda(A) = \inf_{f \geqq \chi_A} l[f] .$$

The function $\lambda(A)$ has the following properties:

1. $\lambda(A)$ is monotonic, and for every sequence A_k, $\lambda(\bigcup A_k) \leqq \sum_k \lambda(A_k)$;

2. if the distance between the sets A_1 and A_2 is positive, that is, if

$$\rho(A_1, A_2) = \inf_{x \in A_1, y \in A_2} \rho(x, y) > 0 ,$$

then

$$\lambda(A_1 \cup A_2) = \lambda(A_1) + \lambda(A_2) .$$

To prove (1), we note that if $\varphi_k(x) \geqq \chi_{A_k}(x)$, then $\sum_k \varphi_k(x) \geqq \chi_{\cup A_k}(x)$.

To prove relation 2, let us choose arbitrary $\alpha > 0$ and a continuous function f such that $\lambda(A_1 \cup A_2) \geqq \lambda[f] - \alpha$ and $f \geqq \chi_{A_1} + \chi_{A_2}$. We let $\varphi_{A_i}^{(\varepsilon)}$ denote the function

$$\varphi_{A_i}^{(\varepsilon)}(x) = \begin{cases} 1 & \text{if} \quad x \in A_i , \\ 1 - \dfrac{1}{\varepsilon}\rho(x, A_i) & \text{if} \quad 0 < \rho(x, A_i) < \varepsilon , \\ 0 & \text{if} \quad \rho(x, A_i) \geqq \varepsilon . \end{cases}$$

Here $\rho(x, y) = \inf_{y \in A} \rho(x, y)$. If $2\varepsilon < \rho(A_1, A_2)$, then $\varphi_1 + \varphi_2 \leqq 1$, $\varphi_{A_1}^{(\varepsilon)} \geqq \chi_{A_1}$, and $\varphi_{A_2}^{(\varepsilon)} \geqq \chi_{A_2}$. Therefore

$$\lambda(A_1 \cup A_2) + \alpha \geqq l[\varphi_{A_1}^{(\varepsilon)} f] + l[\varphi_{A_2}^{(\varepsilon)} f] \geqq \lambda(A_1) + \lambda(A_2) .$$

Relation (2) follows from this inequality by virtue of property (1) and the arbitrariness of $\alpha > 0$. Let $\mathfrak{B}_0$ denote the collection of sets for which $\lambda(A') = 0$, where A' is the boundary of the set A.

Obviously $\mathfrak{B}_0$ is an algebra of sets. Let us show that $\lambda(A)$ is an additive set function on $\mathfrak{B}_0$.

Let A_1 and A_2 denote two disjoint members of $\mathfrak{B}_0$. Let Γ denote the union of the boundaries of the sets A_1 and A_2. Since $\lambda(\Gamma) = 0$, it follows that for every $\varepsilon > 0$, there exists a continuous function g satisfying the condition $g \geqq \chi_\Gamma$ such that $l[g] < \varepsilon$. Define

$$A_i^{(\varepsilon)} = [A_i] \cap \left\{x; g(x) \leqq \frac{1}{2}\right\} ,$$

where $[A]$ denotes the closure of A. The closed sets $A_1^{(\varepsilon)}$ and $A_2^{(\varepsilon)}$

are disjoint and hence are at a positive distance from each other, since X is compact. Define

$$\Gamma_\varepsilon = \left\{x;\, g(x) > \frac{1}{2}\right\} .$$

Then

$$\lambda(\Gamma_\varepsilon) \leqq l[2g] \leqq 2\varepsilon ,$$
$$\lambda(A_i) \leqq \lambda(A_i^{(\varepsilon)}) + \lambda(\Gamma_\varepsilon) \leqq \lambda(A_i^{(\varepsilon)}) + 2\varepsilon .$$

Using the monotonicity of λ and property (2) we obtain

$$\begin{aligned}\lambda(A_1) + \lambda(A_2) &\leqq \lambda(A_1^{(\varepsilon)}) + \lambda(A_2^{(\varepsilon)}) + 4\varepsilon \\ &= \lambda(A_1^{(\varepsilon)} \cup A_2^{(\varepsilon)}) + 4\varepsilon \leqq \lambda(A_1 \cup A_2) + 4\varepsilon .\end{aligned}$$

Taking the limit as $\varepsilon \longrightarrow 0$ and keeping (1) in mind, we see that λ is additive on $\mathfrak{B}_0$.

Thus the conditions of Theorem 2, Section 7, Chapter II are satisfied for the function λ on $\mathfrak{B}_0$. Therefore λ can be extended as a measure μ defined on $\sigma(\mathfrak{B}_0)$. Let us show that $\sigma(\mathfrak{B}_0)$ coincides with $\mathfrak{B}$. Note that for every x there exist only countably many radii r such that $\lambda(S'_r(x)) > 0$, where $S_r(x)$ is the sphere of radius r with center at x and boundary $S'_r(x)$. Therefore $\sigma(\mathfrak{B}_0)$ contains all spheres and hence all Borel subsets of X.

Let us show that for all $f \in C_X$,

$$l[f] = \int f(x)\mu(dx) . \qquad (2)$$

Suppose that f is nonnegative. Let us choose values $0 = c_0 < c_1 < c_1 < \cdots < c_N$ with $c_N > \|f\|$, in such a way that $\lambda\{x;\, f(x) = c_k\} = 0$. Then if $A_k = \{x;\, c_{k-1} \leqq f(x) < c_k\}$, it follows that $A_k \in \mathfrak{B}_0$ and hence $\lambda(A_k) = \mu(A_k)$. Let φ_k for $k = 1, 2, \cdots, N$ denote, a continuous function such that $l[\varphi_k] > \lambda(A_k) + \varepsilon$ and $\varphi_k \geqq \chi_{A_k}$. Since $f \leqq \sum_{k=1}^N c_k\varphi_k$, we have

$$\begin{aligned}l[f] &\leqq \sum c_k l[\varphi_k] \leqq N\varepsilon + \sum_{k=1}^N c_k\lambda(A_k) = N\varepsilon + \sum_{k=1}^N c_k\mu(A_k) \\ &\leqq N\varepsilon + \max_k(c_k - c_{k-1})\,\mu(X) + \int f(x)\mu(dx) .\end{aligned}$$

It follows from this inequality that

$$l[f] \leqq \int f(x)\mu(dx) \qquad (3)$$

for an arbitrary nonnegative function $f \in C_X$. Furthermore,

$$l\left[1 - \frac{f}{\|f\|}\right] \leqq \int\left(1 - \frac{f(x)}{\|f\|}\right)\mu(dx) .$$

Since $l(1) = \int \mu(dx) = \mu(X)$, we have

$$-\,l[f] \leqq -\int f(x)\mu(dx)\,. \tag{4}$$

Comparison of (4) with (3) yields (2) for nonnegative functions and hence for all $f(x) \in C_X$. Thus for arbitrary $f(x) \in C_X$,

$$\lim_{k\to\infty} \int f(x)\mu_{n_k}(dx) = \int f(x)\mu(dx)\,.$$

This completes the proof of the lemma.

We can now prove the theorem.

Proof of the Sufficiency. Let us choose a sequence $\{\varepsilon_m\}$ that approaches 0 as $m \longrightarrow \infty$ and a sequence $\{K^{(m)}\}$ of compact sets such that $K^{(m)} \subset K^{(m+1)}$ and $\sup_n \mu_n(X \backslash K^{(m)}) \leqq \varepsilon_m$. We define

$$\mu_n^{(m)}(A) = \mu_n(A \cap K^{(m)})\,.$$

Let us choose a sequence $\{n_k^{(1)}\}$ such that the sequence of the measures $\mu^{(1)}_{n_k^{(1)}}$ converges weakly to some measure $\mu^{(1)}$. Let us define sequences $\{n_k^j\}$ such that $\{n_k^{(j)}\}$ is a subsequence of $\{n_k^{(j-1)}\}$ and the sequence $\{\mu^{(j)}_{n_k^{(j)}}\}$ converges weakly to some measure $\mu^{(j)}$. Since $\mu^{(j)}$ and $\mu^{(j-1)}$ coincide on $K^{(j-1)}$, it follows that var $|\,\mu^{(j)} - \mu^{(j+p)}\,| \leqq 2\varepsilon_j$. Hence the sequence $\{\mu^{(j)}\}$ converges in variation to some measure μ. Let us show that $\mu_{n_k^{(k)}}$ converges weakly to μ. For every bounded continuous function f,

$$\begin{aligned}
&\varlimsup_{k\to\infty} \left| \int f(x)\mu_{n_k^{(k)}}(dx) - \int f(x)\mu(dx) \right| \\
&\quad \leqq \varlimsup_{k\to\infty} \left| \int_{K^{(m)}} f(x)\mu_{n_k^{(k)}}(dx) - \int_{K^{(m)}} f(x)\mu(dx) \right| \\
&\qquad + \|\,f\,\|\,(\varlimsup_{k\to\infty} \mu_{n_k^{(k)}}(X - K^{(m)}) + \mu(X - K^{(m)})) \leqq 2\|\,f\,\|\varepsilon_m\,.
\end{aligned}$$

This proves the sufficiency of the hypotheses of the theorem.

Proof of the Necessity. If $\{\mu_n\}$ is weakly compact, then $\int 1 \cdot \mu_n(dx)$ is a compact set of real numbers and hence the sequence $\{\mu_n(X)\}$ is bounded.

Let us further suppose that the sequence $\{\mu_n\}$ is weakly compact but that condition b is not satisfied. Note that condition b is equivalent to the condition b′: For all $\varepsilon > 0$ and $\delta > 0$, there exists a compact set K such that $\sup_n \mu_n(X \backslash K_\delta) < \varepsilon$, where K_δ denotes the set of points x whose distance from K does not exceed δ. That b implies b′ is obvious. Conversely, let $K^{(r)}$ denote a compact set such that $\sup_n \mu_n(X \backslash K_{1/r}^{(r)}) \leqq \varepsilon/2^r$. Then the set $\bigcap_r K_{1/r}^{(r)}$

is a compact set for which condition b is satisfied. If b′ were not valid then there would exist an $\varepsilon > 0$ and a $\delta > 0$ such that

$$\sup_n \mu_n(X \backslash K_\delta) > \varepsilon$$

for every compact set K.

Let $K^{(0)}$ denote a compact set such that $\mu_1(X \backslash K^{(0)}) < \varepsilon$. (The existence of such a compact set follows from Theorem 6, Section 7, Chapter II.) Since $\sup_n \mu_n(X \backslash K_\delta^{(0)}) > \varepsilon$, there exists a number n_1 such that $\mu_{n_1}(X \backslash K_\delta^{(0)}) > \varepsilon$; hence, there exists a compact set $K^{(1)}$ such that $\mu_{n_1}(K^{(1)}) > \varepsilon$ and $K^{(1)} \subset X \backslash X_\delta^{(0)}$ (again on the basis of Theorem 6, Section 7, Chapter II). Since $\sup_n \mu_n(X \backslash K_\delta^{(0)} \backslash K_\delta^{(1)}) > \varepsilon$, there exist a number n_2 and a compact set $K^{(2)} \subset X \backslash K_\delta^{(0)} \backslash K_\delta^{(1)}$ such that $\mu_{n_2}(K^{(2)}) > \varepsilon$. Continuing this process, we choose a sequence of numbers n_j and compact sets $K^{(j)}$ such that $\mu_{n_j}(K^{(j)}) > \varepsilon$ and

$$K^{(j)} \subset X \Big\backslash \bigcup_{i=0}^{j-1} K_\delta^{(i)} = X \Big\backslash \Big[\bigcup_{i=0}^{j-1} K^{(i)}\Big]_\delta .$$

Let $\chi_i(x)$ denote a continuous nonnegative function bounded by unity, vanishing on $X \backslash K_{\delta/2}^{(i)}$ and equal to 1 on $K^{(1)}$. Since the distance between any two compact sets of the sequence $K^{(i)}$ is at least δ, the functions $\chi_i(x)$, for distinct values of i, cannot be nonzero simultaneously. Let us choose from the sequence $\{\mu_{n_j}\}$ a weakly convergent subsequence $\{\mu'_k\}$. Suppose that this subsequence converges to μ. Since the measure μ is finite and $\sum_i \chi_i(x)$ is bounded, we have

$$\int \sum_{i=1}^{\infty} \chi_i(x)\, \mu(dx) = \sum_{i=1}^{\infty} \int \chi_i(x) \mu(dx) < \infty$$

and hence

$$\lim_{p\to\infty} \sum_{i=p}^{\infty} \int \chi_i(x) \mu(dx) = 0 .$$

On the other hand,

$$\int \sum_{i=p}^{\infty} \chi_i(x) \mu'_k(dx) \geqq \mu_{n_{p'}}(K^{(p')}) \geqq \varepsilon . \qquad (k = n_{p'})$$

as soon as $k > p$; hence for all p,

$$\sum_{i=p}^{\infty} \int \chi_i(x) \mu(dx) = \lim_{k\to\infty} \sum_{i=p}^{\infty} \int \chi_i(x) \mu'_k(dx) \geqq \varepsilon .$$

This contradiction proves the necessity of condition b. This completes the proof of the theorem.

REMARK 1. The completeness of the space X was used only in proving the necessity of the conditions of the theorem. The conditions of the theorem are sufficient for weak compactness of a sequence of measures in an arbitrary metric space. These conditions are also necessary if the space X can be represented as a Borel subset of some complete separable metric space.

With the aid of Theorem 1 we can establish necessary and sufficient conditions for weak convergence of a sequence of measures in the case of a complete space X. Furthermore, these conditions are sufficient for every metric space.

Theorem 2. *For a sequence of measures μ_n to converge weakly to some measure μ, it is necessary and sufficient that the sequence $\{\mu_n\}$ be weakly compact and that $\mu_n(A) \longrightarrow \mu(A)$ for all A belonging to some algebra $\mathfrak{B}_0$ such that $\sigma(\mathfrak{B}_0) = \mathfrak{B}$.*

Proof of the Necessity. Obviously, every convergent sequence is weakly compact. Let us take an arbitrary set $A \in \mathfrak{B}$. Let $A^{(0)}$ denote its interior (that is, the set of all interior points of A), and let $[A]$ denote its closure. If $\{\mu_n\}$ converges weakly to μ, then by choosing a continuous function $f(x)$ such that $f(x) = 1$ for $x \in [A]$ and $\mu([A]) \geqq \int f(x)\mu(dx) - \varepsilon$, we obtain

$$\mu([A]) \geqq \int f(x)\mu(dx) - \varepsilon = \lim \int f(x)\mu_n(dx) - \varepsilon \geqq \overline{\lim}\, \mu_n(A) - \varepsilon .$$

This means that

$$\overline{\lim}\, \mu_n(A) \leqq \mu([A]) ,$$

$$\overline{\lim}\, \mu_n(X\backslash A) \leqq \mu([X\backslash A]), \quad - \underline{\lim}\, \mu_n(A) \leqq -(A) .$$

Therefore

$$\mu(A^{(0)}) \leqq \underline{\lim}\, \mu(A) \leqq \overline{\lim}\, \mu(A) \leqq \mu([A]) .$$

Thus for all sets A such that $\mu([A]\backslash A^{(0)}) = \mu(A') = 0$, where A' is the boundary of the set A,

$$\lim_{n\to\infty} \mu_n(A) = \mu(A) .$$

Let $\mathfrak{B}_0$ denote the collection of sets A such that $\mu(A') = 0$. Obviously $\mathfrak{B}_0$ is an algebra of sets. Let us show that $\sigma(\mathfrak{B}_0) = \mathfrak{B}$. Note that of all the spheres S_x with center at a given point x, for only countably many of them do we have $\mu(S'_x) > 0$. Consequently $\sigma(\mathfrak{B}_0)$ contains all spheres and hence all Borel sets since the algebra of Borel sets is a minimal σ-algebra containing all spheres. This completes the proof of the necessity.

Proof of the Sufficiency. Let us choose an arbitrary weakly convergent subsequence $\{\mu_{n_k}\}$ of the sequence $\{\mu_n\}$. Let $\bar{\mu}$ denote the limit of this subsequence and let us show that μ and $\bar{\mu}$ coincide. Suppose that $A \in \mathfrak{B}_0$. Then as shown in the proof of the necessity,

$$\bar{\mu}(A^{(0)}) \leqq \underline{\lim}\, \mu_{n_k}(A) \leqq \overline{\lim}\, \mu_{n_k}(A) \leqq \bar{\mu}([A]) .$$

But by hypothesis, $\lim \mu_{n_k}(A) = \mu(A)$. Therefore, for all sets in $\mathfrak{B}_0$,

$$\bar{\mu}(A^{(0)}) \leqq \mu(A) \leqq \bar{\mu}([A]) . \tag{5}$$

Let $\{A_n\}$ denote an arbitrary monotonic sequence of sets that satisfy inequality (5). Then since

$$(\bigcup A_n)^{(0)} = \bigcup A_n^{(0)}, \quad (\bigcap A_n)^{(0)} \subset \bigcap A_n^{(0)} ,$$
$$[\bigcup A_n] \subset \bigcup [A_n], \quad [\bigcap A_n] = \bigcap [A_n] ,$$

we can see that (5) is also satisfied for the limit of the sequence of sets A_n. Thus the collection of sets $\mathfrak{B}_1$ that satisfy inequality (5) is a monotone class containing the algebra $\mathfrak{B}^{(0)}$. This means that this set contains $\sigma(\mathfrak{B}_0) = \mathfrak{B}$ (cf. Chapter II, Section 1, Theorem 3). Hence inequality (5) is valid for all sets $A \in \mathfrak{B}$. Let $\mathfrak{B}_2$ denote the collection of sets A such that $\bar{\mu}(A') = 0$. Then $\mu(A) = \bar{\mu}(A)$ for all $A \in \mathfrak{B}_2$ since $\bar{\mu}(A^{(0)}) = \bar{\mu}([A]) = \mu(A)$.

Obviously, the relation $\mu(A) = \bar{\mu}(A)$ is also satisfied on the minimal σ-algebra $\sigma(\mathfrak{B}_2)$ containing $\mathfrak{B}_2$. As shown in the proof of the necessity of the hypotheses of the theorem, $\sigma(\mathfrak{B}_2) = \mathfrak{B}$. Thus the measures μ and $\bar{\mu}$ coincide. We have shown that the sequence $\{\mu_n\}$ is a weakly compact sequence with a unique limit point that is the measure μ. From this it follows that the sequence $\{\mu_n\}$ converges weakly to μ. This completes the proof of the theorem.

Corollary 1. *If $\{\mu_n\} \Rightarrow \mu$, then $\mu_n(A) \longrightarrow \mu(A)$ for every set A such that $\mu(A^{(0)}) = \mu([A])$.*

The sets A such that $\mu(A^{(0)}) = \mu([A])$ are called *sets of continuity* of the measure μ. This means that $\mu_n(A) \longrightarrow \mu(A)$ for all sets of continuity of the measure μ if $\mu_n \Rightarrow \mu$.

Corollary 2. *Let us assume that the measures μ_n correspond to processes $\xi_n(t)$ and that the measure μ corresponds to the process $\xi(t)$. Then the relation $\mu_n \Rightarrow \mu$ as $n \longrightarrow \infty$ implies that the sequence of distributions of $\{f(\xi_n(t))\}$ converges to the distribution of $f(\xi(t))$ for all $\mathfrak{B}$-measurable functionals f that are almost everywhere continuous with respect to the measure μ.*

Proof. Let A_0 denote the set of points of discontinuity of f. Then $\mu(A_0) = 0$. Let G_α denote the set of those x such that $\{f(x) < \alpha\}$ and let G'_α denote the boundary of the set G_α:

$$G'_\alpha = [\{x;\, f(x) < \alpha\}] \cap [\{x;\, f(x) \geqq \alpha\}] .$$

The intersection of the sets G'_α and G'_{α_1} for $\alpha < \alpha_1$ is contained in the intersection of the sets $[\{x; f(x) < \alpha\}] \cap [\{x; f(x) \geqq \alpha_1\}]$. Therefore the inclusion relation $x \in G'_\alpha \cap G'_{\alpha_1}$ implies that

$$\liminf_{y \to x} f(y) \leqq \alpha, \quad \limsup_{y \to x} f(y) \geqq \alpha_1 ;$$

that is,

$$G'_\alpha \cap G'_{\alpha_1} \subset A_0 .$$

Consequently $\mu(G'_\alpha \cap G'_{\alpha_1}) = 0$. Hence for an arbitrary sequence $\{\alpha_k\}$,

$$\mu(\bigcup_k G'_{\alpha_k}) = \sum_k \mu(G'_{\alpha_k}) .$$

From this it follows that the set of numbers α such that $\mu(G'_\alpha) \neq 0$ is no more than countable. Therefore, for all α except possibly countably many values, the set G_α is a set of continuity of the measure μ, so that $\mu_n(G_\alpha) \longrightarrow \mu(G_\alpha)$ or

$$\mathsf{P}\{f(\xi_n(t)) < \alpha\} \longrightarrow \mathsf{P}\{f(\xi(t)) < \alpha\}$$

as $n \longrightarrow \infty$. This completes the proof.

REMARK 2. As a rule, in considering random processes we assume that the σ-algebra of events of the basic probability space coincides with the minimal σ-algebra containing all events of the form $\{u; \xi(t) \in C\}$, where C is a cylindrical set. This means that weak convergence of the sequences of finite-dimensional distributions of random processes $\xi_n(t)$ to finite-dimensional distributions of the process $\xi(t)$ implies convergence of the sequence of measures μ_n to the measure μ on the algebra $\mathfrak{B}_0$ of all cylindrical sets of continuity of the measure μ, so that in this case, it suffices to verify the conditions ensuring weak compactness of the measures μ.

2. LIMIT THEOREMS FOR CONTINUOUS PROCESSES

In this section we shall assume that the processes $\xi_n(t)$ and $\xi(t)$ are continuous on the interval $[a, b]$. Their sample functions belong, with probability 1, to the complete separable metric space $C[a, b]$ of all functions $x(t)$ on $[a, b]$ with respect to the metric $\rho(x, y) = \sup_{a \leqq t \leqq b} | x(t) - y(t) |$.

We note that in the space $C[a, b]$, the minimal σ-algebra of sets $\mathfrak{U}$ containing all cylindrical sets contains all Borel sets. To see this, if suffices to note that every sphere belongs to $\mathfrak{U}$, since

$$\{x; \sup_t | x(t) - \alpha(t) | \leqq r\} = \bigcap_{k=1}^{\infty} \{x; | x(t_k) - \alpha(t_k) | \leqq r\} ,$$

where $\alpha(t)$ is an arbitrary continuous function and $\{t_k\}$ is an arbitrary sequence everywhere dense on $[a, b]$.

Let H denote a constant and let ω_δ denote a function defined for $\delta > 0$ that approaches 0 from above as δ approaches 0 from above. Let $K(H, \omega_\delta)$ denote the set of functions $x(t)$ such that

$$\sup_{a \leqq t \leqq b} | x(t) | \leqq H, \quad \sup_{|t'-t''| \leqq \delta} | x(t') - x(t'') | \leqq \omega_\delta \quad \text{for all} \quad \delta > 0 .$$

From a theorem of Arzelà (cf. Kolmogorov and Fomin, Heider and Simpson) every compact set in $C[a, b]$ is a closed subset of some set $K(H, \omega_\delta)$ and the latter is also compact. We may now use Theorems 1 and 2 of the preceding section.

Theorem 1. *Suppose that the finite-dimensional distributions oj the processes $\xi_n(t)$ converge to the finite-dimensional distributions of a process $\xi(t)$. Then for the sequence of distributions of $f(\xi_n(t))$ to converge to the distribution of $f(\xi(t))$ for all functionals f that are continuous on $C[a, b]$, it is necessary and sufficient that*

$$\lim_{h \to 0} \sup_n \mathsf{P}\{ \sup_{|t'-t''| \leqq h} | \xi_n(t') - \xi_n(t'') | > \varepsilon\} = 0 \qquad (1)$$

for every $\varepsilon > 0$.

Proof of the Necessity. Suppose that the conclusion of the theorem is satisfied. Then the sequence of measures μ_n corresponding to the processes $\xi_n(t)$ is weakly compact, so that condition (b) of Theorem 1, Section 1 is satisfied. Therefore, for every $\eta > 0$, there exists a compact set $K(H, \omega_\delta)$ such that

$$\sup_n \mu_n(C[a, b] \backslash K(H, \omega_\delta)) = \sup_n \mathsf{P}\{\xi_n(\theta) \notin K(H, \omega_\delta)\} \leqq \eta .$$

Then

$$\mathsf{P}\{ \sup_{|t'-t''| \leqq h} | \xi_n(t') - \xi_n(t'') | > \omega_h\} < \mathsf{P}\{\xi_n(t) \notin K(H, \omega_\delta)\} \leqq \eta .$$

If h is sufficiently small, then $\omega_h < \varepsilon$ and

$$\overline{\lim_{h \to 0}} \sup_n \mathsf{P}\{| \xi_n(t') - \xi_n(t'') | > \varepsilon\} \leqq \eta .$$

In view of the arbitrariness of $\eta > 0$, we obtain (1).

Proof of the Sufficiency. The convergence of the sequence of finite-dimensional distributions implies convergence of the sequence of measures μ_n to the measure μ that corresponds to the process $\xi(t)$ on the algebra $\mathfrak{B}_0$ of all cylindrical sets of continuity of the measure μ. But $\sigma(\mathfrak{B}_0)$ coincides with the minimal σ-algebra containing all cylindrical sets. Hence (as shown earlier in this section) it contains all Borel subsets of the space $C[a, b]$. Therefore, in view of Theorem 2 of Section 1 it will be sufficient for us to prove the weak compactness of the sequence of the measures μ_n. Let us show that for every $\eta > 0$ there exists a compact set $K(H, \omega_\delta)$ such that

$$\sup_n \mathsf{P}\{\xi_n(t) \notin K(H, \omega_\delta)\} \leqq \eta .$$

Since the sequence of the distributions $\xi_n(a)$ converges to the distribution $\xi(a)$, there exists an H such that for all n,

$$\mathsf{P}\{| \xi_n(a) | > H\} \leqq \frac{\eta}{2} .$$

Let us take a sequence $\{\varepsilon_r\}$ that converges to 0 from above. For every ε_r there exists an h_r such that $h_r < h_{r-1}$ and

$$\sup_n \mathsf{P}\{\sup_{|t'-t''|<h_r} | \xi_n(t') - \xi_n(t'') | > \varepsilon_r\} \leqq \frac{\eta}{2^{r+1}} .$$

Let ω_δ denote a nonnegative nonincreasing function such that $\omega_\delta = \varepsilon_r$ for $\delta \in [h_{r+1}, h_r]$. Obviously, $\omega_\delta \downarrow 0$ as $\delta \downarrow 0$. Furthermore,

$$\begin{aligned} &\mathsf{P}\{\xi_n(t) \notin K(H, \omega_\delta)\} \\ &\quad \leqq \mathsf{P}\{| \xi_n(a) | > H\} + \sum_{r=1}^{\infty} \mathsf{P}\{\sup_{|t'-t''| \leqq h_r} | \xi_n(t') - \xi_n(t'') | > \varepsilon_r\} \\ &\quad \leqq \frac{\eta}{2} + \sum_{r=1}^{\infty} \frac{\eta}{2^{r+1}} = \eta . \end{aligned}$$

This completes the proof of the theorem.

REMARK 1. Instead of condition (1), we may require that

$$\lim_{h \to 0} \overline{\lim_{n \to \infty}} \mathsf{P}\{\sup_{|t'-t''| \leqq h} | \xi_n(t') - \xi_n(t'') | > \varepsilon\} = 0 , \tag{2}$$

which is often more convenient to verify. We can use this requirement because it follows from (2) that for every $\eta > 0$ there exists a $\delta > 0$ and an N such that for $n > N$ and $h < \delta$,

$$\mathsf{P}\{\sup_{|t'-t''| \leqq h} | \xi_n(t') - \xi_n(t'') | > \varepsilon\} \leqq \eta . \tag{3}$$

The continuity of the processes $\xi_n(t)$ implies their uniform continuity, so that for every n,

$$\lim_{h \to 0} \mathsf{P}\{\sup_{|t'-t''| \leqq h} | \xi_n(t') - \xi_n(t'') | > \varepsilon\} = 0 .$$

Therefore, there exists a δ such that for $h < \delta$ relation (3) is satisfied for all n.

The next theorem may be more convenient for applications.

Theorem 2. *Suppose that the finite-dimensional distributions of the processes $\xi_n(t)$ converge to the finite-dimensional distributions of the process $\xi(t)$ and that there exist positive numbers α, β, and H such* *that for all t_1, t_2 and for all n,*

$$\mathsf{M} \mid \xi_n(t_1) - \xi_n(t_2) \mid^{\alpha} \leqq H \mid t_1 - t_2 \mid^{1+\beta} . \tag{4}$$

Then for all functionals f that are continuous on $C[a, b]$, the sequence of distributions of $f(\xi_n(t))$ converges to the distribution of $f(\xi(t))$.

Proof. Obviously, because of the continuity of the processes $\xi_n(t)$ we have with probability 1,

$$\sup_{|t_1 - t_2| \leqq h} \mid \xi_n(t_1) - \xi_n(t_2) \mid = \sup_{\substack{|t_1 - t_2| \leqq h, \\ t_1, t_2 \in N}} \mid \xi_n(t_1) - \xi_n(t_2) \mid ,$$

where N is the set of all points of the form $k/2^m$ belonging to $[a, b]$. If $2h < 2^{-k}$, then

$$\sup_{\substack{t', t'' \in N \\ |t' - t''| \leqq h}} \mid \xi_n(t') - \xi_n(t'') \mid \leqq 2 \sup_j \sup_{(j/2^k) < (l/2^m) < \{(j+1)/2^k\}} \left| \xi_n\left(\frac{l}{2^m}\right) - \xi_n\left(\frac{j}{2^k}\right) \right|$$

$$\leqq 2 \sum_{m=k+1}^{\infty} \sup_l \left| \xi_n\left(\frac{l+1}{2^m}\right) - \xi_n\left(\frac{l}{2^m}\right) \right| ,$$

since $(l/2^m) - (j/2^k) = \sum_{r=1}^{s} (1/2^{m_r})$, for $k < m_1 < m_2 < \cdots < m_s \leqq m$. Hence

$$\xi_n\left(\frac{l}{2^m}\right) - \xi_n\left(\frac{j}{2^k}\right) = \sum_{i=1}^{s} \left[\xi_n\left(\frac{j}{2^k} + \sum_{r=1}^{i} \frac{1}{2^{m_r}}\right) - \xi_n\left(\frac{j}{2^k} + \sum_{r=1}^{i-1} \frac{1}{2^{m_r}}\right) \right] .$$

Now we note that

$$\mathsf{P}\left\{ \sup_{a \leqq (i/2^m) < \{(i+1/2^m)\} \leqq b} \left| \xi_n\left(\frac{i+1}{2^m}\right) - \xi_n\left(\frac{i}{2^m}\right) \right| > \frac{1}{m^2} \right\}$$

$$\leqq \sum_i \mathsf{P}\left\{ \left| \xi_n\left(\frac{i+1}{2^m}\right) - \xi_n\left(\frac{i}{2^m}\right) \right| > \frac{1}{m^2} \right\}$$

$$\leqq m^{2\alpha} \sum_i \mathsf{M} \left| \xi_n\left(\frac{i+1}{2^m}\right) - \xi_n\left(\frac{i}{2^m}\right) \right|^{\alpha}$$

$$\leqq m^{2\alpha}(b - a) 2^m H \frac{1}{2^{m(1+\beta)}} = L \frac{m^{2\alpha}}{2^{m\beta}} .$$

Consequently, if $\sum_{m=k+1}^{\infty} (1/m^2) < (\varepsilon/2)$ where $k = \log_2(1/h) + 1$, then

$$\mathsf{P}\{ \sup_{|t_1 - t_2| \leqq h} \mid \xi_n(t_1) - \xi_n(t_2) \mid > \varepsilon \}$$

$$\leqq \sum_{m=k+1}^{\infty} \mathsf{P}\left\{ \sup_i \left| \xi_n\left(\frac{i+1}{2^m}\right) - \xi_n\left(\frac{i}{2^m}\right) \right| > \frac{1}{m^2} \right\}$$

$$\leqq L \sum_{m=[\log_2(1/h)]+2} \frac{m^{2\alpha}}{2^{m\beta}} \longrightarrow 0$$

uniformly with respect to n as $h \longrightarrow 0$. This completes the proof of the theorem.

REMARK 2. It follows from the proof of the theorem that instead of condition (4) we can require that

$$\mathsf{M}\,|\,\xi_n(t_1) - \xi_n(t_2)\,|^\alpha \leqq H\varphi(t_1 - t_2)\ ,$$

where the function $\varphi(t)$ is such that for some $\beta > 0$

$$\sum_{m=1}^{\infty} 2^m m^{\alpha(1+\beta)} \varphi\left(\frac{1}{2^m}\right) < \infty\ .$$

Furthermore, it is sufficient for us to require the convergence of the sequences of finite-dimensional distributions of the processes $\xi_n(t)$ to the finite-dimensional distributions of the processes $\xi(t)$ for t in some set that is dense on $[a, b]$.

3. CONVERGENCE OF SEQUENCES OF SUMS OF INDEPENDENT RANDOM VARIABLES TO PROCESSES OF BROWNIAN MOTION

Consider a double sequence of random variables $\xi_{n1}, \xi_{n2}, \cdots, \xi_{nk_n}$ (independent in each sequence) that satisfy the conditions

1. $\mathsf{M}\xi_{ni} = 0$,
2. $\mathsf{D}\xi_{ni} = b_{ni}$, where $\sum_{i=1}^{k_n} b_{ni} = 1$.

Let us construct a random function $\xi_n(t)$, for $t \in [0, 1]$: We define $S_{nk} = \sum_{i=1}^{k} \xi_{ni}$, $t_{nk} = \sum_{i=1}^{k} b_{ni}$,

$$\xi_n(t) = S_{nk} + \frac{t - t_{nk}}{t_{n,k+1} - t_{nk}} [S_{n,k+1} - S_{nk}]$$

for $t \in [t_{nk}, t_{n,k+1}]$; we also define $S_{n0} = 0$ and $t_{n0} = 0$. The function $\xi_n(t)$ is a random broken line connecting points in the $t\xi$-plane with coordinates (t_{nk}, S_{nk}) for $k = 0, 1, \cdots, k_n$.

In this section we shall study the conditions under which sequences of finite-dimensional distributions of the processes $\xi_n(t)$ and the sequences of the distributions of functionals of these processes converge to the finite-dimensional distributions and the distributions of the corresponding functionals of a process of Brownian motion $w(t)$.

Let $F_{ni}(x)$ denote the distribution function of a random variable ξ_{ni}. Then, if for every $\varepsilon > 0$,

$$\lim_{n\to\infty} \sum_{i=1}^{k_n} \int_{|u|>\varepsilon} u^2 dF_{ni}(u) = 0\ , \tag{1}$$

the random variable ξ_{ni} is said to satisfy *Lindeberg's condition.*

Theorem 1. *Suppose that random variables* ξ_{ni} *satisfy conditions* 1 *and* 2 *and Lindeberg's condition. Then, the finite-dimensional dis-*

tributions of the processes $\xi_n(t)$ converge to the finite-dimensional distributions of the process $w(t)$ and the sequence of distributions of $f(\xi_n(t))$ converges to the distribution of $f(w(t))$ for every functional f that is continuous on $C[0, 1]$.

Proof. Let us first show that the finite-dimensional distributions of the processes $\xi_n(t)$ converge to the finite-dimensional distributions of $w(t)$. Let $\xi'_n(t)$ denote the random process defined by

$$\xi'_n(t) = \sum_{t_{ni}<t} \xi_{ni} \,.$$

Then for every $\alpha > 0$,

$$\mathsf{P}\{|\xi'_n(t) - \xi_n(t)| > \alpha\} \leqq \mathsf{P}\{\sup_i |\xi_{ni}| > \alpha\} \leqq \sum_{i=1}^{k_n} \mathsf{P}\{|\xi_{ni}| > \alpha\}$$
$$= \sum_{i=1}^{k_n} \int_{|u|>\alpha} dF_{ni}(u) \leqq \frac{1}{\alpha^2} \sum_{i=1}^{k_n} \int_{|u|>\alpha} u^2 dF_{ni}(u) \longrightarrow 0 \,.$$

Therefore, to prove the first assertion of the theorem, it will be sufficient to show that the finite-dimensional distributions of the process $\xi'_n(t)$ converge to the finite-dimensional distributions of the process $w(t)$. But since $w(t)$ and $\xi'_n(t)$ are processes with independent increments and $w(0) = \xi'_n(0) = 0$, it will be sufficient to prove that the distributions of $\xi'_n(t'') - \xi'_n(t')$ converge to the distribution $w(t'') - w(t')$ for all $0 \leqq t' \leqq t'' \leqq 1$. We also note that

$$\xi'_n(t'') - \xi'_n(t') = \sum_{t' \leqq t_{ni} < t''} \xi_{ni} \,;$$

that is $\xi'_n(t'') - \xi'_n(t')$ is the sum of independent random variables for which Lindeberg's conditions (cf. Theorem 5, Section 3, Chapter 1) are satisfied. Consequently $\xi'_n(t'') - \xi'_n(t')$ is asymptotically normal with mathematical expectation 0 and variance

$$\lim_{n\to\infty} \sum_{t' \leqq t_{ni} \leqq t''} \mathsf{D}\xi_{ni} = t'' - t'$$

(since

$$\Big| \sum_{t' \leqq t_{ni} \leqq t''} \mathsf{D}\xi_{ni} - (t'' - t') \Big|$$
$$\leqq \max_i b_{ni} \leqq \varepsilon^2 + \max_i \int_{|u|<\varepsilon} u^2 dF_{ni}(u) \leqq \varepsilon^2 + \sum_{i=1}^{k_n} \int_{|u|>\varepsilon} u^2 dF_{ni}(u) \,,$$

and the expression on the right can be made arbitrarily small for sufficiently large n).

This proves the convergence of the finite-dimensional distributions.

To prove that the distributions of $f(\xi_n(t))$ converge to the distributions of $f(w(t))$ for all functionals f that are continuous on $C[0, 1]$, let us show that for arbitrary $\varepsilon > 0$,

$$\lim_{h\to 0}\overline{\lim_{n\to\infty}}\,\mathsf{P}\{\sup_{|t'-t''|\leq h}|\xi_n(t')-\xi_n(t'')|>\varepsilon\}=0\,, \tag{2}$$

and let us use Remark 1 of Section 2. Since

$$\sup_{|t'-t''|\leq h}|\xi_n(t')-\xi_n(t'')|\leqq 2\sup_k\sup_{kh<t\leqq(k+2)h}|\xi_n(t)-\xi_n(kh)|$$
$$\leqq 4\sup_k\sup_{kh<t\leqq(k+1)h}|\xi_n(t)-\xi_n(kh)|\,,$$

we have

$$\mathsf{P}\{\sup_{|t'-t''|\leq h}|\xi_n(t')-\xi_n(t'')|>\varepsilon\}$$
$$\leqq\sum_{kh<1}\mathsf{P}\left\{\sup_{kh<t\leqq(k+1)h}|\xi_n(t)-\xi_n(kh)|>\frac{\varepsilon}{4}\right\}.$$

To find a bound for the probability

$$\mathsf{P}\left\{\sup_{kh<t\leqq(k+1)h}|\xi_n(t)-\xi_n(kh)|>\frac{\varepsilon}{4}\right\},$$

let us use Theorem 2, Section 4, Chapter III.

We note that

$$\sup_{kh<t\leqq(k+1)h}|\xi_n(t)-\xi_n(kh)|\leqq 2\sup_{j_{nk}<r\leqq j_{n,k+1}}\left|\sum_{j=j_{nk}}^{r}\xi_{nj}\right|,$$

where j_{nk} is the greatest of the indices j such that t_{nj} does not exceed kh. Since

$$\mathsf{P}\left\{\left|\sum_{j=j_{nk}}^{j_{n,k+1}}\xi_{nj}\right|>\frac{\varepsilon}{8}\right\}\leqq\frac{64}{\varepsilon^2}\sum\mathsf{D}\xi_{nj}\longrightarrow\frac{64h}{\varepsilon^2}\,,$$

it follows that for sufficiently small h,

$$\overline{\lim_{h\to\infty}}\,\mathsf{P}\left\{\sup_{kh<t\leqq(k+1)h}|\xi_n(t)-\xi_n(kh)|>\frac{\varepsilon}{4}\right\}$$
$$\leqq\frac{1}{1-\dfrac{64h}{\varepsilon^2}}\overline{\lim_{n\to\infty}}\,\mathsf{P}\left\{|\xi_n(t_{nj_{n,k+1}}+1)-\xi_n(t_{nj_{nk}})|>\frac{\varepsilon}{8}\right\}.$$

It follows from the convergence (proved above) of the sequences of finite-dimensional distributions of $\xi_n(t)$ to the finite-dimensional distributions of $w(t)$ that

$$\overline{\lim_{n\to\infty}}\,\mathsf{P}\left\{|\xi_n(t_{nj_{n,k+1}}+1)-\xi_n(t_{nj_{nk}})|>\frac{\varepsilon}{8}\right\}$$
$$=\frac{1}{\sqrt{2\pi h}}\int_{|u|>(\varepsilon/8)}\exp\left(-\frac{u^2}{2h}\right)du\,.$$

Consequently

$$\overline{\lim_{n\to\infty}} \mathsf{P}\{\sup_{|t'-t''|\leq h} | \xi_n(t') - \xi_n(t'') > \varepsilon\}$$
$$\leq \sum_{kh<1} \frac{1}{1-\frac{64h}{\varepsilon^2}} \frac{1}{\sqrt{2\pi}} \int_{|u|>(\varepsilon/8\sqrt{h})} \exp\left(-\frac{u^2}{2}\right)du$$
$$= \frac{1}{\sqrt{2\pi}} \frac{1}{1-\frac{64h}{\varepsilon^2}} \frac{1}{h} \int_{|u|>(\varepsilon/8\sqrt{h})} \exp\left(-\frac{u^2}{2}\right)du .$$

Since

$$\frac{1}{h}\int_{|u|>(c/\sqrt{h})} \exp\left(-\frac{u^2}{2}\right)du \leq \frac{1}{c^2}\int_{|u|>(c/\sqrt{h})} u^2 \exp\left(-\frac{u^2}{2}\right)du \longrightarrow 0 ,$$

it then follows that equation (2) holds. This completes the proof of the theorem.

REMARK 1. It follows from Corollary 2 to Theorem 2, Section 1 that if the conditions of Theorem 1 are satisfied, the distributions of $f(\xi_n(t))$ converge to the distribution of $f(w(t))$ for every functional f defined on $C[0, 1]$ and continuous (with respect to the metric of $C[0, 1]$) almost everywhere (with respect to the measure μ_w corresponding to a process of Brownian motion on $[0, 1]$).

Let $\xi_1, \xi_2, \cdots, \xi_n, \cdots$ denote a sequence of independent identically distributed random variables such that $\mathsf{M}\xi_i = 0$ and $\mathsf{D}\xi_i = 1$. From Theorem 1 we immediately have:

Theorem 2. *Let $\xi_n(t)$ denote a random broken line with vertices $\{(k/n), (1/\sqrt{n})S_k\}$, where $S_k = \xi_1 + \cdots + \xi_k$. Then for every functional f defined and continuous on $C[0, 1]$ almost everywhere with respect to the measure μ_w, the distributions of $f(\xi_n(t))$ converge to the distribution of $f(\xi(t))$.*

Proof. It will be sufficient to show that Lindeberg's condition is satisfied for the variables $\xi_{nk} = (1/\sqrt{n})\xi_k$. If we let $F(x)$ denote the distribution function of the variable ξ_k, then the distribution function of the variables ξ_{nk} is $F(\sqrt{n}x)$ and

$$\sum_{k=1}^{n}\int_{|x|>\varepsilon} x^2dF_{nk}(x) = \sum_{k=1}^{n}\int_{|x|>\varepsilon} x^2dF(\sqrt{n}x)$$
$$= \int_{|x|>\varepsilon} nx^2dF(\sqrt{n}x) = \int_{|u|>\varepsilon\sqrt{n}} u^2dF(u) \longrightarrow 0 .$$

Corollary. *If the conditions of Theorem 2 are satisfied and $\eta_n = \sup_{0\leq k\leq n} S_k$, then*

$$\lim_{n\to\infty} \mathsf{P}\{\eta_n < \alpha\sqrt{n}\} = \frac{2}{\sqrt{2\pi}}\int_0^{\alpha} \exp\left(-\frac{u^2}{2}\right)du .$$

Proof. We have $\eta_n/\sqrt{n} = \sup_{0\le t\le 1} \xi_n(t)$, and since $\sup_{0\le t\le 1} x(t)$ is a continuous functional on $C[0, 1]$, it follows that

$$\lim_{n\to\infty} \mathsf{P}\{\sup_{0\le t\le 1} \xi_n(t) < \alpha\} = \mathsf{P}\{\sup_{0\le t\le 1} w(t) < \alpha\} ,$$

so that it only remains for us to apply Corollary 1 of Section 5, Chapter VI. Analogously, using Corollary 3, Section 5, Chapter VI, we see that

$$\lim_{n\to\infty} \mathsf{P}\{\max_{1\le k\le n} | S_k | < \alpha\sqrt{n}\}$$
$$= \sum_{k=-\infty}^{\infty} (-1)^k \frac{1}{\sqrt{2\pi}} \int_{-\alpha}^{\alpha} \exp\left(-\frac{(u-2k\alpha)^2}{2}\right)du .$$

Theorem 3. *Suppose that a function $\varphi(x)$ is defined for $x \in (-\infty, \infty)$ is Riemann-integrable over every finite interval and that the random variables ξ_k satisfy the conditions of Theorem 2. Then*

$$\lim_{n\to\infty} \mathsf{P}\left\{\frac{1}{n}\sum_{k=1}^{n} \varphi\left(\frac{1}{\sqrt{n}}S_k\right) < a\right\} = \mathsf{P}\left\{\int_0^1 \varphi(w(t))dt < a\right\}$$

for all a such that $\mathsf{P}\left\{\int_0^1 \varphi(w(t))dt = a\right\} = 0.$

Proof. Let us show that the functional

$$f[x(t)] = \int_0^1 \varphi(x(t))dt$$

is continuous μ_w almost-everywhere with respect to the metric of $C[0, 1]$. Suppose that $\{x_n(t)\}$ converges uniformly to $x(t)$ on $[0, 1]$. Then $\varphi(x_n(t)) \to \varphi(x(t))$ for all values of t such that $x(t) \notin \Lambda_\varphi$, where Λ_φ is the set of points of discontinuity of the function φ. Let $\chi_\varphi(x)$ denote the characteristic function of the set Λ_φ. Then the functional $f(x(t))$ is continuous at the point $x(t) \in C[0, 1]$ if for almost all t, we have $x(t) \notin \Lambda_\varphi$, that is, if

$$\int_0^1 \chi_\varphi(x(s))ds = 0 ,$$

so that (in this case) $\varphi(x_n(t)) \to \varphi(x(t))$ for almost all t and the functions $\varphi(x_n(t))$ are bounded by a single constant since $\sup_{n,t} |x_n(t)|$ is finite and $\varphi(x)$ is bounded on every interval. We should note only that

$$\mathsf{M}\int_0^1 \chi_\varphi(w(i))dt = \int_0^1 \mathsf{M}\chi_\varphi(w(t))dt = \int_0^1 dt \frac{1}{\sqrt{2\pi t}} \int_{\Lambda_\varphi} \exp\left(-\frac{u^2}{2t}\right)du = 0 ,$$

since Λ_φ has Lebesgue measure 0 (by virtue of the Riemann-integrability of the function φ). The quantity $\int_0^1 \chi_\varphi(w(t))dt$ is non-negative. Therefore

$$\mathsf{P}\left\{\int_0^1 \chi_\varphi(w(t))dt \neq 0\right\} = 0 .$$

If we let $A \subset C[0, 1]$ denote the set of points of discontinuity of the functional f, then

$$A \subset \left\{x(t); \int_0^1 \chi_\varphi(x(s))ds > 0\right\} .$$

Hence

$$\mu_w(A) \leqq \mathsf{P}\left\{\int_0^1 \chi_\varphi(w(t))dt \neq 0\right\} = 0 .$$

If $\xi_n(t)$ is the process introduced in Theorem 2, then on the basis of that theorem,

$$\lim_{n\to\infty} \mathsf{P}\left\{\int_0^1 \varphi(\xi_n(t))dt < a\right\} = \mathsf{P}\left\{\int_0^1 \varphi(w(t))dt < a\right\} ,$$

if

$$\mathsf{P}\left\{\int_0^1 \varphi(w(t))dt = a\right\} = 0 .$$

Let $\varphi_\varepsilon^+(x)$ and $\varphi_\varepsilon^-(x)$ denote two continuous functions satisfying the inequalities

$$\varphi_\varepsilon^+(x) > \varphi(x) > \varphi_\varepsilon^-(x)$$

and

$$\int_{-\infty}^{\infty} [\varphi_\varepsilon^+(x) - \varphi_\varepsilon^-(x)]dx < \varepsilon .$$

For every continuous function $\bar{\varphi}(x)$,

$$\begin{aligned}
&\left|\int_0^1 \bar{\varphi}(\xi_n(t))dt - \frac{1}{n}\sum_{k=1}^n \bar{\varphi}\left(\frac{1}{\sqrt{n}}S_k\right)\right| \\
&\quad = \left|\sum_{k=1}^n \int_{(k-1)/n}^{k/n} \left[\bar{\varphi}(\xi_n(t)) - \bar{\varphi}\left(\xi_n\left(\frac{k}{n}\right)\right)\right]dt\right| \\
&\quad \leqq \sum_{k=1}^n \int_{(k-1)/n}^{k/n} \left|\bar{\varphi}(\xi_n(t)) - \bar{\varphi}\left(\xi_n\left(\frac{k}{n}\right)\right)\right| dt \\
&\quad \leqq \sup_{\substack{|x-y|\leqq\eta_n \\ |x|\leqq\zeta_n}} |\bar{\varphi}(x) - \bar{\varphi}(y)| ,
\end{aligned}$$

where

$$\eta_n = \sup_k \left|\xi_n\left(\frac{k+1}{n}\right) - \xi_n\left(\frac{k}{n}\right)\right| = \frac{\sup_k \xi_k}{\sqrt{n}} , \quad \zeta_n = \sup_k \frac{|S_k|}{\sqrt{n}} .$$

Therefore

$$\mathsf{P}\left\{\left|\int_0^1 \bar{\varphi}(\xi_n(t))dt - \frac{1}{n}\sum_{k=1}^n \bar{\varphi}\left(\frac{1}{\sqrt{n}}S_k\right)\right| > \varepsilon\right\} \leqq \mathsf{P}\{\eta_n < \delta\} + \mathsf{P}\{\zeta_n > c\} ,$$

where δ and c are chosen in such a way that $|\bar{\varphi}(x) - \bar{\varphi}(y)| < \varepsilon$ whenever $|x - y| \leqq \delta$ and $|x| \leqq C$. Since $\eta_n \to 0$ in probability as $n \to \infty$ (this follows from Lindeberg's condition and was established in the proof of Theorem 1), it follows that $\mathsf{P}\{\eta_n > \delta\} \to 0$ as $n \to \infty$ for every $\delta > 0$. The quantity $\mathsf{P}\{\zeta_n > c\}$ can be made arbitrarily small by choosing sufficiently large c for all n (this follows from Corollary 1). Therefore

$$\left|\int_0^1 \bar{\varphi}(\xi_n(t))dt - \frac{1}{n}\sum_{k=1}^{n} \bar{\varphi}\left(\frac{1}{\sqrt{n}}S_k\right)\right| \to 0$$

in probability, so that

$$\lim_{n\to\infty} \mathsf{P}\left\{\frac{1}{n}\sum_{k=1}^{n} \bar{\varphi}\left(\frac{1}{\sqrt{n}}S_k\right) < a\right\} = \mathsf{P}\left\{\int_0^1 \bar{\varphi}(w(t))dt < a\right\},$$

if $\mathsf{P}\left\{\int_0^1 \bar{\varphi}(w(t))dt = a\right\} = 0.$

Since

$$\mathsf{P}\left\{\frac{1}{n}\sum_{k=1}^{n} \varphi_\varepsilon^-\left(\frac{1}{\sqrt{n}}S_k\right) < a\right\} \geqq \mathsf{P}\left\{\frac{1}{n}\sum_{k=1}^{n} \varphi\left(\frac{1}{\sqrt{n}}S_k\right) < a\right\}$$
$$\geqq \mathsf{P}\left\{\frac{1}{n}\sum_{k=1}^{n} \varphi_\varepsilon^+\left(\frac{1}{\sqrt{n}}S_k\right) < a\right\},$$

by taking the limit as $n \to \infty$ we find that for every $h > 0$,

$$\mathsf{P}\left\{\int_0^1 \varphi_\varepsilon^-(w(t))dt < a + h\right\} \geqq \overline{\lim_{n\to\infty}}\, \mathsf{P}\left\{\frac{1}{n}\sum_{k=1}^{n} \varphi\left(\frac{1}{\sqrt{n}}S_k\right) < a\right\}$$
$$\geqq \underline{\lim_{n\to\infty}}\, \mathsf{P}\left\{\frac{1}{n}\sum_{k=1}^{n} \varphi\left(\frac{1}{\sqrt{n}}S_k\right) < a\right\}$$
$$\geqq \mathsf{P}\left\{\int_0^1 \varphi_\varepsilon^+(w(t))dt < a - h\right\}.$$

But

$$\mathsf{M}\left|\int_0^1 \varphi_\varepsilon^+(w(t))dt - \int_0^1 \varphi(w(t))dt\right|$$
$$\leqq \mathsf{M}\int_0^1 |\varphi_\varepsilon^+(w(t)) - \varphi(w(t))|\, dt$$
$$\leqq \frac{1}{\sqrt{2\pi}}\int_0^1 \frac{dt}{\sqrt{t}}\int_{-\infty}^{\infty} |\varphi_\varepsilon^+(x) - \varphi_\varepsilon^-(x)| \exp\left(-\frac{x^2}{2t}\right)dx$$
$$\leqq \frac{\varepsilon}{\sqrt{2\pi}}\int_0^1 \frac{dt}{\sqrt{t}} = \frac{2\varepsilon}{\sqrt{2\pi}}.$$

This means that the variable distribution of $\int_0^1 \varphi_\varepsilon^+(w(t))dt$ converges to the distribution of $\int_0^1 \varphi(w(t))dt$ as $\varepsilon \to 0$. An analogous assertion is valid for φ_ε^-. Taking the limit as $\varepsilon \to 0$, we see that for all $h > 0$,

$$\mathsf{P}\left\{\int_0^1 \varphi(w(t))dt < a + h\right\} \geqq \overline{\lim_{n\to\infty}} \mathsf{P}\left\{\frac{1}{n}\sum_{k=1}^n \varphi\left(\frac{1}{\sqrt{n}}S_k\right) < a\right\}$$

$$\geqq \underline{\lim_{n\to\infty}} \mathsf{P}\left\{\frac{1}{n}\sum_{k=1}^n \varphi\left(\frac{1}{\sqrt{n}}S_k\right) < a\right\}$$

$$\geqq \mathsf{P}\left\{\int_0^1 \varphi(w(t))dt < a - h\right\}.$$

Taking the limit as $h \to 0$ and remembering that

$$\mathsf{P}\left\{\int_0^1 \varphi(w(t))dt < z\right\}$$

is continuous at $z = a$ if

$$\mathsf{P}\left\{\int_0^1 \varphi(w(t))dt = a\right\} = 0 ,$$

we obtain proof of the theorem.

Corollary. *Let ν_n denote the number of positive sums in the sequence $S_1, S_2, \cdots, S_n$. Then for $0 \leqq a \leqq 1$,*

$$\lim_{n\to\infty} \mathsf{P}\{\nu_n < na\} = \frac{2}{\pi} \arcsin \sqrt{a} .$$

Proof. $\nu_n = \sum g_+\{(1/\sqrt{n})S_k\}$, where $g_+(x) = 1$ for $x > 0$ and $g_+(x) = 0$ for $x \leq 0$. Therefore on the basis of Theorem 3,

$$\lim_{n\to\infty} \mathsf{P}\{\nu_n < na\} = \mathsf{P}\left\{\int_0^1 g_+(w(t))dt < a\right\}$$

for all a such that $\mathsf{P}\left\{\int_0^1 g_+(w(t))dt = a\right\} = 0$. Now we use Theorem 4, Section 5, Chapter VIII, and the proof is complete.

4. CONVERGENCE OF A SEQUENCE OF MARKOV CHAINS TO A DIFFUSION PROCESS

Let us consider a double sequence of random variables $\xi_{n0}, \xi_{n1}, \cdots, \xi_{nk_n}$ that form along in each sequence a Markov chain. We let $p_{nk}(x, A)$ denote the transition probabilities

$$p_{nk}(\xi_{nk}, A) = \mathsf{P}\{\xi_{n,k+1} \in A \mid \xi_{nk}\} \qquad (\text{mod } \mathsf{P}) .$$

Suppose also that $0 = t_{n0} < t_{n1} < \cdots < t_{nk_n} = 1$ is a sequence of partitions of the interval $[0, 1]$. Let us construct a random broken line $\xi_n(t)$ with vertices at the points (t_{nk}, ξ_{nk}). In this section we shall study conditions under which the sequences of finite-dimensional distributions of $\xi_n(t)$ and the sequences of distributions of functionals of $\xi_n(t)$ converge to the corresponding distributions of the Markov process $\xi(t)$ representing the solution of a stochastic equation of the type studied in Chapter VIII.

We define

$$\Delta t_{nk} = t_{n,k+1} - t_{nk} ,$$

$$a_n(t_{nk}, x) = \frac{1}{\Delta t_{nk}} \int (y - x) p_{nk}(x, dy) ,$$

$$b_n(t_{nk}, x) = \sigma_n^2(t_{nk}, x) = \frac{1}{\Delta t_{nk}} \int (y - x)^2 p_{nk}(x, dy) - \Delta t_{nk} a_n^2(t_{nk}, x) .$$

Theorem 1. *Let $\xi(t)$ denote a solution of the stochastic equation*

$$\xi(t) = \xi_0 + \int_0^t a(s, \xi(s)) ds + \int_0^t \sigma(s, \xi(s)) dw(s) ,$$

where ξ_0 is independent of $w(t)$ and where $a(s, x)$ and $\sigma(s, x)$ are functions that are continuous over the set of variables and satisfy a Lipschitz condition with respect to x:

$$| a(s, x) - a(s, y) | + | \sigma(s, x) - \sigma(x, y) | \leqq K | x - y | .$$

For the finite-dimensional distributions of the processes $\xi_n(t)$ to converge to finite-dimensional distributions of the process $\xi(t)$, it is sufficient that the following conditions be satisfied:

a. $\lim_{n\to\infty} \max_k \Delta t_{nk} = 0$,

b. $\lim_{n\to\infty} \sum_{k=1}^{k_n} \mathsf{M}(| a_n(t_{nk}, \xi_{nk}) - a(t_{nk}, \xi_{nk}) |^2 + | \sigma_n(t_{nk}, \xi_{nk}) - \sigma(t_{nk}, \xi_{nk}) |^2) \Delta t_{nk} = 0$,

c. *for some $\delta > 0$*

$$\lim_{n\to\infty} \sum_{k=1}^{k_n} \mathsf{M} | \xi_{n,k+1} - \xi_{nk} |^{2+\delta} = 0$$

and

$$\mathsf{P}\text{-}\lim_{n\to\infty} \sup_{k=1} \sum_{i=k}^{k_n} \mathsf{M}(| \xi_{n,i+1} - \xi_{ni} |^{2+\delta} | \xi_{nk}) = 0 ;$$

d. *the functions $1/\sigma_n(t_{nk}, x)$ and $a_n(t_{nk}, x)/\sigma_n(t_{nk}, x)$ are uniformly bounded with respect to n;*

e. *the limiting distribution of the random variable ξ_{n0} coincides with the distribution of the random variable ξ_0.*

Proof. we set

$$\omega_{nk} = [\xi_{n,k+1} - \xi_{nk} - a_n(t_{nk}, \xi_{nk}) \Delta t_{nk}] (\sigma_n(t_{nk}, \xi_{nk}))^{-1} . \qquad (1)$$

Then

$$\xi_{n,k+1} = \xi_{nk} + a_n(t_{nk}, \xi_{nk}) \Delta t_{nk} + \sigma_n(t_{nk}, \xi_{nk}) \omega_{nk} .$$

Let $\mathfrak{F}_{nk}$ denote the minimal σ-algebra with respect to which the variables $\xi_{n0}, \xi_{n1}, \cdots, \xi_{nk}$ are measurable. The quantity ω_{nk} is measurable with respect to the σ-algebra $\mathfrak{F}_{n,k+1}$ and

$$\mathsf{M}(\omega_{nk} \mid \mathfrak{F}_{nk}) = 0, \quad \mathsf{M}(\omega_{nk}^2 \mid \mathfrak{F}_{nk}) = \Delta t_{nk} . \tag{2}$$

Consider the variables η_{nk} defined by the relations

$$\eta_{n0} = \xi_{n0}, \eta_{n,k+1} = \eta_{nk} + a(t_{nk}, \eta_{nk})\Delta t_{nk} + \sigma(t_{nk}, \eta_{nk})\omega_{nk} .$$

Let us find a bound for $\mathsf{M}(\eta_{nk} - \xi_{nk})^2$. Obviously, the η_{nk} are also measurable with respect to the σ-algebra $\mathfrak{F}_{nk}$. We have

$$\eta_{n,k+1} - \xi_{n,k+1} = \eta_{nk} - \xi_{nk} + [a(t_{nk}, \eta_{nk}) - a(t_{nk}, \xi_{nk})] \times \Delta t_{nk} + [\sigma(t_{nk}, \eta_{nk}) - \sigma(t_{nk}, \xi_{nk})]\omega_{nk} + \varepsilon_{nk}$$

$$\varepsilon_{nk} = [a(t_{nk}, \xi_{nk}) - a_n(t_{nk}, \xi_{nk})]\Delta t_{nk} + [\sigma(t_{nk}, \eta_{nk}) - \sigma_n(t_{nk}, \xi_{nk})]\omega_{nk} .$$

Therefore, by using equations (2) and the Lipschitz conditions for a and σ and also the inequality $2ab \leqq a^2 + b^2$ we obtain

$$\begin{aligned}
\mathsf{M} \mid \eta_{n,k+1} - \xi_{n,k+1} \mid^2 &\leqq \mathsf{M} \mid \eta_{nk} - \xi_{nk} \mid^2 \\
&+ 2\mathsf{M}(\eta_{nk} - \xi_{nk}) \times (a(t_{nk}, \eta_{nk}) - a(t_{nk}, \xi_{nk}))\Delta t_{nk} \\
&+ 2\mathsf{M} \mid \eta_{nk} - \xi_{nk} \mid \times \mid a(t_{nk}, \xi_{nk}) - a_n(t_{nk}, \xi_{nk}) \mid \Delta t_{nk} \\
&+ \mathsf{M}[(a(t_{nk},\eta_{nk}) - a(t_{nk},\xi_{nk}))\Delta t_{nk} + (\sigma(t_{nk},\eta_{nk}) - \sigma(t_{nk},\xi_{nk}))\omega_{nk} + \varepsilon_{nk}]^2 \\
&\leqq \mathsf{M} \mid \eta_{nk} - \xi_{nk} \mid^2(1 + 2K\Delta t_{nk} + \Delta t_{nk})\Delta t_{nk} \\
&+ \mathsf{M} \mid a(t_{nk}, \xi_{nk}) - a_n(t_{nk}, \xi_{nk}) \mid^2 \Delta t_{nk} \\
&+ 2\mathsf{M}(a(t_{nk}, \eta_{nk}) - a(t_{nk}, \xi_{nk}))^2 \Delta t_{nk}^2 \\
&+ \mathsf{M}(\sigma(t_{nk}, \eta_{nk}) - \sigma(t_{nk}, \xi_{nk}))^2 \mathsf{M}(\omega_{nk}^2 \mid \mathfrak{F}_{nk}) \\
&+ 2\mathsf{M}\, \varepsilon_{nk}^2 \leqq \mathsf{M} \mid \eta_{nk} - \xi_{nk} \mid^2(1 + L\Delta t_{nk}) + \alpha_{nk} ,
\end{aligned}$$

where $L = 2K + 1 + 4K^2$ and

$$\begin{aligned}
\alpha_{nk} = \mathsf{M}[a(t_{nk}, \xi_{nk}) - a_n(t_{nk}, \xi_{nk})]^2 (\Delta t_{nk} + 2\Delta t_{nk}^2) \\
+ \mathsf{M}[\sigma(t_{nk}, \xi_{nk}) - \sigma_n(t_{nk}, \xi_{nk})]^2 \Delta t_{nk} .
\end{aligned}$$

Since $\mathsf{M} \mid \eta_{n0} - \xi_{n0} \mid^2 = 0$, we have $\mathsf{M} \mid \xi_{n1} - \eta_{n1} \mid^2 \leqq \alpha_{n0}$,

$$\begin{aligned}
\mathsf{M} \mid \xi_{n2} - \eta_{n2} \mid^2 &\leqq \alpha_{n0}(1 + L\Delta t_{n0}) + \alpha_{n1} \leqq (1 + L\Delta t_{n0}) [\alpha_{n0} + \alpha_{n1}] , \\
\mathsf{M} \mid \xi_{n3} - \eta_{n3} \mid^2 &\leqq [\alpha_{n0} + \alpha_{n1}] (1 + L\Delta t_{n0}) (1 + L\Delta t_{n1}) + \alpha_{n2} \\
&\leqq [\alpha_{n0} + \alpha_{n1} + \alpha_{n2}] (1 + L\Delta t_{n0}) (1 + L\Delta t_{n1}) , \\
\mathsf{M} \mid \xi_{nk} - \eta_{nk} \mid^2 &\leqq \sum_{i=0}^{k-1} \alpha_{ni} \prod_{i=0}^{k-2} (1 + L\Delta t_{ni}) \leqq e^L \sum_{i=0}^{k_n-1} \alpha_{ni} .
\end{aligned}$$

It follows from condition b that $\lim_{n\to\infty} \sum_{i=0}^{k_n-1} \alpha_{ni} = 0$.

Consequently, the finite-dimensional distributions of the process $\xi_n(t)$ converge to the finite-dimensional distributions of the process $\xi(t)$ if the finite-dimensional distributions of the process $\eta_n(t)$, where $\eta_n(t)$ is a random broken line with vertices at the points (t_{nk}, η_{nk})

converge to the finite-dimensional distributions of $\xi(t)$.

To continue the proof, we use:

Lemma 1. *Suppose that* $w_n(t) = \sum_{t_{nk}<t} \omega_{nk}$. *Then the finite-dimensional distributions of the process* $w_n(t)$ *converge to the finite-dimensional distributions of a process* $w(t)$ *of Brownian motion.*

Proof. It will be sufficient here to show that for all λ_{nk} such that $\sup_{n,k} |\lambda_{nk}| < \infty$,

$$\lim_{n\to\infty}\left[\mathsf{M}\exp\left(i\sum_{k=0}^{k_n-1}\lambda_{nk}\omega_{nk}\right) - \mathsf{M}\exp\left(i\sum_{k=0}^{k_n-1}\lambda_{nk}[w(t_{n,k+1}) - w(t_{nk})]\right)\right] = 0\ .$$

We note that the function $\{e^{i\lambda x} - 1 - i\lambda x + (\lambda^2x^2/2)\}/|\lambda|^2|x|^{2+\delta}$ is for $0 < \delta < 1$ bounded on the entire real line. Therefore

$$\left|e^{i\lambda x} - \left(1 + i\lambda x - \frac{\lambda^2x^2}{2}\right)\right| \leqq O(\lambda^2)\,|x|^{2+\delta}\ .$$

Hence

$$\begin{aligned}
&\mathsf{M}\prod_{k=0}^{r} e^{i\lambda_{nk}\omega_{nk}}\\
&= \mathsf{M}\left(\prod_{k=1}^{r-1} e^{i\lambda_{nk}\omega_{nk}}\right)\mathsf{M}[e^{i\lambda_{nr}\omega_{nr}} \mid \mathfrak{F}_{nr}]\\
&= \mathsf{M}\left(\prod_{k=0}^{r-1} e^{i\lambda_{nk}\omega_{nk}}\right)\mathsf{M}\left[1 + i\lambda_{nr}\omega_{nr} - \frac{\lambda_{nr}^2}{2}\omega_{nr} + O(|\omega_{nr}|^{2+\delta}) \mid \mathfrak{F}_{nr}\right]\\
&= \mathsf{M}\left(\prod_{k=0}^{r-1} e^{i\lambda_{nk}\omega_{nk}}\right)\left(1 - \frac{\lambda_{nr}^2}{2}\Delta t_{nr}\right) + O(\mathsf{M}\,|\omega_{nk}|^{2+\delta})\\
&= \prod_{k=0}^{r}\left(1 - \frac{\lambda_{nk}^2}{2}\Delta t_{nk}\right) + O\left(\sum_{k=0}^{r}\mathsf{M}\,|\omega_{nk}|^{2+\delta}\right).
\end{aligned}$$

It follows from formula (1) for ω_{nk} and the boundedness of $1/\sigma_n$ and a_n/σ_n that

$$\mathsf{M}\,|\omega_{nk}|^{2+\delta} \leqq L(\mathsf{M}\,|\xi_{n,k+1} - \xi_{nk}|^{2+\delta} + |\Delta t_{nk}|^{2+\delta})\ .$$

Therefore

$$\lim_{n\to\infty}\sum_{k=0}^{k_n-1}\mathsf{M}\,|\omega_{nk}|^{2+\delta} = 0\ .$$

Furthermore

$$\begin{aligned}
\left|\sum_{k=0}^{k_n-1}\left(1 - \frac{\lambda_{nk}^2}{2}\Delta t_{nk}\right) - \sum_{k=0}^{k_n-1}\exp\left(-\frac{\lambda_{nk}^2}{2}\Delta t_{nk}\right)\right| &\leqq O\left(\sum \Delta t_{nk}^2\right)\\
&= O(\max_k \Delta t_{nk}) \longrightarrow 0\ .
\end{aligned}$$

Consequently

$$\lim_{n\to\infty}\left\{\mathsf{M}\exp\left(i\sum_{k=0}^{k_n-1}\lambda_{nk}\omega_{nk}\right)-\exp\left(-\sum_{k=0}^{k_n-1}\frac{\lambda_{nk}^2}{2}\Delta t_{nk}\right)\right\}=0\,.$$

Since

$$\mathsf{M}e^{i\lambda_{nk}[w(t_{n,k+1})-w(t_{nk})]}=\exp\left(-\frac{\lambda_{nk}^2}{2}\Delta t_{nk}\right),$$

the lemma is proved.

REMARK 1. It follows from Lemma 1 that for every continuous function $\alpha(t, x)$ that is uniformly continuous with respect to x in every finite interval, we have

$$\mathsf{M}\{\exp(i\lambda\sum_{t'\leqq t_{nk}\leqq t''}\alpha(t_{nk}, x)\omega_{nk})\mid\mathfrak{F}_{l_n}\}\to\mathsf{M}\left\{\exp i\lambda\int_{t'}^{t''}\alpha(t, x)dw(t)\right\},$$

if l_n is such that $t_{nl_n}\leqq t'$. This is true because just as in the proof of Lemma 1, we can show that

$$\mathsf{M}\{\exp(i\lambda\sum_{t'\leqq t_{nk}<t''}\alpha(t_{nk}, x)\omega_{nk}\mid\mathfrak{F}l_n\}$$
$$-\exp\left\{\frac{\lambda^2}{2}\sum_{t'\leqq t_{nk}<t''}\alpha(t_{nk}, x)^2\Delta t_{nk}\right\}\to 0$$

uniformly with respect to x in every finite interval (by virtue of the boundedness of $\alpha(t, x)$ on every finite x-interval).

We should note finally that

$$\sum_{t'\leqq t_{nk}<t''}\alpha(t_{nk}, x)^2\Delta t_{nk}\to\int_{t'}^{t''}\alpha(t, x)^2dt$$

uniformly with respect to x in every finite interval, since $\alpha(t, x)$ is uniformly continuous with respect to all the variables in every finite x-interval.

Lemma 2. *If the measurable functions $\varphi_n(x_1, \cdots, x_m)$ are bounded by a single constant, if they converge uniformly on every compact set to a continuous function $\varphi_0(x_1, x_2, \cdots, x_m)$, and if the distribution functions $F_2(x_1,x_2,\cdots,x_m)$ converge weakly to a function $F_0(x_1,x_2,\cdots,x_m)$, then*

$$\lim_{n\to\infty}\int\varphi_n(x_1, \cdots, x_m)dF_n(x_1, \cdots, x_m)$$
$$=\int\varphi_0(x_1, \cdots, x_m)dF_0(x_1, \cdots, x_m)\,.$$

Proof. Since

$$\lim_{n\to\infty}\int\varphi_0(x_1, \cdots, x_m)dF_n(x_1, \cdots, x_m)$$
$$=\int\varphi_0(x_1, \cdots, x_m)dF_0(x_1, \cdots, x_m)\,,$$

by virtue of the weak convergence of $\{F_n(x_1, \cdots, x_m)\}$ to $F_0(x_1, \cdots, x_m)$, it follows that to prove the lemma it is sufficient to show that

$$\overline{\lim_{n\to\infty}} \int | \varphi_n(x_1, \cdots, x_m) - \varphi_0(x_1, \cdots, x_m) | \, dF_n(x_1, \cdots, x_m) = 0 \, .$$

But for every $K > 0$,

$$\int | \varphi_0 - \varphi_n | \, dF_n \leqq \int_{\sum_i |x_i| \leqq K} | \varphi_0 - \varphi_n | \, dF_n + \int_{\sum_i |x_i| > K} | \varphi_0 - \varphi_n | \, dF_n \; ;$$

the first integral approaches 0 as $n \longrightarrow \infty$ since $\{| \varphi_0 - \varphi_n |\}$ converges uniformly to 0 for $\sum | x_i | \leqq K$; the second integral can be made arbitrarily small for all n by choosing K sufficiently large (because of the boundedness of $| \varphi_0 - \varphi_n |$ and the weak convergence of the sequence of the distributions F_n). This completes the proof of the lemma.

Lemma 3. *Let $\{\xi_1^{(n)}\}, \cdots, \{\xi_m^{(n)}\}$ for $n = 0, 1, \cdots,$ denote m sequences of random variables. Let $\Phi_k^{(n)}(\lambda, x_1, \cdots, x_{k-1})$ denote functions such that with probability* 1,

$$\Phi_1^{(n)}(\lambda) = \mathsf{M} e^{i\lambda\xi_1^{(n)}} \, ,$$

$$\Phi_k^{(n)}(\lambda, \xi_1^{(n)}, \cdots, \xi_{k-1}^{(n)}) = \mathsf{M}(e^{i\lambda\xi_k^{(n)}} \mid \xi_1^{(n)}, \cdots, \xi_{k-1}^{(n)}) \, , \qquad k = 2, \cdots m \, .$$

Suppose that for all k the functions $\varphi_k^{(0)}(\lambda, x_1, \cdots, x_{k-1})$ are continuous and that $\{\Phi_k^{(n)}(\lambda_1, x_1, \cdots, x_{k-1})\}$ converges to $\Phi_k^{(0)}(\lambda, x_1, \cdots, x_{k-1})$ for $k = 1, 2, \cdots, m$ uniformly as $n \longrightarrow \infty$ on every compact set. Then the joint distributions of the variables $\xi_1^{(n)}, \cdots, \xi_m^{(n)}$ converge weakly to the joint distribution of the variables $\xi_1^{(0)}, \cdots, \xi_m^{(0)}$.

Proof. Taking $k = 1$, we see that the distributions of the variables $\xi_1^{(n)}$ converge to the distribution of the variable $\xi_1^{(0)}$. Let us suppose that the joint distributions $F_{k-1}^{(n)}$ of the variables $\xi_1^{(n)}, \cdots, \xi_{k-1}^{(n)}$ converge to the joint distribution $F_{k-1}^{(0)}$ of the variables $\xi_1^{(0)}, \cdots, \xi_{k-1}^{(0)}$. Then

$$\begin{aligned}
&\mathsf{M} \exp\left\{i \sum_{j=1}^{k} \lambda_j \xi_j^{(n)}\right\} \\
&\qquad = \int \exp\left\{i \sum_{j=1}^{k-1} \lambda_j x_j\right\} \Phi_k^{(n)}(\lambda_k, x_1, \cdots, x_{k-1}) dF_{k-1}^{(n)}(x_1, \cdots, x_{k-1}) \\
&\longrightarrow \int \exp\left\{i \sum_{j=1}^{k-1} \lambda_j x_j\right\} \Phi_k^{0}(\lambda_k, x_1, \cdots, x_{k-1}) dF_{k-1}^{(0)}(x_1, \cdots, x_{k-1}) \\
&\qquad = \mathsf{M} \exp\,(i\lambda_1\xi_1^{(0)} + i\lambda_2\xi_2^{(0)} + \cdots + i\lambda_k\xi_k^{(0)})
\end{aligned}$$

on the basis of Lemma 2. Therefore, the joint distributions of the variables $\xi_1^{(n)}, \cdots, \xi_k^{(n)}$ also converge to the joint distribution of the variables $\xi_1^{(0)}, \cdots, \xi_k^{(0)}$. We prove the lemma by induction on k.

We now return to the proof of the theorem. Let us choose an arbitrary partition of the interval [0, 1]: $0 = \tau_0 < \tau_1 < \cdots < \tau_N = 1$. Define

$$\eta_n^*(\tau_0) = \xi_{n0} ,$$

$$\eta_n^*(\tau_k) = \eta_n^*(\tau_{k-1}) + \sum_{\tau_{k-1} \leq t_{nr} < \tau_k} a(t_{nr}, \eta_n^*(\tau_{k-1}))\Delta t_{nr} + \sum_{\tau_{k-1} \leq t_{nr} < \tau_k} \sigma(t_{nr}, \eta_n^*(\tau_{k-1}))\omega_{nr} , \qquad k = 1, \cdots, N ,$$

$$\eta_0^*(\tau_0) = \xi_0 ,$$

$$\eta_0^*(\tau_k) = \eta_0^*(\tau_{k-1}) + \int_{\tau_{k-1}}^{\tau_k} a(s, \eta_0^*(\tau_{k-1}))ds + \int_{\tau_{k-1}}^{\tau_k} \sigma(s, \eta_0^*(\tau_{k-1}))dw(s) , \qquad k = 1, \cdots, N ,$$

Obviously,

$$\mathsf{M}(e^{i\lambda\eta_0^*(\tau_k)} \mid \eta_0^*(\tau_0), \cdots, \eta_0^*(\tau_{k-1})) = \Phi_k^{(0)}(\lambda, \eta_0^*(\tau_{k-1})) ,$$

where

$$\Phi_k^{(0)}(\lambda, x) = \exp\left\{i\lambda x + i\lambda \int_{\tau_{k-1}}^{\tau_k} a(s, x)ds - \frac{\lambda^2}{2}\int_{\tau_{k-1}}^{\tau_k} \sigma^2(s, x)ds\right\}$$

is a continuous function of x. Let t_{nl_k} denote the maximum point t_{nr} such that $t_{nr} < \tau_k$. It follows from the remark to Lemma 1 that when we substitute x for $\eta_n^*(\tau_{k-1})$,

$$\begin{aligned} &\mathsf{M}(e^{i\lambda\eta_n(\tau_k)} \mid \eta_n^*(\tau_0), \cdots, \eta_n^*(\tau_{k-1})) \\ &\quad = \mathsf{M}\{\exp(i\lambda\eta_n^*(\tau_{k-1}) + i\lambda \sum_{\tau_{k-1} \leq t_{nr} < \tau_k} a(t_{nr}, \eta_n^*(\tau_{k-1}))\Delta t_{nr}) \\ &\quad\quad \times \mathsf{M}[\exp(i\lambda \sum_{\tau_{k-1} \leq t_{nr} < \tau_k} \sigma(t_{nr}, \eta_n^*(\tau_{k-1}))\omega_{nr}) \mid \mathfrak{F}_{l k-1}] \mid \eta_n^*(\tau_0), \\ &\quad\quad \cdots, \eta_n^*(\tau_{k-1})\} \end{aligned}$$

converges to $\Phi_k^{(0)}(\lambda, x)$ uniformly with respect to x in every finite interval. Therefore, on the basis of Lemma 3 the joint distributions of the variables $\eta_n^*(\tau_0), \cdots, \eta_n^*(\tau_N)$ converge to the joint distribution of the variables $\eta_0^*(\tau_0), \cdots, \eta_0^*(\tau_N)$.

Since

$$\begin{aligned} &\xi(\tau_k) - \eta_0^*(\tau_k) \\ &\quad = \xi(\tau_{k-1}) - \eta_0^*(\tau_{k-1}) + \int_{\tau_{k-1}}^{\tau_k} [a(s, \xi(s)) - a(s, \eta_0^*(\tau_{k-1}))]ds \\ &\quad\quad + \int_{\tau_{k-1}}^{\tau_k} [\sigma(s, \xi(s)) - \sigma(s, \eta_0^*(\tau_{k-1}))]dw(s) , \end{aligned}$$

by using the Lipschitz condition and some simple transformations we obtain

$$\begin{aligned}
\mathsf{M} \mid \xi(\tau_k) - \eta_0^*(\tau_k) \mid^2 &= \mathsf{M} \mid \xi(\tau_k) - \eta_0^*(\tau_{k-1}) \mid^2 \\
&\quad + 2\mathsf{M} \mid \xi(\tau_{k-1}) - \eta_0^*(\tau_{k-1}) \mid \int_{\tau_{k-1}}^{\tau_k} \mid a(s, \xi(s)) - a(s, \eta_0^*(\tau_{k-1})) \mid ds \\
&\quad + \mathsf{M}\left(\int_{\tau_{k-1}}^{\tau_k} (a(s, \xi(s)) - a(s, \eta_0^*(\tau_{k-1}))ds\right)^2 \\
&\quad + \left(\int_{\tau_{k-1}}^{\tau_k} (\sigma(s, \xi(s)) - \sigma(s, \eta_0^*(\tau_{k-1})))dw(s)\right)^2 \\
&\leqq \mathsf{M} \mid \xi(\tau_{k-1}) - \eta_0^*(\tau_{k-1}) \mid^2 \\
&\quad + 2K\mathsf{M} \mid \xi(\tau_{k-1}) - \eta_0^*(\tau_{k-1}) \mid \int_{\tau_{k-1}}^{\tau_k} \mid \xi(s) - \eta_0^*(\tau_{k-1}) \mid ds \\
&\quad + 2\mathsf{M} \int_{\tau_{k-1}}^{\tau_k} ([a(s, \xi(s)) - a(s, \eta_0^*(\tau_{k-1}))]^2 \\
&\quad + [\sigma(s, \xi(s)) - \sigma(s, \eta_0^*(\tau_{k-1}))]^2)ds \\
&\leqq \mathsf{M} \mid \xi(\tau_{k-1}) - \eta_0^*(\tau_{k-1}) \mid^2 (1 + H(\tau_k - \tau_{k-1})) \\
&\quad + H \int_{\tau_{k-1}}^{\tau_k} \mid \xi(s) - \xi(\tau_{k-1}) \mid^2 ds ,
\end{aligned}$$

where $H = 2K + 8K^2$. Consequently, using the bounds that we have already applied, we obtain

$$\mathsf{M}(\xi(\tau_k) - \eta_0^*(\tau_k))^2 \leqq He^H \mathsf{M} \sum_{k=1}^{N} \int_{\tau_{k-1}}^{\tau_k} \mid \xi(s) - \xi(\tau_{k-1}) \mid^2 ds .$$

But on the basis of Lemma 2, Section 3, Chapter VIII,

$$\mathsf{M} \mid \xi(s_1) - \xi(s_2) \mid^2 \leqq C \mid s_1 - s_2 \mid ,$$

so that

$$\mathsf{M}(\xi(\tau_k) - \eta^*(\tau_k))^2 = O[\max_k (\tau_k - \tau_{k-1})] . \tag{3}$$

By analogous reasoning we can show that

$$\overline{\lim_{n\to\infty}} \mathsf{M}[\eta_n(\tau_k) - \eta_n^*(\tau_k)]^2 = O(\max_k (\tau_k - \tau_{k-1})) . \tag{4}$$

We now note that for any real numbers $\lambda_1, \cdots, \lambda_m$ and random variables $\xi_1, \cdots, \xi_m, \eta_1, \cdots, \eta_m$,

$$\begin{aligned}
&\mid \mathsf{M} \exp (i\lambda_1\xi_1 + \cdots + i\lambda_m\xi_m) - \mathsf{M} \exp (i\lambda_1\eta_1 + \cdots + i\lambda_m\eta_m) \mid \\
&\quad \leqq \mathsf{M} \mid \exp \{i\lambda_1(\xi_1 - \eta_1) + \cdots + i\lambda_m(\xi_m - \eta_m)\} - 1 \mid \\
&\quad \leqq \mathsf{M} \mid \lambda_1(\xi_1 - \eta_1) + \cdots + \lambda_m(\xi_m - \eta_m) \mid^2 \\
&\quad \leqq \sqrt{\sum_{k=1}^{m} \lambda_k^2 \sum_{k=1}^{m} \mathsf{M}(\xi_k - \eta_k)^2} .
\end{aligned} \tag{5}$$

Let us choose arbitrary numbers $s_1, \cdots, s_m$ in $[0, 1]$. We may assume that $s_i = \tau_{k_i}$. Then

$$\begin{aligned}
&\overline{\lim_{n\to\infty}} \mid \mathsf{M} \exp\{i \sum \lambda_i \eta_n(\tau_{kj})\} - \mathsf{M} \exp\{i \sum \lambda_j \xi(\tau_k)\} \mid \\
&\quad \leqq \overline{\lim_{n\to\infty}} \mid \mathsf{M} \exp\{i \sum \lambda_j \eta_n(\tau_{kj})\} - \mathsf{M} \exp\{i \sum \lambda_j \eta_n^*(\tau_{kj})\} \mid \\
&\quad\quad + \overline{\lim_{n\to\infty}} \mid \mathsf{M} \exp\{i \sum \lambda_j \eta_n^*(\tau_{kj})\} - \mathsf{M} \exp\{i \sum \lambda_j \eta_0^*(\tau_{kj})\} \mid \\
&\quad\quad + \mid \mathsf{M} \exp\{i \sum \lambda_j \eta_0^*(\tau_{kj})\} - \mathsf{M} \exp\{i \sum \lambda_j \xi(\tau_{kj})\} \mid \\
&\quad \leqq O(\sqrt{m \max_k (\tau_k - \tau_{k-1})}
\end{aligned}$$

by virtue of relations (3) to (5) and the convergence of the joint distributions of the variables $\eta_n^*(s_j)$ to the joint distributions of the variables $\eta_0^*(s_j)$. Since $\max_k (\tau_k - \tau_{k-1})$ is arbitrary, the conclusion of the theorem follows.

Theorem 2. *Under the hypotheses of Theorem* 1, *for any continuous functional f on $C[0, 1]$, the distributions of $f(\xi_n(t))$ converge to the distribution of $f(\xi(t))$.*

Proof. Since the finite-dimensional distributions of the processes $\xi_n(t)$ converge to the finite-dimensional distributions of the process $\xi(t)$, by virtue of Remark 1 of section 2 we can reduce the proof of this theorem to proof of the equation

$$\lim_{h\to 0} \overline{\lim_{n\to\infty}} \mathsf{P}\{ \sup_{|t'-t''|\leqq h} \mid \xi_n(t') - \xi_n(t'') \mid > \varepsilon\} = 0 . \tag{6}$$

Following the reasoning used to prove Theorem 2 of Section 3, we see that

$$\begin{aligned}
&\mathsf{P}\{ \sup_{|t'-t''|\leqq h} \mid \xi_n(t') - \xi_n(t'') \mid > \varepsilon\} \\
&\quad \leqq \sum_{kh<1} \mathsf{P}\left\{ \sup_{kh\leqq t\leqq (k+1)h} \mid \xi_n(t) - \xi_n(kh) \mid > \frac{\varepsilon}{4}\right\} .
\end{aligned}$$

Let s_k denote the greatest of the indices r such that $t_{nr} \leqq kh$. Then

$$\begin{aligned}
&\mathsf{P}\left\{ \sup_{kh\leqq t\leqq (k+1)h} \mid \xi_n(t) - \xi(kh) \mid > \frac{\varepsilon}{4}\right\} \\
&\quad \leqq \mathsf{P}\left\{ \sup_{s_k\leqq j_1\leqq j_2\leqq s_{k+1}+1} \mid \xi_{nj_1} - \xi_{nj_2} \mid > \frac{\varepsilon}{4}\right\} \\
&\quad \leqq \mathsf{P}\left\{ \sup_{s_k\leqq j\leqq s_{k+1}+1} \mid \xi_{nj} - \xi_{nsk} \mid > \frac{\varepsilon}{8}\right\} .
\end{aligned}$$

To find a bound for this probability, we shall need:

Lemma 4. *If $\xi_1, \xi_2, \cdots, \xi_m$ constitute a Markov chain and if*

$$\mathsf{P}\{\mid \xi_m - \xi_k \mid > c \mid \xi_k\} \leqq \alpha ,$$

with probability 1, *where $\alpha < 1$, then*

$$\mathsf{P}\{\sup_k |\xi_k| > 2c\} \leqq \frac{1}{1-\alpha}\mathsf{P}\{|\xi_m| > c\} .$$

Proof.

$$\mathsf{P}\{\sup_k |\xi_k| > 2c, |\xi_m| \leqq c\}$$
$$= \sum_{k=1}^{m} \mathsf{P}\{|\xi_i| \leqq 2c, i \leqq k-1, |\xi_k| > 2c, |\xi_m| \leqq c\}$$
$$\leqq \sum_{k=1}^{m} \mathsf{P}\{|\xi_i| \leqq 2c, i \leqq k-1, |\xi_k| > 2c, |\xi_m - \xi_k| > c\}$$
$$= \sum_{k=1}^{m} \mathsf{M}(\mathsf{P}\{|\xi_i| \leqq 2c, i \leqq k-1, |\xi_k| > 2c \mid \xi_k\}$$
$$\times \mathsf{P}\{|\xi_m - \xi_k| > c \mid \xi_k\})$$
$$\leqq \alpha \sum_{k=1}^{m} \mathsf{MP}\{|\xi_i| \leqq 2c, i \leqq k-1, |\xi_k| > 2c \mid \xi_k\}$$
$$= \alpha\mathsf{P}\{\sup_k |\xi_k| > 2c\} .$$

This means that

$$\mathsf{P}\{\sup_k |\xi_k| > 2c\} \leqq \mathsf{P}\{|\xi_m| > c\} + \mathsf{P}\{\sup_k |\xi_k| > 2c, |\xi_m| \leqq c\}$$
$$\leqq \mathsf{P}\{|\xi_m| > c\} + \alpha\mathsf{P}\{\sup_k |\xi_k| > 2c\} ,$$

from which proof of the lemma follows.

Let us now return to the proof of the theorem. Since

$$\mathsf{P}\{|\xi_{ns_{k+1}+1} - \xi_{nj}| > \delta \mid \xi_{nj}\}$$
$$\leqq \frac{1}{\delta^2}\mathsf{M}([\xi_{ns_{k+1}+1} - \xi_{nj}]^2 \mid \xi_{nj})$$
$$= \frac{1}{\delta^2}\mathsf{M}\left\{\left(\sum_{r=j}^{s_{k+1}} [a_n(t_{nr}, \xi_{nr})\Delta t_{nr} + \sigma_n(t_{nr}, \xi_{nr})\,\omega_{nr}]\right)^2 \middle| \xi_{nj}\right\}$$
$$\leqq \frac{2}{\delta^2}\mathsf{M}\left[\left(\sum_{r=j}^{s_{k+1}} a_n(t_{nr}, \xi_{nr})\,\Delta t_{nr}\right)^2 \middle| \xi_{nj}\right]$$
$$+ \frac{2}{\delta^2}\mathsf{M}\left[\left(\sum_{r=j}^{s_{k+1}} \sigma_n(t_{nr}, \xi_{nr})\omega_{nr}\right)^2 \middle| \xi_{nj}\right]$$
$$\leqq \frac{2}{\delta^2}\mathsf{M}\left[\left(\sum_{r=j}^{s_{k+1}} a_n(l_{nr}, \xi_{nr})\Delta t_{nr}\right)^2 \middle| \xi_{nj}\right]$$
$$+ \frac{2}{\delta^2}\mathsf{M}\left[\sum_{r=j}^{s_{k+1}} \sigma_n^2(t_{nr}, \xi_{nr})\mathsf{M}(\omega_{nr}^2 \mid \xi_{nr}) \mid \xi_{nj}\right]$$
$$+ \frac{4}{\delta^2}\mathsf{M}\Big(\sum_{j=r<l\leqq s_{k+1}} \sigma_n(t_{nr}, \xi_{nr})\,\sigma_n(t_{nl}, \xi_{nl})\omega_{nr}\,\mathsf{M}(\omega_{nl} \mid \xi_{nl}) \mid \xi_{nj}\Big)$$
$$= \mathsf{M}\left[\left(\sum_{r=j}^{s_{k+1}} a_n(t_{nr}, \xi_{nr})\Delta t_{nr}\right)^2 + \sum_{r=j}^{s_{k+1}} \sigma_n^2(t_{nr}, \xi_{nr})\Delta t_{nr} \mid \xi_{nj}\right],$$

since the functions a_n and σ_n are bounded, and since

$$\sum_{r=s_k}^{s_{k+1}} \Delta t_{nj} \leqq h + 2 \max_j \Delta t_{nj} ,$$

it follows that there exists a constant H_1 such that

$$\mathsf{P}\{| \xi_{ns_{k+1}} - \xi_{nj} | > \delta \mid \xi_{nj}\} \leqq \frac{H_1}{\delta^2}(h + 2 \max \Delta t_{nj}) .$$

Consequently, for sufficiently small h and all sufficiently large n,

$$\mathsf{P}\left\{| \xi_{ns_{k+1}+1} - \xi_{nj} | > \frac{\varepsilon}{16}\right\} \leqq \frac{1}{2} .$$

Then on the basis of Lemma 4,

$$\mathsf{P}\left\{\sup_{kh \leqq t \leqq (k+1)h} | \xi_n(t) - \xi_n(kh) | > \frac{\varepsilon}{4}\right\} \leqq 2\mathsf{P}\left\{| \xi_{ns_{k+1}+1} - \xi_{ns_k} | > \frac{\varepsilon}{16}\right\} .$$

It follows from the convergence of the finite-dimensional distributions of $\xi_n(t)$ to the finite-dimensional distributions of $\xi(t)$ that

$$\overline{\lim_{n\to\infty}} \mathsf{P}\left\{| \xi_{ns_{k+1}+1} - \xi_{ns_k} | > \frac{\varepsilon}{16}\right\} \leqq \mathsf{P}\left\{| \xi((k + 1)h) - \xi(kh) | > \frac{\varepsilon}{16}\right\} .$$

This means that

$$\overline{\lim_{n\to\infty}} \mathsf{P}\{\sup_{|t'-t''| \leqq h} | \xi_n(t') - \xi_n(t'') | > \varepsilon\}$$

$$\leqq \sum_{kh<1} \mathsf{P}\left\{| \xi(kh + h) - \xi(kh) | \geqq \frac{\varepsilon}{16}\right\}$$

$$\leqq \sum_{kh<1} \left(\frac{16}{\varepsilon}\right)^4 \mathsf{M} | \xi(kh + h) - \xi(kh) |^4 .$$

But for some L we have $\mathsf{M} | \xi(t + h) - \xi(t) |^4 \leqq Lh^2$ on the basis of Corollary 2, Section 3, Chapter VIII, so that

$$\overline{\lim_{n\to\infty}} \mathsf{P}\{\sup_{|t'-t''| \leqq h} | \xi_n(t') - \xi_n(t'') | > \varepsilon\} \leqq \left(\frac{16}{\varepsilon}\right)^4 Lh .$$

This proves equation (6) and hence the theorem.

5. THE SPACE OF FUNCTIONS WITHOUT DISCONTINUITIES OF THE SECOND KIND

Let $D[0, 1]$ denote the set of real functions $x(t)$ defined on the interval $[0, 1]$ and having right- and left-hand limits at every point. We shall treat two functions that coincide at all points of continuity as the same function. Therefore it is natural to take some sort of standard definition of the values of functions

$x(t)$ at points of discontinuity. In what follows we shall assume that for all functions in $D[0, 1]$,

$$x(t) = x(t + 0) , \qquad x(0) = x(+ 0) , \qquad x(1) = x(1 - 0) . \quad (1)$$

Study of the space $D[0, 1]$ is useful since there are classes of random processes whose sample functions fail, with probability 1, to have discontinuities of the second kind (for example, processes with independent increments, Markov processes under extremely broad conditions). In order to be able to use the results of Section 1 we need to define on $D[0, 1]$ a metric in which $D[0, 1]$ becomes a separable metric space enjoying the property that the minimal σ-algebra containing all cylindrical sets coincides with the σ-algebra of Borel subsets of that space. The metric should be sufficiently "strong" (that is, there should be as few convergent sequences as possible and hence as many continuous functionals in that metric as possible). The uniform-convergence metric

$$\rho_u(x, y) = \sup_{0 \leq t \leq 1} | x(t) - y(t) |$$

is not suitable for this since $D[0, 1]$ is not a separable space in that metric. (The set of functions $x_s(t) = [1 + \operatorname{sgn}(t - s)]/2$ for $0 < s < 1$ has the cardinality of the continuum, but the distance between any two distinct elements of that set is equal to 1.) We introduce into the space $D[0, 1]$ a metric that is somewhat weaker than the uniform-convergence metric.

Let Λ denote the set of all continuous increasing real functions $\lambda(t)$ on $[0, 1]$ such that $\lambda(0) = 0$ and $\lambda(1) = 1$ (that is, λ is a continuous one-to-one mapping of $[0, 1]$ onto itself).

We note that for each $\lambda \in \Lambda$ there exists an inverse function λ^{-1}, also in Λ. If λ_1 and λ_2 belong to Λ, then the composite function $\lambda_1(\lambda_2)$ also belongs to Λ.

Now for every pair $x(t)$ and $y(t)$ in $D[0, 1]$, we define

$$\rho_D(x, y) = \inf_{\lambda \in \Lambda} [\sup_t | x(t) - y(\lambda(t)) | + \sup_t | t - \lambda(t) |] . \quad (2)$$

Let us show that ρ_D defines a metric on $D[0, 1]$. To do this, we need to show that the function ρ_D satisfies the three axioms of a metric: (a) $\rho_D(x, y) \geqq 0$ with equality holding if and only if $x = y$; (b) $\rho_D(x, y) = \rho_D(y, x)$; (c) $\rho_D(x, z) \leqq \rho_D(x, y) + \rho_D(y, x)$ for all $x(t)$, $y(t)$, and $z(t)$ in $D[0, 1]$.

Condition (a) is obvious. Condition (b) follows from the relation

$$\begin{aligned}\rho_D(y, x) &= \inf_{\lambda \in \Lambda} [\sup_t | y(t) - x(\lambda(t)) | + \sup_t | t - \lambda(t) |] \\ &= \inf_{\lambda^{-1} \in \Lambda} [\sup_t | y(\lambda^{-1}(t)) - x(t) | + \sup_t | \lambda^{-1}(t) - t |] = \rho_D(x, y) .\end{aligned}$$

Let us look at condition (c), the triangle inequality. Let $x(t)$, $y(t)$, and $z(t)$ denote functions belonging to $D[0, 1]$. For every $\varepsilon > 0$ there exist functions $\lambda_1(t)$ and $\lambda_2(t)$ such that

$$\left.\begin{aligned} \rho_D(x, y) &\geqq \sup_t | x(t) - y(\lambda_1(t)) | + \sup_t | t - \lambda_1(t) | - \varepsilon , \\ \rho_D(y, z) &\geqq \sup_t | y(t) - z(\lambda_2(t)) | + \sup_t | t - \lambda_2(t) | - \varepsilon . \end{aligned}\right\} \quad (3)$$

Then

$$\begin{aligned} \rho_D(x, z) &\leqq \sup_t | x(t) - z(\lambda_2(\lambda_1(t))) | + \sup_t | t - \lambda_2(\lambda_1(t)) | \\ &\leqq \sup_t | x(t) - y(\lambda_1(t)) | + \sup_t | t - \lambda_1(t) | \\ &\quad + \sup_t | y(\lambda_1(t)) - z(\lambda_2(\lambda_1(t))) | + \sup_t | \lambda_1(t) - \lambda_2(\lambda_1(t)) | \\ &= \sup_t | x(t) - y(\lambda_1(t)) | + \sup_t | t - \lambda_1(t) | \\ &\quad + \sup_t | y(t) - z(\lambda_2(t)) | + \sup_t | t - \lambda_2(t) | , \end{aligned}$$

since $\lambda_1(t)$ ranges over the interval $[0, 1]$ as t ranges over the interval $[0, 1]$. From inequalities (3) we obtain

$$\rho_D(x, z) \leqq \rho_D(x, y) + \rho_D(y, z) + 2\varepsilon ,$$

which, by virtue of the arbitrariness of ε, implies condition (c).

Thus we may take ρ_D as the distance in $D[0, 1]$. To make further study of the properties of the metric ρ_D, we need some auxiliary propositions.

Lemma 1. *Let us define for every function $x(t)$ in $D[0, 1]$,*

$$\begin{aligned} \Delta_c(x) = &\sup_{t-c \leqq t' \leqq t \leqq t'' \leqq t+c} [\min \{| x(t') - x(t) |; | x(t'') - x(t) |\}] \\ &+ \sup_{0 \leqq t \leqq c} |x(t) - x(0) | + \sup_{1-c \leqq t \leqq 1} | x(t) - x(1) | . \end{aligned}$$

Then

$$\lim_{c \to 0} \Delta_c(x) = 0 .$$

Proof. The continuity of $x(t)$ at the points 0 and 1 implies that $\sup_{0 \leqq t \leqq c} | x(t) - x(0) |$ and $\sup_{1-c \leqq t \leqq 1} | x(t) - x(1) |$ approach 0 as $c \to 0$. If we can find an $\varepsilon > 0$ such that for arbitrarily small c,

$$\sup_{t-c \leqq t' \leqq t \leqq t'' \leqq t+c} \min [| x(t') - x(t) |; x(t'') - x(t) |] > \varepsilon ,$$

then we can also find sequences $t'_n < t_n < t''_n$ such that $t'_n - t''_n \to 0$ and

$$| x(t'_n) - x(t_n) | > \varepsilon , \qquad | x(t''_n) - x(t_n) | > \varepsilon . \quad (4)$$

By taking subsequences if necessary, we may assume that $\{t_n\}$ converges to some point t_0 in the interval $[0, 1]$. Then the sequences $\{t'_n\}$ and $\{t''_n\}$ must also converge to t_0. Therefore the quantities $x(t'_n)$, $x(t_n)$, and $x(t''_n)$ converge to one of the numbers $x(t_0 - 0)$ or $x(t_0) = x(t_0 + 0)$, so that at least two of these must have the same limit. It is easy to see that the limit of the sequence $\{x(t_n)\}$ must coincide with the common limit of the sequences $\{x(t'_n)\}$ and $\{x(t''_n)\}$ when these two sequences have the same limit. This means that at least one of the differences $x(t'_n) - x(t_n)$ or $x(t''_n) - x(t_n)$ approaches 0, and this contradicts inequalities (4). This completes the proof of the lemma.

Lemma 2. *Let $x(t)$ denote a function in $D[0, 1]$ and let $[\alpha, \beta]$ denote a subinterval of $[0, 1]$. If $x(t)$ has no jumps exceeding ε in $[\alpha, \beta]$, then the inequality $|t' - t''| \leqq c$ for $t', t'' \in [\alpha, \beta]$ implies*

$$|x(t') - x(t'')| \leqq 2\Delta_c(x) + \varepsilon .$$

Proof. Let us choose arbitrary $\delta \in (0, \varepsilon)$ and a point τ in the interval $[t', t'']$ such that for $t \in [t', \tau]$,

$$|x(t') - x(t)| < \Delta_c(x) + \delta$$

and

$$|x(t') - x(\tau)| \geqq \Delta_c(x) + \delta .$$

If no such point exists, then $|x(t') - x(t'')| < \Delta_c(x) + \delta$. This means that the assertion of the lemma is satisfied. If a point τ does exist then since

$$\min [|x(\tau) - x(t')|; |x(\tau) - x(t'')|] \leqq \Delta_c(x) ,$$

and

$$|x(\tau) - x(t')| \geqq \Delta_c(x) + \delta ,$$

we have

$$|x(\tau) - x(t'')| \leqq \Delta_c(x) .$$

Thus

$$\begin{aligned} &|x(t'') - x(t')| \\ &\quad \leqq |x(t'') - x(\tau)| + |x(\tau) - x(\tau - 0)| + |x(\tau - 0) - x(t')| \\ &\quad \leqq 2\Delta_c(x) + \delta + \varepsilon . \end{aligned}$$

Taking the limit as $\delta \to 0$, we obtain proof of the lemma.

Let $H_{m,n}$ denote the set of functions $x(t)$ in $D[0, 1]$ that are constant on each of the intervals $[(k/n), (k + 1)/n)$ and that assume values that are multiples of $1/m$.

Lemma 3. *For every function $x(t)$ in $D[0, 1]$ there exists a* *function* $x^*(t)$ *in* $H_{m,n}$ *such that*

$$\rho_D(x, x^*) \leqq \frac{1}{n} + \frac{1}{m} 4\Delta_{\frac{2}{n}}(x) .$$

Proof. In each of the intervals $[(k/n), \{(k+1)/n\})$ there can exist no more than a single point with a jump exceeding $2\Delta_{2/n}(x)$. This is true because if τ is one such point, then

$$| x(s) - x(\tau - 0) | = \min (| x(s) - x(\tau - 0) |; | x(\tau) - x(\tau - 0) |)$$

$$\leqq \Delta_{\frac{1}{n}}(x) \quad \text{for} \quad s \in \left[\frac{k}{n}, \tau\right],$$

$$| x(s) - x(\tau) | \leqq \Delta_{\frac{1}{n}}(x) \quad \text{for} \quad s \in \left(\tau, \frac{k+1}{n}\right)$$

and hence $| x(s) - x(s-0) | \leqq 2\Delta_{1/n}(x) \leqq 2\Delta_{2/n}(x)$. Suppose that τ_k is a point of the interval $[k/n, (k+1)/n]$ at which $| x(\tau_k) - x(\tau_k - 0) | \geqq 2\Delta_{2/n}(x)$ if such a point exists in that interval. We let $\lambda(t)$ denote a function in Λ such that $\lambda((k+1)/n) = \tau_k$ and $t - 1/n \leqq \lambda(t) \leqq t$. (Such a function is, for example, the piecewise-linear function defined by the equations $\lambda(0) = 0$, $\lambda((k+1)/n) = \tau_k$, $\lambda(1) = 1$.) Let us define $\bar{x}(t) = x(\lambda(t))$. The function $\bar{x}(t)$ has discontinuities exceeding $2\Delta_{2/n}(x)$ only at points of the form k/n and

$$\rho_D(x, \bar{x}) \leqq \sup_t | \bar{x}(t) - x(\lambda(t)) | + \sup_t | t - \lambda(t) | \leqq \frac{1}{n} .$$

Now define

$$\bar{x}^*(t) = \begin{cases} \bar{x}\left(\frac{k}{n}\right) & \text{for} \quad t \in \left[\frac{k}{n}, \frac{k+1}{n}\right], \; k \leqq n-1 , \\ \bar{x}\left(\frac{n-1}{n}\right) & \text{for} \quad t = 1 . \end{cases}$$

Then

$$\rho_D(\bar{x}, \bar{x}^*) \leqq \sup_t | \bar{x}(t) - \bar{x}^*(t) | \leqq \sup_t \sup_{k/n \leqq t < \{(k+1)/n\}} \left| \bar{x}(t) - \bar{x}\left(\frac{k}{n}\right) \right| .$$

Since the jumps of $\bar{x}(t)$ that exceed $2\Delta_{2/n}(x)$ occur only at points of the form k/n, there are no such jumps in the interval $[k/n, (k+1)/n)$. Hence, according to Lemma 2,

$$\left| \bar{x}\left(\frac{k}{n}\right) - \bar{x}(t) \right| \leqq 2\Delta_{\frac{1}{n}}(\bar{x}) + 2\Delta_{\frac{2}{n}}(x) \quad \text{for} \quad t \in \left[\frac{k}{n}, \frac{k+1}{n}\right) .$$

Let us find a bound for $\Delta_{1/n}(\bar{x})$:

$\Delta_1(\bar{x})$

$$= \sup_{-(1/n) \leqq t' \leqq t \leqq t'' \leqq t+(1/n)} [\min \{| \bar{x}(t') - \bar{x}(t) |; \ | \bar{x}(t) - \bar{x}(t'') |\}]$$

$$+ \sup_{0 \leqq t \leqq 1/n} | \bar{x}(t) - \bar{x}(0) | + \sup_{1-(1/n) \leqq t \leqq 1} | \bar{x}(t) - \bar{x}(1) |$$

$$= \sup_{t-(1/n) \leqq t' \leqq t \leqq t'' \leqq t+(1/n)} [\min\{| x(\lambda(t')) - x(\lambda(t)) |; | x(\lambda(t)) - x(\lambda(t'')) |\}]$$

$$+ \sup_{0 \leqq t \leqq 1/n} | x(\lambda(t)) - x(0) | + \sup_{1-(1/n) \leqq t \leqq 1} | x(\lambda(t)) - x(1) | .$$

We note that, for $t_1 < t_2 < t_1 + (1/n)$, we have

$$t_1 - \frac{1}{n} < \lambda(t_1) < \lambda(t_2) \leqq t_2 < t_1 + \frac{1}{n} ,$$

so that $0 < \lambda(t_2) - \lambda(t_1) \leqq 2/n$. Therefore

$$\Delta_{\frac{1}{n}}(\bar{x}) \leqq \Delta_{\frac{2}{n}}(x) .$$

Hence

$$\rho_D(\bar{x}, \bar{x}^*) \leqq 4\Delta_{\frac{2}{n}}(x) .$$

Finally, let us set $x^*(t) = (1/m)\,[m\bar{x}^*(t)]$, where $[x]$ denotes the integral part of x. Since $| x^*(t) - \bar{x}^*(t) | \leqq 1/m$, we have

$$\rho_D(x, x^*) \leqq \rho_D(x, \bar{x}) + \rho_D(\bar{x}, \bar{x}^*) + \rho_D(\bar{x}^*, x) \leqq \frac{1}{n} + 4\Delta_{\frac{2}{n}}(x) + \frac{1}{m} .$$

This completes the proof of the lemma.

Corollary. *The space $D[0, 1]$ with the given metric is separable.*

This follows from the fact that the countable set $\bigcup_{m,n} H_{m,n}$ is everywhere-dense in $D[0, 1]$ in accordance with Lemma 3.

Theorem 1. *Let L denote a positive constant and let $\omega(\delta)$ denote a function that is defined, continuous, and monotonic for $\delta > 0$. Suppose that $\lim_{\delta \downarrow 0} \omega(\delta) = 0$. Let $K(L, \omega)$ denote the set of functions in $D[0, 1]$ that satisfy the inequalities $| x(t) | \leqq L$ and $\Delta_c(x) \leqq \omega(c)$. Then $K(L, \omega)$ is a compact set in the metric ρ_D.*

Proof. We note first of all that for every $\varepsilon > 0$ the set $K(L, \omega)$ has a finite ε-net consisting of the functions in $H_{m,n}$ that satisfy the relation $| x(t) | \leqq L$ if m and n are chosen so that $1/n + 1/m + 4\omega(2/n) < \varepsilon$. The set $K(L, \omega)$ is closed. It is easy to verify that

$$\Delta_c(x) \leqq \Delta_{c+P_D(x,y)}(y) .$$

Therefore, if $\rho_D(x_n, \bar{x}) \longrightarrow 0$ as $n \longrightarrow \infty$, where each $x_n \in D[0, 1]$, then for every $\alpha > 0$,

$$\Delta_c(\bar{x}) \leqq \overline{\lim_{n\to\infty}} \Delta_{c+\alpha}(x_n) \leqq \omega(c + \alpha) .$$

Because of the continuity of ω, this means that $\Delta_c(\bar{x}) \leqq \omega(c)$. Obviously,

$$\sup_t | x(t) | \leqq \sup_t | y(t) | + \rho_D(x, y) ,$$

so that

$$\sup_t | \bar{x}(t) | \leqq \overline{\lim_{n\to\infty}} \sup_t | x_n(t) | \leqq L .$$

Consequently, the limit of a sequence belonging to $K(L, \omega)$ also belongs to $K(L, \omega)$. We must now show that every fundamental sequence $\{x_n(t)\}$ of elements of $K(L, \omega)$ converges. (Then, we shall show that $K(L, \omega)$ is a complete metric space with finite ε-net for every $\varepsilon > 0$, which will mean that $K(L, \omega)$ is compact.) Let $\{x_n(t)\}$ denote a sequence of functions in $K(L, \omega)$ such that $\rho_D(x_n, x_m) \to 0$ as $n, m \to \infty$ (that is, $\{x_n(t)\}$ is a fundamental sequence). It is sufficient to show that some subsequence $\{x_{n_k}(t)\}$ converges to $\bar{x}(t)$ since we shall then know by virtue of the inequality

$$\rho_D(x_n, \bar{x}) \leqq \rho_D(x_n, x_{n_k}) + \rho_D(x_{n_k}, \bar{x}) ,$$

that $\bar{x}(t)$ is the limit of the sequence $\{x_n(t)\}$. Therefore, we may assume that the sequence $\{x_n(t)\}$ is such that $\rho_D(x_n, x_{n+1}) < 1/2^{n+1}$. Then there exists a sequence $\{\lambda_n(t)\}$ of functions in Λ such that

$$\sup_t | x_n(t) - x_{n+1}(\lambda_{n+1}(t)) | \leqq \frac{1}{2^{n+1}}$$

and

$$\sup_t | t - \lambda_{n+1}(t) | \leqq \frac{1}{2^{n+1}} .$$

Let us define $\mu_1(t) = \lambda_1(t)$ and $\mu_n(t) = \lambda(\mu_{n-1}(t))$. Since

$$\sup_t | \mu_n(t) - \mu_{n-1}(t) | \leqq \sup_t | \lambda_n(t) - t | \leqq \frac{1}{2^n} ,$$

it follows that $\{\mu_n(t)\}$ converges to some nondecreasing continuous function $\mu(t)$ that satisfies the conditions $\mu(0) = 0$ and $\mu(1) = 1$. Furthermore,

$$\sup_t | x_n(\mu_n(t)) - x_{n-1}(\mu_{n-1}(t)) | = \sup_t | x_n(\lambda_n(t)) - x_{n-1}(t) | \leqq \frac{1}{2^n} .$$

Therefore $\{x_n(\mu_n(t))\}$ converges uniformly to some function $x^*(t)$ in $D[0, 1]$.

Let us look at the connection between the functions $x^*(t)$ and $\mu(t)$. Suppose that $\mu(t)$ is constant on some interval $[\alpha, \beta]$. If $x^*(\alpha) = x^*(\beta)$, then $x^*(t)$ is also constant on $[\alpha, \beta]$. On the other hand, if $x^*(\alpha) \neq x^*(\beta)$ there exists a $\gamma \in [\alpha, \beta]$ such that $x^*(t) = x^*(\alpha)$ for $t \in [\alpha, \gamma)$ and $x^*(t) = x^*(\beta)$ for $t \in [\gamma, \beta]$. This is true because otherwise there would be points $t' < t'' < t'''$ belonging to $[\alpha, \beta]$ such that $x^*(t') \neq x^*(t'')$ and $x^*(t'') \neq x^*(t''')$. Then we would have

$$\lim_{n\to\infty} \min [| x_n(\mu_n(t')) - x_n(\mu_n(t'')) | ; \quad | x_n(\mu_n(t'')) - x_n(\mu_n(t''')) |]$$
$$= \min [| x^*(t') - x^*(t'') | ; \quad | x^*(t'') - x^*(t''') |] > 0 ,$$

although $\mu_n(t') < \mu_n(t'') < \mu_n(t''')$ and the sequences $\{\mu_n(t')\}$, $\{\mu_n(t'')\}$, and $\{\mu_n(t''')\}$ all converge to $\mu(\alpha)$. This would contradict the fact that the sequence $\{x_n(t)\}$ belongs to $K(L, \omega)$. Let $\bar{x}(t)$ denote the function in $D[0, 1]$ defined by

$$\bar{x}(t) = x^*(\mu(t)) , \tag{5}$$

which is to be satisfied at all points t at which $\mu(s) > \mu(t)$ for all $s \in (t, 1]$. Equation (5) defines a unique function $\bar{x}(t)$ in $D[0, 1]$. Let us show that this function $\bar{x}(t)$ is the limit of the sequence $\{x_n(t)\}$. To do this, we construct auxiliary functions $\varphi_n(t)$ in Λ. Suppose that $\tau_1, \tau_2, \cdots, \tau_k$ are all points in the interval $[0, 1]$ at which $\bar{x}(t)$ has jumps exceeding $1/n$. We let $[\alpha_i, \beta_i]$ denote the maximum interval on which $\mu(t)$ assumes the value τ_i (this interval may consist of a single point).

Let γ_i denote a point in the interval $[\alpha_i, \beta_i]$ such that $x^*(t) = \bar{x}(\tau_i - 0)$ for $t \in [\alpha_i, \gamma_i)$ and $x^*(t) = \bar{x}(\tau_i)$ for $t \in [\gamma_i, \beta_i]$. If $\alpha_i = \gamma_i$, then $x^*(t)$ assumes the single value $\bar{x}(\tau_i)$ on $[\alpha_i, \beta_i]$.) Let us choose ε_n not exceeding $1/n$ such that $\Delta_{\varepsilon_n}(\bar{x}) < 1/n$. Let $\varphi_n(t)$ denote a function satisfying the equation $\varphi_n(\gamma_i) = \tau_i$ and the inequality $| \varphi_n(t) - \mu(t) | \leqq \varepsilon_n$. Let us find a bound for

$$\sup_t | x^*(t) - \bar{x}(\varphi_n(t)) | .$$

If t does not belong to any of the intervals $[\alpha_i, \beta_i]$, then

$$| x^*(t) - \bar{x}(\varphi_n(t)) | = | \bar{x}(\mu(t)) - \bar{x}(\varphi_n(t)) | \leqq 2\Delta_{\varepsilon_n}(\bar{x}) + \frac{1}{n}$$

by virtue of Lemma 2 since $\bar{x}(t)$ has no jumps exceeding $1/n$ between $\mu(t)$ and $\varphi_n(t)$. If $t \in [\alpha_i, \beta_i)$, then

$$| x^*(t) - \bar{x}(\varphi_n(t)) | \leqq \sup_{s \in [\tau_i - \varepsilon_n, \tau_i)} | \bar{x}(\tau_i - 0) - \bar{x}(s) | \leqq \Delta_{\varepsilon_n}(\bar{x})$$

since $| \bar{x}(\tau_i - 0) - \bar{x}(\tau_i) | > 1/n$. In an analogous manner we can show that $| x^*(t) - \bar{x}(\varphi_n(t)) | \leqq \Delta_{\varepsilon_n}(\bar{x})$ for $t \in [\gamma_i, \beta_i]$. Consequently,

$$\sup_t | x^*(t) - \bar{x}(\varphi_n(t)) | \leqq 2\Delta_{\varepsilon_n}(\bar{x}) + \frac{1}{n} \leqq \frac{3}{n} .$$

Let us now find a bound for $\rho_D(x_n, \bar{x})$. We have

$$\begin{aligned}\rho_D(x_n, \bar{x}) &\leqq \rho_D(x_n(t), x^*(\mu_n^{-1}(t))) + \rho_D(x^*(\mu_n^{-1}(t)), \bar{x}(\varphi_n(\mu_n^{-1}(t))))\\ &\quad + \rho_D(\bar{x}(t), \bar{x}(\varphi_n(\mu_n^{-1}(t))))\\ &\leqq \sup_t |x_n(\mu_n(t)) - x^*(t)| + \sup_t |x^*(t) - \bar{x}(\varphi_n(t))|\\ &\quad + \sup_t |t - \varphi_n(\mu_n^{-1}(t))|\\ &\leqq \frac{1}{2^n} + \frac{3}{n} + \sup_t |\mu_n(t) - \varphi_n(t)|\\ &\leqq \frac{1}{2^n} + \frac{3}{n} + \frac{1}{2^n} + \varepsilon_n .\end{aligned}$$

Thus $\rho_D(x_n, \bar{x}) \to 0$; that is the sequence $\{x_n\}$ converges to the function $\bar{x}(t)$. This completes the proof of the theorem.

Theorem 2. *If the finite-dimensional distributions of the processes $\xi_n(t)$ that have no discontinuities of the second kind converge to the finite-dimensional distributions of the process $\xi(t)$, and if for every $\varepsilon > 0$,*

$$\lim_{c\to 0} \overline{\lim_{n\to\infty}} \mathsf{P}\{\Delta_c(\xi_n(t)) > \varepsilon\} = 0 , \tag{6}$$

then for every functional f defined on $D[0, 1]$ and continuous in the metric ρ_D, the sequence of distributions of $f(\xi_n(t))$ converges to the distribution of $f(\xi(t))$.

Proof. Using Remark 1 of Section 2, we see that condition (6) implies

$$\lim_{c\to 0} \sup_n \mathsf{P}\{\Delta_c(\xi_n(t)) > \varepsilon\} = 0 . \tag{7}$$

Using the sufficiency of Theorem 2 of Section 1 and Theorem 1 of this section, we can see that to prove the theorem, it suffices to show that

$$\lim_{L\to\infty} \sup_n \mathsf{P}\{\sup_t |\xi_n(t)| > L\} = 0 . \tag{8}$$

But for every function $x(t)$ in $D[0, 1]$,

$$\sup_{0\leqq t\leqq 1} |x(t)| \leqq \sup_{0\leqq k\leqq m} \left|x\left(\frac{k}{m}\right)\right| + \Delta_{\frac{1}{m}}(x)$$

since for $t \in [k/m, (k + 1)/m]$, either $|x(t) - x(k/m)| < \Delta_{1/m}(x)$ or $|x(t) - x((k + 1)/m| < \Delta_{1/m}(x)$. Therefore,

$$\mathsf{P}\{\sup_t |\xi_n(t)| > L\} \leqq \mathsf{P}\left\{\sup_{0\leqq k\leqq m} \left|\xi_n\left(\frac{k}{m}\right)\right| > L - \varepsilon\right\} + \mathsf{P}\{\Delta_{\frac{1}{m}}(\xi_n(t)) > \varepsilon\} .$$

The random variable $\sup_{0\leqq k\leqq m} |\xi_n(k/m)|$ is uniformly bounded in probability with respect to n. This follows from the convergence of the finite-dimensional distributions of $\xi_n(t)$ to the finite-dimensional distributions of $\xi(t)$, which implies convergence of the distributions of $\sup_{0 \leq k \leq m} | \xi_n(k/m) |$ to the distribution of $\sup_{0 \leq k \leq m} | \xi(k/m) |$. This means that

$$\overline{\lim_{L \to \infty}} \sup_n \mathsf{P}\{\sup_t | \xi_n(t) | > L\} \leq \sup_n \mathsf{P}\{\Delta_{\frac{1}{m}}(\xi_n(t)) > \varepsilon\} .$$

Taking the limit as $m \longrightarrow \infty$, we see that equation (8) holds. This completes the proof of the theorem.

6. CONVERGENCE OF A SEQUENCE OF SUMS OF IDENTICALLY DISTRIBUTED INDEPENDENT RANDOM VARIABLES TO A HOMOGENEOUS PROCESS WITH INDEPENDENT INCREMENTS

Let $\xi_{n1}, \xi_{n2}, \cdots, \xi_{nn}$ for each n denote independent, identically distributed random variables. We define $S_{nk} = \xi_{n1} + \cdots + \xi_{nk}$. Let us consider the random process $\xi_n(t)$ defined by $\xi_n(t) = S_{nk}$ for $t \in [(k-1)/n, k/n)$ and $\xi_n(1) = S_{nn}$. This process is a process with independent increments and has discontinuities at the points k/n. In this section we shall study the conditions under which the finite-dimensional distributions of the processes $\xi_n(t)$ and the distributions of the functionals of these processes that are continuous in the metric ρ_D converge to finite-dimensional distributions and the distributions of the corresponding functionals of a homogeneous process $\xi(t)$ with independent increments. We shall assume that the sample functions of the process $\xi(t)$ (which we may assume to be stochastically continuous) belong, with probability 1, to $D[0, 1]$. As we know (cf. Chapter I, Section 4), the characteristic function of a homogeneous process with independent increments can be represented in the form

$$\mathsf{M}e^{i\lambda\xi(t)} = \exp\left\{t\left[i\lambda\gamma + \int_{-\infty}^{\infty}\left(e^{i\lambda u} - 1 - \frac{i\lambda u}{1 + u^2}\right)\frac{1 + u^2}{u^2}\, dG(u)\right]\right\} , \quad (1)$$

where G is a monotonic bounded function. Formula (1) completely determines the finite-dimensional distributions of the process $\xi(t)$. Concerning the sequence of sums S_{nk}, we define

$$\gamma_n = n\int_{-\infty}^{\infty} \frac{x}{1 + x^2} dF_n(x) \quad \text{and} \quad G_n(x) = n\int_{-\infty}^{x} \frac{u^2}{1 + u^2} dF_n(u) ,$$

where $F_n(x)$ is the distribution function of the random variables ξ_{ni}.

Theorem 1. *If there exist a number* γ *and a nondecreasing bounded function* $G(x)$ *such that* $\gamma_n \to \gamma$ *and* $G_n(x) \Rightarrow G(x)$ *as* $n \to \infty$, *then the finite-dimensional distributions of the processes* $\xi_n(t)$ *converge to the finite-dimensional distributions of the process* $\xi(t)$ *with characteristic function* (1).

Proof. Since $\xi_n(t)$ is a process with independent increments it is sufficient for us to show that for all $t_1 < t_2$ in $[0, 1]$, the sequence $\{\xi_n(t_2) - \xi_n(t_1)\}$ converges to the distribution $\xi(t_2) - \xi(t_1)$. The characteristic function of the variable $\xi_n(t_2) - \xi_n(t_1)$ takes the form

$$\mathsf{M}e^{i\lambda[\xi_n(t_2)-\xi_n(t_1)]} = [\mathsf{M}e^{i\lambda\xi_{n1}}]^{k_n} = ([\mathsf{M}e^{i\lambda\xi_{n1}}]^n)^{k_n/n} ,$$

where $k_n = [nt_2] - [nt_1]$. Since $k_n/n \to t_2 - t_1$ as $n \to \infty$, the theorem will be proved if we can show that

$$\lim_{n\to\infty} (\mathsf{M}e^{i\lambda\xi_{n1}})^n = \exp\left\{i\lambda\gamma + \int_{-\infty}^{\infty}\left(e^{i\lambda u} - 1 - \frac{i\lambda u}{1+u^2}\right)\frac{1+u^2}{u^2}dG(u)\right\} . \qquad (2)$$

Let us represent the characteristic function of ξ_{n1} in the form

$$\begin{aligned}\mathsf{M}e^{i\lambda\xi_{n1}} &= \int_{-\infty}^{\infty} e^{i\lambda x}dF_n(x)\\ &= 1 + \int_{-\infty}^{\infty}\frac{x}{1+x^2}dF_n(x) + \int_{-\infty}^{\infty}\left(e^{i\lambda x} - 1 - \frac{x}{1+x^2}\right)dF_n(x)\\ &= 1 + \frac{1}{n}\left[\gamma_n + \int_{-\infty}^{\infty}\left(e^{i\lambda x} - 1 - \frac{x}{1+x^2}\right)\frac{1+x^2}{x^2}dG_n(x)\right] .\end{aligned}$$

Then

$$(\mathsf{M}e^{i\lambda\xi_{n1}})^n = \left(1 + \frac{1}{n}\left[\gamma_n + \int_{-\infty}^{\infty}\left(e^{i\lambda x} - 1 - \frac{x}{1+x^2}\right) \times \frac{1+x^2}{x^2}dG_n(x)\right]\right)^n .$$

But $\gamma_n \to \gamma$ as $n \to \infty$ and from the hypothesis of the theorem (we recall that the function $(e^{i\lambda u} - 1 - i\lambda u/(1+u^2))\,(1+u^2)/u^2$ is continuous and equal to $-\lambda^2/2$ at $u = 0$), it follows that

$$\int_{-\infty}^{\infty}\left(e^{i\lambda x} - 1 - \frac{i\lambda x}{1+x^2}\right)\frac{1+x^2}{x^2}dG_n(x) \to \int_{-\infty}^{\infty}\left(e^{i\lambda x} - 1 - \frac{i\lambda x}{1+x^2}\right)\frac{1+x^2}{x^2}dG(x) .$$

The conclusion of the theorem then follows.

We find that we do not need to impose any additional conditions to ensure convergence of a sequence of distributions of functionals that are continuous in the metric ρ_D.

Theorem 2. *Suppose that the hypotheses of Theorem 1 are satisfied. For a functional f that is defined on $D[0, 1]$ and continuous in the metric ρ_D, the sequence of distributions $f(\xi_n(t))$ converges to the distribution $f(\xi(t))$.*

Proof. Since the convergence of the finite-dimensional distributions of the processes $\xi_n(t)$ to the finite-dimensional distributions of the process $\xi(t)$ was shown in Theorem 1, on the basis of Theorem 2 of Section 5 we see that to prove this theorem it suffices to show that for every $\varepsilon > 0$,

$$\lim_{c\to\infty} \overline{\lim_{n\to\infty}} \mathsf{P}\{\Delta_c(\xi_n(t)) > \varepsilon\} = 0 . \tag{3}$$

To prove this equation we use:

Lemma 1. *Suppose that $\xi_1, \xi_2, \cdots, \xi_n$ are independent identically distributed random variables. Then*

$$\mathsf{P}\left\{\sup_{0\leq i<j<l\leq n} \min\left[\left|\sum_{k=i+1}^{j} \xi_k\right|; \left|\sum_{k=j+1}^{l} \xi_k\right|\right] > \varepsilon\right\} \leq \left(\mathsf{P}\left\{\sup_k \left|\sum_{i=1}^{k} \xi_i\right| > \frac{\varepsilon}{2}\right\}\right)^2 . \tag{4}$$

Proof. Suppose that $i < j < l$, $|\sum_{k=i+1}^{j} \xi_k| > \varepsilon$, $|\sum_{k=j+1}^{l} \xi_k| > \varepsilon$. Then either $|\sum_{k=1}^{i} \xi_k| > \varepsilon/2$ or $|\sum_{k=1}^{j} \xi_k| > \varepsilon/2$, and for all $r < j$, either $|\sum_{k=r+1}^{j} \xi_k| > \varepsilon/2$ or $|\sum_{k=r+1}^{l} \xi_k| > \varepsilon/2$. Thus the event

$$\left\{\sup_{i<j<l} \min\left[\left|\sum_{k=i+1}^{j} \xi_k\right|; \left|\sum_{k=j+1}^{l} \xi_k\right| > \varepsilon\right]\right\}$$

implies one of the events $\mathfrak{A}_r$ (for $r = 1, 2, \cdots, n$):

$$\mathfrak{A}_r = \left\{|\xi_1 + \cdots + \xi_k| < \frac{\varepsilon}{2},\ k \leq r - 1 ;\right.$$
$$\left.|\xi_1 + \cdots + \xi_r| > \frac{\varepsilon}{2},\ \sup_{l>r} |\xi_{r+1} + \cdots + \xi_l| > \frac{\varepsilon}{2}\right\} .$$

Since

$$\mathsf{P}\{\mathfrak{A}_r\} = \mathsf{P}\left\{|\xi_1 + \cdots + \xi_k| \leq \frac{\varepsilon}{2},\ k \leq r - 1 ;\right.$$
$$\left.|\xi_1 + \cdots + \xi_r| > \frac{\varepsilon}{2}\right\}\mathsf{P}\left\{\sup_{l>r} |\xi_{r+1} + \cdots + \xi_l| > \frac{\varepsilon}{2}\right\} ,$$

remembering that the variables ξ_i are identically distributed, we obtain

$$\mathsf{P}\left\{\sup_{i<j<l} \min\left[\left|\sum_{k=i+1}^{j} \xi_k\right|; \left|\sum_{k=j+1}^{l} \xi_k\right|\right] > \varepsilon\right\}$$

$$\leqq \sum_r \mathsf{P}\left\{|\xi_1 + \cdots + \xi_k| \leqq \frac{\varepsilon}{2},\ k \leqq r-1;\right.$$

$$\left.|\xi_1 + \xi_2 + \cdots \xi_r| > \frac{\varepsilon}{2}\right\}\mathsf{P}\left\{\sup_{l>r} |\xi_{r+1} + \cdots + \xi_l| > \frac{\varepsilon}{2}\right\}$$

$$\leqq \mathsf{P}\left\{\sup_{1\leqq l\leqq n} |\xi_1 + \cdots + \xi_l| > \frac{\varepsilon}{2}\right\} \sum_r \mathsf{P}\left\{|\xi_1 + \cdots + \xi_k| \leqq \frac{\varepsilon}{2},\right.$$

$$\left.k \leqq r-1;\ |\xi_1 + \cdots + \xi_r| > \frac{\varepsilon}{2}\right\}$$

$$= \left(\mathsf{P}\left\{\sup_{1\leqq l\leqq n} |\xi_1 + \cdots + \xi_l| > \frac{\varepsilon}{2}\right\}\right)^2.$$

This completes the proof of the lemma.

Returning now to the proof of the theorem, we note that

$$\Delta_c(x(t)) \leqq \sup_{0\leqq t\leqq c} |x(t) - x(0)| + \sup_{1-c\leqq t\leqq 1} |x(t) - x(1)|$$

$$+ \max_{0<k<1/c} \sup_{kc\leqq t'<t''<t'''\leqq(k+3)c} \min[|x(t') - x(t'')|;\ |x(t'' - x(t''')|].$$

Therefore

$$\mathsf{P}\{\Delta_c(\xi_n(t)) > \varepsilon\}$$

$$\leqq \mathsf{P}\left\{\sup_{0\leqq t\leqq c} |\xi_n(t) - \xi_n(0)| > \frac{\varepsilon}{4}\right\} + \mathsf{P}\left\{\sup_{1-c\leqq t\leqq 1} |\xi_n(t) - \xi_n(1)| > \frac{\varepsilon}{4}\right\}$$

$$+ \sum_{k<1/c} \mathsf{P}\left\{\sup_{kc\leqq t'<t''<t'''\leqq(k+3)c} \min[|\xi_n(t') - \xi_n(t'')|;\right.$$

$$\left.|\xi_n(t'') - \xi_n(t''')|] > \frac{\varepsilon}{2}\right\}$$

$$\leqq \mathsf{P}\left\{\sup_{0\leqq t\leqq c} |\xi_n(t) - \xi_n(0)| > \frac{\varepsilon}{4}\right\} + \mathsf{P}\left\{\sup_{1-c\leqq t\leqq 1} |\xi_n(t) - \xi_n(1)| > \frac{\varepsilon}{4}\right\}$$

$$+ \sum_{k<1/c} \left(\mathsf{P}\left\{\sup_{kc\leqq t\leqq(k+3)c} |\xi_n(t) - \xi_n(kc)| > \frac{\varepsilon}{4}\right\}\right)^2.$$

If $n \geqq 1/c$, it is easy to see that $\Delta_c(\xi_n(t)) = 0$ since $\xi_n(t)$ is constant on intervals of the form $[i/n, (i+1)/n]$. On the other hand, if $n < 1/c$, then although the number of points of the form i/n in the interval $[kc, (k+3)c]$ varies with varying k, this number still does not exceed the number of these points in the interval $[0, 4c]$, so that we always have

$$\mathsf{P}\{\Delta_c(\xi_n(t)) > \varepsilon\} \leqq 2\mathsf{P}\Big\{\sup_{0\leqq t\leqq 4c} |\xi_n(t) - \xi_n(0)| > \frac{\varepsilon}{4}\Big\} + \Big(1 + \frac{1}{c}\Big)\Big(\mathsf{P}\Big\{\sup_{0\leqq t\leqq 4c} |\xi_n(t) - \xi_n(0)| > \frac{\varepsilon}{4}\Big\}^2\Big).$$

To find a bound for the probability

$$\mathsf{P}\Big\{\sup_{0\leqq t\leqq 4c} |\xi_n(t) - \xi_n(0)| > \frac{\varepsilon}{4}\Big\} = \mathsf{P}\Big\{\sup_{k\leqq 4nc} \Big|\sum_{i=1}^{k} \xi_{ni}\Big| > \frac{\varepsilon}{4}\Big\},$$

we introduce the variable $\xi'_{ni} = \xi_{ni}$ if $|\xi_{ni}| \leqq L$ and $\xi'_{ni} = 0$ if $|\xi_{ni}| > L$, and the variable $\xi''_{ni} = \xi_{ni} - \xi'_{ni}$. Then

$$\mathsf{P}\Big\{\sup_{1\leqq k\leqq N} \Big|\sum_{i=1}^{k} \xi_{ni}\Big| > \frac{\varepsilon}{4}\Big\} \leqq \mathsf{P}\Big\{\sup_{1\leqq k\leqq N} |\xi''_{ni}| > 0\Big\} + \mathsf{P}\Big\{\sup_{1\leqq k\leqq N} \Big|\sum_{i=1}^{k} \xi'_{ni}\Big| > \frac{\varepsilon}{4}\Big\}$$

$$= \sum_{i=1}^{N} \mathsf{P}\{|\xi_{ni}| > L\} + \mathsf{P}\Big\{\sup_{1\leqq k\leqq N} \Big|\sum_{i=1}^{k} (\xi'_{ni} - \mathsf{M}\xi'_{ni})\Big| > \frac{\varepsilon}{4} - \Big|\sum_{i=1}^{N} \mathsf{M}\xi'_{ni}\Big|\Big\}$$

$$\leqq N\cdot\mathsf{P}\{|\xi_{ni}| > L\} + \frac{N\mathsf{D}\xi'_{n1}}{\Big(\dfrac{\varepsilon}{4} - N|\mathsf{M}\xi'_{n1}|\Big)^2}.$$

(Here we used Kolmogorov's inequality, Theorem 1, Section 4, Chapter III.)

If L and $-L$ are points of continuity of the function G, then the following limit relations are valid:

$$\lim_{n\to\infty} n\cdot\mathsf{P}\{|\xi_{ni}| > L\} = \lim_{n\to\infty} \int_{|x|>L} \frac{1 + x^2}{x^2} dG_n(x) = \int_{|x|>L} \frac{1 + x^2}{x_2} dG(x),$$

$$\lim_{n\to\infty} n\cdot\mathsf{D}\xi'_{n1} \leqq \lim_{n\to\infty} n\cdot\mathsf{M}\xi'^2_{n1} \leqq \lim_{n\to\infty} n\cdot\mathsf{M}\frac{1 + L^2}{1 + \xi^2_{n1}} \xi^2_{n1} = (1 + L^2)\int_{-\infty}^{\infty} dG(x),$$

$$\lim_{n\to\infty} n\cdot\mathsf{M}\xi'_{n1} = \lim_{n\to\infty} n\cdot\int_{|x|<L} x dF_n(x)$$

$$= \gamma - \lim_{n\to\infty} n\int_{|x|>L} \frac{x}{1 + x^2} dF_n(x) + \lim_{n\to\infty} n\cdot\int_{|x|\leqq L}\Big(x - \frac{x}{1 + x^2}\Big) dF_n(x)$$

$$= \gamma - \int_{|x|>L} \frac{1}{x} dG(x) + \int_{|x|\leqq L} x dG(x) = \gamma_L.$$

Therefore, if $4c|\gamma_L| < \varepsilon/4$, then

$$\overline{\lim_{n\to\infty}}\, \mathsf{P}\Big\{\sup_{1\leqq k\leqq 4nc} \Big|\sum_{i=1}^{k} \xi_{ni}\Big| > \frac{\varepsilon}{4}\Big\}$$

$$\leqq 4c\int_{|x|>L} \frac{1 + x^2}{x^2} dG(x) + \frac{4c(1 + L^2)\int_{-\infty}^{\infty} dG(x)}{\Big(\dfrac{\varepsilon}{4} - 4c|\gamma_L|\Big)^2}.$$

This means that for sufficiently small c,

$$\overline{\lim_{n\to\infty}} \mathsf{P}\Big\{\sup_{0\leqq t\leqq 4c} |\,\xi_n(t) - \xi_n(0)\,| > \frac{\varepsilon}{4}\Big\} \leqq Kc\ ,$$

were K is some constant such that

$$\overline{\lim_{n\to\infty}} \mathsf{P}\{\Delta_c(\xi_n(t)) > \varepsilon\} \leqq \Big(1 + \frac{1}{c}\Big)(Kc)^2 + 2Kc\ .$$

Equation (3) follows from this inequality. This completes the proof of the theorem.

As examples of corollaries of this general theorem, let us look at a few particular limit theorems.

Theorem 3. *Let $a(t)$ and $b(t) > 0$ denote continuous functions and let α denote a real number. If the hypotheses of Theorem 1 are satisfied, then*

$$\lim_{n\to\infty} \mathsf{P}\Big\{a\Big(\frac{k}{n}\Big) - \alpha b\Big(\frac{k}{n}\Big) < S_{nk} < a\Big(\frac{k}{n}\Big) + \alpha b\Big(\frac{k}{n}\Big), k = 1, 2, \cdots, n\Big\}$$
$$= \mathsf{P}\{a(t) - \alpha b(t) < \xi(t) < a(t) + \alpha b(t),\ 0 \leqq t \leqq 1\}$$

for all positive α such that

$$\mathsf{P}\{a(t) - \alpha b(t) < \xi(t) < a(t) + \alpha b(t),\ 0 \leqq t \leqq 1\}$$

is continuous at α as a function of α.

Proof. Consider the functional defined on $D[0, 1]$ by

$$f(x(t)) = \sup_{0\leqq t\leqq 1} \frac{|\,x(t) + a(t)\,|}{b(t)}\ .$$

It is easy to see that this functional is continuous in the metric ρ_D. Proof of this theorem follows Theorem 2 and the inequality

$$\mathsf{P}\{f(\xi_n(t)) < \alpha - \varepsilon\} < \mathsf{P}\Big\{a\Big(\frac{k}{n}\Big) - \alpha b\Big(\frac{k}{n}\Big) < S_{nk} < a\Big(\frac{k}{n}\Big)$$
$$+ \alpha b\Big(\frac{k}{n}\Big), k = 1, \cdots, n\Big\}$$
$$\leqq \mathsf{P}\{f(\xi_n(t)) < \alpha + \varepsilon\}\ ,$$

which is valid, if $\varepsilon > 0$, for all sufficiently large n.

Theorem 4. *Let $g(x)$ denote a continuous function defined for $x \in (-\infty, \infty)$. Then if the hypotheses of Theorem 1 are satisfied, we have*

$$\lim_{n\to\infty} \mathsf{P}\Big\{\frac{1}{n}\sum_{k=1}^{n} g(S_{nk}) < \alpha\Big\} = \mathsf{P}\Big\{\int_0^1 g(\xi(t))dt < \alpha\Big\}$$

for all α such that $\mathsf{P}\Big\{\int_0^1 g(\xi(t))dt = \alpha\Big\} = 0.$

Proof follows from the fact that the functional

$$f(x(t)) = \int_0^1 g(x(t))dt$$

is continuous in the metric ρ_D, Theorem 2, and the equation

$$\frac{1}{n}\sum_{k=1}^{m} g(S_{nk}) = \int_0^1 g(\xi_n(t))dt \ .$$

Let us define a functional $\gamma_a(x(t))$ that is equal to 0 if $\sup_{0\leqq t\leqq 1} x(t) < a$ and equal to $x(\tau) - a$ if $\sup_{0\leqq t\leqq 1} x(t) \geqq a$, where τ is a point in the interval [0, 1] such that $x(t) < a$ for $t < \tau$ and $x(\tau) \geqq a$. The functional $\gamma_a(x(t))$ is called the time of the first crossing of the function $x(t)$ through the level a. Let us investigate the continuity of $\gamma_a(x(t))$ in the metric ρ_D. One can easily see that for all those $x(t)$ such that $\sup_t x(t) < a$, the functional $\gamma_a(x(t))$ is a continuous functional. It is also continuous on those functions $x(t)$ such that $\gamma_a(x(t)) > 0$.

If $x(\tau) = 0$ but $\sup_{s\in(\tau,\tau+\varepsilon)} x(s) > a$ for every $\varepsilon > 0$, then $\gamma_a(x(t))$ is also continuous at the point $x(t)$. Thus, the only possible points of discontinuity of $\gamma_a(x(t))$ are those functions $x(t)$ for which $x(\tau) = a$ and $\sup_{s\in(\tau,t)} x(s) \leqq a$ for some $t > \tau$. We can now formulate:

Theorem 5. *Let a denote a number such that*

$$\mathsf{P}\{\sup_{0\leqq s\leqq t} x(s) = a\} = 0$$

for $t \in [0, 1]$. Then, the distributions of $\gamma_a(\xi_n(t))$ converge to the distribution of $\gamma_a(\xi(t))$.

The proof follows Theorem 2, Corollary 2 of Section 1, and the fact that under the hypotheses of the theorem, $\gamma_a(x(t))$ is continuous almost everywhere with respect to the measure μ corresponding to the process $\xi(t)$ on $D[0, 1]$.

7. LIMIT THEOREMS FOR FUNCTIONALS OF INTEGRAL FORM

In the preceding sections, by confining ourselves to rather narrow classes of random processes we obtained limit theorems for extremely broad classes of functionals. A characteristic feature of these theorems was the constructive character of the conditions that were imposed in addition to the condition of convergence of sequences of finite-dimensional distributions. The constructive character of these additional conditions permitted convenient checking of the hypotheses in particular cases. As a rule, these conditions reduced to the requirement of some stochastic equicontinuity of the sequence of random processes. In this section we shall consider a comparatively narrow class of functionals of integral form:

$$f(x(t)) = \int_0^1 \cdots \int_0^1 \varphi(s_1, s_2, \cdots s_k, x(s_1), x(s_2), \cdots, x(s_k))ds_1 ds_2 \cdots ds_k \,. \quad (1)$$

Such functionals can be considered for processes of an extremely general form. For example, if φ is a continuous bounded function, then a sufficient condition for the existence of this functional of the process $\xi(t)$ is that the process $\xi(t)$ be measurable. For functionals of the form (1), we can also formulate a limit theorem in terms of some uniform stochastic continuity of the sequence of processes $\xi_n(t)$, while we impose almost no conditions at all on the processes $\xi_n(t)$ themselves (that is, on the form of the sample functions $\xi_n(t)$) other than measurability and existence of certain moments.

We first formulate a very simple theorem, from which we can obtain more general statements.

Theorem 1. *Let $\xi_n(t)$ and $\xi(t)$ denote measurable processes defined on the interval* [0, 1]. *If the sequences of finite-dimensional distributions of the processes $\xi_n(t)$ converge to finite-dimensional distributions of the process $\xi_n(t)$, if*

$$\sup_{t,n} \mathsf{M} \,|\, \xi_n(t) \,|\, \leqq H < \infty$$

and if

$$\lim_{h\to 0} \overline{\lim_{n\to\infty}} \sup_{|t'-t''|\leqq h} \mathsf{M} \,|\, \xi_n(t') - \xi_n(t'') \,| = 0 \,, \quad (2)$$

then the sequence of distributions of the random variable $\int_0^1 \xi_n(t)dt$ converge to the distribution of the random variable $\int_0^1 \xi(t)dt$.

Proof. It follows from condition (2) that for every $\delta > 0$, there exists an m such that for all sufficiently large n,

$$\mathsf{M}\left|\frac{1}{m}\sum_{n=1}^{m} \xi_n\left(\frac{k}{m}\right) - \int_0^1 \xi_n(t)dt\right| \leqq \sup_{|t'-t''|\leqq 1/m} \mathsf{M} \,|\, \xi_n(t') - \xi_n(t'') \,| \leqq \delta \,.$$

An analogous inequality holds for $\xi(t)$, since it follows from equation (2) and Fatou's lemma (Theorem 2, Section 5, Chapter II) that

$$\lim_{n\to 0} \sup_{|t'-t''|\leqq h} \mathsf{M} \,|\, \xi(t') - \xi(t'') \,| = 0 \,.$$

We note that the convergence of the sequences of finite-dimensional distributions of the process $\xi_n(t)$ to finite-dimensional distributions of the process $\xi(t)$ implies convergence of the sequence of distributions $(1/m)\sum_{k=1}^{m}\xi_n(k/m)$ to the distribution $(1/m)\sum_{k=1}^{m}\xi_n(k/m)$. Therefore

$$\overline{\lim_{n\to\infty}}\left|\mathsf{M}\exp i\lambda\int_0^1\xi_n(t)dt - \mathsf{M}\exp i\lambda\int_0^1\xi(t)dt\right|$$

$$\leqq \overline{\lim_{n\to\infty}}\left|\mathsf{M}\exp\left\{i\lambda\int_0^1\xi_n(t)dt\right\} - \mathsf{M}\exp\left\{i\lambda\frac{1}{m}\sum_{k=1}^{m}\xi_n\left(\frac{k}{m}\right)\right\}\right|$$

$$+\left|\mathsf{M}\exp\left\{i\lambda\int_0^1\xi(t)dt\right\} - \mathsf{M}\exp\left\{i\lambda\frac{1}{m}\sum_{k=1}^{m}\xi\left(\frac{k}{m}\right)\right\}\right|$$

$$\leqq |\lambda|\,\mathsf{M}\left|\int_0^1\xi_n(t)dt - \frac{1}{m}\sum_{k=1}^{m}\xi_n\left(\frac{k}{m}\right)\right|$$

$$+|\lambda|\,\mathsf{M}\left|\int_0^1\xi(t)dt - \frac{1}{m}\sum_{k=1}^{m}\xi\left(\frac{k}{m}\right)\right| \leqq 2|\lambda|\delta .$$

This completes the proof of the theorem.

Theorem 2. *Suppose that the sequences of finite-dimensional distributions of the processes $\xi_n(t)$ converge to finite-dimensional distributions of the process $\xi(t)$. Suppose that for every $\varepsilon > 0$*

$$\lim_{h\to\infty}\overline{\lim_{n\to\infty}}\sup_{|t-s|\leqq h}\mathsf{P}\{|\xi_n(t) - \xi_n(s)| > \varepsilon\} = 0 , \tag{3}$$

and that there is a nonnegative function $\psi(x)$ such that $\psi(x)\uparrow\infty$ as $|x|\to+\infty$ and $\sup_n\sup_t \mathsf{M}\psi(\xi_n(t)) = c < \infty$. Then for every continuous function $\varphi(t, x)$ such that

$$\lim_{N\to\infty}\sup_t\sup_{|x|>N}\frac{|\varphi(t,x)|}{\psi(x)} = 0 ,$$

the sequence of distributions of the variable $\int_0^1\varphi(t,\xi_n(t))dt$ converges to the distribution of the variable $\int_0^1\varphi(t,\xi(t))dt$.

Proof. It will be sufficient to show that the hypotheses of Theorem 1 are satisfied for the sequence of processes $\eta_n(t) = \varphi(t, \xi_n(t))$.

Convergence of the finite-dimensional distributions of the processes $\eta_n(t)$ to the finite-dimensional distributions of the process $\eta(t) = \varphi(t, \xi(t))$ follows from the convergence of the finite-dimensional distributions of the processes $\xi_n(t)$ to the finite-dimensional distributions of the process $\xi(t)$ and the continuity of the function $\varphi(t, x)$. Since $|\varphi(t, x)| < K(1 + \psi(x))$ for some K, we have

$$\sup_n\sup_t\mathsf{M}|\eta_n(t)| \leqq K(1 + c) .$$

Finally, let us show that equation (2) is satisfied for the sequence $\{\eta_n(t)\}$. To do this, we define the function $g_N(x)$ that is equal to

x for $|x| < N$ and equal to $N \operatorname{sgn} x$ for $|x| \geqq N$. We define

$$\varepsilon_N = \sup_t \sup_{|x| \geqq N} \frac{|\varphi(t, x)|}{\psi(x)} .$$

Then using the inequality

$$\frac{|\varphi(t, g_N(x)) - \varphi(t, x)|}{\psi(x)} \leqq \varepsilon_N ,$$

we obtain

$$\begin{aligned} \mathsf{M}\,|\eta_n(t_1) - \eta_n(t_2)| &= \mathsf{M}\,|\varphi(t_1, \xi_n(t_1)) - \varphi(t_2, \xi_n(t_2))| \\ &\leqq \mathsf{M}\,|\varphi(t_1, g_N(\xi_n(t_1))) - \varphi(t_2, g^N(t_2)))| \\ &\quad + \varepsilon_N \mathsf{M}\psi(\xi_n(t_1)) + \varepsilon_N \mathsf{M}\psi(\xi_n(t_2)) \\ &\leqq \mathsf{M}\,|\varphi(t_1, g_N(\xi_n(t_1))) - \varphi(t_2, g_N(t_2)))| + 2\varepsilon_N c . \end{aligned} \tag{4}$$

Since $\varphi(t, x)$ is a continuous function, it follows that for every $\varepsilon > 0$ and $L > 0$, there exists a $\delta > 0$ such that

$$|\varphi(t_1, x_1) - \varphi(t_2, x_2)| < \varepsilon$$

whenever

$$|t_1 - t_2| < \delta, \ |x_1 - x_2| < \delta, \ |x_1| \leqq L, \ |x_2| \leqq L .$$

Define $R = \sup_{t, |x| \leqq N} |\varphi(t, x)|$. Then for $|t_1 - t_2| < \delta$,

$$\begin{aligned} &\mathsf{M}\,|\varphi(t_1, g_N(\xi_n(t_1))) - \varphi(t_2, g_N(\xi_n(t_2)))| \\ &\quad \leqq \varepsilon + R[\mathsf{P}\{|\xi_n(t_1) - \xi_n(t_2)| > \delta\} \\ &\qquad + \mathsf{P}\{|\xi_n(t_1)| > L\} + \mathsf{P}\{|\xi_n(t_2)| > L\}] \\ &\quad \leqq \varepsilon + R\left[\mathsf{P}\{|\xi_n(t_1) - \xi_n(t_2)| > \delta\} + 2 \sup_t \frac{\mathsf{M}\psi(\xi_n(t))}{\psi(L)}\right] . \end{aligned}$$

In view of (3) and (4) we obtain

$$\lim_{h \to 0} \varlimsup_{n \to \infty} \sup_{|t_1 - t_2| \leqq h} \mathsf{M}\,|\eta_n(t_1) - \eta_n(t_2)| \leqq \varepsilon + 2\varepsilon_N c + 2Rc[\varphi(L)]^{-1} .$$

The quantities on the right side of this inequality can be made arbitrarily small by the suitable choice of ε, N, and L. Consequently,

$$\lim_{h \to 0} \varlimsup_{n \to \infty} \sup_{|t_1 - t_2| \leqq h} \mathsf{M}\,|\eta_n(t_1) - \eta_n(t_2)| = 0 ;$$

that is, all the conditions of Theorem 1 are satisfied for the sequence of processes $\eta_n(t)$. This completes the proof of the theorem.

Theorem 3. *Suppose that the sequence of processes $\xi_n(t)$ and the function $\psi(x)$ satisfy the conditions of Theorem 2. Then for every continuous function $\varphi(t_1, \cdots, t_k, x_1, \cdots, x_k)$ satisfying the equation*

$$\lim_{|x_1| \to \infty, \cdots, |x_k| \to \infty} \sup_{t_1, \cdots, t_k} \frac{|\varphi(t_1, \cdots, t_k, x_1, \cdots, x_k)|}{\sum_{i=1}^{k} \psi(x_i)} = 0 ,$$

the distributions of

$$\int_0^1 \cdots \int_0^1 \varphi(t_1, \cdots, t_k, \xi_n(t_1), \cdots, \xi_n(t_k))dt_1 \cdots dt_k$$

converge to the distribution of

$$\int_0^1 \cdots \int_0^1 \varphi(t_1, \cdots, t_k, \xi(t_1), \cdots, \xi(t_k))dt_1 \cdots dt_k .$$

Proof. Define

$$\varepsilon_N = \sup_{|x_1| \geqq N, \cdots, |x_k| \geqq N} \sup_{t_1, \cdots, t_k} \frac{|\varphi(t_1, \cdots, t_k, x_1, \cdots, x_k)|}{\sum_{i=1}^{k} \psi(x_i)}$$

$$+ \sup_{|x_1| \leqq N, \cdots, |x_k| \leqq N} \sup_{t_1, \cdots, t_k} \frac{|\varphi(t_1, \cdots, t_k, x_1, \cdots, x_k)|}{k\psi(N)} .$$

It is easy to see that $\varepsilon_N \to 0$ as $N \to \infty$. Let $g_N(x)$ denote a continuous periodic function with period $4N$ that coincides with x for $|x| \leqq N$ and that satisfies the inequality $|g_N(x)| \leqq x$ for $|x| > N$. Then

$$|\varphi(t_1, \cdots, t_k, x_1, \cdots, x_k) - \varphi(t_1, \cdots, t_k, g_N(x_1), \cdots, g_N(x_k))| \leqq \varepsilon_N \sum_{i=1}^{k} \psi(x_i) .$$

Therefore

$$\mathsf{M}\left|\int_0^1 \cdots \int_0^1 \varphi(t_1, \cdots, t_k, \xi_n(t_1), \cdots, \xi_n(t_k)) - \varphi(t_1, \cdots, t_k, g_N(\xi(t_1)), \cdots, g_N(\xi(t_k))dt_1 \cdots dt_k\right| \leqq k\varepsilon_N c .$$

Since on the basis of Fatou's lemma, $\mathsf{M}\psi(\xi(t)) \leqq \lim \mathsf{M}\psi(\xi_n(t)) \leqq c$, we also have

$$\mathsf{M}\left|\int_0^1 \cdots \int_0^1 (\varphi(t_1, \cdots, t_k, \xi(t_1), \cdots, \xi(t_k)) - \varphi(t_1, \cdots, t_k, g_N(\xi(t_1)), \cdots, g_N(\xi(t_k)))dt_1 \cdots dt_k\right| \leqq k\varepsilon_N c .$$

We now note that the function $\varphi(t_1, \cdots, t_k, g_N(x_1), \cdots, g_N(x_k))$ can be extended as a continuous periodic function with period $4N$ with respect to $x_1, \cdots, x_k$ and as a periodic function with period 2 with respect to $t_1, \cdots, t_k$. Therefore, for every $\varepsilon > 0$ there exists a trigonometric polynomial $P(t_1, \cdots, t_k, x_1, \cdots, x_k)$ in the variables $t_1, \cdots, t_k, x_1, \cdots, x_k$ such that

$$|\varphi(t_1, \cdots, t_k, g_N(x_1), \cdots, g_N(x_k)) - P(t_1, \cdots, t_k, x_1, \cdots, x_k)| < \varepsilon .$$

Since

$$\begin{aligned}
&\overline{\lim_{n\to\infty}} \left| \mathsf{M} \exp\left\{i\lambda\int_0^1 \cdots \int_0^1 \varphi(t_1, \cdots, t_k, \xi_n(t_1), \cdots, \xi_n(t_k))dt_1 \cdots dt_k\right\}\right. \\
&\quad \left. - \mathsf{M} \exp\left\{i\lambda\int_0^1 \cdots \int_0^1 \varphi(t_1, \cdots, t_k, \xi(t_1), \cdots, \xi(t_k))dt_1 \cdots dt_k\right\}\right| \\
&\leqq \overline{\lim_{n\to\infty}} \left| \mathsf{M} \exp\left\{i\lambda\int_0^1 \cdots \int_0^1 \varphi(t_1, \cdots, t_k, \xi_n(t_1), \cdots, \xi_n(t_k))dt_1 \cdots dt_k)\right\}\right. \\
&\quad \left. - \mathsf{M} \exp\left\{i\lambda\int_0^1 \cdots \int_0^1 P(t_1, \cdots, t_k, \xi_n(t_1), \cdots, \xi_n(t_k))dt_1 \cdots dt_k\right\}\right| \\
&\quad + \left| \mathsf{M} \exp\left\{i\lambda\int_0^1 \cdots \int_0^1 \varphi(t_1, \cdots, t_k, \xi(t_1), \cdots, \xi(t_k))dt_1 \cdots dt_k\right\}\right. \\
&\quad \left. - \mathsf{M} \exp\left\{i\lambda\int_0^1 \cdots \int_0^1 P(t_1, \cdots, t_k, \xi(t_1), \cdots, \xi(t_k))dt_1 \cdots dt_k\right\}\right| \\
&\quad + \overline{\lim_{n\to\infty}} \left| \mathsf{M} \exp\left\{i\lambda\int_0^1 \cdots \int_0^1 P(t_1, \cdots, t_k, \xi_n(t_1), \cdots \xi_n(t_k))dt_1 \cdots dt_k\right\}\right. \\
&\quad \left. - \mathsf{M} \exp\left\{i\lambda\int_0^1 \cdots \int_0^1 P(t_1, \cdots, t_k, \xi(t_1), \cdots \xi(t_k))dt_1 \cdots dt_k\right\}\right| \\
&\leqq \overline{\lim_{n\to\infty}} \mathsf{M} \left| \lambda\int_0^1 \cdots \int_0^1 [\varphi(t_1, \cdots, t_k, \xi_n(t_1), \cdots, \xi_n(t_k))\right. \\
&\quad \left. - P(t_1, \cdots, t_k, \xi_n(t_1), \xi_n(t_k))]dt_1 \cdots dt_k \right| \\
&\quad + \overline{\lim_{n\to\infty}} \mathsf{M} \left| \lambda\int_0^1 \cdots \int_0^1 [\varphi(t_1, \cdots, t_k, \xi_n(t_1), \cdots, \xi_n(t_k))\right. \\
&\quad \left. - P(t_1, \cdots, t_k, \xi(t_1), \cdots, \xi(t_k))]dt_1 \cdots dt_k \right| \\
&\quad + \overline{\lim_{n\to\infty}} \left| \mathsf{M} \exp\left\{i\lambda\int_0^1 \cdots \int_0^1 P(t_1, \cdots, t_k, \xi_n(t_1), \cdots, \xi_n(t_k))dt_1 \cdots dt_k\right\}\right. \\
&\quad \left. - \mathsf{M} \exp\left\{i\lambda\int_0^1 \cdots \int_0^1 P(t_1, \cdots, t_k, \xi(t_1), \cdots, \xi(t_k))dt_1 \cdots dt_k\right\}\right| \\
&\leqq 2(\varepsilon + k\varepsilon_N c) \\
&\quad + \overline{\lim_{n\to\infty}} \left| \mathsf{M} \exp\left\{i\lambda\int_0^1 \cdots \int_0^1 P(t_1, \cdots, t_k, \xi_n(t_1), \cdots, \xi_n(t_k))dt_1 \cdots dt_k\right\}\right. \\
&\quad \left. - \mathsf{M} \exp\left\{i\lambda\int_0^1 \cdots \int_0^1 P(t_1, \cdots, t_k, \xi(t_1), \cdots, \xi(t_k))dt_1 \cdots dt_k \right| ,\right.
\end{aligned}$$

to prove the theorem, it is sufficient to show that the distributions of

$$\int_0^1 \cdots \int_0^1 P(t_1, \cdots, t_k, \xi_n(t_1), \cdots, \xi_n(t_k))dt_1 \cdots dt_k$$

converge to the distribution of

$$\int_0^1 \cdots \int_0^1 P(t_1, \cdots. t_k, \xi(t_1), \cdots, \xi(t_k))dt_1 \cdots dt_k .$$

But

$$\int_0^1 \cdots \int_0^1 P(t_1, \cdots, t_k, \xi_n(t_1), \cdots, \xi_n(t_k))dt_1 \cdots dt_k$$

is a sum of products of the form

$$\prod_{j=1}^{k} \int_0^1 g_j(m_j t) f_j(n_j \xi_n(t))dt ,$$

where the $g_j(x)$ and the $f_j(x)$ are trigonometric functions ($\sin x$ or $\cos x$) and the m_j and n_j are real numbers. Therefore the conclusion of the theorem follows from the fact that the joint distributions of an arbitrary finite set of variables $\int_0^1 \varphi_j(t, \xi_n(t))dt$, for $j = 1, \cdots, l$ converge to the joint distribution of the variables

$$\int_0^1 \varphi_j(t, \xi(t))dt, \quad \text{for} \quad j = 1, \cdots, l ,$$

so that

$$\begin{aligned}\lim_{n\to\infty} \mathsf{M} \exp\left\{i \sum_{j=1}^{k} \lambda_j \int_0^1 \varphi_j(t, \xi_n(t))dt\right\} &= \lim_{n\to\infty} \mathsf{M} \exp\left\{i\int_0^1 \sum_{j=1}^{k} \lambda_j \varphi_j(t, \xi_n(t)dt\right\} \\ &= \mathsf{M} \exp\left\{i\int_0^1 \sum_{j=1}^{k} \lambda_j \varphi_j(t, \xi(t))dt\right\}\end{aligned}$$

in accordance with Theorem 2.

8. APPLICATION OF LIMIT THEOREMS TO STATISTICAL CRITERIA

One of the problems of mathematical statistics is to determine the distribution function of a random variable from the results of observations.

Suppose that the results of some experiment constitute a random variable with unknown continuous distribution function $F(x)$. How can we approximate the function $F(x)$ if we know n results of independently performed experiments: $\xi_1, \xi_2, \cdots, \xi_n$?

In mathematical statistics we use for this purpose an empirical distribution function $F_n^*(x)$ defined by

$$F_n^*(x) = \frac{\nu_n(x)}{n} ,$$

where $\nu_n(x)$ is the number of variables ξ_k falling in the interval $(-\infty, x)$. It follows from Bernoulli's theorem that the sequence $\{F_n^*(x)\}$ converges in probability to $F(x)$. Thus we can take the function $F_n^*(x)$ as an approximation of $F(x)$. Of course we must consider the error that arises when we do this. However it is often convenient to have an analytic representation for an approximation of $F(x)$. Then we need to answer the question: can a given function $\Phi(x)$ serve as an approximation of $F(x)$ if we know the results of an experiment $\xi_1, \cdots, \xi_n$? In both cases it is important to know the behavior of the difference between the empirical distribution function and the theoretical distribution function $F(x)$. We shall give some results that have applications in statistics.

Theorem 1 (Kolmogorov's criterion). *For all* $\alpha > 0$,

$$\lim_{n\to\infty} \mathsf{P}\{\sqrt{n} \sup_{-\infty<x<\infty} |F(x) - F_n^*(x)| < \alpha\} = \sum_{k=-\infty}^{\infty} (-1)^k e^{-2k^2\alpha^2} .$$

Theorem 2 (Smirnov's criterion). *For all* $\alpha > 0$,

$$\lim_{n=\infty} \mathsf{P}\{\sqrt{n} \sup_{-\infty<x<\infty} (F_n^*(x) - F(x)) < \alpha\} = 1 - e^{-2\alpha^2} .$$

Theorem 3 (the ω^2-criterion). *For all* $\alpha > 0$,

$$\lim_{n\to\infty} \mathsf{P}\left\{n\int [F_n^*(x) - F(x)]^2 dF(x) < \alpha\right\} = G(\alpha) ,$$

where $G(\alpha)$ *is the distribution function whose characteristic function is equal to*

$$\left(\prod_{k=1}^{\infty} \sqrt{1 - \frac{2i\alpha}{k^2\pi^2}}\right)^{-1} .$$

We shall not show how these results can be applied in statistical investigations, but we shall show how they can obtained by means of the limit theorems for random processes.

We introduce the process $\eta_n(x) = \sqrt{n}\,(F_n^*(x) - F(x))$.

Lemma 1. *The processes* $\eta_n(x)$ *converge weakly to a Gaussian process* $\eta(x)$ *with the properties* $\mathsf{M}\eta(x) = 0$ *and*

$$\mathsf{M}(\eta(x)\eta(y)) = F(x)[1 - F(y)]$$

for $-\infty < x \leqq y < \infty$.

Proof. We note that

$$\eta_n(x) = \frac{1}{\sqrt{n}} \sum_{k=1}^{n} [\varepsilon(\xi_k - n) - F(x)] ,$$

where $\varepsilon(z) = 0$ for $z \leqq 0$ and $\varepsilon(z) = 1$ for $z > 0$. Since

$$\mathsf{M}\varepsilon(\xi_k - x) = F(x), \ \mathsf{M}[\varepsilon(\xi_k - x)\varepsilon(\xi_k - y)] = F(x) \quad \text{for} \quad x < y ,$$

and since the processes $\varepsilon(\xi_k - x) - F(x)$ are independent for different values of k, the proof follows from Theorem 4, Section 6, Chapter I.

Corollary. *Let* $F^{-1}(t)$ *denote the inverse of the function* $F(x)$. *We set* $\xi_n(t) = \eta_n(F^{-1}(t))$ *and* $\xi(t) = \eta(F^{-1}(t))$ *for* $0 \leqq t \leqq 1$. *Then the* *sequence of processes* $\xi_n(t)$ *converges weakly to a Gaussian process* $\xi(t)$ *defined for* $t \in [0, 1]$ *with the properties* $\mathsf{M}\xi(t) = 0$ *and*

$$\mathsf{M}(\xi(t)\xi(s)) = t(1 - s) \quad \textit{for} \quad t < s \, .$$

REMARK 1. The process $\xi_n(t)$ can be represented in the form

$$\xi_n(t) = \frac{1}{\sqrt{n}} \sum_{k=1}^{n} [\varepsilon(\eta_k - t) - t] \, ,$$

where the $\eta_k = F^{-1}(\xi_k)$ are independent uniformly distributed random variables on the interval $[0, 1]$.

REMARK 2. The finite-dimensional distributions of the process $\xi(t)$ coincide with the conditional finite-dimensional distributions of a process of Brownian motion $w(t)$ for $0 \leqq t \leqq 1$, with the hypothesis that $w(1) = 0$. Since the conditional distributions of the process $w(t)$ with the hypothesis $w(l) = 0$ are Gaussian distributions, it is sufficient to show that

$$\mathsf{M}(w(t) \mid w(1) = 0) = \mathsf{M}\xi(t), \; \mathsf{M}(w(t)w(s) \mid w(1) = 0) = \mathsf{M}\xi(t)\xi(s) \, .$$

The random variable $\bar{\xi}(t) = w(t) - tw(1)$ is uncorrelated with $w(1)$. Since $\bar{\xi}(t)$ and $w(1)$ have joint Gaussian distributions, the process $\bar{\xi}(t)$ is independent of $w(1)$. Therefore

$$\mathsf{M}(\bar{\xi}(t) \mid w(1)) = \mathsf{M}\bar{\xi}(t) = 0 \, ,$$
$$\mathsf{M}(\bar{\xi}(t)\bar{\xi}(s) \mid w(1)) = \mathsf{M}(\bar{\xi}(t)\bar{\xi}(s)) \, .$$

Using the relation $w(t) = \bar{\xi}(t) + tw(1)$ and the preceding formulas, we obtain

$$\mathsf{M}(w(t) \mid w(1)) = tw(1) \, ,$$
$$\mathsf{M}(w(t)w(s) \mid w(1)) = \mathsf{M}(\bar{\xi}(t)\bar{\xi}(s)) + ts[w(1)]^2$$
$$= \min\,[t, s] - ts + ts[w(s)]^2 \, .$$

Setting $w(1) = 0$, we see the validity of what was said at the beginning of Remark 2.

Lemma 2. *For every functional* f *that is continuous on* $D[0, 1]$, *the sequence of distributions of* $f(\xi_n(t))$ *converges to the distribution of* $f(\xi(t))$.

Proof. We note first of all that the separable process $\xi(t)$ is continuous, so that $\xi(t)$ with probability 1 belongs to $D[0, 1]$. This is true because for $h > 0$, the difference $\xi(t + h) - \xi(t)$ has a Gaussian distribution and

$$\mathsf{M}(\xi(t + h) - \xi(t)) = 0 \, ,$$
$$\mathsf{M}(\xi(t + h) - \xi(t))^2 = (t + h)(1 - t - h)$$
$$+ t(1 - t) - 2t(1 - t - h) = h - h^2 \, .$$

Therefore $\mathsf{M}\,|\,(\xi(t+h)-\xi(t)\,|^4 = O(h^2)$ and the process $\xi(t)$ is continuous by virtue of Theorem 2, Section 5, Chapter IV.

The convergence of the finite-dimensional distributions of the processes $\xi_n(t)$ to the finite-dimensional distributions of $\xi(t)$ has been established. On the basis of Theorem 2 of Section 5, it remains only to show that relation 5 of Section (5) is satisfied. Since

$$\Delta_c(x(t)) \leqq \sup_{|t'-t''|\leqq c} |\,x(t') - x(t'')\,|\,,$$

the lemma will be proved if we can show that

$$\lim_{c\to 0}\varlimsup_{n\to\infty} \mathsf{P}\{\sup_{|t'-t''|\leqq c} |\,\xi_n(t') - \xi_n(t'')\,| > \varepsilon\} = 0 \tag{1}$$

for all $\varepsilon > 0$. The process $\xi_n(t) + \sqrt{n}\,t$ increases monotonically; that is, for $t_1 < t_2 < t_3 < t_4$,

$$\begin{aligned} -\sqrt{n}\,(t_4 - t_1) &\leqq \xi_n(t_3) - \xi_n(t_2) \\ &\leqq \xi_n(t_4) - \xi_n(t_1) + \sqrt{n}\,(t_4 - t_1)\,. \end{aligned}$$

Therefore

$$\begin{aligned} &\sup_{|t'-t''|\leqq c} |\,\xi_n(t') - \xi_n(t'')\,| \\ &\qquad \leqq \sup_{|k_1-k_2|\leqq c\cdot 2^{m+2}} \left|\xi_n\left(\frac{k_1}{2^m}\right) - \xi_n\left(\frac{k_2}{2^m}\right)\right| + \frac{2\sqrt{n}}{2^m}\,. \end{aligned}$$

Suppose that the m_n are chosen so that $\sqrt{n}/2^{m_n} \longrightarrow 0$ as $n \longrightarrow \infty$ and $2^{n/m_n} \geqq 1$. To prove equation (1), it sufficies to show that for every $\varepsilon > 0$,

$$\lim_{c\to 0}\varlimsup_{n\to\infty} \mathsf{P}\left\{\sup_{|k_1/2^{m_n}-k_2/2^{m_n}|\leqq c} \left|\xi_n\left(\frac{k_1}{2^{m_n}}\right) - \pi_n\left(\frac{k_2}{2^{m_n}}\right)\right| > \varepsilon\right\} = 0\,.$$

We note that

$$\sup_{|(k_1/2^{m_n}-k_2/2^{m_n})|\leqq c} \left|\xi_n\left(\frac{k_1}{2^{m_n}}\right) - \xi_n\left(\frac{k_2}{2^{m_n}}\right)\right| \leqq 2\sum_{r=m^{(c)}}^{m_n} \sup_i \left|\xi_n\left(\frac{i+1}{2^r}\right) - \xi_n\left(\frac{i}{2^r}\right)\right|\,,$$

where $m^{(c)}$ is the smallest integer such that $c2^{m^{(c)}} \geqq 1$. (With regard to the last inequality, see the proof of Theorem 2, Section 2.) Let us choose $a < 1$ such that $2a^4 > 1$. Then

$$\begin{aligned} &\mathsf{P}\left\{\sup_{|(k_1/2^{m_n}-k_2/2^{m_n})|\leqq c} \left|\xi_n\left(\frac{k_1}{2^{m_n}}\right) - \xi_n\left(\frac{k_2}{2^{m_n}}\right)\right| > \varepsilon\right\} \\ &\quad \leqq \sum_{r=m^{(c)}}^{m_n} \mathsf{P}\left\{\sup_i \left|\xi_n\left(\frac{i+1}{2^r}\right) - \xi_n\left(\frac{i}{2^r}\right)\right| > \frac{\varepsilon}{2}\cdot\frac{a^{r-m^{(c)}}}{1-a}\right\} \\ &\quad \leqq \sum_{r=m^{(c)}}^{m_n} \sum_{i=0}^{2^r-1} \mathsf{P}\left\{\left|\xi_n\left(\frac{i+1}{2^r}\right) - \xi_n\left(\frac{i}{2^r}\right)\right| > \frac{\varepsilon a^{r-m^{(c)}}}{2(1-a)}\right\} \\ &\quad \leqq \sum_{r=m^{(c)}}^{m_n} \sum_{i=0}^{2^r-1} \mathsf{M}\left|\xi_n\left(\frac{i+1}{2^r}\right) - \xi_n\left(\frac{i}{2^r}\right)\right|^4 \cdot \frac{2^4(1-a)^4}{\varepsilon^4 a^{4(r-m^{(c)})}}\,. \end{aligned}$$

Let μ_n denote the number of variables η_i that fall in the interval $[t, t + h]$. Then

$$\mathsf{P}\{\mu_n = k\} = C_n^k h^k (1 - h)^{n-k}$$

and

$$\xi_n(t + h) - \xi_n(t) = \sqrt{n}\left(\frac{\mu_n}{n} - h\right).$$

Calculations show (cf. Gnedenko, 1963.) That

$$\mathsf{M}(\xi_n(t + h) - \xi_n(t))^4 \leqq 3h^2 + \frac{h}{n} \leqq 3h^2 + \frac{h}{2^{m_n}}.$$

This means that for $h \geqq 1/2^{m_n}$ we have

$$\mathsf{M}(\xi_n(t + h) - \xi_n(t))^4 \leqq 4h^2 .$$

This inequality and the preceding one show that

$$\mathsf{P}\left\{\sup_{|(k_1/2^{m_n} - k_2/2^{m_n})| \leq c} \left|\xi_n\left(\frac{k_1}{2^{m_n}}\right) - \xi_n\left(\frac{k_2}{2^{m_n}}\right)\right| > \varepsilon\right\}$$

$$\leqq \sum_{r=m^{(c)}}^{m_n} \frac{2^4(1 - a)^4}{\varepsilon^4 a^{4(r - m^{(c)})}} \cdot 4 \cdot \frac{1}{2^r}$$

$$\leqq \frac{2^6(1 - a)^4}{\varepsilon^4} 2^{-m^{(c)}} \sum_{r=0}^{\infty} \frac{1}{(2a^4)^r} \leqq L_\varepsilon \frac{1}{2^{m^{(c)}}},$$

where

$$L_\varepsilon = \frac{2^6(1 - a)^4}{\varepsilon^4} \sum_{r=0}^{\infty} \frac{1}{(2a^4)^r}.$$

This completes the proof of the lemma.

Let us now proceed to prove Theorems 1 to 3. We note that the functionals

$$f_1(x(t)) = \sup_{0 \leqq t \leqq 1} | x(t) | ,$$

$$f_2(x(t)) = \sup_{0 \leqq t \leqq 1} x(t) ,$$

$$f_3(x(t)) = \int_0^1 [x(t)]^2 dt$$

are continuous on $D[0, 1]$. On the other hand,

$$\sqrt{n} \sup_{-\infty < x < \infty} | F_n^*(x) - F(x) | = f_1(\xi_n(t)) ,$$

$$\sqrt{n} \sup_{-\infty < x < \infty} (F_n^*(x) - F(x)) = f_2(\xi_n(t)) ,$$

$$n\int_{-\infty}^{\infty} (F_n^*(x) - F(x))^2 dF(x) = n\int_0^1 [F_n^*(F^{-1}(t)) - t]^2 dt = f_3(\xi_n(t)) .$$

Therefore, on the basis of Lemma 2, it is sufficient to show that

$$\mathsf{P}\{f_1(\xi(t)) < \alpha\} = \sum_{k=-\infty}^{\infty} (-1)^k e^{-2k^2\alpha^2} , \tag{2}$$

$$\mathsf{P}\{f_2(\xi(t)) < \alpha\} = 1 - e^{-2\alpha^2}\,, \tag{3}$$

$$\mathsf{M}e^{i\alpha f_3(\xi(t))} = \left(\prod_{k=1}^{\infty} \sqrt{1 - \frac{2i\alpha}{k^2\pi^2}}\right)^{-1}\,. \tag{4}$$

To prove formula (2), we use Theorem 2, Section 5, Chapter VI, from which we easily find the joint distribution of $\sup_{0 \leq t \leq 1} |w(t)|$ and $w(1)$: For $|x| < \alpha$,

$$\mathsf{P}\{\sup_{0 \leq t \leq 1} |w(t)| < \alpha,\ w(1) < x\}$$

$$= \frac{1}{\sqrt{2\pi}} \sum_{k=-\infty}^{\infty} \int_{-\alpha}^{x} \left[\exp\left\{-\frac{1}{2}(u + 4k\alpha)^2\right\}\right.$$

$$\left. - \exp\left\{-\frac{1}{2}(u + 4k\alpha + 2\alpha)^2\right\}\right] du\,.$$

From this formula we find the conditional distribution of $\sup_{0 \leq t \leq 1} |w(t)|$ under the hypothesis that $w(1) = 0$. This conditional distribution coincides with the right-hand member of equation (2). Analogously, by using Theorem 1, Section 5, Chapter VI, we obtain formula (3). To prove formula (4), we note that on the basis of formula (18), Section 2, Chapter V,

$$\xi(t) = \sqrt{2} \sum_{k=1}^{\infty} \xi_k \frac{\sin k\pi t}{k\pi}\,,$$

where the ξ_k are independent normally distributed random variables such that $\mathsf{M}\xi_k = 0$ and $\mathsf{D}\xi_k = 1$. Therefore

$$\int_0^1 \xi^2(t)dt = \sum_{k=1}^{\infty} \frac{\xi_k^2}{k^2\pi^2}\,,$$

$$\mathsf{M} \exp\left\{i\alpha \int_0^1 \xi^2(t)dt\right\}$$

$$= \prod_{k=1}^{\infty} \int_{-\infty}^{\infty} \exp\left\{-\frac{x^2}{2} \frac{k^2\pi^2 - 2i\alpha}{k^2\pi^2}\right\} \frac{dx}{\sqrt{2\pi}} = \prod_{k=1}^{\infty} \sqrt{\frac{k^2\pi^2}{k^2\pi^2 - 2i\alpha}}\,.$$

This proves formula (4).

Bibliographic Notes

These notes contain selected references to the literature on the questions discussed in each chapter. Their purpose is not to give a complete bibliography or to clarify the history of the basic ideas in the theory of random processes. In many cases we refer not to the original works, since they are not always accessible, but instead to later texts and monographs that contain a bibliography on questions we are studying.

CHAPTER I.

§ *1.* A systematic study of the questions of the theory of random processes was begun in the works by Slutskiy (1928) and Kolmogorov (1933 b), (1938 a). A significant role in the construction of the theory of random processes was played by the works of Kolmogorov.

§ *2.* The concept of a stationary process in the broad sense was introduced by Khinchin (1937).

§ *3.* The multidimensional generalization of the central limit theorem was first obtained by Bernstein (1946).

§ *4.* Oscillations with random amplitudes and phases have been studied in many works. See, for example, the appendix by Yu. A. Krutkov in the books by Einstein and Smoluchowski, and Bogolyubov.

§ *5.* The spectral representation of the correlation function of a process that is stationary in the broad sense was obtained by Khinchin (1938). Its multidimensional generalization was produced by Cramér. The representation of a structural function of processes with increments that are stationary in the broad sense was found by Kolmogorov (1940 b). The formula for the correlation function of a homogeneous isotropic random field is contained in the work by Schönberg.

CHAPTER II.

In this chapter the exposition of the theory of measure and the integral was geared to applications in probability theory and the theory of random processes. A more detailed exposition can be found in the books by Saks, Halmos, Kolmogorov and Fomin, and Heider and Simpson.

CHAPTER III.

§ *1.* The set-theoretic axiomatization of probability theory that is generally accepted at the present time was proposed by Kolmogorov in 1929 and was expounded in his monograph (1933 b).

§ *2.* The fundamental theorem (Theorem 3) of this section is from Kolmogorov (1933 b) (in which it is proved for families of random variables). The more general approach to the question has been proposed by Bochner (1947).

§ *3.* In connection with the material of this section see Kolmogorov (1933 b), Halmos and Loève (1963). In Halmos one can find a theorem on the existence of a probability space that is a representation of an infinite sequence of independent elements with values in an arbitrary space X.

§ *4.* The basic ideas and results of this section are from Kolmogorov and Khinchin and Kolmogorov (1929 a). Series of independent random variables are studied in greater detail in Doob, and Loève (1963).

§ *5.* The appearance of ergodic theory is connected with the problems of providing a foundation for statistical mechanics. See Khinchin (1949). The ergodic theorems of von Neumann and Birkhoff served as the starting point of an intensive development of ergodic theory. Khinchin freed the proof of Birkhoff's ergodic theorem from unnecessary assumptions and made possible its application to probability theory problem. A simple proof of the Birkhoff-Khinchin theorem has been presented by Kolmogorov (1938 b). Hopf and Jacobs give good surveys of ergodic theory.

§ *6.* The theory of conditional probabilities and conditional mathematical expectations was constructed in the monograph by Kolmogorov (1933 b). See also Doob, and Loève (1963).

CHAPTER IV.

§ *1.* The problem of constructing a random process that is stochastically equivalent to a given process whose sample functions satisfy specified conditions of regularity was first examined by Ye. Ye. Slutskiy and Kolmogorov (cf. Slutskiy's article [1949]). In the subsequent development of the problems which arose here and the different variations of the axiomatic definition of a random function, many essential results belong to Doob. References to the original works are found in his monograph.

§ *2 and 3.* The basic results of these sections are from Doob.

§ *4.* Theorem 1 was proved by Chentsov and Theorem 2 was proved by Kinney (for Markov processes). The fact that stochastically continuous processes with independent increments have no discontinuities of the second kind was shown by Lévy (1934).

§ *5.* Theorem 2 belongs to Kolmogorov and was first published in the article by Slutskiy (1949). Theorem 4 was proved independently by Dynkin, and Kinney (1952) (for Markov processes).

CHAPTER V.

§ *1.* An introduction to the theory of Hilbert spaces can be found in Kolmogorov and Fomin, and Heider and Simpson. A more complete exposition can be found in Akhiyezer and Glazman.

§ *2.* Cf. Slutskiy (1923) and Loève (1948), (1963).

§ *3.* The theory of stochastic integrals was proposed by Cramér. Kolmogorov first showed the connection between stochastic integrals and the theory of spectral representations of random functions with the theory of Hilbert spaces (1940 a), (1940 b), (1941).

§ *4.* Theorem 1 belongs to Karhunen (1947), Theorem 2 to Cramér, and Theorem 3 to Kolmogorov (1940 b).

§ *5.* By means of the theory of filters, one can easily obtain the spectral decomposition of stationary processes (cf. Blanc-Lapierre and Fortet). A more general theory of linear transformations of random processes can be constructed with the aid of the theory of generalized random processes as proposed by I. M. Gel'fand and K. Ito (cf. Gel'fand and Vilenkin, and Ito (1954)).

§ *6.* The basic results for stationary sequences were obtained by Kolmogorov (1941). Those for processes with continuous time were obtained by Hanner, and Karhunen (1950).

§ *7.* The general formulation of the problem of linear prediction (for stationary sequences), its connection with the geometry of a Hilbert space, and its reduction to a problem in the theory of functions belongs to Kolmogorov (1941). N. Wiener developed methods for the actual solution of problems of linear prediction and filtration for processes with continuous time. A method by A. M. Yaglom with a large number of examples can be found in his survey article.

§ *8.* The theorem on the decomposition of a stationary sequence and the concept of determined and undetermined processes belongs to Wold. The general solution of the problem of the prognosis of a stationary sequence from the entire past was obtained by Kolmogorov (1941) and for processes with continuous time, by Krein (1944), (1954). The problem of prediction of a vector-valued stationary sequence was studied by Rozanov (1958). Details of the prediction of processes with continuous time can be found in Doob, Grenander and Szegö, and Rozanov (1966).

CHAPTER VI.

The decomposition of a process into a continuous and a jump component was used by Lévy (1934) in the derivation of the characteristic function of the process. The idea was presented by Ito (1942).

§ *1.* Measures constructed from the jumps of a process were introduced by Ito (1942) and (1960).

§ *2.* Theorem 2 belongs to Khinchin (1936), chapter I, § 2.

§ *3.* The representation of the discontinuous part of a process by means of stochastic integrals belongs to Ito (1942) and (1960). The general formula for the characteristic function of a process was found by Lévy (1934).

§ *5.* A process of Brownian motion was first considered by Bachelier. Rigorous construction of this process and study of the properties of its sample functions were carried out by Wiener (1923). Theorem 1 is from Bachelier, and Theorem 2 follows from the general results of Petrovskiy.

§ *6.* The growth of homogeneous processes with independent increments was studied by Khinchin (1939) and Gnedenko (1943), (1948). The law of the iterated logarithm for a process of Brownian motion was obtained by Khinchin (1936), Chapter V.

CHAPTER VII.

§ *1.* Markov first studied sequences of associated random variables for which the "future" does not depend on the "past." Kolmogorov's article (1938) served as the basis of the general theory of Markov processes.

§ *2.* Theorem 2 is attributed to Kolmogorov (1951). A more detailed exposition of the questions studied in this section and bibliographic references can be found in the article by Reuter and the monograph by Chung.

§ *3.* Integro-differential equations of jump processes were introduced in Kolmogorov's article (1938). Theorems on the existence of solutions to these equations were considered by Feller (1936), (1940).

§ *4.* Numerous examples of Markov processes in natural science can be found in Feller (1966) and Bartlett.

§ *5.* Branching processes with discrete time were first studied in the article by Watson and Galton. The general definition of a branching process was given in the article by Kolmogorov and Dmitriyev. The survey article by Sevast'yanov was used for the present section.

§ *6.* The definition of a Markov process was analyzed by Doob. For more general definitions and properties of Markov processes see Dynkin (1961).

§ *7.* The structure of sample functions of a homogeneous Markov jump process and also the construction of a process from the function $q(t, x, A)$ were studied by Doob (who gives references to earlier works). A similar investigation of processes with countably many states, including all possible pathological cases, was made by Chung, and a summary of his results are contained in his book. The property proved in Theorem 3 of a process was called the "condition of strict Markovity" by Dynkin. Dynkin (1961) studied general, strictly Markov processes. For the strict Markovity of homogeneous processes with countably many states, see the monograph by Chung. Theorem 5 and 6 were proved in the article by Reuter.

CHAPTER VIII.

A probabilistic treatment of the phenomenon of diffusion was made by Khinchin (1936), Chapter III. Stochastic differential equations for random processes were studied by Bernstein (1934), Gikhman (1950), (1951), and Ito (1946), (1950). Here, we have used the basic terminology and notation of Ito.

§ *1.* Continuous Markov processes and equations for transition probabilities were considered by Kolmogorov (1938). Solutions of Kolmogorov's equations and their generalizations were studied by Feller (1936), (1954, a, b), and Dynkin (1964).

§ *2.* The basic results of this section are from Ito (1944), (1951).

§ *3.* Equations in this form were studied by Ito (1946), (1950), who proved the theorem on existence and uniqueness and showed that the solution is a Markov process.

§ *4.* Differentiability of the solutions of stochastic equations from initial conditions was shown by Gikhman (1951).

§ *5.* The idea of deriving Kolmogorov's equations from initial conditions by using the differentiability of the solution of a stochastic equation is from Gikhman (1951). The derivation of equations for the distribution of an additive functional from a process of Brownian motion belongs to Kac (1949), (1951), and in the general case to Dynkin (1955 b). Theorem 4 was proved by Lévy (1939).

§ *6.* The application of differential equations to random observations in a bounded region was proposed by Petrovskiy. Diffusion processes in bounded regions were studied by Khinchin (1936), Chapters III and IV. A general study of various boundary conditions for homogeneous diffusion processes was made by Feller (1954 a). The distributions of certain functionals connected with the time of reaching the boundaries was studied by Khas'-minskiy.

CHAPTER IX.

The limit theorems for probabilities of events that depend on the entire trajectory of a process (the probabilities that the process remains in a curvilinear strip) were first studied by Kolmogorov (1931), (1933 a), Petrovskiy, and Khinchin (1934). The first general limit theorem for arbitrary functionals that are continuous in the metric of C was obtained by Donsker (1951) (in the case of convergence of sums of identically distributed independent random variables to a process of Brownian motion).

§ *1.* Questions of weak convergence of measures in a metric space were studied by Prokhorov (1953), (1956).

§ *2.* The general limit theorem for continuous processes was obtained by Prokhorov (1953), (1956).

§ *3.* The limit theorems for different particular cases were considered by Kolmogorov (1931), (1933 a), and Kac and Erdös (1946 a), (1946 b). The general theorem belongs to Prokhorov (1953), (1956). A particular case (Theorem 2) was obtained earlier by Donsker (1951).

§ *4.* Specific cases of convergence to diffusion processes were studied by Bernstein (1934), (1946), Khinchin (1936), and Gikhman (1953), (1954). The theorems for C-continuous functionals were studied by Maruyama, Prokhorov (1954), and Skorokhod (1965), (1963).

§ *5.* The convergence studied in this section was introduced by Skorokhod (1956). An interesting limit theorem for processes without discontinuities of the second kind was obtained by Chentsov.

§ *6.* The conditions for convergence of the distributions that we considered in this section follow from the limit theorems for sums of independent random variables. See, for example, Gnedenko and Kolmogorov. The limit theorem for the probability that a sequence of sums will be contained in a curvilinear band was obtained by Gikhman (1958). General theorems for functionals were studied by Skorokhod (1957) and Prokhorov (1956).

§ *7.* With regard to the criteria mentioned, see the survey article by Smirnov. The method given here of proving the criteria of Kolmogorov and Smirnov belongs to Donsker (1952).

Bibliography

Akhiyezer, N. I., and Glazman, I. M.: Theory of Linear Operators in Hilbert Space. New York, Ungar Publ. Co., Vol. I, 1961; Vol. II, 1963.

Bachelier, L.: Théorie de la spéculation. Ann. Sci. Ecol. Norm. Sup., **3**: 21-86, 1900.

Bartlett, M. S.: Introduction to stochastic processes. London, Cambridge University Press, 1955.

Bernstein, S. N.: Principes de la théorie des équations différentiales stochastiques. Trudy Fiz-Mat. in-ta im. Steklova, **5**: 95-124, 1934.

Bernstein, S. N.: Rasprostraneniye predel'noy teoremy teorii veroyatnostey na summy zavisimykh velichin [The extension of a limit theorem in probability theory to the sum of dependent quantities]. Uspekhi Matem. Nauk, **10**: 65-114, 1944.

Bernstein, S. N.: Teoriya Veroyatnostey, [Probability Theory], Izd. 4. Gostekhizdat, 1946.

Blanc-Lapierre, A. and Fortet, R.: Théorie des Fonctions Aléatoires. Paris, 1953.

Bochner, S.: Stochastic processes. Ann. Math., **48**: 1014-1061, 1947.

Bochner, S.: Harmonic analysis and the theory of probability. Berkeley, University of California Press 1955.

Bogolyubov, N. N.: O nekotorykh statisticheskikh metodakh v matematicheskoy fizike [On certain statistical methods in mathematical physics]. Izd-vo Akad. Nauk USSR, 1945.

Chentsov, N. N.: Slavaya skhodimost' sluchaynykh protsessov s trayektoriyami bez razryvov vtorogo roda [Weak convergence of random processes with trajectories without discontinuities of the second kind]. Teoriya Veroyatnostey i Yeye Primeneniya, **1**: 154-161, 1956.

Chung, K. L.: Markov Chains with Stationary Transition Probabilities, Springer-Verlag, 1960.

Cramér, H.: On the theory of random processes. Ann. Math., **41**: 215-230, 1940.

Donsker, M.: An invariance principle for certain probability limit theorems. Mem. Amer. Math. Soc., **6**: 1-12, 1951.

Donsker, M.: Justification and extension of Doob's heuristic approach to the Kolmogoroff-Smirnov theorems. Ann. Math. Stat., **23**: 277-281, 1952.

Doob, J.: Stochastic processes. New York, John Wiley and Sons Inc., 1953.

Dynkin, Ye. B.: Kriteriy nepreryvnosti i otsutstviya razryvov vtorogo roda dlya trayektoriy markovskogo sluchaynogo protsessa [A criterion for continuity and absence of discontinuities of the second kind for the trajectories of a Markov random process]. Isvestiya Akad. Nauk USSR, ser. matem., **16**: 563-572, 1952.

Dynkin, Ye. B:. Nekotoryye predel'nyye teoremy dlya summ nezavisimykh sluchaynykh velichin s beskonechnymi matematicheskimi ozhidaniyami [Certain limit theorems for sums of independent random variables with infinite mathematical expectations]. Izvestiya Akad. Nauk USSR, ser. matem., **19**: 247–266, 1955a.

Dynkin, Ye, B.: Funktsionaly ot trayektoriy markovskikh sluchaynykh protsessov [Functionals of trajectories of Markov random processes]. Doklady Akad. Nauk USSR, **104**: 691–694, 1955b.

Dynkin, Ye, B.: Theory of Markov processes. New York, Pergamon Press Inc., 1961; Englewood Cliffs, New Jersey, Prentice-Hall Inc., 1961.

Dynkin, Ye, B.: Markov processes. New York, Academic Press Inc., 1964.

Einstein, A.: Annalen der Physik, **17**: 549, 1905.

Einstein, A.: Annalen der Physik, **19**: 371, 1906.

Einstein, A. and Smoluchowski, M.: Brownian Motion. ONTI, 1936.

Feller, W.: Zur theorie der stochastischen prozesse. Math. Ann., **113**: 113–160, 1936.

Feller, W.: On integro-differential equations of purely discontinuous Markov processes. Trans. Amer. Math. Soc., **46**: 488–515, 1940.

Feller, W.: Introduction to Probability Theory and its Application. New York, John Wiley and Sons Inc., 1966.

Feller, W.: Diffusion processes in one dimension. Trans. Amer. Math. Soc. **77**: 1–31, 1954a.

Feller, W.: The general diffusion operator and positivity-preserving semigroups in one dimension. Ann. Math. **60**: 417–435, 1954b.

Finetti, B.: Sulle funzioni a incremento aleatorio. Rend. Accad. Naz. Lincei, Cl. Sci. Fis.-Mat. Nat. (6), **10**: 163–168, 1929.

Gel'fand, I. M.: Obobshchennyye sluchaynyye protsessy [Generalized random processes]. Doklady Akad. Nauk USSR, **100**: 853–856, 1955.

Gel'fand, I. M., and Vilenkin, N. Ya.: Generalized functions, Vol. IV. Applications of harmonic analysis. New York, Academic Press Inc., 1964.

Gikhman, I. I.: O nekotorykh differentsial'nykh uravneniyakh so sluchaynymi funktsiyami [On certain differential equations with random functions]. Ukr. Matem. Zhurn., **2**, No. 3: 45–69, 1950.

Gikhman, I. I.: K teorii differential'nykh uravneniy sluchaynykh protsessov [On the theory of differential equations of random processes]. Ukr. Matem. Zhurn., **2** (No. 4): 37–63, 1950; **3**: 317–339, 1951.

Gikhman, I. I.: Ob odnoy teoreme A. N. Kolmogorova [On a theorem by A. N. Kolmogorov]. Nauchn. zap. Kiyevsk, un-ta, Matem. Sb., **7**: 76–94, 1958.

Gikhman, I. I.: O nekotorykh predel'nykh teoremakh dlya uslovnykh raspredeleniy i o svyazannykh s nimi zadachakh statistiki [On certain limit theorems for conditional distributions and on the statistical problems associated with them]. Ukr. Matem. Zhurn., **5**: 413–433, 1953.

Gikhman, I. I.: Protsessy Markova v zadachakh matematicheskoy statistik [Markov processes in problems of mathematical statistics]. Ukr. Matem. Zhurn., **6**: 28–36, 1954.

Gnedenko, B. V.: O roste odnorodnykh sluchaynykh protsessov s nezavisimymi prirashcheniyami [On the growth of homogeneous random processes with independent increments]. Izvestiya Akad. Nauk USSR, ser. Matem., **7**: 89–110, 1943.

Gnedenko, B. V.: K teorii rosta odnorodnykh sluchaynykh protsessov s nezavisimymi prirashcheniyami [On the theory of growth of homogeneous random processes with independent increments]. Sb. Trudov in-ta matem.,

Ukr. Akad. Nauk USSR, **10**: 60-82, 1948.

Gnedenko, B. V.: Theory of probability. New York, Chelsea Publ., Co. 1963.

Gnedenko, B. V. and Kolmogorov, A. N.: Limit distributions for sums of independent random variables. Reading, Massachusetts, Addison-Wesley Publ. Co. Inc., 1954.

Gray, Andrew and Mathews, D. B.: Treatise on Bessel functions and their Applications of Phyysics, 2nd. New York, Dover Publications, Inc.

Grenander, U. and Szegö, G.: Toeplitz forms and their applications. Berkeley, University of California Press, 1958.

Halmos, P. R.: Measure theory. Princeton, Van Nostrand Co. Inc., 1950.

Hanner, O.: Deterministic and nondeterministic stationary random processes. Ark. Mat., **1**: 161-177, 1950.

Hardy, G. H., Littlewood, D. E. and Pólya, G.: Inequalities, 2nd Ed. London, Cambridge University Press, 1952.

Harris, E. T.: Some mathematical models for branching processes. Proc. Sec. Berkeley Symp. on Math., Stat. and Prob., pp. 305-328, 1951.

Heider, L. J. and Simpson, J. E.: Theoretical Analysis. Philadelphia, W. B. Saunders Co., 1967.

Hoffman, K.: Banach spaces of analytic functions. Englewood Cliffs, New Jersey, Prentice-Hall Inc., 1962.

Hopf, E.: Ergodentheorie. Ergebnisse der mathematik und ihrer Genzgebiete, New York, Chelsea Publishing Co., 1948.

Irzhina, M.: Asimptoticheskoye povedeniye vetvyashchikhsya sluchaynykh protsessov [Asymptotic behavior of branching random processes]. Czechosl. Matem. Zurn., **7**: 130-153, 1957.

Ito, K.: On stochastic processes. Jap. J. Math. **18**: 261-301, 1942.

Ito, K.: Stochastic integral. Proc. Imp. Acad. Tokyo, **20**: 519-524, 1944.

Ito, K.: On stochastic integral equations. Proc. Jap. Acad., **1-4**, 32-35 1946.

Ito, K.: Stochastic differential equations in a differentiable manifold. Nagoya Math. J. **1**: 35-47, 1950.

Ito, K.: Stationary random distributions. Mem. Coll. Sci., Univ. Kyoto, **28**: 209-223, 1954.

Ito, K.: On a formula concerning stochastic differentials. Nagoya Math. J. **3**: 55-65, 1951.

Ito, K.: Kakuritsu Katei [Probability Processes]. Tokyo, Iwanami Shoten Publishers. Vol. I, 1960; Vol. II, 1963.

Jacobs, K.: Neuere Methoden und Ergebnisse der Ergodentheorie. Springer-Verlag, 1960.

Kac, M.: On distributions of certain Wiener functionals. Trans. Amer. Math. Soc., **65**: 1-13, 1949.

Kac, M.: On some connections between probability theory and differential and integral equations. Proc. Sec. Berkeley Symp. Math. Stat. and Prob., pp. 189-216, 1951.

Kac, M. and Erdös, P.: On certain limit theorems of the theory of probability. Bull. Amer. Math. Soc., **53**: 292-302, 1946a.

Kac, M. and Erdös, P.: On the number of positive sums of independent random variables. Bull. Amer. Math. Soc., **53**: 1011-1020, 1946b.

Kantorovich, L. V. and Akhilov, G. P.: Functional analysis in normed spaces. New York, Pergamon Press Inc., 1964.

Karhunen, K.: Ueber lineare Methoden in der Wahrscheinlichkeitsrechnung. Ann. Acad. Sci. Fennicae, Ser. A., Math. Phys. **37**: 3-79, 1947.

Karhunen, K.: Ueber die Struktur stationären zufälliger Funktionen. Ark. Math., **1**: 141–160, 1950.

Khas'minskiy, R. Z.: Raspedeleniye veroyatnostey dlya funktsionalov ot trayektorii sluchaynogo protsessa diffuzionnogo tipa [Distribution of probabilities for functionals of the trajectory of a random process of the diffusion type]. Doklady Akad. Nauk USSR, **104**: 22–25, 1955.

Khinchin, A. Ya.: Asimptoticheskiye zakony teorii veroyatnostey [Asymptotic laws in probability theory]. ONTI, 1936.

Khinchin, A. Ya.: Zur Theorie der unbeschränkt teilbaren Verteilungsätze. Matem. Sb., **2**: 79–120, 1937.

Khinchin, A. Ya.: Teoriya korrelyatsii statsionarnykh sluchaynykh protsessov [The theory of correlation of stationary random processes]. Uspekhi Matem. Nauk, **5**: 42–51, 1938.

Kinchin, A. Ya.: O lokal'nom roste stokhasticheskikh protsessov bez posledeystviya [On the local growth of stochastic processes without after-effect]. Izvestiya Akad. Nauk USSR, ser. matem., pp. 487–508, 1939.

Kinchin, A. Ya.: Mathematical foundations of statistical mechanics. New York, Dover Publications Inc., 1949.

Kinney, J. H.: Continuity properties of sample functions of Markov processes. Trans. Amer. Math. Soc., **74**: 280–302, 1953.

Kolmogorov, A. N.: Ueber die Summen durch den Zufall bestimmter unabhängiger Grössen. Math. Ann. **99**: 309–319, 1928, **100**: 484–488, 1929a.

Kolmogorov, A. N.: Obshchaya teoriya mery i ischisleniye veroyatnostey [The general theory of measure and the calculus of probabilities]. Trudy Komm. Akad., Razd. Matem. **1**: 8–21, 1929b.

Kolmogorov, A. N.: Eine verallgemeinerung des Laplace-Liapounoffschen satzes. Izvestiya Akad. Nauk USSR, Otdeleniye Matem. Yestestv. Nauk, pp. 959–962, 1931.

Kolmogorov, A. N.: Sulla forma generale di un processo stocastico omogeneo, Atti Accad. Lincei, **15**: 805–808; 866–869, 1932.

Kolmogorov, A. N.: Ueber die Grenzwertsätze der Wahrscheinlichkeitsrechung, Izvestiya Akad. Nauk USSR, Otdeleniye Matem. Yestestv. Nauk, pp. 363–372, 1933a.

Kolmogorov, A. N.: La transformation de Laplace dans les espaces linéares. Comptes Rendus Acad. Sci., **200**: 1717, 1935.

Kolmogorov, A. N.: Grundbegriffe der Wahrscheinlichkeitsrechnung. Berlin, Springer-Verlag, 1933b.

Kolmogorov, A. N.: Ob analiticheskikh metodakh v teorii veroyatnostey. [On analytical methods in probability theory]. Uspekhi Matem. Nauk, **5**: 5–41, 1938a.

Kolmogorov, A. N.: Uproshchennoye dokazatel'stvo ergodicheskoy teoremy Birkgofa-Khinchina [Simplified proof of the Birkhoff-Khinchin ergodic theorem]. Uspekhi Matem. Nauk, **5**: 52–56, 1938b.

Kolmogorov, A. N.: Krivyye v gil'bertovom prostranstve, invariantnyye po otnosheniyu k odnoparemetricheskoy gruppe dvizheniy [Curves in a Hilbert space that are invariant under a one-parameter group of motions]. Doklady Akad. Nauk USSR, **26**: 6–9, 1940a.

Kolmogorov, A. N.: Spiral' Vinera i nekotoryye drugiye interesnyye krivyye v gil'bertovom prostranstve [Wiener's spiral and other interesting curves in a Hilbert space]. Doklady Akad. Nauk USSR, **26**: 115–118, 1940b.

Kolmogorov, A. N.: Statsionarnyye posledovatel'nosti v gil'bertovom prostranstve

[Stationary sequences in a Hilbert space]. Bull. MGU (Moscow State University), **2** (No. 6): 1-40, 1941.

Kolmogorov, A. N.: K voprosu o differentsiruyemosti perekhodnykh veroyatnostey v odnorodnykh po vremeni protsessakh Markova so schetnym chislom sostoyaniy [On the question of differentiability of transition probabilities in Markov processes that are homogeneous with respect to time and that have countably many states]. Uchen. Zap. MGU (Moscow State University), Ser. Matem., **4** (No. 148): 53-59, 1951.

Kolmogorov, A. N. and Dmitriyev N. A.: Vetvyashchiyesya sluchaynyye protsessy [Branching random processes]. Doklady Akad. Nauk USSR, **56**: 7-10, 1947.

Kolmogorov, A. N. and Fomin, S. V.: Elements of the theory of functions and functional analysis. Albany, New York, Graylock Press Vol. I, 1957, Vol. II, 1961.

Kolmogorov, A. N. and Khinchin, A. Ya.: Ueber konvergenz von reihen deren glieder durch den zufall bestimmt werden. Matem. Sb., **32**: 668-677, 1925.

Krein, M. G.: Ob odnoy interpolyatsionnoy probleme A. N. Kolmogorova [On an interpolational problem of A. N. Kolmogorov]. Doklady Akad. Nauk USSR, **46**: 306-309, 1944.

Krein, M. G.: Ob osnovnoy approksimatsionnoy zadache teorii ekstrapolyatsii i fil'tratsii statsionarnykh sluchaynykh protsessov [On a basic approximation problem in the theory of extrapolation and filtration of stationary random processes]. Doklady Akad. Nauk USSR, **94**: 13-16, 1954.

Lévy, P.: Sur les intégrales dont les éléments sont des variables aléatoires indépendentes. Ann. Scuola Norm. Pisa, **2** (No. 3): 337-366, 1934.

Lévy, P.: Sur certains processus stochastiques homogènes. Comp. Math., **7**: 283-339, 1939.

Lévy, P.: Processus Stochastiques et Mouvement Brownien. Paris, Gauthier-Villars, 1948.

Loève, M. Fonctions aléatoires du second ordre. (Supplement to Lévy), P., 1948.

Loève, M.: Probability theory, 3rd Edition. Princeton, Van Nostrand Co. Inc., 1963.

Markov, A. A.: Rasprostraneniye zakona bol'shikh chisel na velichiny, zavisyashchiye drug ot druga [Extension of the law of large numbers to quantities dependent on each other]. Izvestiya Fiz.-Matem. o-va pro Kazanskom un-te (2) **15**: 135-156, 1906.

Maruyama, G.: Continuous Markov processes and stochastic equations, Rend. Circolo Mat. Palermo, **4**: 1-43, 1955.

Paley, R.E.A.C. and Wiener, N.: Fourier transforms in the complex domain. Providence, Rhode Island, American Mathematical Society, 1934.

Petrovskiy, I. G.: Ueber das irrfahrtproblem. Math. Ann., **109**: 425-444, 1934.

Privalov, I. I.: Granichnyye Svoystva Analiticheskikh Funktsiy [Boundary properties of analytic functions]. Gostekhizdat, 1950.

Prokhorov, Yu. V.: Raspredeleniye veroyatnostey v funktsional'nykh prostranstvakh [Distribution of probabilities in functional spaces]. Uspekhi Matem. Nauk, **8** (No. 3): 165-167, 1953.

Prokhorov, Yu. V.: Metody funktsional'nogo analiza v predelynykh teoremakh teorii veroyatnostey [Methods of functional analysis in the limit theorems of probability theory]. Vestn. Leningr. un-ta **11**: 44, 1954.

Prokhorov, Yu. V.: Skhodimost' sluchaynykh protsessov i predel'nyye teoremy teorii veroyatnostey [The convergence of random processes and limit theorems of probability theory]. Teoriya Veroyatnostey i Yeye primeneiya, **1**: 177–238, 1956.

Pugachev, V. S.: Theory of random functions. Reading, Massachusetts, Addison-Wesley Publ. Co. Inc., 1965.

Reuter, G. E. H.: Denumerable Markov processes and the associated contraction semi-groups on L. Acta Math., **97**: 1–46, 1957.

Rozanov, Yu. A.: Spektral'naya teoriya mnogomernykh statsionarnykh protsessov s diskretnym vremenem [Spectral theory of multidimensional stationary processes with discrete time]. Uspekhi matem. Nauk, **13** (No. 2): 93–142, 1958.

Rozanov, Yu. A.: Stationary random processes. San Francisco, Holden-Day Inc., 1966.

Saks, S.: Theory of the Integral, revised Edition. New York, Dover Publications Inc., 1964.

Schönberg, J. L.: Metric spaces and completely monotone functions. Ann. Math., **39**: 811–841, 1938.

Schwartz, L.: Théorie des distributions. Paris, Hermann, Vol. I, 1950; Vol. II, 1951.

Sevast'yanov, B. A.: Teoriya vetvyashchikhsya sluchaynykh protsessov [Theory of branching random processes]. Uspekhi Matem. Nauk, **6** (No. 6): 47–99 (1951).

Skorokhod, A. V.: Predel'nyye teoremy dlya sluchaynykh protsessov [Limit theorems for random processes]. Teoriya Veroyatnostey i Yeye Primeneniya, **1**: 289–319, 1956.

Skorohod, A. V.: Predel'nyye teoremy dlya sluchaynykh protsessov s nezavisimymi prirashcheniyami [Limit theorems for random processes with independent increments]. Teoriya Veroyatnostey i Yeye Primeneniya, **2**: 145–177, 1957.

Skorokhod, A. V.: Predel'nyye teoremy dlya protessov Markova [Limit theorems for Markov processes]. Teoriya Veroyatnostey i Yeye Primeneniya, **3**: 217–264, 1958.

Skorokhod, A. V.: Studies in the Theory of Random Processes. Reading, Massachusetts, Addison-Wesley, 1965.

Skorokhod, A. V.: Sluchaynyye Protsessy s Nezavisimymi Prirashcheniyami [Random Processes with Independent Increments]. Moscow, Fizmatgiz, 1963.

Slutskiy, Ye. Ye.: Sur les fonctions éventuelles continues, intégrables et dérivables dans le sens stochastique. Comptes Rendus Acad. Sci., **187**: 370–372, 1928.

Slutskiy, Ye. Ye.: Neskol'ko predlozheniy k teorii sluchaynykh funktsiy [Some propositions in the theory of random functions]. Tr. Sr-Az. un-ta, ser. Matem (5), **31**: 3–15, 1949.

Smirnov, N. V.: Priblizheniye zakonov raspredeleniya sluchaynykh velichin po empiricheskim dannym [Approximation of laws of distribution of random variables from empirical data]. Uspekhi Matem. Nauk, **10**: 179–206, 1944.

Smoluchowski, M.: Die Naturwissenschaften, **6**: 253–263, 1918.

Watson, H. M. and Galton, W.: On the probability of the extinction of families. Jour. Antropol. Inst., **4**: 138–144, 1874.

Wiener, N.: Differential space. J. Math. Phys., Mass. Inst. Tech. Press., **2**: 131–174, 1923.

Wiener, N.: Extrapolation, Interpolation and Smoothing of Stationary Time Series with Engineering Applications. Cambridge, Mass., Mass. Inst. Tech. Press, 1949.

Wiener, N. and Masany, P.: Predication theory of multivariate stochastic processes. Acta Math., **98**: 111–150, 1957 and **99**: 93–137, 1958.

Yaglom, A. M.: Vvedeniye v teoriyu statsionarnykh sluchaynykh funktsiy [Introduction to the theory of stationary random functions]. Uspekhi Matem. Nauk, **7** (No. 5): 3–168, 1955.

Index of Symbols

$\forall x X$	for all $x \in X$,
$\exists x X$	there exists an $x \in X$,
$\varnothing$	the empty set
$A \subset B$	B includes A (the event A implies the event B)
$\cup$	union of sets (sum of events)
$\cap$	intersection of sets (coincidence of events)
$A \backslash B$	the difference between the sets A and B
$\bar{A}$	the complement of the set A
$\{x; A\}$	the set of elements x satisfying the relation A
$\mathsf{P}\{A\}$	the probability of the event A
$\mathsf{M}\xi$	the mathematical expectation of the random variable ξ
$\mathsf{D}\xi$	the variance of the random variable ξ
$h \uparrow a$ $(h \downarrow a)$	h approaches a from below (from above)
$(\boldsymbol{x}, \boldsymbol{y})$	the scalar product of vectors $\boldsymbol{x}$ and $\boldsymbol{y}$
δ_{ij}	the Kronecker delta (equal to 1 for $i = j$ and equal to 0 for $i \neq j$)

INDEX